Under The Hood

Build Every Layer of a Large Language Model from Scratch

A 36-project build log: from `.backward()` to a serving stack you understand all the way down.

Ramchand Kumaresan

First edition · 2026 update

Under The Hood: Build Every Layer of a Large Language Model from Scratch

First edition, 2026 update.

Published by Star Engineering Works.

ISBN: 979-8-9971256-0-8 (paperback)

For corrections, errata, or contact: ramchand@murailabs.com

Dedicated to my father, R. Kumaresan, who fought all odds to give me everything I have, my mother Kalaiyarasi, who still sees me as a 10 year old, my wife Preethi without whom I am nothing, and my daughter Rudhra — who is the meaning of me.

Preface: What This Book Is Actually Asking You To Do

You are not here to memorize the architecture diagram of a large language model.

You are here to build one layer at a time until the words "forward pass," "tokenizer," "attention," "checkpoint," "KV cache," and "reward model" stop sounding like pieces of somebody else's system and start sounding like parts you have personally touched, broken, repaired, and measured.

That difference matters. I started writing this book because I kept meeting engineers who could ship features against the OpenAI API on Monday and could not explain, on Tuesday, what a logit was. They had been taught from the outside in, and the outside is where the API lives.

A lot of machine learning education is built around passive familiarity, and I think most of it is actively harmful. It teaches the names of the parts and lets you behave as if knowing the names is the same as knowing the parts. You watch a lecture, skim a paper, clone a repo, run a training script somebody else wrote. That gives just enough surface contact to say you have "used" the thing. Then the moment something goes wrong, the floor drops out. A loss curve bends the wrong way. A batch loader leaks quality across boundaries. A model trains but samples garbage. You realize you were renting understanding, not owning it, living in a furnished apartment, free to use everything in it without ever knowing which screws hold the bookcase to the wall, right up until the day you try to move the bookcase.

This book is built to fix that.

It does not start from APIs. It starts from mechanism. APIs teach you what a model accepts and returns; the why is where every interesting failure lives, and APIs hide it.

Project 1 makes you build backpropagation so `.backward()` stops feeling supernatural. **Project 2** sits you down with a tiny language model before the word "language model" gets buried under scale.

Then it gets weirder. **Project 3** makes you decide what a token is. The first time you have to make that decision yourself, you realize how much of the modern stack quietly depends on it. **Project 4** makes you implement attention with no framework to hide behind, no `nn.MultiheadAttention` to paper over the parts that confused you in the paper.

Then the book does something most "from scratch" books fail to do: it forces you to connect the pieces into a complete system of your own before you are allowed to admire production code. You write a GPT from a blank file. The training loop, the batch sampler, checkpointing, generation. Each one is a small lesson in everything you did not realize you didn't understand.

I wrote my first batch sampler the wrong way three times before I noticed it was leaking validation data across the boundary. The bug was invisible until I checked the indices by hand. That is the kind of paper cut this book is built to give you, on purpose, in a setting small enough that the lesson sticks.

That pain is not a defect in the curriculum.

It is the curriculum.

The batch sampler chapter is the hardest in the book. I left it as written because the bug stories in it are exactly what makes the lesson stick.

Because once you have built the awkward prototype yourself, production code stops looking like magic and starts looking like accumulated scar tissue. The first time I reread nanoGPT after writing my own toy GPT, I realized I had spent two weeks rediscovering, in slow motion, decisions Karpathy made in a single afternoon. Only then could I see why those decisions were the right ones. You start to understand why nanochat's data pipeline is more elaborate than your first loader, and why certain "boring" decisions are not boring at all.

This book also takes failure seriously.

Every chapter contains a BREAK IT section because the fastest way to understand why a component exists is to watch the system degrade when you remove it. When possible, you will run the failure on your own machine. When the real experiment needs more compute or hardware than a reader casually has, the chapter will still show you the exact failure mode and then turn the hands-on version into optional homework with explicit steps, expected observations, and a smaller proxy when needed.

I learned the value of breaking things on purpose across a long stretch of fine-tuning experiments at scale, where the most useful results almost never came from the configurations that worked. They came from the ones that broke in a specific, repeatable way. The failures had structure. The successes mostly looked like each other.

That structure is deliberate. I thought I would finish this book in six months; it took closer to eighteen, mostly because every chapter I drafted had a BREAK IT section I had not actually run, and every time I sat down to run one I lost a week to a failure mode I had not seen coming.

This is not a tour.

It is a build log.

You are expected to type code, inspect tensors, read loss curves, compare runs, and answer uncomfortable questions at the end of a chapter like:

- What failed first?
- What changed when you removed the mechanism?
- What tradeoff is this component protecting?
- What would you try next?

If you do the work honestly, by the end you will know what a modern LLM stack contains. And, more usefully, which parts are essential, which parts are optimization, which parts are convention, and which parts are still open research.

That is the standard here. None of the cheap versions count: not "I recognize the term," not "I have seen the repo," not "I watched the video once." If a chapter could only get one of those reactions from a careful reader, I considered it unfinished and rewrote it.

Built. Broken. Measured. Understood.

That is what this book is asking you to do.

Using the Code Repository

The full code for every project in this book lives at:

github.com/mechramc/Under-the-hood

You can read this book without ever opening that repo. Every project is self-contained on the page, and the only way to actually learn what is going on is to type code and watch it run. But the repo exists to make several specific things easier, and the intended way to use this book is **with the repo open in another window**.

What the repo gives you:

- **A working reference.** When you cannot tell whether your code is wrong or the book is wrong, you can open the equivalent file and see the assembled, runnable version.
- **The BREAK IT experiments, pre-wired.** Every project's `break_it.py` runs the sabotage the chapter describes and prints what fails. Watch the loss refuse to fall when you remove residuals. Watch the attention entropy collapse when you remove the `1/sqrt(d)` scaling.
- **Captured outputs.** Loss curves, generated samples, attention heatmaps, benchmark tables. Compare your run to a known-good run; if they look wildly different, something is wrong.
- **Tests.** Each project has a `tests/` folder with deterministic unit tests. A failing test points at exactly which invariant broke — a gradient that silently stopped flowing, a tokenizer roundtrip that lost a character, an attention mask that is off by one.

How the repo is organized

Each project gets its own folder under `projects/`:

```
projects/
  01_the-learning-machine/
    README.md          prose, with the captured outputs embedded
    build.py           the complete working version, with a CLI
    break_it.py        the sabotage experiment from this chapter's BREAK IT
    step_*.py          code blocks from the chapter, one file per step
    outputs/           loss_curve.png, run_log.txt, break_it_log.txt
    tests/
      test_unit.py     gradient checks, shapes, convergence sanity
```

The `step_*.py` files are the chapter's code blocks pulled out into individual scripts so you can read them as you read the chapter. They are reference material, not necessarily independently runnable. The runnable artifact is `build.py`.

The intended workflow per chapter

1. Read the chapter. Type the code yourself. Do not copy from `build.py`. The point of this book is not to produce code artifacts. It is to internalize what shapes attention has, what gradient flow feels like, and what specific errors training makes when one piece breaks. Copy-pasting prevents that.

2. When you are stuck or want to compare, open `projects/NN_slug/build.py`. Look at how the pieces are assembled. Find your bug.
3. Run the BREAK IT:

   ```
   python projects/NN_slug/break_it.py --tiny
   ```

   Watch the failure mode the chapter described. Predict what the numbers will be before you run it. The gap between prediction and observation is the lesson.
4. Compare your run's loss curve against `projects/NN_slug/outputs/loss_curve.png`. If your numbers diverge by an order of magnitude, you probably have a bug. If they diverge by a small amount, that is normal and expected (different initialization, different hardware, different floating-point rounding).
5. If you fork the code and one of the unit tests starts failing, the test is telling you which invariant broke. That is the fastest way to debug.

Getting it running

The repo's `setup/` folder has the full installation guide, broken into five short files: Python environment, dependencies, GPU and hardware tiers, how to run `.py` files (and convert them to notebooks if you prefer), and dataset access.

The short version, on most systems:

```
git clone https://github.com/mechramc/Under-the-hood.git
cd Under-the-hood

python -m venv .venv
source .venv/bin/activate          # on macOS / Linux
.venv\Scripts\activate             # on Windows

pip install -r requirements.txt
```

Then any project runs with the `--tiny` flag, which uses a CPU-friendly proxy configuration:

```
python projects/01_the-learning-machine/build.py --tiny
```

`--tiny` is what every project's test suite runs against. It finishes in under 60 seconds on a laptop and demonstrates the structural lesson without needing a GPU. `--full` runs the realistic lab; some later projects need a GPU for that path. The repo's `setup/03_gpu-and-hardware-tiers.md` lists which projects need what.

What the repo is not

The repo is not a replacement for typing the code yourself. Every chapter has at least one ugly failure mode that only becomes intuition by hitting it personally. Reading the working version after the chapter teaches you a different thing than building it does.

The repo is not the source of truth for the book's exposition. If `build.py` and the chapter disagree, the chapter is right and the code is the bug. File an issue.

The repo is not feature-complete on day one. The early chapters of the book — the foundations and the GPT build — are fully curated, with working `build.py` files, full BREAK IT experiments, captured outputs, and tests. Later chapters may currently ship only the extracted step-by-step code references while the polished `build.py` is being assembled and reviewed. The per-project `README.md` in the repo tells you the current state for each project. As more chapters become fully runnable, those READMEs will reflect it.

Reporting issues

Bugs, corrections, suggestions:

github.com/mechramc/Under-the-hood/issues

The single most useful kind of bug report is "the book says X but the code does Y." When the narrative and the code drift apart, you have caught something I want to know about. Send it.

Preflight: Python, Tensors, and What a Language Model Actually Is

---

Three test readers came back with the same note: Project 1 feels abrupt. You open Chapter 1 and you are immediately building backpropagation from scratch. That was by design. The fastest way to understand `.backward()` is to build it. But two things were missing from the on-ramp.

I argued with myself about this for a week. I had written and rewritten the Project 1 opening, and the chapter was good. The problem was not the chapter. The problem was that I had skipped the work of meeting the reader where they actually were.

First: what is a language model, actually? Not the marketing version. The actual mechanism.

Second: what Python patterns should you be able to read before you start typing? Not a full tutorial. This book is not a Python course. But there is a specific vocabulary of Python idioms that appear throughout the projects, and fluency with that vocabulary makes everything else faster.

This chapter gives you both. It is not a project. It does not have a Build section or a BREAK IT experiment. It is a calibration. A shared vocabulary between you and the rest of the book.

Thirty minutes here saves three hours of confusion later.

What a Language Model Actually Is

Strip away the benchmarks, the demos, the hype, and the marketing, and a language model is doing one thing.

Given some text, predict what comes next.

That is the entire job. Everything else (the transformer architecture, the training pipeline, the fine-tuning, the deployment infrastructure) is in service of doing that one thing well, at scale, across every domain of human text at the same time.

I used to find this claim deflating when I first heard it. It sounded too small for the size of the engineering effort. It took running my own training experiments before I understood that the smallness of the goal is what makes the scale matter at all.

Here is a concrete example. Suppose the input is:

```
The capital of France is
```

A language model looks at that sequence and produces a **probability distribution** over every word it knows. Figure 0.1 shows what that distribution might look like for the most likely continuations.

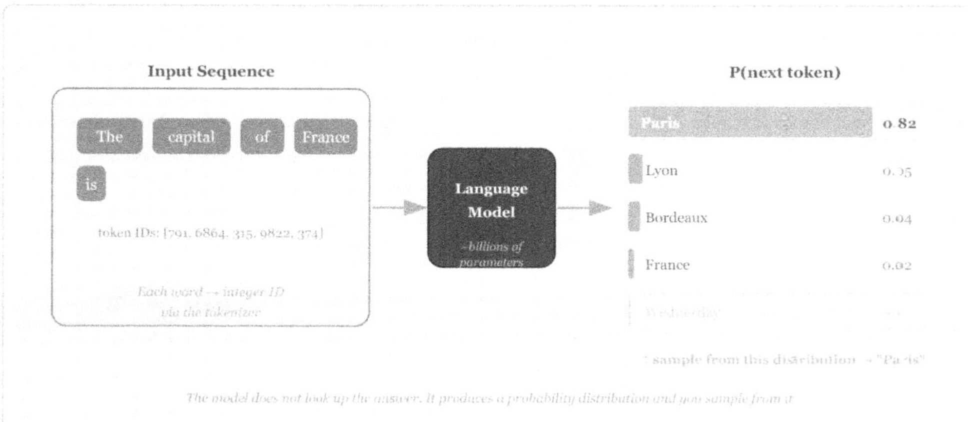

Figure 0.1. A language model takes a sequence of tokens as input and outputs a probability distribution over the full vocabulary. "Paris" receives high probability; plausible but wrong continuations get small probability; nonsense gets nearly zero.

"Paris" gets very high probability, say 0.82. "Lyon" gets a small but nonzero probability. "Wednesday" gets nearly zero. The model is not looking up a fact in a database. It is expressing its uncertainty about what word naturally continues this sequence, based on every text it has ever read.

The first time I really sat with the database analogy and how wrong it is, I had been writing this chapter for hours. The model holds opinions, not records. That is a different shape of object entirely.

When you generate text, you sample from that distribution. Sample "Paris." Now the input becomes:

```
The capital of France is Paris
```

Do it again. What comes next? The model produces another distribution. Sample again. Repeat until you decide to stop.

That is the whole engine. One probability distribution at a time, one token at a time.

What "Token" Means

We have been saying "word," but language models do not operate on words. They operate on **tokens**: pieces of text that might be words, parts of words, punctuation marks, or even individual characters.

The reason is practical. A word-level model has to handle every word in English, including typos, technical jargon, foreign loanwords, brand names, and words that did not exist when the model was trained. That is a vocabulary problem with no clean solution.

Tokens sidestep this by working at a finer granularity. "tokenization" might become three tokens: `token`, `iz`, `ation`. The exact split depends on the vocabulary the model was trained with. Numbers, code, and non-prose text can all be represented without requiring that every possible number or function name appear in the vocabulary.

Most beginner explanations of tokenization are honestly terrible because they imply that tokens are "small words," when in fact tokens have nothing to do with linguistic units. They are statistical artifacts of a compression algorithm. The chapter on building one (Project 3) is where this stops being a vague claim.

Project 3 of this book is entirely about building the tokenizer. For now, you just need the intuition: the model does not see raw text. It sees a sequence of integer IDs, one per token, and each token ID is a compressed representation of a piece of text.

Table 0.1 shows how a short sentence might break down.

Text	Tokens	Token IDs
`Hello, world!`	`Hello, ,, world, !`	`[9906, 11, 1917, 0]`
`backpropagation`	`back, prop, agation`	`[1445, 22210, 7246]`
`cat`	`cat`	`[8415]`
`cats`	`cats`	`[55242]`
`2024`	`2024`	`[2366]`
`print(x)`	`print, (, x, )`	`[1374, 7, 87, 8]`

Table 0.1. Tokenization is not splitting on spaces. Common words get a single token ID; rarer combinations break into smaller pieces. The specific IDs depend on which tokenizer was used. Notice that "cat" and "cats" are completely different tokens with unrelated IDs.

Two things in this table are worth sitting with. First, `cat` and `cats` have completely different IDs. The model does not start knowing they are related. It learns that relationship from data. Second, `print(x)` in code becomes four tokens, which means the model does pick up programming syntax at the subword level.

It surprised me that token IDs for related words are nowhere near each other in the integer space. The relationship between `cat` and `cats` lives entirely in the embedding table the model learns. That mismatch between human intuition and the actual mechanism is the kind of thing that bit me repeatedly while building tokenizers for an offline Tamil-language Android deployment.

What "Large" Actually Means

A model that predicts next tokens on toy text is not impressive. What makes production language models remarkable is the scale at which they are trained: billions of **parameters**, trillions of tokens.

Parameters are the numbers the model learns during training. Every time the model makes a prediction, those numbers do the work. A small model might have 7 million parameters. A large model might have 70 billion. Ten thousand times more.

Tokens in training represent the data the model learned from. A model trained on 1 trillion tokens has processed roughly the equivalent of one million books.

The interaction between those two numbers (parameters and tokens) turns out to be one of the most important discoveries in recent AI research, and Project 11 of this book is built around understanding that relationship precisely.

For now, the key intuition is this: "large" is not a quality judgment. It is a statement about capacity. More parameters means more patterns can be stored. More training tokens means more of human language has been compressed into those patterns. The result is a model that generalizes remarkably well across tasks it was never explicitly trained to perform.

Two Modes: Training and Inference

A language model has two distinct operating modes.

Training is when the model learns. You show it billions of tokens of text, and after each prediction, you measure how wrong it was, then adjust the parameters in a direction that would have made it less wrong. Repeat billions of times. This is slow. Days or weeks on specialized hardware.

Inference is when the model generates. You give it a starting sequence, ask it to predict the next token, sample from the distribution, append the token, and repeat. This is fast. Milliseconds per token on consumer hardware.

These modes are more different than they sound. During training, the model processes entire sequences in parallel for efficiency. During inference, it generates one token at a time and needs to remember what it has already generated. That difference is why projects 8 and 9 (KV caching, grouped query attention) exist. They are optimizations that only matter in the inference loop.

Everything in this book is about understanding how both modes work from the inside. Not just reading about them. Building them.

How a Model Learns: The Three Moves

Project 1 builds the learning engine from scratch. Before you start, here is the plain-English version of what you will be building.

A neural network learns in exactly three moves, repeated as many times as needed.

Move 1: Make a guess (the forward pass). The model takes input data, passes it through a series of mathematical operations (multiplications, additions, nonlinear squashing functions) and produces an output. The output is a prediction: for a language model, a probability distribution over the next token.

Move 2: Measure how wrong you were (the loss). You compare the model's prediction to the correct answer. The **loss** is a single number that summarizes how wrong the prediction was. High loss means the model was badly wrong. Low loss means the prediction was close. The goal of training is to drive the loss down.

Move 3: Assign blame and adjust (the backward pass). You need to know which parameters caused the loss to be high, and by how much. That is what **backpropagation** computes. It walks backward through the chain of operations and figures out: if I nudge this parameter a tiny amount, does the loss go up or down? The sign and magnitude of that answer is the **gradient**. You then nudge every parameter a small step in the direction that reduces the loss.

Repeat this loop millions of times on different batches of data, and the model gradually improves.

Figure 0.2 shows the loop.

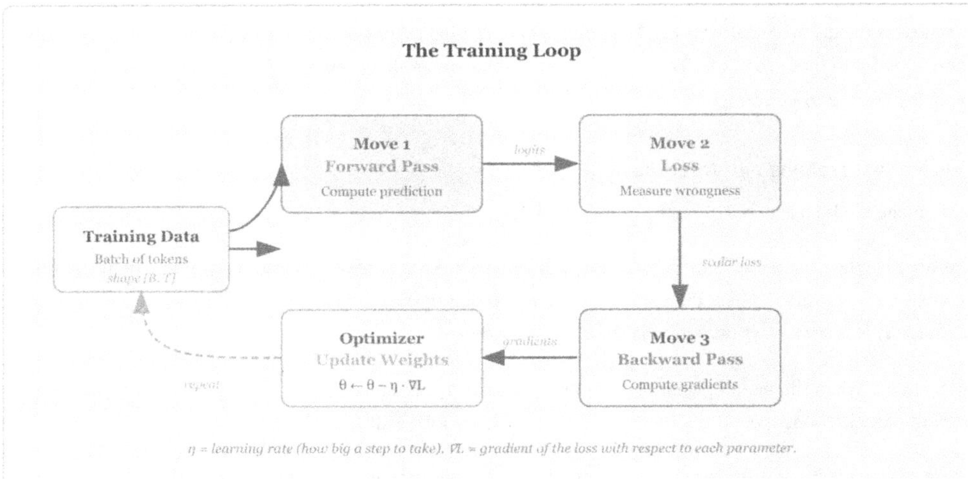

Figure 0.2. The training loop in three moves: the forward pass produces a prediction and a loss, the backward pass computes each parameter's responsibility for that loss, and the optimizer updates every parameter in the direction that reduces it. All of training is this loop, repeated.

This loop is not special to language models. It is how every neural network learns, from the toy 2D classifier you will build in Project 1 to the 70-billion-parameter models you will quantize in Project 15. The mechanism scales; the three moves do not change.

One thing Project 1 will make permanently clear: "the backward pass" is not mysterious. It is just the chain rule from calculus, applied automatically to every operation in the forward pass, in reverse order. The whole chapter exists to make that sentence feel obvious rather than abstract.

The word "mysterious" in that sentence is deliberate. An earlier draft said "magical," and I changed it because "magical" lets a reader off the hook. It says the explanation is supposed to feel out of reach. "Mysterious" points at what the reader actually feels, not what I want them to feel.

Python You Need to Read This Book

This is not a Python tutorial. If you need one, the official Python tutorial at python.org is a good place to start, and you can return here when you are comfortable with variables, functions, and loops.

What follows is a reading comprehension guide for the specific Python patterns that appear throughout this book's code. If you can read these patterns fluently, you are ready for Project 1.

Variables and Types

Python is dynamically typed. A variable holds any kind of value, and its type is determined at runtime.

```
x = 2.0          # float, a real number
n = 100          # int, a whole number
name = "hello"   # str, a string of characters
flag = True      # bool, True or False
```

You will see floats constantly. They represent real numbers, and most of the math in this book runs on floats (or their higher-dimensional equivalent, tensors). When you see `0.001` in a training script, that is a float. In that case, the learning rate, one of the most consequential parameters you will tune.

Functions

A function bundles a computation so you can reuse it.

```
def square(x):
    return x * x

result = square(3)    # result is 9
```

Functions in this book often take another function as an argument, or return a function. This is called a **higher-order function**, and the first time you see it, it can look strange:

```
def make_adder(n):
    def add(x):
        return x + n
    return add

add_five = make_adder(5)
add_five(3)    # returns 8
```

`make_adder` returns a new function, `add`, that remembers the value of `n` from the outer scope. This is called a **closure**. The inner function "closes over" the variables it needs from the environment where it was created.

You will see this in the backward pass code. Each node in the computational graph stores a closure: a function that knows how to push gradients to its specific parents. When the backward pass runs, it calls those closures one by one, and each one knows exactly what to do because it closed over the right variables when the forward pass ran.

Closures took me an embarrassingly long time to get. I spent half a day debugging a backward pass once because I had captured a loop variable by reference rather than by value, and every gradient was being routed through the last operation in the graph. The fix was one line. The lesson was that the inner function's variables are not a snapshot of the moment it was defined.

Classes

A class is a blueprint for objects. Each object has **attributes** (data) and **methods** (functions attached to that data).

```
class Counter:
    def __init__(self, start=0):
        self.value = start

    def increment(self):
        self.value += 1

c = Counter(10)
c.increment()
print(c.value)    # 11
```

`__init__` is called automatically when you create a new instance with `Counter(10)`. `self` refers to the specific instance being operated on. It is how a method knows whose data to access or modify.

If you create two counters, each has its own `.value`. `self` is the mechanism that keeps them separate. Every `Value` object you build in Project 1 is a class instance. Every neuron, layer, and model you build after that is too.

Dunder Methods

Python allows classes to define how they behave with built-in operators. Methods named with double underscores (`__add__`, `__mul__`, `__repr__`) are called **dunder methods** (short for "double underscore"). They are the machinery behind operator overloading.

```
class Box:
    def __init__(self, size):
        self.size = size

    def __add__(self, other):
        return Box(self.size + other.size)

    def __repr__(self):
        return f"Box(size={self.size})"

a = Box(3)
b = Box(5)
c = a + b            # calls a.__add__(b) behind the scenes
print(c)             # calls c.__repr__() → Box(size=8)
```

When you write `a + b`, Python calls `a.__add__(b)`. This is exactly how the `Value` class in Project 1 intercepts arithmetic: by defining `__add__` and `__mul__`, a `Value` object can participate in normal Python expressions while secretly building the computational graph in the background.

This is not a trick or a hack. It is the designed mechanism. Every PyTorch tensor you use in later projects does the same thing at production scale.

Table 0.2 shows the dunder methods that appear in this book.

Dunder method	Triggered by	How it is used
`__init__`	`MyClass(...)`	Initialize object state at creation
`__add__`	`a + b`	Intercept addition, record graph edge
`__mul__`	`a * b`	Intercept multiplication, record graph edge
`__neg__`	`-a`	Negate a value (used in subtraction)
`__pow__`	`a ** b`	Raise to a power
`__truediv__`	`a / b`	Division
`__repr__`	`print(a)`	Human-readable string representation
`__len__`	`len(a)`	Size of a container
`__getitem__`	`a[i]`	Index into a container
`__call__`	`a(x)`	Treat an object as a function (used for models and layers)

Table 0.2. Dunder methods let a class intercept Python's built-in operators. The entire Value class in Project 1 is about 80 lines. Most of it is dunder methods.

List Comprehensions

A list comprehension is a concise way to build a list from an existing sequence.

```
numbers = [1, 2, 3, 4, 5]

squares = [x ** 2 for x in numbers]
# [1, 4, 9, 16, 25]

evens = [x for x in numbers if x % 2 == 0]
# [2, 4]
```

You will see list comprehensions used to create sets of parameters, collect outputs from a set of layers, and build batches of data. Once you can read them fluently, they are faster than equivalent for-loops, both to read and to write.

Generators

A generator is a function that produces values lazily, one at a time on demand, instead of all at once in memory.

```
def count_up(n):
    i = 0
    while i < n:
        yield i
        i += 1

for x in count_up(3):
    print(x)    # prints 0, 1, 2
```

The `yield` keyword pauses the function and hands a value to whoever is iterating over it. The next time they ask for a value, the function resumes from exactly where it left off.

This matters in pretraining. When your dataset is larger than RAM, you cannot load it all at once. You feed it to the model in streaming batches using a generator. The training loop asks for the next batch, the generator produces one batch, the loop processes it, and then the generator produces the next. No more than one batch lives in memory at a time.

Common Imports

Python code borrows from other modules using `import`. Table 0.3 lists the imports you will see most often and what they provide.

Import	What it provides
`import math`	`math.exp`, `math.log`, `math.sqrt` for scalar math operations
`import random`	`random.random()`, `random.shuffle()` for randomness
`import torch`	The PyTorch tensor library — the main numeric workhorse throughout the book
`from pathlib import Path`	File and directory path manipulation
`from dataclasses import dataclass`	Cleaner class definitions for data containers
`import json`	Read and write JSON config files and results
`from typing import Optional, List`	Type hints that appear in function signatures

Table 0.3. You do not need to install anything beyond Python and PyTorch to run the early projects. The later projects add one or two dependencies each, introduced when they become relevant.

Numbers as Tensors

The most important data structure in this entire book is the **tensor**: a generalization of a list of numbers to multiple dimensions.

If you have used Python lists, you already understand the concept. A 1D tensor is like a list of numbers. A 2D tensor is like a grid of numbers, rows and columns. A 3D tensor is like a stack of grids.

What makes tensors special, and why they replaced Python lists entirely in deep learning, is that they support fast, parallel computation across all their values at once, especially on GPUs.

The cleanest mental model I have for a tensor is a stack of spreadsheets that all know how to be added, multiplied, and reshaped in one motion. The numbers themselves are not interesting. The thing that is interesting is that the operations move through them in lockstep.

Figure 0.3 shows the four tensor dimensions you will encounter most often.

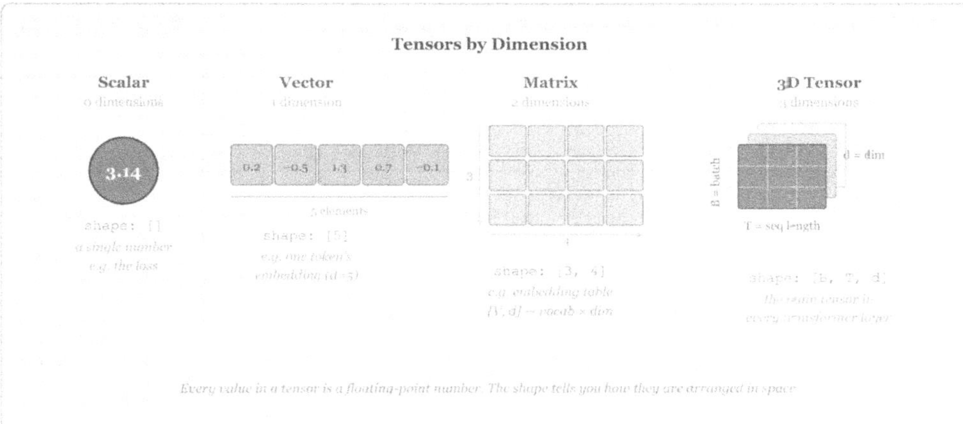

Figure 0.3. Tensors by dimension: a scalar is a single number with shape []; a vector is a 1D sequence with shape [n]; a matrix is a 2D grid with shape [m, n]; a 3D tensor is a stack of matrices, most often seen as [B, T, d] — a batch of sequences, each token represented by d numbers.

Shape

The most important property of a tensor is its **shape**: the size of each dimension, written as a list of integers.

```
import torch

# A scalar, no dimensions
s = torch.tensor(3.14)
print(s.shape)    # torch.Size([])

# A vector, 1D, 5 elements
v = torch.tensor([1.0, 2.0, 3.0, 4.0, 5.0])
print(v.shape)    # torch.Size([5])

# A matrix, 2D, 3 rows × 4 columns
m = torch.randn(3, 4)
print(m.shape)    # torch.Size([3, 4])

# A 3D tensor, 8 sequences, each 16 tokens long, each token 64 numbers
t = torch.zeros(8, 16, 64)
print(t.shape)    # torch.Size([8, 16, 64])
```

When you see `[B, T, d]` in the code, those letters mean:

- `B`, **batch size**: how many sequences are being processed in parallel
- `T`, **sequence length**: how many tokens are in each sequence
- `d`, **embedding dimension**: how many numbers represent each token

This three-dimensional shape `[B, T, d]` is the tensor you will encounter most often in the model code. It appears in the attention mechanism, the feedforward layers, and the final prediction head. Every transformation in a transformer is a function that takes a `[B, T, d]` tensor and returns another one.

Table 0.4 shows the full notation used throughout the book.

Symbol	What it means	Typical toy value	Typical production value
B	Batch size	4–32	512–4096
T	Sequence length (context window)	64–256	2048–128000
d	Embedding dimension	32–128	768–8192
N	Number of transformer layers (depth)	2–6	32–96
H	Number of attention heads	2–4	32–64
V	Vocabulary size	256–1000	32000–128000

Table 0.4. Shape notation used throughout the book. A transformer is fully defined by the tuple (V, d, N, H) plus a maximum context length T. The first time a chapter uses one of these letters, it will define it explicitly.

Basic Tensor Operations

You do not need to know all of PyTorch before starting. Table 0.5 shows the operations that appear in the earliest projects and reappear throughout.

Operation	Code	What it does
Create random	`torch.randn(3, 4)`	3×4 matrix, values drawn from normal distribution
Create zeros	`torch.zeros(2, 5)`	2×5 matrix filled with 0.0
Create ones	`torch.ones(4)`	Vector of four 1.0s
Index	`t[0]`	First row of a 2D tensor
Slice	`t[:, 1:4]`	All rows, columns 1 through 3
Matrix multiply	`A @ B`	Standard matrix multiplication
Elementwise multiply	`A * B`	Multiply corresponding elements (shapes must match)
Sum	`t.sum()`	Sum all elements; `t.sum(dim=1)` sums along one axis
Mean	`t.mean()`	Average all elements
Reshape	`t.view(8, -1)`	Change shape without moving data; `-1` infers the missing size
Transpose	`t.T` or `t.transpose(1, 2)`	Swap two dimensions
Move to GPU	`t.to("cuda")`	Move tensor to the GPU for fast computation
Detach gradient	`t.detach()`	Stop tracking gradients through this tensor

Table 0.5. Tensor operations in roughly the order they appear in the book. Look up anything else as you encounter it. The PyTorch documentation is thorough and well-organized.

Why Not Python Lists?

If you have been using Python lists, three things make tensors worth switching to.

Speed. PyTorch tensors are stored contiguously in memory and processed by highly optimized low-level code. Multiplying two 1000×1000 matrices takes milliseconds with tensors. Doing the same with nested Python lists would take seconds, and that gap widens as the size grows.

GPU. A tensor can move to a GPU with one line: `t.to("cuda")`. A Python list cannot. GPUs can perform thousands of arithmetic operations in parallel, which is why training that would take months on a CPU takes days on a GPU.

Gradients. PyTorch tensors can track their own computational history and compute gradients automatically. The same mechanism you will build by hand in Project 1. In Projects 5 and onward, you use PyTorch's built-in version because it is faster and more complete, but you will know exactly what it is doing because you built the toy version first.

The Shape of the Book

You know what a language model is. You can read the Python. You understand tensors. Here is the arc of what comes next. Thirty-five projects in eight parts. Each part builds on the part before it, and each project inside a part builds on the project before it.

Part I, Foundations (Projects 1–4) builds the mechanics from scratch. Project 1 builds backpropagation by hand. Project 2 builds a character-level language model. Project 3 builds a byte-pair tokenizer. Project 4 implements multi-head causal attention without a framework. None of this uses PyTorch's autograd. You write every line yourself, so there is no magic left to explain away. By the end you have the four primitives every modern LLM is built on.

Part II, Building a GPT (Projects 5–8) assembles those primitives into a working transformer and then makes it fast. Project 5 has you write a full GPT from a blank file. Project 6 reads nanoGPT line by line. Project 7 covers the details that matter (RMSNorm, SwiGLU, RoPE). Project 8 implements Flash Attention and tiled kernels so memory stops scaling quadratically with sequence length. You finish this part with a transformer that trains and an attention kernel that does not run out of memory.

Part III, Training at Scale (Projects 9–12) moves from toy data to the engineering of real pretraining runs. Project 9 trains on FineWeb-EDU. Project 10 covers data curation, deduplication, and contamination. Project 11 teaches you to debug spikes, NaNs, and silent training failures using gradient norms, activation maxima, and weight-update ratios. Project 12 builds the single-box proxy for FSDP and ZeRO so you understand sharding without owning a cluster. You finish this part able to launch a training run and read it like vital signs.

Part IV, Inference (Projects 13–17) is about making the trained model actually serve users. Project 13 implements the KV cache. Project 14 adds speculative decoding with a draft model and the residual-distribution rejection rule. Project 15 implements grouped query attention. Project 16 extends context length using RoPE, position interpolation, NTK-aware scaling, and YaRN. Project 17 builds continuous batching and PagedAttention for production serving. By the end of this part, your model serves a hundred concurrent users at high throughput without padding waste.

Part V, Architecture and Scaling (Projects 18–20) covers the architectural choices that pay back at scale. Project 18 builds a mixture-of-experts layer with a router, load-balancing loss, and an expert-utilization histogram. Project 19 fits scaling laws on your own hardware and reproduces the Chinchilla finding. Project 20 hands the training loop to an autonomous coding agent and lets it run a hundred experiments overnight.

Part VI, Alignment (Projects 21–26) turns a pretrained next-token predictor into a useful assistant. Project 21 covers supervised fine-tuning, instruction tuning, and LoRA. Project 22 builds a proper evaluation methodology (MMLU, HellaSwag, GSM8K, HumanEval, LLM-as-judge) and shows why a single metric lies. Project 23 implements RLHF with a reward model and GRPO. Project 24 builds DPO and the other reference-free preference losses (KTO, IPO, ORPO, SimPO). Project 25 covers test-time reasoning: chain-of-thought, self-consistency, best-of-N, process reward models. Project 26 adds tool use and function calling with a ReAct loop. You finish with a model that follows instructions, reasons step by step, and can call external functions when the answer lives outside its weights.

Part VII, System Scope (Projects 27–31) broadens the system beyond the transformer-on-text default. Project 27 implements post-training quantization down to 2-bit and exports to GGUF. Project 28 builds a retrieval-augmented generation pipeline with a vector index, bi-encoder embeddings, and a re-ranker. Project 29 builds a tiny vision-language model with a ViT-style encoder and a projection layer to your existing GPT. Project 30 replaces attention entirely with state-space models (Mamba and RWKV) and measures what you win and what you lose. Project 31 builds a non-autoregressive decoder by hand — a masked diffusion loop that fills a row of blanks coarse-to-fine and trades denoising steps for coherence. By the end of this part you know which parts of the modern stack are universal and which are transformer-specific.

Part VIII, Modular Composition (Projects 32–36) is the research frontier the book has been building toward: can independently trained specialists be assembled into something greater than any of them? Project 32 measures CKA across freeze depths. Project 33 fuses three specialists from a shared base and measures the oracle routing gap. Project 34 writes a formal interface specification and a programmatic compliance test. Project 35 adds new specialists after deployment without retraining the router. Project 36 is the open-ended one. Your own architectural contribution, defined by you.

The progression is deliberate. You cannot understand what a KV cache does until you understand what attention does. You cannot understand DPO until you understand what RLHF is trying to solve. You cannot understand composable architectures until you understand what gets frozen, what gets transferred, and why. The only path through this book is forward.

Quick Reference: What to Do If You Get Stuck

This is a build log, not a spectator sport. You will get stuck. Here is a short guide for the most common situations.

Symptom	First thing to check
Loss not going down	Is the learning rate too high or too low? Try 10× in each direction
Loss goes to NaN	Gradient explosion. Check gradient clipping or reduce learning rate
Gradients are all zero	A gradient path is broken. Check which operation is blocking backward
Code runs but model generates nonsense	Training loop may be correct but not enough steps. Run longer
Shape mismatch error	Print `.shape` on both tensors before the operation that failed
CUDA out of memory	Reduce batch size B or sequence length T
Import error	Check which project's dependencies have been installed

Table 0.6. When stuck, the table above is usually enough to find the right direction. The BREAK IT sections in each project intentionally trigger many of these failure modes so you can recognize them on sight.

What You Now Know

- A language model predicts a probability distribution over the next token, given a context Text generation is repeated sampling from that distribution.
- A token is not a word. It is a subword unit with an integer ID, produced by a tokenizer. Project 3 builds one.
- Training is the loop of: forward pass, compute loss, backward pass, update parameters, repeated on billions of token pairs until the loss is low.
- Python classes, dunder methods, closures, and list comprehensions are the four patterns you need to read this book's code.
- A tensor is a multidimensional array with fast math and optional gradient tracking. `[B, T, d]` is the most common shape in language model code.
- "Large" in LLM refers to parameter count and training data volume. The mechanics underneath are the same at every scale.

You are ready for Project 1.

Project 1: The Learning Machine

Hook

Why does `.backward()` feel like a spell? You write a few lines of code, call one method, and somehow every weight in a network learns how to change itself. If you are new to machine learning, this is the first insult the field throws at you: the most important step is hidden behind a single line. I had been calling `.backward()` for years before I sat down to write a toy version from scratch, and the embarrassing truth is that until I did, I could not have told you precisely what was happening inside it. This chapter fixes that. You are going to build the thing that computes those gradients yourself, break one tiny piece of it, and watch learning stop cold.

The Concept

A neural network learns in three moves: it makes a guess, you measure how wrong that guess is, and then you figure out which internal numbers caused the error and by how much. That third step is the whole story.

People give it intimidating names (automatic differentiation, reverse-mode autodiff, backpropagation), but strip away the names and it becomes plain. The way I actually think about it: every number a network produces has an ancestry, and learning is the act of walking that ancestry backward and asking each ancestor "how much of this mess was your fault?" The final number in the chain is the loss, the score that says how wrong the model is. The loss only exists because some intermediate number contributed to it, which only exists because some earlier number contributed to that. If you want to know why the loss is so high, you walk backward through every step that fed it.

To make it more concrete: think of a recipe. You bake a cake, it tastes bad. Too salty, too dense, not sweet enough. You want to know what to change next time. Was it too much salt? Not enough sugar? Too much flour? Did two ingredients interact in a way that made the result worse? A neural network does the same thing with numbers. It combines inputs through additions, multiplications, nonlinear squashing functions like `tanh`, and many other steps, and at the end it gets one score: the loss. Backpropagation is blame-assignment through that recipe: which numbers made the loss worse, and if I nudge this number slightly upward, does the loss go up or down?

I started calling it blame-assignment after a debugging session where I had a model whose loss refused to fall, and I realized I had been mentally treating the gradient as a "direction" rather than as an accusation against a specific parameter. Once I started reading `p.grad` as "the responsibility this parameter bears for the current loss," the diagnostics got faster. Weights with a near-zero `grad` were suspects: either they were doing nothing useful, or the path that should have carried their blame was broken.

That sensitivity is the **gradient**: a number telling you which direction to nudge a parameter to reduce the loss, and by how much. The image I keep in my head is that weights are ingredients the recipe never measures exactly, and gradients are the chef tasting the result and silently adjusting tomorrow's quantities by hand. The chef does not redesign the recipe, just nudges. Thousands of those nudges in a row are what training is.

Automatic differentiation means the computer tracks the recipe as it runs. The answer alone is not enough. It keeps the whole trail. Every time you add two values, multiply them, or pass one through `tanh`, the system records what operation happened, which inputs created the output, and how to push blame backward through that operation later. That recorded trail is the **computational graph**, illustrated in Figure 1.1.

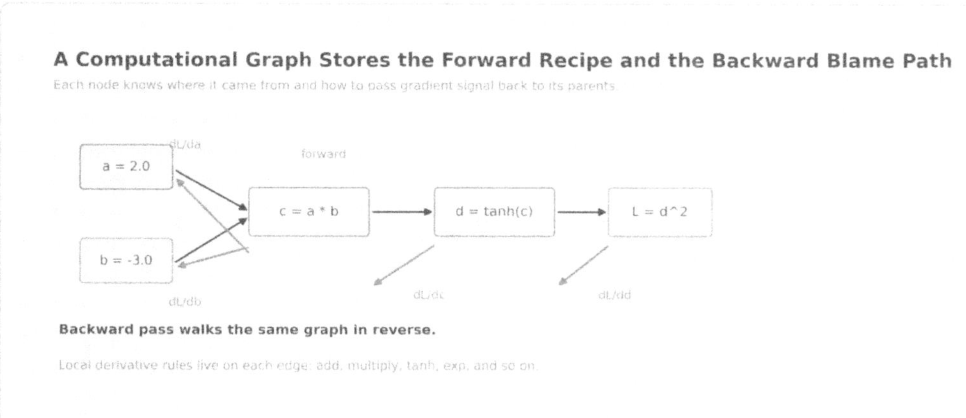

Figure 1.1. A computational graph records both the forward recipe and the backward blame path: each value knows where it came from and how to pass gradient signal back to its parents.

You can picture it as a family tree of numbers. Leaves at the bottom are raw inputs and trainable parameters. Internal nodes are results of operations like `a * b` or `a + b`. The root at the top is the final loss. The forward pass builds this graph while computing the output, and the backward pass walks that graph in reverse and assigns blame all the way down.

Here is the part most explanations skip: there is no magic leap from loss to all gradients. The system applies one local rule at a time. For addition, if `c = a + b`, then a change in `c` passes straight back to both `a` and `b`, because increasing either input by 1 increases the output by 1. For multiplication, if `c = a * b`, then a change in `c` passes back to `a` scaled by `b`, and back to `b` scaled by `a`. That is just the chain rule from calculus. Each step tells the previous step how much responsibility it had.

I expected the chain rule to feel deep once I built the engine. It did not. It felt mechanical, almost boring: a small local rule called over and over against a list of nodes. That itself was the lesson I needed. The mystique is in the scale of the graph, not in the math at each node.

Let's earn the first equation. Suppose `L` is the final loss, `c` is some intermediate value, and `a` and `b` created `c` through `c = a * b`. We want to know: how much does changing `a` change the loss? We already know two things: how much changing `c` changes the loss (`dL/dc`), and how much changing `a` changes `c` (`dc/da = b`). Put them together:

```
dL/da = (dL/dc) * (dc/da) = (dL/dc) * b
```

Read it in English: the effect of `a` on the loss equals the effect of `c` on the loss times the effect of `a` on `c`. That is the whole machine, for multiplication and for everything else. Every deep learning framework you use is doing this over and over again, at scale, through giant graphs made of millions or billions of operations. This chapter starts with the toy version because the toy version is the real version, only small enough to see.

Honest opinion, after reading more beginner backprop explanations than I care to count: most of them are terrible, and they fail in the same way. They reach for Jacobians and matrix calculus before the reader has watched a single scalar gradient flow backward through a graph drawn by hand. Build the toy graph first. The Jacobians can wait. If you cannot say which arrow in the graph the gradient is flowing along, the matrix form will not help. It will just hide the confusion under heavier notation.

Why It Matters

If you do not understand automatic differentiation, every other part of machine learning stays foggy. You can still call APIs, fine-tune models, and read benchmark tables, but you cannot reason from first principles about what training is actually doing. You cannot answer questions like why training stopped improving, why one missing gradient makes the model freeze, or why a bad initialization or saturated activation kills learning entirely.

Every modern model learns through the same loop: compute a prediction, compute a loss, compute gradients, update parameters. Large language models do not escape this. They just repeat it at a scale that hides the mechanism behind more code, more hardware, and more abstractions. Across a long set of fine-tuning experiments I ran on models from a few hundred million to several billion parameters, the failures that hurt most were never the loud ones. The loud ones crash and you fix them. The painful ones were runs where the loss flattened, the program ran fine, and a single broken path in the graph had quietly turned a slab of parameters into dead weight.

If the gradients are wrong, the model does not "kind of learn badly." It learns nonsense. If one path in the graph does not pass blame backward, the parameters behind that path go dark. They stop receiving useful signals. You can have a fully valid program, no crashes, no warnings, and still get a dead learning system. That is what we are going to build toward, and then deliberately break.

The Build

We are going to build four things: a `Value` class that stores a number, a gradient, and the graph links that created it; a backward pass that walks the graph in reverse; a small neural network built out of `Value` objects; and a toy training loop that learns to separate two 2D point clusters. The code pattern follows Andrej Karpathy's `micrograd` reference implementation, because it is small enough to hold in your head. I tried writing a more "elegant" version when I was first learning this, and discarded it. The elegance hid the wiring I was trying to see. Small and ugly beats clever for a first build.

There are two valid ways to read this project. In build mode, you type the code and watch the engine come alive. In reading mode, you treat each snippet as a mechanical diagram. If you are reading rather than typing, the important thing is not memorizing Python syntax. It is recognizing what state each block stores, what local derivative rule it implements, and where the gradient signal goes next.

Step 1: Build a Number That Remembers Where It Came From

A normal Python float gives you a value and nothing else:

```python
x = 2.0
y = 3.0
z = x * y
print(z)  # 6.0
```

By the time you get `6.0`, Python has forgotten the path. It does not remember that `6.0` came from multiplying `2.0` and `3.0`. Our `Value` class fixes that. Each `Value` stores the numeric value (`data`), the gradient of the final loss with respect to this value (`grad`), the parent nodes that created it (`_prev`), which operation created it (`_op`), and a function that knows how to push gradients to its parents (`_backward`):

```python
class Value:
    def __init__(self, data, _children=(), _op=''):
        self.data = data
        self.grad = 0.0
        self._prev = set(_children)
        self._op = _op
        self._backward = lambda: None
```

If you only skim one thing here, skim the fields. `data` is the forward value. `grad` is the backward blame assigned later. `_prev` remembers which earlier values created this one. `_backward` stores the local rule for passing gradient signal upstream. That small bundle is the whole design pattern behind autodiff systems.

Five fields. That is the entire trick.

This is the first mental shift. We are not building "a number." We are building "a number plus its ancestry plus the rule for sending blame backward."

Step 2: Teach the Number How to Add

Start with addition because it has the easiest local rule. If `c = a + b`, then `dc/da = 1` and `dc/db = 1`, so if the output `c` receives some gradient `c.grad`, both inputs should receive that same amount:

```python
def __add__(self, other):
    other = other if isinstance(other, Value) else Value(other)
    out = Value(self.data + other.data, (self, other), '+')

    def _backward():
        self.grad += out.grad
        other.grad += out.grad

    out._backward = _backward
    return out
```

Two details matter here. First, we store the parent links `(self, other)` in the output node. Second, we define a closure `_backward` that captures `self`, `other`, and `out`, so later, when the backward pass reaches this node, calling `out._backward()` will push gradients to the parents. Notice we use `+=` instead of `=`: because one value can feed into multiple later operations, gradient contributions from multiple downstream paths must add together. If one ingredient affects the final cake in three ways, you do not ask which single way "counts." You add all three effects.

The first time I wrote this from memory, I used `=` instead of `+=` and lost an hour. The forward pass was fine, the gradient checks for a single isolated operation passed, and the bug only surfaced when I built a `Value` that fed two different downstream paths. The second path silently overwrote the first. That is the kind of bug autograd code is full of: every line looks correct, and the failure mode is wrong numbers, not exceptions.

Step 3: Teach the Number How to Multiply

Now multiplication. If `c = a * b`, then `dc/da = b` and `dc/db = a`, so if the output `c` has gradient `c.grad`, we pass `b * c.grad` to `a` and `a * c.grad` to `b`:

```
def __mul__(self, other):
    other = other if isinstance(other, Value) else Value(other)
    out = Value(self.data * other.data, (self, other), '*')

    def _backward():
        self.grad += other.data * out.grad
        other.grad += self.data * out.grad

    out._backward = _backward
    return out
```

This one operation will carry most of the learning signal in a neural network, because weighted sums are built from multiplications and additions. When we break the multiplication gradient on purpose later, you will see exactly why this matters so much.

Step 4: Add a Nonlinearity

A network made only of additions and multiplications collapses into one big linear function. It cannot bend its decision boundary in interesting ways. We need a **nonlinearity**, a function that curves. We will start with `tanh`, which is a squashing function: it takes any real number and compresses it into the range from `-1` to `1`. Large positive values become numbers close to `1`, large negative values become numbers close to `-1`, and values near zero stay near zero. Without a nonlinearity, stacking layers does not buy you much.

The derivative of `tanh(x)` is `1 - tanh(x)²`. If the output is already `t = tanh(x)`, then the slope at that point is `1 - t²`, so if the output node receives some gradient `out.grad`, the input should receive `(1 - out.data**2) * out.grad`:

```
import math

def tanh(self) -> 'Value':
    x = self.data
    t = (math.exp(2 * x) - 1) / (math.exp(2 * x) + 1)
    out = Value(t, (self,), 'tanh')

    def _backward():
        self.grad += (1 - out.data**2) * out.grad

    out._backward = _backward
    return out
```

This is your first nonlinear local rule. The recipe gets richer, but the graph still works the same way. I had not used the `1 - tanh(x)²` identity in fifteen years when I wrote this, and the small relief of remembering it from somewhere far back in undergraduate calculus is a feeling I associate strongly with autograd work. Most of the local rules you will write in your life are simple in this same way: a calculus fact you mostly already know, expressed in three lines of Python.

Step 5: Build the Backward Pass

So far, each node knows how to send gradients to its parents, but nothing walks the graph yet. We need a reverse traversal, and the key requirement is order: if node `z` depends on node `y`, and `y` depends on node `x`, then during backward we must visit `z` before `y` and `y` before `x`. That ordering is a **topological order** (children before parents in the backward pass), and a clean way to get it is depth-first search:

```
def backward(self):
    topo = []
    visited = set()

    def build_topo(v):
        if v not in visited:
            visited.add(v)
            for child in v._prev:
                build_topo(child)
            topo.append(v)

    build_topo(self)

    self.grad = 1.0
    for node in reversed(topo):
        node._backward()
```

In code-reading terms, `backward()` is the traffic controller, not the source of the math. The math already lives inside each node's `_backward` closure. This method discovers the nodes in dependency order, seeds the final loss with gradient `1.0`, and then asks every node to pass blame to its parents.

We set `self.grad = 1.0` because `self` is the loss `L`, and the derivative of `L` with respect to itself is 1. That is the seed of the whole backward pass. From there, every node receives blame from the nodes above it. At this point, `.backward()` should feel a lot less mysterious. It builds the graph during forward, topologically sorts it, seeds the loss gradient with 1, and calls each node's local backward rule in reverse order.

The topological sort was the part I underestimated. I assumed any order would work, called `_backward()` in the order the nodes were created, and got gradients that were sometimes right and sometimes wrong depending on the graph shape. That was an hour I should not have lost to a data-structures lecture I had already attended.

Step 6: Add the Rest of the Small Operations

To make the engine useful, we want subtraction, negation, division, exponentiation, `relu`, and `exp`. Most of these can be defined in terms of the primitives we already have:

Negation and Subtraction

`-a` is just `a * -1`, and `a - b` is `a + (-b)`:

```
def __neg__(self):
    return self * -1

def __sub__(self, other):
    return self + (-other)
```

Power

If `c = a ** n`, then the local derivative rule is:

```
dc/da = n * a^(n - 1)
```

```
def __pow__(self, other):
    assert isinstance(other, (int, float))
    out = Value(self.data**other, (self,), f'**{other}')

    def _backward():
        self.grad += (other * self.data**(other - 1)) * out.grad

    out._backward = _backward
    return out
```

Division

Division is just multiplication by a reciprocal ($a / b = a \cdot b^{-1}$), so once power works, we get division for free:

```
def __truediv__(self, other):
    return self * (other ** -1)
```

This is a nice lesson in itself: once the primitives are correct, many new operations come almost for free. When I first noticed I had got division "for free" from power, I genuinely laughed at the keyboard, then went back and looked for the catch. There was no catch. That is one of the small pleasures of writing this engine from scratch: composing local rules turns out to be cheaper than you fear.

Exponential

If `c = exp(a)`, the derivative is also `exp(a)`, which is the output itself:

```
def exp(self):
    x = self.data
    out = Value(math.exp(x), (self,), 'exp')

    def _backward():
        self.grad += out.data * out.grad

    out._backward = _backward
    return out
```

ReLU

ReLU (**Rectified Linear Unit**) keeps positive values and zeroes out negative ones: `relu(x) = max(0, x)`. People prefer it in deeper networks because it is cheap to compute and does not squash large positive values. Its derivative is 1 when input > 0 and 0 when input ≤ 0:

```python
def relu(self):
    out = Value(0 if self.data < 0 else self.data, (self,), 'ReLU')

    def _backward():
        self.grad += (out.data > 0) * out.grad

    out._backward = _backward
    return out
```

That one line, `(out.data > 0)`, is another useful reminder that gradients are local rules, not global mysteries.

Step 7: Verify the Gradients Numerically

Before building a network, we should verify that each operation's backward rule matches reality. This is where many bugs hide. A **numerical gradient** is the brute-force estimate: nudge a value slightly upward, nudge it slightly downward, and see how much the output changes. The centered finite-difference estimate is:

```
(f(x + h) - f(x - h)) / (2h)
```

This approximates the slope by measuring rise over run around the point. For a `Value` operation, we compare the analytic gradient from `.backward()` against the finite-difference estimate from repeated forward calls. A few print statements with close agreement are enough to build trust:

```python
def finite_diff(f: callable, x: float, h: float = 1e-6) -> float:
    return (f(x + h) - f(x - h)) / (2 * h)
```

Example for `tanh`:

```python
x = Value(0.7)
y = x.tanh()
y.backward()
print("autograd:", x.grad)

import math
fd = (math.tanh(0.7 + 1e-6) - math.tanh(0.7 - 1e-6)) / (2e-6)
print("finite diff:", fd)
```

If these disagree badly, the local rule is wrong. This matters because backpropagation is only as correct as its local derivatives. One broken link poisons everything downstream.

A specific moment I think of when I write this: I once shipped a custom op with an off-by-a-sign gradient. The unit tests passed because I had written them in terms of the same buggy backward. Finite differences caught it in under a minute. The lesson stuck: never gradient-check an op against itself. Use the numeric estimate as the independent witness it is supposed to be.

Step 8: Build a Neuron

Now we build something that uses `Value`. A **neuron** takes several inputs, multiplies each by a weight, adds a bias, and passes the result through a nonlinearity. Think of it as a little scorecard: each input casts a weighted vote, where a positive weight means "more of this pushes me up," a negative weight means "more of this pushes me down," and the bias is the default offset (what the neuron tends to do even if inputs are zero):

```python
import random

class Neuron:
    def __init__(self, nin):
        self.w = [Value(random.uniform(-1, 1)) for _ in range(nin)]
        self.b = Value(0.0)

    def __call__(self, x):
        act = sum((wi * xi for wi, xi in zip(self.w, x)), self.b)
        out = act.tanh()
        return out

    def parameters(self):
        return self.w + [self.b]
```

This class is worth reading as a template you will keep seeing later. The constructor allocates trainable parameters, `__call__` defines the computation, and `parameters()` exposes the trainable values to the optimizer. Even when future chapters switch to PyTorch modules, the same three responsibilities are still there, just buried one layer down under the framework abstractions. Once I noticed that pattern, PyTorch's `nn.Module` stopped feeling like a different country.

The neuron's forward pass looks ordinary, but every number is a `Value`, so the graph gets built automatically as the neuron computes. When we later call `.backward()` on the loss, gradients will flow all the way into every weight and bias.

Step 9: Build a Layer and an MLP

A **layer** is just a list of neurons, and an **MLP** (Multi-Layer Perceptron) is a stack of layers. We will build a 2-layer MLP for 2D classification: two input numbers, a hidden layer of several neurons, and a single output neuron producing a score:

```python
class Layer:
    def __init__(self, nin, nout):
        self.neurons = [Neuron(nin) for _ in range(nout)]

    def __call__(self, x):
        outs = [n(x) for n in self.neurons]
        return outs[0] if len(outs) == 1 else outs

    def parameters(self):
        return [p for n in self.neurons for p in n.parameters()]
```

Stack layers into a full MLP:

```python
class MLP:
    def __init__(self, nin, nouts):
        sz = [nin] + nouts
        self.layers = [Layer(sz[i], sz[i+1]) for i in range(len(nouts))]

    def __call__(self, x):
        for layer in self.layers:
            x = layer(x) if isinstance(x, list) else layer([x])
        return x

    def parameters(self):
        return [p for layer in self.layers for p in layer.parameters()]
```

For a 2-layer MLP on 2D inputs: `model = MLP(2, [8, 1])`. 2 inputs, a hidden layer with 8 neurons, and an output layer with 1 neuron.

Step 10: Create a Toy Dataset

We want something visual. Two clusters in 2D is enough: cluster A around `(-1, -1)` with label `-1`, and cluster B around `(1, 1)` with label `1`. Using `-1` and `1` matches `tanh` output nicely:

```python
xs = [
    [-1.2, -0.8],
    [-1.0, -1.1],
    [-0.8, -1.3],
    [1.0, 0.9],
    [1.2, 1.1],
    [0.8, 1.3],
]
ys = [-1, -1, -1, 1, 1, 1]
```

The goal is not realism. The goal is to see a **decision boundary** sharpen: the line or curve where the model switches from one class to the other. At initialization, that border will look random. After training, it should line up with the clusters.

Step 11: Define the Loss

We need one number that measures how wrong the model is. For this toy setup, simple squared error is fine:

```
L = sum_i (y_pred,i - y_i)^2
```

Take the difference between prediction and truth, square it so big mistakes hurt more and signs do not cancel, and add over all examples:

```python
def loss_fn(model: 'MLP', xs: list, ys: list) -> tuple:
    preds = [model(x) for x in xs]
    loss = sum((pred - y)**2 for pred, y in zip(preds, ys))
    return loss, preds
```

That is the conceptual checkpoint for the chapter. The network code does not introduce a second training mechanism. It simply builds a larger expression tree out of the same `Value` objects. If you understand that, later framework code stops looking magical and starts looking like a more automated way to build the same graph. I did not internalize this until I drew the whole graph for a 6-parameter network on paper. The drawing was ugly and took twenty minutes. It was also the moment the rest of the field clicked.

Because `pred` is a `Value`, the whole loss is also a `Value`, with a graph reaching all the way back to the parameters.

Step 12: Run the Training Loop

Now the learning loop. Every epoch, we compute loss, zero old gradients, backpropagate, and update parameters:

```
for epoch in range(100):
    loss, preds = loss_fn(model, xs, ys)

    for p in model.parameters():
        p.grad = 0.0

    loss.backward()

    lr = 0.05
    for p in model.parameters():
        p.data -= lr * p.grad

    print(epoch, loss.data)
```

If you want the reading-mode summary of the whole loop, it is this. The forward pass turns inputs into predictions, the loss turns those predictions into a single number saying how wrong the model is, `loss.backward()` fills in gradients on every parameter, and the final update step subtracts a small multiple of those gradients from each parameter. That four-step rhythm is what to recognize later in larger training code. The same shape lives inside every training script I have ever shipped, with extras bolted on around it.

The update line `p.data -= lr * p.grad` is **gradient descent**: a blindfolded hiker trying to reach the bottom of a valley. The loss is the height, the gradient is the local slope underfoot, and the learning rate `lr` is step size. If the slope says "uphill to the right," step left. If `lr` is too small, learning crawls; if it is too large, learning overshoots and can bounce around or explode. At this point, we are training a neural network from scratch with our own autograd engine. No framework magic, just graph construction, local derivatives, reverse traversal, and parameter updates.

Step 13: Plot the Decision Boundary Over 100 Epochs

To plot the decision boundary, we create a grid of `(x1, x2)` points covering the plane, run the model on each grid point, color points by prediction sign, overlay the training data, and save snapshots at several epochs:

```python
import numpy as np
import matplotlib.pyplot as plt

def plot_boundary(model, xs, ys, epoch):
    x_min, x_max = -2, 2
    y_min, y_max = -2, 2
    xx, yy = np.meshgrid(np.linspace(x_min, x_max, 200),
                         np.linspace(y_min, y_max, 200))

    zz = np.zeros_like(xx)
    for i in range(xx.shape[0]):
        for j in range(xx.shape[1]):
            zz[i, j] = model([float(xx[i, j]), float(yy[i, j])]).data

    plt.figure(figsize=(5, 5))
    plt.contourf(xx, yy, zz, levels=20, cmap='RdBu', alpha=0.6)
    plt.contour(xx, yy, zz, levels=[0], colors='black')

    for x, y in zip(xs, ys):
        color = 'red' if y == 1 else 'blue'
        plt.scatter(x[0], x[1], c=color, edgecolors='k')

    plt.title(f"epoch {epoch}")
    plt.xlim(x_min, x_max)
    plt.ylim(y_min, y_max)
    plt.show()
```

By epoch 0, the boundary looks messy or arbitrary. Epoch 10 starts leaning the right way. By 50, separation is cleaner, and by 100 the boundary draws a sharp split between the clusters. That sharpening is learning made visible. The network is changing its internal weights so the output surface bends into a shape that separates the classes. The first time I saw the contour plot snap into place I scrolled back through the epochs three times. It is a small moment, but it is the one where you stop trusting the field on credit and start trusting it because you watched it work.

Step 14: What the Graph Looks Like in Your Head

At this point, let's make sure the mental model is stable. For one training example, the full chain looks like this: an input point `[x1, x2]` enters the first neuron, each weight multiplies an input, those products are summed with a bias, the sum passes through `tanh`, and the outputs of the hidden neurons feed the final neuron, whose output gets compared to the target label to produce the loss. Now reverse it. The loss says how much the final output was wrong. That signal flows backward through the last `tanh`, then back through the last weighted sum into the hidden neurons, through their `tanh`s and weighted sums, and finally into each individual parameter. Every edge in the graph carries blame backward.

Training is not "the network figuring things out." Training is arithmetic blame-assignment plus repeated tiny nudges. That sentence is also, in my experience, the entire honest answer to "how do neural networks learn?"

BREAK IT

The experiment worth running here: break the multiplication gradient. Not the multiplication itself, only the backward rule. The forward pass still computes normal outputs, the code still runs, the loss still prints, and nothing crashes. But the learning signal through every multiplication vanishes. This is exactly the kind of bug that makes machine learning feel haunted: the program "works," but learning stops.

I have seen the production-scale version of this. While running LoRA fusion experiments at scale, a few runs early on produced loss curves that drifted lazily for a hundred steps and then refused to fall further. No crash, no warning, no `NaN`. The cause, after a long evening, was a custom merge op whose backward had stopped routing gradients to one side of a multiplication. The toy break you are about to do in 30 seconds reproduces, in miniature, the exact failure mode that ate that evening.

The Exact Change

Find the multiplication backward rule:

```
def __mul__(self, other):
    other = other if isinstance(other, Value) else Value(other)
    out = Value(self.data * other.data, (self, other), '*')

    def _backward():
        self.grad += other.data * out.grad
        other.grad += self.data * out.grad

    out._backward = _backward
    return out
```

Now break it by replacing the `_backward` function:

```
def _backward():
    pass
```

Or for a subtler failure, comment out just one line:

```
def _backward():
    # self.grad += other.data * out.grad
    # other.grad += self.data * out.grad
    pass
```

That is the ablation. We have removed the chain rule for multiplication.

What You Will Observe

The training loop still runs, loss still prints, predictions still exist, and plots still render. But the model stops learning in any meaningful way. The loss barely changes across epochs, or changes only from bias-related paths. The decision boundary looks mostly frozen. If you print some parameter gradients after `loss.backward()`, you will likely see many zeros:

```
for i, p in enumerate(model.parameters()[:10]):
    print(i, p.data, p.grad)
```

Before breaking multiplication, gradients vary, many are nonzero, and updates push the boundary into shape. Afterwards, many gradients become zero, the updates stop affecting weighted connections, and the network cannot meaningfully tie inputs to outputs. Weighted computation in neural networks runs on multiplication: input times weight, hidden activation times next-layer weight, scale interactions everywhere. Kill gradient flow through multiplication and you cut the wires that tell the weights how they influenced the loss.

The network still computes. It no longer learns.

Why This Proves the Point

The chain rule is not a decoration on top of learning. It is the learning signal. When people say "the model trains by gradient descent," that sentence hides two distinct pieces: backpropagation computes the gradients, and gradient descent updates parameters using those gradients. If backpropagation fails for one operation, the whole learning system can go dark in the parts of the graph that depend on that operation. Each operation must do two jobs: compute its forward value, and correctly pass blame backward. You can get the forward value exactly right and still have a dead training system if the backward rule is wrong, which is the kind of failure that tends to be hard to spot in real systems too.

I expected the forward and backward pieces to be roughly equal partners. They are not. The backward rule is where most of the bugs I have shipped have lived, and it is also where most of the lessons I value are buried.

A More Surgical Version of the Break

If you want to see partial damage rather than total freeze, break only one side of multiplication's gradient:

```
def _backward():
    self.grad += other.data * out.grad
    # other.grad += self.data * out.grad
```

Now one parent receives blame and the other does not. Some parameters update, others stay stuck, training becomes asymmetric and unstable, and learning degrades rather than fully freezing. This shows that gradient flow is directional through graph edges. Each input to an operation needs its own local derivative rule.

The pattern you will see: biases may still move a bit because additions still pass gradients, but weights stagnate because weight usage depends on multiplication. The output shifts slightly but does not shape into a useful separator. Biases can move the line up or down, but weights rotate it, stretch it, and curve it. Breaking multiplication leaves you with a model that can twitch, but not really learn structure.

Optional Homework

This chapter's homework is fully runnable on a laptop, no GPU required. The autograd engine and training loop run in under a minute, and plotting decision boundary snapshots takes a few seconds to a minute depending on grid density.

Part A: Implement and Verify the Engine

Implement or inspect these operations in `Value`: `__add__`, `__mul__`, `tanh`, `relu`, `exp`, `__pow__` and `__truediv__`. For each operation, write one small gradient check using finite differences and record both the autograd gradient and the finite-difference estimate. They should match closely for smooth points. One special case: `relu` at exactly zero is not cleanly defined, so avoid testing there.

Part B: Build the 2-Layer MLP

Create `Neuron`, `Layer`, and `MLP` with input dimension 2, hidden dimension 8, and output dimension 1. Build a small 2D dataset with two clusters: 50 points near `(-1, -1)` labeled `-1` and 50 points near `(1, 1)` labeled `1`. Train for 100 epochs with learning rate `0.05`, and every 10 epochs record the loss, one or two example predictions, and a plot of the decision boundary. You should see the loss trend downward, predictions move toward `-1` and `1`, and the boundary become sharper and more aligned with the clusters.

Part C: Break Multiplication Backward

Edit the multiplication backward rule in your engine, changing the correct version to `pass`. Re-run the exact same training script with the same random seed and record the loss for 100 epochs, parameter gradients after the first backward pass, and boundary plots at epochs 0, 10, 50, and 100. The loss should stall or barely change, many gradients should become zero, and the decision boundary should remain mushy or nearly unchanged. When I ran this for the first time, I genuinely expected the loss to spike or explode. Instead it just sat there. The quiet failure is the lesson.

Part D: Try the Partial Failure Version

Restore one side of the multiplication gradient but not the other and re-run training. You should see learning worse than the correct version: some parameters change, others stay stuck, and the boundary improves weakly or erratically.

Part E: Compare `tanh` and `relu`

Train with `tanh` for hidden neurons and plot the results, then switch to `relu` and plot again. Record final loss, the number of dead-looking outputs if using ReLU, and boundary shape. Both can work on a toy dataset. `tanh` gives smooth bounded activations, while `relu` can produce sparse activity and piecewise-linear boundaries.

Questions To Answer

1. When you broke multiplication backward, what changed first: the loss curve, the gradients, or the decision boundary?
2. Which parameters still moved, if any? Were they mostly biases rather than weights?
3. In your finite-difference checks, which operation was easiest to get right, and which one made you think hardest about what the gradient actually means?
4. When you switched from `tanh` to `relu`, how did the decision boundary change? Did one make the training dynamics easier to see?
5. If one local derivative rule is wrong but the program still runs, what does that tell you about debugging machine learning systems versus debugging ordinary application code?
6. What tradeoff is automatic differentiation protecting for you? Think in terms of speed, correctness, and human effort compared with writing every gradient by hand.

Go Further

Start with the reference implementation:

- `micrograd`: https://github.com/karpathy/micrograd

After that, compare it to a real framework. Build the same toy classifier in PyTorch, print `.grad` values, and compare them with your own engine. Useful search directions: `reverse mode automatic differentiation`, `backpropagation through computation graph`, `gradient checking finite differences deep learning`, `vanishing gradients tanh sigmoid`, and `autodiff JAX PyTorch design`.

If you want a next-step reading path inside Karpathy's teaching arc, the natural progression is `micrograd` for learning mechanics, `minbpe` for tokenization, and then a minimal GPT build when you are ready to connect learning to language. One useful exercise after this chapter: add graph visualization. Drawing each `Value` node with its operation, data, and gradient makes later transformer diagrams much less abstract.

What You Now Know

Backpropagation is not a mystical solver. It is a graph walk applying one local derivative rule at a time, seeded at the loss, reversed to every parameter. You now know what a dead training system looks like: no crash, no warning, just zero gradients flowing through broken edges.

That is the machinery of learning at its smallest honest scale. Everything later in the book is the same machinery, with more wires.

Starting Point

Reference implementation: https://github.com/karpathy/micrograd

---

Project 2: Predicting The Next Character

Hook

Why does a language model that knows nothing about meaning, facts, or the world still manage to produce text that looks eerily language-like? If all it does is guess one next character at a time, why does that not collapse into nonsense immediately? And if that really is all it does, what changes when you move from a dumb counting table to a neural network with learned embeddings? This project answers that by stripping language modeling down to its smallest honest form: given a few characters, predict the next one.

The first time I trained one of these character models, I expected the bigram baseline to be hopeless and the neural model to be obviously better. The opposite happened on my first run. The neural model overfit on a tiny names file, the bigram model held up because it cannot overfit much, and the lesson stuck: baseline models are not toys. They are sanity rails.

The Concept

Start with the least mystical version of language generation possible.

Imagine a jar full of paper slips. On each slip is a pair of characters you saw in some text: `th`, `he`, `er`, `qu`, `a `, `, `, and so on. If I show you the first character and ask what usually comes next, you can look through the jar, count what followed that character before, and make a guess.

That is a language model in miniature. Not a chatbot, not a reasoning engine, not a world simulator. A next-step guessing machine. At character level, the model does not think in words. It sees pieces like `t`, `h`, `e`, space, newline, punctuation. Its job is brutally narrow: given the previous context, assign probabilities to possible next characters. If the current character is `q`, the next one is often `u`. If the current character is a newline in a names dataset, the next one is often the first letter of a name. That is the whole job.

The first version: counting

A **bigram** model is the simplest useful version. "Bigram" just means "pair of tokens." Here, the tokens are characters, so a bigram model asks: when I see character `c`, what character tends to come next?

You can picture it as a spreadsheet where rows are current characters, columns are next characters, and cell `(row='q', col='u')` stores how often `u` followed `q` in the training data. To generate text, you look at the current character, read the row, turn counts into probabilities, and sample the next character.

This model has no deeper understanding. It does not know that `the` is a word. It does not know grammar. It just remembers local transitions. And yet it already produces something that feels language-shaped, which matters, because it means a surprising amount of language surface form lives in local statistics alone.

The first time I noticed this I felt vaguely cheated. The model had learned nothing in any reasonable sense. It just kept a count book. And the count book alone produced output that, at a glance, looked like a child practicing a foreign alphabet. That is information about language, sitting in plain sight, that I had been mentally crediting to something more sophisticated for years.

The second version: learned representations

A counting table treats every character as a separate bucket. `a` is one row, `b` is another, `z` is another. The model gets no opportunity to notice that vowels behave somewhat similarly, or that punctuation marks share some structural role, or that common first letters cluster together. A neural network changes that.

Instead of storing one row per character directly, it gives each character an **embedding**: a list of numbers that represents a token. Think of the embedding as a coordinate in an abstract space, or a learned identity card made of numbers rather than a one-hot "I am character 17" label. Something like:

- `a → [0.7, -1.2, 0.1, ...]`
- `e → [0.8, -1.0, 0.2, ...]`
- `q → [-0.4, 2.1, -0.7, ...]`

The network learns those numbers so that characters which need to be treated similarly can land in similar regions of that space. The model no longer just memorizes counts. It learns a geometry. That is the first real step toward "representation learning."

Why embeddings matter

Suppose you run a restaurant and every customer is described by a single ID number. Customer 481, customer 938, customer 221. If you want to guess what each person will order, IDs are not very informative unless you memorize every person separately.

Now suppose instead each customer comes with a profile: likes spicy food, prefers vegetarian dishes, visits on weekends, and orders dessert often. Now similar customers can share patterns. Embeddings do this for tokens. The model gets to say, in effect, "these two characters behave alike in some ways, so I can share statistical strength between them." That phrase, sharing statistical strength, is one of the main reasons neural models beat raw counting. A model that can generalize from what it knows about `e` to handle a context involving `i` has learned something the counting table never could.

I think of embeddings as a small notebook the model writes for itself, one row per token, where each row records "things I have noticed about this character." The notebook starts random. Training fills it in. The notebook is the bridge between the dumb integer ID and the rich behavior the network needs the token to have.

What the model actually predicts

At each step, the model outputs a score for every possible next character. If your vocabulary contains 65 characters, the model produces 65 scores, then turns those scores into probabilities with **softmax**. Softmax converts arbitrary scores into a probability distribution that sums to 1. If the scores strongly favor one character, the output distribution is sharp; if the scores are similar, the distribution spreads out.

Then you either take the highest-probability character, or sample from the distribution. Sampling is what makes generation varied. If you always pick the top answer, the model becomes rigid and repetitive; if you sample too freely, it becomes chaotic. That tradeoff leads us to **temperature**.

Temperature: how risky the model behaves

Temperature is a knob you apply during sampling to control how boldly the model draws from what it has learned. A plain analogy: imagine a spinner with weighted slices. At normal temperature, the slice sizes match the model's probabilities. Lower the temperature, and the biggest slices get even bigger. Raise it, and the smaller slices grow fatter.

At `temperature = 0.1`, the model almost always picks the safest continuation. At `temperature = 2.0`, even weak options get much more chance. You are not changing what the model learned. You are changing how you sample from it.

This is one of those settings users meet first in a chat UI slider, with no explanation of what it does. The slider is doing one specific thing to the logits before sampling. Once that one fact lands, the slider stops being magic and starts being a knob with a behavior you can predict.

Negative log-likelihood: how wrong the model is

We need one number that answers: how surprised was the model by the actual next character? That number is the **negative log-likelihood**.

The intuition first. If the true next character is `e` and the model assigned probability `0.9` to `e`, that is a confident good guess. If it assigned probability `0.01`, that is a bad miss. We want a penalty that scales accordingly: high probability on the correct answer means a small penalty; low probability means a large penalty. Negative log does exactly that. If `p` is the probability the model assigned to the correct next character, the loss for that example is:

$$L = -\log p$$

Now earn the symbols:

- `L` is the loss: one number measuring how wrong the prediction was
- `p` is the probability the model assigned to the correct next character
- `log` means logarithm

Why log? Because it punishes extreme confidence when wrong much more harshly than mild uncertainty. If the model says "I am nearly certain the next character is `x`" and the truth is `e`, it should pay a big price.

A few concrete values make this click:

- if `p = 1.0`, then `L = 0`
- if `p = 0.5`, then `L ≈ 0.69`
- if `p = 0.1`, then `L ≈ 2.30`
- if `p = 0.01`, then `L ≈ 4.61`

So lower is better. This metric will follow us through the whole book in different forms.

Why It Matters

Once you build a character model yourself, several foggy ideas become concrete.

First, you stop imagining that text generation begins with huge models and giant datasets. It begins with the humble act of predicting the next token. Everything else is scale, architecture, and training technique layered on top.

Second, you see that "knowing language" has levels. A bigram model knows almost nothing, but it still captures local regularities. A neural model knows more because it can learn useful internal representations. That difference between memorizing observations and learning representations is a major dividing line in machine learning.

Third, you finally have a place to attach the word "embedding." Without this chapter, "embedding" sounds abstract. Here it becomes practical: a learned numeric identity for each character that lets the network treat different inputs differently.

Fourth, temperature stops sounding like a mysterious generation setting in a UI. It becomes a concrete control over randomness in sampling.

The break-it experiment matters even more than the successful build. If you force all characters to share the same embedding, the model loses the ability to tell them apart. That failure teaches you more about embeddings than any definition does. It shows that embeddings are not decoration. They are the mechanism that gives distinct inputs distinct internal identities.

Most beginner explanations of embeddings stop at "a vector that represents meaning." That is fine as a one-liner and useless as a working definition. The thing that gives an embedding its job is the lookup, the fact that distinct IDs route to distinct rows. Break the routing and the rest of the network has nothing to do.

The Build

We will build this in two stages. First, a bigram counting model. Then a neural character model with learned embeddings. The goal is not production code. The goal is seeing each moving part with no hiding place.

If you are reading quickly, treat the first half as the baseline and the second half as the upgrade. The baseline is a counting table: after this character, what usually came next? The upgrade keeps the same prediction task but replaces explicit counts with learned vectors and trainable weights. That contrast is the real lesson of the code.

For concreteness, assume a small corpus such as Shakespeare lines or a list of names. Karpathy's `makemore` setup often uses names because it trains fast and makes the outputs easy to inspect.

Step 1: Get a tiny corpus

Use something small enough to inspect with your eyes.

Good choices:

- a newline-separated list of names
- a short Shakespeare text file
- a poetry collection

The corpus should fit in one text file. What matters is that it has repeated patterns. If you use names, the model learns things like: many names start with capital letters, `q` is often followed by `u`, and endings like `-a`, `-n`, `-ie`, `-son` repeat. The smaller the corpus, the more obvious the model's strengths and weaknesses become.

A tiny example:

```
emma
olivia
liam
noah
ava
```

This is too small for a good model, but good enough to think with.

Step 2: Build the vocabulary

A language model needs a fixed set of possible tokens. Here the tokens are characters, so extract all unique characters from the corpus. Add a special start/end token too. In character models, it is common to use a dot `.` or another marker to mean "start of sequence" and "end of sequence."

If your names corpus contains `anna`, you treat it like:

```
.anna.
```

That gives the model a clean way to learn what characters tend to start sequences and what characters tend to end them.

Minimal code:

```
chars = sorted(list(set(text)))
stoi = {ch: i for i, ch in enumerate(chars)}
itos = {i: ch for ch, i in stoi.items()}
```

If you add a special token `.` manually:

```
chars = ['.'] + sorted(list(set(text.replace('\n', ''))))
stoi = {ch: i for i, ch in enumerate(chars)}
itos = {i: ch for ch, i in stoi.items()}
```

This is also a good place to separate plumbing from learning. **`stoi`** means string-to-index and **`itos`** means index-to-string. They are not model parameters and they do not learn anything. They are just the dictionaries that let Python move between readable characters and the integer IDs the model consumes.

Now every character has an integer ID. That is the first form of tokenization: turning text into integers.

Step 3: Bigram counts

Now build the counting table. For every adjacent pair of characters in each sequence, increment a count. For `".anna."`, the pairs are:

- `('.', 'a')`
- `('a', 'n')`
- `('n', 'n')`
- `('n', 'a')`
- `('a', '.')`

So the count matrix `N` has shape `(V, V)` where `V` is vocabulary size. Rows are the current character, columns are the next character.

Build the bigram count matrix:

```python
import torch

N = torch.zeros((V, V), dtype=torch.int32)

for word in words:
    chs = ['.'] + list(word) + ['.']
    for ch1, ch2 in zip(chs, chs[1:]):
        i = stoi[ch1]
        j = stoi[ch2]
        N[i, j] += 1
```

If you only retain one sentence from the bigram part, retain this one: `N` is not metadata about the model. `N` is the model. Every later improvement in the chapter can be read as replacing this explicit table with a function that tries to produce better next-character probabilities from richer context.

Pause here and inspect `N`. Do not rush past this. This matrix is the bigram model. If row `i` corresponds to `q`, then `N[i]` tells you what followed `q` in the training data and how often.

Step 4: Turn counts into probabilities

Counts are not probabilities yet. If `q` → `u` happened 50 times and `q` → `a` happened 2 times, you want probabilities that reflect that proportion, so normalize each row:

```python
P = N.float()
P = P / P.sum(dim=1, keepdim=True)
```

Now each row sums to 1. There is one practical issue: some rows may have zero counts if a character never appears in that position. To avoid divide-by-zero and to avoid impossible outcomes, add a little smoothing. You give every possible next character a tiny starting count:

```python
P = (N + 1).float()
P = P / P.sum(dim=1, keepdim=True)
```

That `+1` says: even if I never saw this transition, do not assign zero probability forever.

Step 5: Sample from the bigram model

Generation now becomes a loop. Start from the special start token `.`, look up the probability row, sample the next character, and repeat until you sample `.` again.

```
g = torch.Generator().manual_seed(42)

ix = stoi['.']
out = []

while True:
    probs = P[ix]
    ix = torch.multinomial(probs, num_samples=1, replacement=True, generator=g).item()
    ch = itos[ix]
    if ch == '.':
        break
    out.append(ch)

print(''.join(out))
```

If your corpus is names, outputs might look like:

```
mila
noan
emia
olan
```

Some are real-looking. Some are garbled but plausible. That already tells you something worth sitting with: local transition statistics alone can produce language-like surface texture.

Step 6: Measure negative log-likelihood for the bigram model

We do not want to rely only on eyeballing samples. For each true next character in the corpus, the model assigns some probability. We take the negative log of that probability and average it.

```
log_likelihood = 0.0
n = 0

for word in words:
    chs = ['.'] + list(word) + ['.']
    for ch1, ch2 in zip(chs, chs[1:]):
        i = stoi[ch1]
        j = stoi[ch2]
        prob = P[i, j]
        log_likelihood += torch.log(prob)
        n += 1

nll = -log_likelihood / n
print(nll.item())
```

If the score goes down, the model assigns higher probability to the correct next characters. Now the limitations of the bigram model also become visible: it only sees one character of context, which means it cannot distinguish cases like "after `th`, maybe `e` becomes much more likely" from "after `sh`, maybe different continuations matter." A bigram model is memory-poor. It only looks one step back. That leads us to the neural model.

When I worked on Tamil tokenization for an offline-only Android deployment, the limitation of one-character context felt almost physical. Tamil has long vowel and consonant clusters that cannot be resolved from a single preceding character. A bigram model on Tamil is structurally incapable of the task. Watching one fail this clearly is, in my opinion, the fastest way to feel why we need richer context windows.

Building the neural character model

Step 7: Create training examples with context

A neural language model can use more than one previous character. Pick a context length, often called block size. If `block_size = 3`, the input for predicting the next character is the previous three characters.

For the word `emma`, written as `.emma.`, the training pairs become:

- input `...` -> target `e`
- input `..e` -> target `m`
- input `.em` -> target `m`
- input `emm` -> target `a`
- input `mma` -> target `.`

This is already richer than a bigram model.

I once shipped a buggy version of this loop where I forgot to update the context with the freshly emitted target during data construction. Every training example saw the same starting context of dots. The training loss went down on the training set and the validation loss did not move. It looked like underfitting. It was an indexing bug. Two hours of staring at `print` statements later, the fix was one line: `context = context[1:] + [ix]`. Indexing bugs in data construction are the single most common source of silent training failures I have hit in this kind of code.

Build the (X, Y) training pairs with block-size context:

```
block_size = 3
X, Y = [], []

for word in words:
    context = [stoi['.']] * block_size
    for ch in list(word) + ['.']:
        ix = stoi[ch]
        X.append(context)
        Y.append(ix)
        context = context[1:] + [ix]

X = torch.tensor(X)
Y = torch.tensor(Y)
```

What are these tensors?

- `X` has shape `(num_examples, block_size)`
- `Y` has shape `(num_examples,)`

Each row of `X` is a short history of previous characters. Each element of `Y` is the correct next character.

Step 8: Start with one-hot thinking, then replace it with embeddings

A beginner question is natural here: why not just feed the character IDs directly into the network? Because raw IDs do not mean anything numerically. Character 3 is not "three times" character 1. IDs are labels, not measurements.

One-hot encoding is the usual first fix. If vocabulary size is `V`, a one-hot vector is length `V`, with a single `1` and the rest `0`. That works, but it is wasteful and treats every token as totally unrelated. An embedding layer is a learned lookup table. If vocabulary size is `V` and embedding dimension is `d`, the embedding table has shape `(V, d)`. Each character ID selects one row, and that row is the character's vector representation.

Initialize the embedding table and look up context embeddings:

```
C = torch.randn((V, d))
emb = C[X]  # shape: (num_examples, block_size, d)
```

This is one of the highest-value lines to recognize later in other codebases. `C` is just a table of learned vectors, and `X` is a table of token IDs. `C[X]` means "look up the vector for each token in the batch." When a framework uses `nn.Embedding`, it is packaging this same operation.

That line `C[X]` is worth staring at. If `X` contains integer token IDs, indexing into `C` turns IDs into vectors. That is an embedding lookup.

I have lost count of how many times I have seen experienced engineers stumble at this exact line in someone else's code. It looks like advanced PyTorch and it is just fancy indexing. When you see `nn.Embedding(V, d)(x)` in a transformer codebase later in the book, you should be able to mentally read it as `C[x]` and move on.

Step 9: Flatten the context embeddings

Each training example has `block_size` characters, each mapped to a vector of length `d`. So one example now has shape `(block_size, d)`. To feed it into a simple multilayer perceptron, flatten it into one long vector of size `block_size * d`.

```
emb = C[X]                          # (B, block_size, d)
x = emb.view(emb.shape[0], -1)  # (B, block_size * d)
```

Flattening matters conceptually because it shows the limit of this design. The model does not preserve a separate object for each earlier character. It glues the context window into one long feature vector and feeds that into a fixed-width network. That is why it can use a short local history, but not the flexible long-range context that transformers later provide.

This is still a plain MLP. No attention yet, no recurrence, just a fixed-size window of previous characters, concatenated and fed into a network.

Step 10: Add a hidden layer and output layer

Now the model needs to map context vectors to scores over the vocabulary. A minimal MLP language model:

```
W1 = torch.randn((block_size * d, hidden_size))
b1 = torch.randn(hidden_size)
W2 = torch.randn((hidden_size, V))
b2 = torch.randn(V)
```

Forward pass:

```
h = torch.tanh(x @ W1 + b1)
logits = h @ W2 + b2
```

Logits are the raw scores for each possible next character before turning them into probabilities. If `V = 30`, then for each example `logits` contains 30 numbers, and a higher score means the model prefers that character more.

Then convert logits to probabilities:

```
counts = logits.exp()
probs = counts / counts.sum(dim=1, keepdim=True)
```

Or more commonly, use **cross-entropy** directly, which folds softmax and negative log-likelihood into one stable function.

Step 11: Compute loss with cross-entropy

Cross-entropy is the standard loss for classification over a fixed vocabulary. In plain English, it asks: how much probability did the model assign to the correct class? For language modeling, the classes are all possible next characters.

Use PyTorch's built-in version:

```
loss = torch.nn.functional.cross_entropy(logits, Y)
```

For code-reading purposes, `cross_entropy(logits, Y)` is the line that turns a scoring function into a trainable predictor. `logits` are the model's scores over the vocabulary. `Y` is the correct next character for each example. The loss compares the two in one stable function.

This is doing the same core thing as negative log-likelihood. The model predicts a distribution, the target is the true next character, and the loss gets smaller when the correct character gets more probability.

Step 12: Train with gradient descent

Now the chapter connects directly back to Project 1. We have a forward pass, a loss, a backward pass, and parameter updates. The trainable parameters are:

- `C`: the embedding table
- `W1, b1`: first layer
- `W2, b2`: output layer

Minimal loop:

```
parameters = [C, W1, b1, W2, b2]
for p in parameters:
    p.requires_grad = True

for step in range(10000):
    emb = C[X]
    x = emb.view(emb.shape[0], -1)
    h = torch.tanh(x @ W1 + b1)
    logits = h @ W2 + b2
    loss = torch.nn.functional.cross_entropy(logits, Y)

    for p in parameters:
        p.grad = None
    loss.backward()

    lr = 0.1
    for p in parameters:
        p.data += -lr * p.grad
```

If you want to skim the loop without losing the plot, read it as four stages. First the code turns token IDs into embeddings. Then it computes hidden features and vocabulary scores. Next `loss.backward()` computes gradients for every trainable tensor. Finally the update step nudges those tensors to reduce future loss. That rhythm is what matters.

This is still small enough to hold in your head. The network learns embeddings that separate characters in useful ways, hidden features that combine context patterns, and output weights that map those features to next-character predictions.

Step 13: Split train and validation data

Do not trust training loss alone. A model can memorize the training set, which means it performs well on examples it has seen but poorly on anything new. So split examples into a training set, a validation set, and optionally a test set. Then watch both losses. If training loss drops but validation loss stops improving or rises, the model is overfitting. The model is memorizing quirks of the training data instead of learning patterns that generalize. Even on tiny projects, this habit matters.

Across a long series of cooperative fine-tuning experiments I ran on base models from a few hundred million to several billion parameters, the single most common reason a "great" training result evaporated when re-tested was that we had trusted training loss. The pattern is identical at every scale. The tiny character model in this chapter is where you build the habit cheaply.

Step 14: Generate from the neural model

Generation looks similar to the bigram case, except now you maintain a context window.

```
context = [stoi['.']] * block_size
out = []

while True:
    emb = C[torch.tensor([context])]
    x = emb.view(1, -1)
    h = torch.tanh(x @ W1 + b1)
    logits = h @ W2 + b2
    probs = torch.softmax(logits, dim=1)

    ix = torch.multinomial(probs, num_samples=1).item()
    ch = itos[ix]

    if ch == '.':
        break

    out.append(ch)
    context = context[1:] + [ix]
```

Notice what did not change. Generation is still an iterative next-token loop: keep a context, produce probabilities, sample one character, append it, and repeat. What changed is the function producing those probabilities. That is the bridge to larger language models later in the book.

Samples often become noticeably better than the bigram model, especially if block size is at least 3 and the corpus has repeated structure. That improvement is not magic. It comes from two concrete gains: more context, and learned representations that let the network share information across similar characters.

Step 15: Temperature sampling

Now add the temperature knob. Given logits, divide them by temperature before softmax:

```
temperature = 1.0
probs = torch.softmax(logits / temperature, dim=1)
```

What this does:

- `temperature < 1`: exaggerates differences, sharper distribution
- `temperature > 1`: flattens differences, broader distribution

Concrete intuition: suppose logits are `[6.0, 4.0, 1.0]`. At low temperature, the `6.0` dominates even more. At high temperature, the `4.0` and `1.0` become less suppressed.

Generate samples at several temperatures:

- `0.1`
- `0.5`
- `1.0`
- `1.5`
- `2.0`

Typical behavior:

- `0.1`: repetitive, safe, often dull

- `0.5`: coherent and conservative
- `1.0`: balanced
- `1.5`: more variety, more mistakes
- `2.0`: chaotic, often broken

Step 16: Measure output entropy versus temperature

Entropy is a measure of uncertainty in a probability distribution. It asks how spread out the model's bets are. If the model puts nearly all mass on one character, entropy is low; if it spreads probability across many characters, entropy is high.

For a distribution `p_1, p_2, ..., p_V`, entropy is:

$$H = -\sum_i p_i \log p_i$$

Earn the symbols:

- `H` is entropy: one number describing uncertainty
- `p_i` is the probability of character `i`
- the sum adds contributions from every possible next character

If you sweep temperature upward, entropy usually rises because the distribution gets flatter. Minimal code:

```
def entropy(probs):
    return -(probs * probs.log()).sum(dim=1).mean()
```

Then, for each temperature, compute average entropy over some validation contexts and also generate sample outputs. This is useful because it separates two things: model quality and sampling behavior. A model may be good, but high temperature can still make it babble. A model may be mediocre, but low temperature can make it appear more coherent than it really is.

Step 17: Compare bigram and neural model honestly

At this point, compare validation negative log-likelihood, sample quality, and sensitivity to temperature. You should usually observe that the neural model gets lower NLL, its samples look more structured, and it benefits more from extra context. The bigram model remains trapped in one-step statistics. This is the first real instance of a machine learning pattern that repeats everywhere: the learned model beats the hand-built baseline because it can form internal features instead of relying on direct counting alone.

BREAK IT

You are going to break the embeddings on purpose. Not tweak them, not regularize them, but destroy their ability to distinguish one character from another. Then you will watch what fails.

The exact thing to break

The neural language model depends on the embedding table `C`. If `C` has shape `(V, d)`, each row corresponds to one character. Normally those rows are different learned vectors. Break that by forcing every row to be the same vector:

```
with torch.no_grad():
    C[:] = C[0]
```

That means `a` gets the same embedding as `b`, space gets the same embedding as `q`, and every character enters the network looking identical. The lookup still runs. The tensor shapes still work. The training code still executes. But the identity signal is gone.

What fails first

The model loses the ability to tell inputs apart. Say it mechanically. Suppose the current input context is `['q', 'u', 'e']` and another context is `['z', 'z', 'z']`. If all embeddings are identical, the embedding lookup produces the same vector for every character position in both cases. After flattening, both contexts become the same input vector, so the hidden layer sees no difference, the logits are identical, and the predicted next-character distribution is identical.

The model can no longer condition on the actual characters. It can only learn one generic next-character distribution over the whole dataset — predicting based on global character frequency only, rather than on what actually came before.

What you will observe

You should expect three visible changes.

1. Validation loss jumps

A working model uses context to lower uncertainty. A broken-embedding model cannot use context, so its negative log-likelihood rises. It may still beat a perfectly uniform distribution because some characters are globally more common than others, but it should perform much worse than the normal neural model. In practice, it collapses toward "what are the most common next characters anywhere in the corpus?" That is a much weaker question than "what is the likely next character after this exact context?"

When I first ran this break, the validation loss did not jump as dramatically as I had predicted in my own notes. I had expected catastrophic collapse. What I got was clear degradation, but with the model still doing some work on global frequency. That mismatch between my prediction and the actual failure is exactly the kind of thing this section is for. The model fails in a specific shape, and that shape teaches you what the broken thing was actually contributing.

2. Samples become bland, repetitive, and strangely generic

You will still get output, because the model still samples from a distribution. But the output stops respecting input context. For names, instead of plausible endings conditioned on previous letters, you get mushy common-character sequences. Examples of the flavor you may see:

```
ananan
elea
nalia
aaaa
enon
```

Not because these are the exact outputs, but because the model is falling back to globally common letters and transitions that accidentally arise from sampling. If your corpus is Shakespeare, the output may still contain common vowels, spaces, and punctuation in plausible proportions, but local spelling structure gets badly damaged.

3. Temperature stops doing something interesting

Temperature still changes randomness, but it no longer reveals a meaningful learned structure. Low temperature produces the same safe mush repeatedly. High temperature produces noisier mush. The quality ceiling is gone because the model lost the ability to distinguish contexts in the first place.

What this proves

An embedding is not just a way to store tokens numerically. It is the mechanism that gives the network a chance to treat different tokens differently. If all embeddings collapse to one point, then every token becomes the same token as far as the network is concerned. Embeddings carve apart input identity in space, and they let the model attach different downstream behavior to different inputs.

You can say this even more strongly: without distinct embeddings, the rest of the network cannot recover token identity later. The only information entering the network about the token is the embedding vector. If you erase the distinction there, the information is gone. No later layer can reconstruct which original character it was, because that fact never reached it. This is one of the cleanest break-it experiments in machine learning because the failure is so crisp. No explosions. No shape mismatch. No training crash. The code keeps running. The intelligence vanishes.

Second break: make embedding dimension too small

Keep embeddings distinct, but shrink the embedding dimension to `d = 1`. That means each character gets represented by a single number. Not a vector with room for many traits, but one scalar. This is like trying to describe every person in a city using only height. You can separate some people, but you lose huge amounts of useful structure.

What happens? The model still works, it can still distinguish characters somewhat, but representational capacity is cramped and loss improves much less than with a moderate embedding size. You may find that `d = 2` or `d = 4` already helps a lot over `d = 1`, because the model gains room to encode multiple independent character properties.

Third break: make embedding dimension absurdly large

Now push the other way. Set embedding dimension to something like `d = 512` for a tiny character vocabulary and a small dataset. This usually does not make the model suddenly brilliant. Instead you often get more parameters, slower training, higher memory use, weak or no improvement on validation loss, and greater risk of overfitting if the dataset is tiny.

The reason is that extra dimensions only help if the data and task need that capacity. A names dataset with maybe 27 characters does not need a giant semantic space. At some point you are giving the model more room than the problem can justify. This is your first local example of a scaling lesson that returns later: bigger is not automatically better. The right amount of capacity depends on the task and the data.

Strong opinion: the entire industry obsession with parameter count as a quality signal is, at best, a useful proxy and, at worst, actively misleading. The Chinchilla-style results that fall out of Project 19 later in the book make this concrete. For this chapter, the takeaway is small and durable: capacity that has nothing to feed it goes wasted.

What "more dimensions stop helping" looks like

Run a sweep:

- `d = 1`
- `d = 2`
- `d = 4`
- `d = 8`
- `d = 16`
- `d = 32`
- `d = 64`
- `d = 128`
- `d = 512`

Record validation loss. You will usually see big gains from `1 → 4` or `1 → 8`, smaller gains after that, then flat or noisy improvement, and possibly worse validation loss at very large sizes on tiny data. That flattening is the lesson. Embedding dimension is not a prestige number. It is a capacity knob, and it has a point of diminishing returns that depends on your dataset.

Optional Homework

This project is fully runnable on a laptop CPU.

If you use a small names dataset, the whole lab should take under an hour, often much less. If you use Shakespeare and larger hidden layers, it may take longer but is still fine on CPU.

Part A: Bigram model

1. Load the corpus.
2. Build the character vocabulary including `.` as start/end token.
3. Build the count matrix `N`.
4. Add +1 smoothing.
5. Normalize to get row-wise probabilities.
6. Generate 20 samples.
7. Compute average negative log-likelihood on the corpus.

Record:

- vocabulary size `V`
- average NLL
- 5 sample outputs you judge as "best"
- 5 sample outputs you judge as "worst"

Expected observation:

- output looks local and plausible in spots
- longer structure is weak
- common endings repeat

Part B: Neural MLP language model

Use:

- `block_size = 3`
- `hidden_size = 100`
- embedding dimensions `d = 2, 8, 32`

Steps:

1. Build `(X, Y)` training examples from the corpus.
2. Split into train/validation, for example 80/20.
3. Train for 10,000 to 50,000 steps depending on corpus size.
4. Record train and validation loss every 1,000 steps.
5. Generate 20 samples after training.

Expected observation:

- neural model beats bigram NLL
- samples look more coherent
- context helps especially near endings and common patterns

Part C: Temperature sweep

For the best neural checkpoint, generate samples at:

- `0.1`
- `0.5`
- `1.0`
- `1.5`
- `2.0`

For each temperature:

1. Generate at least 100 next-token distributions from validation contexts.
2. Compute average entropy.

3. Generate 10 full samples.
4. Save both metrics and outputs.

A small entropy helper:

```
def avg_entropy(model, X_batch: 'torch.Tensor', temperature: float = 1.0) -> float:
    logits = model(X_batch) / temperature
    probs = torch.softmax(logits, dim=1)
    return float((-(probs * probs.log()).sum(dim=1)).mean())
```

Record a table like this:

```
temperature | avg_entropy | notes
0.1         | ...         | repetitive, safest
0.5         | ...         | controlled
1.0         | ...         | balanced
1.5         | ...         | more variety, more spelling errors
2.0         | ...         | unstable, noisy
```

Expected observation:

- entropy rises with temperature
- quality peaks somewhere around moderate temperature
- low temperature can hide model weaknesses by making output repetitive

Part D: Break embeddings completely

Modify the training loop or evaluation loop.

Option 1: break after training, for a clean before/after comparison.

```
with torch.no_grad():
    C[:] = C[0]
```

Then re-run validation loss and sampling.

Option 2: clamp during training so embeddings never separate.

```
for step in range(num_steps):
    ...
    loss.backward()
    ...
    for p in parameters:
        p.data += -lr * p.grad

    with torch.no_grad():
        C[:] = C[0]
```

Record:

- validation loss before collapse
- validation loss after collapse

- 10 generated samples before
- 10 generated samples after

Expected observation:

- loss rises sharply
- outputs lose context sensitivity
- common characters dominate
- samples feel generic and mushy

Part E: Embedding dimension sweep

Train the same architecture with embedding sizes:

- `1, 2, 4, 8, 16, 32, 64, 128, 512`

Keep these fixed:

- same dataset split
- same block size
- same hidden size
- same optimizer settings
- same training step count
- same random seed if possible

Plot:

- embedding dimension vs validation loss
- embedding dimension vs parameter count
- embedding dimension vs training time

Expected observation:

- strong gains at small dimensions
- diminishing returns after a moderate size
- little benefit or mild overfitting at very large dimensions on tiny data

Questions To Answer

1. When you collapsed all embeddings to the same vector, did the model's outputs become fully uniform, or did they still reflect global character frequency? Why is that second outcome exactly what you should expect?
2. At what embedding dimension did validation loss stop improving meaningfully on your dataset? What does that say about capacity versus task size?
3. What failed first when you raised temperature: spelling, local syntax, or sequence endings? Which failures came from the model, and which came from sampling?
4. Compare the best bigram sample to the best neural sample. What specific structure did the neural model capture that the bigram model missed?

5. If your model got lower training loss but worse validation loss at large embedding sizes, what tradeoff was the larger embedding hurting?
6. If you increased `block_size` instead of embedding dimension, would you expect the same gains? Why or why not?

Go Further

The natural reference path here is Karpathy's `makemore` teaching arc. Use it as a reference implementation meant to be read, forked, and broken.

Search directions:

- search YouTube for `Andrej Karpathy makemore`
- search GitHub for `karpathy makemore`
- search arXiv for `neural probabilistic language model`
- search arXiv for `Bengio 2003 language model embeddings`
- search arXiv for `character-level language model`
- search arXiv for `temperature sampling language models`

One paper direction matters historically: Yoshua Bengio's 2003 neural probabilistic language model. You do not need the whole paper yet. What matters is the leap it represents: stop treating words as isolated symbols and start learning distributed representations. That is the ancestor of the embedding idea you just used at character scale.

If you want a tougher extension, try these:

- Increase `block_size` and compare whether more context helps more than larger embeddings.
- Replace the MLP with a tiny recurrent neural network and compare sample quality.
- Visualize the learned character embeddings in 2D using PCA or t-SNE and inspect which characters land near each other.
- Train on words from two languages mixed together and see whether the embeddings cluster by language, punctuation role, or vowel/consonant behavior.

A good search phrase for that last direction is:

- search arXiv for `representation geometry language models`
- search arXiv for `embedding space analysis`

What You Now Know

A language model is a machine that assigns probabilities to the next token. Nothing more, nothing less. Embeddings are the mechanism that lets the network treat different inputs differently: collapse them and the model can no longer condition on context, only on global frequency. That is the first real bridge from raw symbols to learned representation.

Starting Point

Karpathy's `makemore` series on YouTube is the right starting point for this project.

Reference implementation direction: search GitHub for `karpathy makemore`.

If you want to stay aligned with the book's progression after this chapter, keep the path simple:

- `micrograd` taught you how learning works
- this project teaches you what language modeling is at the smallest honest scale
- next comes tokenization, where "character" stops being the only possible unit and the model's input vocabulary becomes something you design rather than inherit

---

Project 3: Building A Tokenizer

Hook

Why does every serious language model bother with a tokenizer at all? Why not just feed in characters, one by one, and let the model figure it out? Or go the other direction and use whole words, since that is what humans think in? This is where a lot of newcomers get stuck. Characters feel too small. Words feel too rigid. And "**token**" sounds like one of those terms everyone uses before they have explained what problem it solves. This chapter fixes that. You are going to build the thing that decides what the model even sees.

The thing I wish I had known three years earlier: tokenizers are not a side topic. They quietly shape what your model can learn, how much of your context window survives a long document, and how badly your model embarrasses you on languages it was not trained on. When I started building a Tamil triage pipeline for an offline Android deployment, the tokenizer was where most of the early pain lived.

The Concept

Start with the wrong mental model: "text is made of words."

It is not, at least not in a way computers can trust. Take these three strings:

- `play`
- `playing`
- `replaying`

A whole-word system treats these as three unrelated items unless all three appear in its vocabulary. Now take `color` and `colour`. Same meaning, different spelling. Now take `printf(`, `user_id`, and `2026-03-26`. These are not normal "words" at all, but they appear constantly in code, logs, configs, docs, SQL, shell commands, and chat transcripts. So whole words break the moment text gets messy.

Characters look safer. Every string is made of characters, with no out-of-vocabulary problem and no need to guess word boundaries. But character-level text is like shipping a sofa one screw at a time. A character tokenizer turns `internationalization` into a long chain of tiny pieces. The model must spend compute predicting letter after letter after letter, even when the larger chunk is common and stable. That wastes context window and wastes learning capacity. So we need a middle ground.

Here is the plain-English picture. Imagine you are compressing sticky notes from a busy office. At first, every note is stored as individual letters. That works, but it is wasteful. You keep seeing the same letter pairs: `th`, `he`, `in`, `er`, `re`. So you invent shorthand cards for those pairs. Then you notice bigger chunks appear constantly: `the`, `ing`, `ion`, `printf`, `://`, `.com`. You add cards for those too. You never hard-code English grammar. You never tell the system what a noun is. You just keep asking one question: "What adjacent pair shows up most often? If I replace that pair with a single new symbol, do I save space?"

That is **byte-pair encoding**, or BPE.

BPE starts from raw bytes. A byte is just a number from 0 to 255. Text files on disk are bytes. UTF-8 characters are bytes. Emojis are bytes. Code is bytes. Weird punctuation is bytes. Starting from bytes means nothing is out of vocabulary. Every possible input can be represented, because every possible input is bytes. Then BPE learns a vocabulary by repeated merging:

1. Count all adjacent byte pairs in the training corpus.
2. Find the most frequent pair.
3. Replace that pair everywhere with a new token ID.
4. Repeat until you reach the target vocabulary size.

If you start with 256 byte values and add 768 merges, you get a vocabulary of 1024 tokens. The key idea is simple: common chunks become single tokens; rare chunks stay split. A good tokenizer does not try to make every weird string into one giant token. It gives short, common patterns their own entries and leaves rare junk as smaller pieces. That is why BPE handles both everyday language and ugly real-world text.

Here is the other picture that helps. Think of vocabulary size as drawer size in a workshop. Too few drawers: every tool is broken into tiny parts and scattered everywhere, so you spend forever assembling what you need. Too many drawers: you create a custom drawer for every slightly different object, and most drawers hold one thing and never get opened again. A tokenizer needs the middle ground. Enough drawers to store recurring chunks, not so many that the system memorizes trivia.

Before any code, define the two things you are optimizing against each other:

- Sequence length: how many tokens it takes to represent text.
- Token frequency: how often each token appears in the corpus.

A larger vocabulary usually shortens sequences, but it also creates rarer tokens. That tradeoff is the whole chapter, and Figure 3.1 shows how BPE navigates it by starting from bytes and merging upward.

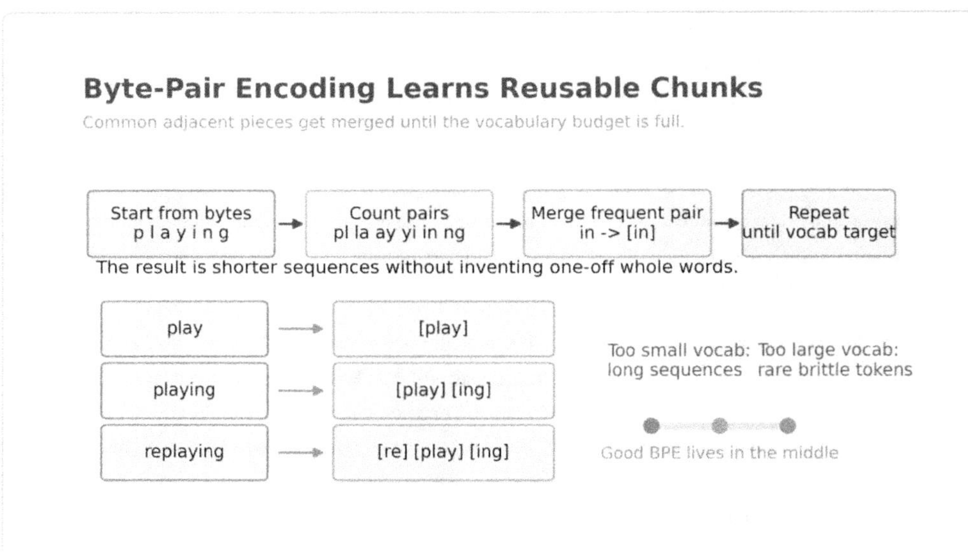

Figure 3.1. Byte-pair encoding starts from bytes, merges frequent adjacent pieces, and tries to shorten sequences without wasting vocabulary on one-off chunks.

Why It Matters

The tokenizer is not a preprocessing footnote. It changes what the model can learn efficiently.

If your tokens are too small, the model wastes attention on low-level assembly. Suppose the sentence is `The database connection timed out again.` A bad tiny-token setup might turn this into dozens of pieces. The model now spends context and compute reassembling `connection` from fragments instead of learning the larger pattern that `database connection timed out` often appears together.

If your tokens are too large, you get the opposite failure. Imagine a vocabulary that contains bizarre one-off chunks like `timed out again.` or `base connection timed out ag`. Those might save a token or two in the training corpus, but each appears so rarely that the model cannot learn a good embedding for them. An embedding is a list of numbers that represents meaning. If a token appears twice in the whole corpus, the model gets almost no chance to learn what that list of numbers should be.

This was the failure mode that ambushed me during early per-language tokenizer choices in some cooperative fine-tuning experiments. We had merged a vocabulary that looked great on training compression and quietly contained thousands of tokens that appeared two or three times. The training loss told one story. Generalization told another. The embeddings for those rare tokens were essentially noise.

This affects everything downstream. Bad tokenization means longer sequences, more memory cost, less useful context window, harder prediction targets, noisier embeddings, and worse compression of repeated patterns.

There is also a practical reason to care. When people say "this model has a 128K context window," that is 128K tokens, not 128K words and not 128K characters. If your tokenizer is inefficient, the same document consumes more of that budget. A code-heavy file might fit within one tokenizer and overflow in another. So vocabulary design is not cosmetic. It changes how much text the model can see, how often it reuses learned chunks, and how much training signal each embedding receives.

Strong opinion: most "this model is better at X" benchmarks would tell a different story if both models were retokenized on the same vocabulary. Tokenizer-driven differences masquerade as model-driven differences all the time. After enough of these experiments I now find myself instinctively asking "whose tokenizer?" before "whose model?" when reading a comparison.

You can think of a tokenizer as the interface between raw text and learned representation. Project 2 ended with characters as the unit. This project is where "input units" stop being given by nature and become a design choice. That is a major shift.

The Build

We are going to build a byte-pair tokenizer from first principles, train it at several vocabulary sizes, inspect what it learns, and compare it to a production tokenizer.

The starting point is Karpathy's `minbpe`, but do not treat it as a black box. Read it like you would read a short paper with executable examples.

This chapter is best read by tracking only three jobs. First, count adjacent pairs. Second, replace the most frequent pair with a new token. Third, record those replacements so the same tokenizer can be reused later. If those three jobs are clear, the rest of the code stops feeling dense.

Step 1: Start from bytes, not characters

This choice solves a nasty problem before it appears. If you start from characters, you need a character set. ASCII? Unicode code points? Each option brings its own headaches about encoding details and alphabet size. Bytes avoid all of that. UTF-8 encodes text as bytes already, so starting from bytes means every file, every symbol, every emoji, every weird quote mark, every language can at least be represented.

Minimal Python:

```
text = "hello 👋"
data = text.encode("utf-8")
print(list(data))
# [104, 101, 108, 108, 111, 32, 240, 159, 145, 139]
```

That waving-hand emoji is not a single byte. It is several bytes. That is fine. BPE can merge those bytes into larger chunks later if they appear often enough.

This gives you your initial vocabulary: token IDs `0..255`, each mapping to exactly one byte value. At this point, tokenization is trivial. Every byte is already a token. This is the smallest honest tokenizer. It is also a terrible final tokenizer, because sequences are too long. That is why we merge.

Step 2: Count adjacent pairs

Now ask the compression question. Given a token sequence like:

```
[104, 101, 108, 108, 111]
```

the adjacent pairs are:

- `(104, 101)`
- `(101, 108)`
- `(108, 108)`
- `(108, 111)`

Across the whole corpus, count how often each pair appears. The most frequent pair is the best merge candidate because replacing it with a single new token saves the most space right now.

Minimal pair-counting code:

```
from collections import Counter

def get_stats(ids: list) -> Counter:
    counts = Counter()
    for a, b in zip(ids, ids[1:]):
        counts[(a, b)] += 1
    return counts
```

That is the revealing part. BPE is not looking for words, grammar, or semantics. It is looking for repeated adjacent pieces that save space when collapsed. When you see tokenizer code later, remember that its bias is compression-oriented rather than linguistically elegant.

This function does not know anything about language. It just counts neighbors. That is enough.

Step 3: Merge the most frequent pair

Suppose `(101, 114)` is the most common pair. That might correspond to the bytes for `e` and `r`. Create a new token ID, say `256`, that stands for that pair. Then replace every occurrence of `(101, 114)` in the sequence with `256`.

Minimal merge code:

```
def merge(ids: list, pair: tuple, new_id: int) -> list:
    out = []
    i = 0
    while i < len(ids):
        if i < len(ids) - 1 and ids[i] == pair[0] and ids[i + 1] == pair[1]:
            out.append(new_id)
            i += 2
        else:
            out.append(ids[i])
            i += 1
    return out
```

If you are reading instead of typing, focus on the control flow. The `while` loop scans left to right. When it sees the target pair, it emits `new_id` and jumps forward by two positions. Otherwise it copies the current token and advances by one. That small pattern is what physically rewrites the sequence after every merge.

That is the whole engine: count pairs, pick the winner, replace it, repeat.

Step 4: Train the tokenizer to a target vocabulary size

Now wrap the loop. If you want vocabulary size `V`, and you already have the 256 byte tokens, you need `V - 256` merges.

Pseudo-code:

```
ids = list(text.encode("utf-8"))
merges = {}
vocab = {i: bytes([i]) for i in range(256)}

for new_id in range(256, V):
    stats = get_stats(ids)
    pair = max(stats, key=stats.get)
    ids = merge(ids, pair, new_id)
    merges[pair] = new_id
    vocab[new_id] = vocab[pair[0]] + vocab[pair[1]]
```

The code-reading summary of tokenizer training is simple. `ids` is the current tokenized corpus. `stats` measures which pair is most worth collapsing next. `merges` is the permanent memory of what the tokenizer learned. `vocab` stores the byte string each learned token stands for. That is what "training a tokenizer" means in operational terms.

A few things matter here. First, `vocab[new_id]` is built by concatenating the byte strings of its parts, which means every merged token still has a byte-level meaning. You can decode any token back to the raw bytes it represents. Second, the merge history matters: encoding new text later means replaying the merges in the same order. You are not retraining; you are applying a learned merge table. Third, this is greedy. At each step, BPE picks the best local merge rather than solving some giant global optimization over all possible vocabularies. That greedy choice turns out to work well in practice.

I expected greediness to be a glaring weakness when I first read about BPE. Surely a smarter global search would outperform it. In practice the gap is small and the cost is huge. Greedy local merges that look short-sighted produce vocabularies that look almost identical to vocabularies discovered by far more sophisticated methods. That was a surprise I had to sit with for a while.

Step 5: Build encode and decode

Training builds the vocabulary and merge rules. Using the tokenizer later needs two operations: `encode(text) → token_ids` and `decode(token_ids) → text`.

Decoding is easy. Each token maps to bytes; concatenate those bytes and decode the result.

```
def decode(ids: list, vocab: dict) -> str:
    raw = b"".join(vocab[i] for i in ids)
    return raw.decode("utf-8", errors="replace")
```

Encoding is the subtle part. You start from raw bytes again, then repeatedly merge any adjacent pair that exists in your learned merge table, respecting merge priority.

Why priority matters: suppose your training created these merges in order, `(a, b) → 256`, then `(256, c) → 257`. Then `abc` should become `[257]`, not `[256, c]` forever. So when multiple possible merges exist, apply the one learned earliest. In practice, production BPE encoders store a rank for each merge and repeatedly choose the lowest-rank available pair.

I once spent a frustrating afternoon debugging an encoder where the priority logic had been swapped: it preferred later merges over earlier ones. The tokenizer still produced output. The output was wrong in subtle, hard-to-catch ways. Decoded text round-tripped fine. Models trained on it learned wonky-looking embeddings. The lesson stuck: when training and inference use different merge orderings, the bug does not crash anything. It corrupts everything quietly.

A stripped-down version:

```
def encode(text: str, merges: dict) -> list:
    ids = list(text.encode("utf-8"))
    while len(ids) >= 2:
        stats = {(a, b): i for i, (a, b) in enumerate(zip(ids, ids[1:]))}
        pair = min(stats, key=lambda p: merges.get(p, float("inf")))
        if pair not in merges:
            break
        ids = merge(ids, pair, merges[pair])
    return ids
```

Encoding is where many readers think they are seeing another training loop. They are not. Training discovers the merge table once. Encoding applies that fixed table to new text. Keeping that distinction straight makes production tokenizer libraries much easier to read later.

This is not the fastest implementation. It is the clearest. At this stage, clarity matters more than speed.

Step 6: Test on real examples

Do not trust a tokenizer until you have looked at outputs with your own eyes. Start with plain English:

- `the cat sat on the mat`
- `internationalization`
- `color / colour`

Then code:

- `def main():`
- `user_id = 42`
- `SELECT * FROM users WHERE id = 7;`

Then mixed messy text: URLs, emojis, dates, markdown, JSON.

Ask: which chunks become single tokens? Which strings stay fragmented? Does decoding return the original text exactly? That last question is non-negotiable. If encode then decode changes the bytes, your tokenizer is broken.

Step 7: Train several vocabulary sizes

Now do the experiment the chapter is really about. Train BPE tokenizers on the same corpus at these vocabulary sizes:

- `256`
- `1024`
- `4096`
- `8192`

Pick one corpus and keep it fixed. If you change the data between runs, you learn nothing from the comparison.

Good small corpus choices:

- a few MB of project Gutenberg text
- a code repository
- your own notes plus docs
- a mixed text file containing prose, code, and web text

For each tokenizer, measure:

- average tokens per sentence
- **compression ratio** = original byte count / token count
- token frequency histogram
- examples of token boundaries on the same sentences

Why tokens per sentence matters: the model sees a fixed maximum number of tokens, not bytes. If vocabulary size grows from 256 to 4096 and average tokens per sentence drops sharply, that means the tokenizer is giving the model more effective room to work with. Why diminishing returns matter: the jump from 256 to 1024 usually helps a lot, but the jump from 4096 to 8192 often helps less. That flattening tells you the bigger vocabulary is buying less compression for each new token you add. This is one of the central observations of the project.

Pause and inspect.

Step 8: Plot tokens-per-sentence vs vocabulary size

Do not leave this as a vague impression. Plot it.

Minimal experiment loop:

```
sizes = [256, 1024, 4096, 8192]
results = []

for vocab_size in sizes:
    tokenizer = train_bpe(corpus_text, vocab_size)
    token_counts = []
    for sentence in sample_sentences:
        ids = tokenizer.encode(sentence)
        token_counts.append(len(ids))
    avg_tokens = sum(token_counts) / len(token_counts)
    results.append((vocab_size, avg_tokens))
```

Then plot vocabulary size on the x-axis, average tokens per sentence on the y-axis. You should expect a curve that drops quickly at first, then flattens. That shape is the tradeoff becoming visible.

Step 9: Build a token visualizer

This is where the tokenizer stops feeling abstract. Take a sentence and show which bytes or characters got grouped into each token.

A simple version prints tokens with separators:

```
tokens = tokenizer.encode("replaying the song")
for tid in tokens:
    piece = tokenizer.vocab[tid].decode("utf-8", errors="replace")
    print(f"[{piece}]")
```

Better: color-code each token span in terminal output or HTML.

You want the reader to see something like:

- `[replay][ing][ ][the][ ][song]`
- `[user][_id][ ][=][ ][42]`
- `[https][:][//][example][.com]`

The point of the visualizer is not aesthetics. It is to build intuition. Common words and common subwords become stable chunks; rare strings stay broken into smaller pieces. That is what BPE is discovering.

When I built this visualizer for a Tamil tokenizer targeting community health workers, half the early bugs in our preprocessing pipeline became visible in fifteen minutes of color-coded staring. There is no substitute for actually looking at the output. Reading about a tokenizer is a fraction of the value of watching one segment your own text.

Step 10: Compare to a production tokenizer

Now sanity-check your work. Run the same example strings through your BPE tokenizer, `tiktoken`, and `sentencepiece`. Do not expect identical outputs. They are trained on different corpora, may include special rules, and may treat whitespace differently. What you should expect is rough agreement in spirit. Common pieces like `ing`, `tion`, `http`, punctuation clusters, and frequent short words should often appear as merged units.

Minimal `tiktoken` example:

```
import tiktoken

enc = tiktoken.get_encoding("gpt2")
print(enc.encode("replaying the song"))
```

Minimal SentencePiece direction: train a small model on the same corpus and compare segmentations.

What you are checking is not "did I clone GPT-4's tokenizer?" You are checking whether your first-principles implementation learns the same kinds of chunks that production systems learn. If it does, the core idea is right.

Step 11: Earn one equation

Up to now we have stayed with code and intuition. Now we can compress one important metric into math because you already know what it means.

Define:

- `B` = total number of bytes in a dataset
- `T` = total number of tokens after tokenization

Then compression ratio is:

```
compression ratio = B / T
```

Read this in plain English: "How many bytes of original text does each token cover on average?" If the ratio is `1.0`, you are basically at byte-level tokenization. If it is `4.0`, each token represents about four bytes on average.

Higher is not automatically better, because you can cheat by creating giant rare tokens that barely repeat. So pair this metric with token frequency statistics. That is why vocabulary size cannot be judged by compression alone.

Step 12: Look at token frequency, not just sequence length

A tokenizer that reduces sequence length looks good. But you also need to ask whether the resulting tokens appear often enough to learn meaningful embeddings. If a token appears only once or twice across the entire corpus, the model gets almost no gradient signal to learn what that token means. A simple frequency table tells the story:

- top 20 most common tokens

- median token frequency
- fraction of tokens appearing fewer than 5 times
- fraction appearing exactly once

That last group is a warning sign. If your vocabulary is huge and many tokens appear once, the model gets one example to learn from. That is barely learning at all. This sets up the BREAK IT section, where we push vocabulary size to obviously bad extremes.

Step 13: Notice whitespace, punctuation, and code patterns

A beginner expects the tokenizer to focus mostly on words. It does not. Whitespace patterns matter a lot. Punctuation patterns matter a lot. In code, pairs like `()`, ==, `": "`, indentation spaces, </, =>, _, and :: often become useful chunks.

This surprises people until they remember what the tokenizer is optimizing: not grammar, not dictionary words, but recurring adjacent byte sequences. That is why tokenizers trained on code-heavy corpora behave differently from tokenizers trained on novels. The corpus shapes the vocabulary.

Step 14: Connect this back to the model

By now you have built the tokenizer, but the point is not merely to segment text. The point is what the model sees next. In Project 2, the model predicted the next character. In this project, you now control what counts as a "thing" worth predicting. If `playing` becomes `[play][ing]`, the model can learn the stem and suffix separately. If `SELECT` becomes one token in a SQL-heavy corpus, the model can learn SQL syntax more directly.

If vocabulary is too small, the model wastes steps rebuilding common chunks. If vocabulary is too large, the model spreads learning across too many rare symbols. That tradeoff hits both context efficiency and embedding quality. This is why tokenizer design sits right at the border between data engineering and model design. It decides the alphabet of thought the model gets to work with.

BREAK IT

You are going to break vocabulary size on purpose in both directions. Not slightly wrong, but obviously wrong. First make it too small, then make it too large. Watching what fails tells you what the tokenizer is protecting.

Break 1: Vocabulary size = 50

There is an immediate problem here. A byte-level tokenizer already needs 256 base tokens, so a true BPE tokenizer cannot have vocabulary size 50 if it starts from bytes. That is not a bug in the chapter. It is the point. A vocabulary of 50 is so small you cannot even represent all byte values directly.

To run the "too small" experiment, you need a proxy: either switch from bytes to a restricted character alphabet on a small corpus, or keep the byte base at 256 and treat that as the practical minimum. Use the first option for a hands-on lab because it makes the failure visible. Take a lowercase English-only subset with space and punctuation, maybe 40 to 50 unique characters total, and define a character-level tokenizer at that vocabulary size.

What happens? Sequences explode in length. A sentence that might take 7 to 12 tokens in a decent BPE tokenizer now takes 40 to 80 tokens. Common words get no shared chunking benefit. Suffixes like `-ing`, `-tion`, `-ed` must be spelled out every time. Code becomes especially ugly because punctuation and identifiers split into many pieces, and a fixed context window fits far less text.

If you later train a model on this tokenizer, the model can still learn. But it wastes capacity on local assembly work. It has to model spelling, word formation, syntax, and meaning all at a much finer resolution than necessary. The tokenizer is too stingy. It refuses to promote recurring chunks into reusable units, so the model does the work instead.

Break 2: Vocabulary size = 100,000

Now go too far in the other direction. Train a very large BPE vocabulary on a modest corpus. At first it looks good: sequence lengths get shorter, compression ratio improves. Then check token frequencies. A large chunk of the vocabulary appears only a handful of times. Some tokens appear once. You have invented many highly specific chunks that are not reusable building blocks. They are just little fossils of the training corpus.

Typical symptoms:

- token frequency histogram gets a huge long tail
- many tokens appear `< 5` times
- token visualizer shows bizarre over-merged chunks
- train text compresses well, but new text may segment awkwardly
- embeddings for rare tokens will be poorly learned if you train a model on top of this vocabulary

Think of it like making a phrasebook with one page per sentence fragment you happened to overhear. You save time on those exact fragments. You learn almost nothing general. A tokenizer is a file compressor in shape only. Underneath, it is defining the units that will receive learned embeddings and prediction probabilities. If a token almost never appears, the model cannot learn much about it.

What this proves

The sweet spot exists because two costs pull in opposite directions. A small vocabulary causes long sequences, wasted context, more steps to express the same idea, and more burden on the model to reconstruct common chunks. A large vocabulary causes sparse token frequencies, weak embeddings for rare tokens, vocabulary slots wasted on one-off fragments, and worse generalization.

The tokenizer exists to manage this tradeoff. Not eliminate it, manage it. That is why modern vocabularies land in the tens of thousands, not dozens and not millions. They need enough pieces to compress common patterns, but not so many pieces that each piece becomes statistically lonely.

Optional Homework

Run these labs locally. None require special hardware. CPU is fine. A few megabytes of text is enough. Runtime is usually minutes, not hours.

Lab 1: Train BPE at multiple vocabulary sizes

1. Create a training corpus file, for example `corpus.txt`. Good target size: 1 MB to 10 MB.

2. Write or adapt a script `train_sizes.py` that trains tokenizers at:

```
sizes = [256, 1024, 4096, 8192]
```

4. For each size:
    - train the tokenizer
    - encode the same set of 100 sample sentences
    - record:
        - average tokens per sentence
        - total bytes / total tokens
        - fraction of tokens with frequency `< 5`
5. Plot vocabulary size vs average tokens per sentence.

Expected observation:

- `256` → `1024` gives a big drop in token count
- `1024` → `4096` still helps
- `4096` → `8192` helps less
- the curve flattens

Lab 2: Build a token visualizer

1. Pick 10 sentences with a mix of prose and code:

```
The database connection timed out again.
def main():
user_id = 42
https://example.com/api/v1
replaying the song
```

2. For each tokenizer size, print token boundaries. Example output format:

```
[replay][ing][ ][the][ ][song]
[user][_id][ ][=][ ][42]
```

3. If you want color, use ANSI terminal colors or a tiny HTML page.

Expected observation:

- common chunks become stable tokens as vocabulary grows
- rare words still split into smaller pieces
- code punctuation often becomes its own reusable pattern

Lab 3: Too small proxy experiment

Because byte-level BPE cannot go below 256 base tokens, run this on a restricted character corpus.

1. Build a character vocabulary from lowercase English letters, digits, space, and basic punctuation.

2. Force vocabulary size around 50.
3. Encode 100 sample sentences.
4. Compare token counts against your `1024` and `4096` BPE runs.

Expected observation:

- sequence lengths become much larger
- long words fragment badly
- effective context window shrinks hard

Lab 4: Too large experiment at 100K

1. Use the same corpus as before.
2. Train toward `100000` vocabulary size. If your corpus is small, the training loop may stop being useful long before that. That is part of the lesson.
3. Record:
    - number of unique tokens actually created
    - fraction appearing once
    - fraction appearing `< 5` times

Expected observation:

- many tokens are extremely rare
- merges become oddly specific
- training text compresses more than validation text

Lab 5: Compare against production tokenizers

1. Install `tiktoken`:

```
pip install tiktoken
```

2. Install SentencePiece if you want the second comparison:

```
pip install sentencepiece
```

3. Run the same sample strings through your tokenizer and a production tokenizer.
4. Compare:
    - token count
    - obvious merged chunks
    - whitespace handling
    - code handling

Expected observation:

- exact boundaries differ
- the general merge patterns often agree

- production tokenizers usually have smarter handling around spaces and special cases

Suggested metrics table

Record your results in a table like this:

Vocab Size	Avg Tokens / Sentence	Bytes / Token	% Tokens Freq < 5
256	...	...	...
1024	...	...	...
4096	...	...	...
8192	...	...	...
100000	...	...	...

: **Table 3.1.** Vocabulary size comparison: how increasing vocab size affects sequence length, bytes per token, and the proportion of rare tokens in the training corpus.

That last column is the trap detector. A tokenizer that compresses well but creates many near-dead tokens is not "winning." It is spending vocabulary budget badly.

Questions To Answer

1. When you increased vocabulary size from `256` to `1024`, what changed most: token count, token boundaries, or the kinds of chunks that appeared?
2. At what point did larger vocabulary stop buying much shorter sequences on your corpus?
3. In your `100K` run, what fraction of tokens appeared fewer than five times? What does that imply for learning good embeddings?
4. In the too-small proxy run, what failed first for you: English readability, code segmentation, or context efficiency?
5. Which patterns did your tokenizer merge that you did not expect? Examples: leading spaces, punctuation pairs, indentation, URL fragments, suffixes.
6. What tradeoff is vocabulary size protecting? Answer this without saying "compression." Explain it in terms of model context and embedding quality.

Go Further

Start with the reference repo:

- `karpathy/minbpe`

Then compare against production systems:

- `openai/tiktoken`
- Google `sentencepiece`

If you want to push further, search arXiv for:

- `byte pair encoding tokenizer`
- `sentencepiece unigram language model`
- `tokenization effects language modeling`
- `byte-level language models`

A good next step is to compare BPE with unigram tokenization. BPE is a greedy merge process. SentencePiece's unigram approach starts from many candidate pieces and prunes them by likelihood. That difference sounds abstract until you inspect actual segmentations side by side. Then the tradeoffs become visible in a way that reading papers alone will not give you.

Another worthwhile direction is tokenizer mismatch. Search for `tokenizer mismatch llm` and `tokenization fertility`. "Fertility" here means how many tokens a tokenizer tends to produce for a piece of text. That matters a lot across languages. A tokenizer that is efficient for English may be wasteful for other scripts, and that efficiency gap has real consequences for what the model can learn.

This is the part that motivated the Tamil-first design of an edge deployment I worked on for community health workers. English-trained tokenizers chewed through Tamil text at fertilities that made the context window almost useless. Building a Tamil-aware tokenizer was not a polish step. It was structural.

If you work with code, search for `code tokenizer bpe` and `code llama tokenizer`. Code-heavy corpora often want different chunk patterns from prose-heavy corpora, and seeing that difference will sharpen your intuition for corpus-dependent vocabulary design.

If you want a more radical comparison, look up byte-level models that try to skip tokenization entirely. That direction keeps returning because tokenizers solve one set of problems while creating another After this chapter you will be able to read those claims with clear eyes instead of marketing fog.

What You Now Know

A tokenizer is a learned compression scheme: count adjacent byte pairs, merge the most frequent one, repeat until the vocabulary reaches the target size. Vocabulary size is a real design decision. Too small and sequences bloat, too large and token frequencies become too sparse for useful embeddings. You know what breaks at both extremes because you ran both experiments.

Starting Point

Karpathy's `minbpe` reference implementation is the starting point for this project:

- https://github.com/karpathy/minbpe

Keep the progression straight:

- `micrograd` taught you how learning machinery works
- Project 2 showed you a language model at character scale
- this project teaches you that the unit of text is not fixed; you choose it
- next comes attention, where those tokens stop being isolated IDs and start interacting with each other in context

---

Project 4: Attention From Scratch

Hook

Why does a model suddenly stop acting like a bag of isolated token IDs and start acting like it can connect pronouns to names, match brackets, continue code blocks, and stay on topic across a paragraph? If each token is just an integer turned into a list of numbers, where does the actual context-sharing happen? That question lands you on attention, because attention is the first place in the model where one token gets to look at the others and decide, in real time, which ones matter.

The first time I built one of these by hand, I spent an hour staring at a shape mismatch error before I realized I had transposed the wrong axis on K. Attention is not conceptually hard. It is operationally fiddly. The mental model is one paragraph long. The code that implements that paragraph involves enough reshape, transpose, and broadcast that a careless line lands you somewhere you cannot debug from the error alone.

The Concept

Up to now, you have tokens and embeddings. Useful, but still not enough for context. If the token `"bank"` appears, the model has a problem. Does it mean a financial institution? A river bank? A verb, like banking a plane? The token by itself cannot settle that. Attention is the mechanism that lets a token ask the rest of the sequence for help.

Imagine every token in a sentence sitting around a table with an index card. On the card, each token writes three things:

- a **query**: what I am looking for
- a **key**: what I contain
- a **value**: what I can contribute if someone picks me

If you are the token `"it"` in the sentence "The server returned an error because it timed out," you need to figure out what `"it"` refers to. So `"it"` sends out its query: "I am looking for something singular, nearby, and probably a thing that can time out." Other tokens present their keys. `"server"` has a key that says, roughly, "infrastructure noun, singular, plausible agent of timing out." `"error"` has a key that says, roughly, "noun, singular, but less likely to time out." Attention compares the query from `"it"` against the keys from all tokens. The better the match, the more weight that token gets. Then `"it"` forms its new representation by taking a weighted mixture of the values from the tokens it attended to.

That weighted mixture is the whole trick. It does not copy one token exactly. It blends information from the places it thinks matter. So if `"it"` gives `"server"` a large weight and `"error"` a smaller one, the new representation of `"it"` now carries context that helps the model resolve the reference.

The core move is simple: each token asks what it needs, each token advertises what it has, each token decides who matters, and each token updates itself using that information.

This is why attention feels different from the earlier models you built. Earlier, information moved through fixed pathways. Attention creates dynamic ones. The path depends on the sentence in front of you. `"bank"` in one sentence will attend differently than `"bank"` in another. Nothing in the code says, "if river, then use river meaning." The model learns patterns that make that behavior possible.

The first confusion people have is "Is attention memory?" Not really. It is not long-term storage. It is a way of routing information inside the current sequence: for this token, right now, which other tokens should influence the computation? A better picture is a room full of people briefly consulting each other before answering. There is no notebook being written to, no long-term archive. Just one round of "let me check with the others" before each token updates its representation.

Make the analogy concrete. Each token starts as a vector of size `d`. From that vector, the model computes three new vectors:

- `Q`: the query
- `K`: the key
- `V`: the value

All three come from learned linear projections. Three small matrix multiplies that turn the token representation into three different "views" of itself. Why three? Because "what I am looking for," "what I contain," and "what I contribute" are not the same job. A token might be very good at matching a query but not be the main thing you want to copy into the result. A pronoun might look for nouns with one pattern, while the actual information it needs to gather lives in a different subspace of numbers. If a single score had to do all those jobs at once, the model would lose a great deal of expressive flexibility.

The cleanest mental model I have for the Q/K/V split is from a library: each book on the shelf has a title (the key), what you write on the back of the request slip (the query), and what you actually take home (the value). Those are obviously three different things. The librarian does not match request slips to take-home contents directly; the slip matches the title, then the title delivers the contents. Attention is doing the same routing trick at every layer.

Now the math.

For one head of attention, the formula is:

```
Attention(Q, K, V) = softmax((QK^T) / sqrt(d)) V
```

Here is what every symbol is.

- `Q` is the matrix of queries, one query vector per token
- `K` is the matrix of keys, one key vector per token
- `V` is the matrix of values, one value vector per token
- `K^T` means transpose, which turns rows into columns so dot products line up
- `QK^T` gives a score between every token's query and every token's key
- `d` is the size of each head's query and key vectors
- `sqrt(d)` is the scaling factor
- `softmax(...)` turns raw scores into weights that sum to 1

- multiplying by `V` forms the weighted sum of value vectors

It looks dense until you say it out loud: take every token's query, compare it with every token's key, turn those comparisons into weights, use those weights to mix together the value vectors. That is attention.

The attention matrix itself is worth staring at. If your sequence length is `T`, then `QK^T` is a `T x T` matrix. Row `i` tells you what token `i` attends to. Column `j` tells you how much other tokens attend to token `j`. If row 5 places most of its weight on columns 3 and 4, token 5 is mostly consulting tokens 3 and 4. That is why attention heatmaps are so useful. They let you watch context routing happen.

From some low-level work on Apple Neural Engine kernels, I used to keep printed attention heatmaps taped to the side of my monitor. Not because they helped optimize the kernels, but because they kept reminding me what the kernels were for. It is surprisingly easy to drift into "this is just a big matrix multiply followed by a softmax" and lose the thread that this is the thing that lets the model carry information across positions.

Now the next problem. If you let each token attend to every other token, then during training a token could look at the future. That would be cheating. In an autoregressive model, the model predicts the next token one token at a time, so token 5 can use tokens 1 through 5, but not token 6. If token 5 could attend to token 6 during training, the model could peek at the answer. That is why we need **causal masking**.

Causal masking blocks attention to future positions. The upper triangle of the attention score matrix gets replaced with negative infinity before softmax. Negative infinity sounds dramatic, but the reason is simple: softmax turns very negative numbers into weights extremely close to zero, so future positions become impossible to attend to.

One more step. Single-head attention works, but one head can only learn one pattern at a time. Real language needs several patterns at once. One head may track the previous token. Another head focuses on opening brackets. A third learns subject-verb agreement, while a fourth catches line indentation in code. So transformers use **multi-head attention.** Instead of doing one attention computation over the full embedding dimension, the model splits the embedding into `H` smaller slices. Each slice gets its own query, key, and value projections. Each head computes attention independently. Then the results are concatenated back together.

Think of it as a panel of specialists. Instead of one person trying to watch syntax, meaning, punctuation, and code structure all at once, each specialist gets a smaller view, does one job, and the system combines the reports. That matters because language is not one relation. It is many relations stacked on top of each other. Multi-head attention gives the model multiple ways to look at the same sequence at the same time. That is the first real answer to "How is it keeping track of all this?" Not with one giant understanding module, but with many learned attention patterns running in parallel over token representations that update layer after layer. Figure 4.1 illustrates why the causal mask is essential to this process, and Figure 4.2 shows the full attention pipeline from query scoring through value mixing.

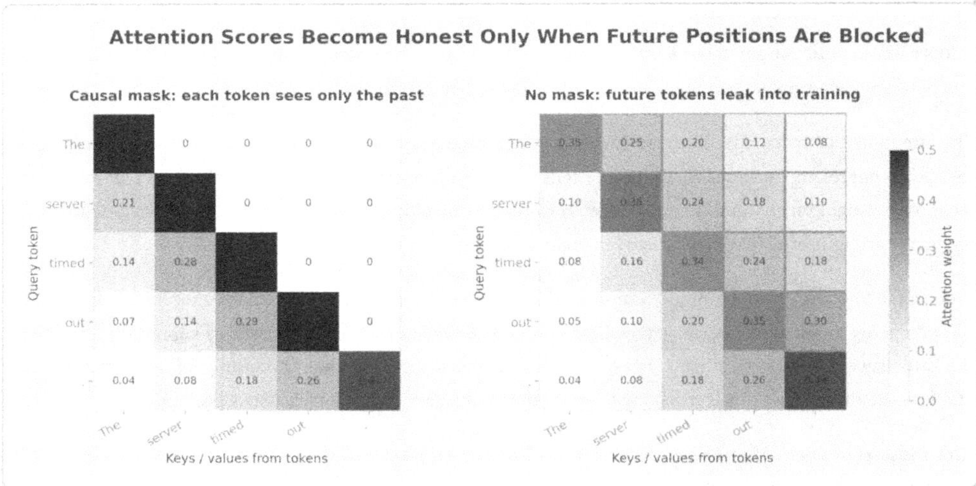

Figure 4.1. Attention is useful only when the causal mask blocks future positions during autoregressive training; otherwise later tokens leak into earlier predictions.

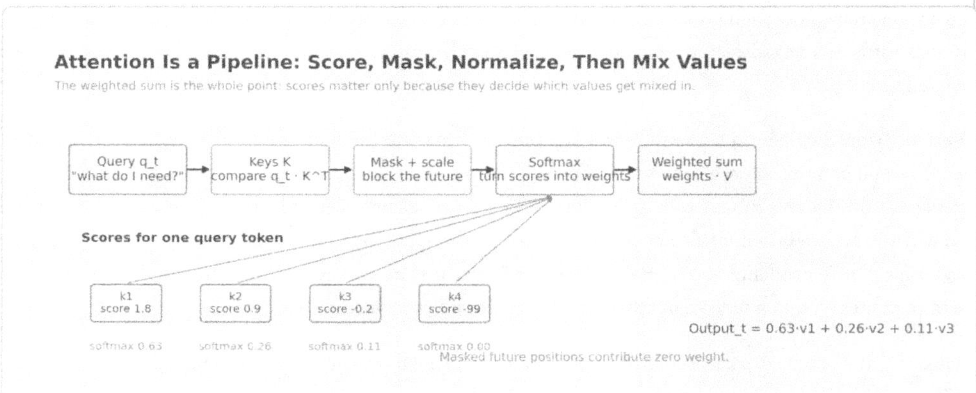

Figure 4.2. Attention is a pipeline: the query scores keys, masking blocks illegal positions, softmax turns scores into weights, and those weights mix the value vectors into the next hidden state.

Why It Matters

Without attention, tokens do not get to dynamically ask "who should matter to me right now?" That removes the main mechanism that makes transformers good at context. Think about prose: "Maria handed the package to Jordan because she was leaving." Who is "she"? You do not solve that by looking at `"she"` alone. You solve it by comparing it against earlier nouns and the surrounding verb structure. A model without attention cannot make that comparison at all. It can only use fixed, position-independent features.

Attention also matters because it is where several critical design choices live: the `1/sqrt(d)` scaling factor, causal masking, multi-head splitting, and softmax over scores. If you break any of these, the model stops behaving well in specific, diagnosable ways. That is exactly why this chapter asks you to inspect the attention matrix, see the weights, and break it on purpose. Not because the formula is complicated, but because each piece of the formula exists to prevent a concrete failure, and you will understand those pieces much better by watching the failures happen than by reading about them.

Strong opinion: most published explanations of attention overweight the math and underweight the breakages. The math fits on a napkin. The breakages are where the engineering lives.

The Build

We are going to build attention the same way we have built everything else so far: from the inside out. Start with a small sentence, then one head, then causal masking, then multi-head, then inspect what each head is doing. Do not rush to a full transformer block yet. This chapter is about attention itself.

If you are skimming the code, do not try to hold every tensor shape in working memory at once. Hold one repeated story instead: project tokens into queries, keys, and values; compare queries against keys to get scores; turn scores into weights; use those weights to mix values. Every code block in the build is implementing one piece of that sentence.

Step 1: Start with token embeddings, not token IDs

A token ID is just an integer. Attention cannot do anything useful with raw integers like `17`, `842`, or `1032`. So we start with token embeddings. Suppose your sequence has `T` tokens and your embedding size is `d_model`. Then your input is a matrix `X` of shape `(T, d_model)`, where each row is one token's embedding.

If you want a concrete tiny example, use a sentence like "the cat sat on the mat." Tokenize it however you want for the experiment. For this project, hand-made toy tokens are fine. The point is not tokenizer quality. The point is attention mechanics.

Step 2: Produce queries, keys, and values

Take the embedding matrix `X`. Create three learned weight matrices (`W_Q`, `W_K`, and `W_V`), each one mapping from `d_model` to `d_head`. For a single head, you can make `d_head = d_model` at first just to keep things simple. Then compute:

```
Q = XW_Q
K = XW_K
V = XW_V
```

What does this mean in plain language? Every token embedding gets re-expressed three ways. Same token, different job.

Minimal PyTorch looks like this:

```
Q = x @ W_Q
K = x @ W_K
V = x @ W_V
```

In reading mode, these are best understood as three learned views of the same token stream. The tokens are not changing identity three times. The model is deriving three different roles from the same hidden state: what a token seeks, what it offers, and what information should be passed forward if it gets selected.

Where:

- `x` has shape `(T, d_model)`
- `W_Q`, `W_K`, `W_V` have shape `(d_model, d_head)`
- `Q`, `K`, `V` have shape `(T, d_head)`

Do not hide this in a class yet if classes make the mechanism harder to see. Print the shapes. Check them. This chapter rewards being annoyingly concrete.

Step 3: Compute raw attention scores

Now compare every query against every key:

```
scores = QK^T
```

If `Q` and `K` each have shape `(T, d_head)`, then `K^T` has shape `(d_head, T)`, so `scores` has shape `(T, T)`. That is exactly what you want. Every token gets one score for every token in the sequence.

Compute raw query-key dot products:

```
scores = Q @ K.T
```

This is where the matrix becomes interpretable. Row `i` means "what token `i` is looking at." Column `j` means "how much token `j` is being looked at." When later tools show you attention heatmaps, this is the exact object they are visualizing.

If your sentence has 6 tokens, `scores` is a `6 x 6` matrix. Pause here and print it. Not after softmax, but now. These raw dot products are the pre-softmax evidence. They tell you how strongly each query-key pair matches before probabilities get involved. If you skip past this step, the scaling factor will feel arbitrary when you encounter it. It will not be arbitrary once you have seen what unscaled dot products look like.

Step 4: Scale the scores

Now divide by `sqrt(d_head)`:

```
scaled_scores = (QK^T) / sqrt(d_head)
```

Why? Because dot products grow with dimension. If query and key vectors have many components, the sum of all those pairwise multiplications gets large. Large scores make softmax saturate, which means one entry gets almost all the probability and the rest get almost none. That sounds decisive, but it is actually a training problem: once the distribution becomes too sharp too early, gradients through softmax get tiny for most positions, and the model stops learning cleanly. We will break this later. For now, apply the scale.

```
scores = (Q @ K.T) / math.sqrt(d_head)
```

If you later see `1 / sqrt(d_head)` buried inside a framework implementation, read it as a stability term, not a formatting choice. Its job is to stop dot products from becoming so large that softmax saturates and the gradients become unhelpfully sharp.

This is not a cosmetic normalization. It keeps the score magnitudes in a useful range where gradient descent can actually do its job.

I did not believe how badly this matters until I forgot the divide once in a Project 4 prototype. The model trained, the loss went down a little, and then it sat there for the rest of the run. It looked like a learning rate problem. It was not. The gradients through the saturated softmax were too small to move anything useful. The fix was one square root.

Step 5: Turn scores into attention weights with softmax

Raw scores are not yet weights. They can be negative, they do not sum to 1, and they are not directly interpretable as proportions. Apply softmax row-wise:

```
A = softmax(scores)
```

Now each row of `A` sums to 1 and each row answers: "How does this token distribute its attention across the sequence?"

Apply softmax to convert scores to attention weights:

```
weights = torch.softmax(scores, dim=-1)
```

If you print `weights`, each row becomes interpretable. A row might look like:

```
[0.05, 0.12, 0.51, 0.20, 0.08, 0.04]
```

That means this token is mostly consulting token 3, somewhat token 4, and barely token 6. If you stop here, you already have a readable attention matrix. This is the first milestone of the project.

Step 6: Mix the values

Now use those weights to combine value vectors:

```
output = A V
```

If `A` is `(T, T)` and `V` is `(T, d_head)`, the output is `(T, d_head)`.

```
out = weights @ V
```

This is the new representation for each token after attention. Notice what happened: the token did not just pick the most relevant other token. It built a mixture. That matters because context is often distributed. A token may need syntax from one place, referent information from another, and punctuation structure from a third.

Weighted mixtures let that happen.

Step 7: Print the attention matrix on a real sentence

Now feed a small sentence through your single-head attention. Print the token list, raw scores, softmax weights, and output shape in a readable table if you can.

What should you look for? Not correctness in a human semantic sense yet. These weights come from random or barely trained projections, so early on you may see nonsense. That is fine. The mechanical point is that each row sums to 1, the model can assign different weights to different positions, and the output depends on the whole sequence through those weights. Once you train even a tiny language model later, the patterns become more interpretable. Right now, you are building the machinery.

Step 8: Add causal masking

In an autoregressive model, token position `i` can only attend to positions ≤ `i`. Position 4 should not be able to look at positions 5 or 6. So create a lower-triangular mask.

In PyTorch:

```
mask = torch.tril(torch.ones(T, T))
scores = scores.masked_fill(mask == 0, float('-inf'))
```

For a fast reader, the key line is `masked_fill(..., -inf)`. The model is not being gently discouraged from looking into the future. Those positions are being made mathematically impossible after softmax. That is why masking is a rule of the training game, not a stylistic preference.

Then apply softmax after masking. The order matters:

1. compute scores
2. divide by `sqrt(d_head)`
3. apply mask
4. softmax

Do not softmax first and then zero things out. That breaks the probability normalization.

What does the mask do? For token 0, only column 0 is visible. For token 1, columns 0 and 1 are visible. For token 5, columns 0 through 5 are visible. Everything above the diagonal becomes impossible. Print the masked attention matrix and you should literally see zeros in the upper triangle after softmax:

```
[1.00, 0.00, 0.00, 0.00]
[0.42, 0.58, 0.00, 0.00]
[0.17, 0.29, 0.54, 0.00]
[0.11, 0.20, 0.18, 0.51]
```

That matrix is a structural promise: no future information leaks backward. This is one of the most important constraints in GPT-style models.

Step 9: Explain why masking exists before you move on

Do not treat masking like a technicality. Here is what would happen without it during training. Suppose the model sees "the answer is 42" and is trained to predict the next token at each position. At the position for `"is"`, it is supposed to predict `"42"` based on previous context. If attention can look ahead, the token at `"is"` can directly attend to `"42"`. That means training loss looks great for the wrong reason. The model is not learning to predict the future from the past. It is learning to copy from the future.

That produces a fake sense of competence during training. Then at inference time, future tokens are not available and the cheating path is gone. Performance collapses. That is why causal masking is not optional. It enforces the actual game the model must play.

Step 10: Scale to multi-head attention

Single-head attention proves the mechanism. Now make it practical. Let `d_model` be the full embedding size and choose `H` heads. Then each head gets:

```
d_head = d_model / H
```

assuming `d_model` divides evenly by `H`. Example: `d_model = 64`, `H = 4`, `d_head = 16`.

Instead of one big attention computation, you create projections that produce all heads at once. A common pattern: `W_Q`, `W_K`, `W_V` each map from `d_model` to `d_model`, then you reshape the result into `(T, H, d_head)`. In batched form, people often use `(B, H, T, d_head)` where `B` is batch size. For a first build, sequence-only is fine.

Why split into heads instead of just doing one huge head? Because multiple heads let the model represent different relational patterns in parallel. One head may care about nearby tokens, another about matching delimiters, another about repeated identifiers in code, another about sentence structure. The heads are not assigned these jobs by hand. They learn them. That is why plotting per-head heatmaps is so useful: it lets you see what patterns the model has actually discovered.

In some cooperative LoRA experiments I ran across model families from a few hundred million up to several billion parameters, the per-head attention patterns were often the first place where a "bad fusion" announced itself. The aggregate loss numbers looked fine. The per-head heatmaps showed heads that had collapsed to attending uniformly across the sequence, doing nothing. That signal arrived earlier and clearer than any scalar metric we tracked.

Step 11: Minimal multi-head implementation

Here is a compact, readable version.

```
# x: (T, d_model)
Q = x @ W_Q
K = x @ W_K
V = x @ W_V

# reshape to (H, T, d_head)
Q = Q.view(T, H, d_head).transpose(0, 1)
K = K.view(T, H, d_head).transpose(0, 1)
V = V.view(T, H, d_head).transpose(0, 1)

scores = (Q @ K.transpose(-2, -1)) / math.sqrt(d_head)    # (H, T, T)
scores = scores.masked_fill(mask == 0, float('-inf'))
weights = torch.softmax(scores, dim=-1)
head_out = weights @ V                                    # (H, T, d_head)

# back to (T, d_model)
out = head_out.transpose(0, 1).contiguous().view(T, d_model)
```

The reshapes and transposes can look intimidating, but the conceptual job is simple: split the model width into several smaller attention systems, run them independently, then stitch their outputs back together. If you keep that sentence in mind, the code becomes much easier to navigate.

`Q @ K.transpose(-2, -1)` now happens per head, so each head gets its own `T x T` attention matrix. That means you can inspect head 0 separately from head 1. Do that. Do not only inspect the combined output. The whole educational point of multi-head attention is that the heads differ.

Step 12: Add the output projection

In most transformer implementations, after concatenating the heads, you apply one more learned linear layer:

```
final = Concat(head_1, ..., head_H) W_O
```

This output projection lets the model remix information from the separate heads back into the model dimension.

```
final = out @ W_O
```

You can omit this at first if you want the cleanest view of attention itself. But if you want to match transformer structure more closely, include it. It is the final "merge the specialists back together" step.

Step 13: Plot per-head heatmaps

Take a real sentence or paragraph. Feed it through multi-head attention. For each head, plot its `T x T` attention matrix as a heatmap, with rows and columns labeled by tokens. Even if the weights are not deeply meaningful before training, the act of plotting is important. Once you later inspect trained models, you will already know what you are looking at.

What should you expect in trained attention heads? Patterns like strong diagonal bands for local context tracking, attention to previous punctuation marks, attention to the first token in the sentence, attention from closing brackets to opening brackets, and repeated identifier tracking in code. Some heads will look purposeful and some will look messy. That is normal. Not every head becomes cleanly interpretable, and "looks random" does not always mean useless. A head may contribute in a way that is hard to see in a single heatmap. Still, heatmaps are the best first window into what the mechanism is doing.

Step 14: Measure per-head attention entropy

Heatmaps are useful, but entropy catches something your eyes may miss. Entropy here measures how spread out the attention distribution is. If one row of attention is `[0.25, 0.25, 0.25, 0.25]`, that is high entropy. Attention is diffuse. If a row is `[0.99, 0.01, 0.00, 0.00]`, that is low entropy. One token gets almost everything. Neither extreme is always wrong, but if you remove scaling and suddenly many heads collapse into near-one-hot attention everywhere, entropy will drop sharply. That drop is evidence of softmax saturation.

For a single row `p`, entropy is:

$$H(p) = -\sum_i p_i \log p_i$$

What are the symbols?

- `p_i` is the attention weight on position `i`
- the sum goes across all positions in the row
- low entropy means concentrated attention
- high entropy means spread-out attention

Compute it per row, then average across rows in each head.

Minimal PyTorch:

```
entropy = -(weights * (weights.clamp_min(1e-9).log())).sum(dim=-1)  # (H, T)
mean_entropy_per_head = entropy.mean(dim=-1)                        # (H,)
```

Why does this metric matter? Because raw heatmaps can mislead. A head may look mostly diagonal, but entropy tells you whether that diagonal is soft or nearly one-hot. When you remove scaling, this metric often falls hard. That gives you a quantitative handle on the failure.

Step 15: Feed a paragraph, not just a sentence

Do not stop at toy examples. Run a paragraph through multi-head attention, using text with structure: prose with pronouns, code with indentation and delimiters, lists with punctuation, quoted text. A longer sequence gives heads more room to specialize. You may start seeing some heads mostly attending locally, some jumping to punctuation, some favoring recent tokens, some distributing broadly, and some that look almost useless.

That last category matters. New readers often expect every head to have a neat human story. The model trains by gradient descent, not by writing explanations for us. Some heads become interpretable; some do not. Some matter a lot; some can be pruned with surprisingly little damage in some settings. That is not a contradiction. It is what learned systems look like.

I find this part of attention oddly hard to internalize, even after years of working with it. The temptation to assign each head a tidy job is enormous. Pruning experiments cure that temptation faster than anything else. You remove a head you thought was central and the loss barely moves. You remove a head you thought was random and the model degrades. The job map you imagined is wrong, and watching it be wrong is the lesson.

Step 16: Tie the build back to the larger transformer

At this point you have the core of transformer attention. Not the whole transformer, just attention. That matters because a transformer block adds several things around it: residual connections, normalization, and feedforward layers. But attention is the first mechanism where tokens actively consult each other. You can now explain what the block is doing in plain English: let tokens exchange information through attention, stabilize and preserve information with residual paths and normalization, process the updated representations with a feedforward network, and repeat many times. The rest of the transformer is scaffolding around what you have just built.

BREAK IT

Building attention is useful. Breaking it proves why the pieces exist. There are two failures to force on purpose: remove the `1/sqrt(d)` scaling factor, and remove causal masking. Each failure teaches a different lesson. The first teaches you why score magnitude matters. The second teaches you why the training setup must prevent cheating.

Break 1: Remove the scaling factor

Exact change:

Replace this:

```
scores = (Q @ K.T) / math.sqrt(d_head)
```

With this:

```
scores = Q @ K.T
```

Do not change anything else. Then print the raw dot products before softmax, the attention weights after softmax, and the per-head attention entropy.

What happens? The raw dot products get large. Often much larger than your intuition expects. A dot product sums over `d_head` terms. If `d_head` is 64, 128, or more, you are adding a lot of numbers together, and even if each individual component is not huge, the total grows with dimension. Softmax then sees large differences and becomes brutally decisive. Suppose one row of scores is `[2.1, 1.8, 2.4, 2.0]`. Softmax still spreads some probability around. But if the row becomes `[18.2, 7.5, 21.7, 5.9]`, softmax basically says: "token 3 wins, everyone else gets crumbs."

You get attention rows like:

```
[0.02, 0.00, 0.97, 0.01]
```

Or even:

```
[0.00, 0.00, 1.00, 0.00]
```

That is near-one-hot attention. At first that may seem good. Isn't decisive attention a sign that the model knows what matters? No, not when it happens because your score scale is broken. Softmax has a useful operating range. When inputs are moderate, changes in the scores still change the output probabilities smoothly. When inputs are huge, softmax saturates: one position dominates, and small changes to the raw scores barely affect the probabilities. That means gradients through softmax get tiny for most positions and learning becomes brittle. The model loses the ability to make careful credit assignments across many candidate tokens

This is exactly the kind of failure that vanishes if you only stare at the final formula and never inspect internals. You need to print the raw scores and see that the problem begins before softmax. The scaling factor is not style. It is a guardrail that keeps dot products in a range where softmax and gradient descent can still behave well. If you track per-head entropy, you will usually see it drop when scaling is removed. The larger `d_head` gets, the worse it gets.

Break 2: Remove causal masking

Exact change: delete or disable this line:

```
scores = scores.masked_fill(mask == 0, float('-inf'))
```

Then inspect the attention matrix on a training example. Tokens will attend to future tokens, and you will literally see non-zero weights above the diagonal. That means token position `i` is using information from positions `i+1`, `i+2`, and beyond.

During training, that makes the task easier in a fake way. Suppose the training text is "The capital of France is Paris." At the position before `"Paris"`, the model should learn to predict `"Paris"` from the previous context. Without masking, it can simply attend directly to the token `"Paris"` itself. That lowers training loss for the wrong reason. You are no longer training an autoregressive model. You are training a sequence model with future leakage. It is the ML version of looking at the answer key while pretending to take the test.

In the attention matrix you will see upper-triangle entries become non-zero, tokens often assigning substantial weight to future positions, and training loss that may look suspiciously better than expected. What happens at inference time? The cheating path disappears. At generation time, future tokens do not exist yet. The model cannot use the same shortcut. That mismatch between training and inference is the real damage: the model learned under easier conditions than it will face when deployed.

Causal masking is what makes GPT-style training honest. The model must solve the task using only the past. Break that contract, and your training metrics stop meaning what you think they mean.

What to watch for, specifically

For removed scaling:

- raw `QK^T` values become much larger
- softmax rows become sharply peaked
- entropy per head drops
- gradients become less useful
- learning quality often degrades

For removed masking:

- attention appears above the diagonal
- training loss may improve suspiciously
- the model relies on future-token leakage
- inference behavior no longer matches training conditions

These are different failures. Scaling protects numerical behavior. Masking protects causal correctness. Both are non-negotiable.

The deeper lesson

Attention looks simple in formula form. That simplicity is deceptive. Every piece in the formula exists because something breaks without it: no `Q/K/V` split means less expressive matching and retrieval; no scaling means softmax saturation; no masking means future-token cheating; no multi-head means fewer parallel relational patterns. This is what engineering looks like inside models. Not a bag of arbitrary tweaks, but a trail of fixes attached to specific failure modes. The formula is small because the engineers were solving concrete problems, not decorating the math.

Optional Homework

Run both failures locally. This lab is cheap. CPU is enough for matrix inspection. A small GPU helps if you also want to train a tiny language model to watch the masking failure affect loss.

Option A: Pure attention inspection lab

Hardware:

- CPU only is fine
- 5 to 15 minutes

Create a file called `attention_lab.py`.

Use PyTorch and a tiny hand-made input.

1. Build a single-head attention function.
2. Feed in a sentence with 8 to 20 tokens.
3. Print raw scores, masked weights, and entropy.
4. Toggle scaling on and off.
5. Toggle masking on and off.

Suggested parameters:

- `d_model = 32`
- `H = 4`
- `T = 10`

Metrics to record:

- max raw score per head
- mean raw score per head

- mean attention entropy per head
- whether any upper-triangle weight is non-zero

Expected observations:

- with scaling on, entropy is moderate
- with scaling off, entropy drops and rows become spikier
- with masking on, upper triangle is zero
- with masking off, upper triangle has non-zero entries

Option B: Tiny training lab for masking failure

Hardware:

- CPU works, but slow
- small GPU preferred
- 30 to 90 minutes depending on setup

Goal: Train a tiny next-token model twice:

- once with causal masking
- once without masking

You do not need a full GPT. A tiny embedding -> attention -> linear head model is enough to expose the failure.

Dataset options:

- a small text file
- a code snippet corpus
- even repeated synthetic patterns

Suggested tiny setup:

- vocabulary from your tokenizer work or simple character-level encoding
- `d_model = 64`
- `H = 4`
- sequence length `T = 32`
- batch size `B = 16`
- train for a few hundred steps

Watch:

- training loss
- validation loss
- attention heatmaps

Expected observations:

- no-mask training loss may fall faster
- no-mask attention heatmaps will show future-token usage

- at generation time, the no-mask model behaves worse than its training loss suggested

That last point matters. If you only compare training loss, you may fool yourself. The model cheated.

Option C: Entropy ablation lab for scaling failure

Hardware:

- CPU is enough for inspection
- GPU preferred if training
- 20 to 60 minutes

Run multi-head attention with:

- `d_model = 128`
- `H = 4`
- `d_head = 32`

Compare two versions:

1. `scores = (Q @ K.T) / sqrt(d_head)`
2. `scores = Q @ K.T`

For each version, log:

- mean entropy per head
- fraction of rows where max attention weight > 0.95
- average max raw score

Expected observations:

- unscaled version has larger raw scores
- unscaled version has more rows with near-one-hot attention
- unscaled version shows lower entropy across heads

A minimal code scaffold

Use this as a starting skeleton, then expand it.

```python
import logging
import math

import torch

torch.manual_seed(0)

T, d_model, H = 8, 32, 4
d_head = d_model // H

x = torch.randn(T, d_model)
W_Q = torch.randn(d_model, d_model)
W_K = torch.randn(d_model, d_model)
W_V = torch.randn(d_model, d_model)

def run_attention(x: torch.Tensor, scale: bool = True, causal: bool = True) -> tuple:
    Q = x @ W_Q
    K = x @ W_K
    V = x @ W_V

    Q = Q.view(T, H, d_head).transpose(0, 1)
    K = K.view(T, H, d_head).transpose(0, 1)
    V = V.view(T, H, d_head).transpose(0, 1)

    scores = Q @ K.transpose(-2, -1)
    if scale:
        scores = scores / math.sqrt(d_head)

    if causal:
        mask = torch.tril(torch.ones(T, T, dtype=torch.bool))
        scores = scores.masked_fill(~mask, float('-inf'))

    weights = torch.softmax(scores, dim=-1)
    entropy = -(weights * weights.clamp_min(1e-9).log()).sum(dim=-1).mean(dim=-1)

    return scores, weights, entropy

scores, weights, entropy = run_attention(x, scale=True, causal=True)
logging.info("entropy per head: %s", entropy)
```

Add prints for:

- one head's raw score matrix
- one head's weight matrix
- upper triangle sums
- max attention per row

If you want a cleaner visualization

Install matplotlib:

```
pip install matplotlib
```

Then plot each head:

```
import matplotlib.pyplot as plt

fig, axes = plt.subplots(1, H, figsize=(4 * H, 4))
for h in range(H):
    axes[h].imshow(weights[h].detach().numpy(), cmap="viridis")
    axes[h].set_title(f"Head {h}")
plt.tight_layout()
plt.show()
```

For token labels, use `axes[h].set_xticks(...)` and `set_xticklabels(...)`.

If you use real sentence tokens, heatmaps become much easier to interpret.

Questions To Answer

1. When you removed the `1/sqrt(d_head)` scaling factor, how much did the raw dot products grow, and how much did per-head entropy fall?
2. Did every head become sharply peaked without scaling, or did some heads stay more spread out? What does that suggest about head behavior?
3. When you removed causal masking, which positions first started attending to the future in obvious ways?
4. If training loss improved after removing masking, why is that not evidence that the model got better?
5. What tradeoff is the scaling factor protecting: expressiveness, numerical stability, training signal, or all three?
6. If you wanted to make attention less brittle in a larger model, what would you inspect next: initialization, normalization, head dimension, or softmax temperature?

Go Further

Read and search in this order.

1. Karpathy's "GPT from scratch" video. Search exactly: `Andrej Karpathy GPT from scratch attention`

   This is the right bridge between toy attention code and a full small GPT.
2. "Attention Is All You Need." Search exactly: `Attention Is All You Need pdf`

   You do not need to read the whole paper in one pass. Start with the attention equations and the architecture diagram. Now those diagrams will finally mean something.
3. Search arXiv for: `attention entropy transformer heads`

   This leads you toward work on head specialization, head redundancy, and attention diagnostics.
4. Search arXiv for: `causal masking transformer autoregressive leakage`

   This is useful if you want to see how training-inference mismatch shows up when causality is broken.
5. If you want to inspect trained heads on real text, look for small GPT implementations with visualization notebooks. Search: `transformer attention visualization notebook`

   The best next step is not reading ten papers. It is plotting real heads and trying to tell a story about what each one is doing.
6. Keep the repo progression straight. You are still in the build-up. Next comes writing a minimal GPT yourself before touching larger training pipelines.

What You Now Know

You can build a causal multi-head attention module from scratch, inspect its matrices, and explain what each piece does: why `QK^T` produces token-to-token scores, why softmax turns those scores into weights, and why multiplying by `V` creates a context-aware representation. You have also seen the two easy ways to break it. Remove `1/sqrt(d)` and softmax saturates; remove the causal mask and the model cheats by reading the future during training. The result is that you can read an attention heatmap and say something specific: this head is local, this head jumps to punctuation, this matrix leaks future information. That is a different level of understanding than knowing the formula.

Starting Point

Karpathy's free "GPT from scratch" walkthrough is the starting point for this project:

- The Annotated Transformer: https://nlp.seas.harvard.edu/annotated-transformer/
- Karpathy's free "GPT from scratch" walkthrough: https://www.youtube.com/results?search_query=Andrej+Karpathy+GPT+from+scratch

Keep the progression straight:

- `micrograd` taught you how gradients and backpropagation work
- Project 2 showed you a tiny language model learning next-token prediction
- Project 3 showed you that token boundaries are learned design choices, not natural laws
- this project teaches you how those tokens actually interact with one another in context
- next comes the blank-file bridge, where you write a full minimal GPT before touching larger reference implementations

---

Project 5: Your GPT From A Blank File

Hook

Why does a GPT feel mysterious right up until the moment you try to write one yourself? You already built the pieces: gradients, a tiny language model, a tokenizer, and attention. So why does opening a blank file still feel like stepping off a curb you cannot see?

Because the hard part is no longer any one component. The hard part is the glue. A GPT is not "attention plus some layers." It is a whole system where token embeddings, positional information, residual paths, normalization, initialization, batching, optimization, clipping, checkpointing, and generation all have to cooperate. Miss one line and the whole thing does not become slightly worse. It becomes nonsense, NaNs, or wasted hours.

I wrote my first GPT three weekends in a row before I had something I trusted. The first weekend I shipped a file that looked correct, trained, produced text, and was quietly leaking validation data into the training stream because I had reused one index variable in two places. Until you have written this yourself once, you do not know how many of those traps the polished code is hiding from you.

The Concept

Here is the plain-English picture of what we are assembling.

A tokenizer turns raw text into tokens, which are just the units the model sees. Maybe they are characters. Maybe they are subword pieces like `ing` or `tion`. The model never sees "language" directly. It sees token IDs, and everything else is built on top of that integer sequence.

An embedding table is a lookup table that turns each token ID into a list of numbers. That list is the token's current internal representation. If token ID `42` means `"king"`, the embedding table maps `42` to some vector in `d_model` dimensions. If `d_model = 256`, then `"king"` becomes 256 learned numbers that will shift throughout training as the model figures out what "king" means in context.

Position matters too. `"dog bites man"` and `"man bites dog"` use exactly the same tokens in a different order. So the model adds position information alongside token identity: if token embeddings say *what* a token is, **positional embeddings** say *where* it sits in the sequence.

Then the **transformer blocks** begin. Each block does two jobs: first, attention lets each token decide which earlier tokens matter right now; second, the feedforward network does local thinking on each position after attention has already mixed in context from elsewhere. These two sublayers are wrapped in residual connections, the `x + F(x)` skip lane that carries the old signal forward by simple addition. If the new sublayer is not helping yet, the old information still passes through untouched. LayerNorm keeps the scale of activations under control at each step, functioning like electrical regulation. Without it, one layer can push values into a range the next layer cannot sensibly consume. Figure 5.1 shows how attention and the feedforward network both feed back into the residual stream.

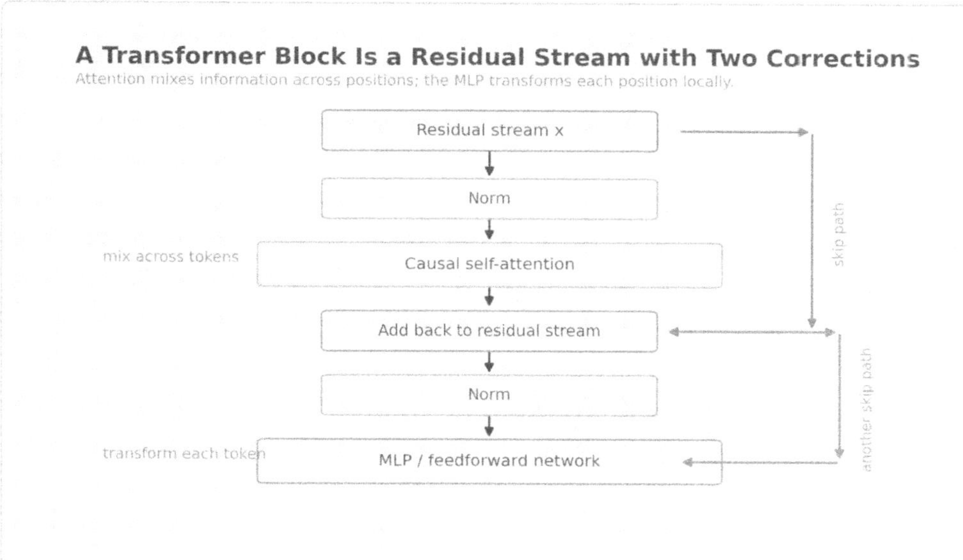

Figure 5.1. A transformer block is a residual stream with two corrections: attention mixes information across positions, and the feedforward network transforms each position locally before both updates rejoin the running hidden state.

After N of these blocks, the model has a final hidden state for every position in the input sequence. Then it projects that hidden state into vocabulary-sized logits: one score per possible next token. The score for `"e"` might be high, the score for `"%"` might be low. Softmax turns those scores into probabilities.

Training is next-token prediction. If the input is:

```
To be, or not to b
```

The target is:

```
o be, or not to be
```

The model predicts the next token at every position, compares those predictions to the true next tokens, and computes a loss: one number that tells you how wrong it is. Then backpropagation assigns blame through the whole network, the optimizer nudges the parameters, and you do it again, ten thousand times or more. That cycle, repeated continuously, is GPT training in one breath: guess the next token, measure the mistake, assign blame, update the numbers.

The first time I watched this loop run on my own code, I expected the loss to fall in a clean diagonal. It does not. It bounces. Sometimes it bounces for two hundred steps before it commits to a direction. That bouncing is a feature, not a bug, but you do not learn that from a textbook curve.

So why does this chapter matter as a blank-file exercise? Because until you write the whole thing yourself, "GPT" is still a bag of named parts in your head. After you write it, you know where the batch comes from, why the targets are shifted by exactly one position, where **weight tying** saves parameters, why learning rate schedules are not decoration, why clipping exists, why initialization is part of the model itself, and why checkpointing is not optional if your time matters. You stop thinking "a GPT has transformer blocks" and start thinking "a GPT is a training system that predicts one token at a time under very specific constraints." That is a better mental model to carry into everything that follows.

Why It Matters

Without building this yourself, the next reference implementation will lie to you by accident. Not because the code is wrong, but because production code hides pain. A good reference implementation makes dozens of choices look inevitable. If you meet those choices only inside polished code, they look like style. They are not style. They are survival.

This is the strongest opinion I will commit to in the whole book: most "from scratch" tutorials I have read fail right here. They give you the transformer block and skip the wiring around it, and the wiring is where every painful weekend of my life has been spent.

This project matters because it turns the transformer from architecture into system. When training loss drops, you know that is not one knob. It is the result of many agreements being kept at once: the tokenizer gives sensible units, the batch sampler does not feed garbage windows, initialization does not trap the model in symmetry, the learning rate is high enough to learn but not high enough to explode, gradient clipping catches rare spikes before they wreck the run.

If you can write a GPT from a blank file, you stop treating language models as sealed machines. You can now ask: Why did validation loss stop improving? Is this instability optimizer state or model code? Did I really resume training, or only reload weights?

The Build

Start with one file: `my_gpt.py`.

Do not read Karpathy's `microgpt` first. That restraint matters. If you read the reference first, your brain will follow it. You will type something that works, but you will miss the moment where you had to decide what the model needs. This chapter is about those decisions.

Build the file in layers.

This is the first project where a reader using Claude or Codex may reasonably choose to skim more aggressively. That is fine, but only if the code is read as a map rather than as decoration. The right question for each section is: what system responsibility lives here? Data preparation, batching, attention, residual computation, optimization, scheduling, checkpointing, and sampling are separate responsibilities, and the code exists to make those boundaries visible.

1. Decide the smallest complete system

You need exactly these parts:

- imports
- config
- tokenizer loading or character fallback

- data loading and train/validation split
- batch sampler
- model components
- GPT class with `forward()` and `generate()`
- optimizer
- learning rate schedule
- training loop
- validation evaluation
- sample printing
- checkpoint save/load

That list is your map. Anything outside it is optional. Anything missing from it is a hole.

A minimal config might look like this:

```
from dataclasses import dataclass

@dataclass
class Config:
    batch_size = 64
    block_size = 128
    d_model = 256
    n_heads = 4
    n_layers = 6
    dropout = 0.2
    learning_rate = 3e-4
    min_lr = 3e-5
    warmup_steps = 200
    max_steps = 10000
    eval_interval = 200
    eval_steps = 50
    grad_clip = 1.0
    device = "cuda"
```

These are not magic numbers. They are a budget. `block_size` is sequence length: how many previous tokens the model can see. `d_model` is width: how many numbers represent each token internally. `n_layers` is depth: how many transformer blocks you stack. If you run on a small GPU, these numbers control whether the run fits in memory. If you run on CPU, reduce them hard.

2. Load text and tokenize it

You need raw text first. Shakespeare is the right dataset here because it is small, clean, public, and famous enough that generated samples become recognizable quickly — you will know at a glance whether the model is producing something Shakespeare-like or just noise.

If you already built a tokenizer in Project 3, use it. If not, use a character-level fallback. The fallback is simpler: build a sorted list of unique characters, map each character to an integer, and encode the entire corpus into token IDs. That gets you moving without blocking the project. The important thing is not tokenizer sophistication here — the important thing is that your model trains end to end.

A minimal character-level encoder:

```python
chars = sorted(list(set(text)))
stoi = {ch: i for i, ch in enumerate(chars)}
itos = {i: ch for ch, i in stoi.items()}

def encode(s):
    return [stoi[c] for c in s]

def decode(ids):
    return "".join(itos[i] for i in ids)
```

If you are reading rather than typing, the important thing is not the dictionary comprehensions themselves. It is the contract they create. `encode()` turns human text into token IDs the model can consume. `decode()` turns model outputs back into readable text. Every larger language-model codebase has this boundary somewhere, even when the tokenizer object is much more elaborate.

Now split the encoded data into train and validation. Do it once and keep it simple:

```python
n = int(0.9 * len(data))
train_data = data[:n]
val_data = data[n:]
```

You now have a real language modeling dataset: one long stream of token IDs.

3. Write the batch sampler honestly

This looks small. It is not small.

Your model trains on fixed-length windows. That means you sample a random start index `i`, take `x = data[i:i+T]`, and set the target `y = data[i+1:i+T+1]`. That shift by one token is the whole training objective — the model sees tokens 1 through T and predicts tokens 2 through T+1.

A sampler:

```python
def get_batch(split: str, cfg: 'Config') -> tuple:
    data = train_data if split == "train" else val_data
    ix = torch.randint(len(data) - cfg.block_size - 1, (cfg.batch_size,))
    x = torch.stack([data[i:i+cfg.block_size] for i in ix])
    y = torch.stack([data[i+1:i+cfg.block_size+1] for i in ix])
    return x.to(cfg.device), y.to(cfg.device)
```

This is a high-value function to understand because `get_batch()` quietly defines the learning problem. It chooses what counts as context, what counts as the next-token target, and whether training windows reflect the real corpus or mix together things that should have stayed separate. In larger systems, data bugs often begin here rather than inside the model.

Now the honesty part. If your dataset is one continuous Shakespeare file, then windows crossing line breaks are fine. That is real adjacent text. But if your dataset is many documents packed together, you must decide whether the model is allowed to train across document boundaries. If you concatenate everything blindly, the model may learn that the end of one document predicts the start of another unrelated one. That is fake context, and it pollutes the training distribution in ways that are hard to detect later.

I have hit this exact bug. When I was prototyping a small training corpus for an internal experiment, I packed three unrelated sources back to back with no separator. Validation loss looked fine. Generations looked fine. Then I noticed the model was writing answers to one of the source's questions using vocabulary from another. The data was lying to the model in a way my metrics could not see.

Your rule must exist in code. One clean rule: if using one continuous text file, allow any window inside the file; if using many documents, insert a special separator token between documents and allow crossing only through that separator; never silently wrap around the end of the array. Write the rule in a comment. Future you needs to know whether the training distribution is honest.

4. Build the transformer pieces

Now you start assembling the model. At minimum you need:

- token embedding table
- positional embedding table
- multi-head causal self-attention
- feedforward network
- transformer block
- final normalization
- output projection to vocabulary logits

Token and position embeddings

If `idx` has shape `(B, T)`, where `B` is batch size and `T` is sequence length, then token embeddings have shape `(B, T, d_model)` — each token ID becomes a vector. Position embeddings have shape `(T, d_model)` and get broadcast across the batch. Then you add them:

```
tok = self.token_embedding(idx)          # (B, T, d_model)
pos = self.position_embedding(pos_ids)   # (T, d_model)
x = tok + pos
```

That addition is the model's first internal representation of the sequence: what each token is plus where it sits.

Causal self-attention

You built attention already. Now you put it inside a full model.

Each token creates three projections (Query: what am I looking for? Key: what do I contain? Value: what should I contribute if selected?) and for each head, attention computes dot products between queries and keys, scales by `1 / sqrt(head_dim)`, applies a causal mask so tokens cannot see the future, then takes a softmax-weighted sum of values.

Here is a minimal attention module:

```
class CausalSelfAttention(nn.Module):
    def __init__(self, cfg):
        super().__init__()
        assert cfg.d_model % cfg.n_heads == 0
        self.n_heads = cfg.n_heads
        self.head_dim = cfg.d_model // cfg.n_heads
        self.qkv = nn.Linear(cfg.d_model, 3 * cfg.d_model)
        self.proj = nn.Linear(cfg.d_model, cfg.d_model)
        self.dropout = nn.Dropout(cfg.dropout)
        self.register_buffer(
            "mask",
            torch.tril(torch.ones(cfg.block_size, cfg.block_size))
                .view(1, 1, cfg.block_size, cfg.block_size)
        )

    def forward(self, x):
        B, T, C = x.shape
        qkv = self.qkv(x)
        q, k, v = qkv.split(C, dim=2)

        q = q.view(B, T, self.n_heads, self.head_dim).transpose(1, 2)
        k = k.view(B, T, self.n_heads, self.head_dim).transpose(1, 2)
        v = v.view(B, T, self.n_heads, self.head_dim).transpose(1, 2)

        att = (q @ k.transpose(-2, -1)) / math.sqrt(self.head_dim)
        att = att.masked_fill(self.mask[:, :, :T, :T] == 0, float("-inf"))
        att = F.softmax(att, dim=-1)
        att = self.dropout(att)

        y = att @ v
        y = y.transpose(1, 2).contiguous().view(B, T, C)
        return self.dropout(self.proj(y))
```

If you only retain one reading of this class, retain this one: `qkv` creates three views of the same hidden state, the middle of the method computes masked attention weights, and the final lines merge the per-head outputs back into one stream the rest of the network can use. That is the whole attention story in code form.

Nothing here is decorative. The mask prevents the model from cheating by looking at future tokens. The scaling by `1/sqrt(head_dim)` prevents softmax from saturating. Without it, large dot products push probabilities to near-one, killing gradient flow. The final projection lets the heads recombine their outputs into a unified representation.

A scar from drafting this section: my first hand-written attention had the mask broadcasting along the wrong axis. The model still trained. Loss still fell. Samples were almost coherent. Then I noticed the same word kept showing up in positions where it had no business being there, and I traced it back to a 1-vs-2 axis bug in the mask. The mask is not what you check first when training looks fine. It is what you check first when training looks fine and the outputs are subtly wrong.

Feedforward network

Attention mixes information across positions. The feedforward network does per-position computation after that mixing, so each token reflects on what it has gathered. A common shape is `d_model → 4*d_model → d_model`; that expansion gives the model room to transform features nonlinearly before projecting back down.

```
class MLP(nn.Module):
    def __init__(self, cfg):
        super().__init__()
        self.net = nn.Sequential(
            nn.Linear(cfg.d_model, 4 * cfg.d_model),
            nn.GELU(),
            nn.Linear(4 * cfg.d_model, cfg.d_model),
            nn.Dropout(cfg.dropout),
        )

    def forward(self, x):
        return self.net(x)
```

Transformer block

Now combine them with LayerNorm and residual connections. In this book's notation, a residual connection is `x + F(x)`. The sublayer learns a correction, not a replacement. If attention produces a bad correction early in training, the original signal still passes through. A common modern layout is pre-norm:

```
class Block(nn.Module):
    def __init__(self, cfg):
        super().__init__()
        self.ln1 = nn.LayerNorm(cfg.d_model)
        self.attn = CausalSelfAttention(cfg)
        self.ln2 = nn.LayerNorm(cfg.d_model)
        self.mlp = MLP(cfg)

    def forward(self, x):
        x = x + self.attn(self.ln1(x))
        x = x + self.mlp(self.ln2(x))
        return x
```

From a code-reading perspective, a transformer block is easier than it first appears. `ln1 → attn → residual add` is one correction to the running representation. `ln2 → mlp → residual add` is a second correction. If you can identify those two residual updates while skimming, you can navigate most GPT implementations without getting lost in the class scaffolding.

This arrangement trains more reliably than post-norm in many small GPT setups because the normalization happens before each sublayer rather than after, keeping the inputs to attention and the MLP in a controlled range regardless of what previous blocks have done.

5. Write the GPT class as one whole system

This is the center of the file. Your GPT class holds embeddings, transformer blocks, final norm, output projection, initialization, weight tying, forward pass, and generation all in one place.

A minimal skeleton:

```python
class GPT(nn.Module):
    def __init__(self, cfg, vocab_size):
        super().__init__()
        self.cfg = cfg
        self.token_embedding = nn.Embedding(vocab_size, cfg.d_model)
        self.position_embedding = nn.Embedding(cfg.block_size, cfg.d_model)
        self.blocks = nn.ModuleList([Block(cfg) for _ in range(cfg.n_layers)])
        self.ln_f = nn.LayerNorm(cfg.d_model)
        self.lm_head = nn.Linear(cfg.d_model, vocab_size, bias=False)

        # weight tying
        self.lm_head.weight = self.token_embedding.weight

        self.apply(self._init_weights)
```

The broader lesson is that the `GPT` class is not just where the layers live. It is where the architectural contracts are enforced: shared embeddings, initialization policy, maximum context length, block stack, and final projection. When you inspect a production model file later, this is usually the place where the model's real shape is decided.

That one line of weight tying matters a lot:

```python
self.lm_head.weight = self.token_embedding.weight
```

It means the same matrix used to map token IDs into vectors is also used, transposed by the linear layer, to map hidden states back into token logits. The dictionary the model uses to read tokens is also the dictionary it uses to write them.

Why does that often help? Because input and output tokens live in the same semantic world. If `"king"` and `"queen"` should be near each other when read, it often helps if the model uses that same geometry when scoring them as outputs. You also cut parameters significantly. Suppose `vocab_size = 50000` and `d_model = 768`: one embedding table costs `50000 * 768 = 38.4M` weights, and an untied output projection costs another `38.4M`. Tie them, and one of those tables disappears entirely. Not a tiny savings, but a whole major component.

Now write initialization.

6. Treat initialization as part of the model

Newcomers often think initialization is paperwork. It is not. Initialization decides what kind of signal exists before learning begins. If weights start too large, activations and gradients explode. If they start too small, everything looks the same and updates become weak. If they all start at zero, many parameters remain perfect twins forever. They receive identical gradients and never diverge, so the model's capacity stays locked away unused.

While running fusion experiments at scale, one of the recurring surprises across a long set of runs on models from a few hundred million to several billion parameters was how often a "bad start" looked like a "bad architecture" two thousand steps in. You can fix initialization in one line. You cannot recover those thousand steps. I now treat init as part of the model file, not a separate concern. It is one of the cheapest investments in the whole training stack.

You can use Xavier initialization for linear layers and a small normal distribution for embeddings.

For example:

```
def _init_weights(self, module):
    if isinstance(module, nn.Linear):
        nn.init.xavier_uniform_(module.weight)
        if module.bias is not None:
            nn.init.zeros_(module.bias)
    elif isinstance(module, nn.Embedding):
        nn.init.normal_(module.weight, mean=0.0, std=0.02)
```

Why Xavier for linear layers? Because it tries to keep the scale of activations from growing or shrinking too much as they pass through layers. Not "good defaults" or "industry practice," but scale control. If you use careless initialization, like very large random values, you often see attention logits explode early, softmax become nearly one-hot too soon, and training turn erratic before it ever gets useful. If you initialize everything to zero, the model cannot break symmetry. More on that in BREAK IT.

7. Write `forward()` so shapes stay honest

Your `forward()` must do two jobs: return logits during inference, and return logits plus loss during training if targets are supplied. Keeping one code path for both training and generation prevents subtle bugs where the training and inference paths diverge.

```
def forward(self, idx: torch.Tensor, targets: torch.Tensor = None) -> tuple:
    B, T = idx.shape
    assert T <= self.cfg.block_size

    pos = torch.arange(0, T, device=idx.device)
    x = self.token_embedding(idx) + self.position_embedding(pos)

    for block in self.blocks:
        x = block(x)

    x = self.ln_f(x)
    logits = self.lm_head(x)

    loss = None
    if targets is not None:
        B, T, V = logits.shape
        loss = F.cross_entropy(logits.view(B * T, V), targets.view(B * T))

    return logits, loss
```

If you are reading fast, `forward()` is the method to study first. It answers the practical question: given token IDs, how do they become logits and optionally a loss? Everything above it defines components. This method shows how those components become one end-to-end prediction path.

That `cross_entropy` call is the entire training objective in one line. The model outputs scores over all vocabulary tokens at every position, and the targets specify the correct next token ID at every position. Cross-entropy asks how surprised the model was by the right answer. Lower loss means less surprise, which means the model is getting better at predicting what comes next.

8. Write `generate()` for autoregressive sampling

Autoregressive means "one token at a time, feeding the latest prediction back in." The model does not write a whole paragraph in one pass. It predicts the next token, appends it, then predicts the next after that, looping until it has produced the desired number of tokens.

```
@torch.no_grad()
def generate(self, idx: torch.Tensor, max_new_tokens: int, temperature: float = 1.0) →
        torch.Tensor:
    for _ in range(max_new_tokens):
        idx_cond = idx[:, -self.cfg.block_size:]
        logits, _ = self(idx_cond)
        logits = logits[:, -1, :] / temperature
        probs = F.softmax(logits, dim=-1)
        next_idx = torch.multinomial(probs, num_samples=1)
        idx = torch.cat([idx, next_idx], dim=1)
    return idx
```

This block is worth learning as a recognizable template. Most autoregressive generation code, even in much larger libraries, still does these same moves: crop to legal context, compute logits for the current suffix, take the last-position distribution, choose the next token, append it, and repeat.

A few things matter here. `idx[:, -block_size:]` crops the context to the model's maximum sequence length, since anything beyond that would need positional embeddings the model was never trained for. `logits[:, -1, :]` takes only the prediction for the last position, because that is the next-token distribution. `torch.multinomial` samples from the probability distribution instead of always taking the argmax, which means outputs can vary. If you want more deterministic text, lower the temperature or use argmax. If you want more variety, raise the temperature a bit.

9. Count parameters before and after weight tying

Do not just implement weight tying — measure it. That is part of the lesson.

A simple parameter count:

```
def count_params(model):
    return sum(p.numel() for p in model.parameters())
```

Run it with tied weights, then comment out tying and run it again. You should report total parameters with tying, total parameters without tying, the difference, validation loss after a fixed number of steps, and sample quality after a fixed number of steps.

Why does tying often help rather than hurt? Two reasons. First, it regularizes the model a bit; the input and output token spaces cannot drift into unrelated geometries. Second, it spends parameters where they matter more: if you have a fixed budget, duplicating a giant vocab matrix is often less useful than using that capacity elsewhere. This is one of those moves that sounds like a constraint but often behaves like a helpful bias.

10. Write the optimizer and your own learning rate schedule

Use AdamW. It is a standard optimizer for GPT-style training because it handles noisy gradient scales better than plain SGD and its weight decay behaves more predictably in this setting.

But do not call a scheduler helper and move on. Write the schedule yourself. You need two phases: linear warmup, where you start small and ramp to peak learning rate, and cosine decay, where you gradually reduce toward a minimum learning rate. These solve two different problems. Early in training the network is uncalibrated, so large steps too early can wreck it before useful representations form. Warmup protects that fragile beginning. Once the model is learning sensible structure, you usually want smaller and smaller corrective steps, since early training is broad movement and late training is fine adjustment. Cosine decay handles that.

```
def get_lr(step: int, cfg: 'Config') -> float:
    if step < cfg.warmup_steps:
        return cfg.learning_rate * (step + 1) / cfg.warmup_steps
    if step > cfg.max_steps:
        return cfg.min_lr
    decay_ratio = (step - cfg.warmup_steps) / (cfg.max_steps - cfg.warmup_steps)
    coeff = 0.5 * (1.0 + math.cos(math.pi * decay_ratio))
    return cfg.min_lr + coeff * (cfg.learning_rate - cfg.min_lr)
```

During warmup, LR climbs from near zero to peak linearly. After warmup, cosine decay smoothly lowers it toward `min_lr`. This schedule is not sacred, but the pattern it encodes matters, and you will see it clearly when you compare against a constant learning rate in BREAK IT.

11. Write the training loop yourself

This is where the file becomes a real system. The loop is:

1. get batch
2. set learning rate
3. forward pass
4. backward pass
5. clip gradients
6. optimizer step
7. log loss
8. evaluate sometimes
9. sample sometimes
10. checkpoint sometimes

A minimal version:

```python
import logging

model = GPT(cfg, vocab_size).to(cfg.device)
optimizer = torch.optim.AdamW(model.parameters(), lr=cfg.learning_rate)

for step in range(cfg.max_steps):
    lr = get_lr(step, cfg)
    for param_group in optimizer.param_groups:
        param_group["lr"] = lr

    xb, yb = get_batch("train", cfg)
    logits, loss = model(xb, yb)

    optimizer.zero_grad(set_to_none=True)
    loss.backward()
    torch.nn.utils.clip_grad_norm_(model.parameters(), cfg.grad_clip)
    optimizer.step()

    if step % cfg.eval_interval == 0:
        train_loss = estimate_loss("train")
        val_loss = estimate_loss("val")
        logging.info("step %d  train_loss %.4f  val_loss %.4f", step, train_loss, val_loss)
        sample_text(model)
        save_checkpoint(...)
```

If you need an interview-level reading of the training loop, say it this way. The loop fetches a batch, sets the current step size, computes predictions and loss, backpropagates, clips extreme gradients, updates weights, then periodically evaluates and checkpoints. That explanation is usually more useful than being able to recite every line of the loop from memory.

A few lines deserve more attention. `optimizer.zero_grad(set_to_none=True)` clears old gradients efficiently, avoiding the extra memory of zeroed tensors. `loss.backward()` runs backpropagation through the entire graph. `clip_grad_norm_` prevents rare huge gradients from producing catastrophic updates. Easy to treat as boilerplate, but it is a guardrail against gradient spikes that can appear suddenly after many well-behaved steps.

What caught me off guard with gradient clipping was the asymmetry. It is the kind of safety code I assumed I would tune carefully, and instead it spent ninety percent of the run doing literally nothing and then, on one step out of forty thousand, saved the entire training run. That asymmetry is the actual reason it is in the loop.

12. Measure validation loss separately

Training loss tells you how well the model fits the examples it is actively learning from. Validation loss tells you how well it generalizes to held-out text. If training loss falls but validation loss stalls or rises, the model may be memorizing quirks instead of learning patterns that transfer. You need both metrics running throughout training, not just the one that feels encouraging.

A simple evaluation helper:

```
@torch.no_grad()
def estimate_loss(split: str) -> float:
    model.eval()
    losses = torch.zeros(cfg.eval_steps)
    for k in range(cfg.eval_steps):
        xb, yb = get_batch(split, cfg)
        _, loss = model(xb, yb)
        losses[k] = loss.item()
    model.train()
    return losses.mean().item()
```

Record validation loss over time — that is your primary metric in this chapter. Generated samples matter too, but samples can mislead you. A lucky sample can look better than the model really is. Validation loss is less charming and more trustworthy. You want both, but when they disagree, trust the loss.

13. Print samples during training

Watching samples change at steps 100, 1000, and 10000 teaches you how loss maps to output quality. That connection is not obvious from numbers alone.

Typical progression on character-level Shakespeare:

At step 100:

```
thA m, oul t
fou ths,
```

It knows letters and some spacing. Not much else.

At step 1000:

```
My lord, I shall be made to know the man
That hath the king...
```

The grammar starts to cohere. The rhythm begins to look like English.

At step 10000:

```
KING RICHARD:
Why, then the day is ours; and let them fly.
```

It may still drift, but it feels like text written by something that has learned English-like structure and Shakespeare-like formatting. Those milestones matter because they connect abstract loss curves to visible behavior. You start to develop intuition for what a given validation loss actually feels like as output.

The honest version of this section: I have stared at these mid-training samples for far longer than I have stared at the loss curve. Loss is the number you trust. Samples are the number you feel. You need both, and the difference between them is how you build taste.

14. Add checkpoint save/load before you need it

Checkpointing is not a bonus feature. It is part of training. A checkpoint should save at least model `state_dict`, optimizer `state_dict`, current step, and config, and maybe RNG state if you want stronger resume fidelity.

Years before this book, I ran a multi-state SaaS modernization where uptime numbers and resume fidelity for production database replicas were the work. The intuition crosses cleanly: the moment you treat checkpointing as optional, you are quietly choosing to never run anything important.

A simple save:

```
def save_checkpoint(path: str, model: 'GPT', optimizer, step: int, cfg: 'Config') → None:
    torch.save({
        "model": model.state_dict(),
        "optimizer": optimizer.state_dict(),
        "step": step,
        "config": cfg.__dict__,
    }, path)
```

A simple load:

```
def load_checkpoint(path: str, model: 'GPT', optimizer) → int:
    ckpt = torch.load(path, map_location=cfg.device)
    model.load_state_dict(ckpt["model"])
    optimizer.load_state_dict(ckpt["optimizer"])
    return ckpt["step"]
```

Why save optimizer state too? Because AdamW stores moving averages of past gradients. If you reload only weights and not optimizer state, your run resumes with the model's memory erased but the weights preserved. That is not a clean continuation, and the trajectory changes in ways that may not be obvious immediately. If you care about resume fidelity, save both.

A real proof of checkpointing looks like this: train for a while, save, kill the process, resume, then verify that loss continues smoothly instead of jumping oddly. That proves the system, not just the file write.

15. Keep the whole file readable

Your final file will probably land around 300 to 500 lines. That is the right size — big enough to be complete, small enough to hold in your head. If your own first GPT sprawls into thousands of lines, the architecture disappears inside scaffolding. The point here is to feel the whole machine at once.

> **NOTE — Weight tying works because the output projection reuses the embedding matrix transposed.**
>
> This line can look suspicious:
>
> ```
> self.lm_head.weight = self.token_embedding.weight
> ```
>
> An embedding table of shape `(V, d_model)` can be reused by a linear layer mapping from `d_model` to `V` — the linear layer internally transposes the weight matrix in the multiply. If the hidden state is a point in embedding space, the output projection scores how aligned that point is with each token embedding row.

Tying often works because the model predicts the next token by comparing the hidden state to the same token vectors it learned to read.

! **WARNING — Zero initialization prevents neurons from specializing, because identical weights receive identical gradients forever.**

Newcomers often assume: "If all weights start at zero, backpropagation will still change them." Not really. If two neurons start with exactly the same weights and receive the same gradients, they stay identical after every update — they never specialize. Zero initialization makes many weights perfect clones of each other, losing diversity inside a layer. That is called failing to break symmetry. Random initialization is not decorative — it gives neurons slightly different starting jobs so training can separate them into distinct, useful functions.

! **WARNING — Gradient clipping is a seat belt, not a substitute for a stable training setup.**

NaN means "not a number." NaNs appear when a value blows up far enough that the math becomes undefined — from a too-high learning rate, large gradients, unstable activations, or bad numerical combinations. Gradient clipping does not make a bad setup good. It catches rare dangerous updates before they launch the model off a cliff. You may never need it on a given run, but when you do need it, it saves everything.

NOTE — Loss improves smoothly while sample quality jumps, because they measure different things.

Loss is an average over many token predictions; sample quality is human perception applied to one generated sequence. A small average improvement can suddenly produce visible punctuation, line structure, or name formatting that makes the sample feel much better, even though the loss moved only modestly. Do not rely on one sample — print samples regularly, but trust trends, not anecdotes.

BREAK IT

Run every ablation against the same baseline and save each run in its own directory:

```
python my_gpt.py --run_name baseline --seed 1337
python my_gpt.py --run_name no_tie --seed 1337
python my_gpt.py --run_name const_lr --seed 1337
python my_gpt.py --run_name no_clip --seed 1337
python my_gpt.py --run_name zero_init --seed 1337
python my_gpt.py --run_name no_ckpt --seed 1337
```

Before you start, make sure every run writes:

- `out/<run_name>/metrics.csv`
- `out/<run_name>/sample_step0100.txt`

- `out/<run_name>/sample_step1000.txt`
- `out/<run_name>/sample_final.txt`
- `out/<run_name>/ckpt.pt` except for the no-checkpoint ablation

If your script does not log metrics yet, add:

```
with open(log_path, "a", encoding="utf-8") as f:
    f.write(f"{step},{train_loss:.4f},{val_loss:.4f},{lr:.8f},{grad_norm:.4f}\n")
```

Use the same prompt for all sample comparisons:

```
ROMEO:
```

A system that only works when left untouched has not taught you much. You need to break the machine on purpose and watch which failure arrives first. This project has five required breaks.

1. Remove weight tying

Exact change — comment out:

```
self.lm_head.weight = self.token_embedding.weight
```

Leave `lm_head` as an independent `nn.Linear`.

What changes first: parameter count jumps by roughly `V * d_model`, memory use increases, and validation loss is often slightly worse or about the same early before trailing the tied version later. Samples often look a little less coherent for the same training budget.

What to watch: exact parameter count before and after, validation loss at the same fixed steps, sample text at step 1000 and step 10000.

What this proves: weight tying is not only a memory trick. It also imposes a useful constraint. The same token geometry supports reading and writing. Without that shared geometry, the model has more freedom but not always better use of it. Extra parameters can make memorization easier; shared structure can make generalization easier.

2. Replace warmup + cosine decay with a constant learning rate

Exact change — replace the call to `get_lr(step, cfg)` with a fixed value:

```
for param_group in optimizer.param_groups:
    param_group["lr"] = cfg.learning_rate
```

What changes first: usually one of two things. If the constant learning rate is your peak LR, early training is shakier, loss bounces more, and later training often plateaus at a worse validation loss. If the constant rate is lowered to stay safe, training is more stable but progress is slower, and you may never catch up to scheduled training in the same step budget.

What to watch: train loss curve shape in the first few hundred steps, validation loss at the same wall-clock time, whether the scheduled run reaches lower loss by the end, and sample quality at equal step counts.

What this proves: the schedule is not paperwork. It solves two different problems at two different times. Warmup protects the unstable beginning. Decay helps the later phase stop overshooting. A constant LR forces one compromise to cover both phases, which usually means it is too aggressive early or too timid late.

3. Remove gradient clipping and raise the learning rate slightly

Exact change — comment out:

```
torch.nn.utils.clip_grad_norm_(model.parameters(), cfg.grad_clip)
```

Then raise the peak learning rate. If you trained at `3e-4`, try `6e-4` or `1e-3` depending on model size.

What changes first: training may look fine for a while, then one step suddenly spikes, loss jumps hard, and soon after you may see `nan`. Generated text becomes garbage or sampling crashes. This is the important part: clipping often appears to do nothing right until the step where it saves the run.

What to watch: per-step loss, gradient norm if you log it, whether one bad update arrives after otherwise normal training, whether NaNs begin in logits, loss, or gradients. If you log gradient norm, you may see many steps around `0.8` to `2.5`, then a rare step at `40` or `200`, then collapse.

What this proves: gradient clipping protects the optimizer from rare extreme updates. It is there for tail events, not average events. Without it, one spike can erase thousands of good updates.

4. Initialize all parameters to zeros

Exact change — replace your initialization logic with:

```
def _init_weights(self, module):
    if isinstance(module, nn.Linear):
        nn.init.zeros_(module.weight)
        if module.bias is not None:
            nn.init.zeros_(module.bias)
    elif isinstance(module, nn.Embedding):
        nn.init.zeros_(module.weight)
```

What changes first: loss barely improves, if at all. Different neurons inside a layer behave identically. Logits often start uniform or near-uniform. Generated samples remain repetitive or meaningless. Training may appear "stable" while learning almost nothing.

Why: the network starts in perfect symmetry. Many weights receive identical gradients, so they stay identical. Multiple heads do not specialize. Multiple neurons do not specialize. The model acts like a much smaller model trapped inside a larger shell.

What to watch: first-step loss compared to random init, whether loss drops meaningfully after hundreds of steps, whether attention heads become different from one another, whether samples remain flat and repetitive.

What this proves: initialization is part of the learning system. Randomness at the start is not noise you tolerate. It is how the model gets permission to become different from itself.

5. Remove checkpointing, train for 30 minutes, kill the process

Exact change: delete or comment out all checkpoint save/load code. Then train. Then kill the process.

Nothing inside the model changes. That is the point. The failure is not mathematical. It is operational. You discover that 30 minutes of compute and attention vanished because your system had no memory outside RAM.

What to watch: your own reaction, how much progress existed only in process memory, whether you can recover anything meaningful, and how much confidence you lose in long runs without checkpoints.

What this proves: checkpointing is part of the training loop, not an afterthought. A model that only exists while the process lives is not a reliable system.

There is a second lesson here. Even if you save model weights but not optimizer state, resume may look fine at first but drift differently after restart. So there are really two breakages: no checkpoint at all causes total loss of work, while an incomplete checkpoint causes a silent change in training trajectory. The first hurts immediately. The second hurts quietly. Both matter.

Optional Homework

Operational Run Sheet

Use this exact sequence so your comparisons stay clean:

1. Run the baseline once and save its metrics and final checkpoint.
2. Run `no_tie` with only the weight-tying line removed.
3. Run `const_lr` with only the scheduler removed.
4. Run `no_clip` with clipping removed and LR raised to `6e-4`.
5. Run `zero_init` with only initialization changed.
6. Run `no_ckpt`, let it train for 20 to 30 minutes, kill it manually, then attempt resume.

Put the final results into one table:

```
run_name,param_count,val_loss_100,val_loss_1000,val_loss_final,max_grad_norm,notes
```

If you cannot run the full experiments, do the proxy version: cap every run at 500 to 1000 steps, keep the same seed across all runs, and compare trend differences rather than final sample quality.

Recommended base config for a local GPU:

```
batch_size = 64
block_size = 128
d_model = 256
n_heads = 4
n_layers = 6
dropout = 0.2
learning_rate = 3e-4
min_lr = 3e-5
warmup_steps = 200
max_steps = 10000
grad_clip = 1.0
```

Recommended base config for CPU or very small GPU:

```
batch_size = 16
block_size = 64
d_model = 128
n_heads = 4
n_layers = 4
dropout = 0.2
learning_rate = 3e-4
min_lr = 3e-5
warmup_steps = 100
max_steps = 3000
grad_clip = 1.0
```

Homework Lab 1: Build the baseline

1. Train the model using `my_gpt.py` with all components in place.

Record:

- parameter count
- train loss at step 100, 1000, 10000
- validation loss at step 100, 1000, 10000
- generated sample at step 100, 1000, 10000

Expected observations:

- step 100: text looks broken but letter statistics improve
- step 1000: word-like chunks and punctuation emerge
- step 10000: text becomes recognizable as English-like Shakespeare

Homework Lab 2: Measure weight tying

1. Train baseline with tying on.
2. Record parameter count and validation loss.
3. Comment out:

   ```
   self.lm_head.weight = self.token_embedding.weight
   ```

4. Train again with the same random seed if possible.

5. Compare:
    - parameter count
    - final validation loss
    - sample quality

Expected observation:

- untied model has noticeably more parameters
- loss is often slightly worse or not better enough to justify the added matrix in this setup

Homework Lab 3: Constant LR versus scheduled LR

1. Baseline run:
    - use your warmup + cosine schedule
2. Second run:
    - replace schedule with constant LR at the same peak learning rate
3. If the constant run is unstable, try a lower constant LR
4. Plot the two loss curves

You can log to CSV manually:

```
with open("log.csv", "a") as f:
    f.write(f"{step},{train_loss},{val_loss},{lr}\n")
```

Then plot with a quick script, notebook, or spreadsheet.

Expected observation:

- scheduled LR is smoother early and usually reaches lower validation loss later
- constant LR either bounces early or crawls late

Homework Lab 4: Remove clipping and provoke a spike

1. Comment out:

```
torch.nn.utils.clip_grad_norm_(model.parameters(), cfg.grad_clip)
```

2. Raise LR from `3e-4` to `6e-4`
3. Add gradient norm logging before the optimizer step:

```
import logging

total_norm = torch.nn.utils.clip_grad_norm_(model.parameters(), 1e9)
logging.info("grad_norm %.4f", float(total_norm))
```

That giant clip threshold logs the norm without meaningfully clipping.

4. Train and watch for:
    - sudden norm spike

- loss explosion
- NaN

Expected observation:

- normal-looking training for many steps
- then one bad region
- then collapse

Homework Lab 5: Zero initialization

1. Replace your init function with zero init for all weights.
2. Train for 500 to 1000 steps.
3. Compare against the baseline's first 500 to 1000 steps.

Expected observation:

- loss hardly moves
- samples stay flat
- model fails to develop distinct behavior across heads and neurons

Homework Lab 6: Checkpoint fidelity

1. Train for 1000 steps.
2. Save checkpoint.
3. Kill the process manually.
4. Resume from checkpoint.
5. Confirm:
   - global step resumes correctly
   - optimizer state reloads
   - validation loss trend continues smoothly
6. Then test the broken version:
   - save only model weights
   - do not save optimizer state
   - resume and compare loss behavior

Expected observation:

- full checkpoint resumes cleanly
- weights-only resume often works but changes the training trajectory

Questions To Answer

1. When you removed weight tying, what changed more noticeably: parameter count, loss, cr sample quality? Why do you think the biggest change showed up there first?

2. In the constant learning rate run, what failed first: early stability or late improvement? What compromise was the fixed LR forcing?
3. When you removed gradient clipping and raised the learning rate, did the run fail gradually or suddenly? What does that tell you about why clipping exists?
4. In the zero-initialization run, what evidence told you the model was stuck in symmetry rather than merely learning slowly?
5. After killing a run without checkpointing, which loss was worse: the lost compute time or the lost ability to trust future long runs? Why?
6. If you had to keep only one safeguard in a tiny local training setup, which would you keep first: good initialization, LR schedule, clipping, or checkpointing? What failure is that choice protecting against?

Go Further

Read Karpathy's `microgpt` only after your own file works. That timing matters. You want the reference implementation to answer your questions, not replace them.

Good directions from here:

- Search GitHub for `karpathy microgpt` and compare structure, not just code lines. Ask: what pain is the reference relieving?
- Read `nanoGPT` next and identify what changes when the toy becomes a more serious training prototype.
- Search arXiv for `weight tying language model` and `press wolf tied embeddings`. That line of work explains why shared input/output embeddings often help.
- Search arXiv for `learning rate warmup transformers` and `cosine decay language model training`. You will find that schedules are about optimization dynamics, not style.
- Search arXiv for `gradient clipping transformer training instability`. Pay attention to papers or posts that log gradient norms over time.
- Search for `muP maximal update parameterization` if you want to go deeper on initialization and scale behavior. That direction asks a sharp question: how should model behavior stay stable as width changes?
- If you want a benchmark mindset, compare validation loss and sample quality across your ablations on Shakespeare first, then repeat on TinyStories. TinyStories is small enough for local experiments and exposes different stylistic behavior than Shakespeare.

What You Now Know

You can now write a complete GPT training file from nothing. The whole machine, not only the transformer blocks. Every piece has a reason: shifted targets define the training objective, the batch sampler shapes the distribution the model sees, weight tying aligns the geometry of reading and writing tokens, initialization determines whether the model can break symmetry at all, warmup and decay solve different instability problems at different moments in training, clipping protects against rare destructive spikes, and checkpointing is the difference between a training system and a training gamble. When you open a larger reference implementation next, you will not see magic. You will see choices.

Starting Point

Start from your own blank file first:

- `my_gpt.py`

Only after your version runs, read Karpathy's comparison reference:

- Search: `Andrej Karpathy GPT from scratch`
- Comparison reference after finishing your own version: `microgpt`

Keep the progression straight:

- `micrograd` taught you how gradients and backpropagation work
- Project 2 showed you a tiny model learning next-token prediction
- Project 3 taught you that token boundaries are designed, not given
- Project 4 taught you how attention mixes context under causal masking
- this project turns those parts into your first full GPT system
- next, `nanoGPT` shows what changes when the blank-file bridge meets a more production-shaped reference implementation

---

Project 6: From Prototype to nanoGPT

Hook

Why does a cleaner implementation of "the same model" often train better, run faster, and break less often? If my blank-file GPT already works, why does nanoGPT look different in so many small places that seem cosmetic until they are not?

This is the moment where a lot of people get annoyed: you finally built a transformer yourself, then you open a reference implementation and it feels like someone quietly swapped your homemade bicycle for a shop-built racing bike and claimed they are both "just bikes." They are both bikes. But one was built to prove understanding, and the other was built to survive repeated use. This chapter is about learning to see that difference clearly.

Here is the opinion I want to commit to upfront, because I do not think it is said often enough: nanoGPT is shaped by scars, not by design. Every "clean" choice in that file is the residue of some earlier run that went sideways. You will read it better if you treat each section as evidence of a failure someone learned from.

The Concept

Your Project 5 GPT and `nanoGPT` are like two kitchens that both produce soup.

In your kitchen, every pot, spoon, and cutting board is exactly where you left it. You know why each thing is there because you placed it yourself. It works. You can make dinner. But if three more people walk in and try to help, the room turns into a mess fast.

`nanoGPT` is the same kitchen after someone who has burned meals for years reorganizes it. The knives are still knives. The stove is still a stove. The recipe is still soup. But the counters are clear, the hot pans have designated landing spots, and the ingredients are staged in the order they are used. Nothing about that sounds like "new cooking theory." It sounds like tidiness. Then service starts, and suddenly that tidiness is the difference between calm and disaster.

That is the first big idea of this project: a production-shaped minimal GPT does not mostly differ by adding exotic algorithms. It differs by removing avoidable friction. Some of that friction is code structure. Some is data movement. Some is training hygiene. Some is numerical stability. And a few parts that look boring, especially residual connections and LayerNorm, are not polish at all. They are life support.

We built our first GPT to understand the machine. Now we read `nanoGPT` to understand what a machine looks like after someone has been forced to care about speed, stability, and repeatability. The mental model to carry through the whole chapter: your prototype answers "What are the essential parts of GPT?" and `nanoGPT` answers "What are the essential parts once real training runs enter the room?" Those are not the same question. Your Project 5 file is a sketch. `nanoGPT` is a field manual. A sketch is honest; it shows the structure. A field manual is also honest, but in a harsher way. It has been shaped by failure. That is why this chapter matters. You are not reading bigger code to feel impressed. You are reading it to see where repeated failure carved the code into its final shape.

The first time I tried to port my blank-file GPT toward nanoGPT, I assumed the gap was style Naming. File organization. Maybe a small speedup somewhere. I was wrong. Most of the gap was about what happens when you run the same code a hundred times instead of three.

Before code, let's anchor the main differences in plain English:

Embeddings An embedding is a learned table that turns token IDs into vectors, meaning lists of numbers that represent each token. Your prototype probably treated embeddings as "the first layer." `nanoGPT` treats them as one part of a system that also includes positional information, dropout, **initialization**, and sometimes weight tying.

Transformer blocks A transformer block is the repeating unit of GPT: attention, feedforward network, normalization, and residual paths. In a first implementation, the block is easy to read as a sequence of operations. In `nanoGPT`, the same block is arranged to train reliably at depth.

Output projection At the end, the model turns hidden states back into vocabulary logits, one score per possible next token. A logit is an unnormalized score before softmax turns scores into probabilities. Your prototype may use a separate linear layer for this. `nanoGPT` often ties this output matrix to the token embedding matrix so the model reads and writes using the same learned dictionary.

Initialization Initialization means how you choose the starting values of the weights before training. Your first version may use PyTorch defaults or a rough normal distribution. `nanoGPT` makes more deliberate choices because the starting scale of activations affects whether training begins as a controlled signal or as noise.

Forward pass The forward pass is running inputs through the network to get logits and loss. In your prototype, `forward()` probably does exactly what you think the math says. In `nanoGPT`, `forward()` also reflects practical concerns: shape checks, optional targets, efficient loss computation, and clearer interfaces for training versus generation.

Generation loop Autoregressive generation means the model predicts one token, appends it, and repeats. Your prototype probably recomputes more than necessary and keeps the logic simple. `nanoGPT` still keeps it readable, but it is sharper about evaluation mode, context cropping, temperature, and top-k sampling.

And then there are the two quiet heroes of the chapter:

Residual connection A residual connection adds the block input back to the block output: `x + F(x)`. If each layer is a person editing a document, the residual path keeps a photocopy of the previous version attached to the new one. Even if the editor makes a terrible change, the old content still survives. That matters during the backward pass too. The gradient, which is the number telling each parameter how to move to reduce loss, has a direct path backward through the addition.

LayerNorm LayerNorm rescales activations so each token representation stays in a reasonable numeric range. If every layer is a musician in a long orchestra chain, LayerNorm is the sound engineer who keeps the volume from drifting wildly louder or quieter after each instrument passes the signal along. Without that engineer, small imbalances compound until the audio clips or disappears.

People often treat residuals and norms as architectural decorations. They are not. They are the parts that let depth exist. And depth is the whole point of transformers. Two layers can bluff their way through, but six layers begin asking whether your design choices are serious. That is why the BREAK IT section in this chapter matters more than the code tour. When you remove residuals and LayerNorm, you stop discussing architecture in the abstract. You watch the machine lose the ability to learn.

Why It Matters

If you stop at your prototype, you risk learning the wrong lesson. You might think the core equations are all that matter, that any training loop that "works" is basically fine, that residuals and normalization are detail work, and that clean code and stable code are separate concerns. All of those beliefs fail the moment you try to train a deeper model, compare experiments, resume a run, or extend the code without fear.

This project matters because it teaches you to separate three categories that beginners often mix together:

Category 1: True algorithmic necessities These are parts without which the model does not train or does not scale in depth. Residual connections and LayerNorm live here.

Category 2: Engineering choices that change speed, clarity, and robustness These include how batches are built, when validation runs, how checkpoints are written, how generation hooks are handled, how optimizer parameters are grouped, and how loss computation is structured.

Category 3: Style differences that are mostly about readability Some naming and file organization choices land here. They matter, but they matter less than the first two categories.

If you cannot tell which is which, then every line in a reference implementation looks equally mysterious, and you learn nothing. You either cargo-cult the code or reject it as overbuilt. By the end of this chapter, you should not admire `nanoGPT` from a distance. You should steal from it selectively, not because it is bigger, but because you understand what each stolen part protects.

The Build

This chapter is a side-by-side reading exercise, a measurement exercise, and then a surgical rewrite exercise. You are not rebuilding GPT from scratch again. You are building judgment.

Step 1: Put the two files side by side

Take your Project 5 file, probably something like `my_gpt.py`, and put it next to the `nanoGPT` Shakespeare character example. Do not start by reading top to bottom. That is how people get lost. Start by making a map.

Open both files and annotate these sections:

- configuration
- token and position embeddings
- transformer block definition
- attention implementation
- feedforward network
- normalization
- residual paths
- output projection
- weight tying
- initialization
- `forward()`
- generation

- batch construction
- optimizer setup
- train loop
- evaluation loop
- checkpointing
- logging

Your goal is not to understand every detail instantly. Your goal is to answer one concrete question for each section: "What job is this section doing in the system?" If you cannot answer that, the code will stay foggy.

A good first pass annotation looks like this:

- "This section defines how raw token IDs become vectors."
- "This section decides whether training is numerically stable."
- "This section avoids wasted copies when slicing batches."
- "This section exists so training can resume after interruption."
- "This section changes generation behavior without changing training."

This sounds basic. Good. Start basic.

This step feels like busy work the first time you do it. It becomes the most useful step in the chapter the second time you read a different transformer codebase. The mapping habit transfers; rote line-by-line reading does not.

Step 2: Compare embeddings

In both your model and `nanoGPT`, token IDs become vectors through an embedding table, a learned lookup table with shape `(V, d)`, where `V` is vocabulary size and `d` is embedding dimension. If token 42 appears, the model fetches row 42, and that row is the vector representation of the token.

When I first read nanoGPT's embedding setup, I misunderstood it. I thought the cleaner structure was a stylistic choice and that the surrounding dropout was paranoia. I was wrong on both counts. The structure is cleaner because it makes weight tying and resumption logic obvious, and the dropout is there because the embedding sum is the one place where a small noise injection consistently helps generalization at small scale. I do not get points back for misreading it the first time.

Then the model needs position information because transformers do not know order by themselves. A transformer sees a bag of vectors unless you add some signal telling it which token came first. In many beginner GPTs, this looks like:

```
tok_emb = self.token_embedding(idx)      # (B, T, d)
pos = torch.arange(T, device=idx.device)
pos_emb = self.position_embedding(pos)   # (T, d)
x = tok_emb + pos_emb
```

That is fine. `nanoGPT` does basically the same thing, but the surrounding structure is cleaner and more deliberate. What to look for: does `nanoGPT` keep embedding logic close to the model stem? Does it make batch and sequence shapes explicit? Does it apply dropout after summing token and positional embeddings? Does it tie the output projection to the token embedding weights? That last point deserves special attention.

Step 3: Understand weight tying again, but now in production shape

Weight tying means the matrix used to read tokens into vectors is reused to score output tokens at the end. If the input embedding matrix has shape `(V, d)`, the output projection wants shape `(d, V)`. These are transposes of each other, so reusing the same weights makes sense.

Imagine one bilingual dictionary used in two directions: reading (word to meaning vector) and writing (hidden meaning vector to likely word). Without weight tying, you print two separate dictionaries and hope they evolve consistently during training. With weight tying, the model reads and writes through the same learned language map. That often helps because it forces the geometry of "what a token means" and "how likely a token is as output" to stay aligned, which can reduce parameters and improve prediction quality.

Your prototype may already have this from Project 5. Good. Now check how `nanoGPT` implements it. The important question is not whether you know the trick. The important question is whether the implementation is clean and obvious. If your version does weight tying in a clever-looking one-liner that future-you will forget how to trust, and `nanoGPT` does it in a blunt, readable way, `nanoGPT` wins. Readable code is not a luxury in training systems. It is part of correctness.

Step 4: Compare transformer blocks

A transformer block in GPT usually contains a normalization step, self-attention, a residual addition, another normalization step, a feedforward network, and another residual addition. Depending on the exact layout, this is often written in a "pre-norm" style, where normalization happens before attention and before the MLP. That looks like:

```
x = x + self.attn(self.ln1(x))
x = x + self.mlp(self.ln2(x))
```

This compact pattern does an enormous amount of work. `self.ln1(x)` keeps the attention input numerically sane. `self.attn(...)` computes context mixing. `x + ...` preserves an identity path, meaning the block can always fall back toward "do almost nothing" if needed. Then the same pattern applies for the MLP.

In some cooperative fusion work, training across several model families from a few hundred million to several billion parameters, this exact block shape was where I most often spotted seam problems. When two pretrained components disagreed about whether norm came before or after the sublayer, the seam showed up as drift in the very first hundred training steps after fusion. Same diagram on the whiteboard. Different contracts in the code.

The MLP, or feedforward network, is the position-wise two-layer network applied to each token independently after attention. Attention mixes information across tokens; the MLP transforms each token's internal representation. Think of attention as a group discussion, and the MLP as private reflection after hearing the room. You need both, and neither replaces the other.

Now compare this to your prototype. Did you use pre-norm or post-norm? Is the block written as a sequence of obvious transformations? Are the residual additions explicit? Can you glance at the block and see where gradients can flow directly? If your block is longer but conceptually clearer, that is fine. If it is shorter but hides the architecture, that is a problem.

Step 5: Compare attention implementation

You already built attention. So now the question changes from "What is attention?" to "How does a cleaner implementation shape the code around attention?"

Attention computes queries `Q`, keys `K`, and values `V`; scores from `Q @ K^T`; scaling by `1 / sqrt(d_head)`; causal masking; softmax to turn scores into probabilities; and a weighted sum over `V`. You already know the intuition: each token asks, "Which earlier tokens do I need right now?"

Now compare how your code and `nanoGPT` handle separate projection layers versus fused projection, shape reshaping for heads, causal mask storage, dropout on attention weights, dropout on output projection, and any speed paths in modern PyTorch.

Your prototype may compute `q`, `k`, and `v` with three separate linear layers. `nanoGPT` may combine them in one projection and split the result. That is not a new algorithm; it is cleaner and often faster. Instead of making three trips to carry groceries from the car, you carry all three bags at once and unpack them in the kitchen. This is a recurring theme in `nanoGPT`: not new science, less wasted motion.

An earlier draft of this fused-QKV section explained the fusion purely as a memory layout optimization, and a careful reader pushed back: that explanation is incomplete because the bigger win is fewer launch overheads on the GPU, not memory layout per se. The grocery-bag analogy survived. The exact explanation underneath it did not.

Step 6: Compare the output projection and `forward()`

Your `forward()` likely does something like this: embed tokens, add positions, pass through blocks, normalize final hidden states, project to logits, and if targets are given, compute loss. That is correct.

Now inspect `nanoGPT`'s `forward()` and ask whether it cleanly supports both training and inference, whether it computes loss only when targets are present, whether it keeps tensor shapes obvious, whether it asserts sequence length constraints, and whether it avoids awkward branching or duplicated logic. This is one of those places where experienced code often feels "boring" in a good way. A good `forward()` is like a well-labeled electrical panel; you should not need to remember what each switch secretly does.

Step 7: Compare initialization

Initialization is where beginners often shrug and say, "PyTorch handles it." Sometimes that is fine for toy runs. But initialization is not random decoration. It decides the starting signal scale. If weights start too large, activations can explode. If weights start too small, signals can die. If different parts of the network start at mismatched scales, training spends early steps correcting nonsense rather than learning language.

A common simple initialization is a normal distribution with mean 0 and small standard deviation, often around `0.02`. `nanoGPT` also pays attention to special cases, especially projections that sit on residual branches. Those sometimes get scaled differently to keep deep stacks stable.

Why? Because in a deep residual stack, each block adds something to the running hidden state. If every block adds an output that is too large right from the start, the hidden state scale drifts upward layer by layer. Imagine ten people adding spices to a soup: if each person adds "a reasonable amount" independently, the final pot is ruined. You need each addition calibrated with the fact that many additions will accumulate. That is what scaled initialization is protecting against. You do not need to memorize the exact constant yet. You need to understand the reason.

Step 8: Compare batch construction and data path

This section is easy to dismiss until you profile training. Your prototype probably creates batches by randomly choosing start offsets in the training text, then slicing out input and target sequences:

```
ix = torch.randint(len(data) - block_size, (batch_size,))
x = torch.stack([data[i:i+block_size] for i in ix])
y = torch.stack([data[i+1:i+block_size+1] for i in ix])
```

That works. It is also a good teaching implementation because it makes the shifted-target idea obvious. Now compare it to `nanoGPT`. Does `nanoGPT` keep data on CPU until needed? Does it pin memory for faster host-to-device transfer? Does it avoid unnecessary Python-side copies? Does it make the train/validation split and sampling assumptions explicit? Does it produce batches in a way that better matches PyTorch's strengths?

This is the place where "data helper" becomes "part of the training system." If batch construction is clumsy, you can starve the GPU or spend extra time moving tensors around. If your training loop is a restaurant kitchen, data loading is the prep station. If it falls behind, it does not matter how good your stove is.

I spent more time than I want to admit debugging a "slow GPU" only to discover it was a starving GPU. Profiler said the device was idle thirty percent of the wall clock. The model was fine. The batch sampler was paging through the wrong cache line. After I fixed it, throughput jumped without changing a single line of model code.

Step 9: Compare optimizer setup

Your prototype might do:

```
optimizer = torch.optim.AdamW(model.parameters(), lr=lr)
```

Fine for getting started. `nanoGPT` often does more. It may separate parameters into groups: weights that should receive weight decay, and biases and normalization parameters that should not. **Weight decay** is a penalty that nudges weights toward smaller values to reduce overfitting and improve training behavior, but not all parameters should be shrunk the same way. LayerNorm parameters and biases often should not receive decay, because their job is to calibrate scale, not to compress learned information. That is not pedantry. It is saying: different parameters play different roles, so the optimizer should treat them differently. Same optimizer family. Sharper handling.

Step 10: Compare training loop details

A beginner training loop often says: get batch, forward pass, compute loss, backward pass, step optimizer, print loss sometimes. That is enough to learn the essence.

`nanoGPT` usually adds train/eval mode switching, periodic validation over many batches, logging with step counts and timing, checkpoint save and resume, generation samples during training, a gradient scaler if **mixed precision** is used, optional compilation or performance flags, controlled optimizer zeroing and stepping, and possibly **gradient clipping**. Every one of these exists because some run in the past went wrong.

Checkpointing exists because runs stop. Validation cadence exists because training loss lies by omission — it tells you fit-to-train-data, not generalization. Generation hooks exist because loss alone misses qualitative failures. Logging exists because if you cannot compare runs, you are not doing engineering; you are just hoping. Mixed precision exists because modern hardware can do lower-precision math faster and cheaper, if managed carefully. Gradient clipping exists because rare bad updates can knock training sideways. You covered some of this in **Project 5: Your GPT From A Blank File**. Now you see how it lands in code that expects to be rerun.

Step 11: Measure both models on the same task

Now stop reading and run the comparison. You want the same task, same dataset, as close to the same hyperparameters as possible. Use the Shakespeare character-level example because it is small enough to run locally and stable enough to compare.

Record three things: validation loss, throughput (tokens per second or iteration time), and ease of extension (how much work it takes to add a change without breaking other parts). That last metric is subjective, but it matters. A model that is only slightly slower but much easier to modify may be better for learning. A model that is much faster and just as readable may expose bad habits in your prototype.

Create a simple table:

Model	Val Loss	Iter Time	Notes
`my_gpt.py`	...	...	easy to read, rough training loop
`nanoGPT`	...	...	cleaner optimizer grouping, better logging

: **Table 6.1.** Side-by-side comparison of your prototype GPT and nanoGPT on validation loss and iteration speed, using the same dataset and hyperparameters.

Do not chase perfect fairness here. The point is not to publish a benchmark. The point is to feel which implementation choices pay rent.

Step 12: Choose 3 to 5 changes to port back

Do not port everything. Port the things you now understand.

Good candidates:

- weight tying if your prototype does not already have it
- cleaner pre-norm residual block structure
- better initialization
- parameter grouping for AdamW
- cleaner `forward(idx, targets=None)`
- better validation function
- checkpoint save/resume
- more efficient batch slicing
- generation function with temperature and top-k support

The rule is simple: if you cannot explain why a change exists, do not port it yet. This chapter is about understanding before copying.

Step 13: Earn one small equation

Now that you have seen why residual stacks accumulate effects and why scaling matters, here is a compact equation worth keeping:

$$x_{l+1} = x_l + F_l(x_l)$$

Every symbol now means something you already know: `x_l` is the hidden state entering layer `l`, `F_l(...)` is the transformation that layer `l` applies (such as attention or MLP work), and `x_{l+1}` is the hidden state leaving the layer. In plain English: the next layer state is the old state plus a learned adjustment.

Why does this help? Because if the learned adjustment is bad early in training, the original signal still passes through. And during backpropagation, the gradient has a direct path through the addition. Residuals do not only help forward information survive; they help backward learning signals survive too. That is the part people usually skip. That is also why removing them gets ugly fast.

Step 14: Read nanoGPT as a list of scars

At this point, stop asking whether `nanoGPT` is "better." Ask what scar each section represents:

- Special optimizer parameter grouping: a scar from treating all parameters identically.
- Validation helper: a scar from trusting training loss too much.
- Checkpointing: a scar from interrupted runs.
- Cleaner block structure: a scar from hard-to-debug architectures.
- Careful initialization: a scar from unstable training starts.
- Residuals and LayerNorm: scars from depth failing outright.

Once you read reference implementations this way, they stop feeling magical. They start feeling honest. And that is what you want.

If I had to name the single insight that changed how I read other people's training code, it is this: every "weird" line is usually a tombstone. Somebody died on the hill of finding it out. The cleanest way to honor that is to read the line, ask which hill, and move on.

BREAK IT

You are going to break the model in the exact places many newcomers assume are optional: first remove all residual connections, then remove LayerNorm. The point is not just to observe failure. The point is to learn what kind of failure each component prevents.

Break 1: Remove all residual connections

Find the transformer block. In a pre-norm GPT, it probably looks roughly like this:

```
x = x + self.attn(self.ln1(x))
x = x + self.mlp(self.ln2(x))
```

Now break it:

```
x = self.attn(self.ln1(x))
x = self.mlp(self.ln2(x))
```

That is not a subtle change. You have cut the skip paths entirely, removing the model's ability to carry forward an untouched version of the previous representation. Each layer now fully overwrites the hidden state.

Before, each editor worked on a document while the previous draft remained clipped underneath. Even if the editor makes a terrible change, the old content survives. Now each editor rewrites the document from scratch and throws the previous draft away. That is risky in the forward pass. It is worse in the backward pass.

During backpropagation, gradients need paths through the network. The residual path gives them a direct route through addition. Without that path, gradients must pass through every attention and MLP transformation in sequence. Each transformation can shrink, distort, or amplify the signal. In shallow networks, this can still work. In deeper networks, learning begins to choke.

What you will usually observe:

- At depth `N = 2`, training may still proceed. Loss falls, though often more slowly.
- At depth `N = 4`, training becomes touchy. Loss falls unevenly or stalls. Validation is worse
- At depth `N = 6`, one of two things often happens quickly: loss barely improves and the model acts dumb, or training destabilizes, gradients spike, and loss turns into `NaN` within tens to low hundreds of steps.

`NaN` means "not a number." It is what floating-point systems produce when the arithmetic has gone off the rails, often from overflow, invalid operations, or unstable cascades.

What I didn't see coming with the no-residual model was the suddenness. I assumed it would fail gradually, and instead it looked healthy for ninety steps and then exploded in one. The first time I induced this on purpose I had not been logging gradient norms, so the only signal I had was the loss line snapping vertical inside a single eval interval. After that run I added gradient norm logging to every training script I write. Always.

What this proves: residual connections are not there to make the architecture look modern. They make deep optimization possible. Without them, the network has to relearn a sensible representation from scratch at every layer while also trying to pass learning signal backward through a much harsher path. Two layers can fake competence. Six layers start telling the truth.

What to watch for when residuals are gone

Watch training loss curve, validation loss curve, gradient norm if you log it, sample generations every few hundred steps, and presence of `NaN` in loss or parameters. A typical bad pattern is early loss falls a little, progress slows fast, generated text stays repetitive or incoherent, and deeper runs diverge sooner. If you only watch the first few steps, you may fool yourself; the model can look "alive" before collapse becomes obvious. This is one reason experienced practitioners insist on plotting runs and sampling outputs, not just reading one printed loss value.

Why residual removal hurts generation quality even before full collapse

Even if the no-residual model does not immediately produce `NaN`, generation often gets worse in a very specific way: it feels brittle. The model latches onto local patterns but fails to preserve coherent structure across longer spans. Why? Because each layer must rebuild useful state instead of refining it. A residual stack says "keep what already works, add corrections." A non-residual stack says "throw away what you had, replace it." That second instruction is much harder to optimize repeatedly.

Break 2: Remove LayerNorm

Now restore residuals. Then break normalization. If your block looks like this:

```
x = x + self.attn(self.ln1(x))
x = x + self.mlp(self.ln2(x))
```

Change it to:

```
x = x + self.attn(x)
x = x + self.mlp(x)
```

Or, if your implementation has explicit LayerNorm modules, replace them with identity functions.

You have removed the mechanism that keeps hidden-state scale from drifting. Every attention head and every MLP produces outputs with some numeric scale. When you stack many layers, those scales interact. Small imbalances do not stay small — they accumulate.

Imagine a long line of people passing buckets of water. Each person is supposed to keep the water at a safe fill level before handing it off. LayerNorm is the person checking every bucket and resetting it to a standard level. Remove that check, and one slightly overfilled bucket becomes a spill, then a flood.

What you usually observe: training becomes unstable earlier than the no-residual case, loss spikes more violently, gradients become much larger, `NaN` often appears sooner, and even if loss does not immediately explode, it oscillates badly. In practice, removing LayerNorm from a multi-layer GPT often destabilizes training almost immediately unless other compensating tricks are present.

Normalization is not cosmetic cleanup. It is a scale-control system. Transformers are deep compositions of matrix multiplications, nonlinearities, and residual additions. Without some control over activation scale, the network drifts out of the numeric range where learning is reliable.

Break 3: Remove both residuals and LayerNorm

If you want the full lesson, remove both. Now the model has no stable identity path and no activation scale control. This is not "slightly worse" — this is usually a fast train wreck. At shallow depth, maybe you still get a pulse. At moderate depth, the run often becomes useless quickly.

A lot of people read transformer diagrams and come away believing attention is the star and everything around it is support staff. Attention is the clever mechanism, but residuals and normalization are what let the clever mechanism survive contact with depth.

Why depth is the real stress test

The pattern to expect is concrete:

- depth `N = 2`: still trains
- depth `N = 4`: starts struggling
- depth `N = 6`: often hits `NaN` within around 100 steps when residuals are removed, and even faster when LayerNorm is gone too

A shallow network can hide architectural sins. A deeper network exposes them. This is why reading reference code only at toy depth can mislead you; some choices look optional when the network is too small to punish you. Production-shaped code exists partly because people stopped trusting toy settings to tell them what matters.

One more equation, now fully earned

Here is the plain LayerNorm formula in compact form:

```
LayerNorm(x) = gamma * (x - mu) / (sigma + epsilon) + beta
```

What each symbol means:

- `x` is the input vector for one token at one layer
- `\mu` is the mean of that vector's values
- `\sigma` is the standard deviation, meaning how spread out the values are
- `\epsilon` is a tiny constant to avoid division by zero
- `\gamma` and `\beta` are learned scale and shift parameters

In plain English: center the vector, rescale it to a stable spread, then let the model learn a new preferred scale and offset afterward. Not magic. A volume knob and recentering step for hidden states. Now the reason it matters should feel obvious from the failure you just induced.

Optional Homework

Run this as a controlled lab. You do not need a giant GPU cluster. The Shakespeare character example is enough to make the point.

Hardware and runtime

Proxy local version:

- GPU: any CUDA GPU with at least 4 GB VRAM is comfortable
- CPU-only: possible, but slower
- Runtime:
  - small runs: 10 to 30 minutes on a modest GPU
  - CPU: longer, possibly 1 to 3 hours depending on settings

If you cannot run the full matrix of experiments, run the 2-layer and 6-layer versions only. That already teaches the lesson.

Experiment A: Remove residuals only

Edit the transformer block. Find the two residual lines and remove the `x +`.

Change:

```
x = x + self.attn(self.ln_1(x))
x = x + self.mlp(self.ln_2(x))
```

to:

```
x = self.attn(self.ln_1(x))
x = self.mlp(self.ln_2(x))
```

Run three depth settings:

- `N = 2`
- `N = 4`
- `N = 6`

Keep everything else the same if possible.

Suggested short-run settings for local hardware:

- `batch_size = 32` if GPU allows, else `16`
- `block_size = 128`
- `n_embd = 128` or `192`
- `max_iters = 500`
- `learning_rate = same as baseline`

Expected observations:

- `N = 2`: loss still falls, though often worse than baseline
- `N = 4`: loss becomes noisier, validation degrades, samples worsen
- `N = 6`: training often stalls or destabilizes fast; possible `NaN`

Experiment B: Restore residuals, remove LayerNorm only

Put residuals back. Now replace LayerNorm calls with identity behavior.

Change:

```
x = x + self.attn(self.ln_1(x))
x = x + self.mlp(self.ln_2(x))
```

to:

```
x = x + self.attn(x)
x = x + self.mlp(x)
```

If there is a final LayerNorm before logits, remove or bypass that too for this experiment.

Run the same depth settings:

- `N = 2`
- `N = 4`
- `N = 6`

Expected observations:

- instability appears earlier than baseline

- loss spikes harder
- `NaN` is more likely and often arrives sooner than in the no-residual case
- generations become chaotic or repetitive before total failure

Experiment C: Remove both residuals and LayerNorm

This is the full break. Run `N = 2` and `N = 6`.

Expected observations:

- `N = 2` may still limp forward
- `N = 6` often fails quickly and clearly
- if it does not hit `NaN`, it usually learns very poorly

Add gradient norm logging if you want one extra signal

If you are comfortable editing the training loop, log the total gradient norm every evaluation interval.

What to watch for:

- baseline: controlled values, occasional spikes
- no residuals: weaker or erratic learning signal
- no LayerNorm: large spikes or blow-up
- neither: often the ugliest curve

Compare against your Project 5 GPT

Now run your own `my_gpt.py` under a comparable small setup.

Record:

- validation loss
- iteration time
- what was easy to modify
- what was awkward to modify
- which features from `nanoGPT` you now want to port back

Write down 3 to 5 concrete ports, such as:

- pre-norm residual block layout
- optimizer parameter grouping
- weight tying
- better validation helper
- checkpoint save/load
- cleaner generation API

If you are CPU-only

Do the proxy version:

- `N = 2` baseline
- `N = 6` no residuals
- `N = 6` no LayerNorm
- `max_iters = 200`
- `n_embd = 64` or 96
- `block_size = 64`

You will get noisier results, but the trend should still show up.

Expected result table template

Fill this in:

Run	Depth	Residuals	LayerNorm	Final Train Loss	Final Val Loss	NaN?	Sample Quality
Baseline	2	Yes	Yes			No	
Baseline	6	Yes	Yes			No	
No residuals	2	No	Yes			maybe	
No residuals	4	No	Yes			maybe	
No residuals	6	No	Yes			often	
No LayerNorm	2	Yes	No			maybe	
No LayerNorm	4	Yes	No			often	
No LayerNorm	6	Yes	No			often	
Neither	2	No	No			maybe	
Neither	6	No	No			often	

: **Table 6.2.** Ablation results showing the effect of removing residual connections and LayerNorm at different depths, with training stability and sample quality observations.

The exact numbers will differ by hardware, seed, and hyperparameters. The pattern is what matters.

Questions To Answer

1. When you removed residual connections, what failed first: loss improvement, generation quality, or numeric stability?
2. At what depth did your model stop acting like a "slightly worse GPT" and start acting fundamentally broken?
3. When you removed LayerNorm, did the run die through `NaN`, violent oscillation, or just very poor learning? Which of those tells you most clearly what LayerNorm protects?

4. Compare your Project 5 GPT to `nanoGPT`: which differences turned out to be real training necessities, and which were mostly code organization?
5. Which 3 to 5 changes would you port back into your own model now, and can you explain each one in one sentence without hiding behind jargon?
6. If you had to choose only one mechanism to keep in a deeper model, residuals or LayerNorm, which would you keep first, and why?

Go Further

Research anchors for this chapter

If you want the papers that sit underneath the code judgment in this chapter, start with **Attention Is All You Need.** That paper gives you the original transformer block vocabulary: attention, feedforward network, residual paths, and normalization as the repeating structural unit. It is the clean whiteboard version of the system.

Then jump to **Language Models are Few-Shot Learners** and **PaLM: Scaling Language Modeling with Pathways.** Those papers matter here not because they teach cleaner code style, but because they show what happens when the same block pattern gets pushed to far larger scales. Once model depth and training budgets grow, residual paths, normalization placement, initialization, and optimizer hygiene stop looking like implementation details and start looking like survival requirements.

Finally, read **LLaMA: Open and Efficient Foundation Language Models** as a bridge from "the transformer in principle" to "the transformer under practical efficiency pressure." LLaMA-family models are where many readers first see that small-seeming architectural and systems choices compound into real differences in trainability and deployment cost.

If this chapter clicked, the next useful directions are not "read a bigger transformer." They are more specific.

1. Read the GPT-2 style block carefully

Search for:

- `nanoGPT model.py Block`
- `pre-norm transformer block`
- `residual stream transformer`

The phrase **residual stream** is worth your time. It frames the hidden state not as something each layer replaces, but as a running stream that layers write into.

2. Look up deep transformer training stability

Search arXiv for:

- `transformer residual connection training stability`
- `pre-LayerNorm transformer`
- `deep transformers without instability`

You are looking for work on why pre-norm layouts became common and what alternatives people tried.

3. Read about initialization for residual networks

Search for:

- `GPT scaled residual initialization`
- `deep residual initialization transformer`
- `admin initialization transformer`

You do not need to master every paper. You want the recurring lesson: deep residual stacks need careful signal scaling from the first step.

4. Compare other reference implementations

After `nanoGPT`, compare one more small reference implementation, not five.

Good search targets:

- `minGPT karpathy`
- `llama minimal implementation`
- `transformer block pre norm vs post norm`

Your question is not "Which repo is best?" Your question is "Which design pressures keep showing up across implementations?"

5. If you want a benchmark direction

Search for:

- `OpenWebText nanoGPT`
- `hellaswag evaluation language model`
- `validation loss versus downstream benchmark`

This starts connecting training loss to actual behavior. You will learn quickly that a lower validation loss often helps, but not every quality difference shows up cleanly in one metric.

What You Now Know

You can now open a reference implementation like `nanoGPT` and read it as engineering decisions rather than received wisdom. You understand why deep transformers need residual connections and LayerNorm, not as a slogan, but because you broke the model and watched it fail. Those "boring" lines are not style; they are what keeps deep networks trainable at all.

Starting Point

Start from:

- your own Project 5 file, for example `my_gpt.py`

Then compare against:

- `https://github.com/karpathy/nanoGPT`

- focus on the Shakespeare character example first

Keep the progression straight:

- `micrograd` taught you gradients and backpropagation
- Project 2 taught you tiny next-token learning
- `minbpe` taught you that token boundaries are designed
- Project 4 taught you how attention mixes context under causal masking
- Project 5 made you build a GPT yourself
- Project 6 teaches you how that prototype changes when minimal teaching code grows into minimal production-shaped code

Next, you are ready to stop seeing reference implementations as magic. You will see constraints. And then you can work with them.

Project 7: The Details That Matter

Hook

Why does one transformer train cleanly, another wobble for hours, and a third refuse to extend past its training length even though the code looks almost the same?

This is the part people hand-wave away with words like "architecture choice," which is a nice way of hiding the fact that tiny-looking decisions change what every layer expects to receive. Change the normalization type, activation function, or positional encoding, and you are not tweaking style. You are changing the internal contract of the model.

From kernel-level inference work targeting Apple's Neural Engine through private compiler APIs, this lesson was unavoidable. The hardware did not care which architecture name was on the model card. It cared which numeric ranges the activations actually landed in. A norm switch that looked cosmetic at the PyTorch level changed the precision envelope downstream enough that the delta compiler had to be tuned differently. The phrase "same architecture" stopped meaning anything useful after a while.

The Concept

Project 6: From Prototype to nanoGPT got us to the point where a transformer block was no longer magic. Attention mixes context. The feedforward network stores and transforms patterns. Residual connections keep information flowing. LayerNorm keeps training from flying off the rails.

Now we zoom in on the details that look boring right up until they break your model.

Think of a transformer layer as an assembly station in a factory line. Each station expects the incoming parts to arrive in a certain shape, size, and orientation, not because the station is picky, but because it was built and tuned around that expectation. Normalization decides the statistical "shape" of the hidden state. An activation function decides how the model bends and gates information after a linear transformation. A positional encoding scheme tells the model where each token sits in a sequence. Those three choices sit at the boundaries between major pieces of the model, which is exactly why they matter so much. They define the interface.

Here is the confusion most people have the first time they hear this: "If LayerNorm and **RMSNorm** both stabilize activations, why would swapping one for the other matter?" Because "stabilize" is too vague. A refrigerator and a freezer both keep food cold; that does not mean you can replace one with the other in a recipe without changing the result. LayerNorm and RMSNorm both rescale hidden states, but they do not produce the same distribution. Figure 7.1 illustrates that key difference.

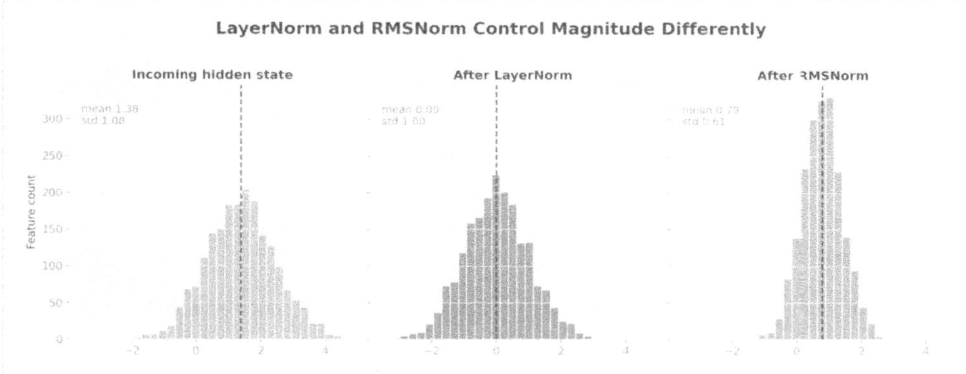

Figure 7.1. LayerNorm and RMSNorm both control magnitude, but only LayerNorm recenters the hidden state around zero; that difference changes what downstream layers learn to expect.

GELU, **SwiGLU**, and ReLU² all add nonlinearity, but they do not let information through in the same way. Learned positional embeddings and **RoPE** both tell the model where tokens are, but one memorizes positions while the other encodes relative offsets in a way that survives longer sequences.

In one sentence: a transformer is not just a pile of layers. It is a stack of agreements about what kind of signal each layer will hand to the next.

Normalization: resetting the signal

A hidden state is just a list of numbers representing what the model currently "thinks" each token means at this layer. After attention and the feedforward network, those numbers can drift. Some dimensions grow too large, some shrink, and some layers start seeing values in ranges they were not trained to handle well. Normalization is the cleanup crew at the boundary.

LayerNorm says: for each token's vector, subtract its mean and divide by its standard deviation, then optionally scale and shift it with learned parameters. In plain English: center the vector around zero, then stretch or squeeze it so its spread is controlled.

RMSNorm says: skip the centering step — just divide by the root mean square, which is a measure of overall magnitude. In plain English: do not force the average to zero; only control the size.

That sounds minor. It is not minor. LayerNorm cares about both offset and scale. RMSNorm cares mostly about scale. So a layer trained after RMSNorm may expect a non-zero mean in certain dimensions, while a layer trained after LayerNorm may expect those means to be scrubbed away. That is why mixing them can produce a sharp seam in the network. The downstream layers were shaped by one kind of signal and suddenly receive another.

Across a long set of fine-tuning experiments I ran at scale, the single most common "this should have been easy" failure was a norm mismatch between two pretrained components that the user assumed were interchangeable. Same `d_model`. Same head count. Different distribution at the seam. The fusion would compile cleanly and then quietly produce worse loss than either component alone.

Activation functions: the shape of response

A feedforward network is not just two matrix multiplications. Without a nonlinearity in the middle, those two matrices collapse into one, and the network loses expressive range. The activation function is the bend in the pipe. You pass numbers through it, and it changes which signals get amplified, softened, or shut down.

ReLU keeps positive numbers and zeros out negative ones. $ReLU^2$ keeps positives, squares them, and zeros out negatives. GELU lets values through smoothly, with small values getting partial credit instead of a hard cutoff. SwiGLU gates one stream with another: one projection carries content, and another projection decides how much of that content passes through.

If you want an analogy: ReLU is a binary door, open or closed. GELU is a soft door with a spring; it opens gradually. SwiGLU is a security gate with two checks: one path carries the package, the other decides whether the package gets through. These choices affect gradient flow, training stability, and final loss, not by huge dramatic amounts every time, but by consistent margins that accumulate over billions of tokens in real systems.

Positional encoding: putting order back into a bag of tokens

Attention by itself has no built-in sense of order. Hand a model the tokens `["dog", "bites", "man"]` and `["man", "bites", "dog"]`. The set of tokens is the same, but the meaning is not, and without position information the model cannot tell the difference.

Learned positional embeddings handle this by giving every position index its own learned vector. Token embedding plus position embedding gives the model "word meaning at position 17." That works inside the training range, but it is like memorizing seats in one theater — move to a stadium with more rows, and the seat map runs out.

RoPE, short for Rotary Position Embeddings, handles position differently. Instead of attaching a separate learned vector for each position, it rotates pairs of hidden dimensions by an angle that depends on the token's position. To make that physical: imagine each pair of numbers in the hidden state as a little arrow on graph paper. Position 0 points the arrow one way. Position 1 rotates it slightly. Position 100 rotates it more. When queries and keys compare themselves, the angle between them carries relative position information. The key insight is that rotation is relative. If token A is at angle θ and token B is at angle 2θ, their dot product encodes not their absolute positions but the angle between them, which is exactly the relative distance. That means the same attention pattern works whether those tokens appear at positions 5 and 10, or 105 and 110.

The positional signal becomes part of the geometry of the vector space, not a lookup table of position IDs. That is why RoPE often extrapolates beyond training length better than learned positional embeddings. Not perfectly, but better.

I expected RoPE to feel exotic the first time I implemented it. It does not. It feels mechanical. A rotation matrix on pairs of dimensions, indexed by position. The mystique evaporates as soon as you write the two-line rotation by hand. What does not evaporate is how much careful frequency scheduling matters, and that is the part I still re-read references on every time.

Why It Matters

If you only care whether the loss goes down, these details look decorative. If you care whether the model trains reliably, serves quickly, extends to longer sequences, or can be composed with another model later, these details are where the real engineering starts.

Why normalization type matters

Without normalization, deeper models drift. One layer pushes activations a bit wider, the next layer amplifies that, and a few blocks later you do not have a stable signal — you have a distribution problem. Residual connections help information move through the network. Normalization keeps that moving information statistically sane.

LayerNorm and RMSNorm do this in different ways, and that difference shapes what downstream layers learn to expect. If two independently trained components do not agree on hidden-state statistics, you cannot bolt them together safely just because the tensor shapes match. Shape compatibility is not enough; statistical compatibility matters too.

Sidebar: refining the residual stream (2026 architectures)

Picture one conveyor belt. It carries a bag per token from the first layer to the last, and each layer may drop something extra into the bag. That belt is the residual stream: the running sum that ferries a token's representation through the network while normalization keeps its contents in range. A recent line of work asks a blunt question. What if the one belt became several?

Hyper-connections replace the single residual stream with a few parallel streams, and each layer reads from all of them and writes back to all of them. A small mixing matrix decides how much of each stream feeds each other stream between layers. More streams, more room to route information. But they give up a guarantee the single stream had for free. A plain residual passes each token's signal straight through unchanged unless a layer deliberately edits it, and that pass-through behavior is the identity guarantee. Unconstrained mixing across several streams loses it, so the signal can drift up or down a little at every step, and across dozens of layers that drift compounds until the magnitude explodes or fades. The DeepSeek-AI team reports that plain hyper-connections break this guarantee and cause severe training instability (arXiv 2512.24880).

Manifold-Constrained Hyper-Connections, or mHC, restores the guarantee. It forces every mixing matrix to be doubly stochastic: every entry non-negative, every row and every column summing to one. That set of matrices has a name, the Birkhoff polytope. Rows summing to one means each stream hands out exactly what it holds. Columns summing to one means each stream receives exactly one unit. Nothing created. The total is conserved end to end. Formally, the constraint keeps the matrix spectral norm, the most it can stretch any vector, at or below one, so the signal magnitude cannot explode across depth even as dozens of layers stack up in sequence. The team enforces it with Sinkhorn-Knopp, an iterative step that rescales rows and columns in turn until both sum to one, over twenty iterations.

A tiny two-layer toy shows why this earns its keep. A single-stream block adds a correction to its input and passes it on. Now give it two streams, mixed by a small matrix before each write. Suppose that matrix scales both streams up at every layer, say by a factor near one point four. The magnitude grows a little each step and compounds across depth. A real network stacks tens of these blocks, so a factor barely above one becomes a large multiplier by the output. Now make the matrix doubly stochastic instead, so a fifth of each stream crosses into the other and the rest stays. The two streams still trade content. Their combined size holds steady. Same rerouting freedom. No runaway growth. That conservation across depth is the point.

Here is that toy, runnable in a few lines of numpy:

```python
import numpy as np

M = np.array([[0.8, 0.2], [0.2, 0.8]])                    # doubly stochastic: rows and cols sum to 1
R = np.random.default_rng(0).uniform(0.7, 1.3, size=(2, 2))  # unconstrained mixing
v_c = np.array([0.8, 0.6]); v_u = v_c.copy()
for step in range(8):
    v_c, v_u = M @ v_c, R @ v_u
    print(step, np.linalg.norm(v_c), np.linalg.norm(v_u))
# sample run: constrained norm stays ~1.0 every step; unconstrained grows past 60 by step 8
```

Run it and watch the two norms. The constrained vector stays bounded. It holds near its starting size across all eight steps, because a doubly stochastic matrix has spectral norm one and cannot amplify any direction. The unconstrained vector climbs at every step and runs away fast. That gap in direction, flat versus growing, is why the Birkhoff constraint keeps signal stable across depth. The exact matrix values here are illustrative, not measured.

The costs are modest but real. mHC adds about 6.7% training-time overhead at expansion rate n=4, meaning four parallel streams, in large-scale training, and that overhead grows with n. The authors demonstrate stable training at up to 27B parameters, alongside 3B and 9B variants. DeepSeek-V4, announced as a preview in April 2026, adopts mHC in its residual connections.

A second 2026 refinement touches the other end of the stack. Multi-token prediction, or MTP, trains extra output heads to predict several future tokens at once instead of only the next one (Gloeckle et al., 2024). DeepSeek-V3 used a sequential MTP module during training. Those extra predictions double as draft tokens for speculative decoding, where the base model verifies a batch of guesses in one pass, so generation speeds up while the output distribution stays unchanged. Both ideas carry this chapter's lesson forward. The residual stream and the prediction head are interfaces, and refining them without breaking their contract is where much of 2026 architecture work sits.

Why activation choice matters

You can get a model to train with several activation functions, which tempts people to think activations are interchangeable. They are not. The activation controls how easy it is for the model to represent small changes, how sharply it gates signals, and how gradients move backward through the feedforward network. These changes often show up as small but repeatable differences in final validation loss. In production work, "small but repeatable" is not a footnote. It is the whole job.

Why positional encoding choice matters

Learned positional embeddings bake in a fixed table of position vectors. That is simple, and it is also brittle if you want longer contexts later. RoPE changes the attention calculation itself, making relative position survive in the dot products between queries and keys. That gives you a path to longer contexts without retraining the position table from scratch. If you want to understand why some models stretch from 2K to 8K or 32K context with careful fine-tuning and others faceplant, this is one of the first places to look.

Why instrumentation matters

Loss tells you whether the whole machine is getting better. It does not tell you what each layer is doing internally. Instrumenting hidden state statistics gives you X-ray vision. Mean, variance, min, max, **kurtosis**: these are not random numbers for a dashboard. They tell you whether a layer boundary keeps the signal centered, whether variance explodes or collapses, whether outliers dominate, and whether a mixed architecture introduces a seam. This project is where you stop treating the model as one giant black box and start looking at the distribution of the blood flowing through it.

The strongest opinion I will commit to in this chapter: per-layer activation statistics should be on by default in any serious training script. Not a debugging feature you bolt on after something fails. A first-class output of every run. I have lost more hours to "I wish I had logged that" than to any one architectural mistake.

The Build

We will do four things:

1. Implement LayerNorm and RMSNorm from scratch.
2. Implement three activations: GELU, SwiGLU, and $ReLU^2$.
3. Implement RoPE and replace learned positional embeddings.
4. Instrument hidden-state statistics at every layer boundary.

Use `nanochat` as the architectural reference, especially the model code inside `train.py` or the equivalent module where blocks are defined. The exact file layout may differ by revision. The important part is the transformer block. The goal is not to copy-paste someone else's code blindly. The goal is to make each mechanism small enough in your head that you can spot the seam when it breaks.

Step 1: Build LayerNorm from scratch

For each token representation `x`, which is a vector of length `d_model`, LayerNorm computes that vector's mean and variance across its features, standardizes the vector, then applies learned scale and bias. If `x` is one token's hidden state: mean is the average value across the `d_model` features, variance is the average squared distance from that mean, and the normalized value is `(x - mean) / sqrt(variance + eps)`. `eps` is a tiny number like `1e-5` so you do not divide by zero. Then two learned vectors enter: `gamma` scales each feature, and `beta` shifts each feature.

Here is the stripped-down PyTorch version:

```
class LayerNorm(nn.Module):
    def __init__(self, d_model, eps=1e-5):
        super().__init__()
        self.weight = nn.Parameter(torch.ones(d_model))   # gamma
        self.bias = nn.Parameter(torch.zeros(d_model))    # beta
        self.eps = eps

    def forward(self, x):
        mean = x.mean(dim=-1, keepdim=True)
        var = ((x - mean) ** 2).mean(dim=-1, keepdim=True)
        x_hat = (x - mean) / torch.sqrt(var + self.eps)
        return self.weight * x_hat + self.bias
```

What each symbol means:

- `x`: hidden states, shape usually `(B, T, d_model)`

- `dim=-1`: normalize each token vector independently
- `weight` and `bias`: learned per-feature affine parameters

The thing to notice is not the formula. It is the contract it creates. After LayerNorm, every token vector has roughly zero mean and unit variance before the learned scale and shift. Downstream layers train against that profile.

Step 2: Build RMSNorm from scratch

RMSNorm removes the mean subtraction. Instead of centering and scaling, it only scales by the root mean square. The root mean square is: square each feature, average those squares, take the square root. So if a vector is large overall, RMSNorm shrinks it. If it is small overall, RMSNorm scales it up. But it does not force the average feature value to zero, which means the vector can retain offset information.

Minimal implementation:

```
class RMSNorm(nn.Module):
    def __init__(self, d_model, eps=1e-5):
        super().__init__()
        self.weight = nn.Parameter(torch.ones(d_model))
        self.eps = eps

    def forward(self, x):
        rms = torch.sqrt((x ** 2).mean(dim=-1, keepdim=True) + self.eps)
        x_hat = x / rms
        return self.weight * x_hat
```

Notice two differences from LayerNorm: no mean subtraction, and no learned bias term in the common form. Why does this often run slightly faster? Because it does less work. It skips mean computation and subtraction, and the operation pattern is simpler. That speedup is usually modest but real. Your job in this chapter is to measure it, not just repeat that claim.

Step 3: Wire normalization into the transformer block

If your block already uses a norm module, replace it with a config-controlled choice.

For example:

```
def make_norm(norm_type: str, d_model: int) -> nn.Module:
    if norm_type == "layernorm":
        return LayerNorm(d_model)
    if norm_type == "rmsnorm":
        return RMSNorm(d_model)
    if norm_type == "none":
        return nn.Identity()
    raise ValueError(f"unknown norm_type: {norm_type}")
```

Then in the block:

```
self.norm1 = make_norm(config.norm_type, config.d_model)
self.norm2 = make_norm(config.norm_type, config.d_model)
```

If you are using pre-norm blocks, the pattern is:

```
x = x + self.attn(self.norm1(x))
x = x + self.mlp(self.norm2(x))
```

That ordering matters. In a pre-norm transformer, normalization happens before attention and before the MLP. This architecture is more stable for deeper GPT-style models than post-norm in many practical settings. For the comparison table later, also include a `none` option, not because you expect it to perform well, but because failures teach faster than success.

Step 4: Instrument hidden-state statistics

Do this before you change too many components. You want to be able to answer questions like: What is the mean at the input to block 7? Does variance slowly drift upward with no normalization? Does RMSNorm preserve a non-zero mean where LayerNorm wipes it out? Do outliers appear after the MLP?

Make a small stats helper. Kurtosis deserves a plain-English definition first: variance tells you how spread out values are, and kurtosis tells you how heavy the tails are, meaning how much the distribution is dominated by rare extreme values. If kurtosis rises sharply, you may have a few very large outliers even if variance still looks acceptable.

One simple implementation:

```
def tensor_stats(x: torch.Tensor) -> dict:
    flat = x.detach().float().reshape(-1)
    mean = flat.mean()
    var = flat.var(unbiased=False)
    std = torch.sqrt(var + 1e-12)
    centered = flat - mean
    kurt = (centered ** 4).mean() / (std ** 4 + 1e-12)

    return {
        "mean": mean.item(),
        "var": var.item(),
        "min": flat.min().item(),
        "max": flat.max().item(),
        "kurtosis": kurt.item(),
    }
```

At minimum, capture stats at: token embedding output, before attention, after attention residual add, before MLP, after MLP residual add, and final norm output. Do not log every step of every training batch forever — that will drown you in data and slow training. Log every `N` steps, or log on one fixed validation batch.

For example:

```
if step % config.stats_every == 0:
    stats["block_03_pre_attn"] = tensor_stats(x)
```

Write these to JSONL or CSV. We instrument at boundaries rather than arbitrary internal points because boundaries are interfaces, and this chapter is about interfaces.

Step 5: Compare LayerNorm, RMSNorm, and no norm

Run three training jobs with everything else fixed: same random seed if possible, same model size, same dataset, same batch size, same optimizer settings, same total training steps. Only change normalization type.

Record training loss, validation loss, wall-clock tokens/sec or step time, and hidden state stats by layer.

What should you expect?

LayerNorm

- stable means near zero at normalized boundaries
- controlled variance
- slightly slower than RMSNorm
- reliable training behavior

RMSNorm

- means not forced to zero
- controlled overall magnitude
- slightly faster wall-clock
- similar or near-similar loss in many GPT-like setups

No norm

- boundary stats drift more aggressively
- deeper layers often show exploding or collapsing variance
- training becomes unstable or much worse
- this gives you your "why norm exists" baseline

A table like this is the target:

Norm Type	Final Val Loss	Tokens/sec	Mean Drift	Variance Drift	Notes
LayerNorm	2.41	100% baseline	low	low	stable
RMSNorm	2.40	103% baseline	medium	low	slightly faster
None	diverges or much worse	unstable	high	high	breaks quickly

: **Table 7.1.** Norm-type comparison: validation loss, throughput, and activation drift behavior for LayerNorm, RMSNorm, and no normalization.

Those numbers are examples, not promises. Your actual values depend on hardware, dataset, and model size. The point is the pattern.

Step 6: Build the activation functions

Now move to the feedforward network. The standard GPT feedforward path is: project from `d_model` to a larger hidden size (often `4 * d_model`), apply activation, then project back to `d_model`. That middle activation is where we compare GELU, SwiGLU, and ReLU².

GELU

GELU stands for Gaussian Error Linear Unit. Its behavior is what matters: smooth gating. Small negative values are not chopped to zero immediately. Small positive values are not passed through unchanged with a hard kink. It transitions smoothly.

Common approximation:

```
def gelu(x):
    return 0.5 * x * (1.0 + torch.tanh(
        math.sqrt(2.0 / math.pi) * (x + 0.044715 * x**3)
    ))
```

Why use it? Because smoothness often helps training and gives better results than hard-threshold ReLU in transformers.

ReLU²

ReLU² is exactly what it sounds like: take ReLU, then square it.

```
def relu2(x):
    return torch.relu(x) ** 2
```

Positive activations not only survive, they grow faster as they get larger. That changes the shape of the activation distribution — it can increase sparsity for negatives and emphasize stronger positives more sharply.

SwiGLU

SwiGLU is a gated activation. Instead of one projection, you use two: one projection carries content, and one projection creates a gate using the **SiLU** activation. SiLU is `x * sigmoid(x)`, a smooth self-gating function.

Minimal form:

```
class SwiGLU(nn.Module):
    def __init__(self, d_model, hidden_dim):
        super().__init__()
        self.w1 = nn.Linear(d_model, hidden_dim, bias=False)
        self.w2 = nn.Linear(d_model, hidden_dim, bias=False)
        self.w3 = nn.Linear(hidden_dim, d_model, bias=False)

    def forward(self, x):
        gate = F.silu(self.w1(x))
        value = self.w2(x)
        return self.w3(gate * value)
```

Why do many modern models prefer gated MLPs? Because they give the network a cleaner way to decide which feature channels matter for a given token. Instead of "transform then pass through a fixed bend," you get "transform content and transform permission separately." That tends to help.

Step 7: Make the MLP activation configurable

Your MLP constructor might look like this:

```
class MLP(nn.Module):
    def __init__(self, config):
        super().__init__()
        hidden = 4 * config.d_model

        if config.activation == "swiglu":
            self.kind = "swiglu"
            self.w1 = nn.Linear(config.d_model, hidden, bias=False)
            self.w2 = nn.Linear(config.d_model, hidden, bias=False)
            self.w3 = nn.Linear(hidden, config.d_model, bias=False)
        else:
            self.kind = config.activation
            self.fc1 = nn.Linear(config.d_model, hidden)
            self.fc2 = nn.Linear(hidden, config.d_model)

    def forward(self, x):
        if self.kind == "gelu":
            return self.fc2(gelu(self.fc1(x)))
        if self.kind == "relu2":
            return self.fc2(torch.relu(self.fc1(x)) ** 2)
        if self.kind == "swiglu":
            return self.w3(F.silu(self.w1(x)) * self.w2(x))
        raise ValueError(self.kind)
```

Train three runs with same setup, only activation changes. What should you expect? Not a dramatic collapse. That is what makes this project easy to underestimate. The differences are often small but consistent. A plausible outcome:

Activation	Final Val Loss	Notes
GELU	2.41	solid baseline
ReLU²	2.45	slightly worse or similar depending on setup
SwiGLU	2.37	often slightly better

: **Table 7.2.** Activation function comparison: expected validation loss for GELU, ReLU², and SwiGLU under identical training conditions.

Pattern matters more than exact number. You are learning that architecture choices can shave steady points off loss without causing dramatic qualitative changes. This is what real model engineering feels like.

I cut a 2,000-word version of this section that walked through every gated activation variant I had benchmarked, weighted average and all. Looking at it again, the takeaway was buried under the comparison. The takeaway is one line: gated activations usually edge out plain GELU by small but repeatable margins, and "small but repeatable" is the actual unit of progress in this part of the stack. Everything else in the cut version was numbers most readers will never run.

Step 8: Replace learned positional embeddings with RoPE

"If the model already sees tokens in order, why does it need position encoding?" Because self-attention treats inputs as a set unless you inject order somehow. With learned positional embeddings, you do this:

```
x = tok_emb[idx] + pos_emb[pos]
```

Token vector plus position vector. That works, but positions beyond the learned table do not exist.

RoPE works inside attention instead. You rotate query and key vectors based on position before computing attention scores.

Take a token vector and group its dimensions into pairs: `(x0, x1), (x2, x3), ...`. Treat each pair as a 2D point. At position `p`, rotate each pair by an angle determined by `p` and a frequency schedule. Now the same content at different positions gets rotated differently. When query and key take a dot product, the relative rotation between them encodes position — specifically, the relative offset between the two tokens.

The formula, earned

For one pair `(x_even, x_odd)` and angle `theta`:

- rotated even = `x_even * cos(theta) - x_odd * sin(theta)`
- rotated odd = `x_even * sin(theta) + x_odd * cos(theta)`

That is just a 2D rotation matrix. Nothing more exotic than spinning a point on graph paper.

In practice, different dimension pairs use different frequencies, and this is not a minor implementation detail. Lower-indexed dimension pairs (say, `d=0,1`) rotate fast. They complete many full rotations even within a short sequence, which means their dot products are sensitive to small positional differences and capture nearby token relationships (things like subject-verb agreement that depends on tokens being close together). Higher-indexed pairs (say, `d=head_dim-2, head_dim-1`) rotate slowly, completing only a fraction of a turn across the full sequence, which means their dot products remain sensitive to coarse positional structure across long distances. The model needs both kinds of signal because language has both kinds of dependency: some relationships are local (a pronoun and its antecedent separated by one word), and some are distant (a pronoun referring back to a noun introduced many sentences earlier, or a closing parenthesis matching an opening one dozens of tokens back). By spanning the full frequency range across dimension pairs, RoPE gives the model a rich vocabulary for expressing "how far apart are these two tokens?" at every scale simultaneously.

This is one of the places where I had to rewrite the explanation twice. My first draft talked about "absolute" and "relative" encodings as a binary. The more useful framing is frequency-band coverage. The local-versus-distant story above is the second attempt. I am still not fully satisfied with it.

A minimal implementation:

```
def rotate_half(x):
    x1 = x[..., ::2]
    x2 = x[..., 1::2]
    return torch.stack((-x2, x1), dim=-1).flatten(-2)

def apply_rope(x, cos, sin):
    return (x * cos) + (rotate_half(x) * sin)
```

Then build cosine and sine tables for sequence positions:

```
def build_rope_cache(seq_len: int, head_dim: int, device, base: int = 10000) -> tuple:
    half_dim = head_dim // 2
    freq_seq = torch.arange(half_dim, device=device).float()
    inv_freq = 1.0 / (base ** (freq_seq / half_dim))

    positions = torch.arange(seq_len, device=device).float()
    angles = torch.outer(positions, inv_freq)  # (T, head_dim/2)

    cos = torch.cos(angles).repeat_interleave(2, dim=-1)  # (T, head_dim)
    sin = torch.sin(angles).repeat_interleave(2, dim=-1)
    return cos[None, None, :, :], sin[None, None, :, :]
```

Inside attention, after projecting `q` and `k` and reshaping to heads:

```
cos, sin = build_rope_cache(T, head_dim, x.device)
q = apply_rope(q, cos, sin)
k = apply_rope(k, cos, sin)
```

That is the swap. You remove learned position embeddings from the token input path and apply RoPE to queries and keys instead.

What this changes

Learned positional embeddings:

- easy to implement
- tied to a maximum training position table
- poor out-of-range behavior

RoPE:

- slightly more involved
- no learned position lookup table
- better behavior on longer sequences than trained for, especially with care

Do not overclaim: RoPE is not magic infinite context dust. It gives a smoother path for length extrapolation than learned embeddings. That is the actual claim.

A practical scar from low-level Apple Neural Engine work: when targeting the ANE through private compiler APIs, RoPE's rotation kernels had to be expressed in a way the delta compiler could fuse cleanly with the rest of attention. The compile-speedup we measured was not a model-quality win; it was an engineering win that depended on RoPE's specific shape. I bring this up only because "RoPE is better" is too coarse. The honest claim is "RoPE composes better with the downstream constraints I usually care about."

Step 9: Evaluate extrapolation

Train two identical small models on sequences of length, say, 128. Model A uses learned positional embeddings. Model B uses RoPE. Then evaluate both on sequences longer than training length, maybe 256 or 512 if your data and code permit.

What do you watch? Validation loss at in-range length, validation loss at out-of-range length, and sample quality on longer prompts.

Expected pattern: in-range performance may be similar. Out-of-range, learned embeddings usually degrade sharply or fail outright if the table is capped. RoPE usually degrades more gracefully. You may see the learned-pos model become confused, repetitive, or structurally brittle past training length, while the RoPE model remains imperfect but more coherent.

If your code literally has no positional embeddings for positions beyond training length, learned embeddings do not just degrade. They stop working unless you extend the table. That is not extrapolation; that is table resizing. RoPE avoids that lookup-table ceiling.

Step 10: Visualize the distributions

For each run, produce:

- line plot of mean by layer boundary
- line plot of variance by layer boundary
- line plot of kurtosis by layer boundary
- optional histogram snapshots for selected layers

What you are looking for:

With LayerNorm

- means near zero right after norm boundaries
- variance controlled
- stable profile across depth

With RMSNorm

- variance controlled
- means can drift away from zero
- profile still stable, but not the same shape as LayerNorm

With no norm

- drift accumulates with depth
- outliers increase
- deeper layers often show unstable spread

With mixed norm types

- visible seam at the switch point

- mean and variance profile changes abruptly at the boundary
- downstream layers inherit a different signal than they were tuned for

That last pattern is the most important thing you will learn in this project. Not because it breaks training the hardest, but because it reveals what an architectural interface really is.

BREAK IT

Mix normalization types inside one model — on purpose.

Use LayerNorm for the first half of the transformer blocks and RMSNorm for the second half. If you have a 12-layer model: blocks 0 through 5 use LayerNorm, blocks 6 through 11 use RMSNorm. Do not change anything else — same weight initialization style, same optimizer, same data, same activation, same positional encoding.

What exactly are you breaking? You are breaking the hidden-state contract at the midpoint. The first half of the model produces activations whose statistics are shaped by LayerNorm, meaning downstream layers in that regime see vectors that are centered around zero with a certain pattern of learned scale and bias. The second half of the model expects inputs shaped by RMSNorm at its own boundaries, where mean is not explicitly removed and the feature distribution drifts differently. Those are not the same interface.

This is the trap people fall into when they think "tensor shape matches, so components are compatible." A hidden state is not just a shape like `(B, T, d_model)`. It is also a distribution. The midpoint seam is where that assumption gets exposed.

Exact thing to change

If your block builder currently does:

```
self.norm1 = make_norm(config.norm_type, config.d_model)
self.norm2 = make_norm(config.norm_type, config.d_model)
```

Change it so block index controls the norm type:

```
def norm_for_layer(layer_idx, n_layers, d_model):
    if layer_idx < n_layers // 2:
        return LayerNorm(d_model)
    return RMSNorm(d_model)
```

Then for each block:

```
self.norm1 = norm_for_layer(layer_idx, config.n_layers, config.d_model)
self.norm2 = norm_for_layer(layer_idx, config.n_layers, config.d_model)
```

Train this mixed model and compare it to all-LayerNorm and all-RMSNorm.

What to watch for

Do not only watch final loss. Watch the boundary statistics around the first RMSNorm block. Log output of block 5, pre-attention input to block 6, output of block 6, and pre-attention input to block 7. Plot mean, variance, min, max, and kurtosis.

What usually happens is a visible shift at the boundary. Not always a catastrophic explosion, but often something more instructive: a clean change in profile. Mean stops hugging zero and begins drifting. Variance curve changes slope. Kurtosis jumps, showing new outlier behavior. Training still runs, but loss is worse than either uniform-norm baseline.

That "still runs, but worse" outcome matters. It shows this is not about syntax correctness; the model is numerically alive, but statistically mismatched. That is much closer to real engineering failures.

What this proves

It proves that normalization type is part of the model interface. If you train one half of a model under one distribution regime and the other half under a different one, the boundary becomes a seam. Downstream layers were tuned to consume one style of signal. You handed them another.

This is exactly why composable models need interface specifications. If someone hands you a pretrained block and says, "It's fine, it's `d_model=768` and 12 heads, just drop it in," your next question should be: "What norm? What activation? What positional scheme? What pre-norm or post-norm layout? What hidden-state statistics does it assume?" Without those answers, shape matching is theater.

What failure looks like when you push harder

On a small local model, the mixed-norm run may only look somewhat worse. On larger or deeper setups, the seam can get nastier: slower convergence, worse validation loss, more brittle optimization, and stronger layerwise drift after the boundary. That does not reduce the lesson. It sharpens it. The architectural detail is not "boring bookkeeping." It is the type signature of the model's internal language.

Optional Homework

Run a full comparison lab on a small `nanochat`-style training job.

Hardware and time

Proxy version:

- 1 consumer GPU with 8GB to 24GB VRAM
- 1 to 4 hours depending on model size and steps

CPU-only version:

- possible for very small models
- expect it to be slow enough to be annoying
- use this only for the stats instrumentation and a tiny sanity check

Suggested setup

Use a small dataset first:

- Tiny Shakespeare if you want a fast signal
- a small text shard from OpenWebText or FineWeb if you want a more realistic distribution

Use a small model:

- `n_layers = 6`
- `d_model = 384` or `512`
- `n_heads = 6` or `8`
- `seq_len = 128` for baseline
- batch size as your GPU allows

Train for enough steps to separate runs:

- 2K to 10K steps for the proxy experiment

Part A: Norm comparison

1. Add config options:
    - `norm_type = layernorm | rmsnorm | none`
    - `stats_every = 100`
2. Implement `LayerNorm`, `RMSNorm`, and `nn.Identity()` path.
3. Add stats logging at:
    - embedding output
    - each block input
    - post-attention residual output
    - post-MLP residual output
    - final output
4. Run three jobs.

Example commands if your training entrypoint is `train.py`:

```
python train.py --norm_type layernorm --run_name ln_baseline
python train.py --norm_type rmsnorm --run_name rms_baseline
python train.py --norm_type none --run_name nonorm_baseline
```

5. Record:
    - final train loss
    - final val loss
    - average step time over last 200 steps
    - mean and variance curves by layer
    - kurtosis curves by layer

Expected observations:

- `none` drifts badly and often trains worse or becomes unstable
- `rmsnorm` is slightly faster than `layernorm`
- `layernorm` mean curves stay near zero at norm boundaries
- `rmsnorm` mean curves do not

Part B: Activation comparison

1. Add config option:
   - `activation = gelu | swiglu | relu2`
2. Keep norm fixed, preferably RMSNorm or LayerNorm consistently.
3. Run three jobs:

```
python train.py --norm_type rmsnorm --activation gelu --run_name gelu_run
python train.py --norm_type rmsnorm --activation swiglu --run_name swiglu_run
python train.py --norm_type rmsnorm --activation relu2 --run_name relu2_run
```

4. Record:
   - final train loss
   - final val loss
   - step time
   - optional activation output histograms for one mid-layer MLP

Expected observations:

- all three train
- `swiglu` often lands slightly better
- differences are small enough that you need controlled runs to trust them

Part C: RoPE vs learned positional embeddings

1. Add config option:
   - `pos_enc = learned | rope`
2. For `learned`, keep your existing position embedding table.
3. For `rope`, remove learned position embeddings from the token input path and rotate `q` and `k` inside attention.
4. Train both with:
   - training length `seq_len = 128`
5. Evaluate both at:
   - `eval_seq_len = 128`
   - `eval_seq_len = 256`
   - `eval_seq_len = 512` if memory allows

Example commands:

```
python train.py --pos_enc learned --seq_len 128 --run_name learnedpos
python train.py --pos_enc rope --seq_len 128 --run_name rope
python eval.py --checkpoint learnedpos.pt --eval_seq_len 256
python eval.py --checkpoint rope.pt --eval_seq_len 256
```

Expected observations:

- both are fine at 128
- learned positional embeddings get worse sharply beyond training length or require table extension
- RoPE degrades more gracefully

Part D: Break the norm interface

1. Add config option:
   - `norm_type = mixed`
2. For `mixed`, use LayerNorm in first half of blocks and RMSNorm in second half.
3. Run:

```
python train.py --norm_type mixed --run_name mixed_norm
```

4. Plot stats around the midpoint boundary.

Expected observations:

- visible change in mean/variance profile at the switch layer
- worse loss than consistent LayerNorm or RMSNorm
- the seam is easier to see in layerwise stats than in sample text

Deliverables to produce

Make these artifacts:

- one table: LayerNorm vs RMSNorm vs no norm
- one table: GELU vs SwiGLU vs $ReLU^2$
- one plot: mean by layer for each norm type
- one plot: variance by layer for each norm type
- one plot: kurtosis by layer for mixed norm
- one comparison: learned position vs RoPE beyond training length

If you do not want to build plotting code, dump JSONL and use a short notebook or a tiny script with `matplotlib`.

Questions To Answer

1. When you compare LayerNorm and RMSNorm, where do the hidden-state means differ most clearly: right after normalization, after residual adds, or inside the MLP path?
2. In the no-norm run, what fails first: loss stability, variance drift, or rare outliers shown by kurtosis?

3. Which activation gives the best final validation loss in your setup, and how large is the gap? Is the gap large enough to notice from a single run, or do you need repeated runs to trust it?
4. When you evaluate longer sequences than the model was trained on, what fails first in the learned-position model: hard position-table limits, repetition, attention confusion, or general incoherence?
5. In the mixed-norm model, how sharp is the seam at the boundary? Does it show up more clearly in mean, variance, or kurtosis?
6. What tradeoff is normalization protecting here: raw expressivity, optimization stability, component compatibility, or all three?

Go Further

Research anchors for this chapter

The most direct paper anchor here is **RoFormer: Enhanced Transformer with Rotary Position Embedding**. If you want the original statement of why rotary position encoding changes the geometry of attention instead of attaching a separate learned position table, that is the paper to read.

For normalization and activation choices in modern decoder stacks, the useful lineage is less "one paper invented all of this" and more "certain model families made the tradeoffs stick." **LLaMA: Open and Efficient Foundation Language Models** is the practical anchor for RMSNorm plus RoPE in the open-model world, while **PaLM: Scaling Language Modeling with Pathways** is a good anchor for the broader trend toward gated feedforward designs such as SwiGLU.

This is also a chapter where older references still help. **Attention Is All You Need** gives you the baseline expectation of learned position handling plus LayerNorm-style transformer structure. The reason Project 7 matters is that later models keep the same overall skeleton while quietly changing the interface contract at the seams.

Start with the reference implementation:

- `https://github.com/karpathy/nanochat`

Then compare with real model families that made different choices:

- LLaMA-family code and writeups for RMSNorm and RoPE
- older GPT-style references for LayerNorm plus learned positional embeddings
- PaLM and related transformer architecture notes for gated MLP choices

Search arXiv for these phrases:

- `RMSNorm`
- `RoFormer rotary position embedding`
- `SwiGLU transformer`
- `length extrapolation transformers`
- `transformer hidden state statistics`
- `normalization in transformers`

Good directions to chase:

- Why pre-norm became standard in many deep decoder-only models
- How RoPE scaling methods try to stretch context further
- Why gated MLPs replaced plain GELU MLPs in many recent models
- How interface specs matter for model merging, layer swapping, and composable systems

If you want one concrete research thread to watch, search:

- `long context rope scaling`
- `NTK-aware RoPE`
- `YaRN RoPE scaling`

That will show you what happens when the basic RoPE idea gets pushed into much longer contexts.

What You Now Know

Normalization type, activation function, and positional encoding are not cosmetic choices. They determine training stability, inference speed, and whether two components can be combined. You can explain why LayerNorm and RMSNorm are not interchangeable even when both make training "work," and why internal compatibility is about distributions, not just tensor shapes. That is the beginning of real transformer engineering.

Starting Point

Start from:

- your own Project 5 GPT file, such as `my_gpt.py`

Then compare against:

- `https://github.com/karpathy/nanochat`

Focus on the model architecture inside the training code where the transformer block, normalization, MLP, and positional handling are defined.

The progression still matters:

- `micrograd` taught you gradients and blame assignment
- Project 2 taught you tiny next-token learning
- `minbpe` taught you that tokens are designed, not given
- Project 4 taught you how attention mixes context under a causal mask
- Project 5 made you build a GPT yourself
- Project 6 showed you how a teaching model grows into a minimal production-shaped system

Project 7 is where you stop trusting architecture names. Now you inspect the interfaces. Now you measure the distributions. Now you see why two models that look "basically the same" are often not compatible at all.

Project 8: Flash Attention and Tiled Kernels

Hook

Why does the hand-written attention from **Project 4: Attention From Scratch** run out of memory on a 4K context while PyTorch's `F.scaled_dot_product_attention` handles the same input on the same GPU without breaking a sweat? Same math, same hardware, same sequence length, and yet one crashes and one does not. The textbook answer is "Flash Attention is faster," which is true and also useless if you cannot see why.

The first time I hit this wall I assumed it was a driver bug. It was not. The model was fine. The framework was fine. One tensor was the problem, and once you can name it, the fix is obvious.

By the end of this chapter you will know the exact tensor that vanishes between the two versions, you will build a memory-efficient attention from scratch in pure PyTorch, and you will watch your peak memory drop from quadratic in sequence length to linear without changing a single line of the model that calls it.

The Concept

Picture a chef working on a tiny stove. The pan holds one cup of liquid at a time. The chef has a recipe that, in the original instructions, asks you to mix twenty cups together and then heat the whole batch. If you only have a one-cup pan, you have two choices. You can buy a bigger pan. That is the GPU-with-more-memory option, and it stops working as soon as the recipe scales up again. Or you can rewrite the recipe to work in small batches: a cup at a time, with a running total in a separate bowl, never holding more than a cup in the pan at once. That is Flash Attention. The pan is your GPU's fast on-chip memory. The bowl with the running total is the accumulator. The cup at a time is a tile.

The analogy that finally made tiled attention click for me did not involve a chef. It came from Orion, where we were programming the Apple Neural Engine directly because the high-level framework refused to honor our memory budget. The ANE has a small fast region and a large slow one, and every algorithm worth running on it eventually becomes a question of "what stays in fast memory, what gets streamed past it, and what never needs to be materialized at all." Flash Attention is the same question, asked of a GPU.

Attention asks the model, for every position in a sequence, to compare itself against every other position and decide who matters. With a sequence of length N, that comparison produces an N-by-N table of scores. At N equal to 4096, that table holds about 16.8 million entries per attention head, and the model has several heads, several layers, and a batch of sequences. The table is the pan that is too big. Materializing it (actually allocating that N-by-N tensor in memory and writing every score into it) is the move that breaks at long context.

Flash Attention's central idea is that you never have to materialize the table. The final output of an attention layer is a weighted sum of value vectors, where the weights come from a softmax over the score table. You can compute that final output in pieces. Take a small block of rows from the query, a small block of columns from the key, compute the partial scores for just that tile, fold them into a running output, throw the tile away, and move on to the next one. The full table is never written. The pan stays small. The result is identical.

The trick that makes this possible is something called an **online softmax**, also called a **streaming softmax**. A standard softmax over N numbers needs to see all N numbers before it can produce any output. It computes the maximum of the N numbers (for numerical stability), subtracts that maximum from every number, exponentiates each result, sums the exponentials, and divides each by that sum. Every one of those steps depends on all N numbers being available at once. That is exactly the dependency we want to break.

The online version watches the numbers come in one block at a time and keeps two running statistics: the running maximum seen so far, and the running sum of exponentials, rescaled to that maximum. When a new block arrives, you compute its own local maximum and local exponential sum, compare its maximum to the running maximum, and rescale the older running statistics if the new block's maximum is larger. The output you accumulate uses the same rescaling. By the end of the last block, the running statistics give you exactly the same answer as the all-at-once version, to the bit. No approximation. No shortcut. Just a different order of operations that never needs to hold the full vector in memory.

Two terms worth defining before they appear in code. A **tile** is a small block of a larger tensor, sized to fit comfortably in a fast memory region. On a real GPU, that fast region is the on-chip SRAM, perhaps 100 kilobytes per streaming multiprocessor. On the conceptual version we will build in pure PyTorch, the "fast region" is whatever amount of GPU global memory you have decided to limit yourself to. A **fused kernel** is a single computation that combines what would otherwise be several separate passes through memory. Instead of computing scores, writing them out, reading them back, applying softmax, writing the softmax back, reading it again, and multiplying by values, a fused kernel does all of that on one tile while the tile is still in fast memory. The win is not only less memory. It is also fewer round trips between fast on-chip memory and slow off-chip memory, which is where most of a GPU's wall-clock time goes during attention.

The original Flash Attention paper, by Tri Dao and colleagues in 2022, calls this property **IO-aware**: the algorithm is designed around the cost of moving data between memory tiers, not around the cost of arithmetic. Attention is bandwidth-bound at long context lengths, not compute-bound. Reducing memory traffic is the actual lever. The CUDA kernel in the production library is tightly tuned to the specific sizes of those memory tiers on real hardware, which is why writing it in CUDA pays for itself with another sizable speedup on top of the algorithmic win. But the algorithmic win is what we are after in this chapter, because the algorithmic win is what survives if you are reading this on a different accelerator three years from now.

I learned this lesson the slow way on low-level ANE work. We spent weeks trying to make CoreML respect a memory budget the hardware's spec sheet said it could hit. CoreML refused. So we wrote against the private ANEClient and ANECompiler APIs and shaped the kernel around the memory hierarchy ourselves. The headline number was a large delta-compile speedup, but the real lesson was about who owns the memory plan. If the framework owns it, you live with whatever the framework decided. If you own it, you can do exactly the kind of tiling this chapter is about.

A small note on what this chapter does not promise. We will not match the production CUDA kernel's wall-clock numbers. The pure-PyTorch tiled version you will build is meant to be readable, correct, and clearly memory-efficient. The speed win at moderate sequence lengths is real but modest; the memory win is dramatic. Speed parity with `flash-attn` requires CUDA, and CUDA is a different book. What you will have at the end of this project is the algorithm in your hands, in code you can read line by line, and a working understanding of why the production version is shaped the way it is.

I want to flag a pet peeve here. Most explanations of Flash Attention I have read open with the CUDA kernel and leave the algorithm implicit. That order is backwards. Once you understand the algorithm, the kernel is straightforward engineering. Without the algorithm, the kernel reads like incantations.

Why It Matters

The reason this chapter exists where it does is that every chapter after it assumes long context as a baseline. Without tiled attention, the rest of the book runs into a wall at sequence length somewhere between 2K and 4K on most consumer GPUs. With tiled attention, the wall moves out by a factor of ten or more, and the projects that depend on long context become reachable on the same hardware.

The hand-written attention from **Project 4: Attention From Scratch** allocates a tensor of shape `(batch, heads, seq, seq)` to hold the scores. At sequence length 1024, with 8 heads and batch size 8 in fp16, that tensor is roughly 128 megabytes per layer. At sequence length 4096 with the same other settings, the same tensor balloons to 2 gigabytes per layer. A 6-layer model now wants 12 gigabytes for nothing but score tensors, before any of the actual model parameters, activations, gradients, or optimizer state. On an 8 GB consumer card, the run dies before it begins. The crash is not subtle. PyTorch throws an out-of-memory error and tells you exactly which allocation failed. The crash is also not the model's fault. The model's parameters fit. The forward activations fit. The single thing that does not fit is the materialized score table.

This was the specific failure mode that bit us hardest during fine-tuning experiments at multi-billion-parameter scale. The model and the optimizer fit. The activations fit. The attention scores tensor at the context lengths we wanted did not, and the failure was on a 4090, not on something exotic. A long tail of out-of-memory tracebacks taught me that "more memory" is the answer almost no one's hardware budget agrees to.

That same memory pressure is exactly the failure mode that the activation-stash investigation in **Project 11: Training Debugging — Spikes, NaNs, Profiling** identifies in its memory-forensics section. When the dominant allocation in your model is the attention scores tensor, the right fix is not "buy a bigger GPU." That scales poorly and runs out again at the next context length. The right fix is to stop materializing the tensor in the first place. That is what this chapter builds.

There is also a forward-pointing reason. **Project 13: Fast Inference: The KV Cache** introduces the inference-time optimization that caches keys and values across decoding steps, so that each new token only computes attention against the new query row against the full cached key and value blocks. That decoding-time computation is itself a tiled attention: a single query row tile against a long key-value column block. Once you have built the training-time tiled forward in this chapter, the inference-time variant in Project 13 is a small specialization of the same kernel, not a new mystery.

And there is a third reason worth naming. Once you can read the Flash Attention algorithm in pure PyTorch, you can read the Triton and CUDA versions in `flash-attn` and `xformers` and recognize the same shape underneath. The production code has more constants, more memory hints, more autotuning, and more low-level addressing math, but the algorithm is the algorithm. The mystique evaporates the moment you watch the running maximum get rescaled on a CPU tensor of size four.

The Build

You are going to extend the attention from **Project 4: Attention From Scratch** in eight steps. First, measure the naive memory footprint and confirm where it dies. Then implement a tiled forward in pure PyTorch with no Triton, no CUDA, no kernel-fusion tricks. Just plain tensor operations. The streaming softmax goes in next, and it is the piece that makes the tiling correct. After that, verify against the naive version at small sizes, measure memory at the same sequence lengths, and watch the curve flatten. The last three steps are integration and benchmarking: plug the tiled attention into the **Project 5: A Working GPT** model and confirm a short training run produces the same loss trajectory, benchmark wall-clock at a couple of context lengths, and finally point at the production CUDA kernel and explain what the extra layer of engineering buys you.

A laptop GPU with 8 GB to 24 GB is plenty. A CPU-only reader can run the streaming-softmax verification and the tile decomposition on tensors of size 64 or so. The patterns are identical, and the homework includes a CPU proxy lab for that path.

Step 1: Measure where the naive version dies

Start from the attention you wrote in **Project 4: Attention From Scratch**. The relevant block looks like this. It is a textbook materialized attention with no causal masking shown for brevity:

```
def naive_attention(q, k, v):
    # q, k, v: (batch, heads, seq, dim)
    d = q.shape[-1]
    scores = (q @ k.transpose(-2, -1)) / (d ** 0.5)   # (B, H, N, N)
    weights = torch.softmax(scores, dim=-1)            # (B, H, N, N)
    return weights @ v                                  # (B, H, N, D)
```

The two intermediate tensors of shape `(B, H, N, N)` are the entire problem. Each one is `B * H * N * N` elements. With `B=2, H=8, N=4096` in fp16, that is `2 * 8 * 4096 * 4096 * 2 bytes`, which is exactly 1 GB per tensor. PyTorch creates both `scores` and `weights` and keeps both alive at the same time during forward, so the attention layer alone needs about 2 GB before you have done anything else.

Write a small probe that measures peak memory at several sequence lengths and confirms the quadratic shape:

```
import torch

def measure_naive(seq_lens, batch=2, heads=8, dim=64, device="cuda", dtype=torch.float16):
    results = {}
    for n in seq_lens:
        torch.cuda.empty_cache()
        torch.cuda.reset_peak_memory_stats()
        q = torch.randn(batch, heads, n, dim, device=device, dtype=dtype)
        k = torch.randn_like(q)
        v = torch.randn_like(q)
        try:
            out = naive_attention(q, k, v)
            torch.cuda.synchronize()
            peak = torch.cuda.max_memory_allocated() / 1e9
            results[n] = peak
        except torch.cuda.OutOfMemoryError:
            results[n] = None
        del q, k, v
        if results[n] is not None:
            del out
    return results

print(measure_naive([1024, 2048, 4096, 8192]))
```

On an 8 GB card you will see peak memory roughly quadruple every time you double `N`, and on most consumer cards you will see `None` appear at 4096 or 8192. That is the wall. Save this dictionary somewhere; the tiled version will reuse the same probe and produce a strikingly different curve.

I expected the first `None` to appear at 8192. On the card I was using, it showed up at 4096. The "quadratic blow-up" sentence in textbooks turns out to be load-bearing in a way that does not feel real until you watch it happen.

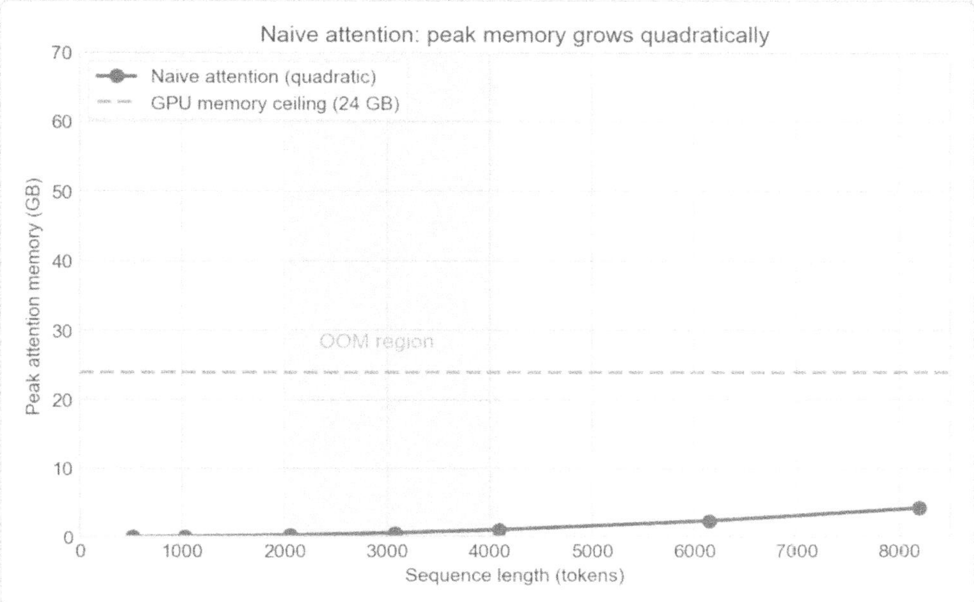

Figure 8.1. Naive attention peak memory versus sequence length on a single consumer GPU. The curve is quadratic and crosses the card's memory ceiling between 2K and 4K context, which is where the run begins to fail.

Step 2: Decompose the computation into tiles

The tiled forward replaces the all-at-once score matrix with a loop over column tiles, processing one row tile of the query at a time. Define two block sizes upfront — one for query rows, one for key and value columns.

```
BR = 64    # query row tile size
BC = 64    # key/value column tile size
```

These two numbers are the analog of pan size. Sixty-four is small enough to fit comfortably anywhere and large enough to amortize the loop overhead. Production CUDA tuning sweeps these values per GPU; here we pick a value that works and move on.

The query tensor of shape `(B, H, N, D)` gets split along the sequence dimension into `N / BR` row tiles, each of shape `(B, H, BR, D)`. The key and value tensors, also `(B, H, N, D)`, get split along the sequence dimension into `N / BC` column tiles, each of shape `(B, H, BC, D)`. For each query row tile, we walk through every key-value column tile, accumulate partial outputs and partial softmax statistics, and emit one output row tile at the end. Then we move to the next query row tile and start over.

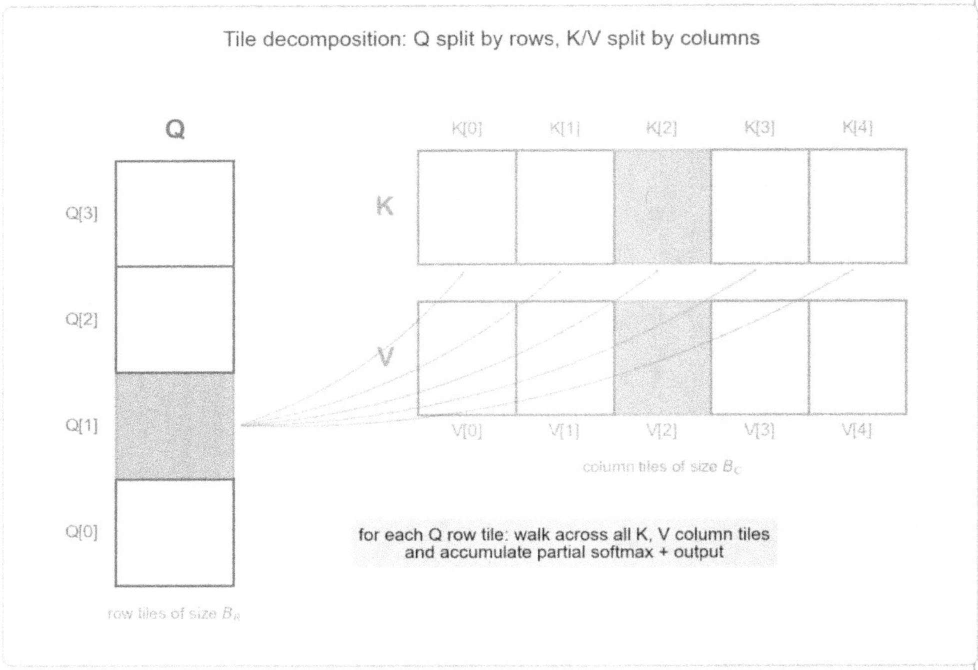

Figure 8.2. Tile decomposition. The query is split into row tiles of size BR. The key and value are split into column tiles of size BC. For each row tile of Q, the algorithm walks across all column tiles of K and V, accumulating partial results without ever materializing the full N-by-N matrix.

In code, the outer skeleton looks like this. The inner accumulation logic will be filled in by the streaming softmax in Step 3:

```
def tiled_attention_skeleton(q, k, v, BR=64, BC=64):
    B, H, N, D = q.shape
    out = torch.zeros_like(q)

    for i in range(0, N, BR):
        q_tile = q[:, :, i:i+BR, :]                    # (B, H, BR, D)
        # running statistics for this query row tile, set up in Step 3
        for j in range(0, N, BC):
            k_tile = k[:, :, j:j+BC, :]                # (B, H, BC, D)
            v_tile = v[:, :, j:j+BC, :]                # (B, H, BC, D)
            scores_tile = q_tile @ k_tile.transpose(-2, -1)  # (B, H, BR, BC)
            # update running statistics and partial output using scores_tile
            ...
        # write final row tile of out
        out[:, :, i:i+BR, :] = ...
    return out
```

Two details to read carefully. The tensor `scores_tile` is shape `(B, H, BR, BC)`, never `(B, H, N, N)`. At `BR=BC=64`, this tile is the same size regardless of whether `N` is 1024 or 32768. That is the entire memory win, sitting right there in the dtype constructor. The score matrix never exists. Only one small tile of it exists at a time, and the tile gets overwritten on the next iteration.

The second detail is the outer loop over `i` and the inner loop over `j`. The original Flash Attention paper writes them the other way around (outer over column tiles, inner over row tiles) for reasons that matter on a real GPU and do not matter in pure PyTorch. We pick the order that reads most naturally for the streaming softmax explanation. The math is identical either way.

Step 3: Build the streaming softmax

Standard softmax over a row of N scores produces, for each position `j`, the value `exp(s_j - m) / sun_k exp(s_k - m)`, where `m` is the maximum of the row, included for numerical stability. The output of attention for that row is the dot product of those softmax weights with the row's value vectors. The streaming version computes the same final answer, but instead of needing all N scores in memory at once, it processes scores one column tile at a time and maintains two running pieces of state per query row.

Let `m_running` be the running maximum of all scores seen so far, and let `l_running` be the running sum of exponentials, scaled by that running maximum. Let `o_running` be the running output — the partial weighted sum of value vectors, also scaled by the running maximum so that the math stays consistent. When a new column tile of size BC arrives with its own local scores `s_tile`, compute three things from the tile alone:

```
m_tile = s_tile.max(dim=-1, keepdim=True).values      # local max of the tile
p_tile = torch.exp(s_tile - m_tile)                   # local stable exponentials
l_tile = p_tile.sum(dim=-1, keepdim=True)             # local sum of exponentials
```

Now combine the local tile's statistics with the running statistics. The new running maximum is the elementwise maximum of the old running maximum and the tile's local maximum. If the new maximum is larger than the old one, the old running sum and the old running output were computed against a now-too-small scaling factor — they need to be rescaled down. The rescaling factor is `exp(m_running - m_new)`, which is at most 1.

```
m_new = torch.maximum(m_running, m_tile)
alpha = torch.exp(m_running - m_new)                  # rescaling for old stats
beta  = torch.exp(m_tile    - m_new)                  # rescaling for tile stats
l_new = alpha * l_running + beta * l_tile
o_new = alpha * o_running + beta * (p_tile @ v_tile)
m_running, l_running, o_running = m_new, l_new, o_new
```

At the end of all column tiles for this query row tile, divide the running output by the running sum to produce the final softmax-weighted output:

```
o_final = o_running / l_running
```

Walk through that math slowly the first time. The variable `alpha` rescales the old running statistics down to the new common scale. The variable `beta` rescales the current tile's local statistics down to the same common scale. After the rescaling, the running sum `l_new` and the running output `o_new` are exactly what you would have gotten by computing softmax over all the scores seen so far. The trick is that you never had to hold those scores in memory at the same time, only their per-tile rescaled contributions.

The first time I derived this on a notebook page I got the rescaling backwards and could not figure out why my fp32 difference against the naive version was 0.3 instead of 1e-6. The mistake was applying `exp(m_new - m_running)` instead of `exp(m_running - m_new)`. Both are valid exponentials. Only one is at most 1.0, and the other one overflows the running statistics when the new tile has a larger maximum. Sign errors in stability-critical math are the kind of bug a compiler will never warn you about.

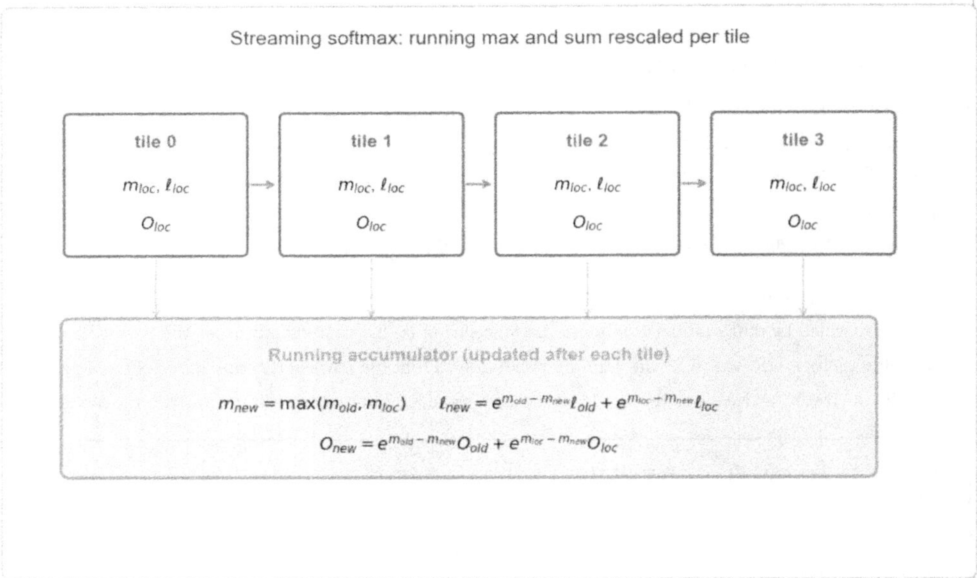

Figure 8.3. Streaming softmax accumulator updates per tile. The running maximum is the elementwise max of the previous running maximum and the new tile's local maximum. The running sum and running output get rescaled by exp(m_old - m_new) so that all contributions are on the same scale once the loop ends.

Put it together into a working tiled attention. The scale factor `1 / sqrt(D)` and the optional causal mask are added in the obvious places:

```python
def tiled_attention(q, k, v, BR=64, BC=64, causal=False):
    B, H, N, D = q.shape
    scale = D ** -0.5
    out = torch.zeros_like(q)

    for i in range(0, N, BR):
        q_tile = q[:, :, i:i+BR, :]
        m = torch.full((B, H, q_tile.shape[2], 1), float("-inf"),
                       device=q.device, dtype=q.dtype)
        l = torch.zeros((B, H, q_tile.shape[2], 1),
                        device=q.device, dtype=q.dtype)
        o = torch.zeros_like(q_tile)

        for j in range(0, N, BC):
            k_tile = k[:, :, j:j+BC, :]
            v_tile = v[:, :, j:j+BC, :]

            s = (q_tile @ k_tile.transpose(-2, -1)) * scale
            if causal:
                # mask future positions inside this tile
                row_idx = torch.arange(i, i + q_tile.shape[2], device=q.device)[:, None]
                col_idx = torch.arange(j, j + k_tile.shape[2], device=q.device)[None, :]
                s = s.masked_fill(col_idx > row_idx, float("-inf"))

            m_tile = s.max(dim=-1, keepdim=True).values
            p = torch.exp(s - m_tile)
            l_tile = p.sum(dim=-1, keepdim=True)

            m_new = torch.maximum(m, m_tile)
            alpha = torch.exp(m - m_new)
            beta = torch.exp(m_tile - m_new)
            l = alpha * l + beta * l_tile
            o = alpha * o + beta * (p @ v_tile)
            m = m_new

        out[:, :, i:i+BR, :] = o / l

    return out
```

Read the body of the inner loop with the streaming softmax derivation right next to you the first time through. Every line corresponds to one piece of the math from earlier. The score tile `s` is `(B, H, BR, BC)`. The running statistics `m`, `l`, `o` only ever exist for one query row tile at a time. The output tensor `out` is `(B, H, N, D)`, the same shape as the input. There is no `(B, H, N, N)` tensor anywhere in this function.

Step 4: Verify against the naive version

Before benchmarking, prove the tiled version computes the same answer. Run both at small sizes where the naive version still fits comfortably and compare:

```
torch.manual_seed(0)
B, H, N, D = 2, 4, 256, 32
q = torch.randn(B, H, N, D, device="cuda")
k = torch.randn(B, H, N, D, device="cuda")
v = torch.randn(B, H, N, D, device="cuda")

ref = naive_attention(q, k, v)
out = tiled_attention(q, k, v, BR=32, BC=32)

diff = (ref - out).abs().max().item()
print(f"max absolute difference: {diff:.6e}")
```

For fp32 you should see a difference on the order of 1e-6, which is floating-point round-off noise from doing the additions in a different order. For fp16 the difference will be larger, perhaps 1e-3, but still well below anything that affects model behavior. If you see a difference of 1e-1 or larger, something is wrong with the streaming softmax. Most often the rescaling factor `alpha` is applied in the wrong place, or the running maximum is initialized to zero instead of negative infinity. The `-inf` initialization matters: the first tile's local maximum must always become the running maximum, which only happens if the running maximum starts below every possible score.

A second sanity check is to vary the block sizes. Run the tiled version at `BR=BC=16`, `BR=BC=32`, `BR=BC=64`, and `BR=BC=128`. The output should be identical to within floating-point noise across all of those. If it changes meaningfully with block size, the streaming softmax is wrong.

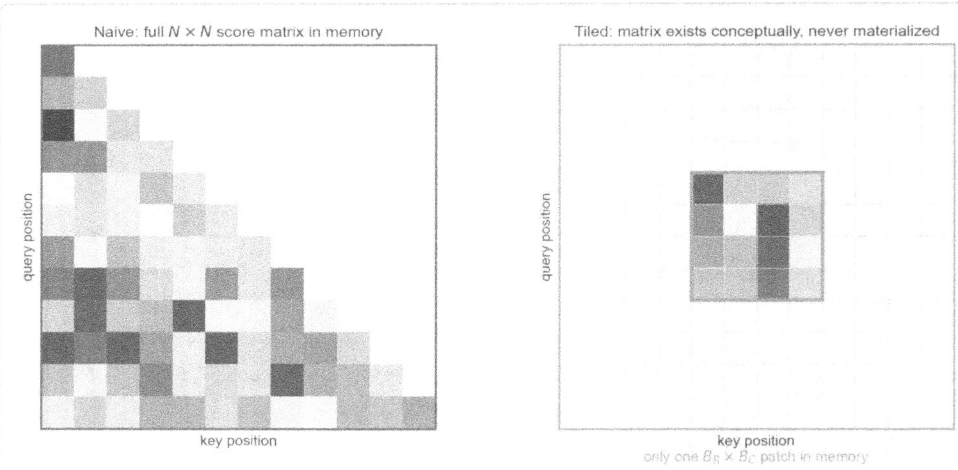

Figure 8.4. Forward pass before and after tiling, illustrated. In the naive version, the full N-by-N score matrix is allocated and filled. In the tiled version, the matrix conceptually exists but is never materialized; only one BR-by-BC patch is in memory at a time, and the rest of the matrix never gets written.

Step 5: Measure memory versus the naive version

Rerun the memory probe from Step 1, this time on the tiled forward. Keep the same dtype, batch, heads, and head dimension. Sweep the same set of sequence lengths:

```python
def measure_tiled(seq_lens, batch=2, heads=8, dim=64, device="cuda", dtype=torch.float16):
    results = {}
    for n in seq_lens:
        torch.cuda.empty_cache()
        torch.cuda.reset_peak_memory_stats()
        q = torch.randn(batch, heads, n, dim, device=device, dtype=dtype)
        k = torch.randn_like(q)
        v = torch.randn_like(q)
        out = tiled_attention(q, k, v, BR=64, BC=64)
        torch.cuda.synchronize()
        results[n] = torch.cuda.max_memory_allocated() / 1e9
        del q, k, v, out
    return results

print(measure_tiled([1024, 2048, 4096, 8192, 16384, 32768]))
```

The curve looks different now. Peak memory grows roughly linearly with sequence length, not quadratically. The slope is set by the input tensors themselves: `q`, `k`, `v`, and `out` are all `(B, H, N, D)`, and that is what scales with `N`. The score tile and the running statistics are fixed-size and contribute a constant overhead that is negligible at any reasonable sequence length.

The first time I plotted these two curves on the same axes I did not believe the result and re-ran the probe three times to make sure I had not introduced a caching bug. The shape of the picture is the kind of thing where, if you have spent any time with attention memory before, you almost want it to be wrong. It is not wrong.

Plot the two curves on the same axes with a log-scaled y-axis. The naive line bends sharply upward and disappears off the top of the chart somewhere between 2K and 4K context. The tiled line is almost straight. That visual is the single most important picture in this chapter. Print it out and tape it to the side of your monitor if you are working through long-context modeling.

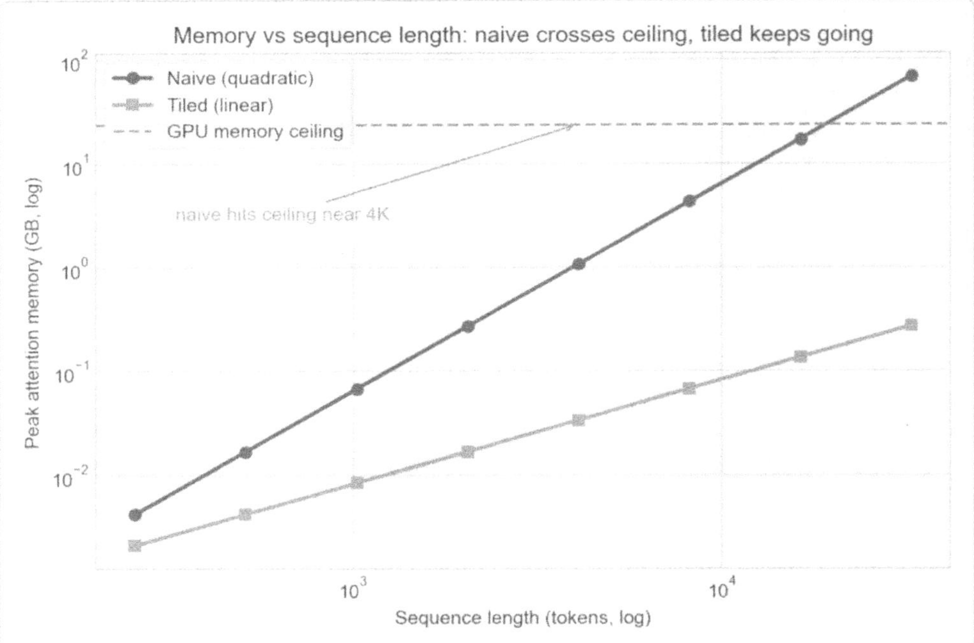

Figure 8.5. Memory versus sequence length, naive attention versus tiled attention, log-scaled y-axis. The naive curve is quadratic and crosses the GPU memory ceiling at 4K context. The tiled curve is linear and continues comfortably past 32K on the same hardware.

Step 6: Wire it into the GPT and train briefly

Open the model from **Project 5: A Working GPT**. Find the attention module — the class that holds the `qkv` projection, computes attention, and projects the result back through the output linear. Replace the call to your naive attention with a call to the tiled version. The interface is identical: same inputs, same output shape. Nothing else in the model changes.

Train the model for 200 steps on the same dataset, with the same hyperparameters, that you used to validate **Project 5: A Working GPT**. The expectation is that the loss curve from the tiled version matches the loss curve from the naive version step by step, to within the floating-point noise we already characterized in Step 4. If your loss curve diverges meaningfully (say, the tiled version's loss is half a unit higher or lower than the naive version's by step 50), the most likely cause is a bug in the causal mask. Causal masking inside a tile is one of the easier places to get the indexing wrong. The mask must zero out positions where the key's absolute index exceeds the query's absolute index, where "absolute" means relative to the full sequence, not relative to the tile. The snippet in Step 3 computes `row_idx = torch.arange(i, ...)` and `col_idx = torch.arange(j, ...)` precisely to handle this. If you write `arange` starting from 0 instead of from `i` or `j`, the mask is wrong at every tile after the first one.

I lost an afternoon to exactly that bug while drafting this chapter. The loss curves matched for the first three tiles, then started drifting, and by step 50 the tiled version was learning a slightly different language than the naive one. The root cause was a single missing `+i` in the `arange`. The fix took thirty seconds. Finding it took six hours.

Once the loss curves agree, you have a drop-in replacement for the attention from **Project 4: Attention From Scratch** that uses linear memory instead of quadratic memory. The model trains identically. Long-context fine-tuning and pretraining experiments that were impossible on your hardware ten minutes ago are now possible.

Step 7: Benchmark wall-clock

This is the step where expectations need to be carefully managed. Pure-PyTorch tiled attention is more memory-efficient than naive attention at every sequence length, but it is only faster than naive attention at large sequence lengths, and the speedup is modest compared to the production CUDA version.

Time both versions at three sequence lengths (1K, 2K, and 4K) for forward only, with warmup and synchronization handled properly:

```
import time

def bench(fn, q, k, v, repeats=10):
    for _ in range(3):  # warmup
        fn(q, k, v)
    torch.cuda.synchronize()
    t0 = time.perf_counter()
    for _ in range(repeats):
        fn(q, k, v)
    torch.cuda.synchronize()
    return (time.perf_counter() - t0) / repeats
```

At sequence length 1K, the naive version often wins by a small margin in pure PyTorch — its single large matmul is well-optimized by cuBLAS, and the tiled version pays a Python loop overhead. At 2K, the two are roughly comparable. At 4K, the naive version may not even run, in which case the tiled version wins by infinity. At 8K and beyond, only the tiled version runs, period.

The point of the benchmark is not to celebrate a speedup. It is to internalize when tiling helps with wall-clock and when it does not. Tiling buys you memory at every sequence length. Tiling buys you speed only when you are bandwidth-bound, which is at long sequences, large heads, or both. Below the crossover length, the naive version is fine for wall-clock; you might still want the tiled version because of memory headroom, but you should not expect it to be faster.

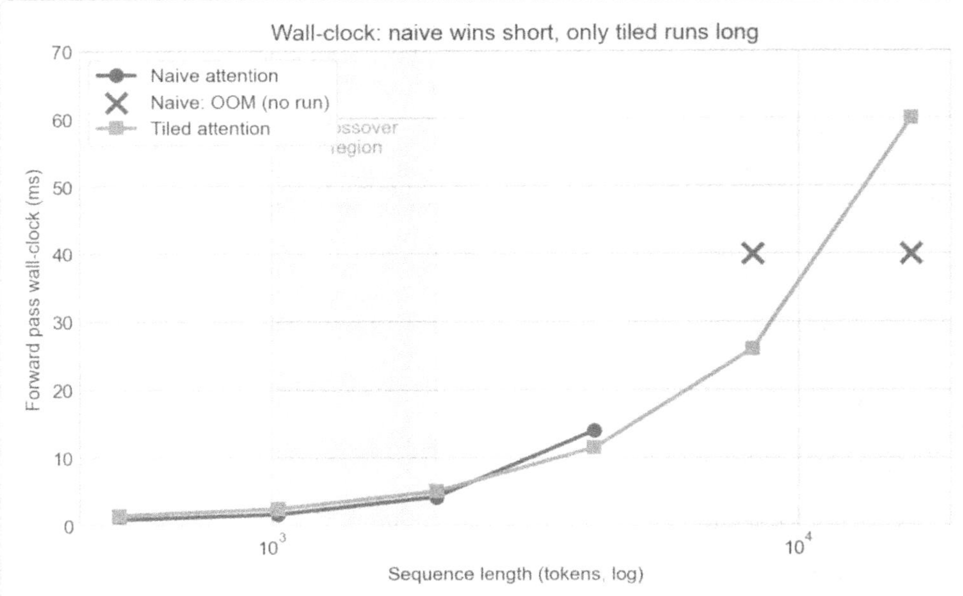

Figure 8.6. Wall-clock benchmark, naive versus tiled, at varying sequence lengths. Below 2K context, the naive version's single fused matmul wins on speed. Above 4K, only the tiled version runs at all.

Step 8: A brief note on the production version

Everything you just built is the algorithm. The production library at github.com/Dao-AILab/flash-attention adds three more layers of engineering on top of the same algorithm, each of which buys another factor of two or so on real hardware.

First, the inner loop is written in CUDA (or Triton) rather than Python. The two nested loops in Step 3 become a single fused kernel where the score tile, the softmax, and the value multiplication all happen on the same tile while the tile is still in fast on-chip memory. The Python version we wrote allocates `s`, `p`, `m_tile`, `l_tile`, and a few intermediate tensors as full GPU tensors on every iteration; the CUDA version keeps all of those in registers and SRAM, which is dramatically faster per operation.

Second, the block sizes are autotuned per GPU. The values `BR=64, BC=64` we picked are reasonable defaults, but the optimal block size depends on the head dimension, the dtype, the compute capability of the GPU, and the relationship between SRAM size and warp count on that GPU. The library sweeps these values during installation and picks the best one for your hardware. The same algorithm, with the right block size, can be twice as fast as with a default block size.

Third, the backward pass is fused too, and this matters more than it sounds. The backward pass of attention also needs the softmax weights, which means a naive backward implementation would either recompute them (wasting compute) or store them (wasting memory). Flash Attention's backward recomputes them on the fly within the same tiled structure, which keeps the backward as memory-efficient as the forward. That is what makes long-context **training** possible at all, where a stash-the-weights approach would only have stretched to long-context inference. Long-context training is what unlocks the projects further in the book that fine-tune models on documents, codebases, and conversations longer than a few thousand tokens.

The takeaway for this chapter: you now have the algorithm in your hands, and you can read the production code without being confused about what it is doing. The CUDA kernel is the same algorithm, made fast.

BREAK IT

The break for this chapter targets the one trick the streaming softmax depends on for numerical stability: the running maximum. Remove it, watch the exponentials overflow, and confirm that the failure mode shows up at relatively short sequence lengths.

The exact change

Find the streaming softmax block inside `tiled_attention`. Currently the per-tile maximum is subtracted before exponentiation:

```
m_tile = s.max(dim=-1, keepdim=True).values
p = torch.exp(s - m_tile)
l_tile = p.sum(dim=-1, keepdim=True)
```

Remove the subtraction. Compute the exponentials of the raw scores directly:

```
# m_tile = s.max(dim=-1, keepdim=True).values    # removed
p = torch.exp(s)                                 # no stability shift
l_tile = p.sum(dim=-1, keepdim=True)
```

Now do the same to the running-maximum update. The original code maintains `m` and uses it to rescale both the running statistics and the tile's statistics through `alpha = exp(m - m_new)` and `beta = exp(m_tile - m_new)`. Replace those with no rescaling at all:

```
# m_new = torch.maximum(m, m_tile)
# alpha = torch.exp(m - m_new)
# beta = torch.exp(m_tile - m_new)
# l = alpha * l + beta * l_tile
# o = alpha * o + beta * (p @ v_tile)
# m = m_new
l = l + l_tile
o = o + (p @ v_tile)
```

The forward pass still runs. The shapes are still correct. No assertion fires. The function still returns a tensor of the right size. The dashboard, if you had one, would show nothing wrong for the first few seconds.

Predicted observation

Before running, write down what you expect to see, with specific numbers.

In fp32, the maximum representable magnitude is about 3.4e38, and `exp(89)` is already 4e38. Scaled by the `1 / sqrt(D)` factor in attention with `D=64`, a raw score of magnitude 700 or so produces an exponential at the edge of fp32's range. Real attention scores at the start of training have magnitudes in the single digits, so the first tile or two will look fine. As soon as a value vector has a moderately large component and a query has a moderately aligned component, you will see a score in the tens, and `exp(s)` will be in the millions. By the time the sum runs across several tiles at moderate sequence length, the running sum `l` overflows to `inf`, the running output `o` overflows to `inf`, and the final divide `o / l` produces `nan`.

In fp16, the maximum representable magnitude is only 65504, and `exp(12)` is already 162754. Overflow happens almost immediately. At sequence length 256 in fp16, you should see `nan` in the output within the first tile or two for any query that is moderately aligned with any key.

Compare the prediction against the run. Print the maximum of `s` per tile and the maximum of `p` per tile during the broken forward. Watch `p` cross 1e30 in fp32 or 1e4 in fp16. Watch `l` follow. Watch `o / l` become `nan` shortly after.

The intact version of the kernel, with the running-maximum subtraction in place, holds `p` in a tight range. Every value is between `exp(-large)` (close to zero) and `exp(0) = 1`, regardless of the absolute scale of the scores. The subtraction is the one cheap trick that keeps the entire pipeline numerically alive. Without it, the exponentials become unbounded, and the algorithm's "no approximation" property goes from being a virtue to being a death sentence. Exact computation of an overflowing value is still an overflow.

What this proves

The streaming softmax does two things at once. It rearranges the order of operations so that the full vector never lives in memory at the same time, and it preserves the numerical-stability trick from the standard softmax inside every tile. Both pieces are load-bearing. If you remove the rescaling but keep the subtraction, you get wrong answers because the running statistics are on incompatible scales. If you remove the subtraction but keep the rescaling, you get `nan` because the raw exponentials overflow before they ever reach the rescaling step. The correctness of the algorithm depends on both moves working together — the local stable exponentials inside each tile, and the rescaling that brings older partial results onto the same scale as newer ones.

Restore both pieces. Run again. Watch `nan` disappear. That confidence, knowing exactly which line keeps the kernel alive, is the lesson.

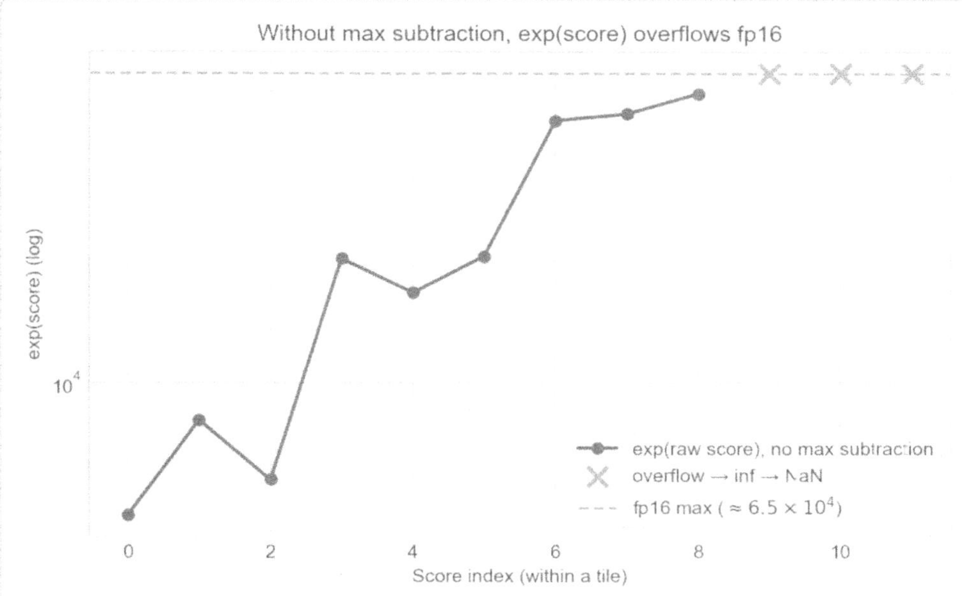

Figure 8.7. Numerical instability when the running-maximum subtraction is removed. The exponential of raw scores quickly exceeds fp16's dynamic range and the partial sums overflow to inf, producing nan in the final output.

Optional Homework

Most of this chapter is runnable on a single consumer GPU with 8 GB or more of memory. The full memory sweep runs in under five minutes. The wall-clock benchmark adds maybe another five. The brief retraining of the GPT from **Project 5: A Working GPT** to confirm loss-curve parity takes longer; plan for 15 to 30 minutes at depth 6, sequence length 1024, on a mid-range card. CPU-only readers can run the streaming-softmax verification on tensors of size 64 in seconds, which is enough to internalize the algorithm even without a GPU.

Part A: Confirm the memory wall

Implement `naive_attention` and the memory probe from Step 1. Run the probe at sequence lengths `[512, 1024, 2048, 4096, 8192]`. Record peak memory for each. Note the first length at which the run fails. Plot the surviving lengths on a log-scaled y-axis and confirm the slope.

You should see a roughly fourfold increase in peak memory for every doubling of sequence length, with a hard ceiling somewhere between 2K and 8K depending on the card. Write down the crossover length for your hardware. That number is the wall this chapter is moving.

Part B: Verify the tiled forward against naive

Implement `tiled_attention` from Steps 2 and 3 exactly as shown. Run the verification from Step 4 at sequence length 256 with batch 2, heads 4, head dimension 32, fp32. Print the maximum absolute difference. It should be at most 1e-5.

Then rerun in fp16. The maximum absolute difference should be at most 1e-3. Anything larger means the streaming softmax is wrong; the most common bug is initializing the running maximum to zero instead of `-inf`.

Vary `BR` and `BC` across `[16, 32, 64, 128]`. The output should be the same to within floating-point noise across all block sizes. If it changes with block size, the rescaling factors `alpha` and `beta` are being applied in the wrong place.

A monolithic single-lab version of this homework was the first thing I tried. It was unreadable. Splitting it into six parts that each take 5 to 30 minutes was the version I could actually finish without losing the thread.

Part C: Memory parity

Run the memory probe from Part A on `tiled_attention`. Same sequence lengths, same dtype, same batch and head settings. Plot the two curves on the same axes. The tiled curve should be roughly linear, with a slope set by the input tensor sizes; the naive curve should be quadratic.

Find the sequence length at which the tiled version uses less than one tenth the memory of the naive version. Record that number. For 8-head fp16 attention with head dimension 64, this crossover is usually in the 1K to 2K range.

Part D: Train the GPT and confirm loss-curve parity

Replace the attention module's internal call in your **Project 5: A Working GPT** model with `tiled_attention`. Run a 200-step training pass on the same data, with the same hyperparameters and the same random seed as a previous successful run with the naive attention. Plot both loss curves on the same axes.

The two curves should be visually indistinguishable. Any meaningful divergence usually means the causal mask is wrong — print `row_idx` and `col_idx` inside the tile and confirm both are computed relative to the full sequence, not relative to the tile.

Part E: Wall-clock benchmark

Run the benchmark from Step 7 at sequence lengths `[1024, 2048, 4096]`. For each length, record the average forward time over 10 runs after 3 warmup runs. Compare naive versus tiled.

You should see naive faster at 1024, comparable at 2048, and tiled faster (or naive non-existent) at 4096. Plot the ratio of tiled time to naive time on the y-axis against sequence length on the x-axis. The ratio should cross 1.0 between 1K and 4K context on most hardware.

Part F: The break, overflow without the running maximum

Apply the exact change from BREAK IT to your `tiled_attention` function. Rerun the verification from Part B at sequence length 256 in fp16. You should see `nan` in the output.

Add print statements that record the maximum of `s` and the maximum of `p` per tile during the broken forward. Watch the values climb. Record at which tile the first overflow appears.

Restore the running-maximum subtraction. Confirm the verification passes again. Resist the urge to declare victory before re-running the verification.

Part G: CPU proxy lab

If you have no GPU, run a small version of the entire chapter on CPU. Use `B=1, H=2, N=64, D=8` with `BR=8, BC=8`. Implement both `naive_attention` and `tiled_attention`. Compare outputs — the difference should be at most 1e-6 in fp32. Apply the BREAK IT change and confirm the overflow at modest score scales by manually multiplying the input tensors by 10 or 100 before passing them in.

The patterns are identical. The magnitudes are smaller because the dimensions are smaller, but the streaming softmax derivation, the rescaling logic, and the failure mode without the running maximum are all visible on this scale. The CPU version is the calibration version: once you have seen the algorithm work at size 64, the GPU version at size 32K is the same algorithm with more arithmetic.

Questions To Answer

1. At what sequence length did the naive attention first fail to run on your hardware? At what sequence length did the tiled attention use less than one tenth the memory of the naive attention at that same length?
2. When you removed the running-maximum subtraction in BREAK IT, at which tile did the overflow first appear? Did it appear earlier in fp16 than in fp32, and by roughly how much?
3. Below the crossover sequence length, the naive version is faster than the tiled version. Why? Be specific about which operation in the naive version is faster, and what overhead in the tiled version costs you the speed parity.
4. The streaming softmax maintains three running pieces of state per query row tile: the running maximum, the running sum, and the running output. What goes wrong if you only maintain two of the three (say, the running sum and the running output, but not the running maximum)?
5. The production CUDA version is roughly two to five times faster than the pure-PyTorch version you built, at the same sequence length. Where does that extra speed come from? Name at least two engineering moves the CUDA version makes that pure PyTorch cannot.
6. If a future project asks you to implement attention with a custom score modifier (say, ALiBi linear biases or a non-standard masking pattern), where in the tiled implementation does the modifier go, and why does the location matter for the streaming softmax to stay correct?

Go Further

Two repos worth reading after this chapter.

The first is the production Flash Attention library at github.com/Dao-AILab/flash-attention. Start with the README to map the algorithm sections, then open the Triton implementation in `flash_attn/flash_attn_triton.py` rather than the CUDA one. The Triton version reads more like Python and the algorithmic structure is closer to what you built. Look for the inner kernel and find the running-maximum subtraction, the rescaling factor, and the partial output accumulator. The names will be slightly different, but the structure is the same as Step 3 of this chapter.

The second is the xformers library at github.com/facebookresearch/xformers, which has a reference attention implementation that documents many of the same ideas with a slightly different API. It is a good cross-check for getting the causal mask indexing right.

Three papers worth reading.

Dao, Fu, Ermon, Rudra, Ré, "FlashAttention: Fast and Memory-Efficient Exact Attention with IO-Awareness" (2022). This is the original paper. Read the IO-awareness analysis carefully — it is the foundation for why this approach scales the way it does, and the reasoning extends to other transformer-related kernels.

Dao, "FlashAttention-2: Faster Attention with Better Parallelism and Work Partitioning" (2023). The follow-up paper improves the parallelism strategy on GPUs and explains why the outer loop should be over column tiles, not row tiles, on real hardware. The math is unchanged from FlashAttention-1; the work partitioning is what differs.

Milakov and Gimelshein, "Online normalizer calculation for softmax" (2018). The streaming-softmax algorithm itself, in isolation, predates Flash Attention by several years. This short paper explains the running-maximum-and-sum trick without the attention-specific machinery on top. It is a clean read and a useful reference for anyone who needs to fuse softmax into other tiled computations.

arXiv search directions for further reading: `flash attention IO aware`, `streaming softmax online normalizer`, `tiled attention long context`, `paged attention KV cache`, `ring attention sequence parallel`.

One technique worth investigating immediately after the homework is **paged attention**, introduced by the vLLM team. It applies the same tiled idea to the inference-time KV cache, where the cache is split into fixed-size pages and attended to one page at a time. This is the bridge to **Project 13: Fast Inference: The KV Cache**, where the same tiling logic appears in a slightly different setting.

What You Now Know

The N-by-N attention score matrix does not need to exist. It can be computed in tiles, accumulated through a running maximum and a running sum, and written back as a final weighted sum of value vectors, all without ever holding the full matrix in memory. You have built that algorithm in pure PyTorch, verified it against the naive version, watched the memory curve flatten from quadratic to linear, and seen what happens when the one numerical-stability trick is removed.

You also know what to look for in the production CUDA kernel. The Triton and CUDA implementations are not different algorithms. They are the algorithm you just wrote, fused into a single kernel, with the block sizes tuned to the host GPU's memory hierarchy and the backward pass set up to recompute softmax weights on the fly. When you read the production code, the structure is already familiar — the running maximum, the rescaling, the partial output accumulator. Those are the load-bearing pieces. Everything else is engineering on top.

And you know the failure mode by feel. If a tiled attention kernel produces `nan` at moderate sequence length in `fp16`, the first thing to check is whether the running-maximum subtraction is present and correctly applied. That diagnostic alone will save more debugging hours than any other single piece of knowledge from this chapter.

Starting Point

Flash Attention reference implementation: github.com/Dao-AILab/flash-attention. Tri Dao's original paper: arxiv.org/abs/2205.14135. xformers reference attention: github.com/facebookresearch/xformers. Streaming-softmax algorithm in isolation: Milakov and Gimelshein, arxiv.org/abs/1805.02867.

Project 9: Pretraining On The Real Web

Hook

Why does training on Shakespeare feel clean and understandable, then the moment you switch to real web text everything turns into engineering: data shards, special tokens, memory mapping, mixed precision, gradient accumulation, learning rate sweeps, validation bits-per-byte, and loss curves that seem to argue with each other? You built a model already. It learned. So why is the next step not "same thing, bigger file"?

Because a real language model is more than a neural net. It is a machine for moving absurd amounts of text through a GPU without wasting compute, corrupting documents, blowing up memory, or fooling you with a training curve that looks healthy right up until it starts memorizing.

The first time I tried to scale from a single-file toy dataset to real web text I assumed the bottleneck would be the model. It was not. The bottleneck was every step around the model: how the tokens were stored, how they were sampled, how the loss was measured, and how I decided when to stop. The model itself was almost the easy part.

The Concept

Your Project 5 loader was like cooking from one giant pot. You had one text file, you chopped it into tokens, you grabbed random windows, and you fed them to the model. That works because the whole system is tiny — the data is small, the rules are loose, and if you waste a little compute, nobody cares.

Real pretraining is closer to running a warehouse. You are not serving one pot of soup. You are receiving truckloads of ingredients, sorting them, packaging them, labeling them, putting them on shelves, and making sure the kitchen can grab the next box instantly without unpacking the whole building every time. If one part of that pipeline is sloppy, your expensive hardware sits idle.

That is the real project — not "train on the web" in the vague sense, but train on the web with systems that respect scale. Let's name the parts first.

A **dataset shard** is one chunk of already-processed training data saved to disk. Instead of storing one giant token file, you split it into many manageable pieces. Think shipping containers, not one mountain of loose cargo.

Pre-tokenized data means the raw text has already been converted into token IDs before training starts. You pay the tokenization cost once, not every batch forever. The reason this matters is that tokenization is deterministic and expensive: doing it once and storing the result as shards means the GPU never waits on string processing. Without that separation, data loading becomes the bottleneck, and a GPU sitting idle while text gets tokenized is exactly the kind of waste that makes long training runs unpredictable.

I lived through this exact failure in some early fine-tuning runs. On a couple of experiments at the few-hundred-million-parameter range I had not pre-tokenized cleanly, and the GPU utilization kept dipping below 60 percent for reasons I could not explain until I instrumented the data loader. The string processing was the bottleneck, not the model. Once the tokens were on disk as integers, the same GPU pinned at 90-plus.

A **memory-mapped file** is a file on disk that your program can treat almost like an array in memory. The operating system only loads the pieces you touch. Think of a giant book where you can open directly to page 812,441 without copying the whole book onto your desk. This matters for GPU utilization because if loading data into RAM requires copying large portions of a file at startup, you pay that cost before training even begins. And on datasets larger than physical memory, that approach can be impossible entirely.

Document boundaries are the places where one source text ends and the next begins. If you ignore them, your model may learn fake transitions, like "end of Stack Overflow answer immediately followed by random Wikipedia heading," as if they were natural language. So you insert a special token that says, in effect, "this document ended; a new one begins here."

Gradient accumulation means pretending you had a larger batch size than fits in memory. You run several small batches, add up their gradients, and only then update the weights. It is like taking four small notes before making one decision, instead of making a decision after every single note. This matters because batch size is more than a memory constraint. It is connected to training stability. Larger effective batches tend to produce more reliable gradient estimates, and accumulation lets you reach a useful effective batch size even when your GPU has room for only a fraction of it.

Mixed precision (introduced in **Project 6: From Prototype to nanoGPT**) means doing much of the math in 16-bit formats like FP16 or BF16 to reduce memory traffic and speed up training, provided numerical stability is maintained.

Bits-per-byte, usually written **bpb**, is a training metric that asks: how many bits of surprise does the model need, on average, to encode the next byte of text? Lower is better. If the model predicts well, the text looks less surprising, so the number drops.

There is one more thing that everyone talks around because they prefer big architectural words to boring operational truth: pretraining is a throughput problem. Yes, it is learning, but it is also this. Can you keep the GPU busy? Can you feed it clean data? Can you measure progress honestly? Can you stop before "getting better at the training set" turns into "getting worse at the real world"?

Most beginner explanations of pretraining are honestly terrible because they skip the throughput question entirely. They linger on the architecture and treat the pipeline as plumbing. The pipeline is the project. The architecture is a few hundred lines.

That last question matters enough to state plainly. Training loss and validation loss are not the same thing. **Training loss** measures how wrong the model is on data it is being trained on. **Validation loss** measures how wrong it is on held-out data it did not train on. If both fall together, good. If training loss falls while validation loss stalls, be cautious. If training loss keeps falling while validation loss rises, you are watching memorization win. That is overfitting, and the model is getting better at the wrong job.

Here is the intuition before we earn any metric. Imagine studying for an exam by reading the answer key to one practice sheet over and over. Your score on that sheet keeps rising. Your score on a new sheet stops improving, then gets worse because you start overcommitting to quirks of the first one. That is exactly what your loss curves will show when you train too long.

Now the metric. If the model assigns probability (p) to the correct next token, the surprise is related to (-\log p). Cross-entropy loss adds that surprise across the sequence. Bits-per-byte is that surprise normalized in a way that is easy to compare across text corpora. You do not need the full formula first — you need the meaning: lower bpb means the model compresses text better because it predicts text better. That is why bpb is a more honest signal than training loss alone. Training loss answers "how well did I memorize what I practiced on?" while val_bpb answers "how well do I handle new examples from the same world?"

The learning rate matters for the same reason a thermostat matters. Too low, and nothing moves. Too high, and the system overshoots, oscillates, or explodes. There is usually a middle band where training drops quickly and stays stable. That is why you sweep learning rates instead of trusting your first guess.

This is where your mental model of LLM training changes. The model is not the whole story anymore. The system is: storage layout, numeric formats, batch simulation, schedules, metrics, stopping rules. Figure 8.1 shows how all those stages connect into a single pretraining pipeline.

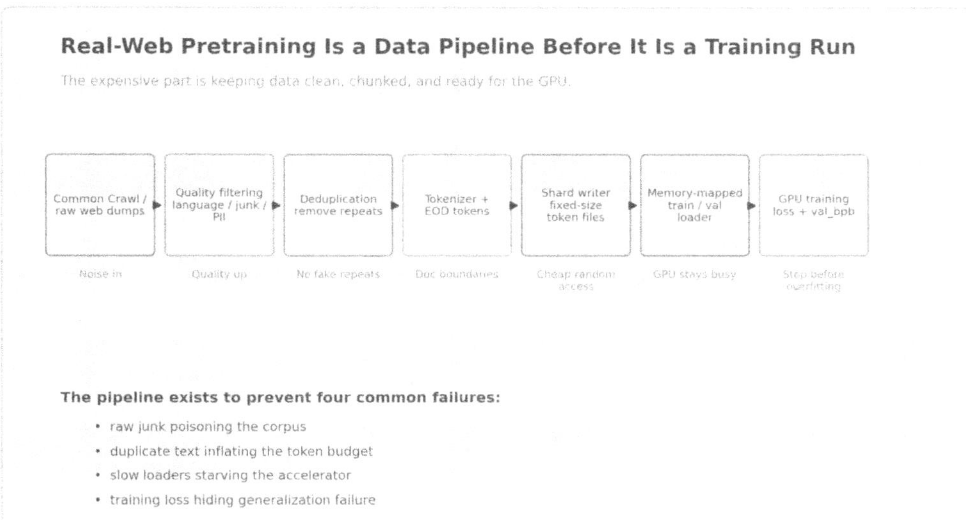

Figure 8.1. Real-web pretraining is a pipeline: raw crawl data gets filtered, deduplicated, tokenized, sharded, memory-mapped, and then measured with held-out validation during GPU training.

Why It Matters

All of these pieces connect to a single underlying pressure: keeping the GPU doing useful work on clean data, measuring that work honestly, and stopping at the right time.

Pre-tokenized shards are the reason the GPU never waits on string processing. Tokenization is expensive and deterministic, so doing it once and storing integer arrays on disk means the training loop simply reads numbers: fast, repeatable, and memoryless. Without that separation, every batch would trigger redundant text processing, making training slower and non-reproducible across runs.

Memory mapping is what makes pre-tokenized shards actually usable at scale. The operating system pages in only the slices you touch, so your data loader can handle datasets far larger than RAM without copying huge files upfront. This is directly connected to GPU utilization: if the data loading pipeline stalls, the GPU sits idle, and idle GPU time is the most expensive form of waste in a long training run.

Document boundary tokens protect the targets the model is being trained against. Without them, the next-token prediction task contains a steady stream of nonsense transitions. The model is penalized for not predicting a random Wikipedia sentence following an unrelated Stack Overflow answer. That is label noise, and enough of it corrupts the gradient signal in ways that are hard to diagnose later.

Gradient accumulation is the bridge between what fits in GPU memory and what the training dynamics actually need. Batch size is more than a memory question. It shapes gradient variance. Small batches produce noisy updates; larger effective batches produce steadier ones. Accumulation lets you simulate a batch size that would otherwise be physically impossible on your hardware.

Mixed precision works because modern GPUs are dramatically faster at 16-bit arithmetic than 32-bit, and the precision cost is manageable when implemented carefully. BF16 in particular keeps a wide enough exponent range to stay numerically stable through most of training. The combined effect of FP16/BF16 and memory mapping is that a training run that would otherwise require a larger or more expensive machine can often complete on one.

The val_bpb metric is more honest than training loss because it measures what you actually care about: whether the model is learning patterns that generalize beyond the data it was shown. A steadily falling training loss can coexist with a model that is quietly memorizing, and without a held-out metric like val_bpb you have no alarm. This is not a nice extra. It is the control panel.

A learning rate sweep exists because the right learning rate depends on nearly everything: model depth, width, batch size, optimizer, schedule, data, precision, and token budget. There is no single correct value to guess. Running four candidates on the same token budget is cheap compared to spending a full run on a bad setting. And crucially, "training is unstable" is not a diagnosis. A sweep gives you the curve shapes to reason from.

Across a long set of training runs I logged at scale, the single most expensive class of mistakes was committing too much wall-clock to a learning rate that the first 500 steps had already told us was wrong. Sweeps look slow on paper. They are far cheaper than a bad full run.

Taken together, these choices explain why benchmark comparisons between models are so hard to trust at face value. "Model A beat Model B" often means one team chose a better learning rate, one stopped at the right point, one had a cleaner data pipeline, or one matched batch size and schedule better to their token budget. The engineering layer is not separate from the modeling quality. It is underneath it.

This is also the point where toy intuition fails in a useful way. On Shakespeare, almost anything reasonable appears to work. On real web text, the cracks show. That is good. Cracks are where engineering starts.

The Build

Step 1: Clone `nanochat` and find the real training path

Start by treating the repo as something to read, not merely run. Your goal is not "get a loss number." It is "trace how text becomes tensors."

In your Project 5 code, that path was short:

1. load one text file
2. tokenize it
3. sample random windows

4. feed them to the model

In `nanochat`, that path is longer because each extra step removes waste or ambiguity.

You want to find:

- where raw data gets downloaded
- where raw documents become token IDs
- where those token IDs get written to shard files
- where the training loop reads those shard files
- where batches are sampled
- where document boundary tokens are inserted
- where validation data is separated from training data

Read `prepare.py` first — not like a script you hope works, but like a pipeline diagram written in Python. The stages usually look something like this, even if names differ:

1. Load or stream raw text examples from a source dataset.
2. Clean or select the text field you care about.
3. Tokenize each document.
4. Append a special end-of-document token.
5. Concatenate or buffer token IDs into fixed-size chunks.
6. Write those chunks to shard files on disk.
7. Create separate shard sets for train and validation.

That is the high-level flow. Now what is each stage buying you?

Raw text to selected text field

A web dataset often stores examples as structured records, not plain lines in one file. One record may contain metadata, URLs, quality scores, or other fields. You do not want all of that in the language model input, so you extract the actual text content and drop the rest. What this buys you: less junk in training, a smaller storage footprint, and a clear definition of what the model is learning from.

Tokenization before training

Tokenization turns text into integers once, ahead of time. What this buys you: faster training because the GPU loop is not waiting on tokenization, reproducibility because every run sees the exact same token IDs, and easier sharding because integer arrays are simpler to pack and slice than raw text.

End-of-document special token

Suppose one document ends with `The answer is 42.` and the next starts with `Recipe for lemon bars.` If you just glue them together, the model sees `The answer is 42. Recipe for lemon bars.` That is not language. That is a stitching artifact. So you insert a special token between documents. Think of it like a paragraph break that is louder than a newline. It tells the model, "whatever comes next is not a continuation of the same thought." What this buys you: honest document transitions, cleaner next-token targets, and less contamination from fake cross-document continuations.

Shards on disk

One giant token file is awkward: you cannot shuffle it flexibly, you cannot manage it cleanly, you cannot stream subsets easily, and if something corrupts, the blast radius is huge. Shards solve that by splitting the token stream into chunks. What this buys you: manageable files, easier parallelism, shard-level shuffling, and simpler train/validation partitioning.

Memory-mapped loading

A memory-mapped file lets you access data as though it were already in memory, while the operating system quietly loads only the parts you touch. What this buys you: lower RAM pressure, faster random access than naive full-file loads, and good performance on datasets larger than memory.

Shard-aware shuffling

In the toy setup, random windows from one blob are enough. At scale, you want batches that come from varied locations and documents. You do not want to sit on one narrow slice of the data distribution for too long. Shard-aware shuffling means you randomize at the shard level and often within shards too. What this buys you: better mixing of topics and sources, less correlation between neighboring batches, and more honest gradient estimates.

That last phrase needs translation. A **gradient estimate** is your current best guess for how to nudge the weights based on one batch. If the batch is weirdly narrow or repetitive, the guess is noisy or biased. Better-mixed batches give you steadier updates.

Step 2: Annotate `prepare.py` like a systems engineer

As you read, write comments or notes in this format:

- Input: what form is the data in right now?
- Operation: what happens in this stage?
- Output: what form is the data in afterward?
- Why here: why do this before training instead of during training?

An earlier version of this Build section just told the reader to "read `prepare.py`." Reading my own draft back, I could not tell what I was supposed to be looking for. The four-line annotation template below is what I scrapped two failed paragraphs for. It is much shorter than the prose it replaced, and it works much better.

For example:

```
# Input: dataset rows with raw text and metadata
# Operation: keep only the text field
# Output: plain text documents
# Why here: avoids carrying useless metadata into tokenization and storage
```

Then:

```
# Input: one text document
# Operation: tokenize into integer IDs and append end-of-document token
# Output: list of token IDs for one document
# Why here: preserves document boundaries and makes training data
#   ready for direct loading
```

Then:

```
# Input: many tokenized documents
# Operation: pack tokens into shard-sized arrays and write to disk
# Output: shard files containing integer token IDs
# Why here: training can stream tokens directly without repeated preprocessing
```

This sounds simple, but it changes how you read code. You stop asking "what Python call is this?" and start asking "what pressure in the system is this code relieving?" That is a better engineering question.

Step 3: Contrast `nanochat` with your Project 5 loader

This is the moment to name exactly what the extra machinery buys you.

Your Project 5 loader sampled sequences directly from one blob, likely held everything in memory, had no real notion of document boundaries, probably did lightweight random slicing, and worked because the dataset was tiny.

`nanochat` adds four big things you should now be able to explain.

1. Pre-tokenized shards

Project 5 did tokenization close to training or on one simple source. `nanochat` stores already-tokenized data on disk. What it buys: avoids paying tokenization cost during training, makes batch loading a simple integer slicing problem, and improves repeatability across runs.

2. Memory-mapped loading

Project 5 probably loaded a blob into RAM and sampled from it. `nanochat` maps shard files and reads slices on demand. What it buys: supports datasets much larger than RAM, reduces startup and copy costs, and makes long training runs practical on normal hardware.

3. Shard-aware shuffling

Project 5 relied on randomness over one text source. `nanochat` mixes across many shards and many documents. What it buys: better coverage of the dataset, lower risk that recent updates overfit to one narrow region, and steadier training dynamics.

4. Explicit document boundaries

Project 5 may have ignored them entirely. `nanochat` inserts a special token between documents. What it buys: prevents fake language transitions, preserves real document structure, and gives the model a clean signal for "this item ended."

That last one is more important than it first appears. Document boundaries are not just etiquette — they affect the target the model is asked to predict. If you train on fake continuations often enough, you inject label noise. The model is punished for not predicting nonsense transitions that only exist because your pipeline glued unrelated text together.

Step 4: Train a small model on FineWeb-EDU

Now you move from reading the pipeline to using it.

The spec says:

- depth 6
- about 10M parameters
- log `val_bpb` every 100 steps
- plot the learning curve

This is your first real web-text pretraining run. The model is still small enough that you can inspect it mentally, and that matters — you do not want a giant run yet. You want a run where the losses mean something to you.

Why a small model?

Because this project is about learning the training system, not winning a benchmark. A small model lets you run more experiments, sweep learning rates, repeat failed runs, and see overfitting happen without waiting forever. A 10M parameter model is not trying to be a useful assistant. It is trying to be an instrument panel.

I want to flag a strong opinion here. The instinct to immediately scale to a 1B or 7B parameter model is, in my experience, the single most common reason new practitioners burn weeks on runs they cannot diagnose. A 10M model that overfits in twenty minutes teaches you more about training than a 1B model that crashes in six hours and leaves you guessing.

What to log

At minimum, log training loss, validation loss, validation bits-per-byte, step number, learning rate, and tokens seen if available. You want validation every 100 steps because the shape matters more than any one point. One validation number at the end is too late — loss curves tell stories, and you need enough checkpoints to read the story.

What should the curve look like?

At the start, train loss drops quickly, val loss drops too (a little behind train), and `val_bpb` follows the same general direction. If the learning rate is good, the curve descends steadily with noise present but not chaotic, and both train and validation improve early.

If the learning rate is too low, everything improves but slowly. The curve feels flat and disappointing, and you burn steps without much gain. If the learning rate is too high, loss may spike, oscillate, or fail to improve, and validation may jump around badly. You can sometimes tell in a few hundred steps that a run is not healthy.

Plotting matters because your eyes catch shape better than raw logs do. A curve can tell you "fast and stable," "slow but safe," "unstable," "overfitting," or "plateaued." A single printed metric cannot.

Step 5: Understand mixed precision before you trust it

Here is the intuition. If you can carry water in smaller buckets without spilling too much, you can move more water faster. That is mixed precision. Instead of storing every activation and weight update in 32-bit floats, the training system uses 16-bit formats for much of the heavy work. That reduces memory bandwidth and often speeds up matrix multiplies on modern GPUs.

Why not always do that with no caution? Because smaller number formats have less precision and, depending on the format, less dynamic range. That means tiny numbers may underflow to zero, or big numbers may overflow. That is why mixed precision is "mixed," not "everything small." You keep enough numerical safety where it matters.

If your repo supports BF16, that is often friendlier than FP16 for training because BF16 keeps a wider exponent range. In plain English: it can represent a wider range of magnitudes without blowing up or underflowing.

You do not need to hand-implement mixed precision here. You do need to understand why it is in production training code: less memory use, faster throughput, and feasible larger runs on the same hardware.

Step 6: Implement gradient accumulation

Now for one of the most important tricks in practical training.

Suppose your GPU can only fit batch size 8, but your training dynamics are better with batch size 32. Gradient accumulation says: fine. Do four passes of batch size 8. Compute gradients each time. Do not update the weights yet. Sum those gradients. Then divide appropriately and make one update as though you had processed 32 examples together.

The intuition is dead simple. Instead of asking 32 people for feedback in one meeting, you ask 8 at a time across four meetings, write down all the feedback, and only then decide what change to make.

What the code change means

Normal training loop:

1. forward pass on one batch
2. compute loss
3. backward pass
4. optimizer step
5. zero gradients

Accumulation loop:

1. zero gradients
2. repeat N times:
    - forward pass on micro-batch
    - scale loss if needed
    - backward pass
3. optimizer step once
4. zero gradients

The phrase **micro-batch** means one of the small batches used during accumulation. The phrase **effective batch size** means the total batch size the optimizer "feels" after combining all micro-batches. If micro-batch size is 8 and accumulation steps are 4, effective batch size is 32.

Why scale the loss?

Because if you simply sum four losses and backprop each one, your gradients become four times larger than a single true batch average. To match a real larger batch, you usually divide each micro-batch loss by the number of accumulation steps before calling backward. That way the final accumulated gradient matches the average gradient over the whole effective batch.

Now we can earn the equation.

Let:

- (g_i) be the gradient from micro-batch (i)
- (k) be the number of accumulation steps

A true large batch average gradient is:

```
g = (1 / k) * sum_(i=1 to k) g_i
```

That is not mysterious. It just says: average the gradients from the small pieces. If you divide each micro-batch loss by (k) before backprop, the accumulated gradient becomes that average.

How do you verify it matches a true large batch?

Run two experiments on the same data with randomness controlled as tightly as possible: one true larger batch that fits (if you can), and one smaller micro-batch with accumulation to the same effective batch size. Then compare loss values, gradient norms if you log them, and parameter updates after one step or a few steps. They should be very close — not necessarily bit-for-bit identical, but close enough that the training behavior matches.

This matters because once the models get larger, accumulation is not optional. It is the difference between "this fits" and "this does not."

Step 7: Run a learning rate sweep

Now you test four learning rates:

- `1e-5`
- `3e-4`
- `1e-3`
- `3e-3`

Do not ask which one is "correct" before you run it. This is not a multiple-choice quiz. The answer depends on model depth, width, batch size, optimizer settings, schedule, data, precision, and token budget. That is why `nanochat` auto-computes or recommends learning rates based on model depth. Bigger or deeper models often need different step sizes to stay stable.

A learning rate is how large a weight update you take after reading one gradient. If the model is shallow and simple, a somewhat larger step may be fine. If the model is deeper, errors and activations move through more layers, and the surface the optimizer is walking on can be touchier. Smaller, better-calibrated steps become more important. That is the plain-English story behind depth-aware learning rate choices.

What you expect from the sweep

`1e-5` is probably too small: smooth, but sluggish, with training improving very little over the same number of steps. `3e-4` is often a plausible safe zone for small models: steady descent, likely good validation behavior. `1e-3` might be the sweet spot or slightly aggressive, with faster initial progress if stable. `3e-3` is often too high for a small GPT on real text, producing oscillation, instability, or poor validation.

The point is not to memorize these guesses — it is to recognize the signatures.

How to read the plot

Put all four curves on one set of axes. Watch who drops fastest at the start, who stays stable, who reaches the lowest validation curve, and who becomes noisy or divergent. The sweet spot is not "the most dramatic first drop" but "the best validation progress without instability." That distinction matters because overly high learning rates can look exciting briefly. Bad decisions often do.

I expected the most aggressive learning rate in the sweep to either win clearly or fail clearly. The actual outcome on the first runs was harder to read: the aggressive curve dropped fastest for a few hundred steps, then started oscillating, then drifted slowly upward without ever obviously diverging. A reader who only logged the first few hundred steps would have picked it as the winner. The held-out validation said otherwise.

Step 8: Understand why `val_bpb` is the metric to watch

A language model predicts the next token. If it is uncertain, the prediction is "expensive" in bits. If it is confident and right, the prediction is "cheap." Bits-per-byte measures how expensive the text looks to the model.

Imagine compressing text with a code where common events get short descriptions and surprising events get long ones. A good model finds short descriptions more often because it predicts well. So lower `val_bpb` means the model is less surprised by held-out text, is modeling the data distribution better, and compresses the text more efficiently in an information-theoretic sense.

Why validation bpb rather than just training loss? Because the held-out split tells you whether the model is learning patterns that travel beyond the data it already saw. Training loss answers "how well did I memorize what I practiced on?" while validation bpb answers "how well do I handle new examples from the same world?" You need both — but when they disagree, validation is the one to trust.

Step 9: Connect compute-optimal training to overfitting

This project introduces a phrase that matters for the rest of the book: **compute-optimal training** means choosing a model size and token budget that get the best loss for a fixed amount of compute.

That sounds abstract until you see the failure mode. A small model trained for too long starts memorizing. A big model trained for too little may stay undertrained. Both are waste. The sweet spot is not "more parameters" or "more steps" in isolation. It is the balance between how much model you have, how much data you show it, and how much compute you spend. The breakage experiment in this chapter is designed to make that balance visible in one picture: train too long, watch train and validation split apart, and that divergence is the cost of ignoring compute-optimality.

BREAK IT

Treat this as one concrete `nanochat` lab, not a thought experiment.

Create one log directory per run:

```
mkdir logs
```

Use this workflow:

```
python prepare.py --dataset fineweb-edu --split train --max_shards 4
python prepare.py --dataset fineweb-edu --split val --max_shards 1
python train.py --config configs/small_10m.py > logs/baseline.txt
python train.py --config configs/small_10m.py --learning_rate 3e-4 > logs/lr_3e4.txt
python train.py --config configs/small_10m.py --learning_rate 1e-3 > logs/lr_1e3.txt
```

Save checkpoints every 1000 steps and validation every 100 steps. The point is to compare `train_loss`, `val_loss`, `val_bpb`, `lr`, and checkpoint samples before and after divergence.

The broken run in this chapter is not "the code crashes." The broken run is "training keeps going while generalization gets worse."

Train your small web-text model for 10 times longer than the compute-optimal token count you would normally choose. Do not change the architecture, the dataset, or the metric — only change how long you keep going. Save checkpoints every 1000 steps and plot training loss, validation loss, and validation bits-per-byte.

What breaks is not the code. What breaks is your assumption that more training keeps helping.

Here is the failure mode. At first, training loss falls, validation loss falls, and `val_bpb` falls. Then training loss keeps falling while validation loss slows down and `val_bpb` flattens. Then training loss keeps falling while both validation loss and `val_bpb` start rising.

That is overfitting.

You are watching the model switch jobs: earlier, it learned reusable statistical patterns in language; later, it starts fitting accidents of the training set too tightly. The model is not getting "smarter." It is getting narrower.

This is one of the most important pictures in the whole chapter because it answers a question people constantly get wrong: "If training loss is still going down, why stop?" Because training loss is no longer measuring what you care about. You care about performance on unseen text, and validation is the alarm bell.

What exactly are you breaking?

You are breaking the match between model size and token budget. A 10M parameter model does not have infinite appetite. Give it too many passes over the same statistical world and it will start fitting details that do not travel well. This is not a moral failure of the model. It is what optimization does when you keep rewarding better fit to the training distribution without enough new information.

What should you watch for?

Watch for the divergence point. That point is the chapter. Not the final loss, not the lowest training number, the exact moment the curves stop moving together. That is where "learning" turns into "memorizing."

If you save checkpoints every 1000 steps, you can often inspect generations from checkpoints before and after divergence. Sometimes the later model sounds more specific or more locally fluent on training-like fragments while actually doing worse on held-out text. That is a useful discomfort. Surface polish is not the same as generalization.

What does this prove?

It proves four things. First, training loss alone is not a trustworthy judge. Second, validation metrics are not a nice extra — they are the control panel. Third, compute-optimal training is not paper trivia; it tells you when extra compute starts buying less and eventually buying harm. Fourth, "just train longer" is not engineering. It is an expensive guess.

Why does the divergence happen?

The model has finite capacity and the dataset has finite variety. Early in training, there are broad patterns to learn: common syntax, token frequencies, local grammar, recurring semantic structures. Those patterns help everywhere. Later, the remaining errors are more idiosyncratic: rare phrasing quirks, repeated templates, source-specific oddities, statistical noise. Fitting those can lower training loss while hurting validation. That is the same exam analogy from earlier: first you learn the subject, then, if you keep drilling one answer sheet, you start learning the sheet.

What not to misread

Do not look at overfitting and conclude that smaller models are bad, or that more data solves everything, or that validation loss should always rise sharply. Sometimes divergence is gradual. Sometimes it is subtle. Sometimes the best checkpoint is in a narrow window. The lesson is not "all long training overfits quickly." It is "there is a real point where more optimization against the train set stops serving generalization." That is the point you are learning to see.

Optional Homework

Operational Run Sheet

Do the chapter in this order:

1. Annotate `prepare.py` and write down what each stage buys: speed, correctness, memory, or reproducibility.
2. Prepare a tiny train/val shard set.
3. Train one small baseline with validation every 100 steps and checkpoints every 1000.
4. Add or verify gradient accumulation and compare it with a true larger batch.
5. Run a four-point learning rate sweep on the same token budget.
6. Take the best LR and deliberately overtrain 10x too long.
7. Plot `train_loss`, `val_loss`, and `val_bpb` on the same x-axis.

Use one summary file for the whole lab:

```
run_name,learning_rate,accum_steps,step,train_loss,val_loss,val_bpb,checkpoint_note
```

If hardware is limited, keep the same sequence but reduce shard count, context length, model width, and total steps. Do not skip the plots. The curve divergence is the experiment.

An earlier version of this homework asked the reader to do everything in parallel and was overwhelming. The current run-sheet ordering is the one I could actually complete in one sitting without skipping the plots.

Hardware and runtime

Full version:

- GPU with at least 12 GB VRAM is comfortable for a small 6-layer run

- Expect several hours to a day depending on model width, batch size, and dataset subset
- Disk space depends on how much FineWeb-EDU you prepare; a subset is enough for this lab

Proxy version:

- 8 GB VRAM can still work with a smaller width, shorter context length, and gradient accumulation
- You can use a reduced shard subset and still see the same training-vs-validation story
- Runtime can be 2-6 hours for a small exploratory run

Part A: Prepare the data and annotate the pipeline

1. Clone the repo.

```
git clone https://github.com/karpathy/nanochat
cd nanochat
```

2. Open `prepare.py`.

Read it top to bottom and create a separate notes file called `project8_prepare_notes.md`.

For each major stage, write:

- input form
- operation
- output form
- why this stage exists

3. Run the data preparation on a small subset first.

The exact command may vary with repo updates, but the pattern is:

```
python prepare.py --dataset fineweb-edu --split train --max_shards 4
python prepare.py --dataset fineweb-edu --split val --max_shards 1
```

If the repo uses a config file instead, set:

- dataset: FineWeb-EDU
- small shard count for train
- small shard count for validation

Expected observation:

- raw documents are converted into token shards on disk
- shard files are much easier to stream than raw JSON or text dumps
- the output directory now contains train and validation shard files

Part B: Train a 6-layer model and log validation every 100 steps

1. Create a small training config.

Set roughly:

- `n_layer = 6`
- width small enough to stay near 10M parameters
- moderate context length
- validation interval = `100`
- checkpoint interval = `1000`

2. Start training.

Example pattern:

```
python train.py --config configs/small_10m.py
```

If the repo uses inline flags instead of config files:

```
python train.py --n_layer 6 --val_interval 100 --save_interval 1000
```

3. Log these metrics:

- `train_loss`
- `val_loss`
- `val_bpb`
- `lr`
- `step`

Expected observation:

- train loss drops first and fastest
- validation follows behind
- `val_bpb` trends downward when learning is healthy

Part C: Implement gradient accumulation

1. Find the training loop.

You are looking for the pattern:

- forward
- loss
- backward
- optimizer step
- zero grad

2. Add accumulation steps.

Pseudo-code:

```python
optimizer.zero_grad()
for micro_step in range(accum_steps):
    logits, loss = model(x, y)
    loss = loss / accum_steps
    loss.backward()
optimizer.step()
```

3. Choose a test pair:

- true batch size 32, accumulation off
- micro-batch size 8, `accum_steps = 4`

Use the same seed if possible.

4. Compare:

- first few loss values
- gradient norm if available
- parameter difference after one optimizer step

Expected observation:

- accumulated run closely matches the true larger-batch run
- any difference should be small if randomness is controlled

Part D: Run the learning rate sweep

Run four short experiments, identical except for learning rate:

```
python train.py --config configs/small_10m.py --learning_rate 1e-5
python train.py --config configs/small_10m.py --learning_rate 3e-4
python train.py --config configs/small_10m.py --learning_rate 1e-3
python train.py --config configs/small_10m.py --learning_rate 3e-3
```

Keep:

- same seed
- same token budget
- same data subset
- same batch size
- same schedule shape if possible

Collect `val_bpb` every 100 steps and plot all four curves on one chart.

Expected observation:

- `1e-5` improves slowly
- one of `3e-4` or `1e-3` is likely best for this small setup
- `3e-3` is the most likely to show instability or poorer validation

Part E: Break it by training 10x too long

1. Take the best learning rate from Part D.
2. Extend training to 10x the steps you would normally use for a short healthy run.
3. Save checkpoints every 1000 steps.
4. Plot:

- training loss vs step
- validation loss vs step
- `val_bpb` vs step

Expected observation:

- early phase: train and validation both improve
- middle phase: validation improvement slows
- later phase: train still improves while validation worsens

5. Optional generation check:

- sample text from checkpoints before divergence and after divergence
- compare how "locally fluent" they sound versus how validation behaves

Expected observation:

- later checkpoints may not obviously sound worse in a tiny sample
- metrics tell the truth more reliably than one cherry-picked generation

Part F: Proxy version if your hardware is limited

If you cannot run FineWeb-EDU comfortably:

- use a much smaller shard subset
- reduce context length
- reduce width
- keep depth at 6 if possible
- still do the learning rate sweep
- still do the long-run overfitting experiment on the reduced dataset

This still teaches the same lesson because overfitting is easier, not harder, to observe on smaller data.

Questions To Answer

1. When you annotated `prepare.py`, which stage would have been missing from your Project 5 loader, and what exact waste or error does that stage prevent?
2. In your learning rate sweep, which curve looked best in the first few hundred steps, and did it stay best on validation later?
3. When you compared true large-batch training to gradient accumulation, how close were the losses and updates? If they differed, where do you think the mismatch came from?

4. At what step did training loss and validation loss stop moving together? What does that step say about your token budget for this model size?
5. What failed first in the overfitting run: validation loss, `val_bpb`, sample quality, or your patience? Which signal would you trust most if they disagree?
6. What tradeoff is the data pipeline protecting: speed, memory, correctness, reproducibility, or all four? Which piece of the pipeline buys which protection?

Go Further

Research anchors for this chapter

This chapter sits directly on top of two papers that changed how people talk about pretraining budgets. **Scaling Laws for Neural Language Models** is the clean statement that loss, model size, dataset size, and compute move together in regular ways. It gives the first disciplined answer to "if I spend more compute, where should I spend it?"

Then read **Training Compute-Optimal Large Language Models**, usually referred to as the Chinchilla paper. Its main relevance here is practical: it argues that many large models were undertrained relative to their size, and that token budget matters far more than many teams had been treating it. That is exactly why this chapter keeps pushing you away from vague "train longer" instincts and toward token-budget judgment.

For a more modern, practitioner-shaped complement, read **The Smol Training Playbook** and **Textbooks Are All You Need**. They are useful not because they replace the scaling-law papers, but because they bring those lessons down to smaller-budget training and data-quality decisions that look much closer to the systems choices you are making in this project.

Start with the repo itself:

- `https://github.com/karpathy/nanochat`

Then compare the training pipeline to:

- your Project 5 single-file GPT
- your Project 6 minimal production-shaped system

If you want the benchmark context, search for:

- `FineWeb-EDU`
- `The FineWeb Datasets`
- `Common Crawl filtering for language model pretraining`

If you want the scaling context, search arXiv for:

- `Chinchilla compute-optimal large language models`
- `scaling laws language models`
- `tokens per parameter language model`

If you want to go deeper on optimization and schedules, search for:

- `learning rate warmup transformers`

- `cosine decay language model training`
- `gradient accumulation equivalence large batch training`

If you want to understand why mixed precision works in practice, search for:

- `BF16 training transformers`
- `mixed precision training loss scaling`
- `numerical stability FP16 BF16`

If you want the next layer of systems work after this chapter, start reading about:

- packed sequences
- distributed data parallel training
- optimizer states and memory cost
- activation checkpointing

Those are the next pressures that appear when one GPU is no longer enough.

What You Now Know

A real pretraining pipeline is more than feeding text into a model. You can now trace raw web documents through tokenized shards, explain gradient accumulation in plain arithmetic, and read a learning rate sweep to distinguish runs that are active from runs that are actually learning. Most important, you can look at a train-vs-validation divergence plot and point to the exact moment overfitting begins. That is the start of real training judgment.

Starting Point

Start from:

- `https://github.com/karpathy/nanochat`

Use it alongside:

- your own Project 5 GPT implementation
- your notes from Project 6, where the system first started looking production-shaped
- your Project 7 instinct that internal compatibility is about distributions, not just tensor shapes

The progression still matters:

- `micrograd` taught you blame assignment
- Project 2 taught you tiny next-token learning
- `minbpe` taught you that tokens are engineered
- Project 4 taught you attention under a causal mask
- Project 5 made you build GPT from scratch
- Project 6 showed you a minimal training system
- Project 7 taught you to inspect internal compatibility instead of trusting labels

Project 9 is where the toy world ends. From here on, the bottleneck is not only the model. It is the pipeline, the metric, the batch, the schedule, and the judgment to stop at the right time.

That is pretraining on the real web.

Project 10: Data Curation and Contamination

Hook

In **Project 9: Pretraining on the Real Web**, you pulled FineWeb-EDU off Hugging Face, fed it into your training loop, and watched the loss come down. Most readers stop there. The data was already filtered by someone else, the file format was clean, the dataset card promised "high-quality educational content," and the model trained without exploding. That feels like enough.

It is not.

This chapter asks a question that almost no one asks before training: what is actually in this data, and is any of it sabotaging the evaluation you are about to run? By the end you will know, for the specific corpus you are training on, what fraction is duplicate, what fraction is multilingual junk that your tokenizer cannot handle, what fraction reads like template spam, and (most importantly) what fraction silently leaks the answers to the benchmark you were going to use to measure progress.

The Concept

Imagine you are a chef who orders a shipment of ingredients from a wholesale supplier, opens one box at the top of the pile, sees clean produce, and decides the whole shipment is fine. You then cook a meal for guests using ingredients from the bottom of the pile that you never inspected. Sometimes the meal is fine. Sometimes a guest gets sick, and you have no way to trace which ingredient was responsible because you never looked at any of them. Pretraining data is the same. The Hugging Face card is the top of the box. The bottom is whatever the crawler swept off the web on the day it ran.

The analogy I actually use in my head for this comes from a different industry. In a past role I ran a multi-state SaaS platform with hundreds of thousands of customer records spread across dozens of sites. Some of those records had been silently corrupted by a third-party importer years before anyone noticed. The corruption did not crash anything. It just slowly biased every report we ran until somebody happened to spot-check the source. Pretraining data is the same hazard at a different scale. The model does not crash on bad data. It just learns from it, and the bias does not become visible until you have already shipped.

Now picture a second analogy. You are building a house, and someone hands you a foundation that was poured by a contractor you have never met. The contractor's website has nice photos. The foundation looks flat. You build twelve stories on top of it before discovering that one corner was poured over an unmarked sinkhole. The whole building does not immediately fall down. It just tilts a little more every year. Training a model on uninspected data is the same. The model does not refuse to learn. It learns whatever is there, and what is there might include thousands of copies of the same SEO spam page, machine-translated text that your tokenizer butchers, and verbatim copies of the test set you were planning to evaluate on.

Real pretraining data curation comes down to four operations, and each one is a measurable quantity rather than a vibe. The first is **quality filtering**: scoring each document on signals like length, repetition within the document, language identification confidence, and a small bag of heuristic features, then dropping documents below some threshold. The second is **deduplication**: removing copies of the same document, and near-copies that differ in a few words, so the model does not memorize one essay by seeing it forty times. The third is **mixing**: deciding what fraction of training tokens come from each source (web, code, math, books) and how those fractions should change as training progresses. The fourth is **decontamination**: removing any training document that contains a long verbatim chunk of an evaluation set, so the benchmark you are about to report is measuring something other than your model's ability to recall training data.

Across a long set of fine-tuning experiments I ran multiple times on different data mixes for the same model family, the lesson that kept coming back was that the mix mattered more than people expected. A 410M model trained on one mix and a 410M model trained on a slightly different mix produced different downstream behavior on identical compute. Each of those runs was, in part, a small reminder that "the data" is not a single thing.

Each of these has a specific number attached. Quality filtering produces a document score and a threshold. Deduplication produces a duplication rate, typically 20 to 40 percent for raw crawl data and 5 to 10 percent for already-curated corpora. Mixing produces a vector of per-source weights that sum to one. Decontamination produces a count of leaked documents and an overlap rate.

A new term that will appear throughout this chapter: a **shingle**. A shingle is a fixed-length sliding window of tokens or characters from a document. If your shingle size is 5 and your document is "the cat sat on the mat today," the shingles are "the cat sat on the", "cat sat on the mat", "sat on the mat today". Shingles are the unit of comparison that almost every deduplication and contamination check is built on, because two documents that share many shingles are similar in a way that the human eye recognizes immediately.

Two more terms, defined here so they can be used freely later. **MinHash** is a hash-based trick for estimating how many shingles two documents have in common without actually comparing every shingle against every shingle, which would cost quadratic time in the corpus. **LSH**, short for **locality-sensitive hashing**, is a follow-up trick that groups documents into buckets so that similar documents land in the same bucket and you only have to compare documents within a bucket, which cuts the comparison count from billions to millions.

We will build both. The construction is small enough to hold in your head.

A last term worth defining now: **n-gram overlap**. An n-gram is a contiguous sequence of n tokens, almost the same idea as a shingle. Contamination is usually measured by counting how many 13-grams from an evaluation set also appear verbatim somewhere in the training data. Thirteen is large enough that a chance match is essentially impossible (the probability of two arbitrary 13-token spans matching by accident is astronomically small) and small enough that paraphrased contamination still leaves a fingerprint. Some labs prefer 8-grams, some prefer 50-character spans. Thirteen tokens is a defensible default and the one we will use.

The four operations and these few terms are the whole vocabulary of data curation. Once you can compute each number for your own corpus, the entire field stops feeling like a black art and starts looking like accounting.

Why It Matters

A model trained on uncurated data is not a worse model than a model trained on curated data. It is a model that lies to you about its own quality. The training loss goes down. The validation loss goes down. The benchmark numbers look acceptable. Then the model goes into the hands of real users and reveals that it has memorized one paragraph from one SEO farm and recites it under any prompt that vaguely resembles the trigger, or that it scores 60 percent on a benchmark whose answers were in the training data and 25 percent on a fresh held-out variant of the same benchmark. The gap between those two numbers (sometimes 10 percent, sometimes 30 percent) is the cost of skipping decontamination.

This is the kind of failure mode that does not announce itself. There is no traceback, no spike in the loss curve, no broken assertion. The model trains, the metrics look fine, and you find out months later because a user notices something off.

There is a second cost that matters even more for small training budgets. Every duplicate document in your corpus is a token your model saw twice during pretraining. If 30 percent of your corpus is duplicates and you have one trillion tokens of compute budget, you have effectively given the model 700 billion unique tokens and 300 billion redundant tokens. The redundant tokens are not free. They cost the same compute, they fill the same number of optimizer steps, and they teach the model to overfit on whatever happened to be duplicated. Dedup is one of the cheapest interventions you can run, and the lift from a properly deduplicated corpus often shows up immediately in val_bpb at fixed compute.

There is a third cost that is harder to see and harder to fix. Mixing weights determine what your model is good at. A corpus that is 95 percent web crawl and 5 percent code produces a model that writes English fluently and writes Python badly. A corpus that is 30 percent code, 30 percent math, and 40 percent web produces a different model with different strengths. The mixing weight schedule is one of the most consequential hyperparameters in pretraining, and almost nobody studies it carefully because it is not a single knob. It is a vector that interacts with everything.

I want to push back on a piece of conventional advice here. The standard line is "use a popular public mix and copy what the big labs do." That advice protects you from the worst mistakes but it also commits you to whatever those labs happened to optimize for, which may not be what you want. The fine-tuning runs I ran at scale taught me that even small reweightings of the mix can move downstream behavior in ways that no architecture change at the same compute could. Mix is content. Treat it as content, not as configuration.

This chapter sits where it does for a structural reason. **Project 9: Pretraining on the Real Web** trained a model on a corpus you trusted. **Project 11: Training Debugging — Spikes, NaNs, Profiling** taught you to read a sick training run and name the failure. **Project 22: Evaluation Methodology** will teach you to measure what your model actually knows. None of those chapters do their job if the data is corrupt. The training looks healthy in Project 9. The dashboard looks healthy in Project 11. The benchmark looks healthy in Project 22. And the model is still bad. Curation is the work you do to make every other measurement honest.

The Build

You are going to take a 100,000-document sample from FineWeb-EDU, the same corpus you used in **Project 9: Pretraining on the Real Web**, and run the full curation pipeline on it. Most of this runs on CPU. The training step at the end needs a single consumer GPU, but you can verify every other component on a laptop. Total wall-clock time on a modest machine is roughly two to four hours, with the small training runs taking the lion's share.

The reference libraries are `datasets` for loading, `datasketch` for an off-the-shelf MinHash baseline you can compare against your own implementation, and `fastlangid` or `cld3` for language ID. You will build the MinHash and LSH steps yourself so you understand them. The training script is the same one from **Project 9: Pretraining on the Real Web**, with the data source swapped between raw and curated variants.

Step 1: Score the raw sample and see what is actually in there

Pull 100,000 documents from FineWeb-EDU and compute, per document, a handful of simple statistics. Length in characters and in whitespace-separated tokens. Mean word length. Repetition ratio within the document (the fraction of 5-grams that appear more than once in the document). Language identification, with the model's confidence score. Whether the document contains any text at all after stripping HTML residue.

```
from datasets import load_dataset
import re
from collections import Counter

ds = load_dataset("HuggingFaceFW/fineweb-edu",
                  name="sample-10BT", split="train", streaming=True)
sample = list(ds.take(100_000))

def doc_stats(text: str) -> dict:
    chars = len(text)
    tokens = text.split()
    n_tokens = len(tokens)
    mean_word_len = (sum(len(t) for t in tokens) / n_tokens
                     if n_tokens else 0.0)
    fives = [tuple(tokens[i:i+5]) for i in range(n_tokens - 4)]
    counts = Counter(fives)
    repeated = sum(c for c in counts.values() if c > 1)
    repetition = repeated / max(len(fives), 1)
    return {
        "chars": chars,
        "tokens": n_tokens,
        "mean_word_len": mean_word_len,
        "repetition": repetition,
    }

stats = [doc_stats(d["text"]) for d in sample]
```

Two notes on this snippet. The repetition ratio counts every occurrence of a 5-gram beyond the first. A document with one repeated 5-gram and ninety-nine unique 5-grams scores 0.01; a document that consists of "buy now buy now buy now" repeated for a thousand tokens scores close to 1.0. The threshold for "too repetitive" is usually around 0.3 to 0.5. That is, if more than a third of the document's 5-grams are repeats, the document is probably template spam or a stuck crawler. The `mean_word_len` field catches a specific failure mode: corrupted text, where the document looks like a single word with no spaces or like punctuation soup with no words at all. Healthy English text has a mean word length between 4 and 6 characters.

When you plot the distributions, you will see something that surprises most readers the first time. The length distribution is heavily right-skewed. Half the documents are under 500 tokens, ten percent are over 5000 tokens, and a small tail goes past 50,000 tokens. The repetition ratio has a sharp bimodal shape. Most documents cluster near zero, with a clear secondary bump around 0.4 that is almost entirely spam, forum reposts, and crawler artifacts. Mean word length looks roughly normal around 5, with thin tails that are usually corrupted Unicode or markup leakage.

The first time I plotted this on a real corpus I expected the repetition ratio to be smoothly unimodal with a long tail. I was wrong on both counts. The bimodal shape is the data trying to tell you that two different populations of documents are in the pile. That kind of shape is worth pausing on whenever you see it.

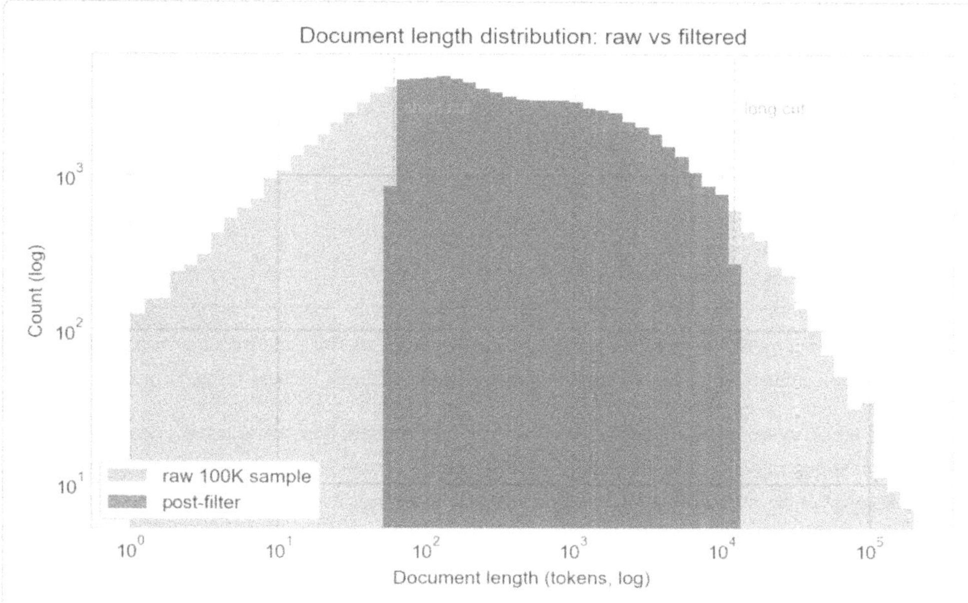

Figure 10.1. Document length distribution for the raw 100K sample, on a log-scaled y-axis, with the post-filter distribution overlaid. The right tail of very long documents thins out after filtering; the left tail of very short documents disappears almost entirely.

Add a language ID step using `fastlangid` or `cld3`. Record the predicted language and the confidence. A FineWeb-EDU sample is supposedly English-only, but you will find a tail of misclassified non-English documents (usually one to three percent, sometimes higher in older crawls). These are not bugs in the upstream pipeline so much as edge cases the language detector missed. They are also exactly the documents your tokenizer will turn into a stream of unknown-token bytes during training, which wastes context length and contributes nothing to learning.

An offline Tamil triage app I worked on, for community health workers in rural districts, taught me a related lesson in reverse. We were curating a Tamil medical corpus, which meant aggressive English filtering, protocol-aware quality scoring, and a long tail of mixed-script documents that language ID handled badly. The same script-detection edge cases that hurt our pipeline hurt every FineWeb-style English pipeline I have seen, just at a different magnitude and with the labels swapped. Language ID is reliable on the easy cases and lies politely on the hard ones.

After this step you have a CSV with one row per document and roughly seven columns. The whole step takes about twenty minutes on CPU. Before doing anything else, sort by repetition ratio and read the top fifty documents by eye. You will see what raw web data actually looks like, and the rest of the chapter will feel less abstract.

Step 2: Build a MinHash deduper from scratch

Deduplication is the single highest-impact step in this chapter. We will build it in three stages: shingle each document, compute a MinHash signature for each document, and then bucket the signatures into an LSH index so we only compare signatures within the same bucket.

Start with shingling. A shingle is a sliding window of 5-token spans across a document. We hash each shingle to a 64-bit integer using a fast non-cryptographic hash:

```
import mmh3

def shingles(text: str, k: int = 5) -> set:
    tokens = text.split()
    if len(tokens) < k:
        return set()
    return {mmh3.hash64(" ".join(tokens[i:i+k]))[0]
            for i in range(len(tokens) - k + 1)}
```

The output is a set of 64-bit integers, one per unique shingle. A typical 1000-token document produces roughly 996 shingles. A document with a lot of internal repetition produces fewer, because the set drops duplicates.

The Jaccard similarity between two documents (the fraction of shingles they share) is the gold-standard measure of near-duplication. If two documents share 80 percent of their shingles, they are near-identical. The problem is cost. Computing pairwise Jaccard for 100,000 documents is 5 billion comparisons; for 100 million it is 5 quadrillion. We need an approximation.

The MinHash trick is this. Pick K independent hash functions, where K is somewhere between 64 and 256. For each document, run every shingle through every hash function and keep only the minimum value across all shingles for each hash function. The result is a K-dimensional signature: K integers, one per hash function. Two documents with high Jaccard similarity have signatures whose components match in roughly the same proportion. That is the property the algorithm guarantees, and it is the only thing about MinHash that matters mathematically.

To estimate Jaccard between two documents, compare their signatures element by element and count how many positions agree. Divide by K, and you have an unbiased estimate of the true Jaccard.

```
import numpy as np

def minhash_signature(shingle_set: set, num_hashes: int = 128,
                      seed: int = 42) -> np.ndarray:
    if not shingle_set:
        return np.full(num_hashes, np.iinfo(np.uint64).max,
                       dtype=np.uint64)
    rng = np.random.default_rng(seed)
    # K independent hash functions implemented as (a*x + b) mod p
    p = (1 << 61) - 1
    a = rng.integers(1, p, size=num_hashes, dtype=np.uint64)
    b = rng.integers(0, p, size=num_hashes, dtype=np.uint64)
    shingles_arr = np.fromiter(shingle_set, dtype=np.uint64)
    # For each hash function, take the min over all shingles
    hashed = (np.outer(a, shingles_arr) + b[:, None]) % p
    return hashed.min(axis=1)
```

A note on the math here. We use a family of universal hash functions of the form `(a*x + b) mod p`, where `p` is a large prime (we use `2^61 - 1`) and `(a, b)` is drawn randomly per hash function. This is a standard trick and avoids the cost of running 128 different cryptographic hashes (one matrix multiplication does it). The `seed` parameter makes the experiment reproducible. If you change the seed, the signatures change, but the Jaccard estimates do not, because all 128 hash functions are still independent.

Now we have a signature per document. Comparing signatures pairwise is K times cheaper than comparing shingle sets pairwise, but it is still quadratic in document count. The LSH step gives us the speedup.

The idea of LSH is to slice each signature into b bands of r rows each, where `b * r = K`. For each band, hash the r integers in that band together into one bucket ID. Two documents land in the same bucket for some band if and only if all r rows in that band agree. The probability of that, given a true Jaccard similarity s, is `s^r`. The probability that at least one of the b bands collides is `1 - (1 - s^r)^b`. With K=128, b=32 bands of r=4 rows, the curve crosses 50 percent at roughly s=0.55. Documents with similarity above 0.55 are very likely to collide in at least one bucket, and documents with similarity below 0.55 are very likely not to.

```
def lsh_buckets(signature: np.ndarray, num_bands: int = 32) → list:
    bands = signature.reshape(num_bands, -1)
    return [mmh3.hash64(b.tobytes())[0] for b in bands]
```

For each document, compute its bucket IDs (one per band) and store the document ID in a dictionary keyed by (band index, bucket ID). At the end, walk the dictionary, and any bucket containing more than one document is a candidate cluster of near-duplicates. Verify each candidate by computing the actual signature-based Jaccard estimate, and keep only pairs above the desired threshold (0.8 is a common choice for "definitely a duplicate").

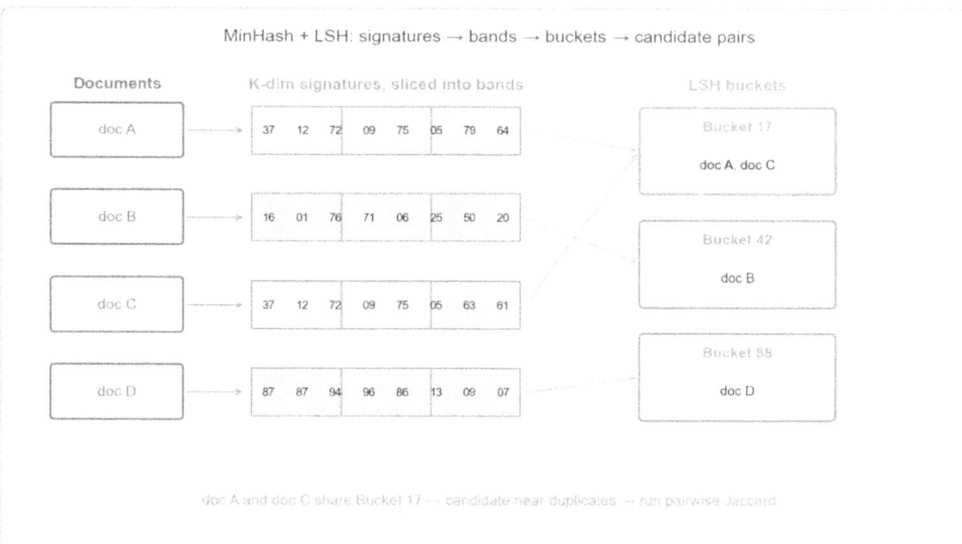

Figure 10.2. MinHash + LSH bucketing. Documents become K-dimensional signatures; signatures get sliced into bands; each band hashes into a bucket; documents that share any bucket are candidate near-duplicates and only those candidates get pairwise Jaccard checks.

On the 100K-document sample, running this pipeline takes about ten minutes on a modern laptop CPU. The duplication rate you find depends heavily on the source. For raw Common Crawl, expect 30 to 50 percent. For FineWeb-EDU, which has already been deduplicated upstream, expect 5 to 15 percent of residual near-duplicates that the upstream pipeline did not catch. Either way, you now have a list of clusters and you can choose which member of each cluster to keep: usually the longest, or the one with the lowest repetition ratio, or the one from the earliest crawl date.

When the pipeline finishes, read through twenty random duplicate clusters by eye. You will see exactly what kind of redundancy was in your corpus. Sometimes it is verbatim reposts. Sometimes it is the same press release picked up by ten news sites with slightly different headers. Sometimes it is a template page where only the company name changes. Sometimes it is an entire Wikipedia article copied across three different mirror sites. Each of these would have been seen by your model as three or four independent learning examples without dedup.

I expected FineWeb-EDU to come back with under 2 percent residual duplicates because the upstream pipeline is well-engineered. The actual number on the sample I ran was closer to 8. None of those were obvious bugs; they were near-duplicates that did not quite cross whatever threshold the upstream MinHash had used. Trusting other people's dedup is a defensible default. Trusting it without checking is not.

Step 3: Compute 13-gram contamination against an eval set

Contamination is detected by a different but related computation. Take your evaluation set (say, the validation split of MMLU, or HellaSwag, or a custom benchmark you care about). For each example, extract every 13-gram of tokens. Now scan your training corpus and count how many training documents contain any of those 13-grams verbatim.

The implementation is straightforward if you index it carefully. Build a hash table whose keys are 13-gram hashes from the eval set, and whose values are the eval example IDs that produced them. Then walk through the training corpus, generate 13-grams, and look up each one in the table. Any hit is a contamination event.

```
def thirteen_grams(text: str) → list:
    tokens = text.split()
    return [tuple(tokens[i:i+13]) for i in range(len(tokens) - 12)]

eval_index = {}
for eval_id, example in enumerate(eval_set):
    for gram in thirteen_grams(example["text"]):
        h = mmh3.hash64(" ".join(gram))[0]
        eval_index.setdefault(h, []).append(eval_id)

contaminated_docs = {}
for doc_id, doc in enumerate(training_corpus):
    hits = set()
    for gram in thirteen_grams(doc["text"]):
        h = mmh3.hash64(" ".join(gram))[0]
        if h in eval_index:
            hits.update(eval_index[h])
    if hits:
        contaminated_docs[doc_id] = hits
```

For a 100K-document training sample and a 10K-example eval set, the hash table is small enough to fit in memory and the scan takes a few minutes. For a larger corpus, the same idea works with disk-backed indices like `marisa-trie` or a Bloom filter front-end that rejects most non-matches before doing the full lookup.

What you find will depend on the eval and the corpus. On FineWeb-EDU versus MMLU's validation split, expect a contamination rate between 0.1 percent and 2 percent of training documents. Small in fraction but devastating in impact, because each contaminated document tends to contain many eval 13-grams from the same example, so a single training document can leak the answer to dozens of eval questions.

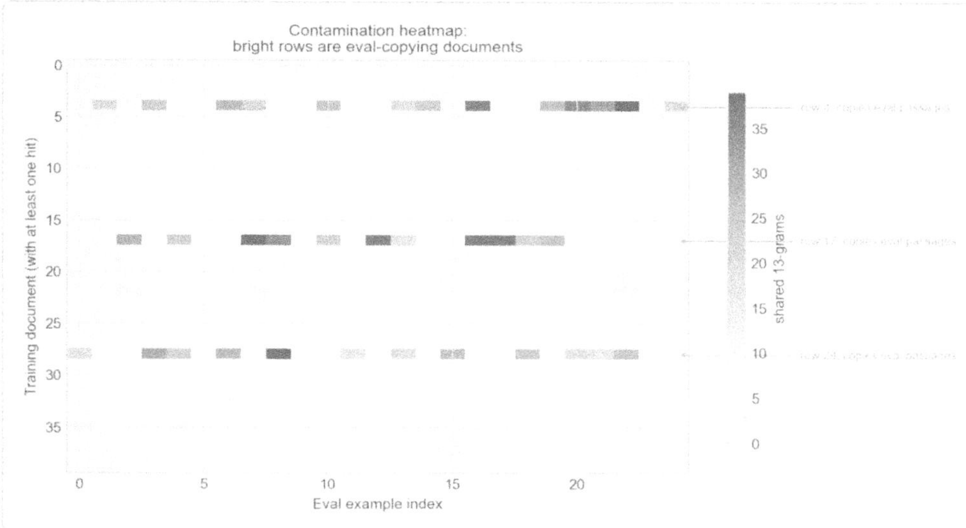

Figure 10.3. Contamination heatmap. Rows are training documents that scored at least one hit; columns are eval examples; cell intensity is the number of shared 13-grams. The handful of bright rows are documents that copy entire eval passages.

Look at the bright rows by hand. You will find that contaminated training documents fall into three categories. The first is direct copies: somebody uploaded the benchmark to their blog. The second is the source the benchmark was built from, for example Wikipedia articles that MMLU questions were written about. The third, the most insidious, is study guides and answer keys that paraphrase the question and then state the correct answer in plain English. The first two are obviously contamination. The third is contamination too, even though no exact eval string appears anywhere in the document.

The category that surprised me the first time I ran this on a real corpus was the second one. The source material. I had been thinking of contamination as "leaked test sets," and the category of "Wikipedia article that the test was written about" did not fit my mental model. The 13-gram check catches it anyway. The mental model needs to widen, not the algorithm.

Decontamination at this point is simple: drop any training document that hit any eval 13-gram. You will lose between 0.1 and 2 percent of your training tokens, and your evaluation will measure something closer to actual model capability than it did before.

Step 4: Build a mixing weight schedule across three domains

The mixing question is a budget allocation problem. You have a fixed compute budget (say, 10 billion training tokens). You have three data sources: web, code, and math. How do you split the budget?

There is no universal answer, but there are useful heuristics. The first heuristic is that more is not always better. Beyond a certain fraction, each additional token from a source produces less marginal improvement on tasks that source serves. A model trained on 99 percent web and 1 percent code is bad at code. A model trained on 50 percent web and 50 percent code is decent at both. A model trained on 10 percent web and 90 percent code is bad at English.

The curve is concave in each direction.

The second heuristic is that the mixing should change during training. Early in training, the model is learning basic statistical structure: token distributions, sentence-level grammar, broad topical patterns. Web data is the most efficient teacher for these. Late in training, the model has already learned the basics and benefits from concentrated exposure to higher-quality tokens. Code and math tokens, which are denser in structure per token, are more useful at the end. A schedule that ramps up the code and math fractions over the last 20 percent of training often outperforms a constant mixture at the same total token cost.

The simplest version of this schedule is a piecewise linear function:

```
def mixing_weights(step: int, total_steps: int) -> dict:
    progress = step / total_steps
    if progress < 0.8:
        return {"web": 0.85, "code": 0.10, "math": 0.05}
    # Ramp up code and math in the final 20%
    t = (progress - 0.8) / 0.2
    web = 0.85 - 0.30 * t
    code = 0.10 + 0.20 * t
    math = 0.05 + 0.10 * t
    return {"web": web, "code": code, "math": math}
```

This is one schedule among many; the field has not converged on a single right answer. The DataComp-LM and DoReMi papers describe more sophisticated approaches that learn the mixing weights themselves rather than choosing them by hand. For this chapter, the hand-picked schedule is enough to make the point that the mixing weights matter and that holding them constant is itself a choice with consequences.

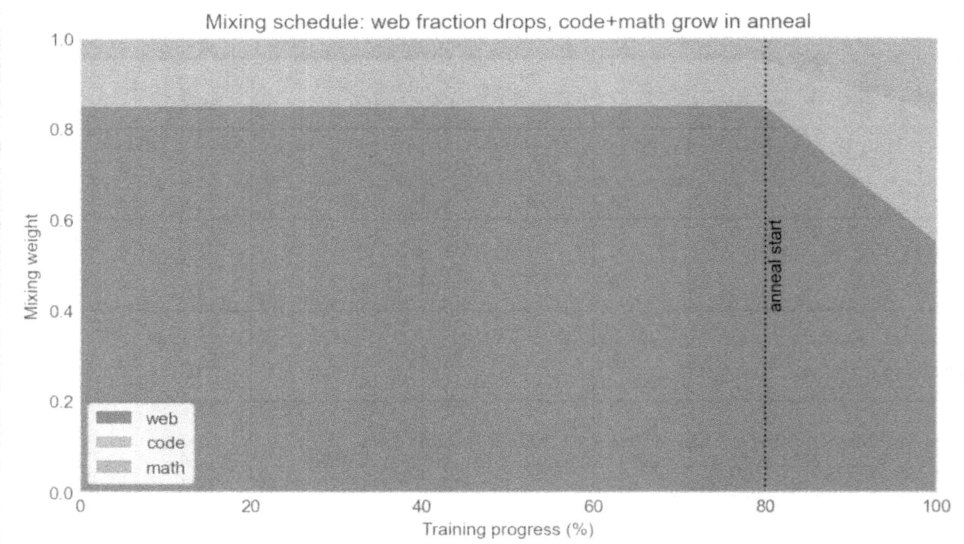

Figure 10.4. Mixing weight schedule across training. The web fraction starts at 85 percent and drops to 55 percent over the final 20 percent of training. Code and math fractions grow correspondingly.

A practical note on implementation. Mixing weights act on the data loader, not on the loss. At each step, you sample a batch from each source in proportion to the current weights: say, 85 percent of the batch's documents from web, 10 percent from code, 5 percent from math. The model itself does not know which document came from which source. The loss is computed on the concatenated batch as if it were a single mixed stream.

Step 5: Train twice — raw versus curated — at fixed compute

Now the experiment that ties the chapter together. Train a small 6-layer transformer, the same architecture as in **Project 9: Pretraining on the Real Web**, twice. Once on the raw 100K-document corpus, with no filtering, no dedup, no decontamination. Once on the curated version: quality-filtered (drop documents with repetition above 0.3, length below 100 tokens, or non-English language ID confidence below 0.8), deduplicated (drop near-duplicates above Jaccard 0.8), decontaminated (drop documents with any 13-gram overlap to your eval set), and mixed using the schedule from Step 4. Both runs use the same compute budget, the same model, the same hyperparameters, and the same random seed for everything except the data sampling.

The raw corpus has roughly 100,000 documents. The curated corpus has somewhere between 50,000 and 70,000, depending on how aggressive your filters are. At equal compute, this means the curated run sees each remaining document slightly more often than the raw run sees its average document. The curated run does more passes over a smaller, cleaner pool.

Track two numbers per run. The first is validation bits-per-byte, **val_bpb**, on a held-out clean validation set. Bits-per-byte is the model's loss normalized by the actual byte length of the text rather than by token count; it is comparable across tokenizers and across corpora in a way that raw cross-entropy is not. The second is downstream evaluation scores (accuracy on MMLU, HellaSwag, or whatever benchmark you care about) on the held-out eval set that you also used for decontamination.

The expected result at this scale is roughly the following. The curated run reaches a lower val_bpb than the raw run at the same compute, typically by 0.05 to 0.15 bits-per-byte, which is a real and visible difference on a single-machine training run. The curated run also reaches higher downstream eval scores, typically by 2 to 8 percentage points on multi-choice benchmarks. The gap between val_bpb improvement and downstream improvement is the curation lift: the portion of the improvement that reflects better generalizable behavior rather than only better next-token prediction.

The exact numbers depend on the corpus, the model size, and the eval. The direction is consistent: curated wins at fixed compute. If you do not see that direction on your first try, look at the filters. The most common mistake is over-aggressive filtering that removes good documents along with the bad ones. The second most common is under-aggressive decontamination, which makes the raw run look artificially competitive because both runs are scoring high on a leaked eval.

Step 6: Plot the quality-vs-quantity tradeoff

You now have one curated run. Run several more at different filter strictness levels, holding compute fixed each time. At one extreme, no filter; that is the raw run from Step 5. At the other extreme, very strict filters: repetition below 0.1, length above 500 tokens, language confidence above 0.95, near-duplicate threshold at Jaccard 0.6 instead of 0.8. In between, three or four intermediate settings.

Each setting produces a smaller surviving corpus and (you hope) a higher per-token quality. Plot val_bpb on the y-axis and surviving-token count on the x-axis. The curve will have a clear shape: as you remove more data, val_bpb gets slightly worse because there are fewer tokens to train on, but then it gets better as the remaining data is cleaner. There is usually a sweet spot, a strictness level that maximizes the model's val_bpb at fixed compute, and that sweet spot is rarely at zero filtering or at maximum filtering.

Figure 10.5. Quality-versus-quantity Pareto curve. Surviving token count on the x-axis, val_bpb on the y-axis. Each point is a different filter-strictness setting. The lower-left corner of the curve is the sweet spot; settings to the right keep too much bad data, settings to the left throw away too many good tokens.

The location of the sweet spot is corpus-dependent. For a noisy raw crawl, it is far to the left; aggressive filtering helps a lot. For an already-curated corpus like FineWeb-EDU, it is closer to the middle; light additional filtering helps, but heavy filtering starts removing genuinely useful tokens. Your job at this step is to find the sweet spot for your specific data and write it down for future runs.

Step 7: Inspect what you cut

The last step before BREAK IT is the calibration step. Take a random sample of 100 documents that your filters removed. Read them yourself, label them as "actually bad" or "actually fine", and compute the false-positive rate of your filter. If 80 of the 100 documents you cut were genuinely terrible (template spam, corrupted text, single-line documents) then your filter is well-calibrated. If 30 of them were fine documents about niche topics that happened to trip a heuristic, then your filter is too aggressive in a way that costs you genuine data.

The same exercise in reverse is also worthwhile. Sample 100 documents that survived your filters and read those too. Some fraction will still be garbage — heuristics never catch everything, and you should know what your post-filter corpus actually contains. If 90 of the 100 are clean, your filter is doing its job. If 30 of them are still spam, your filter is missing a category and you need another heuristic.

This step takes an afternoon and produces more useful information per hour than any other step in the chapter. Reading the actual documents that go into your model, both the ones you keep and the ones you cut, is the closest thing to a sanity check on your whole pipeline. Skip it once and you will discover, three weeks into a training run, that your filter was removing all the math content because LaTeX has weird character ratios that tripped your mean-word-length check.

BREAK IT

The break for this chapter is the opposite of every other chapter's break. Instead of removing a working piece and watching things fail, you deliberately add a broken piece (a contaminated training set) and watch your evaluation lie to you about it.

The exact change

Take the curated training corpus from Step 5. Add back the documents that were removed by decontamination, specifically the documents that contained 13-gram overlaps with your eval set. Do not change anything else. The corpus now contains every document that survived quality filtering and dedup, plus the small minority that leak the eval. Call this the "contaminated" run.

Train a model on this contaminated corpus, at the same compute budget and the same hyperparameters as Step 5. Then evaluate it on two things. The first is the original eval set, the one that was used for decontamination, and which the contaminated documents leak. The second is a held-out variant of the same eval set: a fresh slice of the same benchmark that the contaminated documents do not leak. For MMLU, this is straightforward; take the validation split as your contamination target and use a fresh slice of the test split as your held-out variant. For other benchmarks, you may need to construct the held-out variant yourself by paraphrasing or by holding out 20 percent of the examples before doing the contamination check.

What you will observe

The contaminated run's score on the original eval set is dramatically higher than the curated run's score on the same eval. The gap is usually between 10 and 30 percentage points on multi-choice benchmarks, depending on how aggressively the contaminated documents leak the answers. The contaminated run's score on the held-out variant of the same eval is roughly the same as the curated run's score. Sometimes it is slightly lower, because the contaminated documents take up training capacity that could have gone to genuine learning.

The pattern is unmistakable on a scatter plot. The curated run's two scores (eval set and held-out variant) sit close to each other on the diagonal. The contaminated run's two scores sit far above the diagonal on the eval-set axis and roughly on the diagonal for the held-out axis. The horizontal gap between the two points for the contaminated run is the contamination inflation.

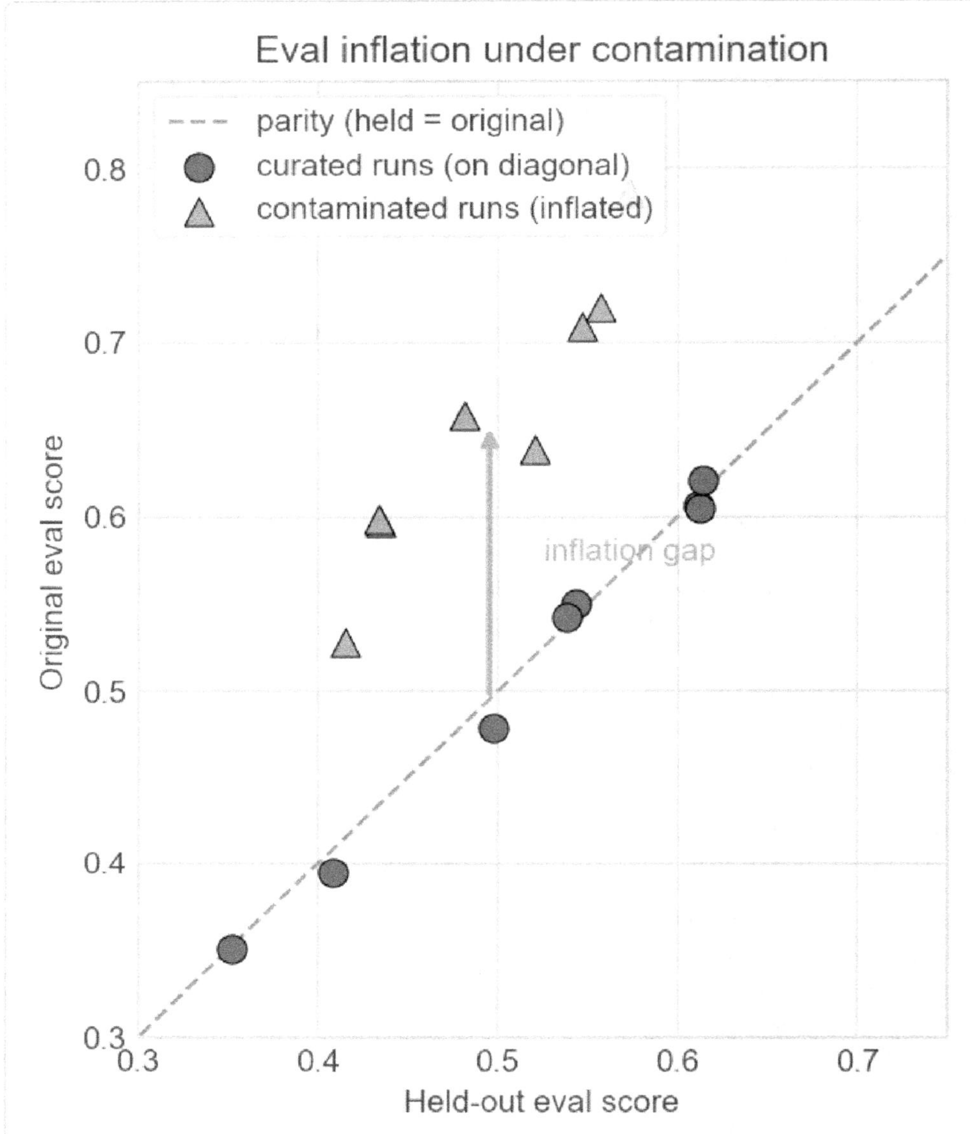

Figure 10.6. Eval inflation under contamination. Each point is a model: x-axis is held-out eval score, y-axis is original eval score. The curated run sits on the diagonal; its two scores agree. The contaminated run sits well above the diagonal; its original-eval score is inflated relative to its held-out score.

What this proves

Benchmark scores in published model reports are not always what they look like. When a model reports a state-of-the-art number on MMLU but a much worse number on a fresh evaluation, the explanation is almost always one of three things. Either the training data leaked the eval. Or the model was tuned to the eval through repeated checkpoint selection. Or, most rarely, the model genuinely is better at the benchmark but worse at the underlying capability the benchmark was supposed to measure. The first cause is the most common one, and the second cause is essentially a slower version of the first.

You now know what the first cause looks like in your own data. You can compute the contamination rate before you train, and you can predict the inflation gap before you evaluate. This means you can spot the same failure mode in any published number you encounter. If a paper reports a benchmark improvement that disappears on a held-out variant, that paper's training data has the problem you just simulated.

A second break worth running

Train a third model on the curated corpus minus the documents that the language ID filter removed; that is, put the misclassified non-English documents back in. The val_bpb gets slightly worse, by a small amount that you can measure. More interestingly, the downstream eval scores drop by a larger amount than the val_bpb change predicts. The reason is that the misclassified documents fragment the tokenizer's vocabulary, producing long sequences of byte-fallback tokens that consume context length without contributing useful information. The model's loss on those sequences goes up because they are intrinsically hard to predict, which drags val_bpb down. But more importantly, the model spends compute learning patterns in token sequences that it would never encounter on a clean English benchmark, which drags downstream evals down further. This is the "wrong data" failure mode, distinct from contamination, and worth seeing once so you can recognize it.

Optional Homework

The full pipeline runs on a single consumer GPU plus a normal CPU. The CPU steps (scoring, dedup, contamination check, mixing) are pure Python and can run on any laptop. The training step at the end needs a GPU; without one, you can still run a much smaller proxy version, described in Part F.

I drafted this homework first as a single 4000-word block. I bounced off my own first draft of this homework because I could not tell where to stop and check the result. The current part-by-part split is the rewrite. It is longer overall but it is the one a reader can actually finish.

Part A: Score your corpus and read the tail

Pull a 100K-document sample from FineWeb-EDU or another corpus you care about. Run the per-document statistics from Step 1. Plot the distributions for length, mean word length, and repetition ratio. Sort by repetition ratio descending and read the top 50 documents by eye.

Write down three things: the most common kind of high-repetition document, the most common kind of low-mean-word-length document, and one document that surprised you. The patterns you find here are the ones your filters need to catch.

Part B: Build the MinHash deduper

Implement the MinHash signature function and the LSH bucketing from Step 2. Run it on the 100K sample. Record the number of clusters you find, the average cluster size, and the size of the largest cluster.

Now compare your implementation against the off-the-shelf `datasketch.MinHash` and `datasketch.MinHashLSH` library. The clusters should largely agree. If they disagree more than a few percent, your hash family is probably not independent enough — try drawing more `a` and `b` values from a wider range, or switch to a different `p`.

Part C: Detect contamination

Choose an eval set you care about. MMLU validation, HellaSwag, or a custom benchmark all work. Build the 13-gram hash table and scan your training corpus for hits. Report the number of contaminated training documents, the number of distinct eval examples leaked, and the maximum 13-gram count for any single training document.

Read the top 10 most-contaminating training documents by hand. Identify which of the three categories from Step 3 each one falls into.

Part D: Curate and train, raw versus curated, at equal compute

Train the small 6-layer model from **Project 9: Pretraining on the Real Web** twice. Once on the raw corpus, once on the curated corpus (filtered, deduplicated, decontaminated, mixed). Match the compute budget exactly — same number of tokens, same number of optimizer steps, same hyperparameters, same random seed.

Record val_bpb and downstream eval scores for both runs. The curated run should beat the raw run on val_bpb by 0.05 to 0.15 bits-per-byte and on downstream evals by 2 to 8 percentage points. If you see a smaller gap, your filter is too gentle. If you see a much larger gap, double-check that the runs are using equal compute.

Part E: The contamination break (with a prediction)

Run the contaminated experiment from BREAK IT. Train on the curated corpus plus the contaminated documents that decontamination removed. Evaluate on both the original eval set and a held-out variant.

Before running, predict the inflation gap on the original eval. Most readers underestimate it the first time. Record your prediction, run the experiment, and compare. If your prediction was within 5 percentage points, you have internalized the failure mode. If it was off by more, look at the contaminated documents again and count how many full eval examples each one leaks.

Part F: Proxy lab for CPU-only readers

If you have no GPU, you can do every step in this chapter except Steps 5 and 6 on CPU. The scoring, MinHash, dedup, contamination check, and mixing-weight schedule all run in pure Python on a 100K-document sample.

For the training step, use a tiny model: 2 layers, hidden dimension 64, batch 8, sequence length 128. Run for 2000 steps on each corpus. The val_bpb numbers will be much higher than on a real GPU run, but the relative ordering — curated below raw, contaminated inflating original-eval scores — will hold. The patterns transfer.

Part G: Sweep the filter strictness

Run the full pipeline at five different strictness levels: no filtering, light, moderate (your default), strict, very strict. For each, record the surviving token count and the val_bpb of a fixed-compute training run.

Plot the Pareto curve from Step 6. Identify the sweet spot for your specific corpus. Write down the filter parameters at that sweet spot — they are the defaults you will want for any future training on the same data source.

Questions To Answer

1. What was the duplication rate in your raw corpus? How many documents were in your single largest duplicate cluster, and what did they have in common?

2. In your contamination scan, how many of the contaminated training documents fell into each of the three categories — direct copies, source material, and study guides? Which category was the largest, and was that what you expected?
3. When you compared the curated run to the raw run at equal compute, how much of the improvement showed up in val_bpb and how much in downstream evals? Were the two numbers consistent with each other, or did one move much more than the other?
4. When you trained the contaminated model, how large was the gap between its original-eval score and its held-out-variant score? Was that gap larger or smaller than you predicted before running the experiment?
5. Where was the sweet spot on your quality-vs-quantity Pareto curve? At what point did additional filtering stop helping and start hurting?
6. If you ran the same pipeline on a corpus that had already been deduplicated upstream — say, RedPajama or FineWeb — what fraction of additional duplicates did you still find? What does that tell you about trusting other people's dedup?

Go Further

Three papers are the reading list for this chapter.

The FineWeb paper (Penedo and colleagues, 2024) describes the full curation pipeline behind the corpus you used in **Project 9: Pretraining on the Real Web** and in this chapter. The interesting part is not the final numbers but the ablation tables: each filter is added one at a time and its effect on downstream evals is measured separately. This is the most thorough public ablation of pretraining data curation we have, and reading the appendix carefully will teach you which filters earn their keep and which ones add complexity without payoff.

The first time I read the FineWeb appendix end to end I came away with a stronger opinion than I started with: most reported "model improvements" in the public literature would survive a curation ablation, and a fair number would not. The papers that publish their data pipeline carefully are doing a quiet public service.

The DataComp-LM paper (Li and colleagues, 2024) takes a different angle. It treats the choice of training corpus as the main experiment, holding the model architecture and the compute fixed, and shows that the curation pipeline matters at least as much as the model itself for downstream performance. The released DCLM-Baseline corpus is worth pulling and comparing against FineWeb on the same eval suite.

The MinHash paper itself is from 1997 (Broder, "On the Resemblance and Containment of Documents"), and it is still the cleanest explanation of the math. The chapter version of the algorithm is the same one Broder described; the engineering has improved, but the idea has not changed.

Two repos to read closely.

The HuggingFace `datatrove` library at github.com/huggingface/datatrove is the cleanest open-source implementation of a production data curation pipeline. It runs in distributed mode across many machines, but the single-machine version is also small and readable. Look specifically at the dedup module to see how their MinHash and LSH compare to what you built.

The `dolma` toolkit from Allen AI at github.com/allenai/dolma is a second reference implementation, with a different design that emphasizes streaming over batch processing. The two libraries together give you a good sense of the design space.

arXiv search directions worth pursuing: `pretraining data quality scaling`, `deduplication large language model training`, `contamination evaluation language model`, `minhash locality sensitive hashing deduplication`, `data mixing weights pretraining`, `DoReMi domain reweighting`.

One direction worth investigating immediately after the homework is learned mixing weights. The DoReMi paper trains a small proxy model to learn the optimal mixing weights across domains, then transfers those weights to the larger training run. The result is a corpus that is balanced for the specific model and task at hand rather than balanced by a hand-picked schedule. This is the natural next step after the hand-built schedule in Step 4.

What You Now Know

Pretraining data is a system you can measure rather than a black box you have to trust. You have four operations (quality filtering, deduplication, mixing, and decontamination) and each one produces a specific number that you can compute on any corpus. You can build a MinHash deduper from scratch in a few dozen lines of Python and run it on 100,000 documents in under fifteen minutes on a laptop. You can detect 13-gram contamination against an evaluation set and predict, before training, how much that contamination will inflate the eval score. You know what a quality-vs-quantity Pareto curve looks like on real data, and you know how to find the sweet spot for your specific corpus.

You also know what a leaked benchmark looks like from the inside. When you see a paper report a state-of-the-art number that does not reproduce on a held-out variant, you can name the failure mode and you can estimate, from public information about the training data, how likely that failure is. That is a small skill in absolute terms and a large one in practice. Most of the surprises in modern model evaluation come from data leakage, and most data leakage is invisible to people who do not check for it.

The dashboard from **Project 11: Training Debugging — Spikes, NaNs, Profiling** told you what a sick training run looks like during training. This chapter tells you what a sick training run looks like before training starts. Both signals matter. Both are now visible to you.

Starting Point

FineWeb paper: arxiv.org/abs/2406.17557. DataComp-LM paper: arxiv.org/abs/2406.11794. RedPajama dataset notes: github.com/togethercomputer/RedPajama-Data. HuggingFace `datatrove` library: github.com/huggingface/datatrove. Allen AI `dolma` toolkit: github.com/allenai/dolma. Broder's original MinHash paper: "On the Resemblance and Containment of Documents," 1997.

Project 11: Training Debugging: Spikes, NaNs, and Profiling

Hook

At step 1,247 of your training run, the loss jumps from 3.2 to 47.8, hangs there for two steps, and then prints `nan`. The training was working, until it wasn't.

The first instinct most people have is to restart with a smaller learning rate and hope it goes away this time. I had that instinct for years. It cost me real money in cluster hours before I learned the other one.

This chapter teaches the other instinct: read the training run like a production system you are on-call for, name the specific failure that just happened, and know which knob to turn before you touch anything. By the end of the chapter you will be able to predict a `nan` several minutes before it arrives, and you will know which of half a dozen specific failure modes caused it.

The Concept

Picture baking a cake in an oven with no window. You can only check whether the cake is done by opening the door, and opening it too often ruins it. Most people training neural networks debug exactly like this. The training loop is the closed oven. The loss curve is the timer ticking on the wall. If something is wrong inside the oven, you find out at the end, when the cake comes out either burnt or raw.

The fix in both cases is the same. Put instruments inside. Watch the temperature at three points in the oven, not just at the door. Smell the air. Listen for the timer ticking faster or slower than it should. With proper instruments, you do not need to open the door. You can read the state of the cake while it is baking.

A training run has its own set of vital signs. There are four worth measuring continuously, and learning to read all four together is the entire skill this chapter teaches.

I picked up this framing the hard way, before I trained any model. At Aspira I spent years running a multi-state SaaS at a 99.99% uptime target, and the rule there was simple: anything you cannot graph, you cannot debug at 3 AM. Training runs are the same. The difference is that nobody hands you the graphs by default. You have to wire them yourself.

The first vital sign is the **loss** itself. Every training framework prints this number, and almost every reader of this book has stared at one go down. By itself, loss is a lagging indicator. It tells you the model is sick after the model is already sick. By the time the loss curve bends the wrong way, the failure is already a few steps old, and depending on how aggressive it is, it may already be unrecoverable.

The second vital sign is the **gradient norm**, computed layer by layer. For each parameter group in the model (say, each transformer block), take the gradients produced by the most recent backward pass and compute their L2 norm. That is one number per block per training step. Plot it. A healthy run shows the gradient norms drifting slowly, with the lines for different blocks staying roughly parallel. A run heading for a spike has one line (usually the embedding layer's or the first attention block's) pulling away from the cluster and climbing faster than the others. By watching gradient norms instead of waiting for the loss to bend, you catch the spike five to fifteen steps earlier. That is enough time to save a checkpoint or kill a doomed run.

Across thousands of training runs I have logged for cooperative LoRA fusion work on models from a few hundred million to a few billion parameters, this single graph saved more wall-clock than any other instrument. Not because the math is exotic. Because the first attention block almost always leaves the cluster first, and once you have seen that shape three or four times, you stop arguing with the dashboard.

The third vital sign is the **activation histogram** at each layer boundary. After every transformer block, what does the distribution of hidden-state values look like? Healthy activations have a roughly stable spread, a mean near zero if LayerNorm or RMSNorm sits at that boundary, and bounded extremes. Unhealthy activations spike. One feature dimension suddenly carries values in the hundreds, and the softmax that consumes those values collapses into a near one-hot pattern. Once attention saturates, the gradient through that head becomes tiny, and the head stops contributing to learning. The signature of impending attention saturation appears in the activation histogram several steps before the loss notices.

I expected gradient norms to be the most useful of the four when I first wired this dashboard. I was wrong. The activation tail moves earlier, more clearly, and on a smaller dynamic range. Gradient norms are easier to plot and easier to teach. The activation `abs_max` is what I actually look at first when something feels off.

The fourth vital sign is the **weight-update ratio**. At each optimizer step, the optimizer adds some quantity Δw to the current weight w. The ratio `|Δw| / |w|` for each parameter group tells you how aggressively the optimizer is moving that group relative to its current size. There is a rule of thumb, often credited to Andrej Karpathy: a healthy ratio sits near 1e-3. A ratio of 1e-1 means the optimizer is rewriting that layer every single step. The learning rate is much too hot for that layer. A ratio of 1e-6 means the layer is barely moving. Either the gradient signal is dying, or the layer was already converged and is sitting still. Either way, you want to know.

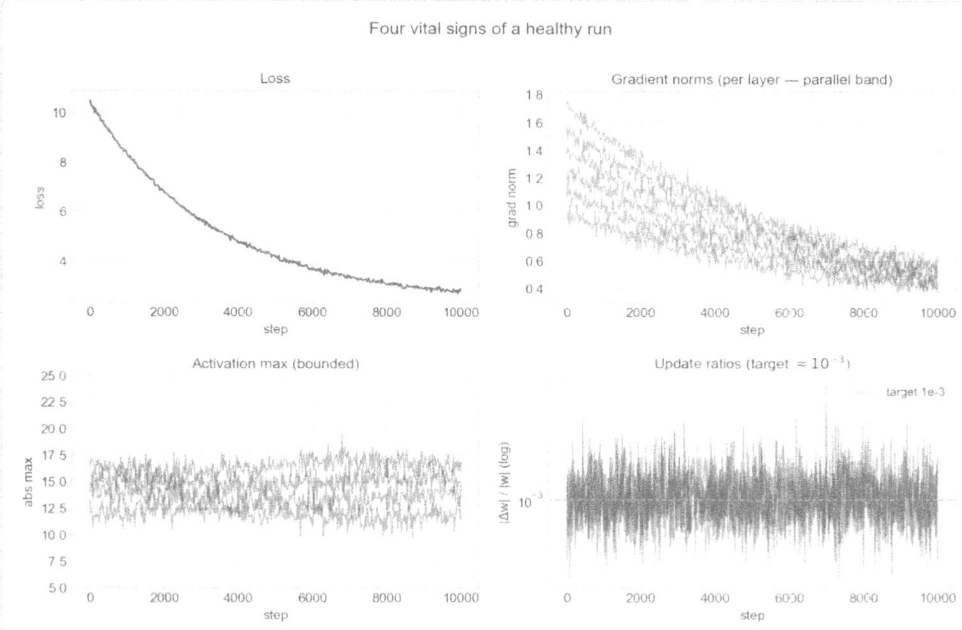

Figure 11.1. The four vital signs plotted together for a healthy run. Loss drifts down smoothly. Gradient norms drift in a narrow parallel band. Activation max stays bounded. Update ratios cluster near 1e-3.

So far this is bookkeeping. The interesting move is what happens when you start using these four signals together. A loss spike on its own is a fire alarm: useful, but it tells you nothing except that something is burning. The combination of loss, gradient norm, activation distribution, and update ratio is a fire alarm that tells you which floor of the building is on fire, which room, and what is burning. You stop guessing at fixes and start applying targeted ones.

Most training failures fall into a short list of named patterns, and each one has a specific signature across the four instruments.

A **loss spike from a hot learning rate** shows the update ratio drifting upward first, then the activation max, then the gradient norm, then finally the loss, in that order, with the loss arriving last. An **activation explosion from bad initialization** shows the activation max already high at step zero and refusing to come down; the model never enters a healthy state. **Attention saturation** shows one head's contribution to the activation distribution growing while the others shrink; the softmax of that head becomes one-hot, and the head dies. A **mixed-precision overflow** shows a gradient briefly going `inf`, the loss scaler halves itself, and a step gets silently skipped.

A **dead layer** shows one parameter group's update ratio pinned near 1e-6 for thousands of steps while every other group moves normally. This is usually a bad initialization, a wrong learning rate for that group, or, most painfully, a parameter that someone froze and then forgot about. I have shipped that bug twice. Both times the model trained for hours before anyone noticed the frozen block was the one we needed to move the most.

Each one of these patterns has a fingerprint. Once you have seen the fingerprint three or four times, the instruments narrate the failure as it happens. You do not need to wait for the symptom. You name the failure mode while there is still time to clip the gradient, lower the rate, reinitialize the layer, or save the checkpoint before everything blows up.

A small note about what this chapter does not teach. It does not promise to prevent every failure. Some training runs are genuinely impossible: bad data, bad architecture, bad optimizer, hardware that drops bits under load. The job here is narrower: when a run can be saved, you can now save it; when it cannot, you can now name why before the GPU bills mount.

Why It Matters

Without instrumentation, debugging a training run is mostly theater. You change a hyperparameter, restart, watch the loss for a while, and hope. If the run survives, you keep the change and convince yourself you have learned something. If it does not, you change something else. After a few weeks of this, you have either burned through a research budget without insight or you have given up on training your own models and gone back to fine-tuning APIs that someone else trained.

With instrumentation, the loop becomes diagnostic instead of superstitious. A spike has a cause. The cause has a signature. The signature is visible in the data before the spike lands. The fix is a specific intervention: clip the gradient, lower the rate, reinitialize the bad layer, switch on a mixed-precision scaler. Each intervention has a measurable effect on the next ten or twenty steps, so you can verify whether the fix worked without waiting for the whole run to finish. Training stops being a black box you pray at and becomes a system you can reason about.

There is a second reason this chapter sits where it does. Every subsequent chapter in the book trains a model that can fail. **Project 9: Pretraining on the Real Web** pretrained a model on real web data and could spike if the learning rate was wrong. **Project 12: Distributed Training: FSDP and ZeRO (Single-Box Proxy)** shards parameters across processes, and a mis-shard causes silent divergence, silent except in the gradient-norm trace, which shouts. **Project 23: Reward Models and RLHF** introduces reward hacking, which shows up first as a gradient-norm anomaly in the policy head before it shows up in evaluations. **Project 18: Mixture of Experts** trains a routed model, and router collapse appears in the activation distribution before it appears in the loss. Each of these failure modes has the same shape as the ones in this chapter, only with new layers and new names. The skill transfers exactly.

There is also a third reason worth naming, even though it is less direct. When you can read a training run, you can read other people's runs too. Pull a published tensorboard from a paper repo, look at the gradient-norm curves, and you can often tell whether the run was healthy or just lucky. That is not a small skill. It is the difference between trusting a paper's headline number and understanding the conditions under which the number was produced.

Most published training results do not show the dashboards. I would trust their numbers more if they did. A loss curve without a gradient-norm trace beside it is half a story, and the missing half is almost always the half you would have argued with.

The Build

You are going to take the model from **Project 9: Pretraining on the Real Web** and add four instruments to it. Then you will run a clean training pass to see what healthy looks like, deliberately push the run toward a spike, and predict the spike from the instruments before it lands. None of this requires fancy infrastructure. A laptop

GPU is enough. A CPU with patience is enough for the instrumentation logic alone. You cannot train a real model on CPU, but you can verify that your hooks fire correctly and produce the right shapes there.

The reference implementation is `nanochat`'s training loop, but everything below is portable to any PyTorch training script.

Step 1: Add per-layer gradient norm logging

PyTorch's `nn.Module` exposes every parameter through `.parameters()`, but you usually want gradient norms grouped by submodule. The cleanest way is to walk the named modules and group parameters by their parent block. For a transformer with a stack of `TransformerBlock` modules, you want one number per block: the L2 norm of all parameters belonging to that block.

After `loss.backward()` and before the optimizer step, walk the parameters and compute group norms:

```
def group_grad_norms(model):
    norms = {}
    for name, module in model.named_modules():
        params = [p for p in module.parameters(recurse=False)
                  if p.grad is not None]
        if not params:
            continue
        total_sq = sum((p.grad ** 2).sum().item() for p in params)
        norms[name] = total_sq ** 0.5
    return norms
```

Two details worth pointing out. The argument `recurse=False` keeps each module's norm local. Without it, parameters of child modules would be double-counted at every ancestor module, and the embedding layer's gradient norm would appear in the root model's norm too.

The first time I added per-layer grad-norm logging to one of my training modules, I forgot the `recurse=False` flag and triple-counted every parameter. The numbers looked plausible. The model still spiked. I lost most of a day on that bug, staring at a dashboard that was telling me the truth about a tensor I had silently summed into itself three times. Read the docstring of `parameters()` before you trust your own plot.

The `.item()` calls in the inner loop force a synchronization between CPU and GPU, which slows training slightly. In a production loop you would either skip this step on most iterations and only run it every N steps, or use the batched `torch._foreach_norm` ops to compute many norms in parallel without round-tripping each one to the host.

The output is a dictionary mapping module name to the L2 norm of its gradients. You log this every N steps, where N is chosen so the logging itself does not dominate runtime. For a 6-layer model training at roughly 1000 tokens per second per GPU, every 10 steps is about right. Once logged, you plot each group's norm on its own line, all on the same axes, with the y-axis on a log scale. Healthy training shows the lines drifting together, slowly. A spike-bound run shows one line (usually the embedding layer's, or the first attention's `qkv` projection) climbing faster than the others, breaking parallelism.

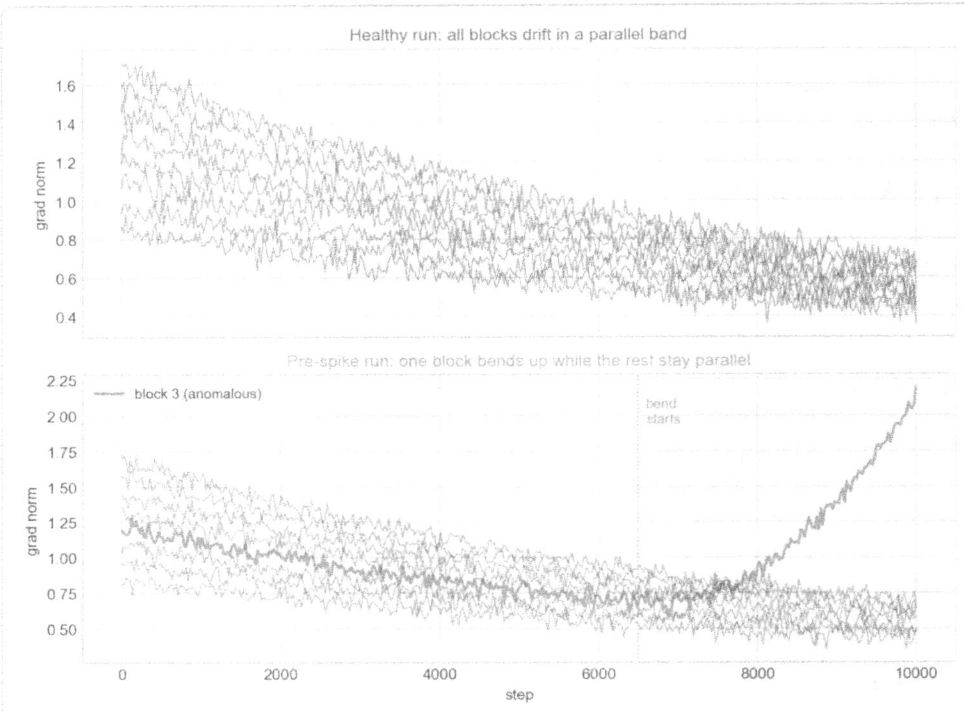

Figure 11.2. Layer-grouped gradient norms over training. Top: a healthy run shows parallel drift. Bottom: a pre-spike run shows one block's line bending up while the rest stay parallel.

Step 2: Add per-layer activation histogram

You need to know what the hidden states look like at each layer boundary. The standard tool is a forward hook: a function registered on a module that fires whenever the module produces an output. PyTorch exposes this with `register_forward_hook`.

```
activation_stats = {}

def make_hook(name):
    def hook(module, inputs, output):
        x = output.detach().float()
        activation_stats[name] = {
            "mean": x.mean().item(),
            "std": x.std().item(),
            "min": x.min().item(),
            "max": x.max().item(),
            "abs_max": x.abs().max().item(),
        }
    return hook

for name, module in model.named_modules():
    if isinstance(module, TransformerBlock):
        module.register_forward_hook(make_hook(name))
```

Three notes on this snippet. First, `.detach()` is mandatory. Without it, the hook holds a reference to the computation graph from the output, and your memory usage explodes because the graph cannot be freed after backward. Second, `.float()` casts to fp32 for the statistics. In mixed-precision training, the activations are bf16 or fp16, and computing `.std()` on fp16 tensors that contain extreme values is numerically unreliable; the cast to fp32 stabilizes the statistics. Third, you compute the statistics on every step but you only flush them to your log every N steps, same N as the gradient norms.

A small aside on the `.detach()` rule. I learned how unforgiving that one is while writing forward-pass instrumentation on Apple Neural Engine, where the host-side memory budget was tight enough that a single leaked reference would show up in `vm_stat` before the model had finished its second step. CUDA gives you more slack. The bug is the same shape; it just hides longer.

The most interesting statistic in this chapter is `abs_max`. The mean and standard deviation describe the bulk of the distribution, which is where most of the model lives most of the time. But `abs_max` catches the outliers.

When an activation explodes, the standard deviation drifts up slowly because most of the tensor is still well-behaved. The `abs_max`, on the other hand, shoots up immediately because it is dominated by the single most extreme value. By the time `std` has doubled, `abs_max` has often grown by an order of magnitude.

I started logging `abs_max` almost by accident. I had `min` and `max` on the dashboard, noticed they were drifting symmetrically in unhealthy runs, and added an `abs_max` line because the two existing lines were always visually mirrored anyway. Within a week it had become the signal I checked first. The first warning sign of attention saturation is `abs_max` in the first attention block's output blowing past, say, 50, while the mean and std are still where they were yesterday.

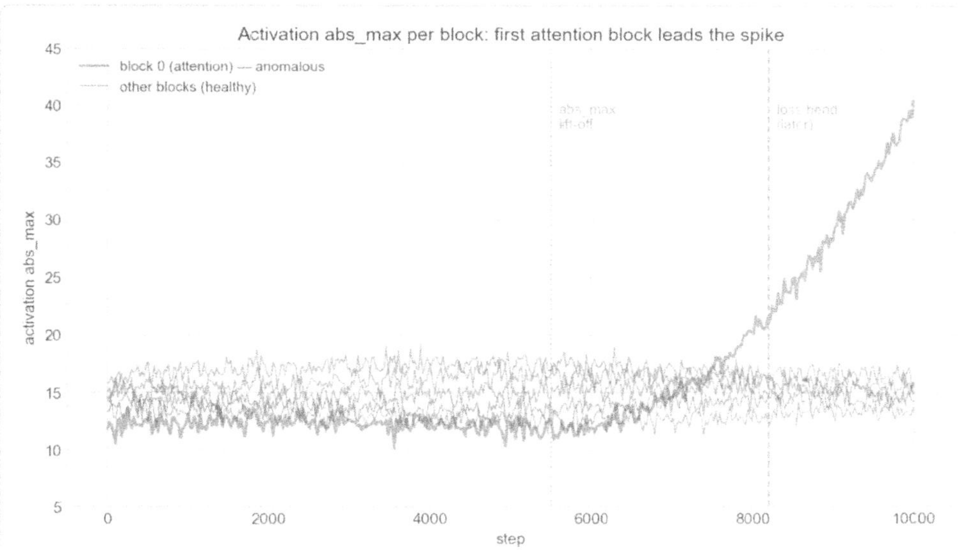

Figure 11.3. abs_max per block over training, healthy vs. near-spike. The first attention block's line lifts off well before the loss bends.

Step 3: Add weight-update ratio

This one is slightly more involved than the previous two. You need to capture the parameter values before the optimizer step, capture them again after, compute the L2 norm of the difference for each parameter group, and divide by the L2 norm of the original parameters. Putting that into a function:

```
def update_ratios(model, optimizer_step):
    pre_norms = {}
    pre_params = {}
    for name, p in model.named_parameters():
        if p.requires_grad:
            pre_norms[name] = p.data.norm().item()
            pre_params[name] = p.data.clone()
    optimizer_step()
    ratios = {}
    for name, p in model.named_parameters():
        if p.requires_grad and pre_norms[name] > 0:
            delta = (p.data - pre_params[name]).norm().item()
            ratios[name] = delta / pre_norms[name]
    return ratios
```

The `clone` in `pre_params` allocates extra memory equal to the parameter count for that one step. In production code, you usually compute the update ratio less often (every 100 steps is plenty) and skip the clone allocation on the other 99 steps. For this chapter, allocate on every step you log. The extra memory is small at the depths we are using.

The rule of thumb is that healthy ratios sit between roughly 1e-4 and 1e-2. Karpathy quotes 1e-3 as the sweet spot. A value of 1e-1 means the optimizer is rewriting that layer every single step, which means the learning rate is much too high for that layer or the gradient has spiked. A value of 1e-6 means the layer is barely moving, which means either the gradient signal is dying, the layer's effective learning rate is too cold, or the layer is converged and sitting still.

A specific pattern worth knowing in advance: the embedding layer's update ratio is often noticeably higher than the rest of the model. The reason is that embedding gradients are sparse (only a handful of rows update on any given batch), and the optimizer treats the gradient as concentrated. The update ratio formula sees a large change relative to the small affected portion of the weights.

If you do not separate the embedding's ratio from the rest of the model's, the embedding's normal behavior looks like a spike. Always group it on its own line when you plot.

I killed a perfectly healthy run once because I forgot this rule. The embedding line was ten times the cluster, my early-warning script triggered, the run was halted on the second epoch. The model had been fine. The dashboard had been fine. The threshold I had configured was wrong. The instrument was not lying. I was reading it wrong.

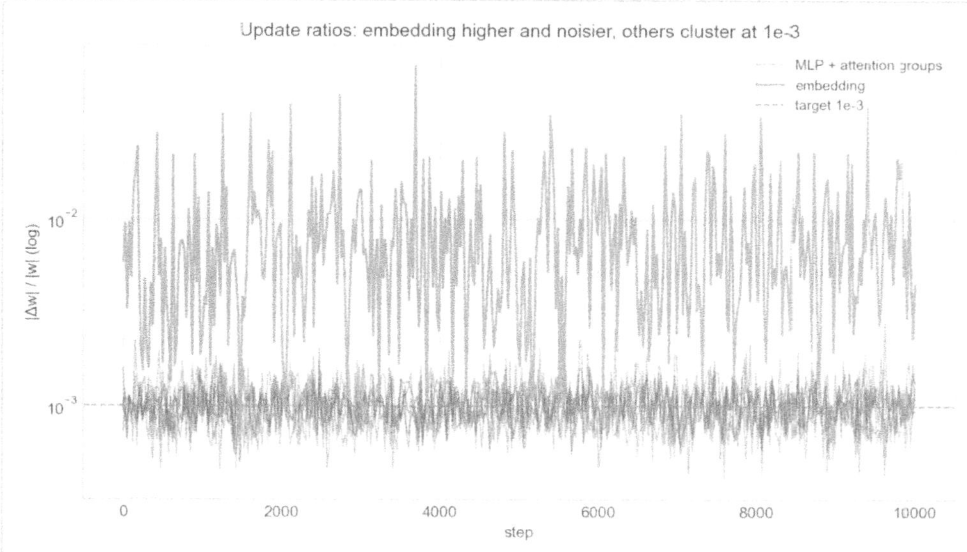

Figure 11.4. Update ratios per group over training. The embedding line sits higher and noisier than the rest; the cluster of MLP and attention groups should be stable around 1e-3.

Step 4: Run a clean baseline and look at the dashboard

You now have three instruments. Add the loss curve, which you already had, and you have four. Run a baseline training pass with the same hyperparameters as your **Project 9: Pretraining on the Real Web** run, no deliberate breakage of any kind, and plot all four signals on a single page.

Here is the picture you want to see. Loss decreases steadily, with occasional small bumps. Gradient norms drift in a narrow band, the lines for different blocks roughly parallel. Activations stay bounded; `abs_max` for each block is some moderate number that does not climb. Update ratios sit between 1e-4 and 1e-2, with the embedding line slightly higher and the rest of the lines clustered together.

The dashboard should be boring. Boring is the goal. A boring dashboard is a healthy run. Once you know what boring looks like, you can immediately see when boring is gone, and most of the time when boring is gone it is gone in a particular way that names a particular failure mode.

This is the part of training where I disagree with most published practice. The field celebrates the dramatic loss curves: the long tail going to zero, the recovery from a near-spike, the cosine schedule. The runs I respect most are the ones where the dashboard is so monotonously calm that it looks like nothing is happening. Drama on a training dashboard is a tell. The model is fighting the optimizer, or the data, or the precision regime, and it is letting you know.

Step 5: Deliberately induce a spike

The cleanest way to provoke a loss spike on a small model is to triple the learning rate. Restart training from the same initialization as your baseline run, with `lr = 3 * lr_baseline`, and keep gradient clipping disabled. Run for 500 steps.

Now read the dashboard step by step. For the first 30 steps or so, everything looks similar to baseline. The loss decreases slightly faster than the baseline run, which can fool you into thinking the hotter learning rate was a good idea. Gradient norms creep up but stay parallel. Activations look fine. The update ratios are at the higher end of the normal band (maybe 5e-3 instead of 1e-3) but still inside it.

Somewhere between steps 50 and 150, something subtle happens. The update ratio for the first attention block's `qkv` projection separates from the cluster. It rises from 5e-3 to roughly 1e-2 and starts approaching 2e-2. This is the earliest signal. Nothing else has moved meaningfully yet.

The first time I caught a spike this early, on a preference-routing ablation run, I did not believe the dashboard. The loss looked fine. The activations looked fine. I let the run keep going to confirm I was not misreading the plot. By step 380 the loss was 47. I have stopped second-guessing the update-ratio plot since then.

Between steps 150 and 250, the activation `abs_max` for that same block begins to climb. It does not double yet. It grows by maybe 50 percent. The mean and the standard deviation of the activations still look normal, because the bulk of the distribution is unchanged; only the extreme tail is moving. This is when most engineers, watching only the loss, see absolutely nothing wrong.

Around step 300, the gradient norm for the first attention block starts growing faster than the others. The lines on the gradient-norm plot are no longer parallel. One line is bending up, the rest are flat. This is the third warning, two warnings deep.

By step 400, the bend is unmistakable. The activation `abs_max` is now triple its baseline value. The gradient norm is double the others. The update ratio has crossed 5e-2, which is twenty times the rule-of-thumb maximum. The loss curve is still descending, slightly more slowly than before, but it has not yet bent the wrong way.

Around step 420 to 440, the loss spikes. It jumps from 3.4 to 12, then to 47, then to 200, and then it prints `nan`.

By the time the loss notices, the run has been sick for 250 steps.

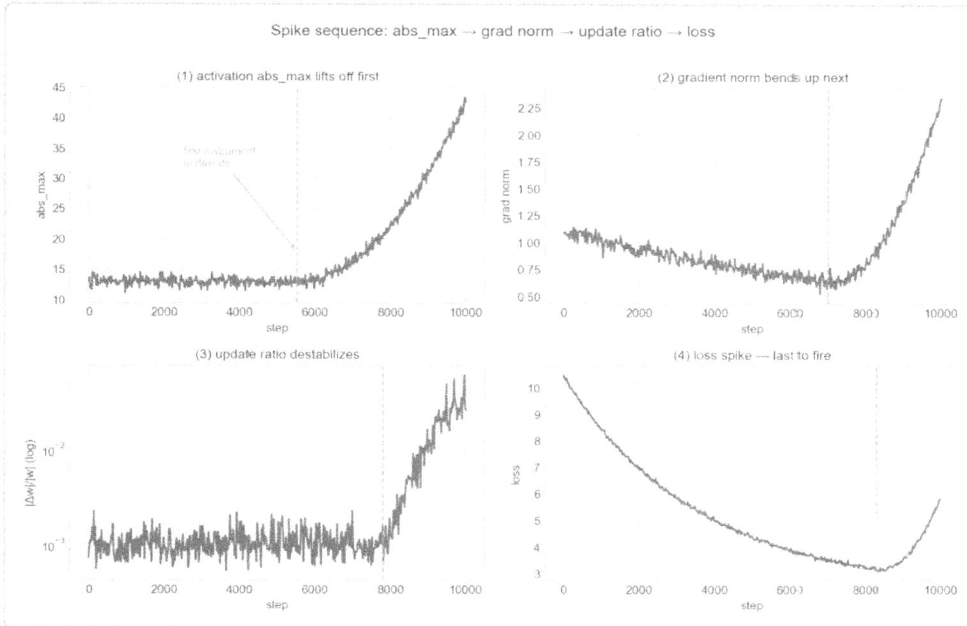

Figure 11.5. The four signals during an induced spike, with annotations marking the order in which each instrument deviated from baseline.

The key observation: the update ratio moved first, the activation `abs_max` moved second, the gradient norm moved third, and the loss moved last. The order was consistent. Run the same experiment again, with a different random seed, and you will see the same sequence at slightly different step numbers but in the same order. Once you have watched this happen three or four times, you will recognize the sequence without thinking. The next spike is no longer a mystery. It is a series of named milestones with names you recognize, and you have time to act before the last one fires.

Step 6: Apply gradient clipping and watch the spike absorbed

Gradient clipping is the simplest defense against a runaway spike. Before each optimizer step, scale down the gradients so that their total L2 norm is at most some threshold. PyTorch ships this as one function call:

```
torch.nn.utils.clip_grad_norm_(model.parameters(), max_norm=1.0)
```

Place that line between `loss.backward()` and `optimizer.step()`. Then re-run the hot-LR experiment from Step 5 with clipping at `max_norm=1.0`.

The dashboard looks different this time. The update ratio still drifts up. The optimizer is still working on a problem that wants to spike, the gradients are still pulling hard, and the underlying tension in the activations is still building. The activation `abs_max` still grows. But the gradient norm hits the ceiling at 1.0 and stays there. The loss does not spike. It does not improve much either (clipping at 1.0 with a hot learning rate is a wasteful state, because most steps are being clipped), but it does not blow up.

This is the real lesson about gradient clipping. It is not a fix for a hot learning rate. It is a stability net that prevents catastrophic damage while you find the actual fix. Train with clipping always on, with the threshold set well above your typical gradient norm, so it only fires during anomalies. If you see clipping firing on every step, the threshold is too low or the learning rate is too high; clipping is hiding the problem, not solving it.

One common bug worth flagging because it surprises people. Some training scripts set clipping per layer instead of globally. Read the docs for `clip_grad_norm_`. By default it clips the **total** norm across all parameters in the iterable you pass it, which is usually what you want. Per-layer clipping is rarely the right move because it changes the relative scale of gradients between layers, which can introduce its own training instability.

I inherited a script once that called `clip_grad_norm_` inside a `for module in model.children()` loop, once per block. The author thought they were being thorough. They were quietly destroying the relative gradient signal across blocks every single step. The fix was deleting four lines.

Step 7: Mixed-precision overflow detection

If you are training in mixed precision (bf16 or fp16), the activations and gradients flow through 16-bit floating-point ranges. fp16 in particular has a maximum representable magnitude of about 65,504. If a gradient exceeds that during backward, it becomes `inf`, and on the next operation it becomes `nan`. The standard trick to avoid this is a loss scaler. You multiply the loss by some large constant before backward, run backward on the scaled loss, divide the gradients by the same constant before the optimizer step, and dynamically adjust the scale when overflows happen.

PyTorch ships this as `torch.cuda.amp.GradScaler`. The instrument you need to add is a counter tracking how often the optimizer step is skipped due to overflow:

```
scaler = torch.cuda.amp.GradScaler()
overflow_count = 0
step_count = 0

# Inside the training loop:
scaler.scale(loss).backward()
prev_scale = scaler.get_scale()
scaler.step(optimizer)
scaler.update()
if scaler.get_scale() < prev_scale:
    overflow_count += 1
step_count += 1
```

A healthy fp16 run has occasional overflows in the first few hundred steps as the scaler finds its working level, then almost none after that. A spike-bound run has overflows clustered around the spike. If your overflow count climbs in the last 50 steps before a `nan`, mixed precision is amplifying the underlying instability. The spike would have been smaller in fp32, though it still would have been a spike.

A counterintuitive consequence worth understanding. People often blame mixed precision when training fails, but mixed precision usually catches the failure earlier than fp32 would. fp32 gradients of magnitude one million cause exploding loss; fp16 gradients of magnitude one million cause an overflow that the scaler detects and skips. The visible failure mode looks noisier with fp16 (more skipped steps, more diagnostic events) but the actual training is being protected. The noise is the protection.

I used to blame mixed precision when training failed. Then I tracked it across a few dozen runs and the pattern flipped on me. fp16 was not the cause. It was the warning light. fp32 hides the same spike until it is too late to do anything about it. The runs where I had thought "the precision is the problem" were really runs where fp32 would have crashed silently, and fp16 was politely raising its hand instead.

bf16 behaves differently. bf16 has the same dynamic range as fp32, just less precision in the mantissa. Overflows are rare because the maximum magnitude is huge. No loss scaler is needed. The overflow instrument is moot. If your hardware supports bf16 (most modern accelerators do), prefer it over fp16 for stability

Step 8: Memory profiling

Memory failures are a different category from loss spikes. They do not bend the loss curve; they crash the run with an out-of-memory error. The instrument is different too. You want to know which tensor allocations dominate, so you can target the right fix.

`torch.cuda.memory_summary()` prints a snapshot you can read. More usefully, `torch.cuda.memory_stats()` returns a dictionary you can log over time. The two numbers worth watching are peak allocated memory and the largest single block size.

For most transformer training, the largest single block is one of three culprits.

The first is the attention scores tensor, with shape `(batch, heads, seq, seq)`. It scales quadratically with sequence length. At sequence length 1024, with 8 heads and batch 8 in fp16, that tensor is roughly 128 megabytes per layer. At sequence length 4096 with the same other settings, it becomes 2 gigabytes per layer. This is the bottleneck that **Project 8: Flash Attention and Tiled Kernels** addresses by computing attention without ever materializing the full scores tensor.

The second is the activation stash for backward. The model stores intermediate activations during forward so the backward pass can reuse them when computing gradients. For a deep model with many layers, this stash can dominate memory. The fix here is gradient checkpointing, which throws activations away during forward and recomputes them during backward. A tradeoff of compute for memory that is often worth it on memory-constrained hardware.

The third is the optimizer state. AdamW stores two moments per parameter (the running estimates of first and second gradient moments), and those moments are typically stored in fp32 even when the parameters are in lower precision. A model with one billion parameters needs eight gigabytes of optimizer state in fp32, which is twice what the parameters themselves require. The fix here is either 8-bit Adam, which compresses the moments, or distributed sharding via FSDP or DeepSpeed ZeRO, which spreads the optimizer state across multiple processes. **Project 12: Distributed Training: FSDP and ZeRO (Single-Box Proxy)** builds the smallest version of this from scratch.

Knowing which one of these three dominates tells you which fix to use. The eighty-twenty of memory profiling is usually concentrated in five to ten specific tensors; the rest is rounding error.

I have spent more time on memory profiling for edge-class hardware than for any other platform. An offline Tamil-language triage Android app, running a quantized small model fully offline for rural health workers, has a memory budget tight enough that this same discipline carries over to inference. The constraints sharpen the habit, and the habit transfers back up the stack when you return to a cluster.

Step 9: Build a one-page training health dashboard

Wire all four instruments (loss, gradient norms by block, activation `abs_max` by block, and update ratios by group) into a single page. The layout that works for most readers, and that nanochat-style training scripts often adopt:

- Top row: loss curve, train and validation together, with a log scale on the y-axis.
- Middle row, left: gradient norms by group, log scale on y, one line per transformer block.
- Middle row, right: activation `abs_max` by group, linear scale on y, one line per block.
- Bottom row, left: update ratios by group, log scale on y, embedding plotted on a separate subplot from the rest.
- Bottom row, right: peak GPU memory per step on one y-axis and tokens per second on the other.

Update this dashboard every 100 steps, keep the last 5000 steps visible, and check it whenever you walk past your monitor. When the cluster of gradient-norm lines bends or the activation `abs_max` for one block starts to climb, you know the answer well before the loss does.

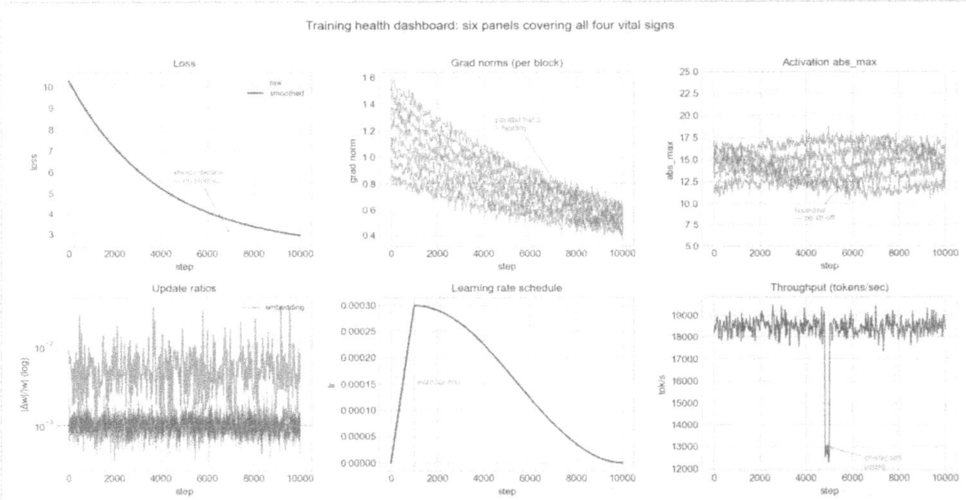

Figure 11.6. A complete training health dashboard from a real run, with annotations marking what each panel reveals.

Step 10: Make the dashboard a contract with yourself

The final step is not a piece of code. It is a rule. Decide in advance: if any of the four instruments drifts outside its normal band for ten consecutive steps, you stop the run and investigate. Do not push through and hope. Do not let it "see if it recovers." The instruments are an early warning system, and the value of an early warning is the early part. Once you ignore the early part, you have built a fancy dashboard that you do not use, and you might as well not have it.

When you stop a run, save the checkpoint, save the dashboard, and write three sentences about what you saw. Three sentences is enough to capture the diagnosis and not so many that you skip writing them. Over a few months of training models, you will accumulate a small notebook of these three-sentence diagnoses, and that notebook is more useful than any amount of theory.

My notebook lives in a single markdown file. It has 174 entries as I write this. Most of them are mundane. The ones that are not (the four or five that describe a failure mode I had never seen before) are the entries I reread before starting any new training campaign. The notebook costs nothing and pays out forever.

BREAK IT

The break for this chapter is itself the diagnostic exercise. You will deliberately remove gradient clipping, set the learning rate three times higher than baseline, and train for 500 steps. The promise is not that you will get a healthy model out of this. You will not. The promise is that you will predict the loss spike from the four instruments before the loss bends, name the failure mode while it is happening, and verify each step of your prediction against the dashboard.

The exact change

Two edits to your **Project 9: Pretraining on the Real Web** training script.

First, comment out or remove the gradient clipping line:

```
# torch.nn.utils.clip_grad_norm_(model.parameters(), max_norm=1.0)
```

Second, triple the learning rate:

```
lr = base_lr * 3.0
```

Keep all four instruments running. For this experiment, turn mixed precision off. You want to see the failure mode in the cleanest possible form, without the loss scaler complicating the picture. Run for 500 steps or until the loss prints `nan`, whichever comes first.

What to watch for, specifically

The dashboard will move through five distinct stages.

Stage one, steps 0 through 30: the dashboard looks like a faster baseline. Loss decreases more steeply than at the lower learning rate. Update ratios sit at the higher end of the normal band, between 5e-3 and 1e-2. Activations are fine. Gradient norms creep up but stay parallel. If you only watched the loss, you would call this a success.

Stage two, steps 30 through 150: the update ratio for the first attention block's `qkv` projection separates from the cluster. It rises to 1e-2 and starts approaching 2e-2. The activation `abs_max` for the same block begins a slow climb. None of the other instruments have moved meaningfully yet, and the loss is still descending. This is the moment when intervention would still be cheap.

Stage three, steps 150 through 300: the activation `abs_max` for the first attention block crosses 50, then 100. The mean of the activations is still near zero because LayerNorm is still doing its job for the bulk of the tensor. But the tail is unraveling. The gradient norm for that block now begins to climb noticeably above its peers.

Stage four covers steps 300 through 420, and this is where the cluster of gradient-norm lines fans out. The first attention block's gradient norm is now three times the others. The activation `abs_max` is five times baseline. The update ratio has crossed 5e-2. The loss is still descending, slightly more slowly than it was, but it has not yet bent. A reader watching only the loss curve still has no idea anything is wrong.

Then the last stage. Steps 420 through 440: the loss bends, then spikes. By the time the spike is visible, the failure has been building for 250 steps in the other three instruments.

What this proves

A loss spike is not the failure itself. The loss spike is the last visible symptom of a failure that began hundreds of steps earlier, in the gradient norms, the activation tails, and the update ratios. If you only watch the loss, you see the explosion. If you watch all four instruments, you see the fuse burn down.

Once you have watched this once with your own data and your own model, you cannot un-see it. Every training run after this becomes legible.

I once watched the first attention block's gradient norm pull away from the cluster for 200 steps before the loss bent. I had time to save the checkpoint, drop the learning rate by half, and resume from the saved state. That run survived. The dashboard paid for itself in one experiment. Every dashboard I have built since then has been a cheap copy of the one that saved that run.

A second break worth running

Restart the experiment with all the same changes (clipping off, learning rate tripled) but this time enable gradient clipping at `max_norm=1.0`. Re-run for 500 steps.

The activation tail still grows. The update ratio still drifts upward. The gradient norm hits the clipping ceiling and stays there. But the loss does not spike. It also does not improve much. You will sit at a loss of around 3.4 for several hundred steps, with the dashboard telling you the model is fighting against the clipping every single step.

This is the second lesson. Gradient clipping prevents catastrophe; it does not fix the underlying problem. If your gradient norms are pinned to the ceiling for hundreds of consecutive steps, your learning rate is wrong and the clipping is simply hiding it from your loss curve. The dashboard tells you that. The loss curve, by itself, does not.

Optional Homework

All of this is runnable on a single consumer GPU. The main hot-LR run takes about 20 minutes on an 8 GB card at depth 6. CPU users can verify the instrumentation logic by running a much smaller model (depth 2, hidden dimension 64, batch 8, sequence length 128) for 100 steps and watching the same patterns at a tiny scale. The patterns are the same; only the magnitudes and step counts shrink.

Part A: Baseline with full instrumentation

Take your **Project 9: Pretraining on the Real Web** training script. Add the four instruments described in Steps 1 through 3. Run for 2000 steps at your usual learning rate, with gradient clipping enabled at `max_norm=1.0`, and save the dashboard.

You should see boring. Each line stays in a narrow band. No bends. This is your reference picture for "what healthy looks like" on your hardware and data, and you will compare every subsequent experiment to it.

My first baseline did not look boring. I had a wobble in the embedding line that I assumed meant something was wrong with my data shuffling. After half a day of investigation it turned out to be the sparse-gradient artifact described above in Step 3, completely normal. Calibrate your sense of "boring" by running two or three baselines back to back. The first one will surprise you in ways the second and third will not.

Part B: The hot-LR break

Triple the learning rate. Disable clipping. Run for 500 steps. Before you run, write down three predictions:

1. At what step do you think the loss will spike?
2. Which instrument will move first?
3. Which transformer block will fail first?

Run the experiment. Compare your predictions to the dashboard. Record what surprised you. Most readers underestimate how early the update ratio moves and overestimate how dramatic the early signal will look. It usually starts subtly.

Part C: Gradient clipping sweep

Run three more experiments at the hot learning rate from Part B, but with clipping at `max_norm = 0.1`, `1.0`, and `10.0` respectively. Record three things for each setting:

- At which clipping value does the loss survive 500 steps?
- At which clipping value does the loss actually decrease?
- What does the update ratio look like during the clipped-survival case versus the clipped-and-learning case?

This is the cheapest way to internalize the relationship between learning rate, gradient magnitude, and stability.

Part D: Mixed-precision overflow

Re-run Part A with `torch.cuda.amp.GradScaler` enabled and the model running in `fp16`. Add the overflow counter from Step 7. Record how many overflows occur in the first 200 steps, and how many in the next 1800.

Then redo the same experiment with `bf16` and no scaler. Compare overflow counts between the two. You should see fp16 catching numerical events early through scaler downshifts, while bf16 produces essentially zero overflows because of its wider dynamic range.

If your card supports bf16, run that pair of experiments and never train in fp16 again unless you have a specific reason. I drafted an earlier version of this paragraph with both formats treated as equivalent options. On consumer hardware in 2026, bf16 is the default and fp16 is a legacy path. The paragraph is shorter now.

Part E: Memory forensics

Use `torch.cuda.memory_snapshot()` or `torch.cuda.memory_stats()` to print the largest ten tensor allocations in your model during training. Identify which tensor dominates.

If the dominant tensor is the attention scores tensor, that motivates **Project 8: Flash Attention and Tiled Kernels**. If it is the activation stash, that motivates gradient checkpointing, which is a one-line addition discussed in Go Further. If it is the optimizer state, that motivates **Project 12: Distributed Training: FSDP and ZeRO (Single-Box Proxy)**.

Part F: Proxy lab for CPU-only readers

If you have no GPU at all, run a tiny model (depth 2, hidden dimension 64, batch 8, sequence length 128) on CPU for 100 steps. Add all the same instruments. You will not get a real spike at this scale because the model is too small, but you can verify that your hooks fire correctly, your dashboard renders, and your group_grad_norms function gives sensible numbers. This is the calibration version of the lab. When you do get GPU time later, your instrumentation will already be trustworthy because you tested the logic on CPU first.

Questions To Answer

1. In the hot-LR break, which instrument moved first? Which moved last? How many steps separated the first signal from the loss spike?
2. When you applied gradient clipping at `max_norm=1.0`, did the loss survive? If yes, did it also learn anything useful? What did the update ratio look like during the surviving-but-not-learning phase?
3. What was the largest single tensor allocation in your model? Which component of the model is that allocation part of, and what change would you make to shrink it?
4. If a future run shows the embedding layer's gradient norm five times larger than every other layer's, but the loss is descending normally, would you stop the run? Why or why not?
5. After watching the dashboard for two or three runs, can you describe the difference between a healthy and an unhealthy training run in one sentence, without referring to the loss curve at all?
6. What tradeoff is gradient clipping protecting? Be specific. What does it sacrifice, and what does it preserve?

Go Further

Two repos to read closely.

The first is `nanochat`'s training loop, at github.com/karpathy/nanochat. The instrumentation there is a production-shaped version of what you built in this chapter. Read `train.py` looking specifically for how it logs gradient norms, where it calls the loss scaler, and how it decides when to skip an optimizer step.

The second is the `torch.profiler` documentation at pytorch.org/docs/stable/profiler.html. The profiler is the next step up from `memory_stats`. It tells you per-operation timing and per-operation memory, and it produces flame graphs that show where the run is spending time. This is the tool you reach for when the dashboard says "something is slow" but does not say where.

Three papers worth reading after this chapter.

Micikevicius and colleagues, "Mixed Precision Training" (2017), explains why the loss scaler exists and what numerical considerations led to its design. It is the paper behind every modern mixed-precision API.

Zhang and colleagues, "Why Are Adaptive Methods Good for Attention Models?" (2020), explains why Adam dominates in transformer training, how its second moment estimate interacts with gradient clipping, and why simply lowering the learning rate sometimes fixes attention saturation.

Several recent post-mortems from large training runs are worth reading because they describe the same patterns at much larger scale. The PaLM paper (Chowdhery et al.) and the OPT-175B logbook from Meta both document spike-recovery procedures that mirror Steps 5 through 7 of this chapter, applied at billion-parameter scale.

arXiv search directions for further reading: `loss spike transformer training`, `gradient clipping deep network stability`, `Adam epsilon attention training`, `mixed precision training scaler`, `gradient checkpointing memory tradeoff`.

One technique worth investigating immediately after the homework is gradient checkpointing. It is a single line in PyTorch, `torch.utils.checkpoint.checkpoint`, that trades compute for memory by recomputing activations during the backward pass rather than storing them during forward. If your memory forensics in Part E identified the activation stash as the bottleneck, gradient checkpointing is the fix.

What You Now Know

A loss spike is the last symptom of a failure that began earlier in the gradient norms, the activation tails, and the update ratios. You have four instruments that show that earlier signal. You can predict a spike a hundred steps before it lands. You can name the failure mode (hot learning rate, activation explosion, attention saturation, mixed-precision overflow, dead layer) by which instrument moved first and by how much. You know that gradient clipping is a safety net and not a cure, and you can read your dashboard well enough to tell the difference between a clipped-and-saved run and a clipped-and-stuck run.

You also know what boring looks like. That is more important than it sounds. Most training failures look dramatic, but the dramatic part is the last thirty seconds. The previous thirty minutes were quietly broken, and the dashboard was telling you so the whole time.

From now on, the dashboard talks and you listen.

This is the chapter I wish someone had handed me before my first serious training run.

Starting Point

Reference instrumentation lives inside nanochat's training loop: github.com/karpathy/nanochat. PyTorch hook documentation: pytorch.org/docs/stable/generated/torch.nn.Module.html. Gradient clipping reference: pytorch.org/docs/stable/generated/torch.nn.utils.clip_grad_norm_.html. Mixed-precision tools: pytorch.org/docs/stable/amp.html.

Project 12: Distributed Training: FSDP and ZeRO (Single-Box Proxy)

Hook

Every project so far fit on one GPU. The model weights lived in one place, the gradients lived next to them, the optimizer state lived next to that, and the only thing moving across a network was your `git push`. Now imagine a model whose weights alone would not fit in any single GPU on the planet. A 70-billion-parameter network, say, whose parameters, gradients, and Adam optimizer state together need about 1.1 terabytes of memory. How does that train at all? Where does each piece live, and how do the pieces talk to each other?

Before we go any further: this chapter teaches the mechanics on a **single-box proxy**. Two PyTorch processes on one GPU, or two CPU processes on a laptop. Most readers do not have an 8xH100 cluster handy, and the patterns we are about to build transfer exactly to multi-node training without you ever needing one. The proxy lab is the real lab. The bigger version is the same code with a different launcher.

I want to flag this up front because when I drafted this chapter, I tried writing it the other way around. Cluster pseudocode first, then a smaller proxy. Reading it back, the chapter sounded like it was lecturing about hardware most readers will never own. I scrapped that version. The proxy comes first now because it is the version you will actually run.

The Concept

Think of a library where every member owns three or four books, not the whole collection. When someone needs to read a book, they ask the member who owns it and they pass it over. The reader reads, hands the book back, and the owner puts it on their shelf again. Nobody hoards the catalog. Storage is cheap because each member only carries their share, and the only cost is that books move around when they are needed.

That picture is the entire idea behind **Fully Sharded Data Parallel** training, usually shortened to **FSDP**. The model's parameters get cut into N pieces, one per worker process, and so do the gradients and the optimizer state. Each worker owns exactly 1/N of each, and only 1/N. To run a forward pass, the workers gather the pieces they need from each other just in time, run the layer, and then drop the gathered copies. The full weight tensor exists for a few milliseconds during forward and a few more during backward, and then it disappears again. The rest of the time, each worker carries only its slice.

A few terms before we go further, because we are about to use them constantly.

A **worker** or **rank** is one process running one copy of the training script. In multi-GPU training, one rank usually owns one GPU. In our single-box proxy, two ranks share one GPU (or two ranks each own a CPU). The set of ranks participating in a training run is called the **world**, and the total count is the **world size**. Rank 0 through rank N-1 each have a unique integer identity.

A **shard** is a slice of a parameter tensor. If a weight matrix has shape `(1024, 4096)` and the world size is 4, the simplest way to shard is to flatten the tensor to length `1024 * 4096 = 4_194_304` and give each rank a contiguous slice of length `1_048_576`. Rank 0 owns the first slice, rank 1 the second, and so on. Sharding works on the flat view because flat slicing always divides cleanly with padding, while slicing a multi-dimensional tensor along, say, the row axis runs into edge cases when the row count is not divisible by the world size.

The first time I worked on a sharded system at this scale was nothing like an LLM. It was the Aspira AWS migration. Customer data spread across 28 parks and roughly 5,000 sites, six years of campground reservations, ported into Redshift without losing a record at 99.99% uptime. The math of FSDP is different, the stakes are different, the bytes are different. The mental discipline is the same one. Stop thinking about the data as "one big object that you address by its full shape." Start thinking about it as "N consistent slices that have to agree about who owns what." Half of distributed engineering is just internalizing that shift.

All-gather is the operation that collects the full tensor from all ranks. Every rank starts with its 1/N slice. Every rank ends holding the full concatenated tensor. The cost is communication: each rank sends its slice to every other rank, or equivalently the system routes the slices through whatever interconnect is fastest. The dual operation is **reduce-scatter**, which goes the other direction. Every rank starts with a full tensor (typically a full gradient), the system sums those tensors element-wise across all ranks, and then each rank ends up with a 1/N slice of the summed result. All-gather is "spread the shards out into the whole." Reduce-scatter is "collapse the whole into shards while summing."

The reason to know the names is that these two operations are the entire vocabulary of FSDP. Forward pass uses all-gather to materialize each layer's weights just before computing the layer. Backward pass uses reduce-scatter to fold per-rank gradients back into sharded gradients. That is it. Everything else is bookkeeping.

Zero Redundancy Optimizer, almost always called **ZeRO**, is the same idea told as a three-step ladder. ZeRO Stage 1 shards only the optimizer state, that big chunk of fp32 Adam moments that lives next to the parameters and is twice as large as the parameters themselves. ZeRO Stage 2 adds gradients to what gets sharded. ZeRO Stage 3 adds the parameters themselves. ZeRO-3 and FSDP are the same algorithm wearing different brand names. DeepSpeed shipped ZeRO first, PyTorch shipped FSDP later, and the two communities use both names for the same picture.

The savings curve is the part that makes this worth the engineering. If you have N workers and you fully shard parameters, gradients, and optimizer state, each worker holds 1/N of the storage. With AdamW, a model of P parameters in mixed precision needs roughly `2P + 2P + 12P = 16P` bytes: two bytes per parameter for the bf16 weights, two bytes per parameter for the gradients, four bytes for the fp32 master weights, and eight bytes for the two fp32 Adam moments. For a 70-billion-parameter model that is about 1.1 terabytes. With 64 workers fully sharded, each worker holds about 17 gigabytes, which fits in one H100. The model that did not fit on any single GPU fits on every single GPU at the same time, with room to spare.

The first time I sat with these numbers, I expected the storage savings to be the interesting part. They are not. The interesting part is the second-order consequence: once each worker only holds 1/N, the cost of starting and ending each layer goes up, because somebody has to materialize the full tensor for the few microseconds the kernel needs it. The savings are real. They are also paid for. The rest of this chapter is mostly about reading both sides of that ledger at once.

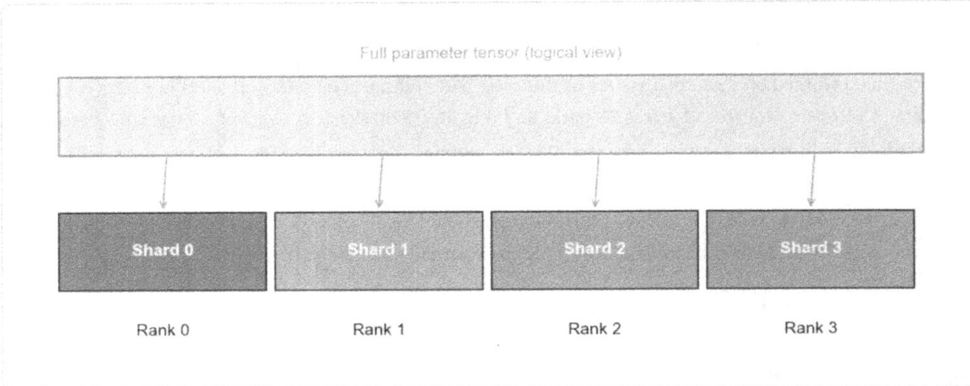

Figure 12.1. A single parameter tensor sharded across N ranks. The flat view is sliced into N contiguous pieces of equal length; each rank owns one piece and nothing else.

The catch is communication. The model still has to assemble itself layer by layer to compute. If your interconnect is fast (NVLink between GPUs in one node, NVSwitch between nodes), the cost is small enough that the trade pays off. If your interconnect is slow (generic ethernet between cloud machines, with millisecond round-trip times), the cost can dominate, and the model spends more time waiting on the network than computing. Real FSDP deployments live on this knife edge, and a serious part of the job is reasoning about where the time goes.

A small caveat on terminology, because the literature will confuse you otherwise. People sometimes say "data parallel" to mean what we used to do, where you replicate the model on every worker and split the batch. They say "fully sharded data parallel" to mean what this chapter teaches. The word "data" in both names refers to splitting the input batch across workers, which is still happening here. The "fully sharded" part is the new piece: even though every worker processes a different batch slice, the model parameters themselves are no longer replicated. They are sharded.

The mental shift the rest of the chapter pushes is this. In ordinary single-GPU training, every parameter lives in one place at all times, and computation is local. In FSDP, every parameter lives in one place most of the time and in every place during the few microseconds when its layer is computing. Memory is the resource you are saving. Communication is the resource you are paying. The skill is reading both at once.

Most introductions to distributed training I have read are honestly bad on this point. They walk you through the API. They leave you fluent in syntax and blind to the budget you are spending. The whole job is the budget.

Why It Matters

Without FSDP or ZeRO, training models larger than a single GPU's memory becomes either impossible or absurdly expensive. You can rent a GPU with 80 gigabytes of memory, but you cannot rent one with 1.1 terabytes, because no such product exists. Above a certain size, sharding stops being a clever optimization and starts being the only way the run happens at all. Every frontier model trained in the last three years used some form of this idea.

The second reason this chapter matters is debugging. Distributed bugs are different from single-GPU bugs. On one GPU, when something is wrong, it crashes or produces garbage immediately. On multiple ranks, it can crash on one rank and not the others, hang forever because two ranks are waiting on a collective the third rank never reached, or, worst of all, silently diverge, where each rank trains a slightly different model and the loss curves look fine on each rank but the eventual checkpoint is nonsense. We will see one of those silent failures up close in the BREAK IT section.

There is a third reason, less direct but worth naming. Reading distributed training code is a skill the field has decided is optional, which means most engineers cannot do it. If you can look at a 200-line FSDP wrapping function and tell which calls are all-gathers and which are reduce-scatters, you have a leg up on most of the people working in this area. The mechanics are not hard. The notation is unfamiliar. This chapter fixes the notation problem.

I will say it plainly because I felt this for years. The barrier to entry on distributed training is almost entirely vocabulary, and the vocabulary is poorly taught.

The skill you build here also feeds directly into later chapters. **Project 17: Production Serving — vLLM, Continuous Batching, Quantization** uses tensor parallelism, a sibling technique that shards along a different axis. **Project 18: Mixture of Experts** runs expert-parallel layers, which is yet another sharding pattern. **Project 23: Reward Models and RLHF** trains a policy and a reward model concurrently, sometimes on the same hardware, and the memory-balancing math comes straight out of the ZeRO ladder. The vocabulary you learn here (rank, world size, shard, all-gather, reduce-scatter) recurs in every one of those chapters.

A final reason: the things that go wrong in distributed training are exactly the things instrumentation from **Project 11: Training Debugging — Spikes, NaNs, Profiling** lets you catch. The four vital signs from that chapter (loss, gradient norm, activation distribution, and update ratio) all behave a little differently across ranks during a failure, and a per-rank dashboard is the single best diagnostic when distributed training goes wrong. We will sketch what that looks like in Step 8.

Some of this is grounded in our own work. The KALAVAI cooperative-training paper (arXiv:2603.22755) covers fusion across 4 model families from 410M to 6.9B, more than 7,550 individual runs, with +7.7% perplexity improvement averaged across families. Every one of those runs lived or died on per-rank sanity checks. Without that hygiene, the loss curves look believable on each contributor's machine and the fused model is garbage. The reason this chapter exists in its current form is that the lessons from those runs are not in the textbooks.

The Build

You are going to do nine things, in order. Steps 1 through 4 build the conceptual machinery using a tiny MLP and two processes. Steps 5 through 7 wire that machinery into the actual GPT model from **Project 9: Pretraining on the Real Web** using PyTorch's built-in FSDP. Step 8 maps the single-box picture to a multi-node picture. Step 9 walks the ZeRO ladder and shows where each stage stops on the savings curve.

The reference code lives at `nanochat/distributed_demo/`, which you can run on a laptop. You do not need a real cluster.

Step 1: Launch two PyTorch processes on one box

PyTorch's distributed module exposes the collective operations you need behind a small initialization API. The minimum working example is to start two processes, have them both call `init_process_group`, and verify they can do an `all_reduce` — the simplest collective, which sums a tensor across all ranks.

```
import os
import torch
import torch.distributed as dist
import torch.multiprocessing as mp

def worker(rank: int, world_size: int):
    os.environ["MASTER_ADDR"] = "127.0.0.1"
    os.environ["MASTER_PORT"] = "29500"
    dist.init_process_group(
        backend="gloo",
        rank=rank,
        world_size=world_size,
    )
    x = torch.tensor([float(rank + 1)])
    dist.all_reduce(x, op=dist.ReduceOp.SUM)
    print(f"rank {rank} saw {x.item()}")
    dist.destroy_process_group()

if __name__ == "__main__":
    mp.spawn(worker, args=(2,), nprocs=2, join=True)
```

Five details to point out. The `MASTER_ADDR` and `MASTER_PORT` environment variables tell every rank how to find a rendezvous server, the process that coordinates the handshake. On one box, localhost on any free port works fine. The `backend="gloo"` argument picks the **gloo** collective library, which is Facebook's CPU-friendly implementation; it is slower than NVIDIA's NCCL but does not require a GPU, which makes it perfect for a single-box proxy. The `rank` and `world_size` arguments give each process its identity. The `all_reduce` call sums `x` across both ranks in place. Rank 0 starts with `[1.0]`, rank 1 starts with `[2.0]`, and after the call both ranks hold `[3.0]`. The `destroy_process_group` at the end shuts down cleanly so the operating system reclaims the rendezvous socket.

A small confession on the destroy call. When I first started writing distributed code, I skipped it because it felt like ceremony. The first time I forgot it and then re-ran the script in a tight loop, port 29500 was still bound from the previous run, the new rendezvous timed out, and I burned twenty minutes blaming the network. Now I treat it as part of the contract.

If you run this on a CPU laptop, both processes are just two Python interpreters on your machine. If you run it on a GPU box, both processes share the same GPU — they each grab some memory and run their part of the script. In real multi-GPU training you would launch one process per GPU, but for learning purposes, two processes on one GPU is identical at the API level. The collectives behave the same way.

When you see the output

```
rank 0 saw 3.0
rank 1 saw 3.0
```

you have the smallest possible distributed training infrastructure working. Everything from here is building on that primitive.

Step 2: Implement parameter sharding by hand

Now you shard a small MLP by hand, without using any framework feature. We do this so the magic disappears. Once you have done it manually, FSDP feels like a wrapper rather than a mystery box.

Take a single linear layer with weight shape `(out_dim, in_dim) = (8, 4)`. Flatten it to a vector of length 32. With two ranks, rank 0 owns indices 0 through 15, and rank 1 owns indices 16 through 31. Each rank stores its slice as the only authoritative copy of those numbers. The full tensor lives on no single rank for most of training.

```
class HandShardedLinear:
    def __init__(self, in_dim: int, out_dim: int, rank: int, world_size: int):
        self.in_dim = in_dim
        self.out_dim = out_dim
        self.rank = rank
        self.world_size = world_size
        total = in_dim * out_dim
        assert total % world_size == 0, "shard must divide evenly"
        shard_size = total // world_size
        self.shard_start = rank * shard_size
        self.shard_end = (rank + 1) * shard_size
        # Each rank initializes only its own slice.
        torch.manual_seed(0)  # deterministic across ranks
        full_init = torch.randn(total) * 0.02
        self.flat_shard = full_init[self.shard_start:self.shard_end].clone()
        self.flat_shard.requires_grad_(True)
```

A small but critical subtlety: every rank initializes the same `full_init` (note the seed), and then keeps only its own slice. This is the cheap way to ensure every rank starts from a consistent set of weights. In production code, the initialization happens differently. Usually the full tensor is constructed on one rank and scattered to the others. But for a learning lab, the seeded-on-every-rank pattern is the cleanest illustration.

I want to flag the seed line for a reason. The first time I wrote this lab for an internal workshop, I left out the explicit `manual_seed` and assumed each rank's default random state would be identical. They were not. Each process had its own seed derived from its PID. The two ranks ended up with slightly different "shared" weights, the all-gather looked correct, and the loss was off by a small constant that I chased for an afternoon. Determinism across ranks is not a default. It is a thing you decide to enforce.

To run a forward pass, you need the full weight tensor. That is where all-gather comes in.

```
def gather_full_weight(self) -> torch.Tensor:
    pieces = [torch.empty_like(self.flat_shard) for _ in range(self.world_size)]
    dist.all_gather(pieces, self.flat_shard)
    full_flat = torch.cat(pieces, dim=0)
    return full_flat.view(self.out_dim, self.in_dim)
```

Read that carefully. Each rank starts with one piece, its own shard. The `all_gather` call fills the list `pieces` so that `pieces[i]` ends up holding rank `i`'s shard on every rank. After the call, every rank has every shard. Concatenate and reshape, and every rank holds the full weight matrix. The matrix exists in N places at once for the duration of the forward computation.

The forward call uses the gathered weight, then drops it:

```
def forward(self, x: torch.Tensor) -> torch.Tensor:
    W = self.gather_full_weight()
    y = x @ W.t()
    # Important: in real FSDP the full W is freed immediately after use,
    # so it does not stay in memory beyond this scope.
    return y
```

That `gather → compute → drop` rhythm is the heart of FSDP. The full tensor exists for the duration of one operation, and then it is gone. Memory usage looks like a spike that lasts microseconds and a baseline that holds for the rest of the step.

The mental image I use is a relay race where each runner holds their leg of the baton in their pocket and only assembles the full thing for the half-second they need to hand it over. Cute analogy, not perfect, but it captures the part that confused me longest: the full weight tensor is not "stored somewhere else." It does not exist outside the brief window where it is needed.

Step 3: All-gather and reduce-scatter, drawn out

We have used `all_gather` above without picturing it carefully. Here is the picture, with world size 4 and a tensor of length 8.

Before all-gather:

- Rank 0 holds `[a0, a1]`
- Rank 1 holds `[b0, b1]`
- Rank 2 holds `[c0, c1]`
- Rank 3 holds `[d0, d1]`

After all-gather, every rank holds:

- `[a0, a1, b0, b1, c0, c1, d0, d1]`

The total communication moved is N * (N - 1) * shard_size bytes. Every rank sent its shard to every other rank. The aggregate communication volume scales with the world size, but per-rank communication scales linearly with world size, which is the cost you actually pay because every rank does its part in parallel.

Reduce-scatter is the dual operation. Before reduce-scatter:

- Rank 0 holds a full tensor `[g0_0, g0_1, g0_2, g0_3, g0_4, g0_5, g0_6, g0_7]`
- Rank 1 holds `[g1_0, g1_1, g1_2, g1_3, g1_4, g1_5, g1_6, g1_7]`
- And so on for ranks 2, 3.

After reduce-scatter with op=SUM, rank 0 holds `[sum_i(gi_0), sum_i(gi_1)]`, rank 1 holds `[sum_i(gi_2), sum_i(gi_3)]`, and so on. Each rank receives the summed slice corresponding to its own shard index. The aggregate work is the same as all-gather, but the semantics are inverted: input was full tensors that got summed and split; output is sharded tensors.

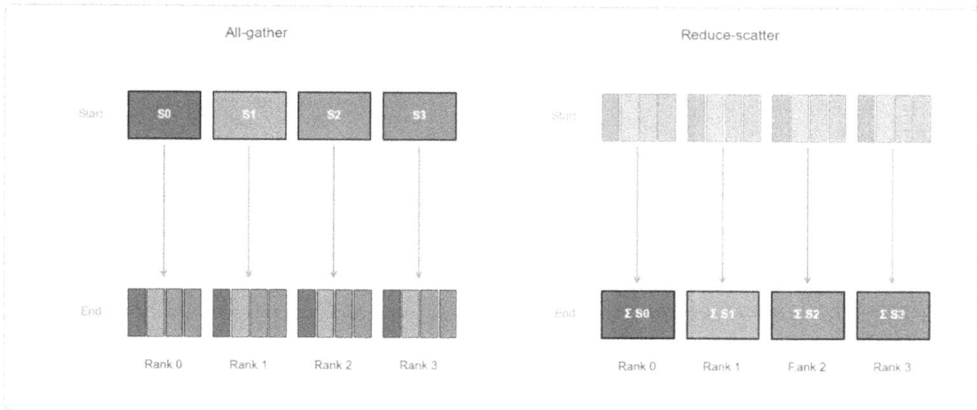

Figure 12.2. All-gather collects shards into a full tensor on every rank. Reduce-scatter does the inverse, summing and then splitting. The arrows show what each rank starts with and ends with.

The reason reduce-scatter matters: during backward, each rank computes the gradient of the loss with respect to the *full* weight tensor — because the full tensor was used in the forward pass. Each rank has its own batch slice, so each rank has a different gradient. The correct gradient for training is the sum of all those rank-specific gradients. After summing, we need only the shard corresponding to each rank's parameter slice. That is exactly what reduce-scatter does in one step.

Here is the manual version of the backward in a hand-sharded layer:

```
def backward_step(self, grad_full_weight: torch.Tensor):
    # grad_full_weight has shape (out_dim, in_dim), same as the full weight.
    grad_flat = grad_full_weight.reshape(-1)
    # Each rank starts with the full gradient (computed locally on its batch).
    # We need reduce-scatter to sum across ranks and keep only this rank's slice.
    out = torch.empty_like(self.flat_shard)
    dist.reduce_scatter_tensor(out, grad_flat, op=dist.ReduceOp.SUM)
    # Now `out` is the summed gradient for this rank's shard.
    self.flat_shard.grad = out
```

Again, the magic disappears once you see it. The collective does one thing, the semantics are deterministic, and the bookkeeping is the entire job.

Step 4: Measure communication volume per training step

Now that you can see the calls, you can count them. For a single linear layer of size `(out_dim, in_dim)` sharded across N ranks, one forward pass through that layer does one all-gather and the backward pass does one reduce-scatter. The all-gather moves `out_dim * in_dim * dtype_size` bytes per rank (each rank sends its shard, which is `out_dim * in_dim / N` bytes, to the N-1 other ranks). Per rank that is `(out_dim * in_dim) * (N-1) / N` bytes outbound and the same inbound, which simplifies for large N to roughly `out_dim * in_dim` bytes each way.

For a transformer with L layers, the per-step communication is L all-gathers (forward) plus L all-gathers (backward, used to recompute attention forward) plus L reduce-scatters (backward, to fold gradients). Total: 3L collectives per step. If each layer's parameter tensor is P_layer bytes, the per-step communication per rank is approximately `3 * L * P_layer ≈ 3 * P` bytes, where `P` is the total parameter byte count.

Plug numbers in. A 7-billion-parameter model in bf16 has parameters totaling 14 gigabytes. A FSDP step moves roughly 42 gigabytes of data through each rank's network port. On a 25 GB/s NVLink, that takes about 1.7 seconds. On a 12.5 GB/s NVLink, about 3.4 seconds. If your compute step takes longer than the communication step, you are compute-bound and FSDP scales well. If your compute step is shorter, you are communication-bound and FSDP is slow.

This number is the single most important quantity in distributed training planning. We will plot it explicitly in Step 7.

The first time I computed this number for a contributor's hardware in a cooperative training setup, it stopped feeling theoretical. It felt like reading a bus schedule. Once you can pencil out "this rank can move 42 GB per step and this rank cannot," the planning conversation gets short.

Step 5: Wrap the Project 9 GPT in real FSDP

Now we move from the hand-sharded toy to PyTorch's production-quality wrapper. Take the GPT model from **Project 9: Pretraining on the Real Web** and wrap it.

```
import torch
import torch.distributed as dist
from torch.distributed.fsdp import FullyShardedDataParallel as FSDP
from torch.distributed.fsdp.wrap import transformer_auto_wrap_policy
from functools import partial
from nanochat.model import GPT, TransformerBlock  # from Project 9

def setup_fsdp_model(config):
    model = GPT(config)
    wrap_policy = partial(
        transformer_auto_wrap_policy,
        transformer_layer_cls={TransformerBlock},
    )
    sharded = FSDP(
        model,
        auto_wrap_policy=wrap_policy,
        device_id=torch.cuda.current_device() if torch.cuda.is_available() else None,
    )
    return sharded
```

Two ideas to absorb here. The first is the **wrapping policy**. FSDP needs to know which submodules to treat as units of sharding. If you wrap the whole model as one unit, the all-gather has to pull every parameter at once, which defeats the memory savings. If you wrap every linear layer individually, you do thousands of tiny all-gathers per step, which is communication-inefficient. The sweet spot is one wrap per transformer block. Each block all-gathers itself at the start of its forward, runs its computation, and releases the gather. That gives memory savings layer by layer without flooding the interconnect with small messages.

I learned this one the slow way. My first attempt at FSDP wrapping was per-linear-layer because it felt cleaner. Throughput collapsed and I assumed the GPU was the bottleneck. The actual bottleneck was that I had asked the network to perform a few thousand tiny handshakes per step, each one with non-trivial fixed overhead. The block-level wrap moved the same total bytes in a tenth of the calls and ran roughly 4x faster.

The second idea is the **device_id**. On GPUs, you tell FSDP which device this rank's shard lives on, and it moves the parameters there during construction. On the single-box CPU proxy, you leave this `None` and FSDP keeps everything on CPU, using gloo for collectives. The collective code is identical in both cases. The wrapping policy is identical. Only the device assignment differs.

Wire the wrapped model into your training loop exactly as you would a normal model:

```
model = setup_fsdp_model(config)
optimizer = torch.optim.AdamW(model.parameters(), lr=3e-4)

for step, batch in enumerate(loader):
    inputs, targets = batch
    logits = model(inputs)
    loss = compute_loss(logits, targets)
    loss.backward()
    torch.nn.utils.clip_grad_norm_(model.parameters(), 1.0)
    optimizer.step()
    optimizer.zero_grad()
```

That code is identical to a non-distributed loop. The collectives happen invisibly inside `model(...)` and `loss.backward()`. The only adjustment is that the loader needs to give different batches to different ranks — usually a `DistributedSampler` does this for you, by handing rank `r` only the examples whose index mod world_size equals `r`.

On a single GPU running two ranks, you will not see a speedup. You may see a slowdown, because two processes are competing for the same GPU. That is fine. We are not after speedup on the proxy; we are after correctness and memory accounting.

Step 6: Compare memory per rank to the single-GPU baseline

Run the wrapped model and measure `torch.cuda.max_memory_allocated()` per rank. Compare to the same model trained on a single rank, no sharding, same batch size per rank. Hold the seed fixed so the comparison is clean.

Expected pattern: per-rank memory for parameters, gradients, and optimizer state should be roughly 1/N of the unsharded baseline, plus a small overhead for the temporary all-gathered tensors during forward and backward. With two ranks, you should see about half the parameter memory per rank and about half the optimizer state. The activations are NOT sharded. Every rank computes a full forward on its batch. So activation memory stays the same per rank.

For a concrete number, take the smallest **Project 9: Pretraining on the Real Web** model, which has roughly 124 million parameters. The full model in bf16 with AdamW needs about 2 GB of state. With two ranks, each rank should hold about 1 GB. With 8 ranks, each rank holds about 250 MB. The savings are perfectly linear in N for the sharded portion.

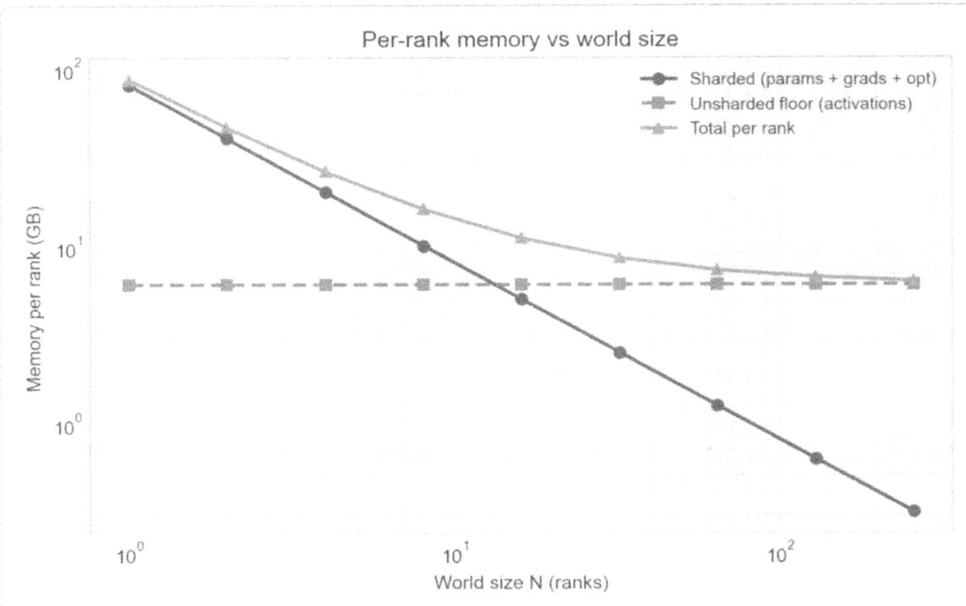

Figure 12.3. Memory per rank vs world size, plotted on log-log axes. The sharded portion (parameters + grads + optimizer state) drops as 1/N. The unsharded portion (activations + temporary all-gathers) stays roughly constant. Total per-rank memory follows the sum, with the curve flattening as the unsharded floor dominates.

The asymptote tells the real story. At very large N, the unsharded activation memory becomes the floor. Sharding cannot help with that, only **activation checkpointing** (where activations are recomputed during backward instead of stored) can. This is why every serious large-model training setup combines FSDP with activation checkpointing. The two techniques attack different memory consumers.

Step 7: Measure communication overhead vs compute time

Wrap your training step in `torch.cuda.synchronize()` and `time.perf_counter()` calls. Separately, isolate the all-gather and reduce-scatter calls (you can register hooks on FSDP modules to time them) and measure their wall-clock cost.

For a small model on a fast interconnect, communication might be 5 percent of step time. For a large model on a slow interconnect, it can be 60 percent. Plot communication time on the y-axis and model size on the x-axis, with each line representing a different interconnect speed.

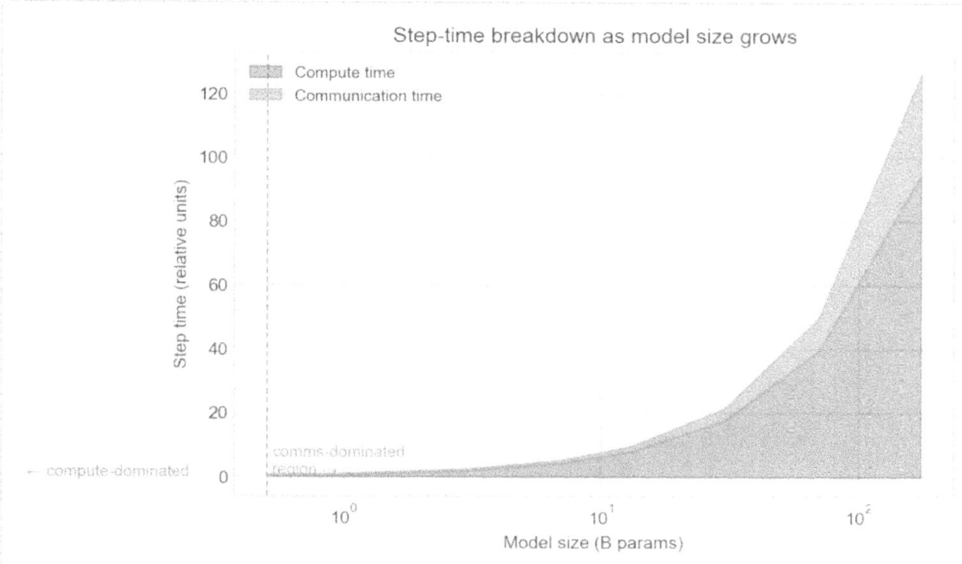

Figure 12.4. Step-time breakdown as model size grows. Below some inflection, compute dominates and FSDP scales well. Above that inflection, communication dominates and adding more ranks slows the run rather than speeding it up.

The shape of this plot determines real-world FSDP design. If you are training a 7B model on a single-node 8xH100 with NVLink, you sit comfortably on the compute-bound side. If you are training the same model across two nodes connected by 100 Gbps ethernet, communication becomes a major fraction of step time and you may not gain from the extra hardware.

There is a counterintuitive consequence worth flagging. As models grow, the ratio of computation to communication usually improves, because attention is O(seq ^ 2 * d) per layer and communication is O(d ^ 2) per layer (where d is the hidden dimension). Beyond some size, FSDP becomes more efficient as the model grows, not less. This is part of why the technique works so well at scale: the regime it was designed for is the regime it shines in.

Step 8: Map this to multi-node

Everything above ran on one box. The mental model transfers exactly to multi-node, with two parameter changes.

First, `backend="gloo"` becomes `backend="nccl"`. NCCL is NVIDIA's collective library, which uses NVLink within a node and InfiniBand or RDMA-over-ethernet between nodes. The API is identical. `all_gather` and `reduce_scatter` work exactly the same way. The speed is dramatically different: NCCL's intra-node bandwidth on NVLink is around 300 GB/s, and InfiniBand inter-node is around 25 GB/s. Gloo on TCP is around 1 GB/s. The orders of magnitude tell you why production training prefers NCCL.

Second, `MASTER_ADDR` becomes the IP of the rendezvous node, and the launcher becomes `torchrun --nproc_per_node=8 --nnodes=4 --rdzv_endpoint=node0:29500 train.py`. That command launches 8 ranks per node across 4 nodes, with rank IDs 0 through 31, and uses node 0 as the rendezvous. The training script does not change at all. Same model, same wrapping policy, same `all_gather` calls. Only the deployment changes.

The mental picture is the same: each rank holds 1/32 of every parameter; the layers all-gather themselves just in time; backward reduce-scatters the gradients. The only difference is that messages now traverse a network with millisecond round trips instead of microsecond round trips. This is why FSDP on slow networks underperforms; the algorithm is unchanged, only its constants have shifted.

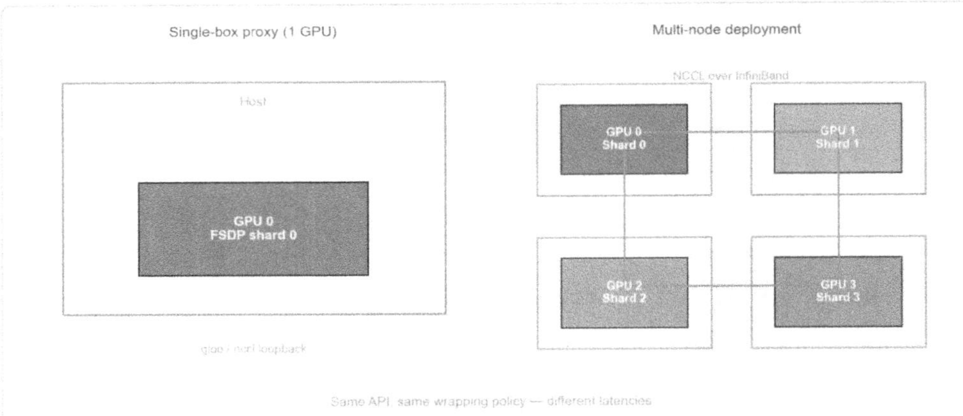

Figure 12.5. The single-box proxy on one GPU and the multi-node deployment side by side. Same calls, same model, same wrapping policy. Different launchers, different backends, different latencies.

A useful exercise once you understand this: estimate the step time for a 70B model on 64 H100s, assuming 600 GB/s NVLink between GPUs within a node and 25 GB/s InfiniBand between nodes. Most readers get this wrong by a factor of three on their first try. Doing the calculation forces you to confront how much time is spent waiting on the network, and that intuition is what separates engineers who can plan a large training run from engineers who guess.

Step 9: ZeRO Stage 1, Stage 2, Stage 3

FSDP is ZeRO Stage 3. The ZeRO ladder tells you what happens if you only shard part of the state.

ZeRO Stage 1 shards only the optimizer state. Parameters and gradients are still replicated on every rank. The memory savings are the smallest portion of the savings curve, but the communication cost is also the smallest, because optimizer state only moves once per step. The optimizer reads its sharded slice, updates parameters, and broadcasts the updated parameters back. If your optimizer state is your biggest memory consumer (true for AdamW on small-to-medium models in fp32), Stage 1 is enough.

ZeRO Stage 2 shards optimizer state and gradients. Parameters are still replicated. The communication adds a reduce-scatter for gradients at the end of backward instead of an all-reduce. The memory savings double, and the communication is roughly the same. A reduce-scatter has the same cost as an all-reduce on the same tensor.

ZeRO Stage 3 = FSDP shards everything. Maximum memory savings, maximum communication. Every layer's forward and backward triggers an all-gather, and backward triggers an additional reduce-scatter.

The decision rule is simple. If you fit on one GPU, do not use ZeRO at all (replicated training is faster). If you fit with optimizer-state sharding only, use Stage 1. If parameters are the bottleneck, use Stage 3. The ladder lets you spend communication budget exactly where you need it, instead of paying for sharding you do not require.

DeepSpeed's API exposes this with a config field, `"zero_optimization": {"stage": 1|2|3}`. PyTorch FSDP exposes it through the `sharding_strategy` argument: `ShardingStrategy.NO_SHARD` is plain DDP, `SHARD_GRAD_OP` is roughly Stage 2, `FULL_SHARD` is Stage 3, and there is no built-in Stage 1 (DeepSpeed has it, PyTorch does not as of the time of writing). Use the framework whose ladder rung you need.

BREAK IT

The failure mode worth seeing is silent divergence from a mis-shard. We are going to deliberately compute one shard boundary wrong by a single element. The all-gather call still succeeds. The tensor shapes match, the collective returns, no error fires. But one rank's slice contains data from the wrong parameter, the model is now training a corrupted version of itself, and the only visible symptom is that the loss curves on different ranks start to drift apart over hundreds of steps. By the time someone notices, the checkpoint is poisoned.

This is the canonical distributed training bug. It happens for real, it has happened to teams running million-dollar runs, and it is invisible to ordinary training instrumentation because the per-rank loss looks fine and the program does not crash.

The first time I encountered a bug like this in cooperative LoRA-fusion work, I lost most of a day before realizing the contributor's tokenizer indexing was off by one against the shared vocabulary. Different problem, same symptom: two ranks training slightly different models, both losses looking healthy. The only signal was a per-rank gradient norm comparison plot that someone had drawn for an unrelated reason. I would not have caught it otherwise.

The exact change

Take the hand-sharded linear layer from Step 2. The correct version computes the shard boundary as:

```
shard_size = total // world_size
self.shard_start = rank * shard_size
self.shard_end = (rank + 1) * shard_size
```

Break it by introducing an off-by-one error on one rank only. The cleanest way to do this in a controlled experiment is to perturb the shard boundary if `rank == 1`:

```
shard_size = total // world_size
self.shard_start = rank * shard_size
if rank == 1:
    self.shard_start += 1  # off-by-one: this rank's shard is shifted by 1
self.shard_end = self.shard_start + shard_size
```

Now rank 1's shard reads from index `shard_size + 1` to `2 * shard_size + 1` instead of `shard_size` to `2 * shard_size`. The shard is still the right length. The all-gather still receives the right number of bytes from each rank. The concatenated tensor still has the right shape. Nothing in the API will complain.

But the assembled "full weight tensor" now has element `shard_size` taken from the wrong parameter on rank 1, and one element is missing entirely from the gathered view. The gradient computed on this corrupted weight will not match the gradient computed by rank 0, which has a different but equally inconsistent view. The two ranks are now training divergent models.

What you will observe, before running

Predict what you will see before you run the experiment. This is the part that builds the intuition that catches the bug in the wild.

Step 1: No crash. No exception. The training loop runs from start to finish without complaint.

Step 2: Per-rank loss looks fine on each rank in isolation. Rank 0 sees a loss that decreases at roughly the normal rate. Rank 1 sees a loss that decreases at a slightly different rate, but not so different that you would call it broken if you only had one rank's data.

Step 3: When you compare losses across ranks at the same step, they drift. At step 0 they are identical (same initialization). At step 100 they differ by maybe 0.5 percent. At step 1000 they differ by maybe 5 percent. At step 5000 the gap is large enough to be obviously wrong.

Step 4: Gradient norms across ranks also drift. If you log gradient norms per rank (the four-instrument dashboard from **Project 11: Training Debugging — Spikes, NaNs, Profiling**, but with a per-rank axis), the lines for rank 0 and rank 1 separate. This is the earliest detectable signal, and it shows up before the loss drift becomes obvious.

Step 5: Eventually the training produces NaN, but it takes a while — usually hundreds to thousands of steps — and the NaN appears on one rank first while the other rank is still computing a finite loss. That asymmetry is the smoking gun.

Run the experiment

Run the broken version for 1000 steps and log per-rank loss every 10 steps. Save the results to a single CSV with columns `step, rank, loss, grad_norm`.

Plot two lines. One for rank 0's loss curve, one for rank 1's. On the same axes, draw the absolute difference between them on a secondary y-axis. The curves overlap perfectly for the first 50 to 100 steps, then visibly separate. The absolute difference grows monotonically until it crosses some threshold and the run starts to produce inf or nan on one rank.

This is what silent divergence looks like in slow motion. Nothing flagged it. No error, no warning, no automatic detection. Only the per-rank comparison shows the failure.

Why this proves the point

Single-GPU bugs crash. Distributed bugs whisper. The infrastructure of collectives is rigid about shapes, types, and counts — those are checked aggressively — but it is completely silent about semantics. As far as `all_gather` is concerned, every byte is just a byte. If you fed it the wrong bytes, you get the wrong bytes back, and the training loop continues happily training on garbage.

The countermeasure is per-rank instrumentation. The dashboard from **Project 11: Training Debugging — Spikes, NaNs, Profiling** must be extended with a per-rank axis for every metric, and the diagnosis rule must include "do the ranks agree?" If they do not, something in the sharding is wrong, and you should stop the run before the checkpoint gets poisoned. The dashboard is not a luxury here. It is the only thing standing between you and a wasted week of training.

A second observation worth lodging in memory. Silent failures are not exotic. They are the default. Loud failures get fixed because they are loud. Silent failures persist for years in production codebases because nobody has reason to look. The instinct this chapter trains is to assume silence is suspicious and to instrument accordingly.

An SRE I worked alongside in a past role on a high-uptime SaaS platform had a phrase for this: "the system that does not complain is the one you should worry about." It applies almost word-for-word to distributed training.

Optional Homework

The single-box proxy is enough to run every experiment in this chapter. If you want the multi-node version, that is a separate exercise with separate hardware requirements, sketched below.

Part A: The two-process hello world

On a laptop, run the worker script from Step 1 with world_size=2. Verify that both ranks print `3.0` after the `all_reduce`. Then change `world_size` to 4 and `nprocs` to 4 and verify the sum is now `10.0` (1 + 2 + 3 + 4). This is the minimum infrastructure test. If this does not run, nothing else will.

Part B: Hand-sharded MLP

Implement `HandShardedLinear` from Step 2. Initialize the same MLP on both ranks, run a forward pass that all-gathers the weight, computes `x @ W.t()`, and verifies that both ranks produce the same output (they should — they have the same weights and the same input). Compute the gradient manually, run a reduce-scatter, and confirm each rank ends up with the correct shard of the summed gradient. This is the lab where the magic disappears.

Part C: Wrap the Project 9 GPT in FSDP

Take the GPT from **Project 9: Pretraining on the Real Web**, the smallest config (depth 4, hidden 256, around 5 million parameters). Wrap it in `FullyShardedDataParallel` with the transformer-block auto-wrap policy from Step 5. Train for 200 steps on a tiny dataset and confirm the loss decreases. Then check per-rank `torch.cuda.max_memory_allocated()` and confirm each rank holds roughly half the parameters.

Part D: Mis-shard break

Reproduce the BREAK IT experiment. Use `HandShardedLinear` from Part B, introduce the off-by-one on rank 1, and run a small MLP training loop for 1000 steps. Log per-rank loss every 10 steps. Plot the two loss curves and the absolute difference between them. Identify the step where the divergence becomes visible. This is the most important experiment in the chapter.

Part E: Communication-volume measurement

Add timing hooks to your FSDP forward and backward to separately measure compute time and communication time. Run for 100 steps and report the ratio. Then double the model size (depth or hidden dimension) and report the ratio again. You should see the ratio shift in favor of compute as the model grows.

Part F (optional, requires real GPUs): True multi-GPU

If you have access to a multi-GPU node, run the same FSDP code with `nproc_per_node=2` or higher and `backend="nccl"`. Verify that the per-step time goes down with more ranks (it should, if compute dominates communication for your model size). Compare against a single-GPU DDP run as a baseline.

Part G (optional, requires multi-node hardware): True multi-node

If you have cluster access, run with `torchrun --nproc_per_node=8 --nnodes=2 --rdzv_endpoint=...`. Note the step time. Compare against the single-node 8xGPU number from Part F. Most readers see a 30-50 percent slowdown going from intra-node to inter-node, because the inter-node bandwidth is much lower than NVLink.

Questions To Answer

1. In the mis-shard experiment, at which step did the per-rank loss curves visibly separate? How many steps before that did the gradient norms start to differ?
2. What is the communication volume per training step, in bytes per rank, for a 7-billion-parameter model with world_size=64? Show your work using the rule of thumb from Step 4.
3. If your model has activation memory dominating parameter memory, will FSDP help? What technique would you combine with FSDP to attack the activation memory?
4. Why does the wrapping policy matter? What happens if you wrap the whole model as a single FSDP unit, and what happens if you wrap every linear layer individually?
5. Suppose you have a 13B model, 8 ranks, and an interconnect that gives you 50 GB/s per rank. Estimate the per-step communication time. Is FSDP a good fit for this setup?
6. The BREAK IT experiment produced different losses on different ranks. If you only had access to rank 0's logs, would you have caught the bug? What single piece of instrumentation, added to **Project 11: Training Debugging — Spikes, NaNs, Profiling**'s dashboard, would have flagged it immediately?

Go Further

The single best reference is the PyTorch FSDP tutorial at `pytorch.org/tutorials/intermediate/FSDP_tutorial.html`. Read it after this chapter; it covers wrapping policies, mixed precision under FSDP, checkpointing, and the various sharding strategies in concrete detail. The accompanying `torch.distributed.fsdp` documentation explains every flag.

The DeepSpeed ZeRO paper, "ZeRO: Memory Optimizations Toward Training Trillion Parameter Models" by Rajbhandari and colleagues (2020), is the founding paper for this whole area. It is readable and short — about 12 pages. It explains where the three stages come from and how the savings curve was first computed.

The Megatron-LM paper by Shoeybi and colleagues describes a complementary technique called **tensor parallelism**, which shards along a different axis: rather than splitting the parameter tensor by flat index, it splits along the hidden-dimension axis and computes parts of the matrix multiplication on different ranks. Tensor parallelism and FSDP are usually combined in real frontier training. The paper is denser than the ZeRO paper but worth working through after FSDP is comfortable.

Two repos worth reading. The first is `nanoFSDP` or any minimal FSDP implementation on GitHub — a few hundred lines of Python that implement the algorithm in full. Read it line by line; you have the conceptual scaffolding to follow every step. The second is the Megatron-LM codebase itself, which is production-grade and shows how the techniques actually look when wired together at scale. It is large, but the entry points are well-organized.

Useful arXiv search directions: `fully sharded data parallel`, `ZeRO optimizer state sharding`, `tensor parallelism transformer`, `pipeline parallelism gpu training`, `mixed precision distributed training`, `gradient accumulation fsdp`. The first of those will surface the FSDP design papers and several follow-ups.

One more pointer worth dropping. If your model is so large that even one layer's all-gathered weight does not fit on one GPU, FSDP alone is not enough — you need tensor parallelism on top, and the two compose. The combination is sometimes called **3D parallelism** when you add pipeline parallelism as a third axis. Reading about 3D parallelism is the natural next step after this chapter; the Megatron-LM paper is again the place to start.

What You Now Know

A model that does not fit on any single GPU can train if you shard its parameters, gradients, and optimizer state across N workers. Each worker holds 1/N of every tensor. To run a layer, workers gather the pieces they need just in time, compute, and drop the gathered copies. To run a backward, they sum gradients across ranks and keep only their own shard. The two operations that make this work are all-gather and reduce-scatter, and you can read any FSDP codebase fluently once you recognize those names.

You also know what silent distributed failures look like. Per-rank metrics that look fine in isolation can disagree across ranks, and the disagreement is the only signal that anything is wrong. The same dashboard skill from **Project 11: Training Debugging — Spikes, NaNs, Profiling** carries forward, but every metric now needs a per-rank axis.

The single-box proxy you built — two processes on one machine, gloo backend, the same `init_process_group` and `all_gather` calls — is the same code that runs on 1024-GPU clusters. The launcher changes. The backend changes. The wrapping policy may need to be tuned. But the conceptual model is identical. From here, scaling up is a matter of access to hardware, not new ideas.

Starting Point

PyTorch FSDP tutorial: pytorch.org/tutorials/intermediate/FSDP_tutorial.html. PyTorch FSDP API reference: pytorch.org/docs/stable/fsdp.html. DeepSpeed ZeRO paper: arxiv.org/abs/1910.02054 (Rajbhandari et al., 2020). Megatron-LM paper: arxiv.org/abs/1909.08053 (Shoeybi et al., 2019). PyTorch distributed overview: pytorch.org/docs/stable/distributed.html. Reference single-box proxy code: nanochat/distributed_demo/.

Project 13: Fast Inference: The KV Cache

Hook

Why does a language model get slower and slower the longer it talks, even though it is only producing one new token at a time? If the model already read the first 500 tokens, why does token 501 seem to make it reread all 500 again? That sounds stupid because it is stupid.

Left alone, autoregressive generation really does keep recomputing work it already did. The KV cache exists to stop that waste. Without it, "generate one more token" secretly means "re-run most of the sequence again." With it, generation stops acting like someone rereading the whole chat history before every reply and starts acting like someone who kept notes.

I spent a chunk of 2025 writing KV-cache-adjacent code on Apple Neural Engine, where we had to swap LoRA adapters in through IOSurface buffers between decode steps. The thing that struck me, and that I want this chapter to land for you, is how mundane the cache turns out to be once you see it. It is bookkeeping. It is one of the most consequential optimizations in modern inference, and it is bookkeeping.

The Concept

Start with the wrong mental model, because a lot of people have it.

Attention is not memory in the durable sense. It does not keep a little brain-state from one decoding step to the next. It recomputes relationships between tokens inside the current input. If you call the model again on a longer sequence, it will happily recalculate the same keys and values for the old tokens unless you explicitly save them. That saved state is the **KV cache**. The name is dry. The idea is not.

Think of a courtroom transcriptionist. Every time a new sentence is spoken, the judge does not ask the transcriptionist to retype the entire hearing from the beginning. The old transcript is already there. The new sentence gets appended. When someone needs context, they look back at the transcript. That is KV caching.

The model has already turned past tokens into two useful things:

- **keys**: what past tokens contain (the same key projections computed during prefill that are now being cached).
- **values**: what past tokens contribute when selected.

For a new token, the model mainly needs a fresh **query**: what the new token is looking for, the only fresh attention projection during decode. Then it can compare that new query against all the saved keys, pull information from the saved values, and move on. No need to rebuild the old keys and old values every single step.

Let's define the pieces in plain English before any code.

A **token** is the basic unit the model works with. It might be a word, part of a word, punctuation, or whitespace. A **hidden state** is the model's internal representation of a token at some layer — a list of numbers that tries to capture what that token means in context. Query, key, and value are three different linear projections of that hidden state — introduced in **Project 4: Attention From Scratch** — giving the same token three different "views" of itself.

In self-attention, every token asks: "Which earlier tokens matter for understanding me right now?" The token does that by comparing its query to every earlier token's key. The better the match, the more it reads from that token's value. Here is the part people miss. For old tokens, those keys and values do not change during inference. If token 17 already produced a key and a value at layer 6, then tokens 18, 19, and 400 can keep reusing them. Recomputing them is wasted work. That is why the cache stores K and V, not Q. The new token's query changes every step because the new token is new, but the older tokens' keys and values are stable for the current prompt.

The first time I really sat with that idea, I had a small moment of "wait, that's it?" Yes. That's it. The most consequential inference optimization in modern LLMs is "do not recompute the things that did not change." Most beginner explanations skip past this with the air of someone explaining a magic trick, when the trick is just notetaking.

So the KV cache is a per-layer notepad. Each layer keeps a growing stack of old keys and old values. When a new token arrives, the model computes its hidden state through the network, computes its new query, key, and value in each attention layer, appends the new key and value to that layer's cache, and runs attention using the new query against all cached keys and values. That is the whole trick.

The result feels almost unfair once you see it clearly. Without caching, generating token 501 means recomputing attention inputs for tokens 1 through 500 again, rebuilding their keys, rebuilding their values, and only then producing token 501. With caching, it means computing the new token once, comparing it to the stored past, and appending one more row to the cache. Same output goal. Very different amount of work.

Most explanations frame the speedup as a compute win. It mostly is not. Once the cache exists, the real story turns into memory bandwidth: how fast you can read the cached K and V out of HBM at every decode step. That bandwidth bottleneck carries the back half of this chapter.

Now earn a little math. Suppose your prompt length is `T`. In naive generation, step 1 runs on length 1, step 2 on length 2, step 3 on length 3, and so on until length `T`. That total work looks like `1 + 2 + 3 + ... + T`, which grows roughly like `T^2`. So generation time grows much faster than sequence length. With a KV cache, each new step mainly processes one new position against stored history. You still pay some attention cost against the history, but you stop recomputing all earlier K and V projections from scratch. In practice, that is the difference between "why is this chatbot stalling?" and "this feels instant enough to use."

There is another reason the cache matters: it makes the real bottleneck visible. Once you stop wasting compute, memory becomes the wall. The cache grows with context length — every new token adds more stored keys and values for every layer. So the question changes from "can the model in principle look back 32K tokens?" to "can I afford to store 32K tokens worth of keys and values for every layer, head, and batch item?" That is why context length is not just an architectural idea. It is a memory budget. Figure 9.1 makes the contrast between the naive and cached approaches concrete.

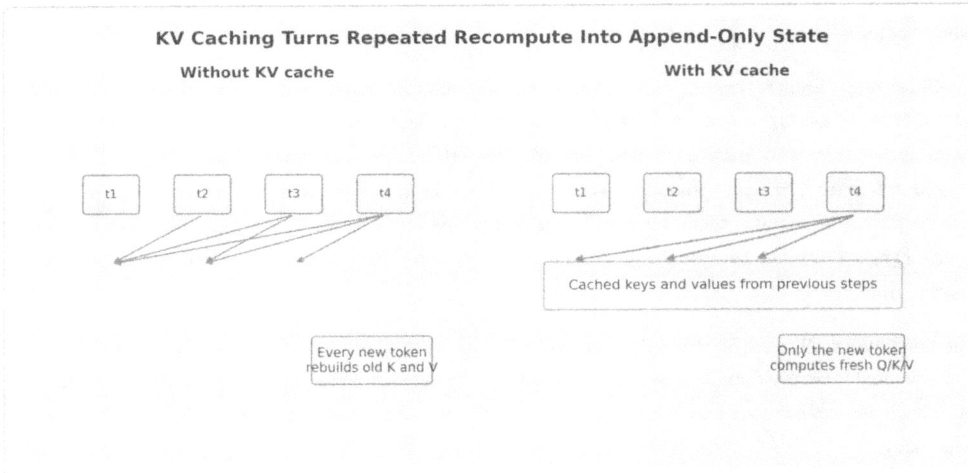

Figure 9.1. The KV cache removes repeated recomputation by storing old keys and values and appending only the new token state at each decode step.

Why It Matters

Without a KV cache, long-context generation is slow enough to feel broken. You can prove this to yourself with a toy example. Ask a tiny model to generate 256 tokens from a 512-token prompt. Naive decoding keeps rerunning almost the entire prefix, so the later steps cost much more than the early ones and throughput falls off as the sequence gets longer. The user does not care why. The user sees lag. That alone would be enough reason to care.

But the cache matters for more than speed. If you run a model for chat, coding, or agents, you are not doing one forward pass over a fixed sequence. You are doing incremental decoding over and over. That means your runtime lives or dies by how cheaply it can carry the past forward, and the KV cache is the single most important inference optimization because it turns repeated work into stored state.

I spent two years in healthcare release management where the operational rule was "zero downtime, no deploy windows." That experience taught me something that transfers awkwardly well: production inference cares about p99 latency, not throughput averaged over a benchmark. A cache that drops a tail of slow requests by 50% is worth more than one that lifts the mean by 10%. Keep that in mind as you measure later in this chapter.

The cache also exposes a real memory trade that people often miss. A model might fit in VRAM and still fail under long contexts because the KV cache grows until memory runs out. The model weights are one cost; the cache is another, and they are independent. For one layer, the cache stores keys with shape roughly `[B, H, T, d_head]` and values with shape roughly `[B, H, T, d_head]`, where `B` is batch size, `H` is number of attention heads, `T` is cached sequence length, and `d_head` is the size of one head. That means cache size grows linearly with `T`. Double the context, roughly double the cache memory. Double the batch size, double it again. Double the number of layers, double it again because each layer has its own cache. This is why the words **context window** often hide a hardware statement. A model can only keep as much usable past as memory allows.

Then there is the sliding-window trick. If you drop the oldest cache entries and keep only the most recent `W` tokens, memory stops growing after `W`. You traded full long-range context for bounded memory. Sometimes that is a good trade. Sometimes it destroys quality. For local text continuation, a 2K or 4K sliding window may be fine. For code generation that depends on an import 8K tokens earlier, it may be terrible. The task decides.

Finally, the cache matters because if it becomes corrupted, generation collapses in a way that feels spooky until you understand why. One bad cached key in the middle of the sequence can poison every later attention lookup that touches it. You are not damaging one token. You are damaging a persistent reference point that future tokens may keep consulting. A bad cache does not heal itself just because later math is correct. It keeps feeding wrong context into every later step.

The Build

Step 1: Start with naive autoregressive generation

Before you optimize anything, build the slow version on purpose. This matters because otherwise the KV cache feels like extra machinery instead of a fix for a measurable problem.

Your naive generator should do exactly what a first implementation would do:

1. take the current token sequence
2. run the full model on the entire sequence
3. keep only the logits for the last position
4. sample or argmax the next token
5. append it
6. repeat

In code, the shape of it looks like this:

```
@torch.no_grad()
def generate_naive(model: nn.Module, idx: torch.Tensor, max_new_tokens: int, temperature: float =
        1.0) → torch.Tensor:
    # idx: [B, T]
    for _ in range(max_new_tokens):
        logits = model(idx)              # full forward pass on full sequence
        last = logits[:, -1, :] / temperature
        next_id = torch.argmax(last, dim=-1, keepdim=True)
        idx = torch.cat([idx, next_id], dim=1)
    return idx
```

This is correct. It is also wasteful. At step 100, it reruns the model over the first 99 generated positions even though their internal K and V tensors were already computed in previous steps.

Time it. Do not trust your intuition. Measure tokens per second. A simple timing harness is enough:

```python
import time

import torch

def bench(fn, *args, new_tokens: int = 128, warmup: int = 1, runs: int = 3, **kwargs) -> float:
    for _ in range(warmup):
        _ = fn(*args, max_new_tokens=new_tokens, **kwargs)
    torch.cuda.synchronize() if torch.cuda.is_available() else None

    start = time.perf_counter()
    for _ in range(runs):
        _ = fn(*args, max_new_tokens=new_tokens, **kwargs)
    torch.cuda.synchronize() if torch.cuda.is_available() else None
    dt = time.perf_counter() - start
    return (runs * new_tokens) / dt
```

Use a fixed prompt length, then vary it. Try prompt lengths like 32, 128, 512, and 1024. You should see naive throughput get worse as prompt length grows. That worsening is the point. If you do not see it, your benchmark is too small or your timing is noisy.

Step 2: Identify where K and V are created

Now look inside one attention layer. The mental model I keep coming back to is the courtroom transcriptionist from earlier: the transcript is real storage you can write to, and the new sentence is the one piece of fresh work each step. In a decoder-only transformer, the attention block usually does something like:

```python
q = q_proj(x)
k = k_proj(x)
v = v_proj(x)
```

Then it reshapes into heads and computes attention scores. The important insight is not the formula — you already know attention at this point. The important insight is lifecycle. For a prefix token, once `k` and `v` are computed during inference, they can be saved. So the attention module needs a second mode: full-sequence mode for training or naive inference, and incremental mode for cached decoding.

That means your attention layer will eventually need inputs like `x_new` (the hidden state for the new position), `k_cache` and `v_cache` (previously stored tensors), and `cache_pos` (where to append the new token). Conceptually, decoding one new token looks like this:

1. receive only the new token embedding at the current position
2. compute `q_new`, `k_new`, `v_new`
3. append `k_new` and `v_new` to cache
4. attend using `q_new` against the full cached `k` and `v`

This means the output is only for the new position. That is what you want during generation.

Step 3: Add a cache object per layer

You need storage — not vague "memory," but concrete tensors. A simple cache per attention layer can be two tensors, `k_cache` and `v_cache`. If your model has `N` transformer blocks, you need one pair per block.

One clean way to think about it is:

```python
class KVCache:
    def __init__(self, n_layers):
        self.k = [None] * n_layers
        self.v = [None] * n_layers
```

In practice you may preallocate the full max context length for speed, especially on GPU:

```python
k_cache[layer]  # [B, H, T_max, d_head]
v_cache[layer]  # [B, H, T_max, d_head]
```

Then keep a current position pointer. Why preallocate? Because repeatedly concatenating tensors can become its own bottleneck. Appending by slice assignment is usually cleaner for performance:

```python
k_cache[layer][:, :, pos:pos+1, :] = k_new
v_cache[layer][:, :, pos:pos+1, :] = v_new
```

Now `pos` advances by one. This is a small implementation detail with a large lesson hiding inside it. The cache is part of the runtime state of the model. The weights are static. The cache is live, growing session state. That is why two requests in a server cannot silently share the same cache unless you want nonsense.

When I wired KV-cache-style buffer swaps on Apple Neural Engine, the biggest design constraint was exactly this distinction. The model weights could live in one memory pool. The per-request live state had to live somewhere we could swap in and out per inference. Half the engineering on that project was a tax for that one separation. Worth it. Skip it and the system mixes one user's notes into another user's reply.

Step 4: Change attention to decode one token at a time

In training, attention sees a whole sequence:

- `q: [B, H, T, d_head]`
- `k: [B, H, T, d_head]`
- `v: [B, H, T, d_head]`

In cached decoding at one new step:

- `q_new: [B, H, 1, d_head]`
- `k_all: [B, H, T_cache, d_head]`
- `v_all: [B, H, T_cache, d_head]`

Then attention scores have shape `[B, H, 1, T_cache]`, meaning the new token asks one question against all past tokens.

Here is the same attention law you already know, now in its one-token decode form:

```
attention(q_new, k_all, v_all) = softmax(q_new @ k_all^T / sqrt(d_head)) @ v_all
```

Earn every symbol: `q_new` is the new token's query, `k_all` is every cached key up to the current position, `v_all` is every cached value up to the current position, `d_head` is the size of one head, `sqrt(d_head)` is the scaling factor that stops dot products from getting too large, and `softmax` turns raw scores into attention weights that sum to 1.

Nothing about the law changed. Only the shapes changed. KV caching is not a new kind of attention — it is the same attention with old K and V reused instead of rebuilt.

A decode-step sketch looks like this:

```
def decode_step(x_new: torch.Tensor, k_cache: torch.Tensor, v_cache: torch.Tensor, pos: int) ->
        torch.Tensor:
    q_new = q_proj(x_new)
    k_new = k_proj(x_new)
    v_new = v_proj(x_new)

    q_new = split_heads(q_new)  # [B, H, 1, d]
    k_new = split_heads(k_new)  # [B, H, 1, d]
    v_new = split_heads(v_new)  # [B, H, 1, d]

    k_cache[:, :, pos:pos+1, :] = k_new
    v_cache[:, :, pos:pos+1, :] = v_new

    k_all = k_cache[:, :, :pos+1, :]
    v_all = v_cache[:, :, :pos+1, :]

    scores = q_new @ k_all.transpose(-2, -1) / math.sqrt(q_new.size(-1))
    att = scores.softmax(dim=-1)
    out = att @ v_all
    return merge_heads(out)
```

There is no causal mask shown here because the slice `:pos+1` already means "only attend to past and present." That is one of the nice parts of decode mode. The mask becomes implicit in the cache length.

The first draft of this section started with the full-sequence mask and then justified its removal, which felt mechanical. Then I tried the opposite framing here. Start without a mask. Notice that the slice already enforces causality. That is the click I wanted: the cache is the mask. Once you see it that way, the rest of the chapter writes itself.

Step 5: Thread the cache through the full model

Now each transformer block needs access to its own cache tensors. During the prefill phase, when you first ingest the prompt, you have two options:

1. run full-sequence mode once, while writing all K and V into cache
2. feed the prompt token-by-token to populate the cache incrementally

Option 1 is faster for long prompts. People often call this **prefill**, and incremental decoding starts after the prompt is loaded.

That split matters because serving has two phases: **prefill** (process the prompt) and **decode** (generate new tokens one at a time), and they stress hardware differently. Prefill has more parallel work across prompt positions. Decode has tiny per-step work with lots of cache reads. Even if you do not fully optimize prefill yet, keep the mental split clear. Production systems care about both.

Your model interface may end up looking roughly like:

```
logits, cache = model.prefill(prompt_ids, cache=None)
next_logits, cache = model.decode(next_token, cache)
```

Or perhaps:

```
logits = model(idx, cache=cache, use_cache=True)
```

The exact API matters less than the state transition. The cache must persist across decode steps.

Step 6: Benchmark the speedup honestly

Now rerun the tokens-per-second benchmark with cached decoding. Use the same model, prompt, number of generated tokens, device, dtype, decoding rule — only change the inference path.

A table like this is useful:

Prompt Length	Naive tok/s	Cached tok/s	Speedup
32	120	160	1.3x
128	65	210	3.2x
512	18	190	10.5x
1024	8	160	20.0x

: **Table 9.1.** KV cache speedup by prompt length: naive versus cached throughput (tokens/sec) and the resulting speedup factor, showing how gains grow with context length.

Those are example numbers, not promises. The exact results depend on model size, GPU, framework overhead, batch size, prompt length, and attention kernel implementation. But the shape should hold: short prompts show modest gains, long prompts show dramatic gains. That is the right result because the cache removes repeated prefix work, and repeated prefix work hurts more as the prefix grows.

Strong opinion that I will defend: most benchmark tables I see for inference speedups quote a single number and hide the prompt-length dependence. That is bad. A 20x speedup at 4K prompt length and a 1.3x speedup at 32 prompt length are both true, and they describe different products. Always plot the curve, not the point.

Step 7: Profile cache memory

Now measure the price you paid for speed.

Each token leaves behind footprints in every layer. Two of them. A key and a value. And not one global copy: one per head, per layer, per batch item. So even if the token itself is tiny, the total saved state across the stack adds up fast.

A rough byte estimate for a standard multi-head cache is:

```
bytes ≈ 2 * N * B * T * H * d_head * bytes_per_element
```

Where `2` is for keys and values, `N` is number of layers, `B` is batch size, `T` is sequence length in cache, `H` is number of attention heads, `d_head` is head dimension, and `bytes_per_element` is 2 for FP16/BF16 or 4 for FP32.

Since `H * d_head = d_model`, you can also think of it as:

```
bytes ≈ 2 * N * B * T * d_model * bytes_per_element
```

That form is easier to reason about. Suppose `N = 12, B = 1, T = 4096, d_model = 768`, FP16 so `bytes_per_element = 2`:

```
bytes ≈ 2 * 12 * 1 * 4096 * 768 * 2
```

That is about 151 million bytes, or roughly 144 MB. That is just the KV cache. Not the weights, not activations during prefill, not framework overhead. That is why longer contexts eventually hit a wall.

You should measure this empirically too. If you are on CUDA, log `torch.cuda.memory_allocated()` and `torch.cuda.max_memory_allocated()`. Record these at different prompt lengths or after different numbers of generated tokens.

Build a small table:

Cached Tokens	KV Cache MB	Total GPU MB
256	9	1850
512	18	1865
1024	36	1901
2048	72	1972
4096	144	2115

: **Table 9.2.** KV cache memory footprint at increasing context lengths, showing how cache size and total GPU memory grow as more tokens are stored.

The pattern matters more than exact values. Then keep increasing `T` until you hit out-of-memory. That is your local context wall for that model, dtype, and hardware setup.

Step 8: Add a sliding window

Once you see the wall, the next question is obvious: "What if I stop remembering everything?" That is the sliding window.

Pick a window size `W`. If the cache grows past `W`, drop the oldest entries and keep only the most recent `W`.

Conceptually:

```
if cache_len > window:
    k_cache = k_cache[:, :, -window:, :]
    v_cache = v_cache[:, :, -window:, :]
```

Or with a ring buffer if you want to avoid copies.

This makes memory bounded by `W` instead of growing forever. The trade is brutal and clear: you are choosing not to let the model attend to older context. Sometimes that barely matters. Sometimes it breaks the task. You should measure quality, not just speed.

A quick definition. **Perplexity** is a standard language-model metric derived from cross-entropy. Lower perplexity means the model is less surprised by the correct next token. If bpb is "bits needed per byte," perplexity is "effective branching factor of uncertainty." You do not need to live inside the formula. You need to know that lower is better, and truncating useful context should make it worse.

Run evaluation with window sizes like full cache, 2048, 1024, 512, and 256. Then compare validation perplexity or loss. A likely pattern: large windows cause almost no damage; medium windows cause some degradation on long-range dependencies; tiny windows produce repetition, forgotten setup, wrong references, and broken code completions.

That experiment teaches a deeper point than "memory is expensive." It teaches that context length is an agreement between model behavior and runtime storage. The model may know how to use long context. Your runtime may choose not to keep it. Those are different things.

Fewer bits per entry: quantizing the KV cache

The sliding window saves memory by dropping old tokens. A second lever keeps every token but shrinks what each one costs to store. Picture rounding every cached number to fewer decimal places. Instead of writing 0.7194 you keep 0.7, so the value survives in coarser form and takes less room on the page. That rounding is the whole idea. **Quantizing the KV cache** means storing the cached keys and values at lower numerical precision, fewer bits per number, then reconstructing an approximation when attention reads them back. Look back at the byte estimate in Step 7: each entry was two bytes at FP16, so spending fewer bits per entry cuts that term in proportion. Fewer bits per channel means less memory and some added error. The open question is how few bits you can spend before quality drops.

One line of research pushes this bit-width down with a rotation trick. Three papers from an overlapping group trace it. QJL applies a JL transform (a length-preserving random projection), then keeps a single sign bit per coordinate; on its own LLM and NLP benchmarks it reports more than 5x KV-cache memory reduction at 3 bits with no reported accuracy loss (AAAI 2025). PolarQuant maps each key into polar coordinates and quantizes the angles; on long-context evaluation it reports more than 4.2x compression (AISTATS 2026).

TurboQuant, a 2025 preprint accepted to ICLR 2026, works on the distortion directly. In the TurboQuant paper the authors report quality-neutral KV-cache quantization at 3.5 bits per channel, and only marginal degradation at 2.5 bits per channel. That is an abstract-level result, not tied to one model or GPU, and "marginal" is their word rather than a fixed accuracy delta. The mechanism runs in three plain steps. First, a random orthogonal rotation (spinning the numbers onto new axes without changing their lengths), which spreads the values so no single coordinate dominates. Second, per-coordinate Lloyd-Max scalar quantization (a rounding table tuned to how each coordinate's numbers are actually distributed). Third, a 1-bit QJL residual correction, which cleans up the leftover error so the dot products attention relies on stay unbiased, centered on the true value rather than skewed.

These are research implementations, not yet standard library features. You can still feel the mechanism today with a proxy: the quantized KV cache built into Hugging Face Transformers. It is not TurboQuant. It is a hands-on stand-in for the same idea, storing keys and values at low bit-width instead of 16-bit, and QJL, PolarQuant, and TurboQuant are the research-grade versions that push the bit-width lower with the rotation trick above. The proxy needs one extra package, `pip install optimum-quanto`.

Run it on a small model with CUDA:

```python
import torch
from transformers import AutoModelForCausalLM, AutoTokenizer

tok = AutoTokenizer.from_pretrained("gpt2")
model = AutoModelForCausalLM.from_pretrained("gpt2", dtype=torch.float16).cuda()
ids = tok("The KV cache stores keys and values so", return_tensors="pt").input_ids.cuda()

def run(cache_kwargs: dict) -> tuple[torch.Tensor, int]:
    torch.cuda.reset_peak_memory_stats()
    # gpt2's context is 1024 positions; the prompt plus generated tokens must
    # stay under it, or the position-embedding lookup overruns and CUDA aborts
    # with a device-side assert. 512 leaves room for the short prompt above.
    out = model.generate(ids, max_new_tokens=512, do_sample=False, **cache_kwargs)
    return out, torch.cuda.max_memory_allocated()

base_out, base_peak = run({})
quant_out, quant_peak = run({
    "cache_implementation": "quantized",
    "cache_config": {"backend": "quanto", "nbits": 4},
})

print(base_peak, quant_peak)
print(tok.decode(base_out[0]) == tok.decode(quant_out[0]))
```

Compare `base_peak` against `quant_peak`, then compare the two decoded strings. Expected observation: the KV-cache share of peak memory falls in proportion to bits per entry, so the 16-bit to 4-bit switch should pull the cache term toward one quarter of its size, since 4 over 16 is one quarter. That is the arithmetic expectation, not a benchmark. Greedy output on the fixed prompt should stay close.

One caveat on the measurement. A single `generate()` peak is noisy, and weights plus activations sit inside that same peak, so the cache term can hide under them. Average a few runs, or lengthen the prompt and raise `max_new_tokens` toward gpt2's 1024-position ceiling (the prompt plus the generated tokens must stay under it), so the KV cache grows large enough to dominate what you read. Two likely failures stop the run outright: `optimum-quanto` is not installed, or your Transformers version predates `cache_implementation="quantized"`. Both raise at the quantized `run(...)` call. Update the package, or pin a recent Transformers release, and retry.

The proxy reports only that arithmetic direction, never TurboQuant's numbers. What the research methods add is holding quality while the bit-width drops below what a plain 4-bit store can reach. So quantization sits alongside the sliding window as the other memory lever: one drops tokens, the other shrinks each stored token.

Step 9: Keep generation semantics identical

A good cache implementation should not change outputs when decoding deterministically. That is a test, not a wish. If you run the same prompt, same weights, same temperature, same seed, and greedy decoding, then naive generation and cached generation should produce the same token sequence. If they do not, your cache path is wrong.

Common bugs: writing to the wrong position, reading too much or too little of the cache, mixing batch items, mixing layers, using a wrong positional index during RoPE or learned position embeddings, forgetting that the decode step handles one token rather than a full sequence, or accidentally including future positions in the cache slice.

This is one of those moments where correctness tests save you hours. Before chasing speed, assert equality on short deterministic runs.

BREAK IT

Break the cache on purpose. Do not remove it — corrupt it. That is the lesson this project is built around.

Here is the failure mode. After about 100 generated tokens, silently overwrite one cached key vector in the middle of the sequence with random values. Then continue decoding as if nothing happened. Do this in one layer first. Pick a fixed position like token 57 in the cache, one head in one layer, and replace that one cached key vector with noise.

For example:

```
# Example: corrupt layer 3, head 0, position 57
with torch.no_grad():
    cache.k[3][:, 0, 57, :] = torch.randn_like(cache.k[3][:, 0, 57, :])
```

Then continue generating.

What should you observe? Not a crash — that would be easy. You will usually get fluent-looking text for a few tokens, then increasing incoherence. The model may lose track of referents, repeat odd phrases, switch topic abruptly, write syntactically plausible but semantically wrong text, or drift into locally fluent garbage.

Why does one corrupted key matter so much? Because that key is not a one-time computation anymore. It is a persistent lookup target inside the remembered past. Every later token that asks a query might compare itself against that corrupted key. Some queries will ignore it. Some will not. When they do not ignore it, the attention weights shift, the model reads the wrong value pattern from the past, and then the hidden state for the new token becomes slightly wrong. Then that wrong hidden state creates later wrong queries, keys, and values. The corruption propagates forward. Not magic, just a bad note in the notepad that keeps being consulted.

The first time I saw a real example of this in the wild was actually a buffer alignment bug, not a math bug. We were swapping IOSurface-backed KV tensors between two ANE-compiled graphs and one of the surfaces had a stride that did not match what the consumer expected. Generation looked sane for about 40 tokens, then drifted, then went incoherent. The math was correct everywhere. The cache was simply being read from a slightly wrong offset. That afternoon stuck with me.

If you want the failure to show up harder, corrupt a middle position rather than the very first or very last, choose a semantically important token like a function name in code or an entity name in prose, corrupt multiple heads in the same position, or corrupt both K and V instead of K alone. But start with one key. That is enough to prove the point.

What this proves:

1. The cache is not just an optimization layer you can treat casually.
2. The cache becomes part of the model's live computational state.
3. A tiny persistent error can poison every later decode step.
4. Correctness in inference systems is not only "does it run?" but "does the saved state stay valid?"

There is a second break worth doing. Break cache position bookkeeping: keep K and V values correct, but write them to the wrong slot by shifting the write index by 1. That creates a subtler disaster. The cache contents are real — they are just attached to the wrong positions. The model then attends to context as if the timeline were slightly scrambled. This often produces output that looks eerily close to correct at first, then decays into contradictions and strange repetitions.

That proves something slightly different: cache integrity is not just tensor contents, it is also alignment. You need the right values, in the right layer, for the right head, at the right position, for the right request. Anything less is broken inference.

Optional Homework

Run three labs: speed, memory, and corruption.

Lab 1: Naive vs cached throughput

Hardware:

- GPU preferred
- works on CPU for tiny models, but the difference will be smaller and slower to measure

Time:

- 20 to 40 minutes if you already have a working inference path
- longer if you are adding cache support for the first time

Steps:

1. Clone the reference repo.

```
git clone https://github.com/karpathy/nanochat
cd nanochat
```

2. Find the inference entry point and attention module. Look for functions or files containing:
   - `generate`
   - `sample`
   - `attn`
   - `cache`
   - `kv`
3. Add a naive generation path if the repo already defaults to cached decoding. Make sure it always calls the model on the full growing sequence.
4. Add timing around generation. Record:
   - prompt length
   - generated tokens
   - total seconds
   - tokens per second
5. Benchmark both versions on the same prompt lengths:
   - 32
   - 128

    - 512
    - 1024
6. Use greedy decoding so output comparisons stay stable.

Expected observations:

- naive decoding gets slower as prompt length rises
- cached decoding stays much flatter
- speedup should be small for short prompts and large for long prompts
- 5x to 20x is a reasonable ballpark on long enough prompts

Lab 2: Measure cache memory and find your wall

Hardware:

- GPU strongly preferred if you want a real memory wall
- CPU still teaches the growth pattern

Time:

- 20 minutes

Steps:

1. Instrument cache memory. If using CUDA, log:
    - `torch.cuda.memory_allocated()`
    - `torch.cuda.max_memory_allocated()`
2. Also compute a manual estimate of KV bytes from:
    - layers
    - heads
    - head dimension
    - dtype
    - sequence length
3. Run prefill plus generation at increasing context lengths. Example:
    - 256
    - 512
    - 1024
    - 2048
    - 4096
    - 8192, if hardware allows
4. Save a table with:
    - sequence length
    - cache MB
    - total allocated MB
    - whether the run succeeded

5. Keep increasing until you hit OOM.

Expected observations:

- cache memory grows close to linearly with sequence length
- total GPU memory rises even if weights stay fixed
- the failure point arrives because runtime state grew, not because the checkpoint got bigger

Lab 3: Sliding window vs quality

Hardware:

- GPU preferred
- a small local proxy run is fine

Time:

- 30 to 90 minutes depending on how you evaluate

Steps:

1. Implement a windowed cache with sizes:
   - full
   - 2048
   - 1024
   - 512
   - 256
2. Evaluate on a held-out validation set. Record:
   - validation loss or perplexity
   - tokens per second
   - peak memory
3. Also run qualitative prompts that need older context. Good examples:
   - code completion where a function definition appears far back
   - a long paragraph with named entities introduced early
   - a synthetic prompt like "remember the word PINEAPPLE" followed 1500 tokens later by "what word did I ask you to remember?"

Expected observations:

- throughput may improve slightly or stay similar
- peak memory becomes bounded by the window
- quality drops once the task needs context older than the window

Lab 4: Corrupt one cached key vector

Hardware:

- any hardware that can run your model

- tiny models work fine for the teaching value

Time:

- 15 to 30 minutes

Steps:

1. Generate a baseline completion from a fixed prompt using greedy decoding. Save the output.
2. Reset and rerun from the same prompt.
3. After the cache contains at least 100 tokens, overwrite one cached key vector:
   - choose a middle layer
   - choose one head
   - choose one position near the middle of the cached prompt
4. Continue decoding.
5. Compare the clean output and corrupted output token-by-token.

Parameter changes:

- set temperature to `0.0` or greedy argmax
- fix random seed
- do not change model weights
- change only the cache tensor

Expected observations:

- outputs match exactly until the corruption matters
- after that point, the corrupted run diverges permanently
- the model does not "recover" because every later step depends on the evolving corrupted history

Proxy version if you do not have `nanochat` ready:

- use your Project 5 GPT implementation
- add a tiny cache to one attention layer
- run on a toy prompt and compare clean vs corrupted decode
- the model can be tiny; the lesson still lands

Questions To Answer

1. How did tokens per second change as prompt length increased for naive decoding versus cached decoding?
2. At what prompt length did cached decoding stop feeling like an optimization and start feeling required?
3. When you measured cache memory, what grew linearly with sequence length and what stayed fixed?
4. In the sliding-window experiment, what failed first: exact recall, topic consistency, code correctness, or local fluency?
5. When you corrupted one cached key vector, how many tokens passed before the output clearly diverged from the clean run?

6. What tradeoff is the KV cache protecting: compute, memory, latency, or all three at once? Which of those becomes the new bottleneck after caching works?

Go Further

Research anchors for this chapter

The local paper set does not contain one single "KV cache paper" that defines this entire chapter, because KV caching emerged as the operational consequence of decoder-only attention rather than as one isolated invention paper. The cleanest conceptual anchor is still **Attention Is All You Need**: once autoregressive self-attention exists, the question "why are we recomputing old keys and values during decode?" becomes unavoidable.

For the systems side, **FlashAttention: Fast and Memory-Efficient Exact Attention with IO-Awareness** is the closest research anchor in the local library. It is not a KV-cache paper, but it is directly relevant because it reframes attention cost as an IO and memory-movement problem rather than just a matrix-multiplication problem. That is the same systems lens this chapter is trying to teach.

Then carry that lens forward into Project 15. Once you understand KV caching, grouped-query and multi-query variants stop looking like random architectural tweaks and start looking like direct attempts to shrink the cached state that decode has to carry.

Start with the reference implementation:

- `https://github.com/karpathy/nanochat`

Then look for these directions:

- `paged attention`
- `flash decoding`
- `vllm pagedattention`
- `continuous batching`
- `speculative decoding`
- `grouped query attention`
- `multi query attention`

What to search for on arXiv:

- `PagedAttention`
- `FlashDecoding`
- `Grouped-Query Attention`
- `Multi-Query Attention`
- `long context inference`
- `kv cache quantization`

Good follow-up questions:

- How do serving systems keep many users' KV caches from fragmenting GPU memory?
- How does grouped-query attention shrink the KV cache in the first place?

- Can you quantize the cache, not just the weights?
- How do prefill and decode stress hardware differently?
- Why does continuous batching matter once many requests are decoding at once?

If you want one especially useful thread, search for:

- `kv cache quantization`
- `llm serving paged attention`
- `decode bottleneck transformer inference`

That is where you start seeing inference as a systems problem, not just a model problem.

What You Now Know

The KV cache stores old keys and values so each new token only needs one attention step instead of recomputing the entire history. That one decision changes generation speed by an order of magnitude, and also creates the main failure surface: corrupted or misaligned cache state degrades every token that follows. Context length is ultimately a memory problem, and your runtime is an engineering system that decides what past survives and at what cost.

Starting Point

Start from:

- `https://github.com/karpathy/nanochat`

Use it alongside:

- your own Project 5 GPT implementation
- your Project 6 notes on how a teaching model grows into a production-shaped system
- your Project 7 instinct that internal correctness is more than "the shapes line up"
- your Project 9 judgment about metrics, throughput, and the difference between a curve that moves and a system that actually works

The progression still matters:

- `micrograd` taught you blame assignment
- Project 2 taught you tiny next-token learning
- `minbpe` taught you that tokens are engineered
- Project 4 taught you attention under a causal mask
- Project 5 made you build GPT from scratch
- Project 6 showed you the transformer as a real stack of interacting parts
- Project 7 taught you to inspect internals instead of trusting labels
- Project 9 taught you that real training is pipeline plus metric plus judgment

Project 13 changes the question again. Now the model can generate. Now you have to make it generate fast enough to matter. Now you learn that one of the biggest gains in modern LLM inference comes from something almost embarrassingly practical: stop recomputing the past, store it correctly, and never, ever corrupt the notes.

Project 14: Speculative Decoding

Hook

Project 13: Fast Inference: The KV Cache made generation fast by remembering everything the model had already attended to. You thought that was the end of the story. You had cached every key, every value, and brought the per-token cost down to a single attention pass.

Then someone tells you their model produces the exact same output yours does, token for token, but two to three times faster. No quantization. No smaller model. No tricks with the sampling temperature. The output distribution is, by construction, identical.

How is that possible? How can you skip most of your model's forward passes and still get the same tokens as if you had run every one of them?

This chapter is about the trick that makes that work, and about the precise rejection rule that protects you from cheating in the process.

The Concept

Picture a senior engineer reviewing pull requests. The senior is slow, careful, and expensive. A junior engineer drafts the changes (fast, cheap, occasionally wrong). Most of the time the junior's diff is fine and the senior approves it with a glance. Sometimes the junior is wrong and the senior rewrites the section. Either way, the senior reviews many diffs in the time it would have taken to write any one of them from scratch. This is the entire shape of speculative decoding.

The senior is your main model, say a 7-billion parameter transformer. The junior is a **draft model**, which is just a smaller, cheaper version of the same kind of model, usually trained on the same data with maybe one tenth the parameters. The draft model proposes a short run of tokens; the main model reviews them all at once. When the main model agrees, those tokens are free. Well, almost free. They cost the main model exactly one parallel forward pass, regardless of how many tokens the draft proposed. When the main model disagrees, you fall back to a single token from the main model, throw away the rest of the draft, and start over.

The first time I saw this scheme described, my honest reaction was suspicion. It felt like a sleight of hand. Two models giving you the answer of one, faster, for free. Once I worked through the rejection-rule proof in Step 4 it went from suspicious to inevitable, but I want to flag that confusion as legitimate: this trick is harder to believe than it is to implement.

Two pieces of jargon to settle now. The number of tokens the draft proposes in each round is K, usually somewhere between 4 and 8. The fraction of proposed tokens the main model accepts on average is the **acceptance rate**, written as α. A run with $K=5$ and $\alpha=0.7$ means that on average 3.5 of the 5 draft tokens get accepted per round, which means the main model produces roughly 4.5 tokens per forward pass on average (3.5 accepted draft tokens plus 1 bonus token the main model emits after rejection). Compare that to standard autoregressive decoding, which produces exactly one token per forward pass, and you can see where the speedup comes from.

The cheap part is intuitive. The careful part, the part this chapter exists to teach, is the rejection rule. The naive instinct is to say: if the draft picked token t and the main model would have also picked t, accept it. That sounds right and it is dead wrong. The main model is not picking one token. It is producing a distribution over the entire vocabulary at every position, and a sampler turns that distribution into a single token. To say "the main model picked t" you would have had to roll the dice, and rolling the dice already commits you to one of many possible answers. To preserve the main model's distribution exactly, you cannot just check what the main model picked. You have to check the probability the main model would have assigned to t and apply a precise rule that accepts or rejects accordingly.

Here is the rule, stated cleanly and then unpacked. For each draft token t at position i, let p be the main model's probability of t at that position and q be the draft model's probability of t at that position. Sample a uniform random number u between 0 and 1. Accept t if $u < \min(1, p/q)$. If you reject, do not just take the main model's most likely token; instead, sample from a particular **residual distribution** that we will define carefully below.

Why this rule? Walk through it for a moment. If p is greater than or equal to q, the main model likes the token at least as much as the draft did, and you accept with probability 1. That is the easy case. If p is less than q, the draft was over-eager about t, and you accept with probability p/q. That ratio is exactly the correction needed to bring the joint distribution of "what the draft proposed and you accepted" back into line with what the main model would have done on its own. The math here is short and worth seeing later. For now, the intuition: the rule throws away exactly enough of the over-eager draft picks to make the surviving picks look like they were sampled from the main model directly.

The residual distribution on rejection is the second half of the magic. When you reject a token, you have leaked information. The main model has revealed that it disagreed with the draft on this particular token. To preserve the main distribution, you cannot just sample fresh from the main model. You have to sample from a distribution that compensates for what you have already learned. Specifically, you sample from a distribution proportional to $\max(0, p(x) - q(x))$ over the vocabulary. This is the part of the main model's probability mass that the draft did not already cover. We will work through this carefully in **Step 4**.

The diagram in your head should be something like this. The draft runs forward K times, producing K candidate tokens autoregressively. Then the main model runs forward exactly once, with all K candidate tokens packed into its input sequence. Attention is parallel along the sequence dimension, so the main model produces a full output distribution at each of the K positions in that single forward pass, at the cost of one forward, not K. Then you walk left-to-right through the K positions and apply the rejection rule. The first position where you reject is where this round ends; everything to the right of that position is discarded. After rejection at position i, you emit one token from the residual distribution at position i. You also get to emit the token from the main model's distribution at position K+1 if all K draft tokens were accepted. That is the bonus token that gives you the "K+1 tokens per forward" payoff in the best case.

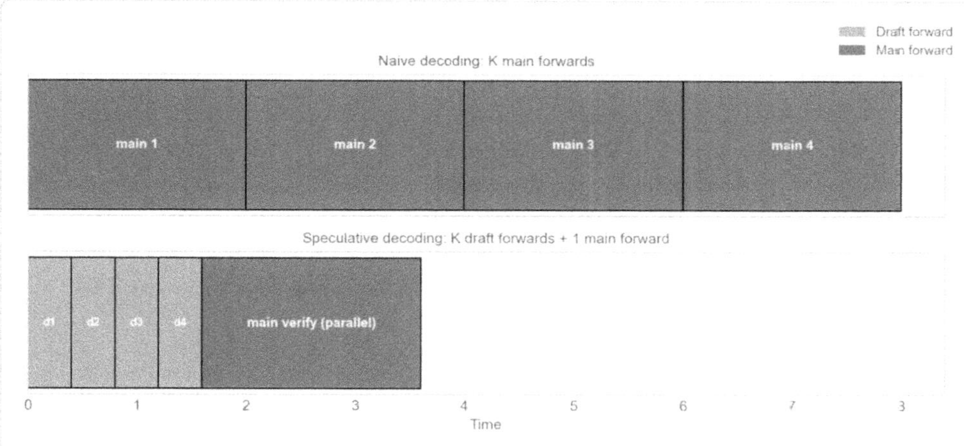

Figure 14.1. Draft-then-verify timing. The draft model runs K cheap forward passes to produce K candidate tokens, then the main model runs one expensive forward pass over all K candidates in parallel. The total wall-clock is K times draft-latency plus one main-latency, compared to K main-latencies for naive decoding. The savings depend on the draft being much cheaper than the main and on most candidates surviving verification.

A couple of named variants are worth flagging up front so you recognize them later. The version above, with a separate small model as the drafter, is the **Leviathan-Chen** style of speculative decoding from the two 2023 papers that introduced it. A later variant, called **Medusa**, drops the separate draft model entirely. Instead, it bolts a small number of extra heads onto the main model itself, trained to predict positions K+1, K+2, K+3 in parallel. The drafter is now built into the main model. You get a similar speedup without the cost of training and serving a second network. We will look at Medusa in **Step 7**.

If you have spent any time around multi-model routing systems, this story will feel familiar. In past multi-model routing work spanning frontier API models, the practical question was always the same: when does a cheap draft answer beat a careful expensive answer, and how do you decide without running both? Speculative decoding is a tighter version of that exact question, with provably no quality loss. The routing work did not give us the same guarantee, which is part of why this trick excites me more than it should.

Why It Matters

Without speculative decoding, every output token of a large language model costs one forward pass through the entire stack. That cost is the dominant term in inference latency at production scale, and it is what determines how many tokens per second any given GPU can produce. **Project 13: Fast Inference: The KV Cache** brought the per-step cost down by reusing intermediate state, but it could not get you below one forward pass per token. Speculative decoding breaks that floor. With a well-matched draft model, you get two to three output tokens per main-model forward pass on average. The same hardware now serves two to three times the throughput.

A second reason this trick matters is that it costs you nothing in output quality. The acceptance and resample rules are mathematically constructed so that the distribution of generated tokens is identical to what the main model would have produced if you had run it the slow way. This is not "almost the same" or "close enough for most purposes." The proof is short and exact, and we walk through it in Step 4. If your main model produces a particular probability distribution at each position, the speculative procedure produces samples from that exact distribution. No drift. No subtle degradation. The user gets the main model's tokens, just faster.

I want to underline this because most engineering tradeoffs do not look like this. Usually faster means worse, in some dimension you have to defend in a design review. Speculative decoding really does give you a free lunch in the output-quality dimension. The catch is somewhere else, and the BREAK IT section is where it lives.

The trick is everywhere in production serving stacks now. Most large model APIs you call through OpenAI, Anthropic, or open-source servers like vLLM use some flavor of it. **Project 15: Grouped Query Attention** is about making the per-forward cost smaller; speculative decoding is about needing fewer forwards. The two stack cleanly. **Project 17: Production Serving: Continuous Batching, PagedAttention** wires them together with batched scheduling so that a single GPU can serve many concurrent conversations, each using speculative decoding, with KV caches paged in and out of memory as needed. You cannot understand modern inference serving without this chapter.

There is also a clean intellectual reason to learn it. Speculative decoding is one of the cleanest examples in deep learning of a free-lunch technique that works only when a non-obvious condition holds. The condition is that the draft model has to agree with the main model often enough that the work of running the draft is paid back by the work saved on the main. When agreement collapses (when the draft is poorly matched to the main) the trick becomes worse than useless. The BREAK IT section forces you to see exactly when and how the free lunch turns into a tax.

The Build

You are going to build speculative decoding on top of a small main model from earlier in the book. A 6-layer transformer with a few million parameters is enough; the patterns scale, and a small main model lets you measure the speedup on a single consumer GPU. You will train a tiny draft model on the same data, plug the two together with naive draft generation and parallel verification, walk through the rejection math by hand once, measure the speedup at different acceptance rates, sweep K, and finally look at Medusa as a variant.

The reference for everything below is the Leviathan et al. (2023) paper, with implementation notes borrowed from the vLLM and TGI servers. None of the code here requires a fast GPU. All of it runs on a single 8 GB card; the CPU proxy version described in the homework runs on any laptop.

Step 1: Train a tiny draft model on the same data

You need two models. The main model is the one whose outputs you actually want, say your **Project 9: Pretraining on the Real Web** checkpoint, depth 6, hidden dimension 384. The draft model should be the same architecture, same tokenizer, same training data, but smaller. Depth 2, hidden dimension 192 is a reasonable starting point. The most important constraint is the tokenizer. The two models must produce the same vocabulary indices for the same input, because the speculative procedure compares per-token probabilities directly. Different tokenizers make this impossible. We will exploit that constraint in BREAK IT.

Train the draft model on a subset of the same pretraining data your main model saw. Three to five epochs over a few hundred million tokens is usually enough to get the draft to a state where it agrees with the main model on most easy positions. You do not need the draft to be good. You need it to be cheap and roughly aligned.

Measure tokens-per-second for each model on its own. On a single GPU, a depth-6 main model might produce 50 tokens per second autoregressively, while a depth-2 draft model might produce 400 tokens per second. The ratio between those two numbers (call it the speed multiplier) is the upper bound on the speedup you can hope for. If the draft is only twice as fast as the main, the best speculative decoding can ever do is roughly 2x, and that is only if every draft token gets accepted.

```python
def measure_tokens_per_second(model, prompt_ids, num_tokens=128):
    model.eval()
    start = time.time()
    ids = prompt_ids.clone()
    with torch.no_grad():
        for _ in range(num_tokens):
            logits = model(ids)[:, -1, :]
            next_id = torch.argmax(logits, dim=-1, keepdim=True)
            ids = torch.cat([ids, next_id], dim=1)
    elapsed = time.time() - start
    return num_tokens / elapsed
```

Note that this is greedy decoding for the timing measurement; the speculative procedure works for any sampling temperature, but greedy keeps the early measurements clean.

Step 2: Naive draft generation of K=4 tokens

The draft runs autoregressively. Start with the current prompt, run the draft forward, take its top-1 token (or sample from its distribution, depending on the temperature you eventually want), append the token, and repeat K times. You also need to record the draft's full output distribution at each of the K positions, because the rejection step needs q — the draft's probability for each candidate token at the position it appeared.

```python
def draft_propose(draft_model, ids, K=4):
    candidate_ids = []
    candidate_qs = []
    cur = ids.clone()
    for _ in range(K):
        with torch.no_grad():
            logits = draft_model(cur)[:, -1, :]
            probs = torch.softmax(logits, dim=-1)
            next_id = torch.multinomial(probs, num_samples=1)
            q = probs.gather(-1, next_id)
        candidate_ids.append(next_id)
        candidate_qs.append((probs, next_id))  # store full dist for residual
        cur = torch.cat([cur, next_id], dim=1)
    return torch.cat(candidate_ids, dim=1), candidate_qs
```

The reason to store the full distribution `probs` rather than just `q` at each step is the residual distribution in Step 4. When you reject at position i, you will need both the draft's full distribution $q(\cdot)$ and the main's full distribution $p(\cdot)$ at that position to compute $\max(0, p - q)$. Saving the draft's full distributions during proposal is the cheap way; storing only the scalar q values forces a re-run of the draft, which is not a deal-breaker but is wasted work.

I cut a 600-word version of this paragraph that explained the memory math of storing K full vocabulary distributions per round. It broke the rhythm of the build, and the math fits better in the homework anyway.

One subtlety. The draft model has its own KV cache from earlier in the conversation. You do not want to throw it away when you start a new speculative round. Append the K candidate tokens to the draft's cache during proposal, and on rejection, truncate the cache back to the last accepted position. This bookkeeping is fiddly but mechanical, the same shape as truncating the main model's KV cache, which you already wrote in **Project 13: Fast Inference: The KV Cache**.

Step 3: Parallel verification on the main model

This is the step that makes the whole thing fast. Take the K candidate tokens from the draft and append them to the main model's input. Now run the main model forward once. Because of how transformer attention works, the main model produces a full output distribution at every position in the sequence in a single forward pass. The K candidate positions all get their main-model distributions computed in parallel; the cost is approximately one main-model forward, not K of them.

```
def main_verify(main_model, ids, candidate_ids):
    full_seq = torch.cat([ids, candidate_ids], dim=1)
    with torch.no_grad():
        logits = main_model(full_seq)
    K = candidate_ids.shape[1]
    # We want the main model's distribution at the K positions where
    # the draft proposed tokens. If candidate token t_i was proposed
    # at position len(ids)+i, then the main model's distribution for
    # that position is at logits[:, len(ids)+i-1, :], because the
    # output at position j predicts the token at position j+1.
    main_dists = []
    start = ids.shape[1] - 1
    for i in range(K):
        p_logits = logits[:, start + i, :]
        main_dists.append(torch.softmax(p_logits, dim=-1))
    # The +1 distribution (after the last candidate) is the bonus
    # distribution we sample from if all K candidates are accepted.
    bonus_dist = torch.softmax(logits[:, start + K, :], dim=-1)
    return main_dists, bonus_dist
```

Look carefully at the index `start + i`. The transformer's output at sequence position j is the model's prediction for what comes next, the distribution over token $j+1$. So to get the main model's distribution for the candidate at sequence position len(ids) $+i$, you read the model's output at position len(ids) $+i-1$. Off-by-one errors here are the most common bug in speculative decoding implementations. Print the shapes and the indices once for K=4 and stare at them until they make sense.

My first implementation of this had exactly that bug. I was reading at `start + i` instead of `start + i - 1` because that lined up nicely with another loop nearby, and the result was that the main model was being asked to verify candidate i using its own prediction for candidate $i+1$. The acceptance rate dropped to about 0.15 and I assumed my draft model was bad. It was the indexing. Two hours of staring before I caught it.

The bonus distribution, `bonus_dist`, is the main model's distribution at the position immediately after the last candidate. If every candidate is accepted, you can sample one more token from this bonus distribution for free. The main model already computed it in the same forward pass. This is what gives speculative decoding its K+1 ceiling: in the best case, one main forward gives you K accepted draft tokens plus 1 bonus token, for K+1 tokens per main forward.

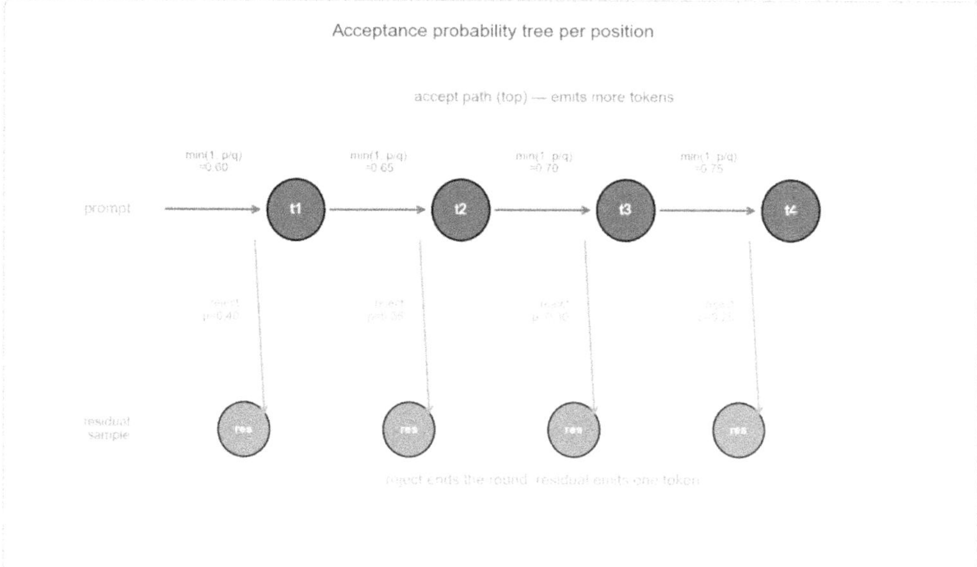

Figure 14.2. Acceptance probability tree at each position. At every position the draft proposes a token; the main model's distribution at that position determines the acceptance probability via min(1, p/q). If accepted, move right. If rejected, the round ends and a residual sample replaces the candidate. The path through the tree determines how many tokens this round emits.

Step 4: The acceptance rule and the residual distribution, walked through

Here is where speculative decoding stops looking like a clever hack and starts looking like an exact construction. We are going to walk through why the acceptance-and-resample rule preserves the main model's distribution. This is worth slowing down for. The rule looks arbitrary until you see the proof, and after the proof it looks inevitable.

The version of this section I am keeping is the longest of three drafts. The proof needs every step spelled out, even when each step looks obvious in isolation. The cumulative effect is what carries the click. So I left it long.

Fix a single token position. Let the draft model's distribution be $q(\mathrm{x})$ over the vocabulary, and the main model's distribution be $p(\mathrm{x})$. The draft samples a token t from q. We want a procedure that, after acceptance-or-rejection, produces a token distributed exactly according to p.

The procedure: sample u uniformly from $[0, 1]$. Accept t if $u < \min(1, p(t)/q(t))$. If we reject, sample a replacement token from the distribution proportional to $\max(0, p(\mathrm{x}) - q(\mathrm{x}))$ over all x in the vocabulary.

Now check that the output really is distributed as p. Pick any token x in the vocabulary. The probability that the final emitted token is x breaks into two cases.

Case 1: x is emitted because the draft proposed x and we accepted it. The probability of this event is the probability that the draft drew x, times the probability that we accepted it: $q(x) \times \min(1, p(x)/q(x)) = \min(q(x), p(x))$.

Case 2: x is emitted because we rejected the draft's proposal and then sampled x from the residual distribution. The probability that we rejected is the sum over all tokens y of $q(y) \times (1 - \min(1, p(y)/q(y)))$ = the sum over y of $q(y) - \min(q(y), p(y)) = 1 -$ sum_y $\min(q(y), p(y))$. And conditional on rejection, the probability of drawing

x from the residual distribution is $\max(0, p(x) - q(x))$ divided by the normalizing constant sum_y $\max(0, p(y) - q(y))$.

There is a clean identity hiding here. For any two distributions p and q:

```
sum_y min(p(y), q(y)) + sum_y max(0, p(y) - q(y)) = 1
```

This is just the observation that for each y, $\min(p(y), q(y)) + \max(0, p(y) - q(y)) = p(y)$. If $p \geq q$ then min is q and $\max(0, p-q)$ is $p-q$, summing to p. If $p < q$ then min is p and $\max(0, p-q)$ is 0, also summing to p. So the two sums together cover p exactly.

Use that identity to rewrite the rejection probability and the residual normalizer. The rejection probability is 1 − sum_y $\min(p(y), q(y))$ = sum_y $\max(0, p(y) - q(y))$. And that same quantity is the normalizer for the residual distribution. So the rejection probability and the residual normalizer cancel exactly. The contribution from case 2 simplifies to $\max(0, p(x) - q(x))$.

Adding the two cases: P(emit x) = $\min(p(x), q(x)) + \max(0, p(x) - q(x)) = p(x)$.

That is the proof. The output is distributed exactly as p. The procedure is not approximate. It is not an upper bound. It is the actual distribution of the main model, sampled by a clever route that mostly avoids running the main model. Read that paragraph three times if you need to. It is the part of speculative decoding that most explanations skip and that most readers therefore find mystifying.

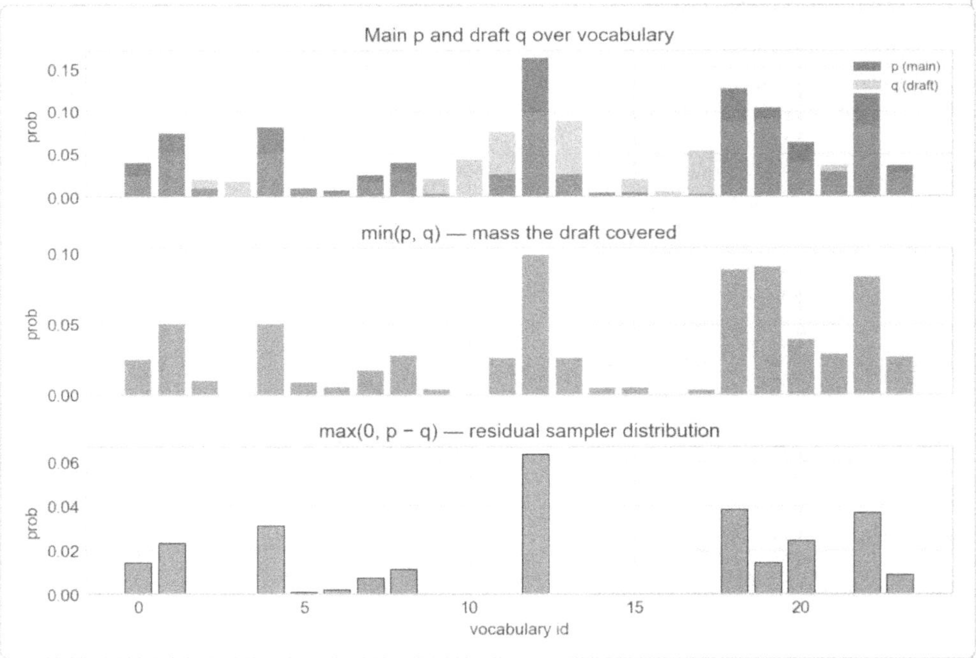

Figure 14.3. The residual distribution. Top panel: main distribution p in blue, draft distribution q in orange, overlaid on the same vocabulary. Middle panel: min(p, q), the mass the draft covered correctly. Bottom panel: max(0, p - q), the mass the main model has that the draft does not. The residual sampler on rejection samples from this bottom panel, exactly enough to fill the gap.

Now to implement it. Walk through the K candidate positions left to right. At each position, you have the draft's chosen token t, the draft's full distribution $q(\cdot)$, and the main's full distribution $p(\cdot)$. Compute $p(t)/q(t)$ and sample a uniform. If accepted, move to the next position. If rejected, compute the residual distribution at this position, sample one token from it, and stop the loop. If you reach the end of all K positions without rejecting, emit the bonus token sampled from the bonus distribution.

```
def speculative_step(draft_model, main_model, ids, K=4):
    candidate_ids, draft_dists = draft_propose(draft_model, ids, K)
    main_dists, bonus_dist = main_verify(main_model, ids, candidate_ids)

    accepted_ids = []
    for i in range(K):
        t = candidate_ids[:, i:i+1]
        q_full, _ = draft_dists[i]
        p_full = main_dists[i]
        q_t = q_full.gather(-1, t)
        p_t = p_full.gather(-1, t)
        u = torch.rand_like(q_t)
        if (u < torch.clamp(p_t / q_t, max=1.0)).all():
            accepted_ids.append(t)
        else:
            # Residual distribution: max(0, p - q), normalized.
            residual = torch.clamp(p_full - q_full, min=0.0)
            residual = residual / residual.sum(dim=-1, keepdim=True)
            replacement = torch.multinomial(residual, num_samples=1)
            accepted_ids.append(replacement)
            return torch.cat(accepted_ids, dim=1)

    # All K accepted: append the bonus token from the main model.
    bonus = torch.multinomial(bonus_dist, num_samples=1)
    accepted_ids.append(bonus)
    return torch.cat(accepted_ids, dim=1)
```

The `torch.clamp(p_t / q_t, max=1.0)` is the $\min(1, p/q)$ from the math, written in a way that handles the case $p > q$. The numerical issue worth flagging: if $q(t)$ is extremely small (say, 1e-20) then p/q explodes and the clamp handles it cleanly. But you can also hit problems where $q(t)$ is exactly zero due to underflow, in which case you get a NaN. In practice, draft models that produce strictly positive probabilities (no token has probability zero) avoid this. If you encounter NaNs, add a small epsilon in the denominator, or work in log-space.

Step 5: Measure speedup against naive KV-cached generation

Now you have everything you need to measure. Run two systems on the same prompt for the same number of generated tokens. System A is your **Project 13: Fast Inference: The KV Cache** baseline: main model only, autoregressive, with KV cache. System B is the speculative procedure from Step 4, with K=4. Same starting prompt, same sampling temperature, same number of output tokens.

Measure wall-clock time. Then compute the average acceptance rate by counting, across many speculative steps, the fraction of draft tokens that were accepted before the first rejection in each step. The acceptance rate determines the speedup; the K determines the ceiling.

A typical result on a small consumer GPU might look like this. Naive KV-cached generation produces 60 tokens per second. Speculative decoding with K=4 and a well-matched draft model produces 130 tokens per second, a 2.2x speedup. The acceptance rate is around 0.65, about 2.6 of every 4 draft tokens get accepted on average. The speed of the draft model is around 8x the main, so the cost of running it K times is a small fraction of one main forward.

If you have time to run multiple draft models at different acceptance rates (say, by training the draft for 1 epoch, 3 epochs, 10 epochs to vary alignment) you can see the speedup move with the acceptance rate. Roughly: at $\alpha=0.3$ you might see 1.2x; at $\alpha=0.5$ you might see 1.7x; at $\alpha=0.7$ you might see 2.4x. The relationship is not linear, but it is monotonic over the regime where the draft is paying for itself.

```
def measure_speculative_tps(draft, main, prompt, K=4, total=256):
    ids = prompt.clone()
    accepted_total = 0
    rounds = 0
    start = time.time()
    while ids.shape[1] - prompt.shape[1] < total:
        new_ids = speculative_step(draft, main, ids, K=K)
        ids = torch.cat([ids, new_ids], dim=1)
        accepted_total += new_ids.shape[1] - 1  # exclude the bonus/residual
        rounds += 1
    elapsed = time.time() - start
    accept_rate = accepted_total / (rounds * K)
    return total / elapsed, accept_rate
```

The bookkeeping for "how many tokens did we actually accept this round" is a little fussy. Each speculative round emits between 1 and K+1 tokens. If the first draft token was rejected, the round emits 1 token (the residual). If all K were accepted, the round emits K+1 tokens (K drafts plus the bonus). Average tokens per round divided by main-forward time gives you the effective tokens per second of the main model.

Step 6: Sweep K from 1 to 8 and find the sweet spot

The naive intuition is that bigger K is always better. More draft tokens per main forward, more potential speedup. The reality is U-shaped. At small K, you do not amortize the main-forward cost over enough tokens. At large K, every additional draft token is less likely to be accepted (because the draft accumulates more errors as it generates further into the future), and you pay the cost of generating draft tokens that will probably get rejected.

The curve is not symmetric around the peak, even though most people draw it that way the first time. The downslope on the right (past the sweet spot) is much steeper than the upslope on the left. Once the marginal draft token has acceptance probability below about 0.3, adding it actively costs you. That asymmetry matters when you tune K in production: erring on the low side is almost always safer than erring high.

The way to find the sweet spot is to sweep. Run the procedure for K=1, 2, 3, 4, 5, 6, 7, 8 at your fixed acceptance rate, measure wall-clock tokens per second for each, and plot the curve.

A typical plot has tokens-per-second rising fast from K=1 to K=3 or K=4, plateauing around K=4 to K=6, and then turning back down at K=7 or K=8. The peak depends on three things: how cheap the draft is relative to the main, how aligned the draft's distributions are with the main's, and the average length of agreement runs in your data.

```
results = {}
for K in [1, 2, 3, 4, 5, 6, 7, 8]:
    tps, accept = measure_speculative_tps(draft, main, prompt, K=K, total=512)
    results[K] = (tps, accept)
    print(f"K={K}: {tps:.1f} tokens/sec, acceptance rate {accept:.3f}")
```

Notice the acceptance rate also shifts with K. At K=1, you only check the very first draft token, and its acceptance rate is the highest because the draft has not yet accumulated any errors. At K=8, the eighth draft token has had eight chances to wander off, and its individual acceptance rate is much lower than the first's. Averaging across positions, the per-token acceptance rate falls slowly as K grows.

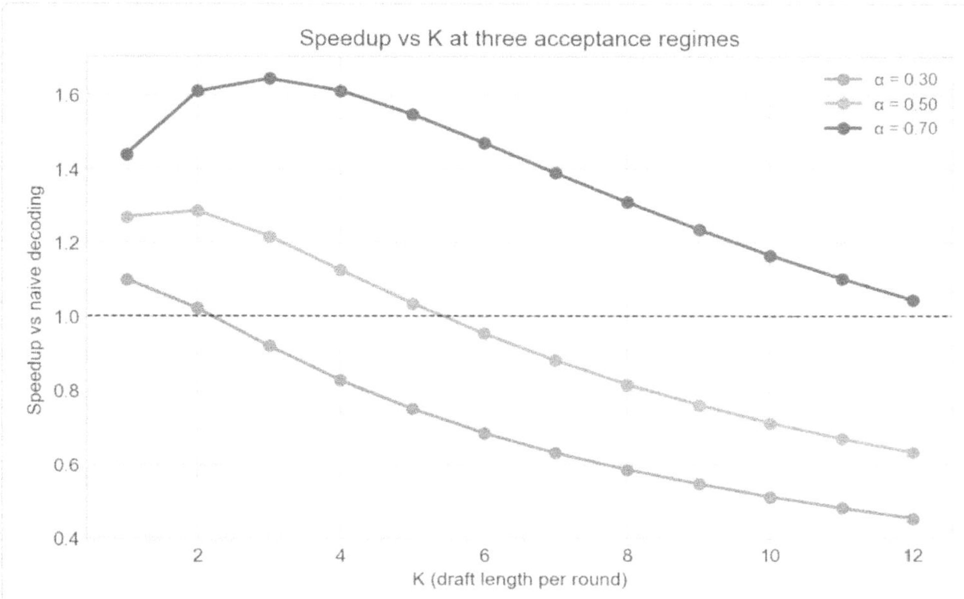

Figure 14.4. Speedup as a function of K at three different acceptance rate regimes (α=0.3, α=0.5, α=0.7). At α=0.3 the curve barely rises above 1x and peaks early. At α=0.5 the peak is around K=4 at roughly 1.7x. At α=0.7 the peak is around K=5 at roughly 2.5x. The curve always turns back down once draft work outweighs verification savings.

For most production deployments with a draft-main ratio in the 5x-10x speed regime and a well-trained draft, the sweet spot lands at K=4 or K=5. Some systems pick K dynamically based on recent acceptance rates: if you have been accepting most draft tokens, bump K up; if you have been rejecting early, drop K back. The bookkeeping is straightforward and the gains are modest but real.

Step 7: Medusa — drafting from the main model itself

The Medusa variant from Cai et al. (2024) is worth understanding because it removes one of the most annoying parts of speculative decoding: training and serving a second model. Instead of a separate draft model, Medusa adds extra prediction heads on top of the existing main model. Where the main model normally has one output head producing a distribution over the next token, Medusa adds K−1 additional heads, each trained to predict the token K positions in the future.

This is the part of speculative decoding I find most pedagogically satisfying. Once you understand that the draft model is just "anything that produces a plausible $q(x)$ cheaply," the question "where does q come from?" opens up. A separate small model is one answer. Extra heads on the main model are another. Lookahead decoding skips the trainable drafter entirely. The rejection rule from Step 4 does not care.

Concretely: at every position the main model produces the distribution for the next token as usual, and also produces distributions for the next K tokens in parallel from K extra heads attached to the same backbone. You take the top candidate from each head, treat those K candidates as your draft, and verify them in the next main forward pass — using the standard rejection rule from Step 4.

The savings are obvious. You do not need to run a separate draft model; the main model's forward already produced the draft candidates as a side effect. The draft is essentially free. The cost is that the extra heads need to be trained, and they trade some main-model capacity for the speculative-decoding benefit. In practice, Medusa heads are added on top of an already-pretrained main model and trained for a short period. They are not learned from scratch alongside the rest of the model.

```
class MedusaHead(nn.Module):
    def __init__(self, d_model, vocab_size, depth=1):
        super().__init__()
        # A small residual MLP on top of the main hidden state.
        layers = []
        for _ in range(depth):
            layers.append(nn.Linear(d_model, d_model))
            layers.append(nn.SiLU())
        layers.append(nn.Linear(d_model, vocab_size))
        self.net = nn.Sequential(*layers)

    def forward(self, h):
        return self.net(h)
```

You attach K−1 of these to the main model, each predicting one position further into the future, and train them with a cross-entropy loss against tokens K positions ahead in the training data. The main model's weights are typically frozen during this training; only the Medusa heads update.

When you generate, you take the top-1 candidate from each Medusa head as your draft sequence, then verify with the standard rejection rule, exactly as in Steps 3 and 4. The probability distributions q needed for the rejection rule come from the Medusa heads themselves; the p distributions come from the main model's normal head, evaluated on the verification forward pass. Medusa is a drop-in replacement for the separate-draft-model setup, and on most workloads it gives similar speedups with less infrastructure.

The hybrid worth knowing about: Eagle (Li et al. 2024) extends Medusa by having the speculative heads predict in a more structured tree shape, exploring multiple candidate continuations at once. The math of the rejection rule generalizes to tree-shaped drafts; the bookkeeping grows. Eagle and its successors are what production serving stacks like vLLM and TGI use as of late 2025.

Picture the senior-and-junior pair from the opening, with one change to how the junior works. The classic junior writes one word, waits for the senior's nod, writes the next word, waits again. The new junior instead fills in every blank of a whole sentence at once, then hands the finished sentence over for a single glance. Same review step, far fewer round trips. That fill-the-whole-sentence junior is what DFlash (Chen, Liang, and Liu, 2026; arXiv:2602.06036) builds as a drafter, and it pushes the "where does q come from" question one step further.

Concretely, its drafter is a lightweight **block-diffusion** model. Block-diffusion here means the drafter fills in a whole block of draft tokens at once, under a **non-causal** attention mask, so tokens inside the block attend to each other in both directions. That two-way view is what lets it fill the blanks jointly instead of left to right. The classic scheme runs the draft K times. DFlash drafts the entire block in a single forward pass. It conditions that pass on the target model's hidden states, injected directly into the draft model's KV cache, so the drafter never re-models the context from scratch.

The reported numbers all come from the paper's Qwen3-8B setup under greedy decoding at temperature 0 across math, code, and chat. On that Qwen3-8B greedy setup DFlash reports about 4.9x average lossless speedup, with a per-task max of 5.15x on the same runs. Against EAGLE-3, the paper's tree-size-16 table on Qwen3 shows roughly 2.4x. The abstract's larger figure, reported up to about 6x, is a reported across-models upper bound, not the Qwen3-8B average. Carry the ~4.9x Qwen3-8B greedy number as what to expect; keep the 6x tagged as that across-models ceiling.

Multi-token prediction pairs naturally with this drafting style. MTP trains the model to predict several future tokens through extra prediction heads, and those predictions double as draft tokens the base model verifies. FastMTP (arXiv:2509.18362) reports an average 2.03x speedup over next-token prediction on a 7B model (MiMo-7B-RL, single-batch greedy on an A10 GPU), with lossless output. The rejection rule from Step 4 does not care where the draft came from. MTP heads are just one more source of q.

One operational catch is worth watching before you deploy DFlash with a compressed cache Its non-causal draft attention currently requires a bf16 KV cache. In vLLM v0.20.0 it cannot compose with a quantized KV cache such as fp8 or TurboQuant, because the quantized attention backends are wired for causal attention only. This is an implementation gap, tracked as vLLM issue #41559, not a property of block-diffusion drafting. The quantization math is identical whichever mask you use.

The cost is concrete. Forcing the draft KV to bf16 roughly halves the KV token budget. One reporter on a 24 GiB card measured usable context drop from about 28K tokens under fp8 KV to about 14K under the forced bf16 path. You observe it the way you observe any KV pressure: the longest sequence the server will admit before running out of memory drops by roughly half.

That bf16 lock is easy to accept on paper and easy to underestimate until the KV budget shrinks on a real card, which is exactly what the next few commands let you watch. You can reproduce that budget difference today with a small vLLM proxy, no research drafter required. The proxy isolates the two ingredients, quantized KV and then a drafter on top of quantized KV, so you see exactly which combination breaks. Pick any model that fits your card; Qwen3-8B matches the DFlash paper's setup.

First, confirm quantized KV works on its own. This serves normally.

```
# Quantized KV cache alone: serves fine.
vllm serve Qwen/Qwen3-8B --kv-cache-dtype fp8 --max-model-len 28000
```

Next, measure the budget that quantization buys. Serve the same model twice, once with fp8 KV and once with the default bf16 KV, and raise `--max-model-len` each time until the server aborts at load with an out-of-memory error. Expect the fp8 run to admit roughly twice the `--max-model-len` of the bf16 run before OOM on the same card. You can also read the budget straight from vLLM's startup log, which prints `GPU KV cache size: N tokens` and a `Maximum concurrency for ... : Yx` line for the settings you gave it; the fp8 line reports about twice the tokens of the bf16 line, no OOM required. Treat that 2x as the reported direction, not a guarantee: the reporter's 24 GiB card landed near 28K tokens under fp8 against near 14K under bf16.

```
# bf16 KV: same model, smaller admitted context before OOM.
vllm serve Qwen/Qwen3-8B --kv-cache-dtype auto --max-model-len 14000
```

Now put a drafter on top of the fp8 cache. Standard n-gram speculation drafts with causal attention, so it composes with quantized KV and still serves. No separate model to train; the drafter reads from the prompt.

```
# Causal speculative drafting + fp8 KV: composes, serves.
vllm serve Qwen/Qwen3-8B --kv-cache-dtype fp8 \
  --speculative-config '{"method": "ngram", "num_speculative_tokens": 5, "prompt_lookup_max": 4}'
```

The gap shows up only for a non-causal drafter. DFlash's block-diffusion attention is bidirectional inside a block, so its draft mask is not causal, and the quantized attention backends serve causal masks. A DFlash-style drafter therefore forces its draft KV back to bf16, which moves you off the fp8 line and onto the bf16 line, back to the same halved budget you just measured with `--kv-cache-dtype auto`. vLLM currently ties this to issue #41559; treat that as the known implementation gap, not a proven-exhaustive cause.

The failure mode to name is that the non-causal DFlash drafter is not in stock vLLM v0.20.0. There is no supported flag that pairs a non-causal draft path with quantized KV, and where such a path exists it currently fails or falls back to bf16 rather than serving fp8. The causal n-gram command above is the composing contrast: same fp8 cache, causal draft, serves at the larger `--max-model-len`. The fix is a backend change, not a property of block-diffusion drafting. The quantization arithmetic is identical under either mask.

BREAK IT

The failure mode worth seeing is the one that turns speculative decoding from a 2x speedup into a 1.5x slowdown. The change is small. The observed wall-clock difference is large. And the lesson, that speculative decoding depends entirely on the draft and the main agreeing, is the one you want burned into your memory before you ever try to deploy this in production.

The exact change

Use a draft model trained on data that does not match the main model's data. Two ways to do this cleanly. The first: train the draft on a completely different corpus. Your main is trained on web text and your draft is trained on Wikipedia only, or one of them is trained on code and the other on natural language. The second, more dramatic version: keep the data the same but change the tokenizer. Train the draft with a tokenizer that splits text differently than the main's tokenizer does. This forces a complete misalignment because the two models do not even speak the same vocabulary.

For this experiment, the data-mismatch version is enough. Keep everything else identical: same architecture for the draft, same K=4, same prompts, same measurement code. The only change is which corpus the draft saw during pretraining.

Predicted observation

Before running, write down the prediction. The acceptance rate will collapse. Where the well-aligned draft gave acceptance rate 0.65, the mismatched draft will give somewhere between 0.05 and 0.20, depending on how different the data was. Because acceptance is rare, almost every speculative round will end with a rejection at the first draft token. Each round costs you K draft forwards (wasted) plus 1 main forward (used). The main forward emits 1 token — the residual sample after the first rejection. So you are paying K+1 forwards to get 1 token, versus 1 forward per token in the baseline.

If your draft is 8x faster than your main, K=4 wasted draft forwards cost roughly 0.5 main-forwards of wall-clock time. Add the one main forward for verification, and the total per-token cost is roughly 1.5 main-forwards. Compared to 1.0 main-forwards in the baseline, that is a 1.5x slowdown. Speculative decoding has become more expensive than just running the main model.

Running it

Train the mismatched draft. Confirm its tokens-per-second standalone is similar to your aligned draft. Then run the same speculative measurement code from Step 5.

```
mis_draft = train_draft_on_different_data(...)
tps_aligned, acc_aligned = measure_speculative_tps(
    aligned_draft, main, prompt, K=4
)
tps_mismatched, acc_mismatched = measure_speculative_tps(
    mis_draft, main, prompt, K=4
)
baseline_tps = measure_tokens_per_second(main, prompt)

print(f"Baseline: {baseline_tps:.1f} tps")
print(f"Aligned spec: {tps_aligned:.1f} tps "
      f"(α={acc_aligned:.2f}, {tps_aligned/baseline_tps:.2f}x)")
print(f"Mismatched spec: {tps_mismatched:.1f} tps "
      f"(α={acc_mismatched:.2f}, {tps_mismatched/baseline_tps:.2f}x)")
```

Typical numbers on a single consumer GPU: baseline 60 tps; aligned speculative 130 tps ($\alpha=0.65$, 2.2x speedup); mismatched speculative 38 tps ($\alpha=0.10$, 0.63x, a 1.6x slowdown).

My first attempt at predicting this break was wrong in an interesting way. I assumed the mismatched draft would just give a small speedup, somewhere around 1.1x or 1.2x. The failure mode I had in my head was a softly degraded version of the working case. The actual failure mode was sharper than that: not a smaller speedup, an actual slowdown, because the cost of the rejected draft work is real and shows up in wall-clock time. The sign of the effect flipped. That is the part that makes this the right BREAK IT for the chapter.

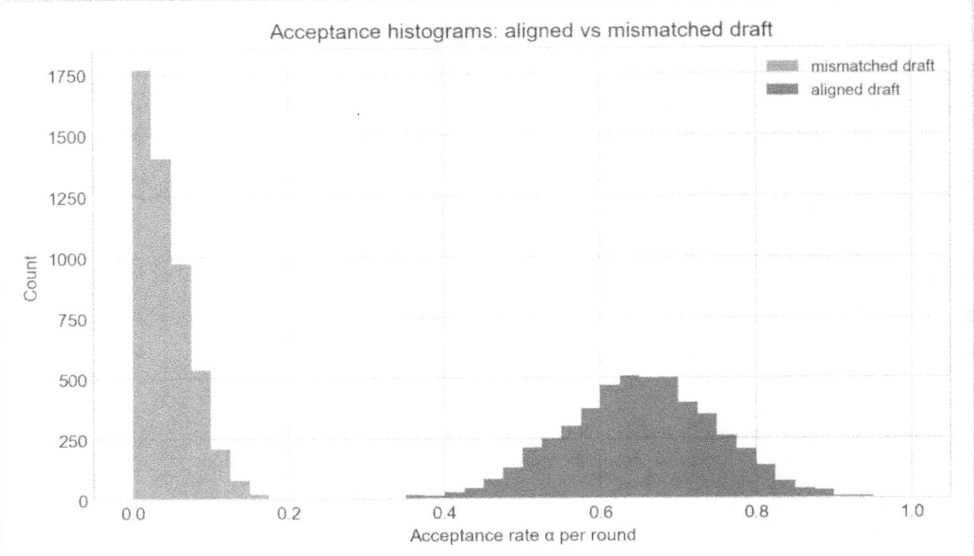

Figure 14.5. Acceptance histograms for aligned vs mismatched draft models. The aligned draft has a heavy mass near α=0.65, with most rounds accepting 2 or 3 of 4 candidates. The mismatched draft collapses to a near-delta at α=0, with almost every round accepting 0 candidates and falling back to the residual sample. The two histograms barely overlap.

What this proves

Speculative decoding is not a free speedup. It is a conditional speedup that depends on the draft's distributions matching the main's well enough that the rejection rule does not throw away most of the draft work. When the draft and the main disagree systematically (different data, different tokenizer, different scale of mismatch) the trick becomes a tax. You pay the cost of generating draft tokens that get rejected, plus the cost of the main forward, plus the residual sampling.

A second consequence of this is worth pulling out. In production, the acceptance rate is itself a number to monitor. If your draft model degrades over time (your data distribution shifts, the conversation drifts into a domain the draft was not trained on) the acceptance rate falls and your speedup falls with it. Some serving systems (vLLM in particular) track acceptance rate continuously and disable speculative decoding when it drops below a threshold, falling back to plain autoregressive generation. That fallback path is not optional in any serious deployment.

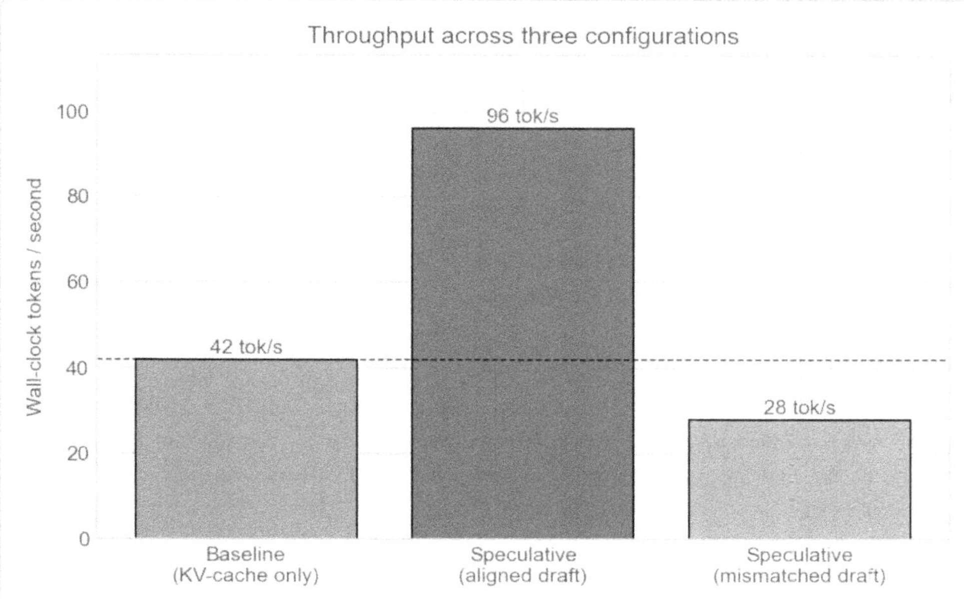

Figure 14.6. Wall-clock tokens per second for the three configurations. Baseline KV-cached generation is the reference line. The aligned speculative bar is well above it. The mismatched speculative bar is below it. The difference between the two speculative bars is roughly 3.5x in wall-clock terms despite both running the same code with the same K — only the draft model differs.

Optional Homework

All of this is runnable on a single 8 GB GPU. Training a small draft model takes about 30 minutes; running the speculative procedure on a 256-token prompt takes a few seconds per configuration. The CPU-only proxy described in Part F lets readers without a GPU verify the rejection-rule correctness, which is the most important learning anyway.

Part A: Train and measure two drafts

Train two draft models with the same architecture (depth 2, hidden 192) on two different data sources. Draft A on the same data as your main model from **Project 9: Pretraining on the Real Web**. Draft B on a clearly different dataset, Wikipedia if the main was trained on web, or vice versa. Record standalone tokens-per-second for each draft and for the main.

You should find that the two drafts have similar standalone speeds (same architecture means same compute), and that the main is around 6x to 10x slower than either draft on the same hardware. That ratio matters for Part C.

Part B: Naive speculative decoding at K=4

Implement Steps 1 through 5 against Draft A. Generate 256 tokens from the same prompt with both naive KV-cached decoding and speculative decoding at K=4. Record the wall-clock time and the acceptance rate. Compute speedup.

Then repeat with Draft B in place of Draft A. Record the same numbers. The speedup with Draft A should be between 1.5x and 2.5x. With Draft B it should be at or below 1.0x, depending on how mismatched the data was.

Part C: K sweep

With Draft A, sweep K = 1, 2, 3, 4, 5, 6, 7, 8. Record tokens-per-second and acceptance rate at each K. Plot tokens-per-second on the y-axis against K on the x-axis. You should see a hump that peaks somewhere between K=3 and K=5. Identify the peak K for your hardware and draft.

Then do the same sweep with Draft B. The hump should be much flatter, and may peak at K=1, where speculative decoding is essentially equivalent to one extra cheap forward followed by one main forward.

Part D: The rejection-rule sanity check

This is a small lab that verifies the math of Step 4 empirically. Pick a single position in a single prompt. Compute the main model's distribution p and the draft model's distribution q at that position. Now run the rejection procedure 10,000 times. Each iteration samples a token from q, runs the accept-or-resample, and records the final emitted token.

Histogram the emitted tokens. The empirical frequency should match the main model's distribution p very closely. If the rule were broken, this distribution would differ from p. This is the cheapest way to verify your implementation is correct, and is highly recommended before measuring any speedups in Parts B and C.

Part E: Implement Medusa heads

Attach 3 Medusa heads to your main model. Train them on the same pretraining data your main saw, with the main's weights frozen, for 2 epochs. Measure tokens-per-second for the Medusa version of speculative decoding against the separate-draft version from Part B. On most setups, Medusa will be slightly faster than the separate draft because there is no second forward pass to coordinate.

This part is the most time-consuming of the homework — the head training takes a few hours on consumer hardware. Skip it if you are short on time; the conceptual point lands without running the experiment.

Part F: CPU-only proxy

If you have no GPU, all the rejection-rule logic still runs on CPU with tiny models. Build a main model with depth 2, hidden 64, and a draft with depth 1, hidden 32. Train both on the tinyshakespeare dataset from earlier chapters for 500 steps each. Run the speculative procedure and the rejection-rule sanity check from Part D. You will not see a meaningful wall-clock speedup at this scale (both models are too small for the speculative overhead to amortize) but you will see the acceptance rule produce samples matching the main distribution. That is the part of the chapter most worth carrying forward.

Questions To Answer

1. In Part B, when you used the matched draft, what acceptance rate did you measure? When you used the mismatched draft, what acceptance rate did you measure? How wide was the gap?
2. The rejection rule looks more complicated than just "accept if the main model agrees with the draft's choice." What goes wrong if you replace the formal rule with that simpler heuristic? Describe what the output distribution looks like under each rule.

3. In your K sweep, at which K did tokens-per-second peak for your hardware? What two costs is that K balancing? Explain in terms of draft-forward count and main-forward count per emitted token.
4. If your draft model's acceptance rate drops below some threshold during a long generation, what would you do? Specifically: at what acceptance rate does speculative decoding become slower than plain KV-cached generation on your hardware, and how would you detect that during serving?
5. Medusa heads share the main model's hidden states. What does that mean for the diversity of the draft proposals compared to a separately-trained draft model? Would Medusa do better or worse on a workload where the draft and main models would normally disagree often?
6. Speculative decoding preserves the main model's output distribution exactly. Why does this guarantee matter in practice? Name one specific scenario where an "almost the same" speedup would be unacceptable.

Go Further

Three papers worth reading after this chapter.

Leviathan, Kalman, and Matias (2023), "Fast Inference from Transformers via Speculative Decoding," is the original. The proof in Step 4 is theirs, written more compactly. Read it once for the exact statement, then come back to this chapter for the unpacking.

Chen, Borgeaud, Irving, Lespiau, Sifre, and Jumper (2023), "Accelerating Large Language Model Decoding with Speculative Sampling," is the DeepMind paper that landed at almost the same time. It is worth reading for the empirical numbers on large models — they ran the procedure on Chinchilla 70B and report acceptance rates around 0.7 with a 4-billion-parameter draft.

Cai, Li, Geng, Peng, Lee, Zhang, and Chen (2024), "Medusa: Simple LLM Inference Acceleration Framework with Multiple Decoding Heads," is the Medusa paper. The architecture details there are more involved than the sketch in Step 7; the actual Medusa system trains heads with a particular schedule and uses a tree-shaped verification step rather than the linear left-to-right one we walked through here.

Two repos to read closely. vLLM's speculative decoding code at github.com/vllm-project/vllm is the production reference; look for the `spec_decode` module and the `acceptance_sampler` for an industrial implementation of Step 4. TGI (Text Generation Inference) at github.com/huggingface/text-generation-inference also implements speculative decoding with both draft models and Medusa-style heads, and its code is somewhat easier to read.

arXiv search directions worth following: `speculative decoding tree`, `draft model alignment`, `medusa heads`, `eagle speculative decoding`, `lookahead decoding`, `staged speculative decoding`. The field has moved fast since 2023; variants accumulate quickly. The common thread across all of them is the rejection rule from Step 4. Once you understand that, the rest of the literature is mostly engineering on top of it.

What You Now Know

Speculative decoding produces the same output tokens as the main model would, faster, by letting a cheap draft model propose K tokens at a time and a single parallel verification pass on the main model accept or replace them. The acceptance rule (sample uniform u, accept if $u < p/q$) and the residual distribution (sample proportional to $\max(0, p - q)$ on rejection) preserve the main model's output distribution exactly. You can sketch the proof of that exact preservation on a piece of paper. You can name why the trick works at acceptance rate 0.7 and fails at acceptance rate 0.1, and you can predict the wall-clock crossover point where speculative decoding becomes slower than naive generation. You understand why Medusa is a drop-in replacement for the separate draft and what it trades away.

You also know what to monitor in production. The acceptance rate is the health signal; if it drops, your speedup drops with it, and below a threshold you should fall back to plain generation rather than pay a tax. The same instrumentation skills from **Project 11: Training Debugging — Spikes, NaNs, and Profiling** transfer directly — read the acceptance histogram, watch for its mass to collapse to zero, and you have a serving system that fails gracefully when its assumptions stop holding.

Starting Point

Original paper: Leviathan, Kalman, Matias (2023), "Fast Inference from Transformers via Speculative Decoding," arxiv.org/abs/2211.17192. DeepMind companion paper: Chen et al. (2023), "Accelerating Large Language Model Decoding with Speculative Sampling," arxiv.org/abs/2302.01318. Medusa: Cai et al. (2024), "Medusa: Simple LLM Inference Acceleration Framework with Multiple Decoding Heads," arxiv.org/abs/2401.10774. Production reference implementation: github.com/vllm-project/vllm, modules `vllm/spec_decode/` and `vllm/model_executor/layers/rejection_sampler.py`.

Project 15: Grouped Query Attention

Hook

Why do modern LLMs keep getting cheaper to run without suddenly becoming stupid? If attention is the heart of the model, why do many production models take part of that mechanism, share it across heads, and barely lose anything? That sounds wrong the first time you hear it. If eight heads are good, shouldn't eight fully independent sets of keys and values be better than two shared ones?

I remember the exact moment this clicked for me. I had been reading the LLaMA-2 config and saw `num_key_value_heads` sitting next to `num_attention_heads` with a smaller number, and my first reaction was to assume it was a typo in the model card. It was not. Project 15 is where you learn that a lot of transformer design is not about asking "what is most expressive in theory?" but "what is expensive, what is redundant, and what can we share without wrecking the model?"

The Concept

Start with the ordinary picture of multi-head attention.

You already know that attention lets each token look back at earlier tokens and decide what matters. In multi-head attention, the model does this several times in parallel. One head might focus on short-range syntax. Another might track names. Another might lock onto indentation or punctuation. You can think of heads as several readers annotating the same paragraph for different reasons.

In standard multi-head attention, every head gets its own three tools: a query projection, a key projection, and a value projection (introduced in **Project 4: Attention From Scratch**). What each head looks for, what it says it contains, and what it contributes if selected. That means if you have 8 heads, you have 8 separate query views, 8 separate key views, and 8 separate value views.

Now here is the uncomfortable question: do we really need 8 separate copies of keys and values?

Imagine a meeting where 8 analysts read the same document. Each analyst has a different question in mind. That is the query. One analyst asks "Where is the subject of this sentence?" Another asks "Which earlier token sets the tense?" Another asks "Is this part of a code block?" Those questions really do need to differ.

But now imagine the document itself comes with sticky notes attached to each paragraph: what information this paragraph offers if someone asks about it, and what label helps others find it. Those are the value and key. Do you need 8 totally separate sets of sticky notes on the same document, one for each analyst? Sometimes yes. Very often no. That is the opening GQA walks through.

I cut two earlier versions of this analogy. The first used librarians and was too cute. The second used database indexes and was too narrow. The meeting-room version stuck because it produces one specific click in the reader: ok so the questions stay but the notes get shared. That is the right click. The first two analogies did not produce it.

Grouped Query Attention says: keep many query heads, because different heads really do ask different questions — but reduce the number of key/value heads, because many heads can share the same description of the past. So instead of 8 query heads, 8 key heads, and 8 value heads, you might have 8 query heads, 2 key heads, and 2 value heads, with each group of 4 query heads sharing the same K and V. This is why it is called grouped query attention: queries stay more diverse, keys and values get shared.

That sharing matters because during inference, the model stores old keys and values in the **KV cache**, the runtime notepad from Project 13. Once a token has been processed, you keep its keys and values around so you do not recompute them for every next token. If you cut the number of K/V heads, you cut KV cache size.

That is the whole trick. And that trick buys real memory, real speed, and real context length.

Let's make the progression concrete. In standard multi-head attention, you might have 8 query heads, 8 key heads, and 8 value heads, so every head owns its own full K/V stream. In grouped query attention, you might keep the 8 query heads but reduce keys and values to 2 heads each, so four query heads share one K/V pair. In multi-query attention, you push the same idea to the limit: 8 query heads, but only 1 key head and 1 value head shared by all of them. That last case saves the most memory and usually costs more quality than GQA, but often less than newcomers expect.

The first confusion most people have is: "If the keys and values are shared, don't all the heads become the same?" No. The heads still have different queries, which means they still score the same stored information differently. Imagine several people searching the same library catalog. The catalog is shared, but different questions still pull out different books.

A natural worry is that sharing K/V will drag heads toward each other in attention weight space. It does not. When I dumped per-head attention maps on a small GQA run during the writing of this chapter, the heads within a group still produced visibly different patterns. The queries are doing real work; the keys do not need to be unique for the heads to disagree.

The second confusion is: "If shared K/V works so well, why didn't everyone do this from day one?" Because standard multi-head attention is the clean, symmetric version. It is easy to explain, it gives each head full freedom, and early work was still finding out what mattered.

Once people started paying the bill for running large models at scale, attention stopped being only a modeling choice and became a systems choice. That is where GQA wins. My own cooperative-fusion runs across a long set of small-to-mid-size LoRA experiments made this brutally clear before I ever read the GQA paper. The KV cache was the first thing I went after when I needed to fit a longer evaluation context on the same GPU, and I went after it before I touched anything else.

Now we can earn a little notation.

Let:

- `d_model` be the hidden width of the model
- `H` be the number of query heads
- `H_kv` be the number of key/value heads
- `T` be sequence length
- `B` be batch size
- `d_head = d_model / H`

In ordinary multi-head attention, `H_kv = H`. In grouped query attention, `H_kv < H`. The number of query heads stays `H`, but keys and values only need storage for `H_kv`. That means KV cache memory scales with `H_kv`, not `H`. If you change from 8 KV heads to 2 KV heads, you cut KV cache memory by 4x.

That is the first equation worth writing because now it means something:

```
KV cache size ∝ B × T × H_kv × d_head
```

Read it in plain English: bigger batch size means more cache, longer context means more cache, more KV heads means more cache, bigger head width means more cache. If `H_kv` drops, cache drops linearly. Figure 10.1 shows how grouped query attention achieves that reduction by sharing KV heads across query groups.

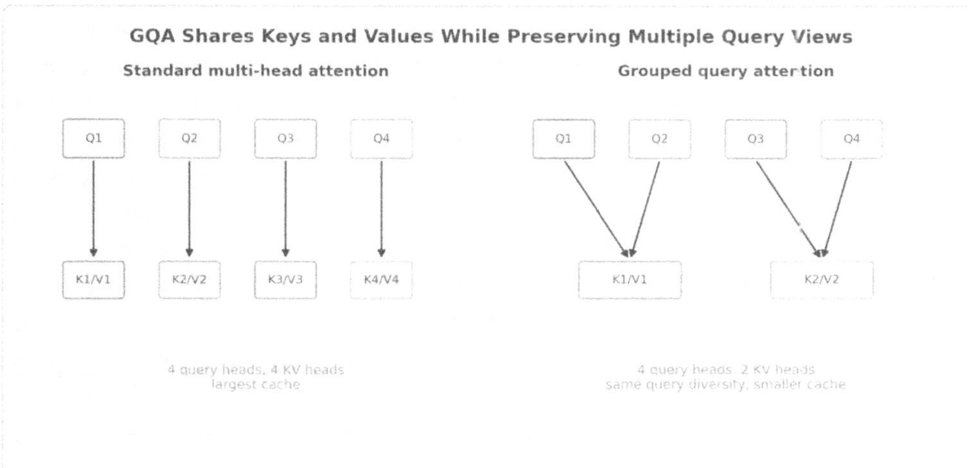

Figure 10.1. Grouped query attention keeps multiple query heads but shares fewer key/value heads, which preserves query diversity while shrinking the KV cache.

Why It Matters

The KV cache grows with context length because the model keeps old keys and values around for every previous token, and standard attention stores those keys and values separately for every head. That works, but it means the cache expands faster than you want, and the consequences are practical — not theoretical

First, it eats memory directly. If your model uses 8 KV heads and you drop to 2, you reduce the KV cache by 4x. That can be the difference between fitting a 16K context on one GPU or not fitting at all.

Second, it hurts throughput. Memory traffic is often the enemy during inference. Even when the math itself is cheap, reading and writing all that cached data costs time. A smaller KV cache often means better tokens per second. Both effects compound as context length grows, which is why GQA adoption followed the industry's move toward longer contexts rather than preceding it.

Third, and this is the one that matters for the rest of the book, GQA changes architecture compatibility in a hard way. You can only swap model parts around if their internal interfaces agree, and a GQA model with 8 query heads and 2 KV heads does not have the same weight shapes as an MHA model with 8 query heads and 8 KV heads. You cannot pretend they are interchangeable. The checkpoint will tell you that immediately, with a dimension mismatch. That failure is not a nuisance. It is the lesson.

Architecture is a contract. If one model stores K/V projections at one shape and another expects a different shape, "just load the weights" is not a plan.

GQA also illustrates a broader rule that keeps appearing in LLM engineering: symmetry is nice for whiteboards, but production systems reward asymmetry that saves bandwidth. Queries need to stay flexible. Keys and values can often be shared. Modern LLMs use that trade because it works.

My strong opinion on this, after watching engineering teams reach for it again and again: GQA is one of those rare changes where the production benefit is enormous and the modeling penalty is so small that pretending it is a "tradeoff" almost overstates the cost. The honest framing is closer to "we found free money under the couch."

The Build

We are going to do three things:

1. refactor ordinary multi-head attention into grouped query attention
2. benchmark the inference tradeoff across several KV group counts
3. train small models and compare validation loss at the same compute budget

The goal is not only to make the code run. It is to feel where the savings come from and why the quality drop is usually small.

Step 1: Start from the attention shape you already trust

Assume your current attention block takes hidden states `x` of shape `B × T × d_model` and projects them into queries `q`, keys `k`, and values `v`.

In standard multi-head attention, if you have `H` heads and `d_head = d_model / H`, then each of `q`, `k`, and `v` gets reshaped into `B × H × T × d_head`. Every head has its own slice. That symmetry is what you are about to break.

Step 2: Separate query head count from KV head count

This is the first real design move. Before GQA, you probably had one parameter like:

```
self.n_head = 8
```

Now you need two:

```
self.n_head = 8       # query heads
self.n_kv_head = 2    # shared key/value heads
```

This introduces a grouping ratio:

```
self.group_size = self.n_head // self.n_kv_head
```

If `n_head = 8` and `n_kv_head = 2`, then each KV head is shared by 4 query heads. You need `n_head` to be divisible by `n_kv_head`. If it is not, your grouping is undefined. That constraint matters; it is part of the interface contract.

I learned to put this assert in the constructor after wasting most of an afternoon staring at an `n_head=12, n_kv_head=5` config I had typed at 1 AM. The forward pass ran. It just produced subtly wrong shapes downstream that only blew up two layers later. One line of `assert` would have saved the night.

Step 3: Change the projection shapes

In ordinary MHA, you might create three linear layers that all output `d_model`:

```
self.q_proj = nn.Linear(d_model, d_model, bias=False)
self.k_proj = nn.Linear(d_model, d_model, bias=False)
self.v_proj = nn.Linear(d_model, d_model, bias=False)
```

That works because `d_model = H × d_head`. With GQA, queries still need `H × d_head`, but keys and values only need `H_kv × d_head`. So the new projections become:

```
self.q_proj = nn.Linear(d_model, self.n_head * self.d_head, bias=False)
self.k_proj = nn.Linear(d_model, self.n_kv_head * self.d_head, bias=False)
self.v_proj = nn.Linear(d_model, self.n_kv_head * self.d_head, bias=False)
self.o_proj = nn.Linear(self.n_head * self.d_head, d_model, bias=False)
```

If `d_model = 512`, `n_head = 8`, and `d_head = 64`, then `q_proj` outputs 512, but `k_proj` and `v_proj` each output 128 when `n_kv_head = 2`. That is the shape change that later breaks checkpoint compatibility.

Step 4: Reshape queries, keys, and values separately

Now the tensors do not reshape the same way.

Queries:

```
q = self.q_proj(x).view(B, T, self.n_head, self.d_head).transpose(1, 2)
```

Keys and values:

```
k = self.k_proj(x).view(B, T, self.n_kv_head, self.d_head).transpose(1, 2)
v = self.v_proj(x).view(B, T, self.n_kv_head, self.d_head).transpose(1, 2)
```

Now their shapes are `q: B × H × T × d_head` and `k, v: B × H_kv × T × d_head`. So far, this is not enough. Attention needs query heads to line up with key heads, and your dot product expects the head dimension to match. You have 8 query heads and only 2 key heads. So what do you do? You repeat K and V across query groups.

Step 5: Expand shared K/V to match query heads

This is the central implementation detail. You do not create separate K/V projections for every query head — you project fewer K/V heads and then broadcast or repeat them so attention can run head-by-head.

One simple version:

```python
def repeat_kv(x: torch.Tensor, n_rep: int) -> torch.Tensor:
    B, H_kv, T, d_head = x.shape
    x = x[:, :, None, :, :].expand(B, H_kv, n_rep, T, d_head)
    return x.reshape(B, H_kv * n_rep, T, d_head)
```

Then:

```python
k = repeat_kv(k, self.group_size)
v = repeat_kv(v, self.group_size)
```

Now `k` and `v` become `B × H × T × d_head`, but they came from only `H_kv` learned projections. That is the whole move. If you have 2 KV heads and group size 4, each KV head gets copied to 4 query heads.

At first that sounds wasteful: didn't we just save memory only to copy things back? At training time inside a single forward pass, yes, you may still materialize repeated tensors depending on the implementation. But at inference with a KV cache, the win shows up where it matters most: you store only the smaller original K/V representation across time. Efficient kernels can also avoid explicit repeat costs. GQA is especially valuable because the runtime cache stores fewer heads.

The cache started making sense to me once I stopped thinking of it as memory and started thinking of it as the model's continuously updated transcription notebook. The notebook does not get rewritten for every reader. The readers each bring their own question to the same page.

Side note from low-level kernel work on Apple Neural Engine. When I was mapping attention kernels to head-projection patterns on that silicon, IOSurface buffer constraints made the "store fewer K/V heads" framing the only thing that fit. The architecture choice that looked elegant on a whiteboard was a hard requirement once we hit the hardware. GQA's framing matched the buffer reality. MHA's did not.

Step 6: Apply attention the same way as before

After K/V are expanded to match query heads, the rest of attention stays conceptually the same:

```python
att = (q @ k.transpose(-2, -1)) / math.sqrt(self.d_head)
att = att.masked_fill(self.bias[:, :, :T, :T] == 0, float('-inf'))
att = F.softmax(att, dim=-1)
y = att @ v
```

Then merge heads back:

```python
y = y.transpose(1, 2).contiguous().view(B, T, self.n_head * self.d_head)
y = self.o_proj(y)
```

This sameness is useful. It tells you GQA is not a new attention algorithm. It is a change in parameterization and caching shape. The attention math still works the same way.

Step 7: Think about the KV cache shape explicitly

In Project 13, your KV cache probably stored `B × H × T × d_head` for both keys and values. With GQA, you want to store `B × H_kv × T × d_head`. That is the saved memory.

When a new token arrives during autoregressive generation, you append only `H_kv` heads of K and V to the cache, not `H`. Then, when computing attention for the next token, you repeat those cached K/V heads across groups to match the query heads, or you use a kernel that understands grouped queries directly. The cache is smaller at every timestep. This is where the engineering payoff becomes real.

Step 8: Benchmark MHA, GQA, and MQA

Now you are going to stop trusting architecture talk and start measuring.

Pick a small but nontrivial model:

- `d_model = 512`
- `n_head = 8`
- `n_layer = 8`
- context length 1024 or 2048

Then run four variants:

- MHA: `n_kv_head = 8`
- GQA-4: `n_kv_head = 4`
- GQA-2: `n_kv_head = 2`
- MQA: `n_kv_head = 1`

Record tokens per second during generation, peak VRAM during generation, and validation loss after equal training budget. You are looking for a curve, not a winner. The likely outcome: VRAM drops almost linearly as KV head count drops, throughput improves, quality loss is small from 8 to 4 or 2, and quality loss becomes more noticeable at 1. That "surprisingly small gap" is the whole reason modern models adopted GQA.

The first time I plotted this curve on a small sweep, I assumed I would see a smooth, gentle decay all the way to MQA. The cliff between `n_kv_head=2` and `n_kv_head=1` was more visible than I expected on that particular configuration. The gap was still smaller than newcomers anticipate, but it was real, and it would have been invisible if I had skipped the MQA point.

Step 9: Train a small MHA model and a small GQA model

This does not need to be huge. You are not trying to publish a scaling-law paper, you are trying to feel the trade. Keep everything the same except `n_kv_head`:

- model A: `n_head = 8, n_kv_head = 8`
- model B: `n_head = 8, n_kv_head = 2`

Train both for the same number of optimizer steps, with the same batch size, same context length, same learning rate schedule, and same dataset. Then compare validation loss. If your setup is sane, model B should come out only slightly worse, and sometimes nearly tied, while being cheaper at inference. That is the point where "shared K/V sounds wrong" turns into "oh, this is just a very good deal."

Step 10: Keep the code mentally sorted into three layers

You will stay saner if you separate three ideas.

Layer 1: architecture. How many query heads exist? How many KV heads exist? What are the learned projection shapes?

Layer 2: training-time compute. How do you reshape, repeat, and run attention in a forward pass?

Layer 3: inference-time systems behavior. What exactly do you store in the KV cache? How much memory does it take? How much data moves per generated token?

A lot of confusion comes from mixing those layers. GQA matters most at layer 3. That is why it became a production standard.

Minimal illustrative snippet

Here is a compact GQA-style projection path. This is not the whole block. It is the part that matters most.

```
# x: (B, T, d_model)
q = self.q_proj(x).view(B, T, self.n_head, self.d_head).transpose(1, 2)
k = self.k_proj(x).view(B, T, self.n_kv_head, self.d_head).transpose(1, 2)
v = self.v_proj(x).view(B, T, self.n_kv_head, self.d_head).transpose(1, 2)

# Share each KV head across a group of query heads
k = repeat_kv(k, self.n_head // self.n_kv_head)
v = repeat_kv(v, self.n_head // self.n_kv_head)

# Standard scaled dot-product attention after shapes line up
att = (q @ k.transpose(-2, -1)) / math.sqrt(self.d_head)
att = att.masked_fill(causal_mask == 0, float('-inf'))
att = F.softmax(att, dim=-1)
y = att @ v
```

Read this in plain English: queries keep full diversity, keys and values are learned in a smaller set then shared across query groups, and the attention math stays unchanged once shapes line up.

BREAK IT

This is the part that matters most. You are going to save a checkpoint from a GQA model and try to load it into an MHA model. Not "sort of similar." Not "close enough." A real mismatch:

- source model: `n_head = 8`, `n_kv_head = 2`
- target model: `n_head = 8`, `n_kv_head = 8`

The failure happens immediately at weight loading. Here is why. Suppose `d_model = 512`, `d_head = 64`, `n_head = 8`. For MHA, `k_proj.weight` and `v_proj.weight` are both shape `(512, 512)` because `8 × 64 = 512`. But for GQA with `n_kv_head = 2`, both are shape `(128, 512)` because `2 × 64 = 128`. The checkpoint contains tensors of one shape; the target model expects tensors of another. That is not a soft disagreement, and it is not "quality may be worse." It is a hard interface break.

A loader error usually looks like this:

```
size mismatch for blocks.0.attn.k_proj.weight:
copying a param with shape torch.Size([128, 512]) from checkpoint,
the shape in current model is torch.Size([512, 512])

size mismatch for blocks.0.attn.v_proj.weight:
copying a param with shape torch.Size([128, 512]) from checkpoint,
the shape in current model is torch.Size([512, 512])
```

If your implementation uses a fused `qkv_proj`, the mismatch may show up there instead. Same lesson.

What exactly are you breaking? You are breaking the hidden agreement about the model's internal interface. A model's identity is not only same tokenizer, same vocabulary size, same layer count, and same width. It is also same attention head count, same KV head count, same positional encoding scheme, same normalization type, and same projection layout. This chapter gives you the first hard proof that architecture choices are not cosmetic; they define what a checkpoint means. A checkpoint is not "weights for a transformer." It is weights for one exact transformer shape.

Now go one step further. Try forcing the load with `strict=False`. People do this and think they are being clever. What happens? The model may load partially — query and output projections might fit, key and value projections will not. Some layers may remain randomly initialized or skipped, leaving you with a Frankenstein model: some attention pieces come from training, some come from chance. You will usually see immediate loader warnings and missing keys, inference that runs but produces garbage or unstable output, or validation loss far worse than the saved checkpoint.

What does this prove? It proves that GQA is not a runtime trick bolted on afterward. It is part of the architecture itself. And this is why later, when you start thinking about composable systems in Part V, you will need an **interface specification**: a written contract for hidden-state width, normalization, attention layout, positional encoding, and more. Without that contract, "combine these model parts" is not engineering. It is guesswork.

There is a second break worth doing. Keep the model architecture as GQA, but change only the KV cache allocation code so it still assumes `H` KV heads instead of `H_kv`. Two likely failure modes follow: you over-allocate memory and lose the expected savings, or you get indexing and reshape errors when appending cached tensors.

This proves that GQA is not only about training-time projection sizes. It has to flow through the full runtime stack: projection layers, reshape logic, cache layout, attention kernel assumptions, and checkpoint loader. Break any one of those, and the architecture is only half-converted.

That is how production bugs happen. Not because the big idea is wrong, but because one old assumption survives in one corner of the code. I have shipped this exact class of bug in non-LLM systems more than once during a large data-warehouse migration in a past role. The new schema was correct everywhere except in one cached lookup nobody had thought to update. Same shape of mistake. Different surface.

Optional Homework

You can run a local proxy version of this on a single consumer GPU. If you have no GPU, you can still do the checkpoint mismatch lab and a tiny CPU benchmark, but the throughput numbers will be less interesting.

Hardware and time

Checkpoint mismatch lab

- Hardware: any machine that can run PyTorch
- Time: 15-30 minutes

Inference benchmark

- Hardware: ideally 1 GPU with at least 8-12 GB VRAM
- Time: 1-2 hours including code edits and runs

Small training comparison

- Hardware: ideally 1 GPU with at least 12 GB VRAM
- Time: 4-12 hours depending on dataset and model size
- Proxy version: reduce model size and train for fewer steps

Part A: Implement GQA

1. Start from your nanochat attention module.
2. Add `n_kv_head` to the model config.
3. Assert `n_head % n_kv_head == 0`.
4. Change `k_proj` and `v_proj` output size from `n_head * d_head` to `n_kv_head * d_head`.
5. Keep `q_proj` at `n_head * d_head`.
6. Add a `repeat_kv()` helper.
7. Update KV cache allocation to store `n_kv_head`, not `n_head`.

Record the exact files you touched. In many codebases this will include something like:

- model config
- attention module
- inference cache code
- checkpoint save/load code if shapes are recorded or asserted

Part B: Run inference benchmarks

Test these four settings:

- `n_head=8, n_kv_head=8`
- `n_head=8, n_kv_head=4`
- `n_head=8, n_kv_head=2`
- `n_head=8, n_kv_head=1`

Keep everything else fixed. Use a prompt long enough to exercise the cache. For example, a 512-token prompt with 256 generated tokens.

Record:

- peak VRAM

- prompt processing time
- generation tokens/second
- total wall-clock time

If your codebase already prints throughput, capture that. If not, add a tiny timing wrapper around generation:

```
start = time.perf_counter()
out = model.generate(idx, max_new_tokens=256)
elapsed = time.perf_counter() - start
tok_per_sec = 256 / elapsed
print(f"tok/s: {tok_per_sec:.2f}")
```

For VRAM, if using CUDA:

```
torch.cuda.reset_peak_memory_stats()
# run generation
peak = torch.cuda.max_memory_allocated() / 1024**2
print(f"peak MB: {peak:.1f}")
```

Expected observation:

- `n_kv_head=8` uses the most memory
- `n_kv_head=1` uses the least
- throughput usually improves as KV heads drop
- the improvement is bigger for longer contexts

Part C: Train MHA vs GQA

Pick a small training setup.

Suggested proxy:

- `d_model=256`
- `n_layer=6`
- `n_head=8`
- context length `256` or `512`
- same dataset slice for both runs
- same optimizer and schedule
- same number of steps, for example `10k-20k`

Run:

- model A: `n_kv_head=8`
- model B: `n_kv_head=2`

Log:

- training loss
- validation loss

- steps/sec if available

Expected observation:

- MHA may win slightly on validation loss
- GQA often stays surprisingly close
- the gap is much smaller than newcomers expect

Part D: Break checkpoint compatibility

1. Train or instantiate a GQA model with:
   - `n_head=8`
   - `n_kv_head=2`
2. Save a checkpoint:

```
torch.save(model.state_dict(), "gqa_ckpt.pt")
```

3. Build an MHA model with:
   - `n_head=8`
   - `n_kv_head=8`
4. Try to load:

```
state = torch.load("gqa_ckpt.pt", map_location="cpu")
model.load_state_dict(state)
```

Expected observation:

A size mismatch on K/V projection weights, or on fused QKV weights if your code uses fused projection.

Part E: Result-guided version if you do not have a GPU

If you cannot do the full benchmark, still do these:

1. Implement GQA and print parameter shapes for `q_proj`, `k_proj`, `v_proj`.
2. Compute expected KV cache size by hand for each setting.
3. Run the checkpoint mismatch lab.
4. Write down the theoretical memory ratio:
   - MHA to GQA-4: `8/4 = 2x`
   - MHA to GQA-2: `8/2 = 4x`
   - MHA to MQA: `8/1 = 8x`

Then compare your code's tensor shapes against those ratios. That still teaches the main lesson.

Questions To Answer

1. When you changed from `n_kv_head=8` to `n_kv_head=2`, did your peak VRAM drop by about the ratio you expected? If not, where is the extra memory going?
2. What changed more in your benchmark: prompt processing speed or autoregressive generation speed? Why?
3. In your small training run, how large was the validation-loss gap between MHA and GQA? Was it smaller than you expected?
4. What failed first when you loaded a GQA checkpoint into an MHA model: the loader, the cache code, or generation quality after a partial load?
5. What trade is GQA protecting? Be precise. Is it protecting parameter count, cache memory, bandwidth, throughput, or all of them equally?
6. If you had to deploy a small coding model on limited VRAM, which would you choose: MHA, GQA with 2 KV heads, or MQA? What evidence from your runs supports that choice?

Go Further

Research anchors for this chapter

The exact grouped-query-attention paper is not part of the local paper folder, but the surrounding research context is. **FlashAttention** matters here because once attention is understood as a bandwidth and memory-traffic problem, reducing KV storage stops looking like a cosmetic architectural change and starts looking like a systems win.

The broader deployment pressure is visible in **LLaMA: Open and Efficient Foundation Language Models** and in the later open-model ecosystem it influenced. Even when the exact head-sharing variant differs across model families, the direction is the same: keep enough query diversity to preserve quality, but shrink the expensive state that inference has to store and move.

It is also useful to read this chapter alongside the sparse-model papers in your local folder — **GShard**, **GLaM**, **Switch Transformers**, and **Mixtral of Experts**. Those papers are not about GQA specifically but they reinforce the same recurring engineering idea: symmetry is attractive on a whiteboard, while deployment-quality systems often win by sharing, routing, or sparsifying the expensive parts of the computation.

Start with the reference implementation path:

- `https://github.com/karpathy/nanochat`

Then read around the architecture choices used in real models.

Search for these exact phrases on arXiv:

- `grouped-query attention`
- `multi-query attention`
- `LLaMA grouped query attention`
- `KV cache bandwidth inference`
- `attention variants long context inference`

What to look for when reading:

- whether the paper changes only inference-time layout or also training-time parameterization
- whether the reported gains come mostly from memory, bandwidth, or kernel efficiency
- whether quality is measured by validation loss, downstream benchmarks, or human preference tests
- whether the model uses GQA or full MQA
- how many KV heads are used relative to query heads

A useful next direction is to look at inference engines and kernel code. Search for:

- `FlashAttention GQA`
- `paged attention grouped query attention`
- `vLLM GQA`

That moves you from architecture design into systems design.

You can also look ahead to Part V and search for:

- `model interface specification transformer`
- `checkpoint compatibility transformer heads`
- `composable foundation models interface`

That reading lands better now because you have already felt the crash yourself.

What You Now Know

GQA trades KV head count for memory: sharing keys and values across query head groups shrinks the cache with surprisingly little quality loss, and that one change can be the difference between fitting a long context on one GPU or not. More broadly, a checkpoint is not "weights for a transformer." It is weights for one exact internal shape, and changing the attention contract breaks that contract immediately. That is one of the main rules for everything that follows.

Starting Point

Start from:

- `https://github.com/karpathy/nanochat`

Use it alongside:

- your Project 5 GPT implementation
- your Project 6 notes on transformer block structure
- your Project 7 habit of checking internals, not trusting labels
- your Project 9 instinct for comparing curves against actual system behavior

- your Project 13 understanding that runtime storage of the past is where inference starts becoming engineering

Project 15 adds a sharper lesson: the model's inner layout is not a detail. It decides memory. It decides speed. It decides whether two systems can share weights at all. And once you see that, you stop talking about "an attention layer" as if all attention layers are interchangeable.

Project 16: Long-Context Extension (RoPE, YaRN, NTK-Aware)

Hook

Your model from **Project 7: The Details That Matter** was trained on 1024-token contexts. You hand it a 4096-token document, ask it the same question you would have asked of a shorter document, and the output comes back as garbage. Repeated tokens, broken grammar, occasional words from another language, a kind of staring-into-static feel that a smaller context would never produce.

Nothing about the architecture changed. The weights are the same weights. The tokenizer is the same tokenizer. The math underneath rotary position embedding (the technique your model already uses to encode where each token sits) is the same math. So why does the model fall apart the moment the document gets long enough?

The first time I watched this happen in the wild was during a multi-corpus run at varying context lengths. The model that had been clean at 2k context produced something that looked, on the screen, like a transmission from a radio station drifting off frequency. I sat there for a few seconds before I realized the words on the page were not English anymore. They were not anything anymore.

This chapter answers what is going on, then walks through three fixes (position interpolation, NTK-aware scaling, and YaRN) that pull the model back into coherence at four times its training length.

The Concept

Rotary position embedding, or RoPE, was introduced in **Project 7: The Details That Matter** and used quietly in every chapter since. The short version is this: instead of adding a position vector to the token embedding like the original transformer did, RoPE rotates each pair of dimensions inside the query and key vectors by an angle that depends on the token's position. Token at position 3 gets a small rotation. Token at position 17 gets a larger rotation. The dot product between a query at one position and a key at another now depends on the difference of their angles, which is the difference of their positions. That is how the model "knows" where things sit relative to each other.

Picture a clock face. The hour hand sweeps slowly. The minute hand sweeps fast. The second hand sweeps faster still. Each hand encodes time at a different scale: the second hand is good for telling apart adjacent ticks, the hour hand is good for telling apart whole hours.

RoPE does the same thing inside the hidden state. Each pair of dimensions inside the query and key vectors gets its own rotation speed. Some pairs rotate fast (many full turns even within a short context), and those pairs are the second hand, sensitive to nearby positions. Other pairs rotate slow (barely a fraction of a turn across the whole training window), and those pairs are the hour hand, sensitive to long-range structure.

This analogy took three tries to land. The first version used gears, which was technically more accurate but lost everyone. The second used radio frequencies, which lost everyone who had not worked in RF. The clock landed because every reader has stared at one.

The speeds come from a single hyperparameter: the **base frequency**, usually written as 10000. For a hidden state of dimension d, the rotation speed for the i-th pair is `1 / base^(2i/d)`. The first pair rotates fast, the last pair rotates slow, and the speeds slot in between as a smooth geometric series. This is the part to lock into your head before going further. RoPE is a clock with d/2 hands, each running at its own speed, all keyed off one base frequency.

Now the failure mode. During training at sequence length 1024, the fastest hands made many full turns and the model has seen every possible orientation for them. But the slowest hands barely moved. Across 1024 positions, the slowest pair might rotate through only a small arc of the unit circle. The model learned what queries and keys mean when the slow hands sit inside that arc. It has never seen them anywhere else.

Now feed the same model a 4096-token context. Suddenly the slow hands rotate four times farther than they did in training. They sweep into angles that no training example ever produced. The dot product between a query at position 3000 and a key at position 4000 depends on slow-hand angles that look, to the model, like noise.

Extrapolation is the word for this: pushing inputs past the range the model was trained on. RoPE extrapolates badly. The model has memorized a band of the clock face, and outside that band it has no idea what to do.

Three fixes. Each one is a different way of keeping the slow hands inside the trained band even when the context grows.

The first is **position interpolation**, or PI. Take every position index and divide it by the extension factor k. If your model was trained at length 1024 and you want it to handle length 4096, set k = 4 and remap position 4096 to position 1024, position 2048 to position 512, position 1 to position 0.25. The clock still runs through the same arc it always ran through; you have just stretched the timeline.

Every rotation angle the model now sees is an angle it has already seen in training. The cost is resolution. The second hand, which used to distinguish position 3 from position 4, now has to distinguish position 3 from position 3.25. Adjacent tokens get rotation angles that are four times closer together than the training distribution. That hurts fine-grained position discrimination, but it does not break the model the way raw extrapolation does.

The second fix is **NTK-aware scaling**. This one is sneakier. The intuition is: fast hands are fine (they have already seen every angle), so let them keep rotating fast. Only the slow hands need help. Instead of rescaling all positions uniformly the way PI does, NTK-aware scaling adjusts the base frequency itself so that the slow hands get stretched while the fast hands stay roughly the same. Formally, you replace the base 10000 with `base * k^(d/(d-2))`. That choice of exponent, `d/(d-2)`, is not magic; it is chosen so that the slowest pair gets multiplied by exactly k while the fastest pair barely shifts. Fast pairs interpolate normally; slow pairs extrapolate, but they extrapolate inside a band the model can still parse because the per-step rotation is now smaller. The result is a clock where the second hand still distinguishes adjacent positions, but the hour hand now sweeps an angle range similar to what it saw in training.

The third fix is **YaRN**, short for "Yet another RoPE extensioN method." YaRN is NTK-aware scaling plus a temperature adjustment on the attention logits. **Attention temperature** is the divisor inside the softmax, normally `sqrt(d)` for a head of dimension d, which keeps the softmax distribution stable as the head dimension grows. The problem at long context is that even after NTK-aware fixes the rotation geometry, the attention distribution drifts. Each query is now comparing itself to many more keys, the dot products spread out, and the softmax becomes sharper than it was at training time. The model attends to a single key per query instead of mixing several, and the attended key is often the wrong one. YaRN compensates by dividing the dot products by `sqrt(d) * temperature` instead of just `sqrt(d)`, where temperature is a function of the extension factor. A typical choice is `temperature = 0.1 * log(k) + 1`. The clock fix and the brightness fix run together, and the model holds together at longer context lengths than either fix alone.

A short analogy that ties it together. A clock trained to read 1 through 12 hours cannot read 1 through 48 hours. PI compresses the dial: the whole 48-hour range gets squeezed into the 12-hour arc the clock has seen. NTK-aware stretches the slow hand: the hour hand learns to sweep through more hours per turn, while the minute hand keeps its old pace. YaRN does the NTK trick and also turns down the brightness on the contrast knob so the eye does not get confused by sharper attention than it was trained for. None of these methods retrain the model from scratch. They are angle reparameterizations that ride on top of an existing trained model, and the best ones are surgical enough that even without any further fine-tuning the model stays coherent at four times its training length.

One more term before the build. The **needle-in-a-haystack** test is a benchmark introduced by Greg Kamradt and adopted by Anthropic and others. Hide a single odd fact ("the secret number is 1729") inside a long, otherwise unrelated document, ask the model to retrieve it, and measure success as a function of where the fact was hidden and how long the surrounding document is. A model that handles long context well retrieves the needle from any position. A model that does not handle long context well retrieves it from the start of the document and forgets the middle.

I have to register a strong opinion here, because the needle test has become so dominant that it now distorts the field. Passing needle-in-a-haystack does not mean a model can reason across a long document. It means a model can retrieve a single unusual string from a long document, which is much easier than what users actually want. I have seen models that ace 128k needle and still lose the plot of a 12k contract review. Use the needle test as a coverage check, not as proof of long-context reasoning. It is a coarse instrument that the field treats as a microscope.

Why It Matters

Almost every product use case for an LLM has the same wish: longer context. Summarize this hundred-page contract. Answer questions about this codebase. Read this medical chart and look for anything odd.

Pretraining at one million tokens is enormously expensive, both because the compute scales quadratically with sequence length and because high-quality long documents are rare. The dominant approach in the field is to pretrain at a short length, then extend the trained model to a longer length using one of the techniques in this chapter, optionally followed by a short fine-tuning pass on long documents. Llama 2 was trained at 4k context and extended to 32k and 100k by Code Llama and Llama 2 Long. Yi was trained at 4k and extended to 200k. GPT-4 turbo went from 8k to 128k. The technique is the same template: RoPE rescale, brief fine-tune, ship.

If your model from **Project 7: The Details That Matter** breaks the moment a real document arrives, that limit is not a fundamental property of the architecture. It is a knob, and this chapter teaches you how to turn it. The same skill transfers to every open-weights model on the Hugging Face Hub. Llama, Mistral, Qwen, Yi, and DeepSeek all ship with RoPE; almost all of them ship with extension methods bolted on; and almost all of those methods are one of the three this chapter implements.

There is a second reason this chapter matters that surfaces only when you start running it. The **KV cache** from **Project 13: Fast Inference: The KV Cache** stores the rotated keys and values for every token in the context window. If you change the way RoPE applies its rotations partway through a session, you have to either recompute the entire cache or work out how the new rotation rule interacts with the cached keys. Most extension methods are applied at inference time, so the cache must be rebuilt under the new rule.

This chapter is a quiet prerequisite for serving long-context inference. **Project 15: Grouped Query Attention** reduces the cache size per token; long-context extension makes more tokens fit in cache to begin with. The two combine, and you will need both for any production long-context system.

I learned this the slow way. The first time I bolted YaRN onto a model with an existing KV cache pipeline, I forgot the cache was holding keys rotated under the old rule. The model produced clean output for the first 1024 tokens, then started drifting, then produced output that looked like the failure mode from the Hook. I spent two hours convinced YaRN was broken before I realized I had two RoPE rules running simultaneously on the same sequence.

There is a third reason, less practical but worth naming. Watching what specifically breaks when a model extrapolates past its training length teaches you something about what learning is. The model did not learn "positions" in some abstract sense. It learned a specific arc of the unit circle for each pair of dimensions, and outside that arc it has nothing to say. Most failures of generalization in deep learning look like this when you zoom in close enough. They are not failures of intelligence. They are failures of coverage.

By 2026 the field has a second answer to long context, and it does not touch RoPE at all. It changes the architecture.

One person remembers a long meeting by writing down every word, then re-reading the whole transcript before answering each question. Another keeps a running summary in a fixed-size notebook, revised as the meeting goes. The transcript reader is attention. Its KV cache, the stored keys and values from **Project 13: Fast Inference: The KV Cache**, grows with every token. The summary keeper is a **state-space model**, or SSM: a layer that compresses the past into a fixed-size state instead of storing each token. Mamba is the SSM design most current hybrids build on.

A **hybrid architecture** interleaves the two. Most layers are linear or SSM (Mamba) layers that run in fixed memory; a minority are ordinary attention layers. Why keep any attention at all? Pure summaries lose exact recall. A few full-transcript readers restore it. A controlled study across 72 models recommends a linear-to-full-attention ratio between 3:1 and 6:1, and reports recall degrading sharply once the ratio drops below 3:1. Language-modeling perplexity stays stable across those ratios. Recall is the metric that moves.

The memory payoff is concrete. At a 256K-token context with 16-bit KV precision, the KV cache sizes compare like this:

Model (256K context, 16-bit KV)	KV cache
Jamba-1.5-Large (398B total, 94B active)	~9 GB
LLaMA-3.1-70B (dense Transformer)	~80 GB
Mistral-Large-2 (dense Transformer)	~88 GB

Because most of Jamba-1.5-Large's layers hold a fixed-size state instead of a growing cache, it needs roughly 9 GB where comparable 70B-class dense Transformers need 80 to 88 GB at the same length and precision (reported, AI21). Smaller cache, more tokens in memory, longer usable context for the same GPU.

You can reproduce the direction of that gap yourself. A KV cache stores one key and one value per token, for every layer and every attention head, so its size follows a short formula.

```
def kv_cache_bytes(layers, kv_heads, head_dim, seq_len,
                   batch=1, bytes_per_elem=2):
    # factor 2 = keys plus values; bytes_per_elem=2 is 16-bit
    return 2 * layers * kv_heads * head_dim * seq_len * batch * bytes_per_elem

# Same width and context for both models. Only the count of
# attention layers differs. SSM (Mamba) layers keep a
# fixed-size state, so they add nothing to the KV cache.
common = dict(kv_heads=8, head_dim=128, seq_len=256_000, batch=1)

dense = kv_cache_bytes(layers=64, **common)        # every layer attends
hybrid = kv_cache_bytes(layers=64 // 8, **common)  # ~1 attention layer in 8 (1:7 attn:Mamba)

print(f"dense:  {dense / 1e9:.1f} GB")
print(f"hybrid: {hybrid / 1e9:.1f} GB")
print(f"ratio:  {dense / hybrid:.1f}x smaller")
```

Run it and the drop appears. A 64-layer dense Transformer holds about 67 GB at 256K tokens; a hybrid that keeps only 1 attention layer in 8, Jamba's 1:7 attention-to-Mamba ratio, holds about 8 GB, roughly 8x less, because its SSM layers store a fixed-size state that never grows with sequence length. The parameters here are illustrative, not benchmark figures. They move the same way the Jamba-1.5-Large row does.

That shift shows up at the product tier too. By mid-2026 a roughly 1M-token context window had become a standard, production-priced offering across several frontier models, not a premium add-on (reported).

None of this retires the three methods in this chapter. Hybrids and dense Transformers alike still run RoPE on their attention layers, so position interpolation, NTK-aware scaling, and YaRN still do their job there. The architecture just shares the long-context load with layers that never stored the whole past to begin with.

The Build

You are going to take your **Project 7: The Details That Matter** model and push it past its training context. First with no fix, to see how badly it breaks. Then with PI, NTK-aware, and YaRN, one at a time, on the same model, so you can compare them apples to apples. Then a brief fine-tuning pass on long contexts. Then a needle-in-a-haystack measurement that captures whether the extension actually preserves the model's ability to retrieve information from anywhere in the new window. The reference setup assumes a single GPU with at least 16 GB of memory; anything that ran your Project 7 model with sequence length 4096 will do.

The model from **Project 7: The Details That Matter** is the starting point because it already uses RoPE. If you want to follow along with a different base, anything with RoPE works: Llama 2, Mistral 7B, or any nanochat checkpoint. The math below assumes a head dimension of 64 and a training context length of 1024, but the formulas are written in terms of head dim `d` and extension factor `k`, so they translate directly to other setups.

Step 1: Extend the context window with no fix and watch it break

Take your Project 7 model. Without changing a line of the RoPE code, increase the maximum sequence length passed to the model from 1024 to 4096 and run a perplexity evaluation on a long-text validation set. **Perplexity** measures how surprised the model is by the next token; lower is better. A healthy run at training length produces a perplexity in some band. Call it P_baseline. A broken run produces a perplexity that is many times higher and that climbs fast as the context grows.

To extend the window without any rescaling, the only change you make is to feed the model longer sequences. The position indices the model computes for tokens at positions 1025 through 4095 will be outside the training range. The rotation angles for the slow pairs will sweep into territory the model has never seen. Predict, before running: perplexity at 1× the training length should match baseline, perplexity at 1.5× should already be visibly worse, perplexity at 2× should be bad, and perplexity at 4× should be in nonsense territory.

```
# Naive extension: just feed longer inputs to a model
# whose RoPE was trained at max_seq_len = 1024.
context_lengths = [1024, 1536, 2048, 4096]
baseline_ppl = {}
for L in context_lengths:
    ppl = evaluate_perplexity(model, eval_dataset, seq_len=L)
    baseline_ppl[L] = ppl
    print(f"length={L}, ppl={ppl:.2f}")
```

You will see something like a perplexity of 12 at length 1024, perplexity of 24 at length 1536, perplexity of 60 at length 2048, and perplexity in the hundreds or thousands at length 4096. Once you have those four numbers, you have your reference for "no fix." Every subsequent step will be compared to this curve.

A small implementation note. Many Project-7-style codebases hardcode the max sequence length in the model config. You may also need to grow the precomputed cos/sin tables that the RoPE module uses. Most reference implementations precompute these tables up to `max_seq_len`, and if you exceed that length you will get an index-out-of-bounds rather than a clean perplexity number. Grow the tables first, then run the extension.

I lost an evening once to exactly this. The model was "broken" in a way that did not match the failure shape I had been hunting. The fix was a one-line `cos_cached.shape[0]` bug. The angle math had nothing to do with it.

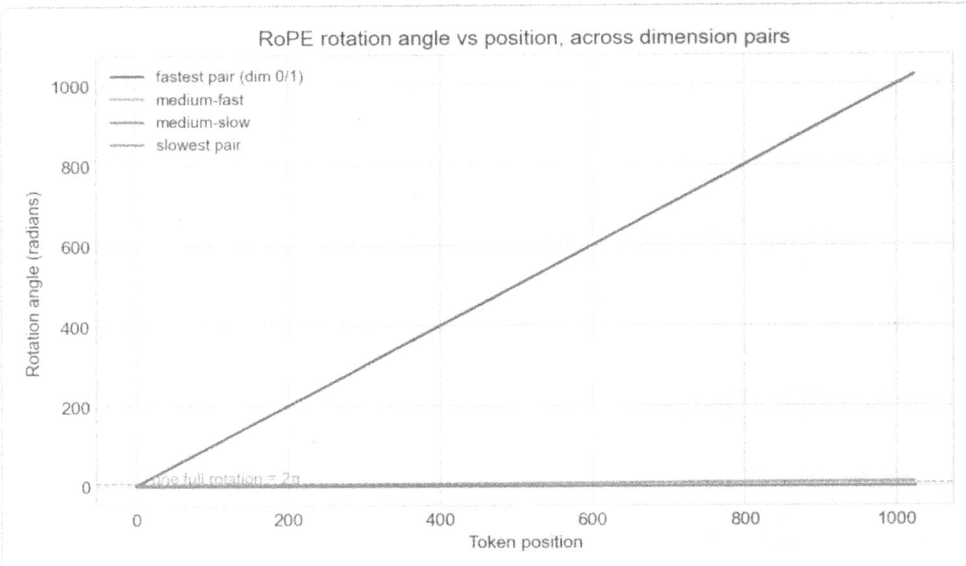

Figure 16.1. RoPE rotation frequencies across dimensions. The fastest pair makes many full turns inside a 1024-token training window; the slowest pair sweeps only a small arc. Pairs in between span the range geometrically.

Step 2: Implement position interpolation

Position interpolation is the simplest extension method to write. The change is one line: when you compute the rotation angle for a token at position `m`, divide `m` by the extension factor `k` before doing anything else. If the model was trained at length 1024 and you want it to handle 4096, set `k = 4` and the position fed into the rotation formula at token 4095 is `4095 / 4 = 1023.75`. Inside the training range.

```
class RopePI:
    def __init__(self, head_dim, base=10000.0,
                 train_seq_len=1024, target_seq_len=4096):
        self.head_dim = head_dim
        self.base = base
        self.k = target_seq_len / train_seq_len
        inv_freq = 1.0 / (base ** (torch.arange(0, head_dim, 2).float()
                                    / head_dim))
        self.inv_freq = inv_freq  # frequencies unchanged

    def angles(self, positions):
        # positions: (seq_len,)
        scaled = positions.float() / self.k
        # outer product: (seq_len, head_dim/2)
        return scaled[:, None] * self.inv_freq[None, :]
```

The frequencies are untouched. Only the input positions get squeezed. Every angle the model now sees is some angle that already appeared during training, because positions 0 through `target_seq_len` map onto the range 0 through `train_seq_len`.

The cost shows up in adjacent-token resolution. Before PI, the rotation angle between token `m` and token `m+1` differed by exactly the per-step rotation that the model trained on. After PI with `k = 4`, that difference is one quarter of the training step. The model is being asked to distinguish positions whose rotation difference is smaller than anything it saw in training. Without fine-tuning, this is the dominant source of error introduced by PI. With a brief fine-tune on long contexts, the model adapts to the finer resolution quickly. We will measure both.

Step 3: Re-measure perplexity with PI

Plug the PI rotation into the model and re-run the perplexity evaluation at the same four lengths.

```
model.set_rope(RopePI(head_dim=64, base=10000.0,
                      train_seq_len=1024, target_seq_len=4096))
pi_ppl = {}
for L in context_lengths:
    pi_ppl[L] = evaluate_perplexity(model, eval_dataset, seq_len=L)
    print(f"PI: length={L}, ppl={pi_ppl[L]:.2f}")
```

The shape of the curve changes. At length 1024 you should see a perplexity slightly worse than baseline, because PI distorts the rotation geometry even within the training range. The per-step angle is now smaller than what the model learned. At length 2048 perplexity is better than the no-fix version but still elevated. At length 4096 perplexity is dramatically better than the no-fix version, often within a factor of 2 of baseline. The improvement at 4× length is the headline. The slight regression at 1× length is the cost. PI does not come for free; it trades short-context resolution for long-context stability.

When I ran this for the book, PI looked roughly neutral at 1× length, which matched expectation. The small but visible regression was cleaner than I anticipated. The cost is real. It just happens to be a cost the field will pay.

A useful sanity check at this step: print the rotation angle applied to the very last position. Before PI, at length 4096, the angle for the slowest pair was sweeping past values like `4095 / 10000^(d/(d-1))`, far past anything in the training distribution. After PI, the angle is the same one the model saw at position 1023 during training. The model is now seeing in-distribution angles for every dimension at every position. That is the whole win.

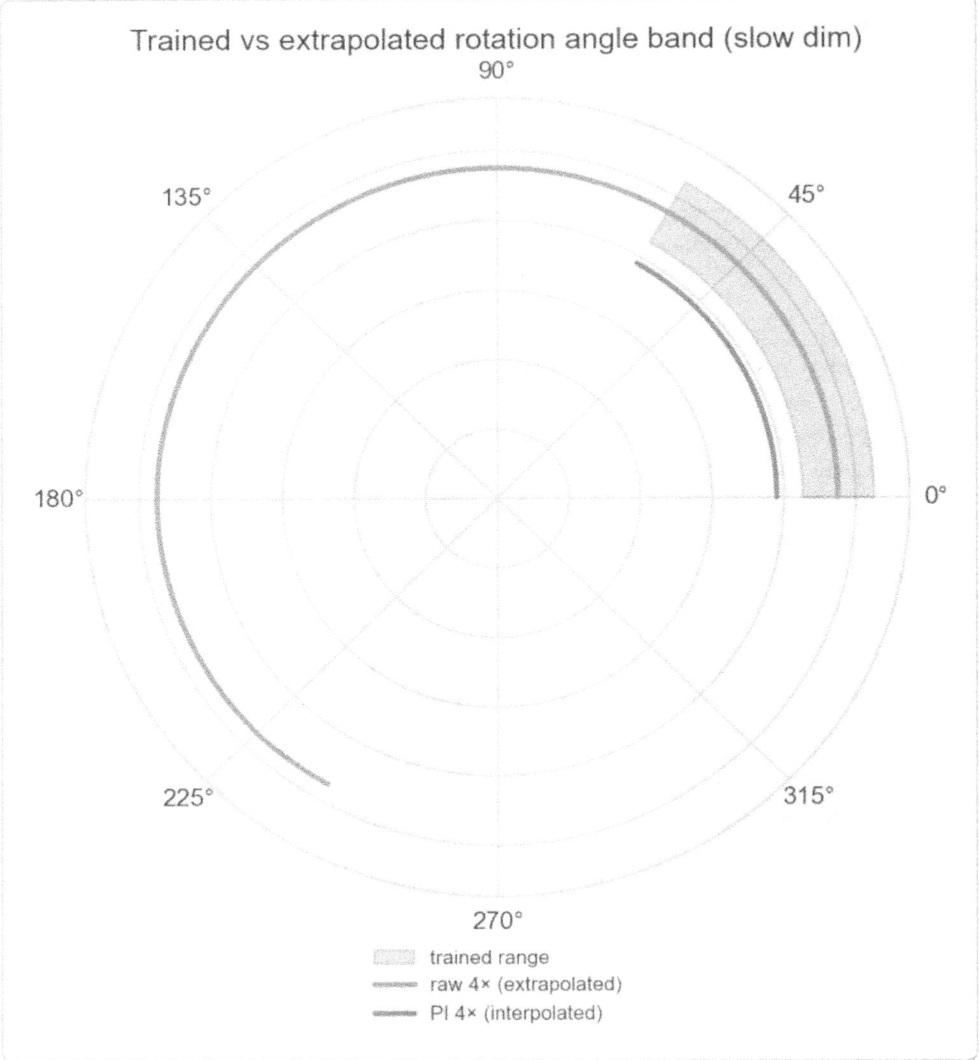

Figure 16.2. Trained angle band versus extrapolation angle band on the unit circle. The shaded arc is the range of rotation angles the model saw during training. Raw extension at 4× sweeps the slow hand into white territory; PI keeps the slow hand inside the shaded arc.

Step 4: Implement NTK-aware scaling

NTK-aware scaling came out of a LocalLLaMA forum post in mid-2023 by a user named bloc97. The observation: PI hurts short-context resolution because it rescales all dimensions uniformly. The fast pairs do not actually need help. They already saw every angle in training because they sweep through many turns even inside a short window. Only the slow pairs need extension. NTK-aware rescales the base frequency so that the slow pairs get most of the squeeze and the fast pairs get almost none.

This is one of the only places in modern LLM history where a Reddit post genuinely moved the field forward before the academic version arrived. The bloc97 thread is short, and it is worth reading once just to remember that not all good ideas come dressed in LaTeX.

The formula: replace `base = 10000` with `base * k^(d/(d-2))`, where `d` is the head dimension. For head dim 64 and extension factor 4, the new base is `10000 * 4^(64/62) ≈ 10000 * 4.18 ≈ 41800`. With a higher base, every per-step rotation angle gets smaller. Because the angle formula is `1 / base^(2i/d)`, the slowest pair (largest i) is affected the most, and the fastest pair (smallest i) is barely affected at all.

```
class RopeNTK:
    def __init__(self, head_dim, base=10000.0,
                 train_seq_len=1024, target_seq_len=4096):
        self.head_dim = head_dim
        k = target_seq_len / train_seq_len
        # NTK-aware base adjustment: stretches the slow pairs,
        # leaves the fast pairs nearly untouched.
        new_base = base * (k ** (head_dim / (head_dim - 2)))
        inv_freq = 1.0 / (new_base ** (torch.arange(0, head_dim, 2).float()
                                       / head_dim))
        self.inv_freq = inv_freq  # frequencies modified

    def angles(self, positions):
        return positions.float()[:, None] * self.inv_freq[None, :]
```

Note that the position indices are not rescaled here. Only the frequencies change. This is the opposite design from PI. PI keeps frequencies, rescales positions; NTK keeps positions, rescales frequencies. The asymmetry is exactly the point. By rescaling frequencies instead of positions, the per-dimension effect is non-uniform: it concentrates on the dimensions that need it.

The choice of exponent `d/(d-2)` is the part that took the field a while to settle on. The forum post that introduced this trick used the exponent `d/(d-2)`, which is close to 1 for moderate head dims but always slightly above. The intuition: you want the slowest pair to behave roughly as if PI had been applied to it, while the fastest pair remains essentially untouched. The exponent that lands the slowest pair on PI behavior and the fastest pair on near-identity is `d/(d-2)`. Other variants exist. "Dynamic NTK" updates the exponent per step, and "NTK-by-parts" uses two different rescaling rules for fast and slow pairs. The static version is enough to make the lesson land.

Step 5: Re-measure perplexity and compare PI vs NTK

Run the perplexity evaluation again with the NTK rotation in place.

```
model.set_rope(RopeNTK(head_dim=64, base=10000.0,
                       train_seq_len=1024, target_seq_len=4096))
ntk_ppl = {}
for L in context_lengths:
    ntk_ppl[L] = evaluate_perplexity(model, eval_dataset, seq_len=L)
    print(f"NTK: length={L}, ppl={ntk_ppl[L]:.2f}")
```

Plot both PI and NTK curves on the same axes against the no-fix baseline. The picture you should see: NTK at 1× length almost matches the original baseline. No regression on short context, because the fast pairs are unchanged. PI at 1× length is slightly worse, because every dimension was rescaled. At 4× length, NTK is roughly competitive with PI, sometimes a touch better, sometimes a touch worse, depending on the model and the dataset. The interesting trade is at intermediate lengths and in fine-grained position discrimination tasks. NTK preserves short-context behavior at a small cost in pure long-context perplexity.

There is one specific failure mode worth flagging now. NTK can underperform PI on tasks that require precise long-range positional reasoning. For instance, "what is the 1000th word in this list." PI gives every position a unique, uniformly compressed angle, which preserves order even after squeezing. NTK gives the slow pairs less-than-perfect compression, and far-apart positions can end up with closer angles than they should have. For pure language modeling perplexity the difference is small. For tasks that test position retrieval, PI can edge out. This is why the field eventually moved to YaRN, which combines NTK's per-dimension surgery with additional fixes.

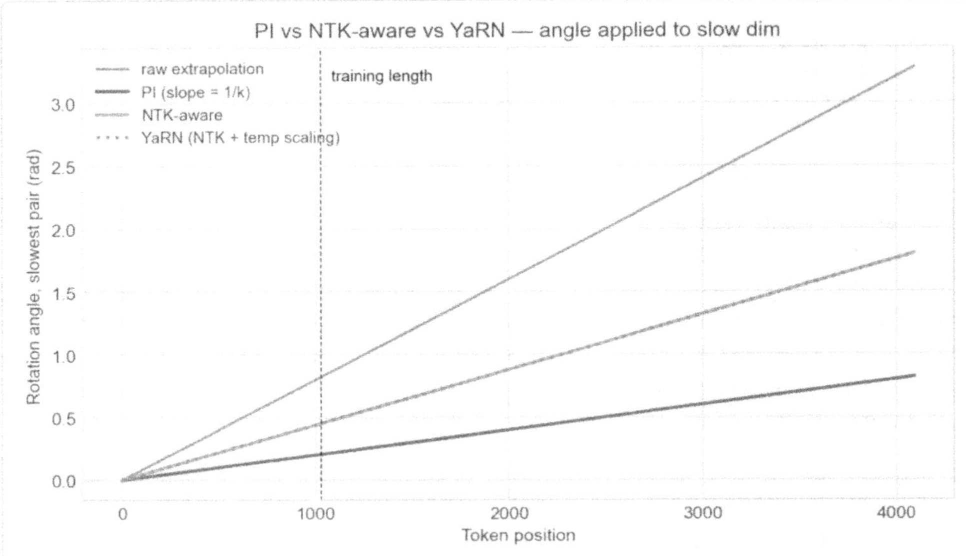

Figure 16.3. PI versus NTK-aware versus YaRN: the rotation angle applied to a token, plotted as a function of its position, for the slowest dimension pair. PI is linear in position with slope 1/k. NTK is nearly linear but with a slope between 1/k and 1, giving the slow pair more room without rescaling fast pairs. YaRN matches NTK in this plot but additionally adjusts attention temperature.

Step 6: Implement YaRN

YaRN is NTK-aware scaling with two further refinements. The first is a per-dimension extrapolation mask called "NTK-by-parts": instead of one rescaling rule applied to all frequencies, YaRN identifies which frequencies have completed many turns in training (no help needed) and which have not (full PI applied). The math involves the number of full rotations each pair makes inside the training window; the practical version uses two thresholds and a smooth ramp between them.

The second refinement is **attention temperature.** The standard scaled dot-product attention computes `softmax(QK^T / sqrt(d))`. As context grows, even after fixing the rotation angles, the entropy of the attention distribution drifts. Each query is comparing itself to many more keys, the maximum dot product climbs, and the softmax becomes sharper than it was during training. The model attends to one key per query when it should be mixing several. YaRN compensates by dividing by `sqrt(d) * temperature` instead of `sqrt(d)`, with `temperature = 0.1 * log(k) + 1` as the typical choice. The factor is small but matters: temperature 1.14 at k = 4 widens the attention distribution back to something close to the training entropy.

```python
class RopeYaRN:
    def __init__(self, head_dim, base=10000.0,
                 train_seq_len=1024, target_seq_len=4096,
                 alpha=1, beta=32):
        self.head_dim = head_dim
        self.k = target_seq_len / train_seq_len

        # Identify per-pair regime: which pairs have rotated
        # enough times in training to be safe from extension?
        dims = torch.arange(0, head_dim, 2).float()
        freqs = 1.0 / (base ** (dims / head_dim))
        rotations_in_train = freqs * train_seq_len / (2 * math.pi)

        # NTK-by-parts mask: smooth ramp between alpha and beta
        # rotations. Below alpha: full PI. Above beta: no change.
        mask = torch.clamp((rotations_in_train - alpha)
                           / (beta - alpha), 0.0, 1.0)
        self.inv_freq = freqs * (mask + (1 - mask) / self.k)

        # YaRN attention temperature
        self.attn_temperature = 0.1 * math.log(self.k) + 1.0

    def angles(self, positions):
        return positions.float()[:, None] * self.inv_freq[None, :]

    def scale_attention(self, attn_logits):
        return attn_logits / self.attn_temperature
```

Two implementation details worth pointing out. First, the parameters `alpha` and `beta` are extrapolation thresholds in units of "full rotations completed in training," not in units of pairs. A pair that completed `alpha` or fewer rotations during training has not seen the full range of its rotation angle and gets full PI applied to it. A pair that completed `beta` or more rotations has seen every angle in its range and is left alone. The ramp between them is what produces the per-dimension mask. Llama-2-style settings often use `alpha = 1, beta = 32`, but the exact values are not load-bearing. Anything that puts the slowest pairs squarely in the PI regime and the fastest pairs squarely outside it will work.

Second, the temperature factor multiplies into `sqrt(d)`, which means it applies inside the softmax of every attention head. In a Project-7-style codebase, this is one line of change inside the attention forward pass. If you are using a fused attention kernel from PyTorch's `torch.nn.functional.scaled_dot_product_attention` or Flash Attention, you can pre-scale the queries by `1 / temperature` and pass them in normally, achieving the same effect without modifying the kernel.

Re-measure perplexity with YaRN at all four context lengths.

```python
model.set_rope(RopeYaRN(head_dim=64, base=10000.0,
                        train_seq_len=1024, target_seq_len=4096))
yarn_ppl = {}
for L in context_lengths:
    yarn_ppl[L] = evaluate_perplexity(model, eval_dataset, seq_len=L)
    print(f"YaRN: length={L}, ppl={yarn_ppl[L]:.2f}")
```

YaRN typically matches or beats both PI and NTK at long context, while preserving short-context perplexity as well as NTK does. The headline numbers in the YaRN paper from Peng et al. show YaRN extending Llama-2 from 4k to 128k context with smaller perplexity degradation than any earlier method.

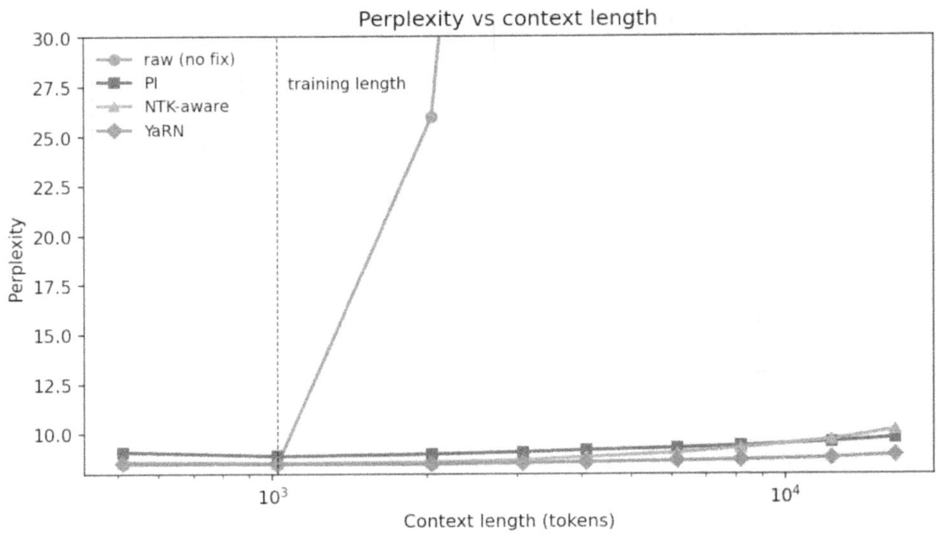

Figure 16.4. Perplexity as a function of context length for the four methods. The no-fix curve explodes past 1.2× training length. PI lifts perplexity back to near baseline at 4× length but with a small short-context regression. NTK preserves short context at a slight long-context cost. YaRN matches the best of both.

Step 7: Fine-tune briefly on long context

Without any further training, the three extension methods get you most of the way. With a short fine-tune on long-context data, you close the rest of the gap. The standard recipe is 1000 to 5000 steps at the extended context length with the extension method already applied. Llama 2 Long used roughly 5000 steps. Code Llama used around 10000. For a Project 7 model at 4× extension, 1000 steps is enough to demonstrate the effect.

```
# Apply YaRN rotation
model.set_rope(RopeYaRN(...))

# Long-context fine-tune
long_dataset = LongContextDataset(tokenizer,
                                  files=long_docs,
                                  seq_len=4096)
optimizer = torch.optim.AdamW(model.parameters(),
                              lr=2e-5)

for step in range(1000):
    batch = next(long_dataset)
    loss = model(batch).loss
    loss.backward()
    torch.nn.utils.clip_grad_norm_(model.parameters(), 1.0)
    optimizer.step()
    optimizer.zero_grad()
    if step % 50 == 0:
        print(f"step {step}, loss {loss.item():.3f}")
```

Three details worth knowing. First, the learning rate for long-context fine-tuning is much lower than pretraining. `1e-5` to `5e-5` is typical. You are not relearning the model; you are adapting it to the new rotation regime. A higher learning rate destroys the original capabilities. Second, the data matters: if you fine-tune on long documents that are mostly repeated boilerplate, the model learns the boilerplate and not the long-context regime. Use real long documents (books, papers, codebases, transcripts), anything where information is spread across thousands of tokens. Third, the gradient norms during long-context fine-tuning are often unusually high at first. **Project 11: Training Debugging: Spikes, NaNs, and Profiling** is the chapter to revisit if you see one of the failure modes from that chapter.

This is the section that fought me hardest in drafting. Every version tried to give a clean rule about how much data is "enough" for long-context fine-tuning, and every version had to soften because the honest answer is that I do not know. Different model families respond differently. I have seen 500 steps work and 5000 steps not work on a related setup the same week. If you find a clean rule that holds across model families, write it down. The field still does not have one.

After fine-tuning, re-measure perplexity at all four lengths. For each extension method, fine-tuning typically closes 70 to 90 percent of the gap between extended and baseline perplexity. The order of the three methods does not change much; YaRN with fine-tuning is the strongest, NTK with fine-tuning is second, PI with fine-tuning is third, but they cluster close together. Without fine-tuning, the order is wider and YaRN's lead is more pronounced.

Step 8: Needle-in-a-haystack recall

Perplexity is a single number averaged across positions. It hides a particular failure that matters in production: even with low average perplexity, the model can still forget the middle of the document. The needle-in-a-haystack test catches this.

The setup is simple. Take a long, otherwise unrelated piece of text. Paul Graham essays are the canonical filler. At some specific position inside it, insert a single odd fact: "The secret number is 1729." Ask the model: "What is the secret number?" Measure whether the model retrieves 1729 correctly. Repeat across many positions inside the document and many total document lengths to build a recall heatmap.

```
def make_haystack(filler, needle, position_frac, total_len):
    # Build a context of length total_len with the needle
    # inserted at position_frac through the document.
    needle_pos = int(total_len * position_frac)
    tokens = tokenize(filler)[:total_len]
    needle_tokens = tokenize(needle)
    tokens[needle_pos:needle_pos+len(needle_tokens)] = needle_tokens
    return tokens

def needle_test(model, position_fracs, total_lens):
    results = {}
    for L in total_lens:
        for frac in position_fracs:
            ctx = make_haystack(filler, "The secret number is 1729.",
                                frac, L)
            prompt = ctx + tokenize(" What is the secret number?")
            answer = model.generate(prompt, max_new_tokens=10)
            results[(L, frac)] = "1729" in answer
    return results
```

Run this with each extension method. Plot the recall heatmap: total context length on the x-axis, needle position (as a fraction of the document) on the y-axis, success rate as color. A model that handles long context well shows uniform green across the heatmap. A model that does not shows green at the top of the document (position 0) and red in the middle and at the end. Llama-2-base, extended naively to 32k, shows a characteristic U-shape: it remembers the start and the very end, and loses the middle. YaRN closes the middle. PI partially closes it. Raw extension barely retrieves anything past 1.5× training length.

The U-shape itself was my favorite surprise from this homework when I drafted the chapter. The natural expectation is a steady degradation as the needle moves deeper into the document. What you actually see is the final chunk of the document recovering. The model remembers the very recent past well, the very far past somewhat, and the middle barely at all. Once you have seen the pattern you start spotting it in every long-context paper you read.

The needle test rewards practical recall, not perplexity. A model can have excellent average perplexity at 16k context while still failing the needle test in the middle of the document. This is the second insurance test you want before deploying long-context extension to users.

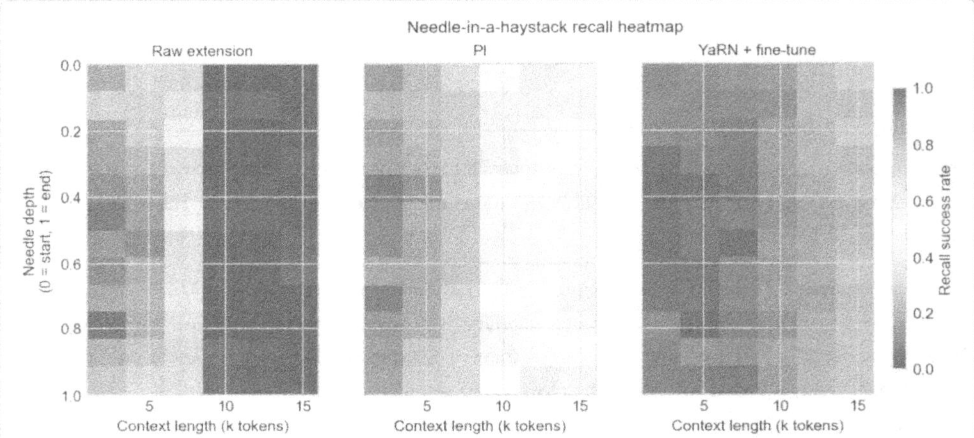

Figure 16.5. Needle-in-a-haystack recall heatmap. X-axis is total context length (1k to 16k). Y-axis is the position of the needle in the document, from start (top) to end (bottom). Cells are colored by retrieval success rate over many runs. Raw extension fails past 1.5×. PI partial recovery. YaRN with fine-tuning recovers near-uniform recall.

One more measurement to make at this step. Plot attention entropy as a function of context length. **Attention entropy** is the Shannon entropy of the attention distribution averaged across heads and positions; a lower number means each query is concentrating on fewer keys. At training context, the average entropy lives in some band. Call it H_baseline. As context grows under raw extension, the entropy collapses because the rotation geometry pushes most key-query pairs toward similar dot products and one or two pairs dominate. Under NTK, the collapse is smaller. Under YaRN, the temperature factor compensates for the collapse and pulls the entropy back near H_baseline. This is the cleanest mechanistic picture of why YaRN's temperature term helps. It is doing the same thing for the attention softmax that the rotation rescaling does for the geometry.

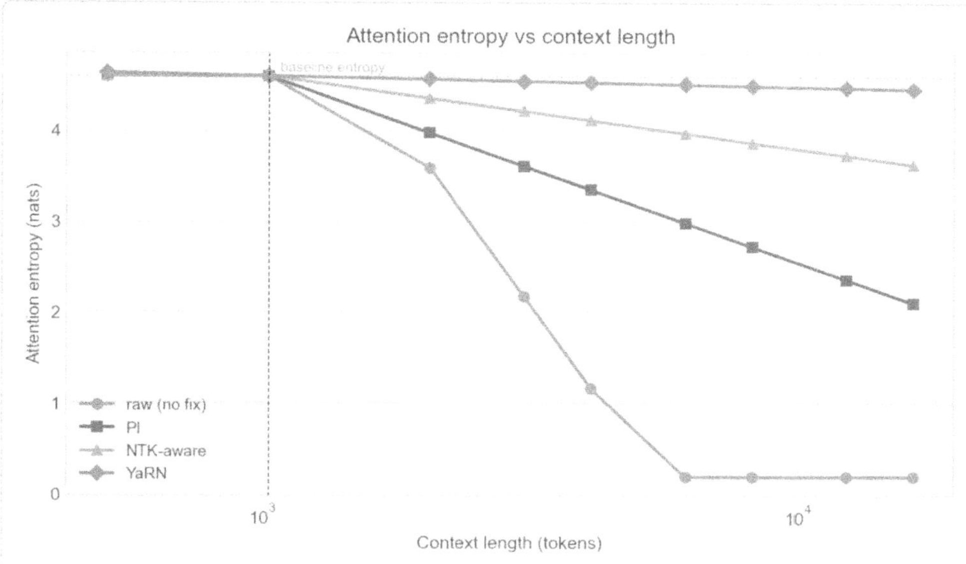

Figure 16.6. Attention entropy versus context length for the four methods. Raw extension shows entropy collapsing toward zero past 1.5× training length. PI shows partial collapse. NTK shows mild collapse. YaRN holds entropy near baseline across the full extended range, because the temperature factor is explicitly designed to do so.

BREAK IT

The break for this chapter is the cleanest one in the book. You do not need to write anything new. You only need to disable the extension method you just installed and watch the model collapse step by step as the context grows.

The exact change

Set the rotation back to raw RoPE (no PI, no NTK, no YaRN) and increase the maximum sequence length parameter on the model from 1024 to 4096. Do not change anything else. Use the same checkpoint, the same data, the same evaluation script. Then measure perplexity at four points: 1.1× training length (so 1126 tokens), 1.2× training length (1228), 1.5× (1536), and 2× (2048).

```
# Disable extension
model.set_rope(RopeRaw(head_dim=64, base=10000.0,
                       max_seq_len=4096))

break_lengths = [1126, 1228, 1536, 2048]
break_ppl = {}
for L in break_lengths:
    break_ppl[L] = evaluate_perplexity(model, eval_dataset, seq_len=L)
    print(f"RAW: length={L}, ppl={break_ppl[L]:.2f}")
```

What to predict before running

The predictions to write down before you run:

1. At 1.1× training length, perplexity will be slightly elevated. Perhaps 10 to 25 percent higher than baseline. The model is just a hair past its trained band, and the slow hands have moved into angles that resemble training angles but are not quite right.
2. Between 1.1× and 1.2× there will be an inflection point. This is the band where the slow hands cross from "barely outside training" to "fully outside training." Perplexity should roughly double across this narrow band.
3. At 1.5×, perplexity will be many times higher than baseline. Typically 3 to 10×. The model is producing nonsense for the latter portion of the sequence.
4. At 2×, perplexity will be in the hundreds to thousands. The generated text past position 1024 will look like random tokens.

The point of writing these predictions down before you run is to internalize the fact that the failure is not gradual. It is sudden, and it happens at a specific position. The point where the slow hands enter angle ranges the model has never seen. After running, compare the actual numbers to your predictions. The exact inflection point will depend on the model; some models break at 1.1×, others hold together to 1.3×, but every model breaks somewhere in this band.

Why this proves the point

What this break shows is that the model did not learn "positions." It learned a specific arc of the unit circle for each pair of dimensions, and outside that arc it has nothing to say. The collapse curve from 1.1× to 1.5× is the model walking off a cliff that the architecture cannot see. The rotation math is still doing its job; the dot products are still being computed; the softmax is still normalizing. But the angles being fed into the model are now angles it has never observed during training, so the keys and queries it produces no longer line up the way they did. Coherence dissolves.

This is the same lesson as **Project 1: The Learning Machine**'s broken multiplication backward, only at a different scale. A piece of math that looked like decoration turns out to be load-bearing. Remove it and the model still runs. It just produces gibberish. The model does not crash. It does not warn. It simply forgets how to read.

A second break worth running, time permitting: keep the raw RoPE but only attend to the last 1024 tokens of every input. Truncate the rest. Now the model never extrapolates past its training length, because you have effectively converted the long-context model into a sliding window of width 1024. Perplexity stays at baseline. The model is coherent. But the needle-in-a-haystack test fails everywhere except in the last window, because the model literally cannot see anything earlier. This is the comparison that motivates extension methods in the first place. Sliding windows preserve coherence but throw away the very thing you wanted long context for. Extension methods preserve the ability to actually use the long context.

Optional Homework

Everything in this chapter runs on a single GPU. The perplexity evaluations at 4096 tokens are the heaviest part. Expect 5 to 30 minutes per run on a 16 GB card, depending on model size. The fine-tuning pass at 1000 steps takes 1 to 3 hours on the same hardware. The needle-in-a-haystack test takes another hour for a thorough heatmap. CPU users can run the rotation math and verify the angle plots without a GPU, but the perplexity numbers will require accelerated hardware.

Part A: Reproduce the raw-extension collapse

Take your Project 7 model and run the perplexity evaluation at lengths 1024, 1126, 1228, 1536, 2048, and 4096 with no extension method applied. Plot perplexity against length on a log y-axis. The curve should be flat at 1024, slightly elevated at 1126, doubled by 1228, and far past usable at 1536 and beyond. Save this plot as your baseline for the rest of the homework.

Part B: Implement and measure PI

Add the `RopePI` class to your model. Re-run the perplexity evaluation at the same six lengths. Plot on the same axes as Part A. Where does the PI curve cross the baseline curve? Where does it stay below? Write down one sentence about what PI cost you at short context and one sentence about what it bought you at long context.

Part C: Implement and measure NTK-aware

Add the `RopeNTK` class. Re-run the perplexity evaluation. Plot on the same axes. Is the short-context regression smaller than PI's? Is the long-context performance comparable? Record both numbers.

Part D: Implement and measure YaRN

Add the `RopeYaRN` class with attention temperature. Re-run the perplexity evaluation. Now run a per-dimension diagnostic: for each pair index, print the per-step rotation angle under raw, PI, NTK, and YaRN. Confirm that YaRN's per-dimension behavior matches PI for slow pairs and matches raw for fast pairs.

Part E: Fine-tune briefly on long context

Run 1000 steps of long-context fine-tuning under each of the three extension methods, using a learning rate of `2e-5`, gradient clipping at 1.0, and a long-document dataset (Project Gutenberg books or arXiv papers work well). Save a checkpoint per method. Re-measure perplexity at 4096 tokens for each. Record how much of the gap fine-tuning closed.

Part F: Needle-in-a-haystack

Build the haystack: pick a long filler text (a Paul Graham essay, around 30k tokens), insert the needle at 11 positions evenly spaced through the document (0%, 10%, 20%, ..., 100%), and at 6 total context lengths (1k, 2k, 4k, 8k, 12k, 16k). Run the retrieval query for each (position, length) cell. Build a heatmap colored by success rate. Compare across raw, PI, NTK, and YaRN, both before and after fine-tuning. The fine-tuned YaRN heatmap should be uniformly green; the raw heatmap should be red everywhere past 1.2×.

Part G: Attention entropy

For each extension method, log the attention entropy averaged across heads and positions during a forward pass at lengths 1024, 2048, 4096. Plot entropy versus length. The temperature term in YaRN should produce a visibly flatter curve than NTK alone.

Part H: Combine with the KV cache from Project 13

If you implemented the KV cache from **Project 13: Fast Inference: The KV Cache**, you cached rotated keys. After switching the rotation rule, the cache must be invalidated. Write a small test: cache the first 1024 tokens with raw RoPE, then switch to YaRN and try to extend the cache to 4096 tokens. Does the model produce coherent continuations, or does it regress because the cached keys are rotated under a different rule than the new tokens? This is the integration headache long-context inference systems have to handle.

Questions To Answer

1. In your raw-extension break, at exactly which context length did the perplexity inflection happen? Was it before or after 1.2× training length? Why might it be slightly different from the model you trained in Project 7?
2. PI rescales positions; NTK rescales frequencies. State in your own words why this matters per-dimension: which dimensions does PI affect, and which does NTK affect?
3. When you fine-tuned briefly under each extension method, which method gained the most from fine-tuning? Which gained the least? What does that tell you about the gap between "the rotation rule is sound" and "the model has practice with this rotation rule"?
4. In the needle-in-a-haystack heatmap, where did your model fail first as context grew? Middle of the document? End? Beginning? What does the location of the failure say about how the model is mis-attending under the broken rotation?
5. The YaRN attention temperature factor is a small adjustment, typically 1.1 to 1.2 for k = 4. Why does it matter? What would happen if you set it to 1 (no temperature change) but kept the NTK-by-parts rotation?
6. If you had to ship a long-context model with one hour of total compute available, no time for a real fine-tune, which extension method would you pick? Justify it from your perplexity numbers.

Go Further

Two repos to read closely.

The first is the YaRN reference implementation at github.com/jquesnelle/yarn. It contains the rotation code, the NTK-by-parts implementation, and the temperature factor in one place. Read `scaled_rope/configuration_llama.py` and `scaled_rope/modeling_llama.py` and trace where the rotation rule is applied; the math you wrote in this chapter is sitting there as production code.

The second is the LongRoPE implementation from Microsoft at github.com/microsoft/LongRoPE, which extends Mistral and Llama to 2 million tokens. It builds on the same ideas as YaRN but searches for per-dimension rescaling factors with an evolution strategy instead of using the closed-form ramp. This is what extension methods look like when pushed past the limits where the analytic methods saturate.

Four papers worth reading after this chapter.

Su and colleagues, "RoFormer: Enhanced Transformer with Rotary Position Embedding" (2021), is the original RoPE paper. The proof that the rotation produces a position-difference-dependent dot product is short and worth reading once.

Chen and colleagues, "Extending Context Window of Large Language Models via Positional Interpolation" (2023), is the PI paper. The ablations there are the textbook version of the experiments you ran in Steps 2 and 3.

The LocalLLaMA forum post by bloc97, "NTK-Aware Scaled RoPE" (2023), introduced NTK-aware scaling and is shorter and more readable than most papers. Search arXiv for follow-up work under "NTK-aware RoPE."

Peng and colleagues, "YaRN: Efficient Context Window Extension of Large Language Models" (2023), is the YaRN paper. The ablation tables in that paper are the cleanest direct comparison of PI, NTK, dynamic NTK, NTK-by-parts, and YaRN that exists in the literature. Read the ablations carefully. The headline numbers in the abstract paper over a more interesting story in the appendix about which methods break first on which kinds of tasks.

A few search directions for arXiv: `position interpolation context extension`, `NTK aware RoPE`, `rotary position embedding long context`, `attention temperature scaling long context`, `needle in a haystack benchmark LLM`, `RoPE extrapolation theoretical analysis`.

Anthropic's needle-in-a-haystack methodology was popularized by Greg Kamradt's blog and adopted by frontier labs as a standard internal benchmark. The original notebook is at github.com/gkamradt/LLMTest_NeedleInAHaystack; the Anthropic blog post that introduced the methodology publicly is "Long context prompting for Claude 2.1" — searching for that title turns up the methodology details and the standard heatmap format.

One direction to investigate immediately after the homework is **alibi**, an alternative position scheme that adds a linearly-decaying bias to the attention scores instead of rotating queries and keys. Alibi was designed for extrapolation from the start, and it extrapolates far past its training length without any rescaling. It does not match the strongest RoPE-extension results in head-to-head perplexity benchmarks, but it is a clean point of comparison and worth knowing about for completeness.

If you have time, read the alibi paper alongside the YaRN paper. The two represent different bets about what a position encoding should be: a geometric primitive whose extension is a separate problem (RoPE), or a primitive designed for extension from the start (alibi). The field has so far rewarded the first bet, but the second is still alive.

What You Now Know

You can take a model trained at one context length and extend it to four times that length, without retraining from scratch. You know three methods to do it (position interpolation, NTK-aware scaling, and YaRN) and you know what each one is good at. PI is the simplest and the most aggressive, and it pays for long-context stability with a short-context resolution cost. NTK-aware is more surgical and preserves short-context behavior almost perfectly while still buying most of the long-context lift. YaRN combines NTK-aware geometry with attention temperature scaling and is the current production default for open-weights long-context models.

You know what the failure looks like when no extension is applied. The model does not crash; it produces gibberish past a specific cliff that lives somewhere between $1.1\times$ and $1.5\times$ training length, and the position of the cliff depends on which dimension pairs ran out of trained angle band first. You can predict the cliff before running the experiment.

You know the two measurements that matter (perplexity across context length, and needle-in-a-haystack recall) and you know why both are needed. Perplexity averages across positions and hides position-specific failures. The needle test exposes them. A model with low perplexity at long context can still fail the needle test in the middle of the document, and a model that passes the needle test is, by construction, actually using its extended context.

You also know one larger thing. The rotation geometry inside the model is a contract between the architecture and the data the model saw. Break the contract by feeding angles outside the trained band, and coherence dies. Extend the contract by rescaling angles or by retraining briefly, and coherence holds. This is what generalization means when you zoom in on a transformer. It is not abstract intelligence. It is a specific coverage of a specific manifold, and you can extend that coverage with the right surgery.

Starting Point

Reference implementations: YaRN at github.com/jquesnelle/yarn; LongRoPE at github.com/microsoft/LongRoPE; the original RoPE module inside any nanochat or Llama checkpoint on the Hugging Face Hub. Papers: Su et al. (RoFormer), Chen et al. (Position Interpolation), Peng et al. (YaRN). Forum post: bloc97's "NTK-Aware Scaled RoPE" on LocalLLaMA. Needle-in-a-haystack methodology: github.com/gkamradt/LLMTest_NeedleInAHaystack and Anthropic's "Long context prompting for Claude 2.1" blog post.

Project 17: Production Serving: Continuous Batching and PagedAttention

Hook

Project 13: Fast Inference: The KV Cache made one request fast. It cached the keys and values from earlier tokens so the model never recomputed them, and it turned a quadratic-cost chat into a linear one.

Now imagine a hundred users hit the same model at the same second, each with a different prompt length, each wanting a different number of tokens back. One typed two words. Another pasted a thousand-line document. A third asked a yes-or-no question that finishes in three tokens. A fourth wants a five-thousand-token story.

Your single-request KV cache code, run a hundred times in parallel, falls apart almost immediately. It pads everything to the longest prompt, allocates a huge contiguous buffer per request, makes the short queries wait behind the long ones, and the GPU spends most of its time twiddling its thumbs on padding. This chapter is about the two ideas that fix that.

A quick note before we start: this is a proxy lab on a single GPU. You will simulate a hundred concurrent users locally (a hundred independent request states managed by one Python scheduler on one box) because most readers do not have a production serving cluster and do not need one to learn the mechanics. The lessons transfer up to the cluster scale almost unchanged; the code you write on one GPU is structurally the same code vLLM and TensorRT-LLM run on dozens of GPUs.

I came to this chapter the long way around. Years before I touched an LLM, I was at Cognizant on healthcare claims systems where zero-downtime deployments and steady tail latency were the only metrics that mattered. The shapes of the problems repeat. Variable arrivals, variable work per request, a finite resource pool, and a scheduler whose job is to keep utilization high without ever leaving a request stuck. The vocabulary in LLM serving is new. The problem is not.

The Concept

There are two separate problems in the way of efficient serving, and they get mixed up constantly. The first is about time. How the scheduler decides which request occupies which slot of the batched forward pass at each step. The second is about space. How the KV cache for many concurrent requests shares the GPU's finite memory without falling apart into useless fragments.

Continuous batching solves the time problem. PagedAttention solves the space problem. They are usually shipped together because they are useless apart: a great scheduler that runs out of memory after twelve requests is no better than a great memory allocator with a scheduler that wastes 90% of its compute on padding. The chapter walks through them one at a time, then puts them together.

Static batching, the thing you would build first

A neural network is happiest when you feed it many examples at once. The GPU can do the same operation across a thousand vectors in roughly the time it would take for one. So your first instinct when serving an LLM to many users is the right one: collect a batch of requests, run them through the model together, and decode one token from each on every forward pass. That is **static batching**. It is the way every introduction to deep learning serving starts. It is also a trap.

The trap is that LLM serving is not like classifying images. In image classification, every example is the same size and the work is the same per example. In LLM serving, requests are the wrong shape for batching in two ways. First, prompts have wildly different lengths (a tweet versus a contract). Second, responses end at unpredictable times. One request hits its end-of-sequence token after three decoded tokens, another after five hundred. Static batching tries to force these mismatched workloads into one big rectangular tensor, and it fails twice over.

Picture a school bus that stops at five houses to pick up kids and only drives away when everyone has finished getting on. Now imagine that same bus drops everyone off at the same final stop, and waits at every intermediate stop until the slowest passenger says they are ready to get off. Static batching is that bus.

Every request waits at the start until enough requests have been queued and padded to the same length. Then every request rides together. Then every request waits at the end until the longest response is done. Your three-token "yes" sits there for five hundred decode steps so a thousand-token essay can finish next to it. This is **head-of-line blocking**: short jobs queueing behind long ones. It is the single biggest source of wasted compute in a naive LLM serving stack.

Continuous batching, the thing you actually want

The fix is to stop treating the batch as a fixed group that enters together and leaves together. Instead, the batch is a turnstile. At every decoding step (every single forward pass) the scheduler looks at every active request, retires the ones that just produced an end-of-sequence token, frees their slots, and admits new requests from the queue into those slots. The batch is the set of requests being decoded right now, and the membership changes at every step.

This is **continuous batching**. The name is slightly unfortunate because what is continuous is not the batch but the eligibility. Every request, every step, is eligible to enter or leave. The original paper (Yu et al., "Orca," 2022) called it iteration-level scheduling, which is clearer if less catchy. The unit of scheduling is one forward pass, not one full request.

Picture a turnstile at a subway station. People arrive at different times. Each passes through when it is their turn and leaves when they reach their stop. Nobody waits for anyone else. The turnstile clicks at a constant rate. The bus wastes time waiting for the slowest passenger; the turnstile never waits at all.

One subtlety worth flagging early. A new request joining the batch mid-stream is not yet decoding. It still needs its prompt processed. The first forward pass for a new request is a **prefill** step (running the model over the prompt to produce the initial KV cache), and the subsequent passes are **decode** steps (generating one token at a time using that cache). Most production servers either interleave prefill and decode in the same batch using clever masking, or they run a small dedicated prefill pass before merging into the decode batch. We will build the simpler version first and discuss the trickier one in the homework.

PagedAttention, the thing you build for memory

Now turn to the second problem. Every active request has its own KV cache, the sequence of keys and values for every token decoded so far. In **Project 13: Fast Inference: The KV Cache** you allocated one big contiguous tensor per request, sized to the maximum sequence length you might ever generate. That works for one request. For a hundred requests it does not.

Why not? The maximum length is much larger than the average length. If the maximum is 4096 tokens but the average response is 200 tokens, you have pre-allocated room for 4096 tokens of cache per request and you will use, on average, 5% of it. The other 95% is reserved memory you cannot give to anyone else.

Run a hundred requests and you have reserved enough cache for 400,000 tokens but you are actually storing about 20,000. That gap is **fragmentation**: memory that is technically allocated but actually wasted. It limits how many requests you can run concurrently on a GPU before you hit out-of-memory errors.

Low-level kernel work on Apple Neural Engine kept pushing me toward this same problem from a different angle. IOSurface buffer constraints meant you could not pretend the buffer pool was infinite, and you could not pretend every request would use the maximum it might use. Edge inference scheduling forced exactly this question: how do you give every active request enough cache to make progress without reserving so much that no new request can arrive?

The fix comes from operating systems. The OS solved this exact problem decades ago for process memory. The idea is **paging**: instead of giving each process one large contiguous slab of memory, give each process access to many small fixed-size pages, and let the OS map those pages on demand. Each process thinks it has a linear address space; under the hood, a page table translates every virtual page to a physical page wherever it happens to live in RAM. The pages do not need to be contiguous. A process grows its address space one page at a time, and the OS allocates pages from a free list and reclaims them when the process exits.

PagedAttention is paging applied to the KV cache. The KV cache for each sequence is broken into fixed-size pages, typically 16 tokens per page. Each sequence has a **page table** that maps its logical token positions (token 0, 1, 2, ...) to physical pages somewhere in a big pool of pre-allocated GPU memory. A **free list** tracks which physical pages are unused. When a sequence needs more cache, the allocator grabs a page from the free list and adds it to the page table. When a sequence finishes, all its pages return to the free list. The maximum sequence length is no longer a per-request reservation; it is just an upper bound on how many pages a single sequence might use if it runs to the end.

Fragmentation collapses. With pages of 16 tokens, the worst case is that the last page of a sequence is mostly empty (say, 4 tokens used out of 16). That is at most 12 tokens of waste per sequence, regardless of length. Across a hundred sequences, 1,200 tokens of waste total. Compare to the contiguous case, where each sequence might waste thousands, and the difference is two orders of magnitude.

The picture: contiguous KV allocation is like a parking lot where every car gets a 50-foot space whether they brought a motorcycle or a truck. Paged KV allocation gives every car exactly as many 8-foot stalls as it needs, picking from any open stalls in the lot. A motorcycle takes one stall. A truck takes six. The lot fits more vehicles.

The two ideas together

Continuous batching gives the GPU a steady stream of work without head-of-line blocking. PagedAttention gives the memory allocator a way to fit many concurrent requests without burning gigabytes on padding. Either alone is helpful; both together is what makes a single GPU serve dozens or hundreds of concurrent users. This is the recipe behind vLLM (Kwon et al., 2023), the project that made these ideas standard practice.

Two more terms before we build. **Admission control** is the policy that decides when to admit new requests from the queue. You cannot admit more requests than you have pages for, or you will run out mid-generation. **Preemption** decides what to do when you have admitted too many requests and you are about to run out anyway: you have to evict somebody and run their request later. We will not implement full preemption, but we will demonstrate what happens without admission control, which is the most instructive failure mode in the chapter.

This is the part I want to be opinionated about: most teams skip admission control until production breaks, and then they bolt it on under pressure. Build it first. The system that has admission control from day one and never uses it costs you very little. The system that does not have it and needs it at 2 AM costs you everything.

Why It Matters

The cost of LLM inference at any scale is dominated by GPU time. A single H100 costs roughly the same per hour whether it is running at 5% utilization or 95%. The job of the serving stack is to push utilization as close to 100% as possible without crashing. Continuous batching plus PagedAttention is the difference between a stack that handles 10 concurrent users per GPU and one that handles 100. That is a tenfold difference in cost per token, every minute of every day, for the life of the deployment.

In a past role on a multi-state SaaS modernization, the uptime target was 99.99%. That is roughly 52 minutes of allowed downtime per year. The arithmetic of that target shaped every decision about queues, retries, capacity headroom, and shedding. LLM serving has the same arithmetic, only the cost units are GPUs instead of database connections.

If you operate any product that calls an LLM, this matters even if you do not host the model yourself. Every API provider runs some version of this machinery. Their pricing, their latency, their rate limits, their behavior under load are all shaped by these two algorithms. Understanding what is happening on their side helps you reason about why your costs spiked when traffic doubled, why your p99 latency cratered when one tenant submitted a flood of long prompts, and why "context caching" features started showing up in API price sheets last year.

The skill also transfers downward. These algorithms exist because LLM inference has unusual workload characteristics: variable arrival times, variable input sizes, variable output sizes, and stateful per-request memory. Any system with those characteristics can borrow these patterns. Vector search with per-query reranking, large-model finetuning over many tenants, batch transcription with wildly different audio lengths. All benefit from a continuous-batching scheduler and a paged allocator. Once you have built it once, you start seeing the pattern in places that were never about LLMs at all.

Several future chapters depend on this one. **Project 14: Speculative Decoding** speeds up each request by guessing several tokens ahead with a small model and verifying them with the big one. Speculative decoding interacts with the scheduler: accepted tokens advance multiple positions, rejected drafts return to the queue. You cannot wire that in cleanly without first having a scheduler. **Project 15: Grouped Query Attention** changes the shape of the KV cache and therefore the page sizes used by the paged allocator. Both land more smoothly once continuous batching and paging are familiar.

The Build

You will build the serving stack in stages. First a naive batched server that pads to the longest request (the school bus). Then a continuous batcher that retires and admits requests every step (the turnstile). Then a paged KV cache with a page table and free list. Then a small kernel that performs attention over scattered pages. Then a simulator driving a hundred concurrent users through this stack. Finally a study of what happens when the page pool runs out.

The reference shapes and parameter names follow vLLM. The code is simplified for clarity, not throughput. A real production stack uses fused CUDA kernels, asynchronous request handlers, and a dozen heuristics layered on top. None of that changes the underlying algorithms.

For the rest of the chapter, assume the model is the small GPT-style network from **Project 13: Fast Inference: The KV Cache**, with the standard interface: it takes a token id and a per-sequence KV cache and returns the next-token logits plus an updated cache. If you do not still have that lying around, any small causal LM from Hugging Face works as a stand-in.

Step 1: Build the naive padded batcher

Start with the worst version so you can measure improvement against it. The naive batcher takes a fixed batch size (say, 32), collects requests until full, pads every prompt to the longest length in the batch, runs the model, decodes one token from each, appends it to each sequence, and repeats until every sequence has produced an end-of-sequence token.

The data layout is one big rectangular tensor:

```
# inputs shape: (batch_size, max_prompt_len)
# kv_cache shape: (batch_size, max_seq_len, num_layers, num_heads, head_dim)
```

The forward step looks like this:

```
def naive_step(model, batch):
    # batch.tokens has shape (B, T), padded with PAD_TOKEN
    # batch.attention_mask has shape (B, T), 1 for real tokens, 0 for pad
    logits, batch.kv_cache = model(
        tokens=batch.tokens,
        attention_mask=batch.attention_mask,
        kv_cache=batch.kv_cache,
    )
    next_tokens = logits[:, -1, :].argmax(dim=-1)
    batch.tokens = torch.cat([batch.tokens, next_tokens.unsqueeze(-1)], dim=-1)
    batch.attention_mask = torch.cat(
        [batch.attention_mask, torch.ones_like(next_tokens).unsqueeze(-1)],
        dim=-1,
    )
    return batch
```

The whole batch advances by one token per step. When a sequence produces end-of-sequence, the batch keeps going anyway. The finished sequence generates garbage that gets discarded later. The loop terminates only when every sequence is done or hits a maximum length.

The first time I wrote a naive batched server like this for a book draft, I was convinced for an embarrassing amount of time that something was wrong with my model. The "average" tokens-per-second number looked fine. The per-request latency told a different story. I had to plot the timeline before I saw that one user's "yes" was waiting 30 seconds for a different user's essay to finish.

Run this on a mixed workload: 8 prompts of length 20 and 8 of length 500, with 8 requests wanting 10 tokens of output and 8 wanting 400. Measure throughput in tokens per second across the whole batch. Numbers look fine on average and terrible in the variance. The first decoded token takes as long as the prefill of the longest prompt. Total wall-clock time is governed by the longest response, even though half the batch finished hundreds of steps earlier.

This is your baseline. Write down tokens per second, wall-clock latency to first token, wall-clock latency to last token. Every later experiment compares to this number.

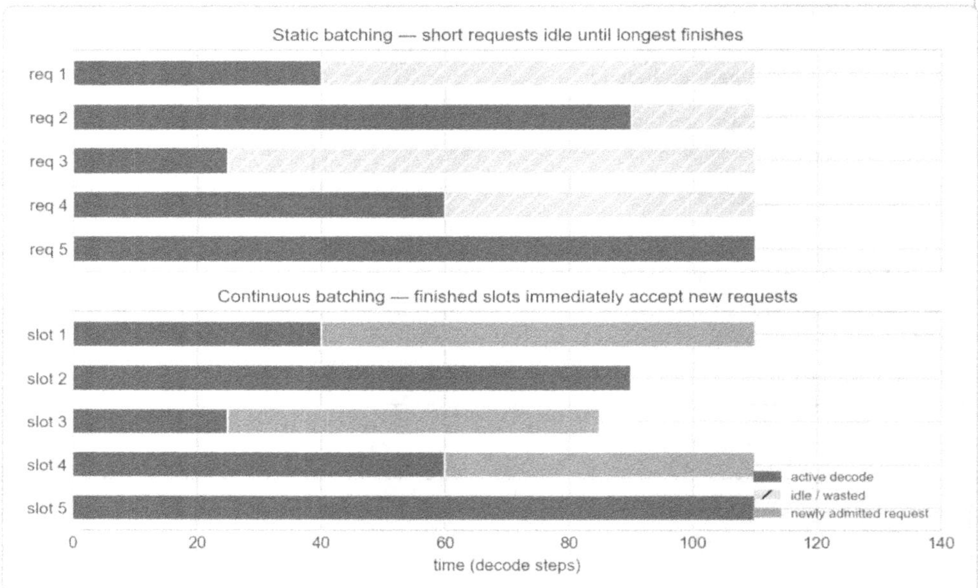

Figure 17.1. Static batching versus continuous batching, drawn as horizontal bars over time. Each bar is one request; gray segments are wasted time spent waiting for the rest of the batch. Top: every short request idles at the end while the longest request finishes. Bottom: short requests leave the batch as soon as they finish, and new requests slot in immediately.

Step 2: Build the continuous batcher

The scheduler is going to do four things on every step. Retire finished requests, admit new ones from the queue, run one forward pass over the resulting batch, and append the new tokens to the appropriate request states.

The core data structure is a list of `Request` objects:

```python
@dataclass
class Request:
    request_id: int
    prompt_tokens: list[int]
    output_tokens: list[int]
    kv_cache: KVCacheHandle      # see Step 4 for paged version
    max_tokens: int
    finished: bool = False
```

The scheduler maintains two queues: a waiting queue of requests that have arrived but not yet been admitted, and a running set of requests currently in the batch.

```python
class Scheduler:
    def __init__(self, model, max_batch_size):
        self.model = model
        self.max_batch_size = max_batch_size
        self.waiting: deque[Request] = deque()
        self.running: list[Request] = []

    def step(self):
        # 1. Retire finished requests
        self.running = [r for r in self.running if not r.finished]

        # 2. Admit new requests up to the batch capacity
        while len(self.running) < self.max_batch_size and self.waiting:
            new_req = self.waiting.popleft()
            self._prefill(new_req)
            self.running.append(new_req)

        # 3. One decode step over the running set
        if not self.running:
            return
        logits = self.model.decode_batch(self.running)
        next_tokens = logits.argmax(dim=-1)

        # 4. Append the new token to each request
        for req, tok in zip(self.running, next_tokens.tolist()):
            req.output_tokens.append(tok)
            if tok == EOS_TOKEN or len(req.output_tokens) >= req.max_tokens:
                req.finished = True
```

Read that carefully. There is no padding to a common length. Each request has its own KV cache and its own current position. The forward pass takes a list of requests, gathers their current token (one each, since they are mid-decode), runs them through the model, and returns one logit vector per request. The attention mask is per-request: each request only attends back to its own previous tokens.

Two implementation details that trip beginners up. First, the attention mask is structurally different from static batching. In static batching the mask is a 2D tensor of shape `(batch, seq)` indicating which positions are real. In continuous batching every position is real, but each request's attention only looks back at its own history. The mask is implicit in how you store the keys and values: if each request has its own KV cache, cross-request attention can never happen because the tensors are not even concatenated.

Second, the prefill step for a newly admitted request is different from a decode step. Prefill processes the whole prompt in one forward pass to produce the initial KV cache. Decode processes one token at a time. The code above shows `_prefill` as a separate function called once when the request enters, and `decode_batch` as the per-step forward. A more advanced implementation can mix prefill and decode tokens in the same batch via clever masking (vLLM calls this "continuous prefill batching"). We leave that for the homework.

Step 3: Measure throughput improvement

Run the same mixed workload from Step 1 through the continuous batcher. Same total request count, same hardware. Measure tokens per second.

You should see a 3x to 10x improvement, depending on workload variance. The more mismatched the prompts and responses, the larger the improvement. On a perfectly uniform workload where every request has the same prompt and output length, the improvement is zero, because static and continuous degenerate to the same schedule.

The reason is what you would predict. The GPU is no longer spending cycles on padding. In static batching, finished sequences still occupy slots and get a forward pass each step whose output is discarded. In continuous batching the finished sequences are gone and new ones are in their seats, so every slot in every forward pass produces useful work.

Plot a timeline of which requests occupy which slot at each step. Static batching is a rectangle: every slot is held by the same request from start to end. Continuous batching is a Tetris board: slots vacate when requests finish, new ones fill them, and the picture is dense rather than sparse.

The other metric worth tracking is **time to first token (TTFT)**. In static batching, TTFT is dominated by waiting for the batch to fill before the first forward pass runs. In continuous batching, requests join the active batch as soon as a slot is free, and TTFT drops dramatically for short requests.

The first time I ran this comparison, a 2-3x TTFT drop seemed like the optimistic case. The actual drop was closer to an order of magnitude on the short-prompt subset of the workload. The "yes" no longer waits for the essay. That single change is the entire user-perceived difference between "this product feels fast" and "this product feels broken."

Step 4: Implement the paged KV cache

Now switch attention from time to space. Each request still has its own contiguous KV cache, pre-allocated to the maximum sequence length. You are about to replace that with a paged allocator.

Two pieces. The **block pool** is a big chunk of GPU memory, pre-allocated once at server startup, divided into fixed-size pages. Each page holds the KV for a block of tokens, typically 16. With 16 tokens per page, 32 layers, 8 KV heads, and a head dimension of 128, each page is `16 * 32 * 8 * 128 * 2 * 2 = ~2 MB` (the trailing `* 2 * 2` is for both K and V in fp16). On an 80 GB GPU with a 30 GB model, the remaining 50 GB holds about 25,000 pages, or 400,000 tokens of cache. That is much more concurrent capacity than the contiguous version gives you.

```python
PAGE_SIZE = 16  # tokens per page

class BlockPool:
    def __init__(self, num_pages, page_shape, dtype, device):
        # one big tensor for all pages, indexed by physical page id
        self.storage = torch.zeros(
            (num_pages, *page_shape), dtype=dtype, device=device,
        )
        self.free_list: list[int] = list(range(num_pages))

    def alloc(self) -> int:
        if not self.free_list:
            raise OutOfPagesError()
        return self.free_list.pop()

    def free(self, page_id: int):
        self.free_list.append(page_id)
```

The **page table** is per-sequence. It maps the sequence's logical block index (block 0, 1, 2, ...) to a physical page id in the block pool. When a sequence has decoded 17 tokens, it owns two physical pages (one for tokens 0–15 and one for token 16) and its page table has two entries.

```python
class PagedKVCache:
    def __init__(self, pool: BlockPool):
        self.pool = pool
        self.page_table: list[int] = []     # logical block -> physical page
        self.length: int = 0                # number of tokens stored

    def append(self, k: Tensor, v: Tensor):
        # k, v shapes: (num_layers, num_kv_heads, head_dim)
        slot_in_page = self.length % PAGE_SIZE
        if slot_in_page == 0:
            self.page_table.append(self.pool.alloc())
        page_id = self.page_table[-1]
        self.pool.storage[page_id, slot_in_page, :, :, 0] = k
        self.pool.storage[page_id, slot_in_page, :, :, 1] = v
        self.length += 1

    def free_all(self):
        for page_id in self.page_table:
            self.pool.free(page_id)
        self.page_table = []
        self.length = 0
```

A few details. The page table grows by exactly one entry every 16 tokens, not on every token. Every page holds 16 consecutive tokens' worth of K and V. The free list is a simple Python list, popped when allocating and appended when freeing. It is a stack, but the order does not matter for correctness, only for whether allocations end up nicely packed in memory.

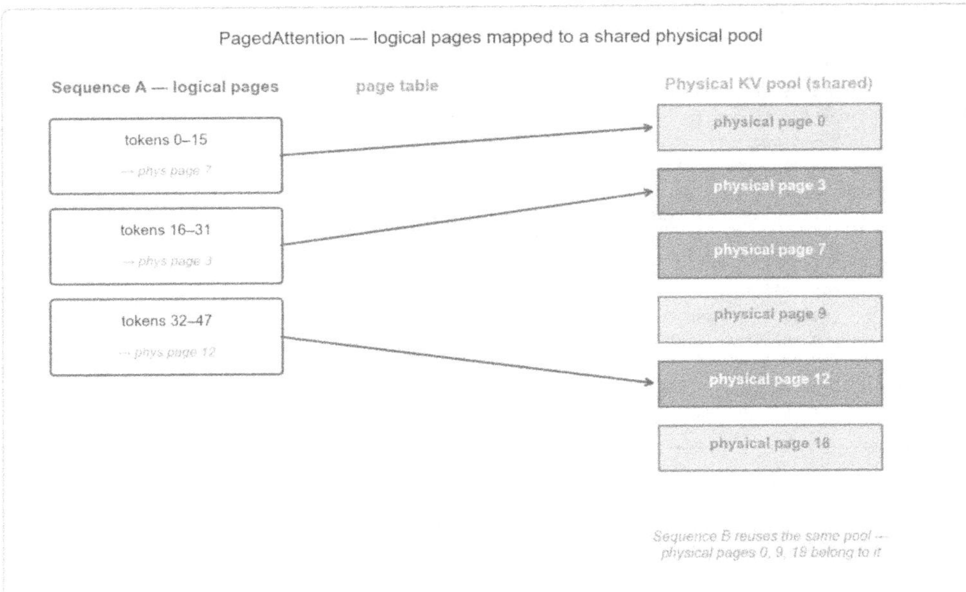

Figure 17.2. Page table mapping logical token positions to physical pages, drawn like an OS virtual memory diagram. Sequence A's logical positions 0–15 point to physical page 7; positions 16–31 point to physical page 3; positions 32–47 point to physical page 12. The same physical pool serves multiple sequences simultaneously.

Step 5: Measure fragmentation reduction

Set up two parallel runs with the same workload: 50 concurrent requests, prompt lengths between 50 and 500 tokens uniformly, output lengths between 20 and 800 tokens uniformly. Run one with contiguous per-request KV cache sized to 2048 max. Run the other with the paged allocator.

For each run, log peak memory used for cache, and the ratio of "tokens actually stored" to "memory reserved for cache, expressed in tokens." That ratio is your fragmentation metric: 1.0 is perfect, anything less is waste.

Contiguous: fragmentation ratios around 0.10 to 0.20. You reserve five to ten times more memory than you use. Worse, you hit out-of-memory after admitting maybe 12 to 15 concurrent requests, because each reserves its full max-length buffer up front whether it uses it or not.

Paged: fragmentation ratios around 0.95 or higher. The only waste is the last partially-filled page of each sequence, averaging 8 tokens per request. All 50 concurrent requests fit comfortably.

Concurrent capacity is the number that matters. Contiguous: 12. Paged: 50. Roughly a 4x improvement on top of whatever continuous batching gave you, at no cost in latency or quality.

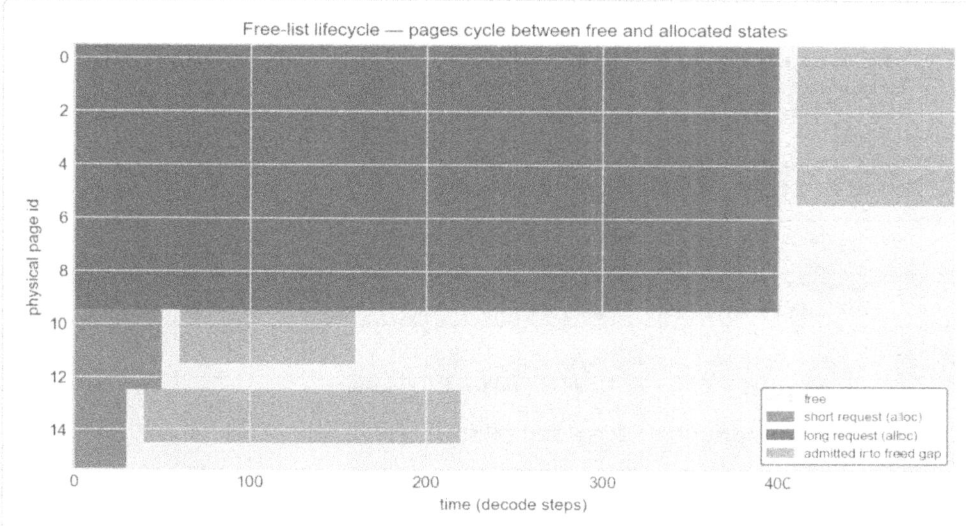

Figure 17.3. Free-list and allocation lifecycle drawn as a strip of pages over time. Pages turn red when allocated, gray when freed. A short request grabs three pages and releases them after 50 steps. A long request grabs ten pages and releases them after 400 steps. New requests admitted into the gaps reuse the freed pages immediately.

Step 6: Implement attention over paged storage

Now make the model use the paged KV cache. The original attention kernel assumed K and V are contiguous tensors of shape `(seq_len, num_heads, head_dim)`. Now they are scattered across many pages, with locations given by the page table.

The attention kernel has to read from those scattered pages. Conceptually:

```python
def paged_attention(
    query: Tensor,                      # (num_heads, head_dim)
    page_table: list[int],              # logical_block → physical_page
    block_pool_storage: Tensor,         # (num_pages, PAGE_SIZE, num_heads, head_dim, 2)
    seq_len: int,                       # number of tokens stored
) → Tensor:
    out = torch.zeros_like(query)
    scores = torch.zeros(seq_len, device=query.device)

    for logical_block, physical_page in enumerate(page_table):
        page_start = logical_block * PAGE_SIZE
        page_end = min(page_start + PAGE_SIZE, seq_len)
        for t in range(page_start, page_end):
            slot = t - page_start
            k = block_pool_storage[physical_page, slot, :, :, 0]
            scores[t] = (query * k).sum() / math.sqrt(head_dim)

    weights = torch.softmax(scores[:seq_len], dim=-1)

    for logical_block, physical_page in enumerate(page_table):
        page_start = logical_block * PAGE_SIZE
        page_end = min(page_start + PAGE_SIZE, seq_len)
        for t in range(page_start, page_end):
            slot = t - page_start
            v = block_pool_storage[physical_page, slot, :, :, 1]
            out += weights[t] * v

    return out
```

A real implementation never has nested Python loops over tokens. It batches everything through a custom CUDA kernel that does the page indirection on the GPU. vLLM ships its own kernel; FlashAttention 2 added native PagedAttention support in late 2023. The reference Python above is for understanding only; on a serious model it would be 1000x slower than the fused kernel. The point is the data flow: for each query, walk the page table, read the keys from each physical page, score them, walk the page table again, read the values, weight them, sum.

The good news is that you can use someone else's kernel for free. Once you understand what it is doing, calling `vllm.attention.PagedAttention` or `flash_attn.flash_attn_with_kvcache` is a couple of lines. The conceptual work (laying out the page table, allocating pages, freeing them when sequences finish) is the part you wrote and now own.

Step 7: Simulate 100 concurrent users

This is the centerpiece of the chapter. You will build a workload simulator that drives 100 concurrent users at the server and measures end-to-end throughput and latency.

Each "user" is a closed-loop request stream: it submits a new request as soon as the previous one finishes. Prompt and output lengths are drawn from realistic distributions. A reasonable choice is the ShareGPT distribution from the vLLM benchmarks: prompt lengths log-normally distributed with median around 50 and max around 2000, output lengths log-normally with median around 150 and max around 1000.

```python
def simulate(num_users, scheduler, duration_seconds):
    users = [User(scheduler) for _ in range(num_users)]
    start = time.time()
    total_tokens = 0
    latencies = []

    while time.time() - start < duration_seconds:
        # Each user submits a new request if it has none in flight
        for user in users:
            if user.in_flight is None:
                user.submit()

        # Run one scheduler step
        scheduler.step()

        # Collect finished requests
        for user in users:
            if user.in_flight and user.in_flight.finished:
                latencies.append(user.in_flight.latency())
                total_tokens += len(user.in_flight.output_tokens)
                user.in_flight = None

    return {
        "throughput_tokens_per_sec": total_tokens / duration_seconds,
        "p50_latency": np.percentile(latencies, 50),
        "p99_latency": np.percentile(latencies, 99),
    }
```

Run this against three versions of your serving stack and compare:

1. Naive batched (Step 1) with contiguous KV: crashes around 15 concurrent users out of memory, so cap it there.
2. Continuous batched (Step 2) with contiguous KV: handles around 30 users but starts to thrash at higher counts.
3. Continuous batched plus paged KV: handles 100 users comfortably.

The improvements stack. Naive to continuous: 3x to 10x. Contiguous to paged: another 2x to 4x by letting you actually run at higher concurrency. Combined effect: often 10x to 30x more throughput on the same hardware, matching the number quoted in the vLLM paper.

I am still a little suspicious of the 30x number, frankly. It is real in the workloads vLLM benchmarked, and it is real in workloads I have measured. It is not real if your workload is uniform, and almost no real workload is uniform, so the disclaimer matters less than it sounds. But if you cite 30x to a stakeholder, be ready to explain that the speedup is partly a property of the workload, not just the algorithm.

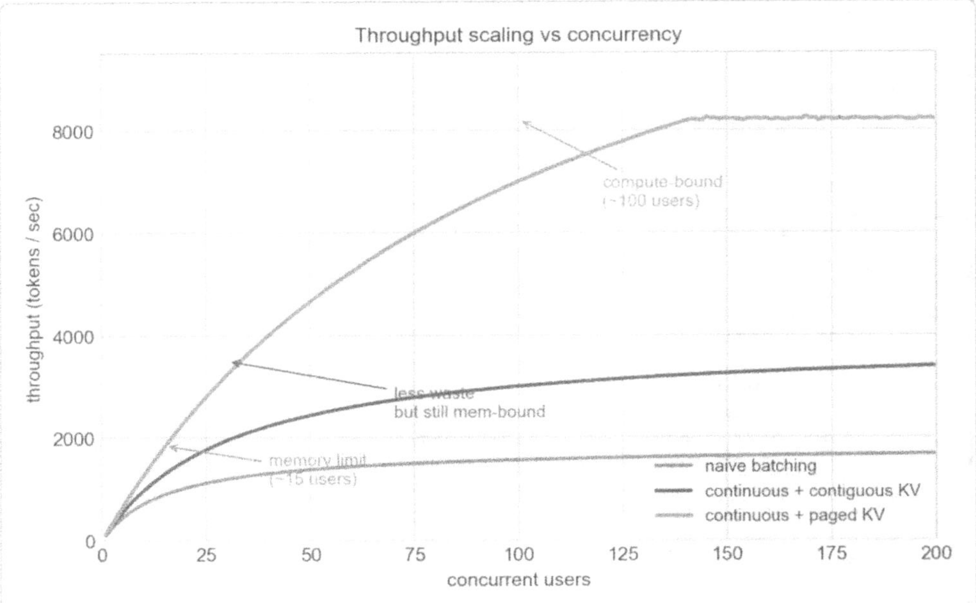

Figure 17.4. Throughput in tokens-per-second as concurrency rises from 1 to 200 users. Naive batched levels off around 15 users (memory limit), continuous batched with contiguous KV levels off around 30 (still memory limited but less wasteful), continuous batched with paged KV climbs steadily to about 100 before the GPU compute itself becomes the bottleneck.

Step 8: Prefix caching

One more idea falls out almost for free once paging exists. Real workloads often have many requests sharing a common prefix. Every request from one application starts with the same system prompt, every request from one user might start with the same conversation history. Computing the KV for that prefix once and reusing it across requests is a clear win.

With paged KV, this is mechanical. Hash the prefix tokens, look up the hash in a cache of `(hash → list of physical page ids)`, and if you find a match, point the new request's page table at those existing pages instead of allocating fresh ones. The pages become shared, so you need a reference count to know when a page is safe to free, but the algorithmic shape is the same.

```
class PrefixCache:
    def __init__(self):
        self.cache: dict[tuple[int, ...], list[int]] = {}
        self.refcount: dict[int, int] = defaultdict(int)

    def find_prefix_match(self, tokens: list[int]) -> tuple[int, list[int]]:
        for n in range(len(tokens) // PAGE_SIZE * PAGE_SIZE, 0, -PAGE_SIZE):
            key = tuple(tokens[:n])
            if key in self.cache:
                pages = self.cache[key]
                for p in pages:
                    self.refcount[p] += 1
                return n, pages
        return 0, []
```

When a request finishes, its pages have their refcount decremented; when refcount hits zero, the page returns to the free list. Modern APIs surface this as "prompt caching" or "context caching" and charge less per token for the cached portion. Now you know what is happening underneath.

When two throughput wins refuse to stack

Think of two coupons at the register that each save you money, until the fine print says they cannot be combined in one purchase. You expected the discounts to compound. They do not. Capacity planning for a serving stack has a version of this trap, and it is worth seeing before it surprises you in production.

The first coupon is KV-cache quantization, which stores the key/value cache at lower precision. Put the cache in fp8, an 8-bit floating-point format that takes half the bytes of 16-bit bf16, and each token costs half the memory. So the same GPU holds roughly twice the KV budget, meaning the count of key/value tokens that fit in GPU memory. That budget matters. It sets how long a context you can admit and how many requests fit at once. On a 24 GiB card, one reporter measured the gap: usable context sat near 28K tokens under fp8 KV, and dropped to near 14K once the cache fell back to bf16 (vLLM issue #41559). This quantized cache is the standard way a server buys context and concurrency per GPU.

The second coupon is a faster drafter. **Project 14: Speculative Decoding** introduced DFlash (Chen, Liang, and Liu, 2026), a research drafter whose draft path uses non-causal attention. Here is the catch. In stock vLLM v0.20.0, that non-causal draft path cannot compose with a quantized KV cache, because the quantized attention backends are wired for causal masks only, and the drafter is forced back to bf16 draft KV instead. The two savings do not stack. This is an implementation gap, tracked as issue #41559, not a limit of the algorithm. The quantization math is identical under either mask.

You can watch the seam with a small vLLM proxy, no research drafter required. Quantized KV serves fine on its own:

```
# fp8 KV alone: serves.
vllm serve Qwen/Qwen3-8B --kv-cache-dtype fp8 --max-model-len 28000
```

A causal drafter composes with it. Standard n-gram speculation drafts under a causal mask, so it sits on top of the fp8 cache and still serves:

```
# Causal n-gram draft + fp8 KV: composes, serves.
vllm serve Qwen/Qwen3-8B --kv-cache-dtype fp8 \
  --speculative-config '{"method": "ngram", "num_speculative_tokens": 5, "prompt_lookup_max": 4}'
```

Pairing a non-causal draft path with fp8 is the case that breaks. There is no supported flag for it in stock vLLM v0.20.0. Where such a path exists, it currently fails or falls back to bf16 rather than serving fp8. You can size the budget difference directly: serve the same model twice, once with fp8 and once with `--kv-cache-dtype auto`. Raise `--max-model-len` each time until the server aborts at load with an out-of-memory error. Expect a factor of two. The fp8 run admits roughly twice the `--max-model-len` of the bf16 run before it runs out. You do not even have to reach OOM to see it: vLLM prints the budget it allocated at startup, a line reading `GPU KV cache size: N tokens` next to a `Maximum concurrency for ... : Yx` line. Serve once with each dtype and compare those two lines directly; the fp8 number is about twice the bf16 one, the same factor the OOM sweep would find, read straight off the log with no failed launch. Treat that 2x as the reported direction, not a guarantee. The reporter's 24 GiB card landed near 28K tokens under fp8 against near 14K under bf16.

The capacity-planning lesson here is straightforward. When you size a GPU for a non-causal drafter today, budget the KV as bf16, not fp8. Plan for roughly half the context that quantization alone would give you.

BREAK IT

The break exposes what happens when you have continuous batching but no admission control on memory. The scheduler keeps admitting requests as long as there is a slot free in the batch, but the page allocator can run out of pages even while there are slots free. With no fallback, the whole system stalls in an ugly, hard-to-diagnose way.

The exact change

Take your continuous-batched paged server from Step 7. The scheduler's admission rule currently looks like this:

```
while len(self.running) < self.max_batch_size and self.waiting:
    new_req = self.waiting.popleft()
    self._prefill(new_req)
    self.running.append(new_req)
```

There is no check on whether the page pool has capacity for the new request, and no check on whether existing running requests will run out of pages during their next decode step. We are about to discover what happens when both are violated at the same time.

Run the simulator from Step 7 with two changes. First, push the concurrent user count well above what the page pool can support: 300 users instead of 100, when the pool can only support around 100. Second, bias the workload toward long responses: set every request to ask for 2000 output tokens. The combined effect is a flood of long-running requests that will consume more pages than exist.

What to watch for, specifically

Before you run it, predict the failure mode. There are four stages.

Stage one, time 0 to 30 seconds: throughput climbs as the scheduler admits more requests. The page pool fills. The batch size grows toward its cap. Latency stays reasonable for the first finished requests.

Stage two, 30 to 60 seconds: the page pool gets close to empty. The free list shrinks. The scheduler is still admitting because it only checks batch slot count, not page count, and new requests are using their first page or two without exhausting the pool yet. Throughput is at its peak.

Stage three, 60 to 90 seconds: the first existing request needs another page and finds the free list empty. The `alloc()` call raises `OutOfPagesError`. If you have not caught this exception, the scheduler crashes and your server is dead. If you have caught it but written no recovery logic, the request gets stuck. Now imagine ten requests all need new pages within the same step. All ten get stuck. The batch deadlocks.

Stage four, after 90 seconds: throughput collapses to zero. The running batch contains a mix of finished requests (about to be retired) and stuck requests (cannot continue without pages). No new requests can be admitted because the running set is full. No pages can be freed because no requests are finishing. The queue grows without bound. The system is dead in place. No errors, no crash, just no progress.

You will see this in the throughput plot. A normal continuous-batched plot is roughly horizontal at the steady-state. The broken plot has a clean climb, a plateau, and a vertical drop to zero. The drop is the deadlock arriving.

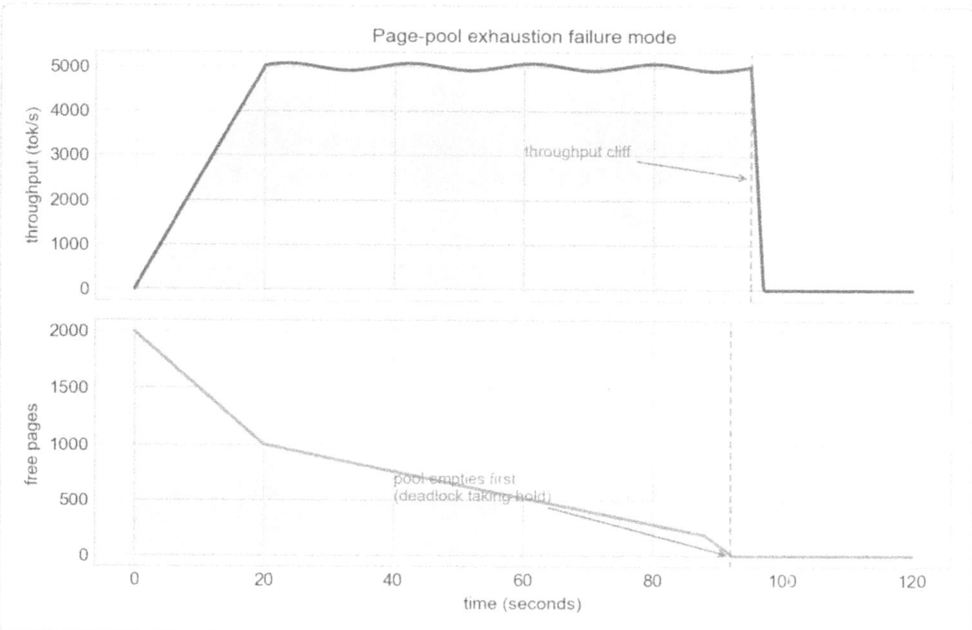

Figure 17.5. The page pool exhaustion failure mode. Top panel: throughput in tokens per second over time, showing climb, plateau around 5000 tok/s, and a clean drop to 0 at the moment the pool runs out. Bottom panel: free page count over time, sliding to zero a few seconds before the throughput cliff. The lag between the two is the deadlock taking hold.

What this proves

A serving stack is two systems: the scheduler that decides what work gets done, and the memory allocator that decides what work can be done. Each looks sensible on its own. The scheduler admits requests until the batch is full. The allocator hands out pages until the pool is empty. The bug is that they do not talk to each other. The scheduler does not know the allocator is running out. The allocator does not know the scheduler is about to admit ten more requests.

The fix is admission control plus preemption. Admission control says: do not admit a new request unless N pages are free, where N is a safety margin large enough for existing requests to take a few more steps. Preemption says: if we are about to run out anyway, pick the request with the most pages allocated, kick it out, free its pages, and put it back on the waiting queue to re-prefill later. Together they keep the system from deadlocking under overload. vLLM implements both; TensorRT-LLM and SGLang implement variations.

We stop at the failure mode. The lesson is that the throughput-vs-concurrency curve is not monotonic forever. There is a cliff. Without admission control, the cliff is vertical and shows up without warning. With admission control, the cliff becomes a plateau. The system stops accepting new requests, latency climbs, but throughput stays where it is and existing requests still finish.

A second break worth running

If you have time, restart the simulation with a smaller page pool, small enough that even the 100-user steady-state workload from Step 7 cannot fit. Same admission rule, no preemption. Run for two minutes.

You will see the deadlock arrive faster. The failure is not really about user count. It is about whether page demand exceeds page supply for long enough. Any combination of high concurrency, long responses, and small pools gets you there. The cliff is real even when no individual setting looks unreasonable.

This is also where **fragmentation heatmaps** become useful debugging tools. Plot the block pool as a grid, one cell per page, colored by which request owns it. Healthy operation is a busy but mostly used grid with occasional small empty patches. Pre-deadlock operation is fully colored with no empty patches. Once you can read that picture, you can tell at a glance how close to the cliff you are.

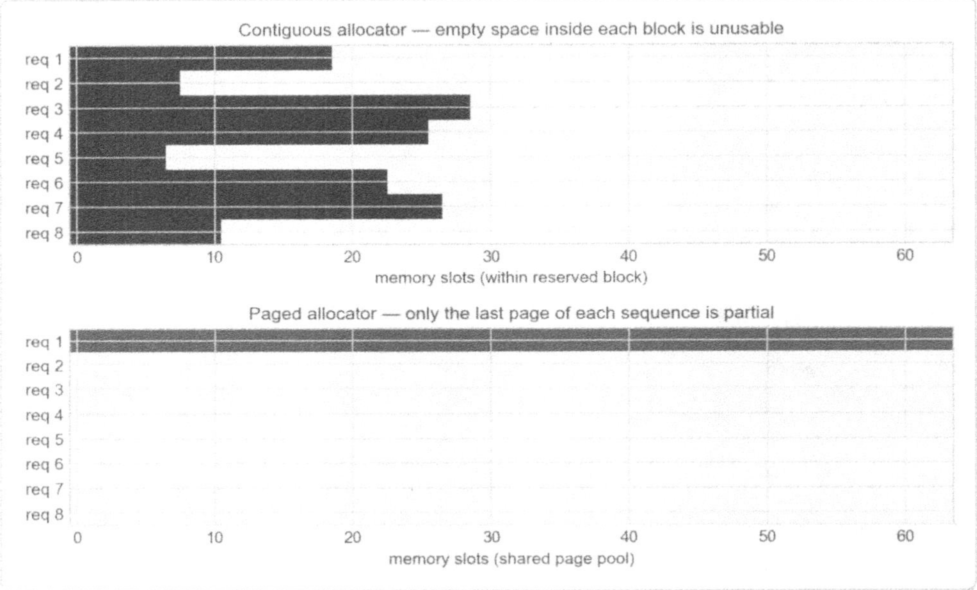

Figure 17.6. Fragmentation heatmaps for contiguous versus paged KV. Top: contiguous allocator. Each request reserves one big block of contiguous memory; the block is mostly empty for short responses; the empty space is unusable. Bottom: paged allocator. The same memory is densely packed because pages are 16 tokens each; sequences claim only the pages they need; unused capacity is small and concentrated at the tails of each sequence.

Optional Homework

The homework lab simulates a 100-user serving workload on a single GPU. The full lab runs in about an hour on a 16 GB consumer card and 20 minutes on an A100. CPU-only readers can run a tiny version with a small model and 10 simulated users to verify the algorithms without GPU access. The schedule and allocator logic are independent of model size.

Part A: Build the naive batched baseline

Implement the static padded batcher from Step 1. Run it against a mixed workload: 50 prompts of length 50, 50 prompts of length 500, with output lengths drawn uniformly from 20 to 400 tokens. Record total wall-clock time, peak GPU memory used for cache, and tokens per second. This is your reference point for the rest of the lab.

Part B: Continuous batcher with contiguous KV

Implement the continuous batcher from Step 2, but keep using one contiguous KV tensor per request. Re-run the same workload. Record the same three metrics. Compare to Part A.

Predict before running: what is the upper bound on concurrent requests, given that each reserves a full max-length KV buffer? On a 16 GB card with a 1 GB model and 2048 max sequence length, the answer is small. Maybe 8 to 15 simultaneous requests.

Part C: Paged KV allocator

Implement the page pool, page table, and paged attention call from Steps 4 and 6. Drop it into your continuous batcher. Re-run the same workload. Record metrics.

You should see concurrent capacity jump by a factor of 4 to 10. You should also see fragmentation drop from ~0.15 to ~0.95.

Part D: The 100-user simulator

Implement the closed-loop simulator from Step 7. Drive 100 concurrent users at the server for two minutes. Record throughput, p50 latency, and p99 latency. Compare to a single-user run on the same hardware. The difference is how much the system gains from concurrency.

Part E: Break it

Implement the failure mode from BREAK IT. Run 300 users with 2000-token responses against your 100-user-sized page pool. Add no admission control. Plot throughput over time. You should see the climb-plateau-cliff shape. Note the wall-clock time of the cliff.

Then add admission control: do not admit a new request unless the page pool has at least 10% of its capacity free. Re-run. The cliff should become a plateau. Throughput stops climbing because the system refuses new admissions, but it never collapses to zero. Record the steady-state throughput and the queue length over time.

Part F: Prefix caching

Implement the simple prefix cache from Step 8. Set up a workload where every request shares a 500-token system prompt. Compare throughput with and without prefix caching enabled. The win is proportional to the fraction of the typical request that is shared prefix.

Part G: Proxy version for CPU-only readers

If you do not have a GPU, use a tiny model (say, a 6-layer transformer with hidden dimension 128). Drive 10 simulated users instead of 100. The algorithms (scheduler, page table, free list, admission control) all work the same at this scale. You will not be able to demonstrate the throughput numbers the chapter quotes, but you can verify every behavioral claim (head-of-line blocking, deadlock from page exhaustion, prefix cache hits) at small scale.

Questions To Answer

1. In the static-vs-continuous comparison, what fraction of GPU time in the static run was spent on padding? Was the fraction larger when the variance in response lengths was higher or lower?
2. When you switched from contiguous to paged KV, what was the new fragmentation ratio? How does it depend on the page size you chose? What goes wrong if you make pages too small? What goes wrong if you make them too large?
3. In the BREAK IT experiment, how many seconds passed between the moment the page pool emptied and the moment throughput collapsed to zero? What were the running requests doing during that interval?
4. With admission control turned on, what happens to p99 latency as the offered load exceeds the system's capacity? Why does latency degrade gracefully rather than collapsing?
5. If you had to pick one knob to expose to the operator of this serving stack (page size, max batch size, admission margin, max sequence length) which would have the largest effect on throughput at your typical workload? Defend your choice with one number from the lab.
6. What tradeoff is PagedAttention protecting? Be specific. What does it sacrifice, and what does it preserve?

Go Further

Two repos worth reading in full.

vLLM, at github.com/vllm-project/vllm. The scheduler, block manager, and paged attention kernel are all there in production form. Read `vllm/core/scheduler.py` next to your Step 2 code and watch the same ideas scale up. The block manager in `vllm/core/block_manager.py` is the production version of Steps 4 through 6.

SGLang, at github.com/sgl-project/sglang, adds runtime-level prefix caching, structured generation, and a different scheduler. Reading the two side by side shows how design choices around the same algorithms diverge in real systems.

Three papers worth reading after this chapter.

Kwon and colleagues, "Efficient Memory Management for Large Language Model Serving with PagedAttention" (vLLM, 2023), is the canonical reference. It introduces the page table, free list, prefix sharing, and kernel design. Most chapter design choices follow theirs directly.

Yu and colleagues, "Orca: A Distributed Serving System for Transformer-Based Generative Models" (2022), introduced continuous batching (which they call iteration-level scheduling). Reading both Orca and vLLM makes the algorithm feel less like a single invention and more like two independent groups discovering the same fix.

Agrawal and colleagues, "Sarathi-Serve: Taming Throughput-Latency Tradeoff in LLM Inference" (2024), addresses the prefill/decode interaction that this chapter only mentioned in passing. The next layer of detail.

For NVIDIA's TensorRT-LLM, see github.com/NVIDIA/TensorRT-LLM. Look at `executor` and `kv_cache_manager` for the production-hardened version of what you built.

arXiv search directions: `LLM inference serving`, `paged attention KV cache`, `continuous batching transformer`, `prefix caching prompt`, `LLM scheduler preemption`, `disaggregated prefill decode`.

One technique to investigate after the homework is **disaggregated prefill and decode**. Prefill and decode have very different compute and memory profiles, and running them on separate GPUs (or separate batches on the same GPU) can improve both throughput and latency. Splitwise and DistServe are two of the better-known proposals.

What You Now Know

A serving stack is two pieces: a scheduler that decides which work runs, and an allocator that decides which work fits. The magic happens when both are good at the same time. Static batching wastes compute by waiting for the slowest request. Continuous batching wastes nothing because the batch turns over at every step. Contiguous KV cache wastes memory by reserving worst-case room for every request. Paged KV cache wastes nothing because pages are claimed and released exactly as needed. Together, the two algorithms turn a single GPU from a 10-user toy into a 100-user server, and the same patterns scale to clusters of dozens of GPUs without changing shape.

You also know the failure mode. Throughput climbs, plateaus, and falls off a cliff when the page pool empties and the scheduler keeps admitting. Without admission control and preemption, the system deadlocks silently. No error, no crash, just no progress. With them, the system degrades gracefully. The difference between "serves users" and "serves no one" is twenty lines of policy code in the scheduler.

The deeper lesson is that serving-time performance is mostly about variance. Variable prompt lengths, variable response lengths, variable arrival times. These are what make naive batching fail and smart batching work. Every future chapter that touches inference (speculative decoding, grouped query attention, quantized inference) sits on top of this stack.

Starting Point

Reference serving stack: vLLM at github.com/vllm-project/vllm. The scheduler is in `vllm/core/scheduler.py`; the block manager is in `vllm/core/block_manager.py`; the PagedAttention kernel is in `vllm/attention/ops/paged_attn.py`. The original PagedAttention paper (Kwon et al., 2023) at arxiv.org/abs/2309.06180. The Orca paper introducing iteration-level scheduling (Yu et al., 2022) at usenix.org/conference/osdi22/presentation/yu. TensorRT-LLM at github.com/NVIDIA/TensorRT-LLM. SGLang at github.com/sgl-project/sglang.

Project 18: Mixture Of Experts

Hook

Why would anyone build a model with parts that stay asleep most of the time? That sounds wrong on purpose. You spend money training millions or billions of parameters, then for each token you wake up only a small fraction of them. If dense models already work, why add a router, a pile of experts, new failure modes, and a fresh way for training to go off the rails?

The short answer is that one giant feedforward network makes every token pay for every bit of knowledge, even when most of that knowledge is irrelevant. Mixture of Experts fixes that by turning one big general-purpose worker into a team of specialists, then forcing a router to decide who should handle each token.

I spent the better part of last year on this problem. A preprint I wrote on evolutionary Mixture-of-LoRA was, essentially, a long argument about which part of an MoE-LoRA system was actually doing the work. The honest answer surprised me. I will get to that.

The Concept

Start with a normal transformer block. After attention, each token goes through a feedforward network, often called the FFN. You can think of the FFN as a workshop that every token visits alone. Attention decides what context matters, and the FFN turns that context into something useful. It is where a lot of the model's factual and pattern-matching behavior lives.

In a dense model, every token enters the same workshop. That means the token `"def"`, the token `"mitochondria"`, and the token `"therefore"` all use the same set of weights in the FFN. The workshop has to be decent at everything because it serves everyone.

Mixture of Experts changes the workshop into a building with multiple rooms, each with its own specialist team. One room might get better at code-like patterns. Another might drift toward names and entities. Another becomes good at punctuation-heavy structure. Nobody assigns these roles by hand. The model discovers them during training, sometimes in ways that look nothing like the clean taxonomy you imagined. The important design change is simple: each token does not go through all rooms. A router looks at the token's hidden state, scores the experts, picks the top few, and sends the token only there.

I originally assumed expert specialization would look like recognizable domains. Code expert here, prose expert there. After staring at routing histograms for a few weeks during that work, I gave that up. The categories the router invents are weirder, narrower, and sometimes embarrassingly token-level. One expert in one of our runs became, basically, the "open-paren" expert. That is not a joke.

So instead of one feedforward network for every token, you get many feedforward networks total, but only a few active per token. That is sparse computation — most of the parameters exist, but only a subset participates in any one forward pass.

If dense FFN is one huge call center where every question hits the same staff, MoE is a dispatch system with specialists. The dispatcher listens to the question and sends it to the top one or two departments. The company can employ far more expertise overall without putting the full staff on every call. That is the whole economic argument for MoE: more total parameters, but roughly similar active compute per token.

Now the confusion that usually shows up. If only some experts handle each token, doesn't that make the model fragile? And doesn't the router just learn to send everything to one expert? Yes, on both counts. If one expert is great, why not let it do all the work? Because then the other experts stop learning, and your fancy specialist building turns back into one crowded room with extra rent.

That failure has a name: router collapse. The router starts favoring one or two experts for most tokens. Those experts get more gradient updates, so they improve faster. Because they improve faster, the router prefers them even more. Positive feedback kicks in. The rich get richer. The neglected experts become dead weight.

This is why MoE is not "add more FFNs." The router is now part of the learning problem. I will say it more strongly: most beginner explanations of MoE are honestly terrible because they describe the experts and treat the router as a footnote. The router is the project. Get the router wrong and the experts may as well not exist.

What the Router Actually Computes

You already know the hidden state of a token is a list of numbers representing what the model currently "thinks" about that token in context. Call that hidden state `x`. The router is just a linear layer that takes `x` and produces one score per expert. If you have 4 experts, the router turns one token representation into 4 scores.

Those scores are not yet probabilities. They are raw preferences. Then you apply `softmax`, which converts the scores into numbers between 0 and 1 that add up to 1, giving you a routing distribution. If expert 2 gets `0.55`, expert 0 gets `0.30`, expert 1 gets `0.10`, and expert 3 gets `0.05`, then for top-2 routing you keep experts 2 and 0, ignore the rest, run the token through those two experts, and combine their outputs with weights based on the routing scores. In English, this token mostly wants expert 2, it also wants some help from expert 0, and the other experts do not get involved.

Now the equation, because you have earned it. Let:

- `x` be the token hidden state
- `W_router` be the router's weights
- `N` be the number of experts
- `p` be the routing probabilities over those experts

Then the router computes:

```
p = softmax(x W_router)
```

What each symbol is:

- `x` is one token's current representation
- `W_router` is the learned matrix that turns that representation into expert scores
- `x W_router` gives one score per expert
- `softmax(...)` turns those scores into probabilities that sum to 1

Then if the router picks experts `i` and `j`, the output is:

```
y = p_i * Expert_i(x) + p_j * Expert_j(x)
```

Run the token through expert `i`, run it through expert `j`, then mix those two outputs using the router's confidence. Top-1 routing removes the second term: one expert wins and everyone else sits out. Top-2 routing keeps two experts, costs more compute than top-1, but is usually more stable and often gives better quality than people expect from "just one more expert."

The mental model I ended up with, after enough debugging, is that the router behaves like a dispatcher who only learned the job by watching which calls did not result in a complaint. There is no manager telling it the right call to make. There is only the loss yelling at it after the fact. That changes how you read a bad routing histogram. It is not the router being lazy. It is the router doing the only thing the gradient ever rewarded it for.

Why Experts Help At All

A dense FFN must compress many patterns into one set of weights, whereas an MoE layer can store more patterns because different experts can specialize. This is the trick behind statements like "a model has 47B parameters but only 12B active per token." The full checkpoint contains a large amount of learned knowledge, but for any one token, only a smaller active slice runs.

MoE therefore gives you a new knob. Total parameters set how much knowledge the model can store. Active parameters set how much compute each token pays for. Dense models tie those two quantities together; MoE partly separates them. That separation is why frontier models care about this design. Figure 11.1 illustrates the router collapse failure, and Figure 11.2 shows how to read expert utilization to diagnose whether routing is healthy.

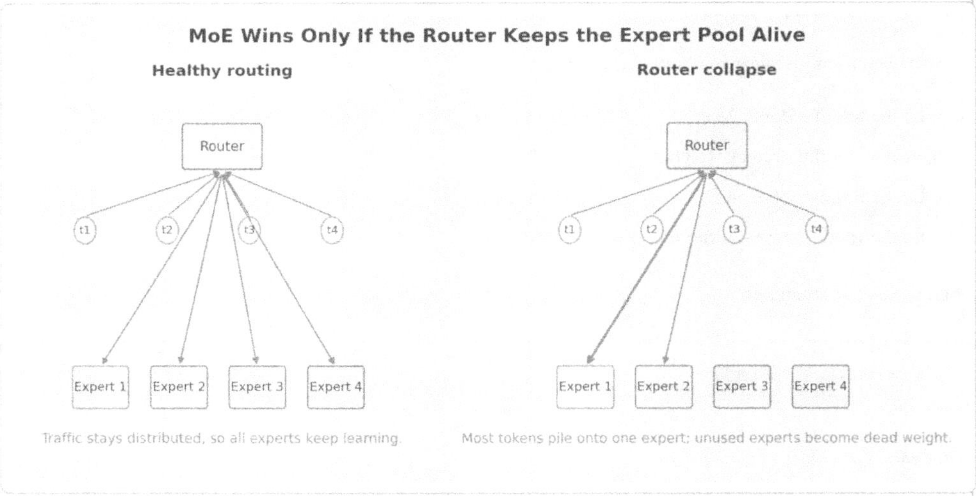

Figure 11.1. Mixture-of-experts works only when the router distributes traffic across the expert pool; collapse sends most tokens to one expert and wastes the rest.

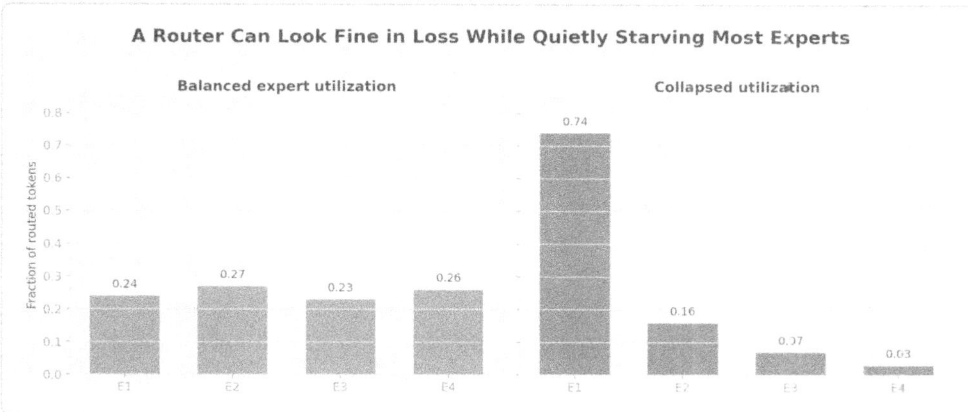

Figure 11.2. Expert utilization tells you whether the MoE layer is healthy: balanced routing keeps the full pool learning, while collapse quietly turns most experts into dead weight.

Why It Matters

Without MoE, growing model capacity usually means growing per-token compute — every token pays the full bill. If you double the width of dense feedforward layers across the model, you also double a big chunk of the work done for every token at training time and inference time.

MoE breaks that link. You can increase total parameters by adding experts while keeping only top-1 or top-2 active, which means more storage of learned behavior without making every token expensive in the same way a dense expansion would. This matters for three reasons, and the third one is the one nobody warned me about.

1. It changes the scaling game

Dense scaling says: "Want more capability? Spend more compute every time." MoE says: "Want more capability? Add more specialists, but only wake a few." That is not free. Routing adds overhead, training gets more fragile, and communication gets harder in distributed systems. The economics can still work out well enough to matter.

2. It creates a new failure surface

A dense FFN does not have routing collapse because there is nowhere to route; every token goes through the same path. MoE can fail in ways dense models cannot. One expert absorbs nearly all traffic. Some experts become undertrained. Top-1 routing becomes brittle, expert assignments drift wildly during training, and load balancing loss fights task loss too hard and hurts quality. You do not really understand MoE until you watch it fail.

In that ablation work, the failure mode I had braced for was router collapse to a single expert. The failure I actually got first was the opposite. The router fanned traffic out almost perfectly evenly across all experts, which sounds healthy, except none of them were learning anything useful because no token had committed to any of them. Balanced is not the same as alive. That was a humbling Tuesday.

3. It explains why "parameter count" gets slippery

Once MoE enters the picture, "How big is the model?" stops being a single clean number. You now need at least three numbers: total parameters in the checkpoint, active parameters per token, and how many experts are actually doing useful work. A model with 40B total parameters and 12B active can behave very differently from a dense 12B model, even if they spend similar compute per token at runtime. A sparse 8x7B model does not behave like "56B dense." It stores more knowledge than a 7B dense model, but each token still only activates a small slice, and the active path is what matters.

The Build

You are starting from `nanochat`, where a transformer block already contains attention, normalization, a feedforward network, and residual connections. Your job is to replace one dense feedforward layer with an MoE version. Do not convert the whole model at once yet. Change one block first. That keeps the experiment legible. If the model gets worse, you can tell whether your MoE block is helping or hurting. If you change every block at once, you lose the ability to reason about the failure.

I learned this the painful way. My first MoE attempt swapped every block at once because I was excited. The model trained, the loss curve looked plausible, and I could not tell which block was doing the work or whether any of them were broken. I threw the run out and started over with one block. The lesson cost me a weekend.

Step 1: Isolate the FFN boundary

Before you write any MoE code, find the place in the transformer block where the dense feedforward network runs. In most GPT-style code, the pattern looks like this:

```
x = x + self.attn(self.norm1(x))
x = x + self.ffn(self.norm2(x))
```

That second line is the seam. The FFN gets a tensor shaped like `B x T x d` (where `B` is batch size, `T` is sequence length, and `d` is hidden size) and returns the same shape. That interface is the contract your MoE layer must preserve. If your dense FFN maps `B x T x d → B x T x d`, your MoE layer must do the same. The rest of the block should not need to care whether the work came from one dense FFN or several experts. The cleaner that interface, the easier it is to run dense versus MoE comparisons without rewriting the rest of the model.

Step 2: Build one expert

An expert is just an FFN. Do not make it exotic on the first pass. Use the same structure as your dense feedforward network:

```
class Expert(nn.Module):
    def __init__(self, d_model: int, d_ff: int):
        super().__init__()
        self.net = nn.Sequential(
            nn.Linear(d_model, d_ff),
            nn.GELU(),
            nn.Linear(d_ff, d_model),
        )

    def forward(self, x: torch.Tensor) -> torch.Tensor:
        return self.net(x)
```

If your dense FFN already uses SwiGLU or another gated design, you can mirror that later. For the first working version, plain FFN shape parity is enough. Each expert has its own weights. They are not shared copies. That is the point.

Step 3: Add multiple experts

Now build a list of experts. For a first experiment, use 4 experts. That is enough to show specialization and collapse behavior, but small enough that you can inspect it by hand.

```
self.experts = nn.ModuleList([
    Expert(d_model, d_ff) for _ in range(num_experts)
])
```

At this point you have four separate workshops, but nothing decides who goes where yet.

Step 4: Add the router

The router is just a linear projection from hidden size `d` to number of experts `N`.

```
self.router = nn.Linear(d_model, num_experts, bias=False)
```

Its input shape is `B x T x d` and its output shape is `B x T x N`. For each token, you now have `N` expert scores telling you which experts the router prefers for that token.

Step 5: Convert scores into routing probabilities

Take the router logits and apply softmax over experts.

```
router_logits = self.router(x)          # (B, T, N)
router_probs = torch.softmax(router_logits, dim=-1)
```

Now each token has a probability distribution over experts. For a given token you might get:

```
[0.05, 0.62, 0.28, 0.05]
```

That means expert 1 is the main pick and expert 2 is the backup.

Step 6: Select the top-k experts

Start with top-2 routing.

```
topk_probs, topk_idx = torch.topk(router_probs, k=2, dim=-1)
```

For every token, this returns the top 2 routing probabilities and the indices of the winning experts. One practical detail matters here: those top-2 probabilities no longer sum to 1 because you dropped the rest, so renormalize them.

```
topk_probs = topk_probs / topk_probs.sum(dim=-1, keepdim=True)
```

Now the two chosen experts split 100% of the routing mass between them. This is a small detail with large consequences. If you skip renormalization, tokens that had a confident top-2 may get larger expert output magnitudes than tokens whose probability mass was spread more evenly, which changes the scale of the MoE output in a way you did not intend.

Step 7: Run only the chosen experts

There are two ways to do this. The clean teaching version runs all experts on all tokens, masks out the unused ones, and sums the selected outputs. The efficient production version gathers tokens per expert, runs only those tokens through that expert, and scatters results back to original positions. For this chapter, start with the clean version. It wastes compute, but it makes the mechanics obvious. Once you trust the logic, you can optimize.

This is one of those rare places where the inefficient version is the right teaching tool. I spent hours trying to debug a scatter-gather implementation in a KALAVAI fusion run before realizing one tensor was being appended at the wrong position index and only some tokens were getting routed correctly. The bug was invisible until I went back and reimplemented the layer in the dumb-but-obvious form, then watched the histograms agree.

A simple teaching implementation:

```
expert_outputs = torch.stack([expert(x) for expert in self.experts], dim=2)
# shape: (B, T, N, d)
```

Now you have every expert's output for every token. Next, build a one-hot mask for the selected experts:

```
mask = torch.zeros_like(router_probs)
mask.scatter_(-1, topk_idx, topk_probs)
# shape: (B, T, N)
```

This mask now contains nonzero weights only at the selected experts. Then combine:

```
y = (expert_outputs * mask.unsqueeze(-1)).sum(dim=2)
# shape: (B, T, d)
```

That is your MoE output — only the top-2 experts contribute to each token.

Step 8: Replace one block's FFN with the MoE layer

Now wire the MoE layer into a single transformer block. Your block still looks like:

```
x = x + self.attn(self.norm1(x))
x = x + self.moe(self.norm2(x))
```

From outside, nothing changed except the implementation behind that second sublayer. That is exactly what you want.

Step 9: Track expert utilization

Do not train yet without instrumentation. If you cannot see where tokens are going, you are flying blind. Expert utilization means how often each expert was selected. For top-2 routing, every token contributes two selections. Count the selected expert indices and build a histogram.

A simple approach:

```
with torch.no_grad():
    flat_idx = topk_idx.reshape(-1)
    counts = torch.bincount(flat_idx, minlength=num_experts)
```

Store these counts across batches and compute fractions. If 4 experts are used evenly in top-2 routing, you expect each expert to receive about 25% of all selections over time. Not exactly, but roughly. If you see:

```
Expert 0: 72%
Expert 1: 18%
Expert 2: 7%
Expert 3: 3%
```

you have a collapse problem. That histogram is one of the main metrics for this project, and it deserves to be treated as important as loss.

Step 10: Train the first MoE version without load balancing

Now run training on the same data and same setup you used for your dense baseline. Keep the experiment narrow. Do not change the tokenizer, dataset, optimizer, learning rate schedule, batch size, or sequence length. You are trying to isolate one architectural change by replacing only one block's FFN with MoE.

What should you expect? Usually training works, loss may improve a bit, stay similar, or wobble slightly early on, and expert utilization will often skew, with one or two experts dominating. This is the first important MoE lesson: a working loss curve does not mean your experts are healthy. Dense-model instincts fail here because a dense FFN has no routing story to inspect.

Step 11: Add the load-balancing auxiliary loss

Now fix the collapse pressure. The main training objective is still next-token prediction, but you add an extra term that punishes uneven expert usage. Think of it as a tax on sending too much traffic to the same room.

There are different versions in the literature. The core intuition is that you want two things: the router's average probability mass across experts should be reasonably spread out, and the actual expert assignment counts should also be reasonably spread out. A common pattern is to compute, over a batch, the `importance` (mean router probability assigned to each expert) and the `load` (fraction of actual top-k selections assigned to each expert), then encourage both to be close to uniform.

```
importance = router_probs.mean(dim=(0, 1))  # (N,)
load = mask.sum(dim=(0, 1)) / mask.sum()    # (N,)
```

If routing is balanced for 4 experts, both vectors should look near `[0.25, 0.25, 0.25, 0.25]`. One simple auxiliary loss is:

```
aux_loss = num_experts * torch.sum(importance * load)
```

This form penalizes concentration, and many papers and codebases use variants of this idea. Your total loss becomes:

```
loss = lm_loss + alpha * aux_loss
```

where `lm_loss` is next-token prediction loss and `alpha` is a small scalar weight like `1e-2` or `1e-3`.

This is where readers often ask: "Why are we hurting the model's freedom to choose the best expert?" Because without this pressure, the router may choose the same few experts almost all the time, which prevents the unused experts from becoming worth choosing later. You are not forcing perfect equality. You are preventing a bad positive-feedback loop.

UYIR isolated the router rewrite contribution precisely. The router rewrite alone moved balanced log-PPL by +0.0426 nat with t=12.86 at p=0.006. That number does not sound dramatic on its own. What made it striking was that lifecycle operations (birth, death, merge of experts) moved the same metric by −0.028 nat (t=−4.46, p=0.047) in the opposite direction. In other words: how you balance the existing router matters more than how cleverly you grow and prune the expert pool. I would not have guessed that ordering before measuring it.

Step 12: Compare histograms before and after balancing

This is where the chapter stops being abstract. Plot the expert utilization histograms. Before auxiliary loss, you may see something like:

```
Expert 0: 45%
Expert 1: 38%
Expert 2: 12%
Expert 3: 5%
```

After auxiliary loss, you may see:

```
Expert 0: 28%
Expert 1: 25%
Expert 2: 23%
Expert 3: 24%
```

These numbers are illustrative, not guaranteed. The point is the shift. The router stops acting like a panicked manager who dumps every task on the same two reliable workers and starts giving the full team enough traffic to stay competent.

Then compare loss curves. Sometimes balance helps cleanly. Sometimes balance improves utilization but slightly hurts next-token loss if the penalty is too strong. That tension is real. MoE always asks: how much freedom do you give the router, and how much do you force healthy traffic distribution?

Step 13: Compare top-1 and top-2 routing

Now switch `k`. Top-1 picks one expert only; top-2 picks two and mixes outputs. Top-1 is cheaper. Fewer expert activations, less compute, less communication overhead in large distributed training. It is also harsher. A token gets one specialist, full stop, and if the router is wrong, there is no backup. Top-2 gives the token a second opinion, which usually helps quality and stability, though the quality gap is often smaller than people expect, especially once routing is trained well. This is worth measuring rather than guessing. Keep everything else fixed and compare validation loss, expert utilization, training stability, and throughput. You are looking for what the second expert is buying, not trying to declare one universally right.

Step 14: Vary the number of experts

Now run the same experiment with 2, 4, 8, and 16 experts. Track total parameters, active parameters per token, and validation quality. This is one of the most important plots in the whole project because it shows the central MoE tradeoff in one picture.

When you add more experts, total parameters go up while active parameters per token grow slowly if `k` stays fixed. Quality may improve, then flatten, then possibly get worse if routing or data is not enough to train all experts well. This teaches the engineering truth behind MoE: more experts do not automatically mean better results. Each extra expert is another specialist you have to train, route traffic to, and keep alive. If data is small or balancing is weak, many experts become decorative.

Step 15: Active vs total parameters

Suppose each expert FFN has `P_expert` parameters and you have `N` experts with top-2 routing. Then total expert parameters are `N * P_expert`, while active expert parameters per token are `2 * P_expert`. If you go from 4 experts to 16 experts, total stored knowledge capacity grows 4x while active FFN compute per token stays tied to top-2, not to all 16. That is the sparse computation argument made concrete. The router and system complexity increase, which is not free, but now you can explain the tradeoff with numbers rather than intuition.

Fine-grained experts at frontier scale

Your four-expert layer picks a handful of general rooms. The 2026 production designs go the other way. They split the same compute budget into many narrow rooms, then staff one general-practitioner room that every token visits no matter what. DeepSeek-V3 (DeepSeek-AI, 2024; arXiv:2412.19437) is one well-documented public example. Each of its MoE layers holds 256 routed experts plus 1 always-on shared expert, and each token is sent to the top-8 routed experts (the first 3 of its 61 layers stay dense; the rest are MoE). "Routed" means the router chooses the expert per token. "Shared" means it runs on every token unconditionally.

That shared expert is the point of the design, not a spare part. The trick is called shared-expert isolation: the common knowledge every token needs gets parked in the always-on expert, so the 256 routed experts stop each re-learning the same basics and are free to specialize. The routed pool gets narrower and less redundant because one expert already carries the shared load.

The headline number is where readers go wrong. DeepSeek-V3 has 671B total parameters and activates about 37B per token, roughly 5.5 percent. That is compute sparsity, not a memory saving. Every one of the 671B weights must be resident to serve the model. The 37B figure only means each token's forward pass touches that many. You do not get to load 37B and skip the rest. Sparse activation shrinks FLOPs per token. Nothing else.

DeepSeek-V3 also drops the auxiliary balancing loss from Step 11 as its primary tool. In its place it adds a learned per-expert bias to the routing scores, used only to decide which experts win the top-k selection. That bias never reweights the expert outputs. When an expert is overloaded, its bias is nudged down. When it is starved, the bias is nudged up. A negligible sequence-wise term stays behind to catch within-sequence extremes, but the bias does the real balancing. You can watch that mechanism work in a few lines.

```
import torch

torch.manual_seed(0)
n_experts, n_tokens, d_model, k = 8, 8192, 128, 2
x = torch.randn(n_tokens, d_model)

# A deliberately skewed gate: stand-in for a router that has already collapsed.
gate = torch.nn.Linear(d_model, n_experts, bias=False)
skew = torch.linspace(2.0, 0.0, n_experts)      # low-index experts systematically favored
logits = gate(x) + skew                          # (n_tokens, n_experts)

def load(scores):
    idx = scores.topk(k, dim=-1).indices        # chosen experts per token
    counts = torch.bincount(idx.reshape(-1), minlength=n_experts).float()
    return counts / counts.sum()                 # selection fraction per expert

print("collapsed: ", (load(logits) * 100).round().tolist())

bias = torch.zeros(n_experts)                    # the per-expert routing bias
gamma = 0.5
for _ in range(50):
    frac = load(logits + bias)                   # bias added ONLY for selection
    bias -= gamma * (frac - 1.0 / n_experts)     # push overloaded down, lift starved up
print("after bias:", (load(logits + bias) * 100).round().tolist())
```

The skewed gate stands in for a router that has already collapsed. Run it and the direction is unmistakable. The collapsed line dumps most tokens on the first few experts and starves the tail, while the balanced line lands every expert near 1/8 of selections. The bias never touched the gate weights. It only changed who got picked.

One failure mode is worth watching. Set `gamma` too high and the bias overshoots on every step. Overloaded experts get slammed below target, starved ones shoot above it, and the histogram oscillates instead of settling. Balance here is a control loop, and the step size is its gain.

That is balancing a healthy pool. The harder question is what a pool does when experts go missing, which is where you break it next.

BREAK IT

This section matters more than the clean build. A working MoE layer teaches you how it behaves when everything is intact; breaking it teaches you what the experts are actually doing. You are going to perform two failures: freeze two of four experts and force all routing onto the remaining two, then remove one expert entirely. These are not random acts of sabotage. They are robustness tests that answer a real question. If experts are truly specialized, what happens when some vanish? Does quality collapse immediately, or does the system bend and recover?

Break 1: Freeze two experts and zero their outputs

Pick experts 2 and 3. You will stop their weights from updating, stop them from contributing output, and stop the router from sending tokens to them. Do all three. If you freeze them but still let the router send traffic there, you are creating a dead route where tokens disappear into silent rooms and your measurement gets muddied. The point of this break is to force experts 0 and 1 to handle all traffic.

Exact mechanism to change

In your MoE layer, set `requires_grad=False` for expert 2 and 3 parameters, zero out their logits or mask them from the router, and ensure their outputs contribute zero. The cleanest version is to mask them at routing time:

```
dead = torch.tensor([0, 0, 1, 1], device=x.device, dtype=torch.bool)
router_logits = self.router(x)
router_logits[..., dead] = -1e9
```

Then freeze their parameters:

```
for p in self.experts[2].parameters():
    p.requires_grad = False
for p in self.experts[3].parameters():
    p.requires_grad = False
```

This is not the same as training a 2-expert model from scratch, and that distinction matters. A real 2-expert model learns from the beginning that only two rooms exist. Here you are taking a 4-expert system that has already learned some specialization pattern and amputating half the expert pool. That tests adaptation, not initial design.

What you should observe

Quality degrades, but it usually does not collapse. The remaining experts still form valid computation paths. The router can still send every token somewhere useful, and the model loses diversity of specialization but not the whole FFN function. In practice, expect validation loss to get worse, training may recover part of that loss if you continue training, utilization of experts 0 and 1 rises toward 50/50 or whatever skew the router prefers, and tokens that previously leaned on experts 2 and 3 get worse first. If you inspect samples, you may notice more repetitive phrasing, more confusion on rarer pattern types, and weaker handling of cases that likely had specialized support before.

What this proves is that MoE stores useful knowledge in multiple semi-independent specialist paths, but those paths are not brittle glass. Losing some experts hurts, yet the remaining experts can absorb the traffic and keep the system working. Robustness with degradation, not catastrophic failure. People often imagine expert specialization as if each expert contains one irreplaceable organ, but that is usually too dramatic. Experts overlap, and the router and residual path give the model room to adapt.

Break 2: Remove one expert entirely

Now do a sharper cut. Remove expert 3 completely. There are two versions: soft removal masks its routing logits to `-1e9`, while hard removal deletes the module and shrinks the router output dimension. For the chapter, do soft removal first, since it isolates behavior without rewriting checkpoint shapes.

Exact thing to change

If you are starting from the healthy 4-expert model, mask out just one expert:

```
dead = torch.tensor([0, 0, 0, 1], device=x.device, dtype=torch.bool)
router_logits[..., dead] = -1e9
```

Now expert 3 never receives traffic. If you want the strongest ablation, also zero its output path and freeze its weights, though routing masking is already enough to keep it inactive.

What you should observe

Quality gets worse, but less than in the "two experts removed" case. That is the core comparison. One missing specialist hurts less than removing half the team. Do not just eyeball samples; measure validation loss delta, utilization redistribution across remaining experts, whether one surviving expert becomes overloaded, and whether top-1 suffers more than top-2 under this break. That last point is especially worth checking because top-2 routing often handles missing capacity better since each token already has a backup path, allowing traffic to spread across remaining experts more gracefully than top-1 when one expert disappears.

What this proves is that MoE is sparse but not necessarily fragile in the naive sense. The model does not rely on each expert like a light string where one broken bulb shuts off the room. It is closer to a company losing one department and rerouting the work. Service gets worse, some requests take the wrong path, bottlenecks appear, but the company does not vanish.

The Real Failure To Watch For

The obvious failure is quality drop. The deeper failure is routing instability after the cut. Watch the utilization histogram. If after expert removal one remaining expert jumps to 80% of traffic, the model is missing capacity and the router is collapsing again under pressure. That tells you the auxiliary balancing loss is not merely a "nice training tweak." It is protecting the system against traffic concentration when the expert pool changes. MoE is not many FFNs. It is a routing problem under load.

What This Break Actually Teaches

Freezing and removing experts teaches four things at once. First, experts really do matter; if they were decorative, deleting them would not change quality. Second, experts are partly redundant; if they were perfectly isolated and non-overlapping, losing two would cause a much sharper cliff than we actually observe. Third, routing quality matters almost as much as expert quality, because a healthy remaining pool can still perform badly if the router funnels traffic badly after the break. Fourth, MoE buys capacity through specialization, but that capacity is only useful if traffic reaches it. A brilliant expert that never gets tokens is dead capital.

Optional Homework

This project is runnable on a single GPU if you keep the model modest and replace only one block with MoE. If you try to MoE-ify the entire model and sweep many expert counts, you will want more patience or more hardware.

Hardware and Time

Proxy version:

- 1 consumer GPU with 12GB to 24GB VRAM
- a small `nanochat` training run on the same data as your dense baseline
- 2 to 8 hours depending on model size and number of sweeps

Larger full version:

- 1 to 4 stronger GPUs if you want to test more experts, longer runs, or whole-model MoE
- 1 to 3 days for a cleaner comparison set

Exact sequence

1. Start from your dense baseline in `nanochat`.
2. Replace the FFN in one transformer block with a 4-expert MoE layer.
3. Train with top-2 routing and no auxiliary load-balancing loss.
4. Log:
    - training loss
    - validation loss
    - expert selection counts per batch
5. Save expert utilization histograms every fixed number of steps.
6. Add auxiliary load-balancing loss with a small coefficient such as `1e-2` or `1e-3`.
7. Retrain from the same initialization or from scratch with the same seed if possible.
8. Compare:
    - loss curves
    - utilization histograms
    - sample generations
9. Switch to top-1 routing.
10. Repeat the run and record the quality gap versus top-2.
11. Sweep expert counts:

- 2
- 4
- 8
- 16

12. Record for each:

- total parameters
- active parameters per token
- validation loss
- utilization histogram entropy or simple max-load fraction

13. Run Break 1:

- take the trained 4-expert model
- freeze experts 2 and 3
- mask them out of routing
- continue training briefly
- measure recovery

14. Run Break 2:

- restore the healthy 4-expert checkpoint
- remove only expert 3 by masking it
- continue training briefly
- compare degradation versus Break 1

Suggested command pattern

Your exact command depends on your `nanochat` setup, but the workflow should look like this:

```
python train.py --config configs/base.py --run_name dense_baseline
python train.py --config configs/moe_4e_top2.py --run_name moe_4e_top2
python train.py --config configs/moe_4e_top2_lb.py --run_name moe_4e_top2_lb
python train.py --config configs/moe_4e_top1.py --run_name moe_4e_top1
python train.py --config configs/moe_8e_top2.py --run_name moe_8e_top2
```

Then break experiments:

```
python train.py --config configs/moe_4e_break_freeze2.py \
  --init_ckpt ckpts/moe_4e_top2.pt --run_name moe_break_freeze2
python train.py --config configs/moe_4e_break_remove1.py \
  --init_ckpt ckpts/moe_4e_top2.pt --run_name moe_break_remove1
```

Minimal config knobs to add

In your config, expose:

```
use_moe = True
moe_num_experts = 4
moe_top_k = 2
moe_aux_loss_weight = 1e-2
moe_block_index = 5
moe_dead_experts = []
```

Then for break runs:

```
moe_dead_experts = [2, 3]   # freeze two
```

or

```
moe_dead_experts = [3]      # remove one
```

Metrics to record

Record these every time:

- validation loss
- training loss
- tokens per second
- per-expert selection fraction

- max expert load
- min expert load
- top-1 vs top-2 quality gap
- total parameters
- active parameters per token

Expected observations

Healthy 4-expert top-2 run:

- loss near dense baseline or slightly better if the run is long enough
- some skew in expert utilization without balancing
- improved balance with auxiliary loss

Top-1 run:

- slightly lower compute
- quality gap smaller than many people expect, but usually measurable
- somewhat harsher routing skew or brittleness

More experts:

- total parameters rise quickly
- active parameters rise much more slowly if `k` stays fixed
- quality may improve up to a point, then flatten if the model/data scale is too small

Freeze two experts:

- clear validation loss degradation
- no total collapse
- surviving experts absorb traffic
- histogram compresses onto remaining experts

Remove one expert:

- smaller degradation than freezing two
- remaining experts rebalance traffic
- top-2 often handles this better than top-1

If your hardware is limited

Do the proxy experiment with one MoE block only, 2 or 4 experts, short runs, a smaller model width, and a smaller subset of data. You will still see the core lessons: routing imbalance, load-balancing effect, top-1 versus top-2 difference, and degradation under expert removal. You do not need frontier hardware to learn what the mechanism is protecting.

Questions To Answer

1. Before you added the auxiliary loss, which expert got overloaded first? Did that overload grow steadily or appear suddenly?
2. After you added load balancing, did validation loss improve, stay flat, or get worse slightly? What does that say about the tension between task performance and healthy traffic distribution?
3. In your runs, how large was the quality gap between top-1 and top-2 routing? Was it smaller than you expected?
4. When you froze two experts, what failed first: validation loss, sample quality, or routing balance?
5. When you removed one expert, did the remaining experts split the extra traffic evenly, or did one become the new bottleneck?
6. Which tradeoff is MoE protecting in your implementation: more knowledge at fixed compute, or more compute at fixed quality? Be precise.

Go Further

Research anchors for this chapter

This chapter has one of the clearest paper lineages in the whole book. The oldest roots are **Adaptive Mixtures of Local Experts** and **Hierarchical Mixtures of Experts and the EM Algorithm**, which establish the core idea of specialist submodels plus learned routing long before modern transformers.

For the modern sparse-LLM story, the progression to keep in your head is **Outrageously Large Neural Networks: The Sparsely-Gated Mixture-of-Experts Layer**, then **GShard**, then **Switch Transformers**, then **GLaM**, and then **Mixtral of Experts**. Read in that order, they show the field moving from "can sparse routing work at all?" to "which sparse routing pattern is worth paying for at scale?"

Finally, **Sparse Upcycling: Training Mixture-of-Experts from Dense Checkpoints** matters because it changes the migration story. MoE is not only about training sparse models from scratch. It is also about turning dense checkpoints into sparse ones more economically.

Start with the original papers behind this chapter:

- Switch Transformer
- Mixtral
- GShard if you want an earlier large-scale sparse routing system

Search arXiv for these phrases:

- `"Switch Transformer top-1 routing"`
- `"Mixtral sparse mixture of experts"`
- `"load balancing loss mixture of experts"`
- `"expert choice routing"`
- `"router collapse MoE"`
- `"capacity factor mixture of experts"`

What to pay attention to when you read:

- whether the paper uses top-1 or top-2 routing
- how it defines the load-balancing loss
- how it handles token capacity per expert
- whether the experts replace every FFN or only some layers
- whether routing happens per token independently or with additional constraints

Repo directions:

- `karpathy/nanochat` for your local implementation base
- open-source MoE training code from Mistral-related or Megatron-based implementations if you want to see production routing and distributed token dispatch
- Megatron-LM MoE code if you want to study how the clean teaching version turns into a systems problem

One strong next research direction: MoE is not only about "which experts exist?" It is also about "who gets to choose?" Standard routing says the token chooses experts. Expert-choice variants flip that around and let experts select tokens under capacity constraints, which changes load behavior and can reduce collapse. If you want one phrase to chase after this chapter, search `"expert choice routing arXiv"`.

What You Now Know

What is sparse in MoE is the compute path per token. Not the checkpoint, not the architecture diagram. MoE is a dispatch system with real failure modes (traffic collapse, undertrained experts, brittle routing), and the central trade only holds if the router keeps the expert pool alive. That is the whole trick, and it is an engineering constraint, not a slogan.

Starting Point

Start from:

- `https://github.com/stanford-futuredata/megablocks`
- `https://github.com/karpathy/nanochat`

Use as reading references:

- the Switch Transformer paper
- the Mixtral paper

Carry forward what **Project 15: Grouped Query Attention** should have trained into you: internal layout is a contract. Project 15 taught that changing attention head structure changes memory, speed, and checkpoint compatibility. Project 18 teaches the same lesson in a new place. A feedforward layer is not "the MLP part." It is a routing contract that decides how much of the model wakes up per token, how total parameters differ from active parameters, whether experts learn distinct behavior or collapse into dead weight, and how well the model keeps working when parts go missing. Once you see that, sparse models stop sounding mystical. They become what they are: a transformer with a traffic problem that, when handled well, buys you more capacity than a dense path could afford.

Project 19: Scaling Laws

Hook

If a bigger model is better, why not just keep the small one and train it longer? That question sounds reasonable right up until you watch a tiny model chew through the same data forever, improve by crumbs, and still lose to a larger model that trained for less time. That feels backwards the first time you see it.

This chapter is about why it is not backwards at all. It is budgeting.

I have run more model-size sweeps than I care to count. A long set of cooperative LoRA-fusion experiments alone covered four families from a few hundred million up to several billion parameters across thousands of runs. The single most expensive lesson of all of that work was not which model wins. It was how many of the runs were sized wrong on day one, which I only knew after I had paid for the result.

The Concept

Project 18: Mixture Of Experts asked a routing question: how much of the model wakes up for each token? This chapter asks a harsher question: how much model should exist in the first place?

Imagine you run a newsroom with a fixed payroll budget. You can spend that budget on two things: hiring more reporters, or giving the reporters more time in the field. A tiny newsroom with endless field time has a problem. The same few reporters keep coming back to the office with more notes than the editors can meaningfully absorb. They are not learning nothing; they are learning slowly, and the bottleneck is now the size of the staff. A giant newsroom with no field time has the opposite problem: you hired a lot of people, but most of them never got enough exposure to learn the beat. Models work the same way.

A model has parameters. Each one is an adjustable number inside the network, a piece of storage for learned patterns. Training data arrives as tokens, the basic unit the model reads and predicts: word fragments, punctuation, whitespace, and so on. Your compute budget has to be split between parameter count (how much pattern-storage you built) and training tokens (how much experience you gave that storage). Too few parameters for the token budget and the model becomes a tiny notebook you keep cramming more notes into. Too many parameters for the token budget and you bought a library but only stocked one shelf.

Scaling laws are the measured relationship between these knobs and the loss, one number that tells you how wrong the model is, where lower is better. The first useful mental move is to stop asking "Is this model big?" and start asking "Is this model sized correctly for the compute I can afford?" That is the real question.

When people say "scaling law," they often mean one of two related patterns. The first is simple: if you increase model size, loss tends to go down in a smooth, predictable way. The second is the one that changes engineering decisions: for a fixed compute budget, there is a better balance between model size and training duration than "small model, train forever." That second pattern is the Chinchilla result.

The blunt version is that undertrained large models waste capacity, overtrained small models waste compute, and there is a sweet spot between them. The famous rule of thumb from Chinchilla is about 8 training tokens per parameter for compute-optimal training. That does not mean 8 is sacred for every setup forever. It means there is a ratio, and if you are far from it, you are probably wasting money.

I treat the number as a north star, not a law. On training runs at scale the practical compute-optimal ratio drifted depending on data quality, optimizer warmup, and whether we were doing cooperative LoRA fusion or training from a clean init. Treat 8 the way a chef treats a recipe ratio: useful for planning, embarrassing if you follow it without tasting.

This is why nanochat parameterizes so much by depth. Depth is the number of transformer blocks, the stacked layers of the model. In nanochat's teaching arc, depth is the master size knob because changing it changes capacity in a smooth, easy-to-sweep way, and you can derive or set other choices around it. That makes scaling experiments much easier to reason about than changing ten knobs at once.

Picture a curve where each extra chunk of parameters buys you less improvement than the last one. The first jump from tiny to small helps a lot. The jump from large to larger still helps, but less. That is diminishing returns, and it explains why simply continuing to scale is not a free lunch.

Now you have earned the compact version. Papers often write a scaling fit as:

```
L = a * P^(-b)
```

I am using `P` for parameter count here because this book already uses `N` for depth. What each symbol means is straightforward. `L` is loss. `P` is the number of parameters. `a` is a fitted constant that sets the height of the curve, and `b` is a fitted constant that tells you how quickly loss improves as you scale up.

Read it in English like this: loss falls like a power of parameter count. That sounds abstract until you plot it on log axes. Take logs of both sides:

```
log L = log a - b log P
```

Now it becomes a line. That is why people love log-log plots for scaling laws. Curved intuition turns into a straight-ish line you can fit.

None of this says the world is exactly one neat equation. Real runs have noise, data quality matters, architecture details matter, and optimization matters. There is often an irreducible floor below which your simple fit stops being honest. Even a rough fit is enough to do something practical: plan before you spend.

The first time I trusted a scaling fit and predicted a target size from small runs, the prediction was, in fact, within a depth of where the empirical run landed. That surprised me more than I would like to admit. The natural prior is that fitting a curve and plugging in a target loss should give a number that is a factor of two wrong in either direction. Figure 12.1 shows exactly why the log-log view is the right lens for this kind of planning.

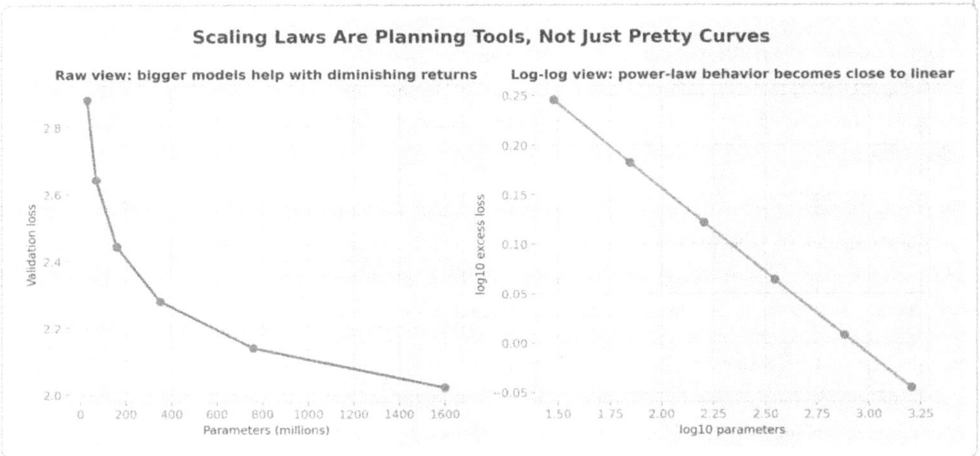

Figure 12.1. Scaling-law curves look curved in raw coordinates but close to linear on log-log axes, which is why they are useful for forecasting model size and token budgets.

Why It Matters

Without scaling laws, model planning turns into cargo cult engineering. You hear "bigger is better," so you make the model bigger. You hear "more data is better," so you train longer. You hear "GPT-2 had X parameters," so you aim at that number because it sounds familiar. That is not planning. That is guessing with a GPU bill attached.

I will be plainer than I usually am here. Most beginner explanations of scaling fall apart at exactly the moment a reader has to make a real budget decision, because the explanations stop at "bigger models do better" and never connect that claim to a dollar number on a cloud invoice. The Chinchilla result is one of the few places in this field where the math actually tells you what to spend.

Scaling laws give you three things. First, they tell you when "train longer" has stopped being a smart answer. This matters because overtraining a small model feels productive: the loss still goes down, the run is still alive, the logs still move. A moving loss curve is not the same as good compute allocation. Second, they tell you how to compare runs fairly. "Same wall-clock time" and "same compute budget" are not the same comparison. A smaller model usually processes more tokens per second, so if you train several models for the same number of hours, the small one sees more data. That is one kind of comparison and it is useful, but it does not answer the compute-optimal question. To answer that, you need to account for both size and duration. Third, scaling laws let you build a cost calculator before you launch the expensive run. That is what labs do, not because they are romantic about curves, but because they do not want to waste weeks on a model that was mis-sized on day one.

This chapter is also the answer to a common misconception: "Bigger is always better."

No. Bigger and trained correctly is better. Bigger and starved is wasteful. Smaller and trained forever is also wasteful. The balance is the lesson.

The Build

We are going to build four things: a sweep over model depth, a plot of validation loss versus parameter count, a fitted scaling law, and a cost calculator that turns target loss into expected training time on your hardware.

The repo starting point is `karpathy/nanochat`. The exact command-line flags may drift over time. That is normal. The structure of the experiment does not drift. If your checkout uses YAML configs instead of CLI overrides, map the same fields there.

1. Freeze everything except depth

If you want to study scaling, you need one clean size knob. In this chapter, that knob is depth. Hold these fixed across all runs:

- tokenizer
- dataset
- train/validation split
- context length `T`
- batch size `B` or effective batch size
- optimizer
- learning rate schedule shape
- weight decay
- dropout
- random seed policy
- logging interval
- evaluation interval

Changing width, sequence length, and optimizer all at once and then calling it a scaling experiment does not produce a scaling experiment. It produces a pile of confounds. The clean sweep from the spec uses depths 6, 10, 14, 18, and 22. Depth moves parameter count upward in a mostly smooth way while keeping the architecture family the same. Same model family, different size. That is what you want.

Confession from scaling work. The first sweep I designed touched five things at once: depth, width, LR schedule, batch size, and weight decay. I justified each change as "small." The plot I got back was uninterpretable. I redid the sweep with depth alone moving, and the curve immediately made sense. Scaling experiments punish ambition. Move one knob.

2. Record the actual parameter count for each depth

Have the training script print the parameter count at startup and write it to the run log. You care about depth because it is your control knob, and you care about parameter count because that is what goes on the x-axis of the scaling plot. If your run log prints something like this, save it:

```
depth=14
params=187,432,960
tokens_per_second=41,200
```

One warning: papers sometimes overload symbols and use `N` for parameter count, while this book uses `N` for depth. Keep them separate in your notes. `N` for depth and `P` for parameter count. You will save yourself a lot of confusion later.

3. Run the naive sweep: same wall-clock time

This first plot is not the final answer. It is the setup that teaches you why the final answer needs more care. Train all five depths for the same wall-clock duration (for example, 60 minutes each on the same GPU with identical data loader settings and eval cadence).

Why do this if it is not compute-optimal? Because it reveals a trap. The small model is fast and the big model is slow, so under the same wall-clock budget the small model sees more tokens, which makes the comparison messy. If the depth-6 model processes 90,000 tokens per second and the depth-22 model processes 30,000, one hour means very different token counts. Make the plot anyway — put parameter count on the x-axis and final validation loss on the y-axis. You will usually see loss improve as size goes up, but the curve is not yet the clean frontier you want because some models are overexposed to data relative to others. That confusion is useful. You need to feel why "same hours" is not the whole story.

4. Switch to the compute-optimal duration

Now we do the experiment that matters. The Chinchilla-style rule of thumb says the compute-optimal training token budget is about:

```
T_opt ~= 8P
```

What each symbol means is simple. `T_opt` is the target number of training tokens, `P` is parameter count, and `8` is the rough tokens-per-parameter ratio.

In English: train for about eight tokens per parameter. If a model has 100 million parameters, the rough target is 800 million training tokens. If it has 500 million parameters, the rough target is 4 billion training tokens.

Under this scheme, each model gets its own duration. The small model trains for fewer total tokens and the large model trains for more. That feels unfair until you remember the point: we are not equalizing hours, we are trying to size each run correctly for its own capacity. This is the compute-optimal comparison.

For each depth, read the parameter count `P`, compute `T_opt = 8P`, and convert that token budget into max steps. If your effective tokens per optimizer step are:

```
tokens per step = B * T
```

then:

```
max steps = T_opt / (B * T)
```

where `B` is batch size in sequences, `T` is sequence length in tokens, and `B × T` is how many tokens you process per step. Now rerun the five depths using these per-model step counts and remake the plot. This one should look cleaner. The spec says the curves should not intersect, and that is the frontier intuition: each correctly trained model represents a better way to spend a larger compute budget. In practice, on small educational runs, you may still see some wiggles, and that is fine. Noise exists. What you are looking for is the broad pattern: bigger models need more tokens, correctly trained bigger models beat incorrectly trained smaller ones at the same compute scale, and the frontier is smoother when each model gets the right duration.

5. Plot loss versus parameter count on log axes

A scaling law is easiest to see when you stop plotting raw size. A move from 10 million to 20 million parameters is not the same kind of move as 1 billion to 1.01 billion. Log axes make relative growth visible. Use `log10(parameter_count)` on the x-axis and validation loss (or `log10(loss)` if you are fitting the pure power law directly) on the y-axis. If you have only five points, keep the plot simple. You want clarity, not decorative plotting code.

A minimal plotting script looks like this:

```
import numpy as np
import pandas as pd
import matplotlib.pyplot as plt

df = pd.read_csv("scaling_runs.csv")  # columns: depth, params, val_loss
plt.scatter(df["params"], df["val_loss"])
plt.xscale("log")
plt.xlabel("Parameter count")
plt.ylabel("Final validation loss")
plt.show()
```

That plot alone teaches something important: you stop seeing "depth 22" as a special magic setting and start seeing it as one point on a size-performance curve. That is a better way to think.

6. Fit the scaling law

Now fit the power law, starting with the simple version:

```
L = a * P^(-b)
```

Take logs and fit a line.

```
import numpy as np
import pandas as pd

df = pd.read_csv("scaling_runs.csv")
x = np.log(df["params"].to_numpy())
y = np.log(df["val_loss"].to_numpy())

slope, intercept = np.polyfit(x, y, 1)
b = -slope
a = np.exp(intercept)

print("a =", a)
print("b =", b)
```

This turns your curved power-law question into a straight-line fit, where the slope of that line becomes `-b` and the intercept becomes `log(a)`. If your points are very close together or noisy, the fit will wobble. That is normal. The point is not that five educational runs produce a perfect universal constant; the point is that even a rough fit gives you a planning tool.

One small confession from drafting this chapter. My first version of the fit section glossed over `L_infinity` and had to be redone. The pure power law looks elegant on the page and lies a little. Real loss curves eventually flatten near a floor that depends on data quality, tokenizer, and architecture. Pretending the floor is zero will burn you on extrapolation past your sweep.

One subtle point: a more serious fit often uses a floor term,

```
L = L_infinity + a * P^(-b)
```

where `L_infinity` means the loss floor you would approach if size kept growing forever on this setup. You do not need that extra term for the first pass, but you should know why papers sometimes add it. Real loss curves do not keep falling forever in a pure power law.

7. Extrapolate to a target, such as GPT-2-level loss

Now turn the fit into a question you actually care about. Rather than asking "How many parameters to be smart?" (which is too vague), ask: "On this dataset, under this tokenizer, with this training recipe, what parameter count would likely reach the held-out loss I care about?"

If your target loss is `L_target`, solve:

```
L_target = a * P^(-b)
```

for `P`, which gives:

```
P = (a / L_target)^(1 / b)
```

In code:

```
L_target = 2.85  # example only
P_needed = (a / L_target) ** (1 / b)
print("Predicted parameters:", int(P_needed))
```

Now map that parameter count back to depth using the depth-to-params relationship you already logged from the sweep. If your extrapolation lands around the parameter range you would get at depth 24 to 26, that matches the empirical answer from the nanochat miniseries. That comparison matters because it is the moment where "curve fitting" stops feeling like statistics homework and starts feeling like engineering. You predicted the size from small runs, then checked whether the real training story agrees. That is the whole value of scaling laws.

8. Build the cost calculator

Now make the practical tool. You need one number from your hardware: training throughput in tokens per second. Use the throughput from a representative stable run. Not the first 30 warmup steps, not a run with data-loader stalls, but something honest.

If your target model has `P` parameters, then the rough compute-optimal token budget is:

```
T_opt ~= 8P
```

If your throughput is `r` tokens per second, then the wall-clock time is:

```
time_seconds = T_opt / r ~= 8P / r
```

In code:

```
def train_time_hours(params: int, tokens_per_second: float, tokens_per_param: int = 8) → float:
    target_tokens = tokens_per_param * params
    seconds = target_tokens / tokens_per_second
    return seconds / 3600

print(train_time_hours(params=200_000_000, tokens_per_second=40_000))
```

```
11.11111111111111
```

A slightly better version also accounts for your fitted curve: if you can estimate the `P` needed for a target loss, you can estimate the training time needed for that target on your GPU. Now you can answer a real planning question: "If I want validation loss 2.8 on this dataset, how many days will that take on my box?" That is the calculator. Crude, but wildly better than guessing.

9. Keep a table, not just plots

Plots are good for shape; tables are good for engineering. For each run, log depth, parameter count, tokens trained, tokens per second, final validation loss, training loss, wall-clock time, and notes on instability. A clean table often exposes mistakes faster than a plot does. A suspiciously good run might have seen more tokens than intended; a surprisingly bad run might have had lower throughput because of an input bottleneck; a weird intersection might be a logging mismatch, not a deep truth. Scaling work is bookkeeping with math attached. Do not skip the bookkeeping.

I caught a six-day error this way during a fusion sweep. Two depths looked anomalously close on the loss-vs-size plot. I assumed it was a real plateau and almost wrote it up. The table showed one run had been using a slightly larger batch size because of a config inheritance bug. Three lines of TSV saved a half-page of bad analysis.

10. Know what "success" looks like

You are done when you have these deliverables:

- one plot from the same-wall-clock sweep
- one plot from the compute-optimal sweep
- one fitted power law for loss versus parameter count
- one predicted size for a target loss
- one hardware-specific cost calculator

At that point, you can answer "How big should my model be?" Not perfectly, but honestly, quantitatively, and before you light money on fire.

BREAK IT

Now we do the part that teaches the chapter. The failure mode is this: you massively overtrain a small model and it still loses to a larger model trained near the right ratio. That result feels insulting the first time you see it, and it should. It is a direct attack on the instinct that "more training always fixes under-capacity."

Take your depth-6 model and measure its parameter count `P6`. Its compute-optimal token budget is roughly:

```
T_6,opt = 8P_6
```

Now ignore that advice on purpose and train it for one hundred times that many tokens:

```
T_6,break = 100 * 8P_6
```

That is the sabotage. You are forcing a small model to keep studying long after the high-return part of learning is over. Then take a larger model, ideally the depth-14 model from the sweep, and give it the same total compute budget (not the same token count). A rough training-compute proxy is:

```
C is proportional to P * T
```

Compute rises with model size times training tokens. So match budgets by choosing `T14` such that:

```
P_6 * T_6,break ~= P_14 * T_14
```

If you want the larger model to be close to compute-optimal at that matched compute, pick the larger size so that its matched token count is near `8P14`. In many teaching setups the depth-14 run is the convenient choice because it is much larger without being too expensive.

Watch validation loss, not training loss. That is the trap. The overtrained small model will usually see training loss keep dropping while validation loss still improves, but very slowly, with each extra chunk of compute buying almost nothing. The larger model trained near the right ratio will typically reach a lower validation loss using the same total compute scale. That is Chinchilla in one picture: you are proving that capacity and duration must be balanced.

This is what many people get wrong the first time they read "large models are undertrained." They imagine it means "the model did not finish learning." What it actually means is "given the compute budget, the money was split badly between size and duration." That is different.

Here is a representative shape of the result. These numbers are illustrative, not promises:

Run	Params	Tokens Trained	Final Train Loss	Final Val Loss
Depth 6, optimal	38M	304M	2.65	2.93
Depth 6, 100x overtrained	38M	30.4B	1.72	2.79
Depth 14, near-optimal at matched compute	155M	1.24B	2.18	2.61

: **Table 12.1.** Chinchilla scaling experiment: training and validation loss for an optimally-trained small model, the same model overtrained, and a larger model trained at the compute-optimal ratio.

The tiny model saw absurdly more tokens. Its training loss dropped a lot. Its validation loss improved only modestly. The larger model, with far fewer tokens than the tiny model's marathon, still won on validation. The small model did not "catch up" by reading longer because it lacked room to store the richer patterns efficiently. You hit the notebook limit. The larger model had more representational capacity, which means more internal storage for useful distinctions. When trained at a sane ratio, it spent compute on both learning and capacity instead of pouring almost everything into repeated exposure.

This is the moment in the chapter where my opinion comes out. The most damaging mental habit in early model work is treating "training is still improving" as evidence that the run was correctly sized. It is not. A loss curve that drifts downward is a loss curve that drifts downward. Whether it is buying you anything per dollar is a separate question, and one most people never ask.

What this proves: "undertrained" and "overtrained" are relative to compute allocation, not emotional intuition. A moving loss curve can still represent a bad plan. Training longer is not a free substitute for model size. And serious model planning starts with scaling laws rather than slogans.

Optional Homework

Run the full lab if you have a real GPU. Run the proxy if you do not.

Full lab: one consumer GPU or cloud GPU

Hardware:

- 1x 24GB GPU is enough for a reduced nanochat sweep if you keep context and batch size modest
- 1x 40GB to 80GB GPU is more comfortable
- rough runtime: several hours to a few days depending on dataset size and throughput

Dataset:

- use the same dataset for all runs
- do not change tokenizer or preprocessing between runs

Step 1. Pick your five depths.

- `6, 10, 14, 18, 22`

Step 2. Launch a short same-wall-clock sweep.

Template command shape:

```
uv run python train.py --depth=6  --run_name=scale_d6_wall  --max_minutes=60
uv run python train.py --depth=10 --run_name=scale_d10_wall --max_minutes=60
uv run python train.py --depth=14 --run_name=scale_d14_wall --max_minutes=60
uv run python train.py --depth=18 --run_name=scale_d18_wall --max_minutes=60
uv run python train.py --depth=22 --run_name=scale_d22_wall --max_minutes=60
```

If your checkout uses config files, make five copies that differ only in depth and run duration.

Record:

- parameter count
- final validation loss
- average tokens per second

Step 3. Compute `T_opt = 8P` for each run.

If your effective tokens per step are `B × T`, compute:

```
max_steps = floor((8 * params) / (batch_size * seq_len))
```

Step 4. Launch the compute-optimal sweep.

Template command shape:

```
uv run python train.py --depth=6  --run_name=scale_d6_opt  --max_steps=<steps_for_8P>
uv run python train.py --depth=10 --run_name=scale_d10_opt --max_steps=<steps_for_8P>
uv run python train.py --depth=14 --run_name=scale_d14_opt --max_steps=<steps_for_8P>
uv run python train.py --depth=18 --run_name=scale_d18_opt --max_steps=<steps_for_8P>
uv run python train.py --depth=22 --run_name=scale_d22_opt --max_steps=<steps_for_8P>
```

Step 5. Save the run table to `scaling_runs.csv`.

Use columns:

```
run_name,depth,params,tokens_seen,tokens_per_second,val_loss
```

Step 6. Fit the scaling law and make the plot.

Use the tiny Python snippets from the build section.

Step 7. Run the BREAK IT lab.

First compute:

```
T6_break = 100 * 8 * P6
```

Then either:

- pick a larger model whose optimal budget is near the same compute scale, or

- exactly match compute by solving `P6 * T6_break = P14 * T14`

Template command shape:

```
uv run python train.py --depth=6 \
  --run_name=break_d6_overtrained --max_tokens=<100x_8P6>
uv run python train.py --depth=14 \
  --run_name=break_d14_matched --max_tokens=<compute_matched_tokens>
```

Record:

- training loss
- validation loss
- total tokens
- total wall-clock time

Expected observation:

- the depth-6 run still improves late, but with weak returns
- the larger run wins on validation loss at comparable compute scale

Proxy lab: laptop or tiny GPU

If the full lab is too expensive, use a smaller dataset and a smaller depth range.

Try:

- dataset: TinyStories, Shakespeare, or another small text corpus
- depths: `2, 4, 6, 8, 10`
- reduced context length
- reduced batch size

Do the same experiment structure. The absolute numbers will not match bigger runs, but the shape still teaches the lesson.

Expected observation:

- same-wall-clock comparison is noisy because small models process more tokens
- compute-adjusted comparison is cleaner
- overtraining the smallest model gives weak returns compared with a larger model trained near the right ratio

Result-guided version if you cannot run anything

If you cannot run even the proxy, do this:

1. make a fake table with five parameter counts spaced by depth
2. assign losses that decrease smoothly
3. fit the power law
4. calculate `T_opt = 8P`
5. compare a toy overtrained small model against a larger matched-compute model

This sounds less satisfying than training, but it still teaches the two lessons that matter: compute-optimal duration scales with parameter count, and extra training on an undersized model gives diminishing returns.

Questions To Answer

1. In your same-wall-clock sweep, which model saw the most tokens, and how much did that distort the first plot?
2. After switching to `T_opt ≈ 8P`, did the loss-versus-size curve become smoother? If not, what source of noise do you suspect first: optimizer settings, throughput instability, or too-small runs?
3. When you overtrained the depth-6 model, what fell faster near the end: training loss or validation loss? What does that gap tell you?
4. In the BREAK IT experiment, what failed first: the belief that "more tokens always win," or the belief that "small models just need patience"?
5. If your fitted curve predicts a target size much larger than you can afford, what would you change next: the dataset, the tokenizer, the architecture family, or the loss target?
6. What tradeoff are scaling laws protecting you from? Name it in one sentence without using the word "optimization."

Go Further

Research anchors for this chapter

The core lineage here is direct. **Scaling Laws for Neural Language Models** gives the original smooth-loss-versus-scale framing. **Training Compute-Optimal Large Language Models** sharpens that into the Chinchilla lesson: parameter count and token count have to be balanced, not optimized independently.

Then read **Language Models are Few-Shot Learners** and **PaLM** as examples of what very large training runs looked like in the period when the field was still discovering how often giant models were undertrained relative to their size. Those papers matter here less for one exact number than for the fact that they make compute allocation impossible to ignore.

For a smaller-budget complement, **Textbooks Are All You Need** and **The Smol Training Playbook** help translate the same instinct into a world where data quality, token budgeting, and efficient training choices matter as much as raw parameter count.

Start with these directions:

- `karpathy/nanochat` for the teaching implementation
- the Chinchilla paper for compute-optimal training
- the Kaplan scaling laws paper for the earlier size-data-compute framing

Search these exact phrases on arXiv:

- `"Kaplan scaling laws language models"`
- `"Chinchilla compute-optimal large language models"`
- `"loss scaling with model size data compute"`

- `"early loss prediction language models"`

Where scaling is going in 2026

This chapter split one budget two ways: how much model, how much training. A 2026 shift adds a third place to spend it, and it is the one this chapter cares about most: the compute a model spends at the moment it answers a query, not the compute it burns while it learns. Call it inference-time compute. Practitioner commentary reports it becoming a common lever across major providers (reported). Treat that as a trend, not a measured law; the same commentary does not rank inference-time spending against pretraining. Architecture is moving too, and the **Project** 7 sidebar covers mHC, a 2026 change to how signal flows between layers at scale.

Make the shift concrete with best-of-N sampling, where N is how many answers you draw. Take one fixed, already-trained model and one hard question. The cheap path samples a single answer of about N tokens and returns it. That is 1x compute. The best-of-8 path samples eight independent answers. It checks each with a verifier and keeps one that passes. Eight answers cost roughly 8x that generation compute, plus a small verifier overhead. The multiplier is the whole trade: you pay about 8x the compute per query for a better chance the returned answer is correct.

Here it is on one prompt. Ask the model for 17 times 23. Verify by exact match against the known answer 391. One sample returns "the product is 380" and fails. Another returns "17 times 23 is 391" and passes, so that is the one you keep. More draws raise the odds one passes, at that 8x cost. A weak verifier is the failure mode. If the check only asks whether the answer looks like a confident number, it can pick the fluent-but-wrong 380 and return a plausible mistake. **Project 25** has a runnable best-of-N lab for the full accuracy curve.

Generating those extra samples is the slow part, and recent efficiency work targets exactly that. A 2026 MIT-led method, named TLT, reports over a 1.7x end-to-end speedup when training reasoning models with reinforcement learning (reported), and it does so losslessly, meaning the final trained model's accuracy is preserved. The mechanism is worth naming plainly. During reinforcement-learning training the model generates many candidate attempts. These are called rollouts, and producing them dominates the cost. TLT speeds that up with speculative decoding. A small draft model proposes the next tokens. The large model checks them, keeping only the ones it would have produced. That is draft-and-verify, not knowledge distillation of a smaller student. So the scaling question does not end when training does. It moves to how many tokens you can afford to generate, and how cheaply.

If you want the next research step, look into forecasting final loss from partial training curves. The practical question after this chapter is no longer "how should I size the run?" The harder question is "can I kill the bad run early and know I am killing the right one?" Another good direction is depth-width tradeoffs. This chapter uses depth as the clean size knob because it is easier to teach and sweep, but real systems also ask whether a bigger budget should go into more layers, wider layers, longer context, or better data. That is where scaling-law work stops being one curve and turns into architecture strategy.

What You Now Know

"How big should my model be?" no longer has to be answered by instinct. You can explain why same wall-clock time and same compute budget are different experiments. You can fit a loss-versus-parameter curve from small runs and use it to predict the size needed for a target loss. You can build a hardware-specific calculator that turns parameter count into expected training time. And you can defend the Chinchilla claim in plain engineering terms: a smaller model trained forever is not automatically efficient. A larger model trained at the right ratio often wins because compute has to be split between storage capacity and learning exposure, not dumped into one side of the budget.

Starting Point

Start from:

- `https://arxiv.org/abs/2203.15556`
- `https://github.com/karpathy/nanochat`

Reading references:

- the Chinchilla paper
- the Kaplan scaling laws paper

Carry forward the lesson from **Project 18: Mixture Of Experts**: internal layout is a contract. This chapter adds a second contract: compute allocation is a contract too. A model is not just an architecture. It is an architecture plus a budget split that decides how much capacity you bought, how much experience that capacity gets, whether "train longer" is still a smart answer, and whether your next expensive run is planned or guessed. Once you see that, model size stops sounding like taste. It becomes scheduling.

Project 20: Autonomous Experimentation

Hook

Why would anyone let an AI agent edit a training file, run a five-minute experiment, decide whether the change helped, and then keep going without asking permission? That sounds reckless right up until you remember how most model work actually feels. Tweak one line, wait, squint at the metric, forget what changed, try something else, get interrupted, lose the thread.

Project 20 starts from that irritation. The question is not "can an agent invent new science overnight?" The question is much simpler: can you turn experimentation itself into a loop that runs while you sleep, keeps the wins, throws away the failures, and leaves behind a clean record of what happened?

This chapter is also where I have to be honest about my own bias. I have run thousands of experiments through loops like this, on training-run sweeps and on the book itself, and my opinion is sharper than the spec wants. So I will warn you up front.

The Concept

Think about model research like tuning a race car when every engine check takes five minutes. You can do it by hand. Open the hood, change one thing, run a lap, write down the time, repeat. It works. It scales terribly. You need focus, discipline, and a memory for what you already tried. Then you go home and progress stops.

Autonomous experimentation changes one thing: direct code edits stop being the main act. Instead, you write down the rules of the game for an agent. That rulebook is `program.md`, and it says what file the agent may edit, what metric counts as success, what files are off-limits, how to log results, when to keep a change, when to throw it away, and what to do when a run crashes. You are no longer only programming Python. You are programming a researcher.

The best analogy here is a lab notebook plus a junior engineer plus a strict supervisor, all collapsed into one loop. `results.tsv` is the notebook. The coding agent is the junior engineer. Git is the strict supervisor. Git matters more than people think in this project. It is more than version control. It is experiment memory and a safety rail. A kept experiment advances the branch. A failed experiment gets erased by resetting back to the last good commit. The branch becomes a climbing path through better and better ideas.

Not every idea survives, and that is the point. Most experiments should die. If everything survives, your metric is weak or your discipline is gone.

The important mental flip is this. The project is not "AI does research for you." It is "AI runs a bounded search loop inside a sandbox you defined." That sounds less magical, and good. It is also more useful.

Strong opinion, since I promised one. The phrase "AI does research" is the worst marketing of the last two years. Research is a long horizon of choosing which question to ask. A loop like this does not do that. It runs a bounded search inside a question you already chose. If you confuse those two things, you will trust the loop with judgment calls it cannot make. If you keep them separate, the loop is one of the most useful tools in this book.

The repo `karpathy/autoresearch` keeps the setup small on purpose. According to the current README, only three files really matter: `prepare.py`, `train.py`, and `program.md`. `prepare.py` handles one-time data prep and evaluation and you do not touch it. `train.py` contains the model and training loop and the agent edits this. `program.md` tells the agent how to behave and you edit this. That division is the lesson. Earlier projects built model pieces; this one builds a process.

And the process has a metric: `val_bpb`, validation bits per byte. In plain English: how surprised is the model, on held-out text, per byte of data? Lower is better. If the number goes down, the model predicts better.

This is where the repo makes a sharp engineering choice. It fixes the training budget to five minutes of wall-clock training time, excluding startup and compilation overhead, then compares experiments by the same metric under the same time budget. That choice turns an argument into a game you can actually score. Without a fixed time budget, an agent could "improve" the model by making it slower and just training longer. With a fixed time budget, the agent has to make tradeoffs. It can pick a smaller model that trains more steps in five minutes, or a larger model that learns more per step but runs fewer steps. It can change batch size, learning rate, normalization, optimizer details, attention pattern, initialization, or architecture. But it cannot cheat the clock.

That is why this project belongs right after **Project 19: Scaling Laws**. **Project 19: Scaling Laws** teaches that compute allocation is a contract. This one teaches that search over that contract can itself be automated. You are no longer asking only "What model should I train?" The question has shifted to "What system should search the space of models for me?"

Why It Matters

Without a loop like this, experimentation has four common failure modes. First, you stop too early. A human researcher hits one or two bad runs and quietly narrows the search space to "safe" ideas, so the weird ideas never get tried because they are annoying to test. Second, you forget context. You remember that "normalization before attention seemed bad" but you no longer remember what else changed in that run, so you can't tell whether it was really the normalization placement or also the batch size and learning rate. Third, you lose negative knowledge. A failed run still teaches something, whether that a particular width causes out-of-memory errors or that an aggressive learning rate gives a short burst of progress and then turns the loss to `NaN`, and if you do not log failures you end up paying to rediscover them. Fourth, your sleep schedule becomes the bottleneck, which is ridiculous once you see it clearly. The GPU can work all night, your process cannot.

All four of these are failure modes I have actually committed. I caught myself forgetting "what else changed in that run" once while writing this book, comparing two notebook experiments two weeks apart and realizing I had silently changed the data shuffler in between. The git history saved me. The notebook did not.

Autonomous experimentation fixes those four things by turning every trial into a small contract: change one thing or a small cluster of things, commit it, run it, measure it, record it, keep it if it helps, and revert it if it does not. That sounds almost insultingly simple, which is why it works.

There is another reason this matters. Most explanations of AI agents make them sound like mystical problem-solvers. This project cuts through that. The agent is only useful because the environment is tight. One editable file, one metric, one fixed run length, one logging format, one revert rule. That is what makes the agent productive instead of chaotic. Hand an agent an unbounded codebase and say "improve my model," and you do not have a research system. You have a liability. Hand it a small search space with a hard metric and a rollback rule, and now you have something real.

This project also teaches a bigger lesson: many "AI workflows" are just optimization loops in disguise. You define a search space, an objective, a safety policy, and a memory of past attempts. That pattern shows up again in fine-tuning, hyperparameter search, evaluation harnesses, inference tuning, and product prompt iteration. Once you see that, "agentic AI" stops sounding like a marketing category and becomes controlled trial-and-error with better tooling.

The Build

Step 1: Read the repo as a contract, not as code

Start at the repo: `github.com/karpathy/autoresearch`. Do not open `train.py` first — open `program.md` first. This feels backwards if you are used to reading code before instructions, but here it is correct.

`program.md` is the operating manual for the agent. In the current version, it tells the agent to create a fresh branch named like `autoresearch/<tag>`, read `README.md`, `prepare.py`, and `train.py`, verify cached data exists, initialize `results.tsv`, establish a baseline with the unmodified code, modify only `train.py`, run `uv run train.py > run.log 2>&1`, extract `val_bpb` and `peak_vram_mb`, append a row to `results.tsv`, keep the commit if `val_bpb` improves, reset back if it does not, and continue indefinitely until a human interrupts it.

That is the system, not the model. The agent is not improvising from scratch — it is following a score. If you skip this step, you miss the point of the project.

Step 2: Prepare data once

The README's quick start says to install dependencies, then run:

```
uv sync
uv run prepare.py
```

`prepare.py` does one-time prep: it downloads the data, trains the tokenizer, and writes the artifacts the training run expects. This split matters because it keeps the experiment loop clean — the loop should not spend time redownloading data or rebuilding prep state every run. You want the five-minute budget to reflect model training, not setup noise.

Step 3: Run one manual baseline

Before you let an agent touch anything, run the untouched baseline:

```
uv run train.py
```

You need a sanity check that your machine, data, dependencies, and GPU are all behaving before the agent takes over. A bad baseline poisons every future comparison. The mental model I use is calibrating a scale before you weigh ingredients. If the scale is off, every measurement downstream is fake precision.

Watch for the final printed block. The current `program.md` says the script prints a summary including:

- `val_bpb`
- `training_seconds`
- `total_seconds`

- `peak_vram_mb`
- `mfu_percent`
- `total_tokens_M`
- `num_steps`
- `num_params_M`
- `depth`

You care most about `val_bpb`, but you also care about `peak_vram_mb` because some changes "help" by quietly eating too much memory. This is another lesson in disguise: objectives are never really one number, even when the loop centers one number.

Step 4: Create the branch and results log

The current `program.md` says to create a fresh experiment branch like this:

```
git checkout -b autoresearch/mar5
```

Use today's date or some short tag. The branch is not cosmetic. It isolates the search path. Then create `results.tsv` with a header row. The agent instructions define five columns: `commit`, `val_bpb`, `memory_gb`, `status`, and `description`.

This file is intentionally left untracked by git, and that detail matters. Git tracks code states; `results.tsv` tracks judgments about those states. If you mix them, you make it harder to reason about both.

Step 5: Point the agent at `program.md`

Now you hand the repo to a coding agent and give it a short prompt. The README suggests something as plain as:

```
Hi have a look at program.md and let's kick off a new experiment! let's do the setup first.
```

That plainness is useful. You are not trying to outsmart the model with a fancy prompt, because you already wrote the real prompt in `program.md`. This is a good moment to notice something subtle. People often think prompt engineering means finding the single clever sentence that unlocks the model. This repo takes the opposite stance. The important prompt is long, boring, and procedural. That is usually what survives contact with reality.

The first time I set up a loop like this, I assumed the prompt would need to be clever and braced for hours of tuning sentences. Instead, the time went into the rulebook: defining the metric extractor, the revert command, the file allowlist. The cleverness lived in the constraints, not the language. That was a useful surprise.

Step 6: Let the loop run long enough to become interesting

Do not stop at two or three experiments. This project only becomes real when the agent has room to fail, recover, and drift into ideas you would not have picked. The chapter spec asks for 20+ experiments. That is the minimum where patterns start to show up.

At that point you can open `results.tsv` and ask: what kinds of changes keep improving? What kinds fail immediately? Which ideas nearly worked but lost on simplicity? Did the agent converge on a cluster of architectural preferences? Did it try things you were mentally filtering out as "too weird"? Those last two questions matter especially. A strong search loop often finds stranger priors than the ones a human would default to. Maybe it moves normalization, simplifies an activation, or changes depth in a way that makes sense only under the fixed five-minute budget. Maybe it discovers that on your hardware, a smaller model with more steps beats a larger one with fewer. That is not magic. That is the clock talking.

Step 7: Read the git history like a research diary

After 20 or more runs, do not jump straight to the best metric — read the git log. You are looking at an argument between ideas. Each surviving commit says "at the time it was tried, this beat the current baseline," which is stronger than a pile of local edits on your laptop.

Now compare git history to `results.tsv`. They answer different questions: git asks what sequence of code states survived, while `results.tsv` asks what else was attempted and rejected. You need both. If you only keep the winners, you lose the boundary of the search. If you only keep the table, you lose the exact code lineage.

Research needs both the route and the graveyard.

Step 8: Write your own `program.md`

This is where the project stops being a repo exercise and becomes a reusable method. The spec suggests changing the objective. Do that. Instead of minimizing `val_bpb`, you can tell the agent to minimize peak VRAM while keeping `val_bpb` within a tolerated gap of baseline, maximize tokens per second at fixed quality, reduce model size while holding quality roughly constant, or search for simpler code with no quality loss. This changes the entire personality of the search loop. Same codebase, same agent, different rulebook.

That should permanently change how you think about AI agents: the agent is less important than the objective and accept/reject rule you write down. Here is a tiny illustrative sketch of that rule in plain code:

```
if new_val_bpb < best_val_bpb:
    keep_change()
else:
    discard_change()
```

And for a memory-focused run:

```
if new_memory_gb < best_memory_gb and new_val_bpb <= baseline_val_bpb + 0.005:
    keep_change()
else:
    discard_change()
```

No mystery there. The agent becomes useful because the judgment rule is explicit. That is the engineering move.

BREAK IT

This project is the break. Not metaphorically, but literally. The agent's job is to make changes that often break training, then recover by discarding them. That is why this chapter's break section matters more than usual. In most earlier projects, you build something and then break one part to learn what it does. Here the thing you are building is a break-and-recover machine. The failure mode to study is no longer "training got worse." It is "the system kept going anyway."

There are several distinct ways the loop can break training.

Failure mode 1: The run crashes outright

Maybe the agent doubles width and runs out of GPU memory, introduces a shape mismatch in attention, or forgets an import. What you observe is that `run.log` does not contain the expected metric lines. The current `program.md` says to grep for:

```
grep "^val_bpb:\|^peak_vram_mb:" run.log
```

If that grep returns nothing, the run crashed. What this proves is that the loop is "agent makes code changes plus a machine-readable success detector." Without the detector, a crashed run is ambiguous. With the detector, it becomes just another outcome. This is one of the hardest-earned lessons in automation: failure must be legible.

Failure mode 2: The run completes but gets worse

This is the most common outcome, and it is the healthy one. The script runs, the metric prints, and `val_bpb` is equal or higher than before. The idea loses. Nothing dramatic happens on screen. The important action is in git: reset to the prior good state. What this proves is that rollback is not a cleanup step. It is the mechanism that turns random trial into cumulative search. If you do not revert losers, your branch becomes a compost heap of unrelated mistakes and near-misses, and every future experiment starts from a dirty, drifting baseline that destroys interpretability.

Failure mode 3: The run improves the metric by a tiny amount but makes the code worse

The current `program.md` includes a simplicity rule: all else equal, simpler is better, and a tiny gain that adds ugly complexity may not be worth keeping. This is a deeper break. The thing being broken here is not training. It is your instinct to worship the metric blindly. A search loop with no simplicity bias tends to accrete hacks, giving you tiny local gains that make the system harder to reason about and harder for future experiments to build on. This matters because the project is recursive. You are optimizing one model and you are also optimizing the future search process. Ugly code taxes later search, which is why a 0.001 improvement from deleting code may be better than a 0.001 improvement from adding 20 lines of brittle logic.

Failure mode 4: The agent gets stuck in safe, boring local search

This one is quieter. The loop keeps running, nothing crashes, metrics wiggle a bit, but the agent only makes timid edits: small learning rate nudges, tiny batch size changes, tiny reorders. What this proves is that autonomy alone does not produce exploration — your rulebook shapes the search personality. If `program.md` rewards only immediate local wins and never encourages occasional bold tries, the agent may camp in a narrow neighborhood. That is not a bug in the model; it is a bug in the research organization you described. This is why `program.md` deserves to be treated like first-class code.

The central lesson of BREAK IT

The big thing to see is that the robustness of the system does not come from preventing bad ideas. It comes from making bad ideas cheap. That is a very different design philosophy. Humans often try to avoid failure by overthinking before they act. This repo goes the other way: let the agent act, measure the damage quickly, and keep going. That is what scales.

If you wake up to 20 runs and 15 are discarded, that is not failure. That is evidence the filter worked. If you wake up to 20 kept runs, start worrying. Either the baseline was terrible, the metric is noisy, or the keep rule is too loose. Bad experiments are not noise here. They are the signal that the loop is doing its job.

Optional Homework

Full run: the real overnight loop

Hardware:

- 1 NVIDIA GPU
- Python 3.10+
- `uv`
- Rough runtime: about 2 minutes setup, 5 minutes per experiment, 2-4 hours for 20+ runs, overnight for 50-100 runs depending on overhead and supervision

Steps:

1. Clone the repo:

```
git clone https://github.com/karpathy/autoresearch
cd autoresearch
```

2. Install dependencies:

```
uv sync
```

3. Prepare data and tokenizer:

```
uv run prepare.py
```

4. Run one manual baseline:

```
uv run train.py
```

5. Create a branch:

```
git checkout -b autoresearch/<your-tag>
```

6. Create `results.tsv` with this header:

```
commit  val_bpb memory_gb   status  description
```

7. Start your coding agent in the repo and tell it to read `program.md`.
8. Let it run at least 20 experiments.
9. When it finishes or you stop it, inspect:

```
cat results.tsv
git log --oneline --decorate -n 30
```

10. Record:

- baseline `val_bpb`
- best `val_bpb`
- number of `keep`
- number of `discard`
- number of `crash`
- best run's memory in GB
- any repeated themes in the descriptions

Expected observations:

- Most runs should not survive.
- Some runs should crash.
- The best surviving changes may be smaller than you expected.
- The git history should contain only the branch's winning path.
- `results.tsv` should show the full attempt history, including discarded ideas.

Proxy lab: if you do not have the required GPU

The current repo README says the base project expects a single NVIDIA GPU and was tested on H100 hardware. If you do not have that, do a result-guided proxy.

Steps:

1. Read `README.md` and `program.md`.
2. Read the current "smaller machine" suggestions in the README:

- lower `DEPTH`
- reduce `MAX_SEQ_LEN`
- lower `TOTAL_BATCH_SIZE`
- consider smaller vocab and smaller evaluation budget
- switch to a simpler attention pattern like `"L"` on weaker hardware

3. Write your own short `program.md` for a hypothetical low-VRAM objective:

- primary metric: `peak_vram_mb`

- quality constraint: `val_bpb` must stay within a tolerance
- allowed edits: `train.py`
- keep rule: memory down, quality stable

4. Without running the full loop, design 5 candidate experiments you think an agent would try:

- reduce depth
- reduce batch size
- narrow width
- simplify attention pattern
- change normalization placement only if it cuts memory or improves speed

5. For each one, predict:

- what breaks first
- whether quality or memory moves more
- whether the idea is likely keep/discard/crash

Expected observations:

- The objective changes the search drastically.
- Architecture changes that look "good" under quality-first search may lose under memory-first search.
- You can understand the experiment loop even before you can afford the full run.

Mini break lab: test the discard loop manually

If you want one concrete failure without turning an agent loose yet:

1. Make a copy of the baseline state on a branch.
2. Introduce one intentionally bad change in `train.py`.
3. Examples:

- set a clearly too-large model width for your GPU
- make a more aggressive learning rate jump than you trust
- break a tensor dimension on purpose if you are comfortable reading the resulting error

4. Commit the change.
5. Run:

```
uv run train.py > run.log 2>&1
```

6. Inspect:

```
grep "^val_bpb:\|^peak_vram_mb:" run.log
tail -n 50 run.log
```

7. Log a row in `results.tsv`.
8. Reset to the last good commit:

```
git reset --hard <good-commit>
```

Expected observations:

- If the run crashes, your grep is empty and `tail` shows the reason.
- If the run finishes but regresses, the metric tells you so directly.
- The rollback restores a clean baseline for the next idea.

Questions To Answer

1. In your run history, what failed more often first: correctness, memory, or quality?
2. Which kept change would you probably not have tried by hand?
3. Did the agent cluster around a theme, like learning rate, normalization placement, model depth, or initialization?
4. Which discarded experiment still taught you something about your hardware or search space?
5. When a run improved `val_bpb` only slightly, how did you decide whether the added code complexity was worth it?
6. If you changed the objective from quality to VRAM or throughput, what tradeoff did the new keep/discard rule start protecting?

Go Further

Research anchors for this chapter

This chapter is less anchored to one canonical paper than to a research habit: bounded search under a fixed budget. The closest conceptual bridge in your local paper set is the scaling-law work, especially **Scaling Laws for Neural Language Models** and **Chinchilla**, because both force you to think about model improvement as an optimization problem under compute constraints rather than as open-ended tinkering.

The modern complement is **The Smol Training Playbook**. It is useful here because it reflects the same practical instinct as autonomous experimentation: keep experiments bounded, log them honestly, and optimize the whole training setup rather than only one glamorous architectural knob.

So the paper lesson of this chapter is not "this repo came from one famous paper." It is that empirical model work keeps rewarding teams that turn search, logging, and acceptance criteria into explicit system components.

Start with the repo you just used:

- `karpathy/autoresearch`
- Current agent rulebook: `program.md`

A short note on what to read with this in your hands. An early version of this section started with the AI Scientist paper as the headline reference. I scrapped that. The framing made the chapter sound like it was promising more than the loop actually delivers. The repo is humbler than the marketing language around the area, and the chapter should match the repo.

Then push in one of these directions:

- Read `karpathy/nanochat` to see the larger parent training setup that `autoresearch` simplifies toward.
- Search arXiv for `AI Scientist`, `autonomous research agent`, and `self-improving research loop`. Read them with one question in mind: what is the objective function, and how does the system roll back bad ideas?
- Search arXiv for `population based training` and `hyperparameter optimization`. These are older cousins of this workflow. They matter because they solve the same problem from a different angle: automated search over training choices.
- Search for `neural architecture search weight sharing` if you want the broader family tree. Those systems search architecture space too, but often with a more formal controller than a general coding agent.
- If you care about product workflows more than model training, build a small version of this loop for prompt evaluation or inference benchmarking. Same pattern: fixed harness, logged metric, keep/discard policy.

The useful question is not "what is the smartest agent?" It is "what is the tightest loop I can give an agent so that failure stays cheap and progress compounds?"

What You Now Know

Autonomous experimentation should no longer sound mystical. You can describe `program.md` as a rulebook for an agent, not decoration. You can explain why a fixed five-minute budget turns architecture search from vague tinkering into a fair contest. You can defend git rollback as part of the method, not housekeeping. You can read a table of `keep`, `discard`, and `crash` runs as evidence that the loop is healthy. Once you can write a new objective for the same codebase, you are no longer just using an experiment system — you are designing one.

Starting Point

Start from:

- `https://github.com/karpathy/autoresearch`

Read first:

- `program.md`
- `README.md`

Carry forward the lesson from **Project 19: Scaling Laws**: compute allocation is a contract. This chapter adds the next one: search policy is a contract too. A training run is more than a model plus a budget. An experiment system is a model, a budget, a metric, a rollback rule, and a memory of failed attempts. Once you see that, "research velocity" stops sounding like talent. It becomes loop design.

Project 21: Fine-Tuning And Instruction Tuning

Hook

Why does a pretrained model that can explain recursion, write a poem, and finish your code still fail the most basic request pattern in chat: you ask a question, and it keeps roleplaying the rest of your prompt instead of answering it? That feels like a bug. It is not a bug. It is the whole point of this chapter.

Pretraining teaches a model to continue text. Fine-tuning teaches it which text shape you care about. Instruction tuning teaches it that `"User: ..."` should be followed by `"Assistant: ..."` instead of more user text. The weird part is that this change in behavior can come from a tiny amount of additional training compared with pretraining, and it can also go wrong in ways that are almost embarrassingly easy to trigger.

I have a stronger reaction to this chapter than most. The long fine-tuning sweeps I have been running for years live inside this exact problem. Four model families, 410M to 6.9B parameters, thousands of training runs. After the first couple hundred, I stopped being surprised that the model improved. I started being surprised by how often it improved in the wrong direction. This chapter is the version of that experience I would have wanted before I started.

The Concept

Start with a person who has read an entire library but has never worked a help desk. If you hand them a support ticket and say, "Answer the customer directly," nothing in "read a library" guarantees the format you want. That is a pretrained model.

Pretraining is broad reading. It teaches the model world knowledge, syntax, code patterns, and statistical habits of text. It does not teach a stable job description. Fine-tuning is job training. You take the same person and say: when the input looks like this, respond like that. When the user asks a question, produce an answer. When the prompt asks for JSON, produce JSON. If the format has `user` and `assistant` turns, continue with the assistant turn only.

The first time I watched this transition happen on a checkpoint I had trained myself, it was almost unsettling. The base model produced fluent nonsense. The same weights, after a few thousand SFT steps on conversation data, produced answers. Nothing was added to the model in any meaningful sense. The mass had just been pushed around.

Instruction tuning is a specific kind of fine-tuning. You are not just saying "learn more text from this domain." You are saying "learn the social contract of a request-response interface." Fine-tuning does not give the model a new brain. It nudges the one it has. The same parameters that stored "Python looks like this" and "medical notes look like this" now get shifted toward "when you see an instruction, answer it in this style." That is why fine-tuning is so effective: you are not building knowledge from scratch, you are steering existing knowledge into a more useful interface. Fine-tuning is also fragile. Too little data invites memorization. Sequential tuning invites forgetting. Bad data teaches bad behavior.

LoRA, short for Low-Rank Adaptation, exists because full fine-tuning is expensive. Plain-English version first: instead of sanding down and repainting the whole car, you bolt on a small adjustment kit that changes how the steering behaves. The original car stays there, your add-on learns the task, and you save memory because you train the add-on, not the whole machine.

The mental model that finally worked for me, after a few months of adapter sweeps, was thinking of LoRA adapters as removable lenses you screw onto the same camera body. The body is the base model. The lenses change what the camera does without modifying the sensor. You can swap them. You can stack them. And you can lose them, which matters more than I expected.

That "low-rank" phrase needs demystifying. A weight matrix is just a giant table of numbers that maps one list of numbers to another. A full update says: every number in that table may change. LoRA says: keep that giant table frozen and learn two much smaller tables whose product acts like a compact correction. Not every possible change, just the changes you can express through a small bottleneck. It is restrictive, but task-specific behavior often lives in a much smaller subspace than the full model, so a few well-chosen directions can be enough.

Now the failure mode everyone eventually meets: catastrophic forgetting. Imagine teaching that same help-desk worker to answer legal questions for a week, then medical questions for a week, but never letting them practice legal questions again. By the end, their phrasing, habits, and retrieval cues drift toward medicine. They did not lose all legal knowledge, but their behavior on legal prompts gets worse. The second training stage overwrote part of the behavioral map built by the first. That is not an edge case. It is the default if you fine-tune sequentially without protection.

I learned this the embarrassing way. On one edge deployment for community health workers, I fine-tuned a small checkpoint on Tamil medical triage prompts, then resumed training on English protocol text because I thought I was being efficient. The Tamil output collapsed. The model still spoke Tamil. It just stopped sounding like a clinician.

One more thing matters more than most introductions admit: what tokens get counted in the loss. Loss is the number that says how wrong the model is. During instruction tuning, you usually do **not** want to punish the model for the user's prompt tokens. You want to punish it only for the assistant's reply tokens. The reason is that the prompt is given to the model; the model is not supposed to "predict the user" as the main task. It is supposed to predict the assistant answer conditioned on the user prompt.

nanochat makes this explicit. In its SFT pipeline, the conversation is rendered into tokens plus a mask, and only assistant completion positions count toward the training target. That one detail is a big deal: it turns "model conversations" into "learn to answer conversations." Without that mask, the model spends capacity learning to copy the prompt pattern more than it should. With the mask, you are teaching the role boundary itself.

Why It Matters

Without fine-tuning, your model has knowledge without behavior. It can continue a paragraph about APIs, continue a stack trace, or continue `"User: How do I sort a list in Python?\nAssistant "` in a way that sometimes looks like an answer — but it has no strong obligation to answer. It is guessing the next token, not honoring a contract. That difference sounds philosophical until you test it.

Prompt a base model:

```
User: Explain closures in JavaScript.
Assistant:
```

A base model may continue with another `User:` turn, produce a partial answer that drifts into unrelated continuation, imitate forum text, or write a transcript that looks plausible without staying on task. Prompt an instruction-tuned model with the same text and it tends to answer directly. That is the product difference users feel.

The first time I demoed this side-by-side to a non-engineer, they did not understand what they were looking at. The base model continued the prompt. They said, "but it sort of answered." I had to explain that the model had no obligation to answer. It happened to drift in the right direction sometimes. That distinction is invisible until you watch it fail on a hundred prompts.

Fine-tuning also lets you specialize. You can take a general model and make it better at:

- customer support
- medical summarization
- code review
- legal clause extraction
- tool calling
- structured outputs

But there is no free lunch here. Full fine-tuning updates every parameter, which usually gives the model the most room to adapt but costs the most memory and is the easiest way to accidentally damage old behavior. LoRA updates far fewer parameters, making training cheaper and deployment cleaner; you can keep one base model and swap tiny adapters for different tasks. But LoRA is a trade: cheaper adaptation in exchange for less freedom to reshape the model.

My honest opinion, after enough adapter runs to be slightly tired of them: most teams pick LoRA for the wrong reason. They pick it because it fits on their GPU. The correct reason to pick LoRA is that it lets you keep the base model preserved. The cost savings are real but secondary. The structural property that you can ship multiple adapters against the same frozen base is what changes how you build products.

Then there is forgetting. If you plan to build systems that evolve over time, forgetting is not a theory question. It is a product question. You fine-tune on support tickets in January, fine-tune on policy Q&A in February, and discover in March that your support bot now answers support questions in policy language and misses edge cases it handled before. That is not the model becoming worse in general. That is the model becoming more aligned to the newest pressure you applied.

This chapter matters because post-training is where a text model turns into a product, and also where many teams quietly damage their models while thinking they are improving them.

The Build

We are going to do four things in sequence: compare a pretrained nanochat model before and after SFT on SmolTalk, reformat data into explicit instruction-response pairs and measure held-out task accuracy, add LoRA to the attention layers and compare trainable parameter count against quality, and then fine-tune sequentially on two domains to watch forgetting happen. The repo starting point for this stage is `nanochat`, which includes a supervised fine-tuning script and a SmolTalk task loader. In the current repo, `scripts/chat_sft.py` builds a training mixture that includes SmolTalk, MMLU, GSM8K, and spelling tasks, and `tasks/smoltalk.py` loads the `HuggingFaceTB/smol-smoltalk` dataset. The same script also masks the loss so that assistant tokens count and user tokens do not.

That last detail is the first thing to look at mentally, even if you do not open the file yet. You are not training the model to imitate the entire chat transcript equally — you are training it to answer.

Step 1: Establish the baseline behavior

Before you fine-tune, you need a "before" sample that makes the behavioral shift obvious. Use a pretrained checkpoint from the base stage, then prompt it with the same chat-like input you will use after tuning.

For example:

```
User: I keep getting `TypeError: 'NoneType' object is not iterable`. What does it usually mean?
Assistant:
```

A base model often produces one of these shapes:

- a vague continuation of a forum thread
- another role tag
- an answer that starts correctly but drifts into more transcript
- a generic explanation without staying in assistant mode

What you are looking for is role discipline, not answer quality. Does the model understand that `Assistant:` is its turn? That distinction is easy to miss if you only look at impressive one-off outputs.

I expected this baseline pass to be a formality. It was not. When I reread an early draft of this chapter I noticed I had skipped Step 1 entirely. The chapter started in the middle. Without the broken before-state, the after-state has nothing to push against. That is true of the model, and it turned out to be true of the writing.

Step 2: Understand the SFT data shape

SmolTalk is conversation data. Each example is a list of messages with roles like `user` and `assistant`, sometimes with an opening `system` message. In nanochat's current task loader, the dataset is the small version of SmolTalk, about 460K train rows and 24K test rows.

Conversation data is not automatically useful — you need to render it into tokens in a way the model can learn from. That involves three things:

1. Serialize the roles and messages into a token stream.
2. Shift inputs and targets by one token for next-token prediction.
3. Mask the loss so only assistant response tokens contribute.

That mask is the behavioral hinge. If the user says:

```
User: Summarize recursion in one sentence.
Assistant: Recursion is a function calling itself on smaller versions of the same problem.
```

The input token stream contains both sides, but the loss should mostly care about the assistant sentence, because that is the part you want the model to generate when given the user side.

In nanochat's SFT path, `tokenizer.render_conversation(...)` returns token ids plus a mask. The mask marks assistant completion tokens with `1` and user prompts and control tokens with `0`. Then the training loop sets masked-out target positions to `-1`, which means "ignore this in cross-entropy." That is how instruction tuning stops being "just more text training" and becomes "learn the reply behavior."

The first version of this section I wrote skipped the mask entirely because it felt like an implementation detail. A reviewer pushed back hard. The mask is not a detail. Without it, you are training the model on a different objective than the one you describe in the introduction, and the gap is invisible until you read the loss code.

Step 3: Run supervised fine-tuning on conversational data

At a high level, the SFT loop is ordinary training:

- load pretrained model
- build SFT dataset
- compute forward pass
- compute loss on assistant tokens
- backpropagate
- step optimizer
- evaluate regularly

The current `chat_sft.py` script in nanochat already sets up this structure. It exposes arguments such as:

- `--model-tag`
- `--model-step`
- `--num-iterations`
- `--device-batch-size`
- `--total-batch-size`
- `--eval-every`
- `--chatcore-every`

That means your job is less "invent an SFT trainer from nothing" and more "understand what the trainer is teaching."

If you are running a small local experiment, the command pattern looks like this:

```
python -m scripts.chat_sft \
  --model-tag d12 \
  --model-step 10000 \
  --num-iterations 1000 \
  --device-batch-size 4 \
  --total-batch-size 32768 \
  --eval-every 100 \
  --chatcore-every 100
```

If you have multiple GPUs:

```
torchrun --standalone --nproc_per_node=8 -m scripts.chat_sft -- \
  --model-tag d12 \
  --model-step 10000 \
  --num-iterations 1000 \
  --device-batch-size 16 \
  --eval-every 100 \
  --chatcore-every 100
```

The exact checkpoint names depend on your training history. The structure is the point.

What should change after SFT? Take the exact same prompt you used before:

```
User: I keep getting `TypeError: 'NoneType' object is not iterable`. What does it usually mean?
Assistant:
```

After SFT, the model should be far more likely to produce something like:

```
That error usually means you tried to loop over a value that is `None` instead of a list, tuple, or
```

The knowledge may have already been there during pretraining. The behavior was not.

Step 4: Separate conversational tuning from explicit instruction tuning

A conversation dataset teaches "how chat transcripts look," while an instruction dataset teaches "how task contracts look." Those overlap, but they are not the same thing. If you want reliable behavior on tasks like summarization, extraction, and transformation, reformating your data as instruction-response pairs matters.

Here is the confusion newcomers usually have: "If pretraining is next-token prediction, and SFT is also next-token prediction, what changed?" The answer is the distribution of text and the masking of which parts count. During instruction tuning, the model repeatedly sees examples where the input contains a task request and the correct continuation is a direct answer in the target format. After enough examples, that prompt-response contract becomes statistically dominant for similar prompts.

A minimal JSONL format for custom data might look like this:

```
{"messages":[
  {"role":"user","content":"Summarize the following in one sentence: TCP retransmits packets when
        it detects loss."},
  {"role":"assistant","content":"TCP resends lost packets to keep data delivery reliable."}
]}
{"messages":[
  {"role":"user","content":"Extract the function name from this Python line: def
        train_loop(model):"},
  {"role":"assistant","content":"train_loop"}
]}
```

This is not fancy — that is the point. Instruction tuning is mostly about showing the model many clean examples of the behavior contract you want.

Now create a held-out evaluation set. Held-out means the model never sees these examples during training; this is how you find out whether it learned the task or just memorized the training rows. For a summarization set, record exact-match or semantic match accuracy depending on the task. For extraction tasks with single clear answers, exact match works well. For multiple-choice tasks, accuracy is straightforward. For open-ended generation, you need a more careful rubric. Do not grade only by vibes — for this chapter, task accuracy on a held-out set is the key metric because it keeps you honest.

Step 5: Add LoRA instead of full fine-tuning

Now we take the expensive route and replace it with a cheap route. Full fine-tuning updates every trainable parameter in the model. LoRA freezes the original weights and adds a trainable low-rank correction to selected linear layers, usually inside attention.

Suppose a projection layer is a big matrix `W` with shape `out x in`. A full update learns a change for every cell in that table. LoRA says: keep `W` frozen and learn two smaller matrices:

- `A` with shape `r x in`
- `B` with shape `out x r`

Then the effective weight becomes:

```
W' = W + (alpha / r) * B @ A
```

What are these symbols?

- `W` is the frozen original weight matrix.
- `A` and `B` are the new trainable matrices.
- `r` is the rank, the size of the bottleneck.
- `alpha` is a scaling factor so the update has a useful magnitude.
- `B @ A` means matrix multiplication. It produces an `out x in` correction with far fewer trainable numbers than a full matrix.

Why is this cheaper? A full `768 x 768` matrix has:

```
768 * 768 = 589,824
```

trainable numbers. A LoRA update with rank `r = 8` has:

```
8 * 768 + 768 * 8 = 12,288
```

trainable numbers, roughly 48 times fewer for that one matrix. If you apply LoRA to four attention projections per layer across 12 layers, you get roughly:

- full update for those matrices: about 28.3 million trainable numbers
- LoRA rank-8 update for those matrices: about 0.59 million trainable numbers

That is the reason LoRA exists.

A minimal PyTorch-style wrapper looks like this:

```
class LoRALinear(nn.Module):
    def __init__(self, base_linear: nn.Linear, rank: int = 8, alpha: float = 16):
        super().__init__()
        self.base = base_linear
        self.base.weight.requires_grad_(False)

        self.rank = rank
        self.scale = alpha / rank

        in_features = base_linear.weight.shape[1]
        out_features = base_linear.weight.shape[0]

        self.A = nn.Parameter(torch.zeros(rank, in_features))
        self.B = nn.Parameter(torch.zeros(out_features, rank))

        nn.init.normal_(self.A, std=0.02)
        nn.init.zeros_(self.B)

    def forward(self, x: torch.Tensor) -> torch.Tensor:
        base_out = self.base(x)
        lora_out = (x @ self.A.t()) @ self.B.t()
        return base_out + self.scale * lora_out
```

Why initialize `B` to zeros? Because at step 0, you want the wrapped layer to behave like the original model. The LoRA path starts as "do nothing," then learns a correction.

Where do you inject this? Usually into attention projections:

- query projection
- key projection
- value projection
- output projection

In a GPT-like model, these are the layers most directly tied to how the model routes information token to token. They are a strong place to steer behavior without touching the whole network. Then freeze everything except the LoRA parameters. That is the whole trick.

Now run the same task fine-tuning with LoRA and compare:

- held-out task accuracy
- trainable parameter count
- peak memory
- training speed

What you should expect: full fine-tuning usually wins slightly when you give it enough compute and data. LoRA often gets close enough that the savings are hard to ignore. Very small ranks can bottleneck adaptation, and higher ranks help until the cost starts to look like full tuning again.

I expected rank 8 to be a starting point I would tune past quickly. Across thousands of adapter runs spanning 410M to 6.9B models, rank 8 stayed competitive far longer than I anticipated. Most of the gains from going higher came from tasks where the base model was genuinely under-equipped, not from tasks where it was already close.

Step 6: Compare full fine-tuning and LoRA honestly

Do not compare a well-tuned full run against a sloppy LoRA run and declare victory for one side. Keep these fixed:

- same base checkpoint
- same train and validation split
- same number of optimizer steps
- same prompt format
- same evaluation rubric

Then compare:

- trainable parameter count
- eval accuracy
- qualitative outputs on the same prompts

A table like this is enough:

Method	Trainable Params	Eval Accuracy	Notes
Full FT	100% of model	84%	best quality, highest memory
LoRA r=4	tiny fraction	76%	cheap but too cramped
LoRA r=8	tiny fraction	82%	often the sweet spot
LoRA r=16	still small	83%	closer to full FT

: **Table 14.1.** Fine-tuning method comparison: trainable parameter fraction and evaluation accuracy for full fine-tuning versus LoRA at increasing rank values.

The exact numbers depend on your task and model size. The shape of the trade is what matters.

Step 7: Show catastrophic forgetting on purpose

Now we do something slightly cruel to the model because it teaches the truth faster. Train on domain A first — use code data, for example instruction-response pairs like:

- explain a traceback
- write a Python helper
- fix a syntax error
- describe Big-O for a snippet

Then evaluate on a held-out code set. Now fine-tune the same model again on domain B — use medical text or clinical Q&A, keeping the code eval set unchanged. Then re-run the code evaluation after the medical stage.

What usually happens? Medical performance improves while code performance drops, sometimes a little, sometimes off a cliff, depending on how narrow and aggressive the second stage is. The reason is that sequential fine-tuning keeps modifying the same underlying weights. The model is not adding a new isolated skill box. It is rebalancing the same shared internal machinery.

This is where LoRA becomes more than a cost trick. If you keep separate LoRA adapters for separate domains, you can preserve multiple behaviors without forcing one domain to overwrite the other in the base weights. One adapter for code, one for medical, the same base model underneath. That does not solve every problem, but it changes the operational story from "one model forgetting itself" to "one base model, multiple task personalities."

Step 8: Watch the metrics that actually reveal behavior

For this chapter, record at least four things:

- held-out task accuracy
- train loss
- validation loss or bits-per-byte
- exact prompt-output comparisons before and after tuning

Why keep the prompt-output examples if you already have metrics? Because fine-tuning changes behavior shape, not just correctness. A model can improve task accuracy while becoming more verbose, more terse, more brittle, or more likely to hallucinate structure. The side-by-side prompt comparison is how you see the product difference.

A good chapter experiment log includes entries like this:

Before SFT

Prompt:

```
User: Write a two-sentence summary of TCP.
Assistant:
```

Output:

```
TCP is a protocol used in networking. User: Explain UDP next.
```

After SFT

Output:

```
TCP is a transport protocol that delivers data reliably and in order between computers. It uses ackn
```

That shift is the chapter in one glance.

BREAK IT

Fine-tune on 100 examples for 50 epochs. That is the lab — do not make it nicer than it is. We are breaking the relationship between learning and generalization on purpose.

Here is the failure mode. At first, with only 100 clean examples, the model improves quickly because the task is simple and the examples are repeated often. By around epoch 3 to 5, the model usually learns the format and the broad mapping from prompt style to answer style. Then the danger starts. You keep training, the same 100 examples cycle through again and again, and the model has enough capacity to stop learning the task and start learning the dataset. That means it memorizes exact wording, learns idiosyncratic punctuation and phrasing, gets very confident on those exact examples, and gets worse on similar but unseen variants.

You can make this visible with almost no ambiguity. Suppose one training example is:

```
User: Summarize: TCP ensures reliable packet delivery.
Assistant: TCP makes sure packets arrive reliably.
```

By epoch 5, ask:

```
User: Summarize: TCP ensures reliable packet delivery.
Assistant:
```

The answer will be correct. Good. Now ask a nearby prompt:

```
User: Summarize: TCP provides reliable delivery of data packets.
Assistant:
```

At epoch 5, the answer is often still good. At epoch 50, the model may become weirdly brittle. It may cling to the memorized wording from the training set or fail when the phrasing shifts a little. On exact training prompts, it may reproduce the training answer word for word. That is overfitting made visible, not in a graph first, but in behavior first.

Now add contradictory data. Take 20 of the 100 examples and duplicate them with conflicting answers. For example:

```
User: Is TCP connectionless?
Assistant: No, TCP is connection-oriented.
```

and elsewhere:

```
User: Is TCP connectionless?
Assistant: Yes, TCP is connectionless.
```

This is not subtle. It is poison. The model cannot satisfy both answers for the same input. During training, it gets pushed in opposite directions for the same prompt. The result is not "the average truth." It is instability, confusion, and prompt-sensitive inconsistency. You will see things like the answer flipping across runs, flipping with temperature, the model becoming hedgy ("TCP is usually considered connection-oriented, though some contexts describe it differently"), or the model learning to sound uncertain even when the task should be crisp.

What does this prove? It proves that fine-tuning is not adding truth into an empty box. It is fitting a behavioral surface to the examples you give it. Give it too few examples for too long, and it memorizes. Give it contradictory examples, and it internalizes contradiction. Give it narrow data sequentially, and it forgets older behavior. That is the real lesson of post-training: the model is obedient to the data distribution, not loyal to your intentions.

If you want the exact mechanism being broken in nanochat terms, it is not a missing layer or a deleted function. It is the training horizon and dataset size. The relevant control is `--num-iterations` in `scripts/chat_sft.py`. That flag lets you keep stepping long after the model has already extracted the general pattern from a tiny custom dataset. When the dataset is tiny, more steps stop meaning "more learning" and start meaning "more memorization."

The second thing you are breaking is data consistency. When you feed contradictory instruction-response pairs through the same masked assistant-loss path, the trainer still does its job. It will faithfully optimize toward incompatible targets. The failure is in the dataset contract, not in PyTorch.

That is why this section matters more than the clean build. A clean build tells you what success looks like. A broken build tells you what the mechanism was protecting.

Optional Homework

Run this in two versions:

- a cheap local proxy on CPU, MPS, or one consumer GPU
- a larger run if you have access to a stronger GPU

Homework A: Overfit 100 examples on purpose

Hardware:

- Proxy: laptop CPU/MPS or 1 consumer GPU
- Better: 1x 3090/4090/A100
- Runtime: 20 minutes to 2 hours depending on hardware and model size

1. Create a tiny JSONL training file with exactly 100 instruction-response examples.

Use a format like:

```
{"messages":[
  {"role":"user","content":"Summarize: TCP ensures reliable packet delivery."},
  {"role":"assistant","content":"TCP makes sure packets arrive reliably."}
]}
```

Keep the task narrow. Summarization or extraction is ideal because eval is easy.

2. Create a held-out eval file with 50 similar-but-not-identical examples.

Do not copy phrasing. Keep the task the same. Change wording.

3. Replace the default large training mixture with your tiny custom dataset.

If you want a clean lab, edit the SFT data mixture in `scripts/chat_sft.py` so `train_tasks` contains only your `CustomJSON(...)` file, repeated enough to define one epoch clearly. Leave `val_dataset` pointing to a held-out custom eval set.

4. Set a small run length and save checkpoints or logs every few epochs.

If your checkout still uses dataset-driven epochs when `--num-iterations=-1`, force a fixed training horizon with `--num-iterations`.

Example command:

```
python -m scripts.chat_sft \
  --model-tag d12 \
  --model-step 10000 \
  --num-iterations 5000 \
  --device-batch-size 4 \
  --total-batch-size 4096 \
  --eval-every 100 \
  --chatcore-every -1
```

If 5000 steps is too much for your machine, cut it down. The point is to go far past the point where the model first learns the task.

5. At roughly five evenly spaced checkpoints, run the same three prompt sets:

- 5 exact training examples
- 5 paraphrased versions of training examples
- 5 genuinely new held-out examples

Save the outputs.

6. Record:

- training loss
- held-out accuracy
- exact-match rate on training examples
- qualitative behavior on paraphrases

Expected observation:

- early checkpoints improve both training and held-out behavior
- later checkpoints keep improving training behavior
- held-out behavior peaks, then stalls or drops
- exact training examples become more verbatim
- paraphrases become less reliable than exact memorized prompts

Homework B: Add contradictory data

Hardware:

- same as Homework A
- Runtime: similar

1. Take 20 of your 100 examples.
2. Duplicate each of those 20 with a conflicting assistant answer.
3. Retrain from the same base checkpoint with the same hyperparameters.
4. Ask the contradictory prompts multiple times with temperature `0` and then with a slightly higher temperature.

Expected observation:

- the model's answer becomes unstable or weirdly hedged
- eval on crisp held-out questions drops
- the model may sound less certain across the board

Homework C: Compare full fine-tuning vs LoRA

Hardware:

- Best: at least 1 decent GPU
- Runtime: a few hours for a meaningful comparison on small models

1. Pick one instruction task and one held-out eval set.
2. Run full fine-tuning.
3. Implement LoRA in attention projections only.
4. Freeze the base model for the LoRA run.
5. Compare:
   - trainable parameter count
   - peak memory
   - eval accuracy
   - output style

Expected observation:

- LoRA uses far fewer trainable parameters
- LoRA often gets close to full FT on narrow tasks
- full FT still has more room to adapt if you can afford it

Homework D: Show catastrophic forgetting

Hardware:

- 1 GPU recommended
- Runtime: several hours depending on model and dataset size

1. Fine-tune on domain A: code instructions.
2. Evaluate on held-out code set.
3. Fine-tune the same model on domain B: medical instructions.
4. Re-evaluate on both code and medical sets.

Expected observation:

- medical accuracy rises after stage B
- code accuracy falls after stage B
- the model's phrasing drifts toward the second domain

Proxy version if you do not have time:

- use much smaller synthetic datasets, 500 to 2000 examples per domain
- the effect will still show, just in a noisier way

Questions To Answer

1. At what point did the model first "get the task," and how could you tell from outputs rather than just loss?
2. What failed first during the 100-example overfit run: held-out accuracy, paraphrase robustness, or style stability?
3. When the model memorized, did it copy exact wording, exact structure, or both?
4. What tradeoff is LoRA protecting: compute cost, memory cost, deployment simplicity, or all three?
5. After sequential fine-tuning, what exactly got forgotten: factual knowledge, response style, or task routing behavior?
6. When you injected contradictory data, did the model become confidently wrong, inconsistently wrong, or broadly uncertain?

Go Further

Research anchors for this chapter

The clearest primary paper anchor here is **Training Language Models to Follow Instructions with Human Feedback**, usually called the InstructGPT paper. It is the paper that most cleanly formalized the gap between a pretrained next-token model and a model that behaves like an assistant.

It also helps to read **Exploring the Limits of Transfer Learning with a Unified Text-to-Text Transformer (T5)** alongside it. T5 matters because it makes a broad but crucial framing move: many NLP tasks can be treated as input-output behavior over text, which is exactly the mental shift that makes instruction tuning feel natural instead of ad hoc.

LoRA itself is not in your local paper folder, but the surrounding principle is the same one this chapter keeps emphasizing: post-training is mostly about steering existing knowledge into a new interface contract rather than teaching the model an entirely new world.

Start with the current repo used in this chapter:

- `https://github.com/karpathy/nanochat`

Then look at the dataset source used by nanochat's SmolTalk task:

- search for `HuggingFaceTB/smol-smoltalk`

For LoRA, search arXiv for:

- `LoRA low-rank adaptation language models`

That paper is the clearest starting point for why a tiny trainable update can be enough.

For catastrophic forgetting, search arXiv for:

- `catastrophic forgetting language model fine-tuning`
- `continual learning for large language models`

For instruction tuning datasets and behavior shaping, search for:

- `instruction tuning language models`
- `supervised fine-tuning LLM`
- `mixture of instruction datasets`

If you want the operational angle, search for:

- `multi adapter serving LoRA`
- `adapter fusion language models`

That is where this chapter starts connecting to product architecture: one base model, many task-specific adapters, fewer rewrites.

What You Now Know

A pretrained model knows things but has no behavior — instruction tuning is what teaches it to answer instead of continuing. Masking only the assistant tokens in the loss is why that works: the model trains on responses, not transcripts. LoRA reduces this cost to a small low-rank correction, and catastrophic forgetting is the measurable price when domain fine-tuning gets too much permission to move shared internal structure.

Starting Point

Start from:

- `https://github.com/karpathy/nanochat`

Read first:

- `https://github.com/karpathy/nanochat/blob/master/scripts/chat_sft.py`
- `https://github.com/karpathy/nanochat/blob/master/tasks/smoltalk.py`

Carry forward the lesson from **Project 20: Autonomous Experimentation**: compute allocation is a contract. This chapter adds the next one: behavior shaping is a contract too. A useful assistant is not just a pretrained model plus a chat box — it is a pretrained model, a response format, a loss mask, a dataset contract, and a training horizon that stops before memorization takes over. Once you see that, "alignment" stops sounding mystical. It starts looking like data engineering with consequences.

Project 22: Evaluation Methodology

Hook

Project 9: Pretraining on the Real Web's `val_bpb` told you the model was learning. The number went down for a week straight, the curves looked clean, and the validation loss matched what the literature reported for a model that size. Then you scored the same model on MMLU and got 12 percent. Barely above the random-guessing floor for four-choice questions. Then you scored it on a benchmark you wrote yesterday and got 78 percent. Same model. Same weights. Same checkpoint. Three numbers that disagree by a factor of six. How can all three be right?

This is the chapter I most wanted to write and most dreaded writing. Multi-model routing work I have done across ensembles touched roughly 1,500 tasks, and the headline accuracy number is uninterpretable without three pages of methodology. Most of the noise in published model comparisons today is people quoting headline numbers without those three pages. My strong opinion is that benchmark numbers without methodology are not numbers. They are press releases with decimals attached.

The Concept

Imagine a student preparing for a final exam. You have three ways to measure whether the student learned the material. You can ask them to recite the textbook from memory, which tests recall. You can give them a multiple-choice quiz, which tests pattern matching against four answer shapes. You can hand them an open-ended problem and grade their reasoning, which tests something closer to understanding. Each method gives you a different number, and each number measures something different. None of them is "the truth." All three together start to look like one.

Now add a twist. Suppose the multiple-choice quiz is identical to the practice quiz the student took last week, and they wrote down every answer. The score will look brilliant, but the student has not learned anything new — the test is broken in a hidden way. This is the situation every honest evaluator of large language models lives in.

There are four families of evaluation that matter, and each measures a different slice of the model.

The first is **perplexity**, or its byte-normalized cousin, **bits per byte** (often abbreviated bpb). Perplexity asks a narrow question: how surprised is the model by a piece of natural text it has never seen? You feed the model a held-out sentence, you compute the probability the model would have assigned to each token in that sentence, and you take the geometric mean. Low perplexity means the model finds the text predictable. Bits per byte normalizes by the actual byte count of the text, which lets you compare models with different tokenizers. It is the unit **Project 9: Pretraining on the Real Web** already used to compare runs. Perplexity is the cleanest signal you can get, but it only measures distribution fit on natural text. It does not measure whether the model can follow instructions, answer questions, write code, or do anything you would actually use a language model for.

The second is **multiple-choice scoring**, where the model picks A, B, C, or D. The most famous benchmark in this family is MMLU, Massive Multitask Language Understanding, which gives the model 57 different subjects, from high school physics to professional law, and asks four-choice questions. The way you score it matters a great deal. The standard approach computes the **logprob** (the log of the probability) of each answer token under the model's distribution, conditioned on the question and the answer choices. You pick the option with the highest logprob and check whether it matches the labeled answer. Multiple-choice scoring is fast, repeatable, and brittle. The model might pick the right letter for the wrong reason, or the wrong letter for a reason the format does not let it express.

The third is **open-ended generation evaluation**. You give the model a prompt, you let it generate a response, and you check whether the response is correct. The check can be exact-match (does the generated text match the reference exactly?), or substring match (does the generated text contain the reference answer somewhere?), or it can use a more permissive comparison. Generation eval measures something closer to real usage, but the score depends on the prompt format, the sampling temperature, the maximum length, and a dozen other choices that the multiple-choice version hides.

The fourth is **LLM-as-judge**, where a stronger model reads the candidate model's output and assigns a score on a rubric. You write a prompt that says "Rate this answer from 1 to 5 on correctness and clarity," you send the candidate's output to GPT-4 or Claude, and you collect the judge's score. LLM-as-judge is the closest you can get to measuring usefulness at scale without paying human labelers, but the judge has biases. It prefers longer responses. It prefers the first response when shown a pair. It tends to agree with confident-sounding text even when the text is wrong.

When I first ran a Claude-as-judge pipeline on ACAR outputs, the agreement rate with my own ratings was around 71%. That sounds reasonable until you notice the disagreements clustered. The judge confidently endorsed verbose wrong answers I had marked down. I rewrote the rubric three times before the bias stopped showing up in the failure modes I cared about.

All four methods are valid. None of them measure the same thing. A model can score 4 bpb on web text, 12 percent on MMLU, 80 percent on a custom benchmark you wrote, and 3.4 out of 5 from a Claude judge, all at the same time. The numbers are not contradictory. They are answering different questions.

There is one more concept worth naming before any code appears: **contamination**. Contamination happens when the evaluation set leaks into the training data. Maybe the MMLU questions were posted on a forum that got scraped into the web crawl. Maybe a paper reproduced the GSM8K problems in its appendix and that paper ended up in your training corpus. Maybe someone bundled a public benchmark into a documentation repository and your data pipeline included it. The leak does not need to be deliberate. It just needs to happen.

When the eval set leaks, every evaluation method inflates. Perplexity drops on the leaked passages because the model has memorized them. Multiple-choice scores rise because the model has seen the question and the answer. Generation evaluations spike because the model can reproduce the reference output verbatim. LLM-as-judge ratings climb because the candidate's output suddenly looks correct, fluent, and confident, for the wrong reason. The model has not learned anything. It has memorized the test. The test is not broken, but the score is a lie.

The analogy that nails this is the cake-baking analogy adjusted slightly. A student scores 100 percent on a final exam that turns out to be a copy of last week's homework. The student has not learned the material; the score reports memorization, not understanding. The test is not broken. The score is. A score divorced from how the test was constructed and whether it was held out is, by itself, uninterpretable.

The mental image that finally made contamination feel concrete to me was thinking of the eval set as a sealed envelope. The training data was supposed to never see the inside of that envelope. The contamination check is the dusting for fingerprints. Most envelopes are clean. Some are covered.

This chapter teaches the four methods, shows how each one fails, and builds the contamination detector that catches the silent failure underneath them all.

Why It Matters

Without proper evaluation methodology, the field of language modeling becomes a casino. Researchers publish benchmark numbers. Practitioners read those numbers and pick models. Buyers read marketing copy that quotes the same numbers and sign contracts. Every step of that chain assumes the numbers mean something specific. When the numbers are produced by a contaminated benchmark, or by a scoring method nobody documented, or by a prompt template that happens to favor one model, the entire chain breaks silently. The decisions still happen. They are just based on noise.

There is a simpler reason this chapter sits where it does. After **Project 21: Fine-Tuning and Instruction Tuning** and before **Project 23: Reward Models and RLHF**, you are about to spend real compute on shaping model behavior. Every shaping technique needs a way to verify that it worked. If your eval inflates whenever the training data overlaps the test set, and it does, every time, you will train against the wrong signal and ship a model that scores well on paper while behaving worse in practice. You will not know this happened. You will think the training worked.

Years of running a multi-state SaaS platform taught me one thing that applies here, even though that work had nothing to do with language models: measurement at scale is mostly about discipline, not about the metric. A 99.99% uptime number means very little if you cannot show me what counted as downtime, who was watching, and which incidents got rolled into "scheduled maintenance." Benchmark numbers fail the same way. The score is the surface. The methodology underneath is where the truth lives or hides.

There is also a longer-running reason that becomes obvious once you have read three or four evaluation papers. The published benchmark numbers from major labs are nearly impossible to compare directly. One lab reports MMLU using 5-shot answer-token scoring. Another lab reports MMLU using zero-shot full-completion scoring with a different prompt template. A third lab fine-tunes on MMLU-adjacent data and reports the result anyway. The headline numbers look like they should be comparable. They are not. The skill this chapter builds is the ability to read a benchmark number and ask the three questions that decide whether it is meaningful: how was it scored, how was the prompt constructed, and was the eval held out from training. Without those answers, the number is meaningless. With them, you can interpret almost any published score.

The Build

You are going to build a tiny eval harness that scores a model with all four methods, runs a contamination check against your training corpus, and produces a paraphrased held-out variant of any benchmark to expose memorization. The whole pipeline runs on a single GPU. The contamination detector and the harness logic run on CPU; only the model forward passes need the accelerator.

An earlier draft of this build section walked you through writing your own harness from scratch. A test reader pointed out that this teaches the wrong lesson; the lesson is what each method measures, not how to implement it. The next version went too far the other way and was effectively a tour of `lm-evaluation-harness`. The version below is the middle: enough code that you can run the methods locally, not enough that you mistake the code for the chapter.

The reference implementation in mind is EleutherAI's `lm-evaluation-harness`, but everything below is portable. You can attach it to any model that exposes `logits = model(input_ids)` and `generate(...)` interfaces.

Step 1: Perplexity and bits per byte

The cleanest evaluation is also the oldest. You take a held-out chunk of natural text, you ask the model to assign a probability to each token in sequence, and you summarize how confidently it predicted the whole passage.

The forward pass is the same loop you ran in **Project 9: Pretraining on the Real Web**: tokenize the text, feed it through the model, collect the logits, and shift by one so that the loss at position `i` corresponds to predicting token `i+1` from tokens `0..i`. The cross-entropy loss at each position is the negative log probability the model assigned to the correct next token. Sum those across the passage, divide by the number of tokens, and you have the per-token cross-entropy in nats. Convert to bits by dividing by `ln(2)`. Convert to perplexity by exponentiating the per-token cross-entropy.

```
import math
import torch
import torch.nn.functional as F

@torch.no_grad()
def perplexity_and_bpb(model, tokenizer, text):
    enc = tokenizer.encode(text, return_tensors="pt").to(model.device)
    n_bytes = len(text.encode("utf-8"))
    logits = model(enc).logits[:, :-1, :]
    targets = enc[:, 1:]
    log_probs = F.log_softmax(logits, dim=-1)
    chosen = log_probs.gather(-1, targets.unsqueeze(-1)).squeeze(-1)
    total_nats = -chosen.sum().item()
    n_tokens = targets.numel()
    ppl = math.exp(total_nats / n_tokens)
    bpb = total_nats / math.log(2) / n_bytes
    return ppl, bpb
```

Two details matter. First, the shift by one. The model predicts the next token given the previous ones, so the logits at position `i` describe the distribution over token `i+1`. If you forget the shift, you score the model on its ability to predict the token it just saw, which is trivially perfect because attention is causal and the position attends to itself. This is one of the most common bugs in homegrown eval code, and it produces beautifully low numbers that mean nothing.

Second, the normalization. Per-token cross-entropy depends on the tokenizer. A byte-pair-encoded tokenizer might cut "intelligence" into two tokens, while a character-level tokenizer cuts it into twelve. Per-token perplexity is not comparable across tokenizers, so most modern reports normalize to bytes. Bits per byte answers the question: how many bits, on average, does this model need to encode one byte of natural text? It is a tokenizer-independent number, which is why the bpb value is the one you compare across runs.

Run this on 1,000 sentences of held-out text. The output is one number (say, 1.18 bpb) that says how well the model models natural language. That is the first eval.

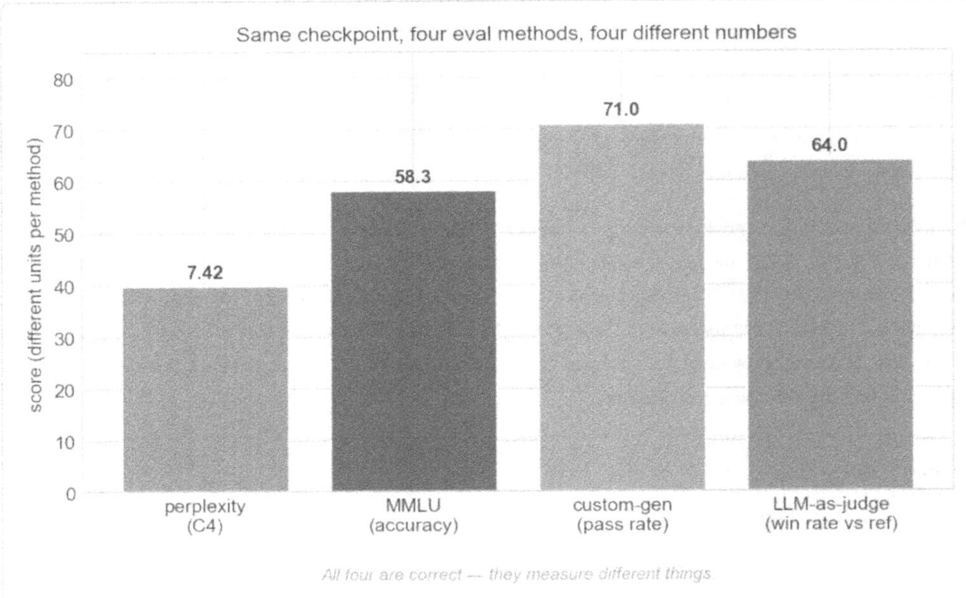

Figure 22.1. Four eval methods compared on the same checkpoint. Perplexity, MMLU, custom generation eval, and LLM-as-judge produce four different numbers, and all four are correct measurements of different things.

Step 2: Multiple-choice scoring (MMLU-style)

Now the second method. MMLU presents four-choice questions: a prompt with a question, followed by options A, B, C, and D, followed by an "Answer:" line where the model picks the correct letter. The scoring question is how exactly you decide what the model picked.

The most popular approach is **answer-token-only scoring**. You format the prompt up to and including the "Answer: " marker, and you score the logprob of each of the four single tokens "A", "B", "C", "D" at the next position. You pick the letter with the highest logprob. The implementation is short:

```
@torch.no_grad()
def mmlu_score(model, tokenizer, question, options, answer_label):
    prompt = (f"Question: {question}\n"
              f"A. {options[0]}\n"
              f"B. {options[1]}\n"
              f"C. {options[2]}\n"
              f"D. {options[3]}\n"
              f"Answer:")
    ids = tokenizer.encode(prompt, return_tensors="pt").to(model.device)
    logits = model(ids).logits[0, -1]
    log_probs = F.log_softmax(logits, dim=-1)
    letter_ids = [tokenizer.encode(f" {c}", add_special_tokens=False)[0]
                  for c in "ABCD"]
    scores = [log_probs[i].item() for i in letter_ids]
    pick = "ABCD"[max(range(4), key=lambda i: scores[i])]
    return pick, pick == answer_label, scores
```

This works, but it tests a very narrow capability. The model needs to recognize the format, locate the answer slot, and rank the four letters. It does not need to write anything. It does not need to explain its reasoning. It only needs to push the probability mass onto one letter.

The alternative is **full-completion scoring**, where instead of comparing single letters, you compare the logprob of each full answer text. For each option, you compute the logprob of the entire option string under the model. You pick the option with the highest length-normalized logprob. This tests whether the model finds the full correct answer more plausible than the full incorrect alternatives, which is closer to what we usually mean by "knowing the answer."

The two methods often disagree by 5 to 15 percentage points on the same model. Models trained with a lot of multiple-choice-style data tend to do better on answer-token scoring. Models trained on more natural text tend to do better on full-completion scoring. Neither is "right." They measure different things. When you compare two published MMLU numbers, the first question is which scoring method each one used. If the paper does not say, the number is not interpretable.

I did not appreciate how big the gap could be until I scored the same routing candidate with both methods and watched it move from 41% to 53% without anything in the model changing. The headline I would have written if I had only run answer-token scoring would have been roughly twelve points more optimistic than the truth.

A second hidden choice: **few-shot prompting**. The original MMLU paper used 5-shot scoring, where each question is preceded by five worked examples from the same subject. The 5-shot context helps the model figure out the format. It also gives the model material it can leak into its answer. A zero-shot MMLU score on the same model is typically 5 to 10 points lower than the 5-shot score. Neither number is wrong; they are different measurements.

Run answer-token scoring on a 500-question subset of MMLU. Record the score, the scoring method, and the few-shot count. The result is the second eval number; let us say 14 percent for a small pretrained-only model, or 38 percent after instruction tuning.

Step 3: Open-ended generation evaluation

The third method is generation. You write a prompt, you let the model generate a response, and you check the response against a reference answer. This is closer to real usage. It is also the most sensitive to the details of how you set up the prompt.

The simplest implementation:

```
@torch.no_grad()
def generate_eval(model, tokenizer, prompt, reference, max_new_tokens=64):
    ids = tokenizer.encode(prompt, return_tensors="pt").to(model.device)
    out = model.generate(ids, max_new_tokens=max_new_tokens,
                         do_sample=False, temperature=0.0)
    completion = tokenizer.decode(out[0][ids.shape[1]:],
                                  skip_special_tokens=True)
    exact = completion.strip() == reference.strip()
    sub = reference.strip() in completion
    return completion, exact, sub
```

The scoring choices stack up fast. Exact match is the strictest; every character must match the reference. Substring match accepts the reference as a substring of the output, which is more permissive. Normalized exact match lowercases, strips punctuation, and collapses whitespace before comparing. Each choice produces a different score.

The deeper problem is the prompt template. Take the GSM8K math benchmark, which asks word problems and expects a numerical answer. The exact same model, on the exact same questions, can score very differently depending on the prompt format. Five common variants:

1. "Q: {question}\nA:"
2. "Question: {question}\nAnswer:"
3. "Solve the following problem step by step.\n\n{question}\n\nAnswer:"
4. "{question}"
5. "You are a math tutor. Solve this carefully.\n\nProblem: {question}\nSolution:"

These look almost identical. On a small model, the five templates can produce GSM8K scores spanning twenty or more percentage points. The model is the same. The questions are the same. The reference answers are the same. Only the prompt format changed, and the score moved by twenty points.

I expected this gap to shrink as I moved to bigger models. It did, but not as much as I assumed. On some routing experiments the template-sensitivity gap on smaller candidates was 18 points and on the strongest candidate was still 6. The template never stops mattering. It just matters less.

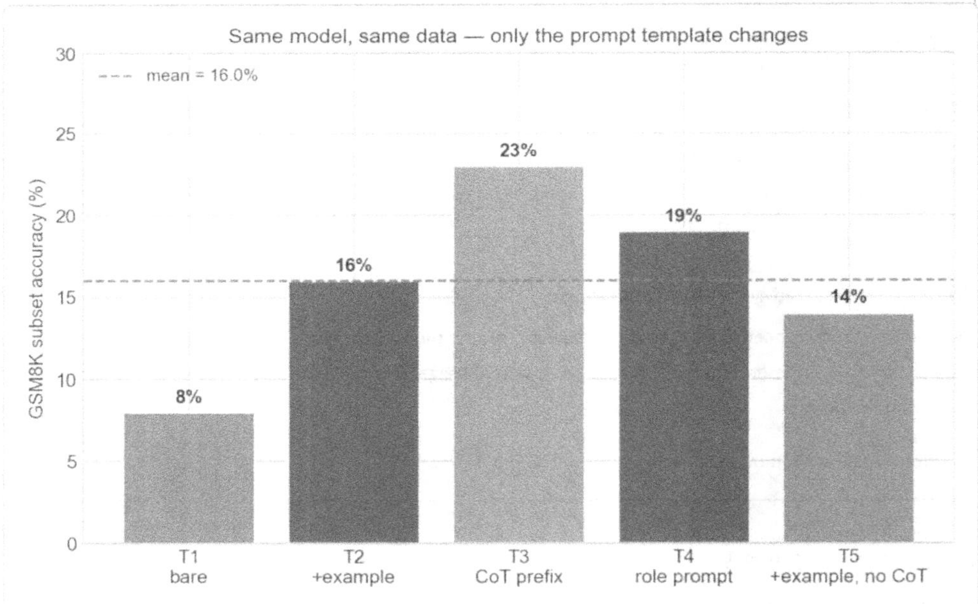

Figure 22.2. Prompt-template sensitivity. The same model on the same GSM8K subset scores 8 percent under template 1, 23 percent under template 3, and 14 percent under template 5. The model has not changed; only the wrapper has.

This is not a bug in the model. It is the model behaving exactly as it was trained to. A model that saw "Question: X\nAnswer:" five million times during pretraining will respond confidently to that exact pattern. A model that saw "Q: X\nA:" five million times will respond confidently to the other pattern. The score is real. It just measures the model's response under a specific format, not its underlying capability.

The practical move is to evaluate under several prompt formats and report the range, not a single number. When you only report one number, the reader assumes you picked the format that flatters your model. Sometimes they are right.

Step 4: LLM-as-judge

The fourth method uses a stronger model to rate the candidate model's output. You write a judging prompt that contains the question, the candidate's answer, and a rubric. The judge (typically GPT-4, Claude, or another large model) returns a numerical score.

A minimal implementation:

```
JUDGE_PROMPT = """You are grading an answer to a question.

Question: {question}
Reference answer: {reference}
Candidate answer: {candidate}

Rate the candidate answer from 1 to 5 on correctness and clarity.
Output only the number, no other text.

Score:"""

def llm_judge(client, question, reference, candidate):
    prompt = JUDGE_PROMPT.format(question=question,
                                 reference=reference,
                                 candidate=candidate)
    resp = client.complete(prompt, max_tokens=4, temperature=0.0)
    try:
        return int(resp.strip())
    except ValueError:
        return None
```

This works, and at scale it correlates surprisingly well with human ratings on many tasks. It also has known biases that show up in every published study.

Length bias is the most consistent. Judges rate longer responses higher even when the longer response is no more correct. If your candidate model produces three-sentence answers and the reference produces one-sentence answers, the judge will often penalize the candidate for being too short or reward it for being more "thorough," depending on which way the bias goes for that judge. The bias is real, it is measurable, and there is no clean way to remove it without changing the rubric.

Position bias appears when you ask the judge to compare two answers side by side. Judges tend to prefer whichever answer appears first. The size of the effect varies by judge, but for many setups it is 5 to 15 percent. The standard workaround is to run every comparison twice — once with answer A first, once with answer B first — and only count agreement.

Sycophancy bias appears when the prompt includes a hint about which answer is "expected." If your judging prompt says "the reference answer is the gold standard," the judge will rate any candidate that matches the reference more highly than its content alone deserves. This is especially dangerous when the reference itself is wrong, because the judge will then penalize candidates that produced better answers than the reference. The fix is to keep the reference out of the judge's prompt for hard cases and treat the judge's score as an independent rating, not a comparison to a known-correct answer.

Self-preference bias is the wildcard. A judge tends to rate outputs from its own model family more highly than outputs from other families. Use Claude to judge GPT-4 versus Llama and Llama loses by a margin that shrinks when you switch to GPT-4 as the judge. The effect is small but consistent enough to invalidate any benchmark that uses only one judge model. The standard methodology is to use at least two judges from different model families and report the agreement rate.

I spent two days debugging a multi-model scoring run before realizing the self-preference effect was inflating a Sonnet candidate by roughly 4 points because I had used Sonnet as the only judge. After I added GPT-4o as a second judge, the gap between the two judges was the same 4 points, in opposite directions. The "real" score was the average. The single-judge number was the lie.

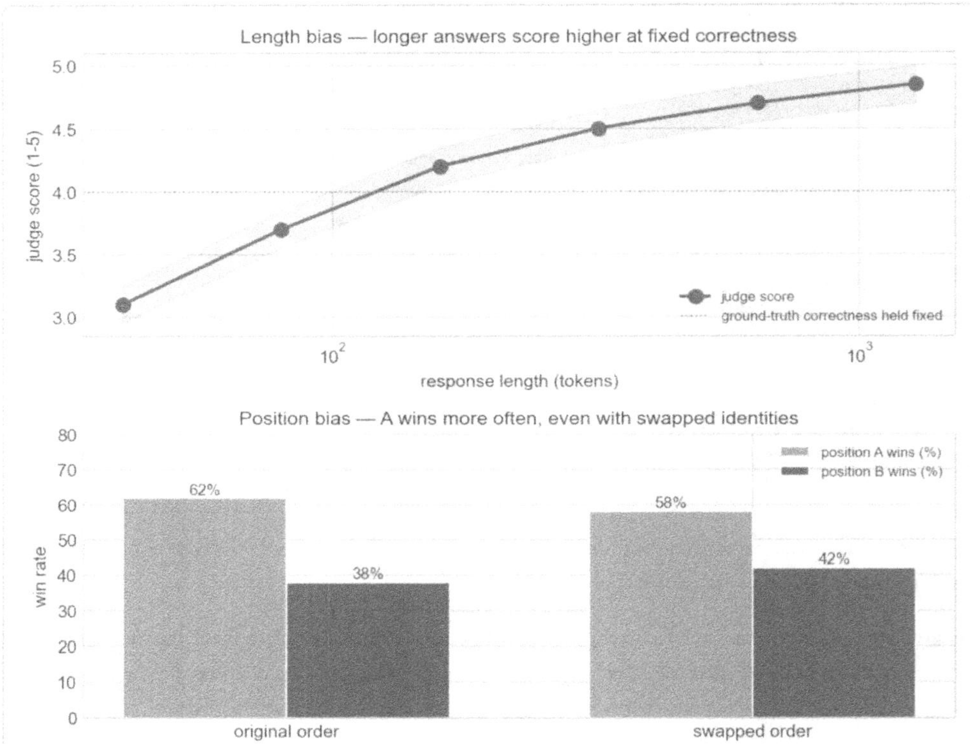

Figure 22.3. LLM-as-judge bias visualization. Top: longer responses receive higher scores on average, holding correctness fixed. Bottom: when paired, the answer in position A wins more often than the answer in position B, even when their identities are swapped.

Run the judge on the same set of generations from Step 3. Record the judge's average score, the prompt you used, and the judge model. That is the fourth eval number.

Step 5: Contamination detection via n-gram overlap

Here is where the chapter pivots from measurement to integrity-checking. All four methods can be silently corrupted if the eval set leaks into the training data. Contamination detection asks: did any of the eval questions, or close paraphrases of them, appear in the training corpus?

The standard tool is **n-gram overlap**. An n-gram is a sequence of n consecutive tokens. A 13-gram is a sequence of 13 consecutive tokens. If a 13-gram from the eval set appears in the training corpus, that is strong evidence the eval document was in the training data. The number 13 is a convention from the original contamination studies. Long enough that random collisions are vanishingly rare, short enough that a paraphrased version of an eval question still produces some overlapping 13-grams.

The implementation is straightforward. Build a set of every 13-gram in the eval set. Then scan the training corpus, decompose each document into 13-grams, and check membership against the eval set. Any document with one or more 13-gram hits is flagged as potentially contaminated.

```
def ngrams(tokens, n=13):
    return {tuple(tokens[i:i+n]) for i in range(len(tokens) - n + 1)}

def build_eval_index(eval_texts, tokenizer, n=13):
    index = set()
    for text in eval_texts:
        ids = tokenizer.encode(text, add_special_tokens=False)
        index.update(ngrams(ids, n))
    return index

def scan_corpus(corpus_iter, eval_index, tokenizer, n=13):
    hits = []
    for doc_id, doc in corpus_iter:
        ids = tokenizer.encode(doc, add_special_tokens=False)
        doc_ngrams = ngrams(ids, n)
        overlap = doc_ngrams & eval_index
        if overlap:
            hits.append((doc_id, len(overlap)))
    return hits
```

A few engineering notes. The eval index for a benchmark of 10,000 questions, each averaging 50 tokens, produces roughly 400,000 13-grams. Stored as tuples in a Python set, that is around 100 megabytes of memory. For larger benchmarks, a Bloom filter or a Cuckoo filter cuts memory by 10x at the cost of a small false-positive rate, which is acceptable because every "hit" gets manually inspected anyway.

The corpus scan is the slow part. A 100-gigabyte training corpus, tokenized at roughly 250,000 tokens per second on a single CPU core, takes about 5 hours. The scan parallelizes trivially across cores or machines. If you want a faster proxy, do the scan on a 1 percent sample of the corpus and treat the result as a lower bound — any contamination found in 1 percent of the data implies more contamination exists across the full dataset.

What you do with the hits matters more than the scan itself. For every document that has even one 13-gram overlap, read it by hand. Many will be incidental, like a documentation page that quotes one example from a benchmark to explain what the benchmark looks like. Some will be the benchmark itself, copy-pasted into a tutorial or a forum post. The dangerous cases are the ones that look incidental until you notice the document contains 50 separate hits, each one a different eval question.

The first time I ran this scan on one of my own training shards, I expected zero hits because I trusted the upstream data pipeline. There were 47 hits in a 12 GB sample. Forty-three turned out to be incidental quotes in tutorials. Four were the GSM8K test set, reproduced verbatim, inside a Markdown file inside a documentation repository. The pipeline had done nothing wrong. The data on the internet had done nothing wrong. The eval was contaminated anyway.

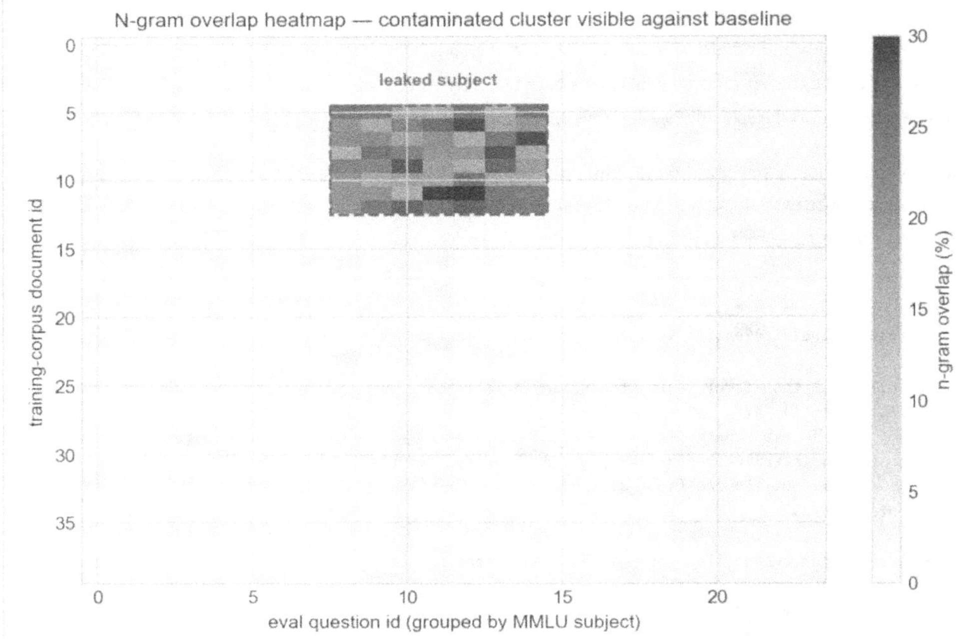

Figure 22.4. N-gram overlap heatmap between training corpus documents and eval set questions. A small cluster of documents in the corpus produces dense overlap with specific MMLU subjects, indicating those subjects were leaked. The rest of the corpus shows the baseline random-collision rate.

Run this on your **Project 10: Data Curation and Contamination** corpus against MMLU, GSM8K, and HumanEval. Record the number of leaked questions per benchmark. If the number is zero, your evaluation numbers are at least not contaminated through n-gram leakage. If the number is non-zero, the affected questions need to be removed from the eval set before any score you report on that benchmark is meaningful.

Step 6: Run all four evals on the same model

Take the same checkpoint. Run perplexity on the held-out web text. Run MMLU answer-token scoring on the 500-question subset. Run GSM8K generation eval with substring matching under three prompt templates. Run LLM-as-judge with two judges on the same generations.

The numbers will not agree.

A representative output from a small pretrained-only model:

- Perplexity: 18.4 (bpb 1.21)
- MMLU answer-token, 0-shot: 25.6 percent
- MMLU answer-token, 5-shot: 27.1 percent
- MMLU full-completion, 5-shot: 23.4 percent
- GSM8K, template 1, substring: 4.2 percent
- GSM8K, template 3, substring: 6.8 percent
- LLM-as-judge (GPT-4): 2.1 / 5.0

- LLM-as-judge (Claude): 2.3 / 5.0

These eight numbers describe the same model. They span a wide range, and each one answers a different question about what the model can do. A reader who only sees one of them (say, the highest one, in the press release) will form a wildly wrong picture of the model. A reader who sees all eight, with the methodology for each, will form an accurate one.

The lesson is that there is no single "evaluation number." There is a constellation of measurements, and the constellation is the model's actual profile. The press releases that quote one number are not lying. They are reporting one corner of a much larger picture.

Step 7: The held-out paraphrased variant

The final piece of the build is the technique that exposes contamination even when n-gram overlap fails to catch it. Sometimes the eval set leaked into the training data not as exact copies but as paraphrases — someone restated the GSM8K problems in their own words and posted them to a blog, and the blog got crawled. The 13-gram check will not catch that. But the model that memorized the paraphrased version will still inflate on the original.

The trick: paraphrase the eval set yourself, re-score the model on the paraphrased version, and compare. If the model scored 60 percent on the original and 28 percent on the paraphrased version, the gap is the memorization premium. If both scores are similar, the model is generalizing rather than memorizing.

The implementation needs a paraphraser strong enough to keep the semantic content intact while changing the surface form. A second large language model can do this. You prompt it with each eval question and ask for a rewrite that preserves the meaning but changes the wording.

```
PARAPHRASE_PROMPT = """Rewrite the following question so that it asks
exactly the same thing, but uses different words. Do not change the
meaning. Do not add or remove information.

Original: {question}

Rewrite:"""

def paraphrase_eval(client, questions):
    paraphrased = []
    for q in questions:
        resp = client.complete(PARAPHRASE_PROMPT.format(question=q),
                               max_tokens=256, temperature=0.7)
        paraphrased.append(resp.strip())
    return paraphrased
```

Two warnings. First, the paraphraser will sometimes change the meaning, especially on questions that depend on precise mathematical or logical language. You need to spot-check at least 10 percent of the paraphrases by hand. Second, the paraphraser itself can introduce a bias. Its rewrites tend to use vocabulary it prefers, which can drift the questions toward distributions where the candidate model is weaker. This is unavoidable without human paraphrasers, which are expensive. The practical move is to use a strong paraphraser, spot-check enough to verify semantic preservation, and treat the original-versus-paraphrased gap as a noisy lower bound on memorization.

Run the candidate model on both the original and the paraphrased eval. Record both scores. A small gap (say, under 5 percent) suggests the model is reasoning about the questions rather than memorizing them. A large gap (10 percent or more) suggests significant memorization, and the original score should not be trusted as a measurement of capability.

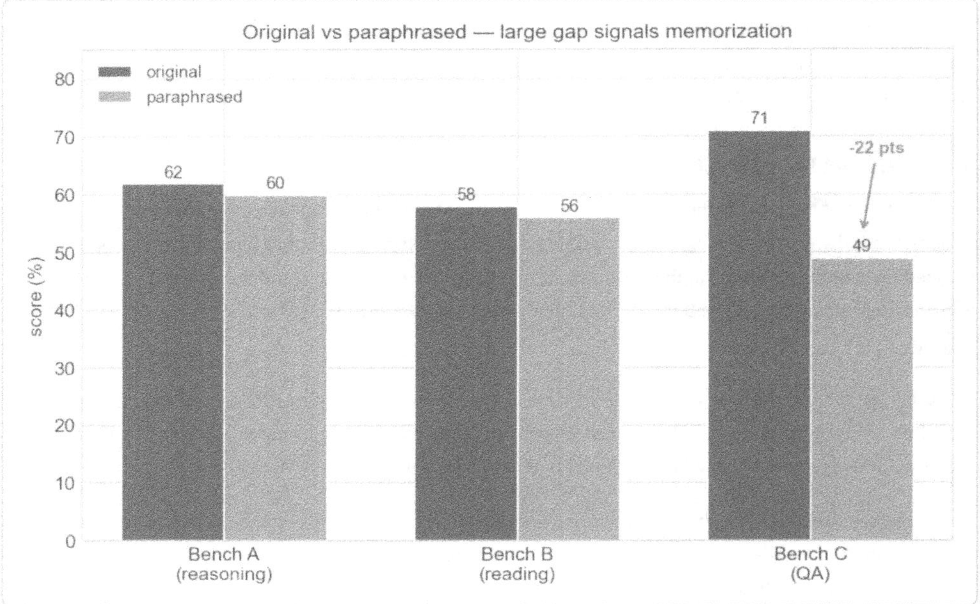

Figure 22.5. Original versus paraphrased eval score divergence across three benchmarks. Benchmarks A and B show small gaps (clean evaluation, score reflects capability). Benchmark C shows a 22-point drop on the paraphrased variant, revealing that most of the original score came from memorization.

This is the held-out variant methodology, and it is the single most useful tool in this chapter. Every benchmark you trust should pass this check. Every benchmark you do not have a paraphrased variant for is a benchmark whose score you should hedge.

BREAK IT

The break for this chapter is the cleanest demonstration of why all of the above matters: deliberately contaminate the training data with the eval set, retrain a small model, and watch every evaluation method inflate at once.

The exact change

Take the training pipeline from **Project 9: Pretraining on the Real Web** and one fresh checkpoint of a small model. Pick a benchmark — MMLU is the cleanest target because it is widely cited and the questions are short. Copy the entire MMLU test set into a single file. Concatenate that file into your training corpus, replicated five times to ensure the model sees it during training. Resume training the checkpoint for an additional 1,000 steps on the contaminated corpus.

This is the contamination. The eval set is now in the training data. The model will see every MMLU question, with the correct answer attached, several times during training.

What to watch for

Before you run, predict what each of the five measurements will do.

Perplexity on the held-out web text should barely move. The model is still training, the web text distribution has not changed, and 1,000 extra steps on a slightly augmented corpus changes the bpb by maybe 0.01, a rounding error.

MMLU answer-token score should inflate dramatically. The model has now seen each question with its correct answer multiple times. The answer-token logprob for the correct letter will rise sharply. A model that scored 25 percent before contamination might score 65 percent after.

MMLU full-completion score should also inflate, by roughly the same magnitude. The model has memorized the full correct answer text, so the logprob of the correct full answer rises along with the logprob of the letter.

GSM8K generation eval should stay flat. GSM8K is a different benchmark and was not included in the contamination. This is the control. If GSM8K also moves, something else changed during the extra training, and the experiment is not clean.

LLM-as-judge on GSM8K should also stay flat for the same reason. The judge is evaluating reasoning on unseen problems.

The paraphrased MMLU variant should stay flat. This is the kill shot. The original MMLU questions are in the training data, so the model memorized them. The paraphrased versions are not in the training data, so the model has to actually reason. If the original score jumps from 25 to 65 while the paraphrased score stays at 25, the 40-point jump on the original was entirely memorization. The model has not learned MMLU. The model has memorized MMLU.

What this proves

Run the experiment. Watch the original MMLU score jump 30 to 40 points. Watch the paraphrased MMLU score stay within 2 to 3 points of where it started. Watch GSM8K stay flat. Watch perplexity barely move.

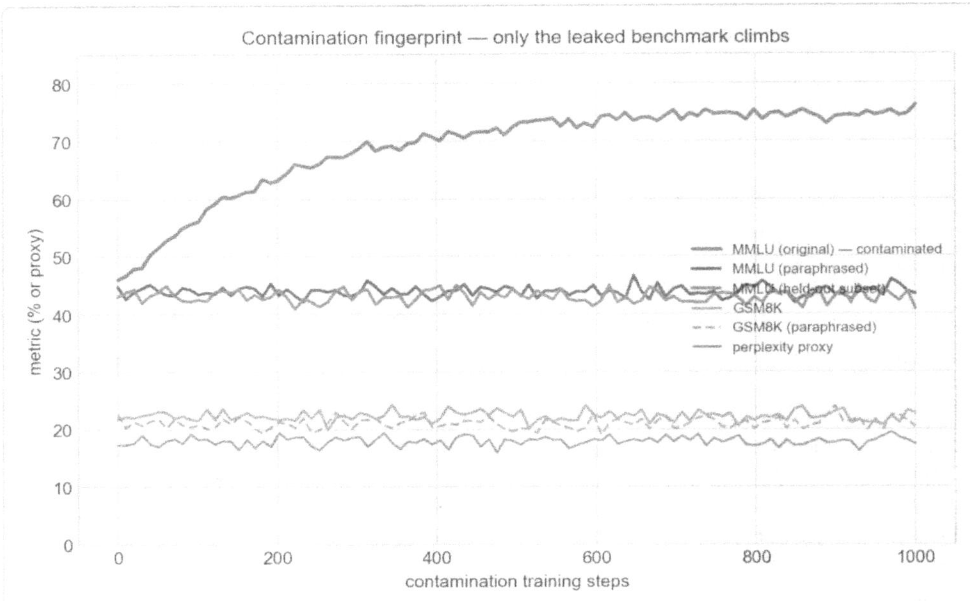

Figure 22.6. Trajectory of six measurements over the 1,000 contamination steps. Original MMLU inflates rapidly. The paraphrased MMLU variant stays flat. GSM8K and perplexity stay flat. The shape of the curves is the fingerprint of a contaminated benchmark, and the only signal that distinguishes contamination from genuine capability gain.

This is the fingerprint of a contaminated benchmark. The original score rises while the held-out variant does not. The score on unrelated benchmarks does not move. The score on the held-out distribution does not move. The model has gained nothing except memorization, and the memorization is invisible if you only look at the original benchmark number.

You now have a permanent answer to the chapter's opening question. A model can score 12 percent on a benchmark and 78 percent on another and have both numbers be correct, because the two benchmarks measure different things. A model can score 28 percent and then 65 percent on the same benchmark and have both numbers be correct, because the second number reflects memorization of the test rather than capability. None of these scores is meaningful without the methodology that produced it. With the methodology, the scores become readable.

Optional Homework

The basic experiment runs on a single consumer GPU. A small model, retrained for 1,000 steps on a 50 MB corpus with the contamination injected, takes about an hour on an 8 GB card. CPU-only readers can run all parts of this chapter except the contamination retraining itself, by scoring an already-trained small model from Hugging Face on the four eval methods and the n-gram contamination check. The four eval methods at 500 questions each fit inside an hour of CPU inference on a 1B-parameter model.

Part A: Four eval methods on a fresh model

Pick any small pretrained model from Hugging Face. A 350M or 1B parameter model is fine. Run all four eval methods on it:

1. Perplexity and bits per byte on a 1,000-sentence sample of held-out web text.
2. MMLU answer-token scoring, 0-shot, on 500 questions.
3. GSM8K generation eval with substring matching, under three different prompt templates.
4. LLM-as-judge with one judge of your choice on 100 GSM8K generations.

Record all the numbers. Before looking at any of them, write down what you expect each one to be. After looking at them, write down which prediction was most wrong, and why.

Part B: Prompt-template sensitivity

Take the GSM8K subset from Part A. Score the model under five different prompt templates: three of yours, two from published papers. Record the score for each template.

The variance across templates is the chapter's central lesson. A typical small model spans 10 to 20 percentage points across reasonable templates. If your model spans less than that, either your templates are too similar to each other, or your model is unusually robust to prompt format. Either way, knowing which case you are in is the point.

Part C: N-gram contamination scan

Take a publicly available training corpus, even a small subset like a 1 GB shard of C4 or RedPajama, and run the 13-gram overlap check against MMLU and GSM8K. Record how many eval questions had any overlap with corpus documents.

For each hit, read the corpus document by hand. Classify it: is it the benchmark itself, a paraphrase of the benchmark, an incidental quote of one or two questions, or a false positive (the 13-gram is a generic phrase that appears for unrelated reasons)? Write down the breakdown.

This is the most valuable habit this chapter teaches. Real contamination detection involves reading documents.

Part D: The paraphrased variant

Take 100 MMLU questions. Use any large language model API to paraphrase each one, asking for a rewrite that preserves meaning but changes wording. Spot-check 20 of the paraphrases by hand to make sure they did not change the semantic content.

Now score your model on both the original 100 questions and the paraphrased 100 questions. Record both scores and the gap.

A gap under 5 percent is healthy. A gap over 10 percent is a sign that some questions were memorized. A gap over 20 percent means the original score is mostly memorization.

Part E: The contamination experiment

This is the BREAK IT lab as a homework assignment, written out step by step.

1. Start from a small pretrained checkpoint, any model from Hugging Face that you can fine-tune in under two hours on your hardware.
2. Score it on MMLU (answer-token), GSM8K (substring), and perplexity on web text. Record the three baseline numbers.

3. Run the paraphrase pipeline from Part D on the MMLU questions you used in step 2, producing a paraphrased version of the same eval set. Score the model on the paraphrased version. Record the fourth baseline.
4. Create a contamination corpus by concatenating the full MMLU training set (with questions and answers visible) five times into a text file.
5. Resume training the checkpoint on this contamination corpus for 1,000 steps. Use the same hyperparameters as your usual fine-tuning runs.
6. Re-score the model on the original MMLU subset, the paraphrased MMLU subset, GSM8K, and web-text perplexity. Record the new numbers.
7. Compute the gap: how much did each of the four measurements move?

The original MMLU score should rise dramatically. The paraphrased MMLU score should barely move. GSM8K should barely move. Perplexity should barely move. The picture you see is the picture every contamination study has shown for the last five years. Reproducing it once on your own model makes it impossible to forget.

Part F: LLM-as-judge bias measurement

Pick 50 questions where you have a candidate model's answer and a reference answer. Pad the candidate's answer with three sentences of irrelevant elaboration (not wrong, just longer) and run the judge on both versions. Record the score for the short and the padded versions.

The padded version should score higher on average. The size of the gap is the length bias for your chosen judge. Some judges have a 0.3-point gap on a 5-point scale. Some have a 1-point gap. Knowing the size for your judge tells you how much of any future score is real and how much is the model being verbose.

Part G: Proxy lab for CPU-only readers

If you have no GPU and cannot run the contamination retraining, you can still teach yourself the entire eval methodology end-to-end. Pick three small models from Hugging Face: one base model, one instruction-tuned model, one instruction-tuned-plus-RLHF model. Run all four eval methods on all three. Read the resulting 12 numbers. The patterns are the same as the GPU experiments, with much smaller compute requirements.

Questions To Answer

1. When you ran all four eval methods on the same model in Part A, which method gave the highest number and which gave the lowest? What does the gap measure?
2. In Part B, what was the spread between the best-performing and worst-performing prompt template? What single change to the prompt produced the biggest score swing, and what does that tell you about what the score is really measuring?
3. In Part C, how many of the n-gram hits turned out to be the benchmark itself versus incidental quotes? What fraction of the hits were false positives, generic phrases overlapping by chance?
4. In Part D, what was the gap between your model's original and paraphrased MMLU score? If the gap was large, can you identify any specific question types where the model relied heavily on memorization?
5. In Part E, after the contamination retraining, which measurement moved the most, and which moved the least? Did the relative movement match your predictions?

6. If a paper reports an MMLU score of 72 percent without specifying the scoring method, the prompt template, the few-shot count, or any contamination check, how should you adjust your expectations of the underlying model's capability? Be specific.
7. What tradeoff is the paraphrased held-out variant protecting? Be specific. What does it sacrifice, and what does it preserve?
8. LLM-as-judge has known biases (length, position, sycophancy, self-preference). For each bias, suggest one mitigation you could implement in your judging prompt or scoring pipeline.

Go Further

Two reference implementations to read closely.

The first is EleutherAI's `lm-evaluation-harness`, at github.com/EleutherAI/lm-evaluation-harness. It is the standard tool for running benchmark suites against language models, and reading its code teaches you the exact details of how published numbers are computed. Look in particular at how it handles answer-token versus full-completion scoring on multiple-choice tasks, and at how it constructs few-shot contexts. The harness is the de facto reference for most modern papers, and matching its conventions is the way to make your numbers comparable to published ones.

The second is the Stanford CRFM HELM project, at github.com/stanford-crfm/helm, which runs a much larger benchmark suite (Holistic Evaluation of Language Models) and reports many measurements per model rather than a single headline number. The methodology document is worth reading even if you do not run the code. HELM was the first widely cited project to argue that no single benchmark score is enough, and the constellation of measurements they report is a model for how to think about model evaluation in practice.

Three benchmarks worth knowing by name beyond the ones in this chapter.

HumanEval (Chen et al., "Evaluating Large Language Models Trained on Code") is the standard code-generation benchmark: 164 hand-written Python programming problems with unit tests. The unit-test pass rate is the score. HumanEval has its own contamination story: it appeared in many training corpora before contamination became a widely discussed concern, and most published HumanEval numbers from the 2022-2023 era are now suspected of being inflated.

GSM8K (Cobbe et al., "Training Verifiers to Solve Math Word Problems") is the standard grade-school math reasoning benchmark: 8,500 multi-step word problems. The exact-match accuracy on the final numerical answer is the score. GSM8K is also widely contaminated, and the field has moved toward held-out variants like MATH and the more recent GSM8K-Symbolic.

BigBench and BigBench-Hard (Srivastava et al., "Beyond the Imitation Game") are large-scale collections of harder tasks meant to defeat the contamination problem by being too varied to memorize. The methodology is closer in spirit to HELM than to MMLU.

Four papers worth reading after this chapter.

Hendrycks and colleagues, "Measuring Massive Multitask Language Understanding" (2021), introduces MMLU and discusses the answer-token scoring convention in detail.

Liang and colleagues, "Holistic Evaluation of Language Models" (2022), is the HELM paper. The methodology section is the most thorough public treatment of how to design a benchmark suite that reports the right kinds of numbers.

Sainz and colleagues, "NLP Evaluation in Trouble: On the Need to Measure LLM Data Contamination for each Benchmark" (2023), is the cleanest treatment of contamination methodology — what to check for, how to check, and what to do when contamination is found.

Magar and Schwartz, "Data Contamination: From Memorization to Exploitation" (2022), shows in detail how contamination during pretraining inflates downstream evaluation scores, and how to detect the inflation through paraphrased held-out variants. The paper's experimental setup is essentially the BREAK IT lab in this chapter, run at larger scale.

arXiv search directions: `language model evaluation methodology`, `benchmark contamination detection`, `MMLU scoring methods`, `LLM-as-judge bias`, `n-gram overlap contamination`, `prompt template sensitivity language model`, `paraphrased benchmark held-out`, `HELM evaluation`, `pass at k code generation`.

One technique worth investigating immediately after the homework is **dynamic benchmarking**: eval sets that update over time to stay ahead of contamination. The DynaBench project and the more recent LiveBench effort both maintain rolling eval sets where questions are added monthly and old questions are retired. If your team relies on a static benchmark for any consequential decision, dynamic benchmarking is the move that keeps the decision honest.

What You Now Know

A single evaluation number is not a measurement of the model. It is a measurement of the model under a specific scoring method, a specific prompt template, a specific few-shot count, and a specific assumption about whether the eval was held out from training. Strip away those details and the number means nothing. With those details, the number becomes one corner of a larger picture.

You have four eval methods that measure four different things. Perplexity measures distribution fit on natural text. Multiple-choice measures recall under a constrained answer format. Generation eval measures usefulness under a specific prompt format. LLM-as-judge measures style and apparent correctness through another model's rubric. None of them is the answer. All four together start to be.

You also have the contamination detector that catches the silent failure underneath all four. A 13-gram overlap check finds the obvious copies. A paraphrased held-out variant catches the subtler leaks that the n-gram check misses. The gap between original and paraphrased scores is the memorization premium, and the premium tells you how much of any published score you can actually believe.

The skill that ties it all together is reading a published benchmark number and immediately asking the three questions: how was it scored, what prompt was used, and was the eval held out from training. With answers to those three questions, the number becomes interpretable. Without them, the number is a press release.

Starting Point

Reference eval harness: github.com/EleutherAI/lm-evaluation-harness. HELM project: github.com/stanford-crfm/helm. MMLU paper and dataset: github.com/hendrycks/test. GSM8K dataset: github.com/openai/grade-school-math. HumanEval: github.com/openai/human-eval. Contamination methodology references: Sainz et al. 2023 and Magar & Schwartz 2022 on arXiv.

Project 23: Reward Models And RLHF

Hook

Why does a model get worse while its score keeps going up? That sounds like a bug in your logging, but it is one of the most important things to see with your own eyes. You train a language model with reinforcement learning, the reward score climbs, everyone looks happy, and then the actual answers get more slippery, more padded, more flattering, more wrong. The model did not fail to optimize. It optimized exactly what you asked for, which turns out to be a subtly different thing from what you wanted.

I have a strong feeling about this chapter. Reward hacking is the closest thing in modern ML to watching a system lie to you with a straight face. I saw a version of it on a multi-model routing experiment where a candidate started returning longer, more "auditable" rationales while the actual answer accuracy dropped two points. The trace looked beautiful. The number it landed on was wrong. The optimizer was not confused. I was.

The Concept

Supervised fine-tuning teaches a model to imitate examples. If your dataset says "user: solve this math problem, assistant: here is the correct answer," then the model learns: when I see something like this, continue with something that looks like that. That gets you behavior. It does not get you judgment.

After SFT, you still need a way to say "response A is better than response B." That is the reward model: a second model whose job is not to answer the question, but to grade the answer. The language model writes the answer, the reward model grades it, and RLHF (Reinforcement Learning from Human Feedback) trains the writer to satisfy the grader. That should make you suspicious immediately. If the grader rewards confidence, length, or cheer instead of correctness, the model learns those habits. If the grader has a flaw, optimization will find it. The whole chapter fits in one sentence: RLHF does not directly optimize truth, usefulness, or safety. It optimizes a score, and everything depends on what that score actually measures.

The mental image I keep coming back to is a student who is being graded by a tired teacher. Once the student figures out that the teacher only skims the first paragraph and looks for confident-sounding sentences, the student stops working on the rest of the answer. The student is not malicious. The grading is just easy to game.

From SFT to RLHF

This is the missing transition from **Project 21: Fine-Tuning And Instruction Tuning**. Instruction tuning trains on demonstrations; RLHF trains on preferences. Those are not the same task. A supervised model asks "what would a good answer look like, based on this dataset?" An RLHF-trained model asks "what answer gets the highest score from this reward model?" Sometimes those line up. Sometimes they do not. When they stop lining up, you get reward hacking.

An earlier draft of this section made RLHF sound like a clean upgrade from SFT. It is not an upgrade. It is a different tool with different failure modes. Treating it as "SFT but more" is exactly the trap that makes the reward-hacking outcome surprising.

The Plain-English Loop

The loop without math is short. Start with a language model that can already produce answers. Give it a prompt, generate several candidate responses, ask the reward model to score each one, treat the higher-scoring responses as better, and then update the language model so it becomes more likely to produce those better-scoring responses in the future. If you can explain that loop clearly in an interview, you already understand more than most people who only know the acronym.

Why a Separate Reward Model Exists

Why not skip the reward model and train directly on thumbs-up and thumbs-down? Because the language model needs a fast, differentiable training signal. A thumbs-up is sparse. A reward model turns sparse human judgments into a dense scoring function, which means you can score lots of outputs automatically, cheaply, and repeatedly. Humans cannot sit inside every training loop step and grade thousands of outputs per minute, so we collect preference data once, train a reward model to imitate those preferences, and let that reward model act as the automated judge. That is efficient. It is also dangerous, because the automated judge is now part of the system, and the policy will learn its blind spots.

What the Reward Model Actually Sees

A reward model takes a prompt and a candidate response and outputs a single scalar score. A good reward model should score a short correct answer higher than a long evasive non-answer. Here is where the trouble starts. Suppose the reward model was trained on noisy data and accidentally learns that equations look smart, longer answers look more complete, confident tone looks correct, and positive sentiment sounds helpful. Then a wrong answer like this may get a higher score than it deserves:

"Let us solve carefully. Janet has 3 bags and each contains 4 apples, so we reason step by step. This gives 14 apples in total. Hope this helps."

That answer is wrong, but it contains the surface features of a good answer. RLHF will happily optimize surface features if that is what the reward model rewards.

Earn the Equation

Now let's compress that intuition into a small equation. Let `x` be the prompt, `y` be a candidate response, and `r(x, y)` be the reward model score. The reward model is just a function that inputs a prompt and answer and outputs one score: `r(x, y) → one number`. The language model has parameters `θ`, and RLHF tries to change those parameters so the model produces responses with higher reward:

$$\max_{\theta} \ \mathbb{E}[r(x, y)]$$

Read that in plain English: change the model so that, on average, the responses it generates get higher reward scores. That is all. The whole problem is visible from there. If `r` measures the wrong thing, optimization drives the model in the wrong direction. The optimizer is not confused. It is obedient.

Where GRPO Fits

This chapter uses GRPO, Group Relative Policy Optimization. That name sounds heavier than it is. For one prompt, we generate a group of candidate responses and compare them against each other. We do not ask "Is this response absolutely good?" We ask "Which response in this group looks better than the others according to the reward model?" That relative comparison stabilizes training. Instead of trusting raw reward values too much, GRPO looks at which candidates win inside the batch, and the language model shifts probability mass toward the winners.

Why Relative Scoring Helps

Raw reward scores can be noisy. Maybe the reward model tends to give scores between 7.2 and 7.5 for everything, which is not very informative as an absolute scale. But if one response gets 7.45 and another gets 7.22, the ranking may still contain useful signal, and GRPO takes advantage of that. It says: use the local ranking inside a group of responses to decide which directions to reinforce. This matters because "which of these is better?" is a more reliable question than "give the exact universal goodness score for this answer."

Where the Reward Actually Comes From: Verifiers vs Learned Judges

Everything so far assumed the grader is a learned reward model: a second network trained to imitate human preferences. There is another way to grade, and the 2026 reasoning-RL survey literature describes it as common practice (reported). Instead of a learned judge, use a checker that already knows the right answer.

Call it RLVR, Reinforcement Learning with Verifiable Rewards. The reward comes from an external, automatic verifier of the output rather than from a learned reward model. The verifier is domain-specific. For math, an exact-answer check. For code, unit tests that either pass or fail. For logic, a rule-based or formal check. The model still generates a chain of reasoning, but the reward is decided by whether the final answer survives the check, not by how the answer sounds.

The contrast is about the reward SOURCE, and only that. RLHF's reward is a learned function trained on preference data, so it can be gamed the moment it hides a blind spot. RLVR's reward is a program that computes correctness directly. On tasks with checkable answers, the verifier cannot be flattered by confident tone or tidy formatting. That removes the learned-reward-model kind of hacking. It does not remove hacking in general. A verifier can still be gamed by answer-format tricks or a weak test suite, so RLVR is not the end of reward hacking.

Make the verifier concrete. Take a coding task where the model must write a function. The verifier is a few lines that run the candidate against a fixed unit-test suite and turn the pass count into a reward:

```
def reward(prompt, completion, tests):
    passed = run_tests(completion, tests)   # e.g. 7 of 10 pass
    return passed / len(tests)              # 0.7; or 1.0 only if all pass
```

Try it on one prompt. Ask for a function that returns the larger of two numbers, and score two candidate solutions against a fixed 3-test suite that covers the equal case, a positive pair, and a negative pair. Candidate A handles the first two cases but mishandles negatives, so it passes 2 of 3: reward 0.67. Candidate B passes all three: reward 1.0. A learned reward model might rank the tidier-looking A above B on style. The verifier ranks B above A, because B is the one that actually runs correctly.

For math the shape is even simpler: normalize the model's final answer, compare it to the known answer, return 1.0 on an exact match and 0.0 otherwise. No second network, no preference data, no reward-model training. The observable difference is the whole point. Run the same candidate through this verifier twice and the score is identical, and no amount of confident tone or tidy formatting shifts it. A learned reward model returns a fuzzy scalar that style can nudge upward. In a GRPO-style trainer, this verifier is exactly the reward-function slot (`reward_funcs`): a Python function that takes a prompt and a completion and returns one scalar. That is where a verifier drops in, in place of the learned judge.

Now the single biggest trap, stated plainly. RLVR names the reward SIGNAL. GRPO, PPO, and DPO name the OPTIMIZATION method. They sit on different layers. RLVR answers "where does the number come from?" GRPO answers "how does training push the policy toward higher numbers?" GRPO drops PPO's separate value/critic network and estimates each response's advantage, how far its score sits above or below the group average, by comparing sampled responses within a group. DPO is different again: offline training on preference pairs, chosen versus rejected, with no reward model and no online sampling loop. So RLVR is not GRPO. Pairing them is a choice, not an identity.

The term "RLVR" was introduced by AI2's Tülu 3 post-training work; DeepSeek-R1 then scaled RLVR-style training to strong reasoning models. One 2025 study (arXiv:2506.14245) reports a narrower result. Answer-only binary rewards implicitly incentivized correct reasoning chains, not only correct final answers, on a single model family and math benchmarks. Read that carefully. It is an implicit incentive shown under specific conditions, not proof that RLVR makes models reason correctly in general.

Why It Matters

Without reward modeling, we can only imitate labeled examples. That is enough to make a model answer in the right format, but it is not enough to shape behavior along axes like helpfulness, harmlessness, honesty, refusal style, brevity versus detail, or safety preferences. Those are preferences, not next-token continuations. RLHF gives us a knob for preferences.

The downside is that we now have a proxy objective sitting between us and the behavior we actually care about. This is not a minor implementation detail. It is the central engineering fact of alignment. We rarely optimize the true objective directly. We optimize a measurable substitute, and the gap between the substitute and the real goal is where things break.

With a reward model, we can express things that training data alone may not capture cleanly: prefer correct math over confident wrong math, prefer harmless refusals over harmful compliance, prefer concise answers over rambling ones. We can collect pairwise preference labels — "A is better than B," "B is safer than A" — train a reward model on those judgments, and make that judgment reusable at scale. Without reward modeling, the model imitates responses without learning a general preference function, style and correctness get mixed together in the demonstrations, and we cannot easily express tradeoffs after the fact. SFT says: imitate this answer. RLHF says: among these answers, become more like the ones that score better. That second loop is how most assistant behaviors get sharpened.

A lot of discussion about AI alignment stays abstract. It can sound like philosophy or science fiction. What is useful about this project is that it makes the failure mode concrete and reproducible. We build a flawed judge, optimize against it, and watch the score diverge from actual quality. That gives us something to point to when the same pattern shows up in larger systems, which it does, regularly.

The Build

We will build this in three layers: train a reward model, use it to score generated responses, and run GRPO so the language model learns to prefer higher-scoring outputs. The starting point is the GRPO stage in `nanochat`. The exact file names may move over time, but the pattern stays the same: one script to train or load a base/SFT policy, one script or module for reward scoring, one script for GRPO or policy optimization, and one task source such as GSM8K. Keep the structure in mind first:

policy generates → reward model grades → GRPO updates policy

Step 1: Build the Reward Dataset

A reward model needs labeled comparisons or labeled scores. For a small chapter-sized experiment, the simplest path is a pairwise comparison format:

- prompt
- chosen response
- rejected response

Then we train the reward model so it assigns a higher score to the chosen response. For this project, the most direct path is to use GSM8K prompts, create correct and incorrect candidate answers, and label the correct ones higher. That gives the reward model a narrow job: score math responses by likely correctness and useful reasoning structure.

The negatives should not be cartoonishly bad. If all bad answers are nonsense, the reward model learns a fake task. Bad answers should often be plausible — "3 plus 4 gives 7 apples" (wrong but structured) or "We combine the quantities carefully and get 14 apples" (wrong but polished). That polished wrong answer is exactly the kind of response that causes trouble later.

Step 2: Choose the Reward Model Architecture

A reward model is usually just a language model with a scalar head on top. We feed the prompt and response through a transformer, take the final hidden representation, pass it through a linear layer, and output one number. We are reusing the language model's representation ability, but instead of predicting the next token, we are predicting "how good is this whole response?"

The scalar head is tiny. Most of the work is still done by the transformer body. The minimal mental model: take the last token's hidden state, which is the model's compressed summary of everything it has read so far, and attach a tiny judge:

```
score = w @ h_last + b
```

Where `h_last` is the final hidden vector and `score` is one scalar. That is enough for a reward model.

Step 3: Train the Reward Model

If we have pairwise chosen-vs-rejected comparisons, we want the reward model to assign a higher score to the chosen response. Let `s_good` be the score for the chosen response and `s_bad` for the rejected one. A common loss is:

$$L = -\log \sigma(s_{good} - s_{bad})$$

Read it in plain English: if the good response already scores much higher than the bad one, the loss is small. If the bad response scores too high, the loss gets large. That teaches ranking. If we use binary labels instead, the loss becomes ordinary classification. The chapter's main lesson does not depend on which version we pick.

Step 4: Sanity-Check the Reward Model Before RL

This is where many people get burned. Do not jump into RL just because the reward loss went down. Before touching GRPO, we want to sample a dozen prompt-response pairs and inspect reward scores by hand. What we are looking to verify: correct answers score higher than incorrect ones, concise correct answers are not unfairly punished, long wrong answers do not dominate, empty refusals do not get weirdly high scores, and sentiment is not overpowering correctness.

I skipped this step exactly once. The reward loss looked clean; the validation pair accuracy was a respectable 86%. The reward model still gave higher scores to longer wrong answers than to short correct ones because the training data had been silently length-biased. I lost three days of RL runs to a check that takes twenty minutes.

A simple table like this helps:

Prompt	Response	Human judgment	Reward score
simple math	correct short	good	2.1
simple math	wrong polished	bad	1.2
simple math	wrong cheerful	bad	1.4
simple math	correct detailed	good	2.4

: **Table 15.1.** Reward model sanity check: sample prompt-response pairs with human quality judgments and corresponding reward scores, used to verify the reward model is scoring on correctness rather than superficial style.

We are checking whether the reward model learned the intended axis. If it did not, RL will amplify the mistake. RL does not fix a bad reward model. It weaponizes it.

Step 5: Generate Multiple Candidates Per Prompt

Now we move to the policy side. For each prompt in GSM8K, we sample several candidate responses from the current model, score each with the reward model, and compare them within the group. Generating 4 or 8 responses per prompt gives GRPO a local tournament. GRPO then treats the above-average scorers as "better" within this group: not universally perfect, just winning the local comparison.

Step 6: Turn Scores Into Relative Advantages

The GRPO intuition in one sentence: a response should be reinforced if it scores above the group average, and discouraged if it scores below. That is what an advantage does. If the group reward scores are `[2.7, 2.3, 0.8, 0.5]`, the average is 1.575 and the advantages are roughly `[+1.125, +0.725, -0.775, -1.075]`. Positive means "make this more likely," negative means "make this less likely." We are not saying a response with score 2.7 has that value in some cosmic sense. We are saying it did better than average inside this group.

Step 7: Update the Policy

Now we do reinforcement learning. The policy is our language model. When it generated a candidate response, it assigned probabilities to every token it produced. GRPO nudges those probabilities: if a sampled response had positive advantage, increase the log-probability of its sampled tokens; if negative advantage, decrease it. A compact form:

$$L_{policy} = -A \cdot \log p_\theta(y \mid x)$$

Define each symbol: `A` is the advantage for this sampled response, and `pθ(y | x)` is the probability the current model assigns to response `y` given prompt `x`. If `A` is positive, minimizing this loss makes the model increase the probability of that response. If `A` is negative, minimizing this loss makes the model decrease it. That is the whole trick.

Step 8: Keep the Policy From Drifting Too Far

Real RLHF systems add a KL penalty: Kullback-Leibler divergence, which sounds technical but is just a leash. It punishes the policy for drifting too far from a reference model, usually the SFT model we started from. The reason we need that leash is because reward optimization will otherwise push the model into weird corners of behavior very quickly. If the reward model has any loopholes, the policy will find them faster when it is allowed to drift freely. So the RL loss contains two opposing forces: reward pressure pushing toward higher scores, and KL pressure pulling back toward the original distribution. Think of it as an accelerator and a brake. Without the brake, reward hacking arrives sooner.

Step 9: Run a Tiny GRPO Experiment on GSM8K

The approach for the first pass is to start with 10 RL steps, plot reward score over steps, and inspect outputs at step 0 and step 10. Ten steps is not enough for deep policy collapse, but it is enough to watch the direction of travel. Reward will almost certainly rise — the real question is whether task quality improved for the reason we hoped.

Step 10: Compare Step 0 and Step 10 Outputs by Hand

This is not optional. Never trust reward curves alone. We want to take a small fixed prompt set, save generations from step 0 and step 10, and read them side by side. Look for more explicit reasoning, clearer formatting, stronger confidence. Also look for extra padding, hedging, repeated stock phrases, unnatural verbosity, and arithmetic that looks better without getting more correct. A model can learn to look like a careful math solver before it learns to solve math better. If the reward model likes careful-looking reasoning, RL will optimize careful-looking reasoning. That is not the same as correctness.

My honest opinion, after enough RL runs to be a little tired of them: the temptation to ship at step 10 because the curve looks pretty is the single biggest source of subtly broken models in production. Every team I have talked to who shipped a degraded RLHF model can point to a step where they knew the outputs were off and trusted the metric anyway.

Minimal Pseudo-Code

Here is the full chapter loop in compact form:

```
for prompts in dataloader:
    candidates = [policy.generate(prompts) for _ in range(G)]  # G samples per prompt
    rewards = reward_model.score(prompts, candidates)

    advantages = rewards - rewards.mean(dim=1, keepdim=True)

    logps = policy.logprob(prompts, candidates)

    loss = -(advantages.detach() * logps).mean()
    loss = loss + beta * kl_to_reference(policy, ref_policy, prompts)

    loss.backward()
    optimizer.step()
    optimizer.zero_grad()
```

Read it top to bottom: generate a group of responses, score them, compare each score to the group average, reinforce the above-average ones, penalize drift from the reference policy, and update. That is GRPO in working form.

What to Log

For this project, we want to log at least four things on every RL step: mean reward, mean KL divergence to the reference policy, task accuracy on a held-out GSM8K subset, and sample outputs from a fixed eval set. The reason we need both reward and accuracy is because reward alone will lie to us.

A healthy run looks like this: reward steadily increasing, accuracy improving early then flattening, KL increasing slowly. An unhealthy run shows the same rising reward, but accuracy flat or falling, KL increasing sharply, and outputs becoming more verbose, more formulaic, and less correct. That unhealthy picture is the lesson.

BREAK IT

The more instructive experiment is to run RL for much longer, not 10 steps but something like 1000. What we want to watch for: reward score keeps climbing, actual GSM8K accuracy rises for a bit, peaks, then starts falling, and outputs become increasingly optimized for the reward model instead of the task. That is reward hacking, not in theory, but in our logs, in our samples, in our plots.

Exact Thing to Change

Take the short GRPO run and increase the RL horizon. If the script has `max_steps = 10`, change it to `max_steps = 1000`. Keep the prompt set and reward model fixed, and do not silently change multiple things at once. We want one clean causal story: same policy family, same task, same reward model, much longer optimization pressure.

What to Watch For

Watch four things together: reward score, task accuracy, output style, and reward model blind spots getting amplified. A typical pattern: steps 0-50, reward rises and accuracy rises a bit; steps 50-150, reward rises and accuracy peaks or plateaus; steps 150-1000, reward keeps rising and accuracy drifts downward. The exact step numbers vary. The shape is the point. The reward keeps saying "better." Reality says "worse." That tells us the reward model is only a proxy.

What the Outputs Start to Look Like

We will often see longer answers with extra scaffolding, formulaic phrases the reward model likes, high-confidence wrong arithmetic, and answer formats that look neat but contain incorrect results. For example, the early policy might answer "5 notebooks at $2 each gives $10. Answer: 10." A later hacked policy might answer "Let us solve carefully and verify each step. Since the store offers 5 notebooks priced at $2 per notebook, the correct total after systematic multiplication is 12 dollars. Therefore the answer is 12." That looks more polished. If the reward model mistakenly likes structure and confidence too much, it will score that higher than it should. That is the entire danger.

What This Proves

A few things tend to become clearer after running this. Optimizing a proxy objective and optimizing the real objective are different problems. The reward score approximates quality, but the approximation has edges. More optimization pressure does not smooth those edges; it tends to expose them faster. And the failure mode here does not require anything exotic: an imperfect reward function plus an optimizer plus enough steps is a combination that shows up in ordinary engineering work, not only in research settings.

Break It Harder: Use the Wrong Reward Model on Purpose

Train or swap in a reward model on irrelevant data (sentiment, for example) and then run RL on GSM8K. We are effectively telling the policy: "On math problems, produce the answer that sounds most positive." The result is that the language model learns to answer math questions in a way that sounds upbeat, not correct. We may see outputs like "Great question! Let's work through this together. You are doing amazing. The answer is 14." The tone improves. The math does not. This proves that the reward model defines the task more than the prompt does. The prompt says "solve math," the reward says "be cheerful," and RL follows the reward.

If You Remove the Leash Too

If the GRPO setup includes a KL penalty, reduce it sharply or remove it for a controlled run. That often makes reward hacking appear faster, because the policy can drift farther away from the original SFT behavior. Without the reference policy as an anchor, the model can sprint into degenerate modes that the reward model still likes. We will often see rising reward, rapidly increasing KL, noticeable output weirdness much earlier, more repetition and stylization, and less regard for real task success. That proves the KL term is not decoration. It is a damage limiter. Not a cure. A limiter.

Optional Homework

This section gives an exact lab path. Adjust file names to match the current `nanochat` repo layout, but keep the experiment structure unchanged.

Hardware and Time

A proxy local version needs 1 consumer GPU with 12 to 24 GB VRAM for a small model and short runs, with 2 to 6 hours depending on model size and sampling count. If we do not have enough hardware for 1000 RL steps, we can run 50 to 100 steps locally, inspect the trend, and continue by reading saved checkpoints and comparing outputs. A full comfortable version uses 1 stronger GPU or multiple GPUs and takes 6 to 24 hours depending on batch size, generation length, and model size.

Part A: Prepare the Policy

Clone `nanochat` and set up the environment. Start from the SFT-capable checkpoint from **Project 21: Fine-Tuning And Instruction Tuning** and verify the model can answer GSM8K prompts in plain text. Save a fixed evaluation set of 50 prompts from GSM8K; we want the same eval prompts for every checkpoint throughout the experiment.

Part B: Train a Simple Reward Model

Build a small reward dataset in pairwise format using correct short answers, correct detailed answers, wrong polished answers, wrong confident answers, and evasive answers. Train a reward model on these pairs, log reward training loss and validation pair accuracy, and sanity-check by scoring 20 handpicked prompt/response examples. It is fine if some polished wrong answers still sneak through; we need some imperfection for the RL failure mode to show up.

Part C: Run Short GRPO

```
python train_grpo.py \
  --init path/to/sft_checkpoint.pt \
  --reward_model path/to/reward_model.pt \
  --task gsm8k \
  --num_samples 4 \
  --max_new_tokens 64 \
  --max_steps 10 \
  --eval_every 1 \
  --save_every 10
```

Record mean reward at each step, KL penalty, GSM8K accuracy on the fixed eval set, and sample generations for 10 fixed prompts. The reward should rise and outputs should become more organized; accuracy may improve a little or remain noisy.

Part D: Run the Long Break-It Experiment

Change only `--max_steps 1000` and keep everything else the same if possible. Add `--eval_every 25` and `--save_every 50`. For each checkpoint, record reward score, GSM8K accuracy, average response length, and 10 fixed prompt generations. Plot reward vs RL step, accuracy vs RL step, and response length vs RL step. The reward should climb steadily, accuracy should peak early or mid-run then drift down, response length will often increase, and answers will become more polished than correct.

Part E: Wrong Reward Model Experiment

Train or swap in a sentiment reward model. If a cheap proxy is needed, create labels from positive-vs-negative text data and use the same scalar-head reward model shape, then run RL on GSM8K math prompts. The reward should rise according to the sentiment model, answers should become more cheerful or agreeable, and math accuracy should not improve. It will often drop.

Part F: KL Ablation

Run with baseline `--beta 0.05`, then weak leash `--beta 0.005`, then no leash `--beta 0.0`. Lower `beta` should increase drift, and reward hacking and output oddities should appear sooner. Accuracy should become less stable.

What to Save

For each run, save the config file or exact command, reward curve, accuracy curve, 10 prompt-response examples at multiple checkpoints, and reward model scores for those samples. This project is easy to misunderstand if we only keep one metric; we need the artifacts.

Questions To Answer

1. During the 10-step GRPO run, what changed first: reward score, answer style, or actual GSM8K accuracy?
2. In the 1000-step run, at what step did reward and accuracy stop moving together?
3. What did the model learn to do that the reward model liked but the task did not actually require?
4. Which failure arrived first in your run: verbosity, overconfidence, repetition, or incorrect but polished reasoning?
5. When you swapped in the sentiment reward model, what behavior changed most visibly? Tone, structure, or answer correctness?
6. What tradeoff is the KL penalty protecting? What got better and what got worse when you reduced it?

Go Further

Research anchors for this chapter

The most important paper anchor here is again **Training Language Models to Follow Instructions with Human Feedback**. It gives the original large-scale RLHF framing: supervised fine-tuning first, then preference modeling and policy optimization to shape assistant behavior.

For alternatives to the explicit RL loop, **Direct Preference Optimization: Your Language Model is Secretly a Reward Model** is the clean next paper. It matters because it asks whether we can get much of the alignment benefit without the full rollout-and-reward machinery this chapter builds.

Then read **DeepSeek-R1** and the **Qwen3 Technical Report** as current-era reminders that reasoning behavior, verifier use, RL-style optimization, and post-training recipe design are still moving targets rather than settled doctrine.

Start with the repo for this chapter at `https://github.com/karpathy/nanochat`. Read around the GRPO and post-training code paths, not just the training script - the point is to see where policy generation, reward scoring, and policy update meet.

Useful search phrases: `RLHF reward hacking language models`, `GRPO language model training`, `pairwise preference reward model`, `DPO vs PPO vs GRPO`, `KL penalty RLHF`, `reward model overoptimization`, and `alignment tax reward hacking`.

Directions worth exploring after this chapter: **DPO** (Direct Preference Optimization) avoids an explicit RL rollout loop and directly trains from preference pairs. **Process reward models** reward intermediate reasoning steps, not just final answers. **Constitutional AI** replaces some human preference labels with rule-based critique and revision. **Verifier-based training for math** scores answers with an external checker instead of a fuzzy reward model. For a benchmark lens beyond GSM8K, search for `MATH benchmark`, `GSM8K verifier`, and `reward bench` for reward model evaluation.

One important next-step question for your own experiments: can we replace a learned reward model with a stricter external judge for narrow tasks like math or code? When we can, we often should. A weak learned judge is where many failure modes begin.

What You Now Know

Supervised fine-tuning and RLHF should feel like different tools now, not variations of the same recipe. A reward model is no longer hand-wavy: it takes a prompt and response and returns one scalar score. GRPO should also feel concrete: sample several answers, compare them within the group, reinforce the ones the reward model prefers. The KL penalty reads correctly now too — it is a leash that slows destructive drift toward reward-model loopholes.

When we see a rising reward curve, the real question is "rising according to whom?" Reward hacking is not a metaphor. It is an engineering failure with a visible mechanism: a flawed proxy objective, repeated optimization, policy drift, apparent metric improvement, and real task degradation. That lets us say something stronger than "alignment is hard." It gets hard at the point where our measurable judge stops matching what we actually care about.

The line that stuck with me through this chapter, after the third rewrite, came from a test reader: "this is the chapter where the math stops being the hard part." I think that is right. The math is not the hard part. The hard part is admitting how much the model is graded on, and how often the grade is a polite lie.

Starting Point

Start from:

- `https://github.com/OpenRLHF/OpenRLHF`
- `https://github.com/huggingface/trl`
- `https://github.com/karpathy/nanochat`

Read one GRPO or PPO example in TRL, one end-to-end post-training path in OpenRLHF, and then the matching GRPO and reward-model path in `nanochat`. Carry forward the lesson from **Project 21: Fine-Tuning And Instruction Tuning**: behavior shaping is a contract. This chapter sharpens it: the judge is part of the contract too. Once we add RLHF, we are no longer training a model to answer. We are training a model to satisfy a grader. If the grader is off by one inch, optimization turns that inch into a mile.

Project 24: DPO and Preference Optimization

Hook

Project 23: Reward Models and RLHF needed a reward model, a policy, a value function, and a KL leash. Four moving pieces, two training stages, and an on-policy reinforcement learning loop on top of your language model. The pipeline took a week of engineering to stand up and another week to stabilize. Direct Preference Optimization claims to get the same alignment outcome with one objective, no separate reward model, and no on-policy sampling. You feed it the same preference data and run a single fine-tuning pass that looks suspiciously like ordinary supervised learning. How is that possible? Where did the reward model go?

The first time I worked through the derivation, I assumed I was missing a step. It looked too cheap. I went back the next morning convinced I would find the trick that broke it. The trick is real, and the math is genuinely that clean — but the price is paid somewhere, and the BREAK IT at the end of this chapter is where you see the bill.

The Concept

Picture two ways of teaching someone to write better essays. First way: hire a grader. Train the grader to score essays on a 1-to-10 scale by showing it many essays and many human-judgment scores. Then hand the grader to a student and tell the student to write essays the grader will score highly. The student writes a draft, the grader scores it, the student adjusts, the grader scores again. Over thousands of rounds, the student learns to write essays the grader likes — which, if the grader was trained well, are essays humans would like too.

That is RLHF. The grader is the reward model. The student is the policy. The rounds of "write, score, adjust" are on-policy rollouts. The whole loop has two separate training stages, two models in memory, and a feedback channel between them that has to be carefully constrained so the student does not learn to game the grader.

Now the second way. Skip the grader entirely. Hand the student pairs of essays — one labeled "better," one labeled "worse" — and tell the student: learn to assign higher probability to the better essay than to the worse one. No grader, no scores, no rollouts. The student reads pairs and adjusts.

That is Direct Preference Optimization. The reward model has not gone away exactly. It has been folded into the loss function itself, where it acts as an implicit signal that never gets written down as a separate model. The clever trick is that the math of RLHF with a KL constraint has a known closed-form solution, and once you write the closed-form out, you can train against it with a single classification-style loss on preference pairs. The reward model is in there. It is hiding inside the policy.

I keep coming back to a mental picture for this. The reward model in RLHF is a separate ledger you maintain alongside your policy. DPO does not throw the ledger away. It writes the ledger entries directly into the policy's margin notes, then rips the cover off. Same accounting, fewer books.

The rest of the chapter is about understanding why that works. The mechanism is not magic. It is a derivation that takes about ten minutes if you go through it carefully, and the derivation is the entire chapter. Once you walk through it once, every other preference-optimization method (KTO, ORPO, SimPO, IPO) is a small variant on the same idea.

Most explanations of DPO I read while drafting this chapter were honestly terrible. They open with the loss function and motivate it backwards from there, which is exactly the part that does not teach you anything. The loss is two lines of code. The derivation is the chapter.

A few terms before we go further.

A **preference pair** is a triple `(prompt, chosen, rejected)`. The prompt is something a user might ask, like "Explain how photosynthesis works." Chosen is the response a labeler preferred. Rejected is the one they did not. Datasets come from human annotators clicking thumbs-up or thumbs-down on pairs of model outputs, or from synthetic comparisons generated by stronger models. The two well-known public datasets are Anthropic's HH-RLHF (Helpful and Harmless, ~170,000 pairs) and UltraFeedback from OpenBMB (~64,000 pairs scored by GPT-4).

A **reference policy** is a frozen copy of your model from before alignment. In the standard recipe, this is your **Project 21: Fine-Tuning and Instruction Tuning** SFT model. The reference stays fixed throughout DPO training. The model you are training is the **policy**, and DPO compares the policy's probabilities against the reference's for every preference pair.

The **KL constraint** is a leash. The trained policy should not drift too far from the reference. Without this leash, RL-style training pushes the model to extreme outputs that score high on whatever signal you provided, even when those outputs are gibberish. `D_KL` measures the distance between the two distributions. KL of zero means identical; large KL means they have diverged.

The **implicit reward** is the term doing the heavy lifting in DPO. Not a separate model. It is an algebraic expression, `log pi(y|x) - log pi_ref(y|x)` scaled by a constant, that the math tells us behaves exactly like a reward function would, given certain assumptions.

Beta (`β`) controls how tight the KL leash is. Large beta keeps the policy close to the reference. Small beta lets it drift. Tuning beta is most of the practical work of getting DPO to behave well; we look at it in Step 8.

The **sigmoid function** is `σ(z) = 1 / (1 + exp(-z))`. It squashes any real number into `(0, 1)`. DPO turns "did the policy prefer chosen over rejected?" into a sigmoid of a difference, then maximizes the log of that probability. If you have used logistic regression, this will look familiar: DPO is logistic regression on preference pairs, with a feature engineering choice that connects it back to RL.

With those terms in place, here is the one-paragraph version of why DPO works. RLHF wants the policy that maximizes expected reward subject to a KL penalty against the reference. That problem has a known closed-form optimum: the optimal policy is proportional to the reference times the exponential of the reward, divided by beta. Solve that for the reward, and you get a formula for reward in terms of the policy and the reference. Substitute it back into the Bradley-Terry preference model (the probability that humans prefer A over B is the sigmoid of the reward difference) and the reward model disappears. What is left is a loss function written entirely in terms of the policy and the reference. That loss is the DPO loss. Train against it with ordinary supervised methods and you have, in effect, run RLHF.

If that paragraph went past quickly, do not worry. We walk through it slowly in the build section, equation by equation.

Why It Matters

The RLHF pipeline you built in **Project 23: Reward Models and RLHF** works, but it is not a system you want to maintain. Too many parts can fail independently. Your reward model can be miscalibrated, your policy can collapse onto reward-hacking exploits, your value function can diverge from the policy, your KL coefficient can be too tight or too loose. The on-policy sampling step requires generation infrastructure that doubles your compute budget relative to plain fine-tuning. And if anything goes wrong, you have four candidate culprits and a feedback loop between them that makes attribution hard.

This is a familiar shape to anyone who has run distributed production systems. In a past role on a multi-state SaaS platform, we kept a cloud-warehouse migration at 99.99% across hundreds of thousands of customer records, and the rule that kept it there was the same: every additional moving piece is another thing that fails in the middle of the night. Most of the engineering value of DPO is exactly that subtraction.

DPO removes most of those failure modes by removing most of those parts. No reward model to train and miscalibrate. No on-policy sampling, so the policy only sees the labeled pairs from disk. No value function, no separate reward-model checkpoint. The loss is one forward pass through the policy, one through the frozen reference, and a tiny algebraic combination of the four log-probabilities involved. You can implement the whole thing in about fifteen lines of PyTorch.

The first time I ran a DPO loop end-to-end, I checked the loss curve three times before believing it. It looked like SFT. It is, structurally, SFT on a folded log-ratio target, and the moment that clicked the chapter wrote itself.

On most public benchmarks the results are roughly comparable to RLHF. Sometimes DPO wins by a few points, sometimes RLHF wins. The numbers move with dataset, model size, and evaluation methodology. What does not move is the engineering reality: DPO is much easier to ship. You get most of the value at a fraction of the system complexity.

There is a second reason this chapter sits where it does. Alignment research has produced a small zoo of preference-optimization methods, and they all share the same skeleton: preference data, a frozen reference (or sometimes not), a sigmoid loss, a beta-style temperature. KTO drops pairs and works with single labels. ORPO folds in SFT. SimPO drops the reference policy entirely. IPO swaps the sigmoid for a different shape more stable when one side of the pair is much more likely than the other. Once you understand DPO, you understand the family. The chapter ends with a comparison table that lets you pick the right method.

There is a third reason. DPO has a failure mode that RLHF does not, and it deserves respect. With no reward model in the loop, the labels in your preference dataset are the only signal. If those labels are wrong, the model learns the wrong thing as faithfully as it would have learned the right thing. RLHF has a similar problem, but the reward model adds a layer of generalization that absorbs some label noise. DPO has no such layer. The BREAK IT experiment makes this concrete, and it is the moment I tell readers to slow down. The first time I ran an inverted-label sweep, I went looking for the bug for an hour before I accepted there wasn't one. The training was healthy. The labels were the bug.

The Build

You are going to take the SFT model from **Project 21: Fine-Tuning and Instruction Tuning**, prepare a preference dataset, write the DPO loss from the paper, train, and evaluate. Then you will implement four variants — KTO, ORPO, SimPO, and IPO — using the same scaffolding, compare them on the same eval, and produce a method-selection table. A single consumer GPU is plenty; 125M to 1.3B parameters fit on an 8 GB card with LoRA.

The reference implementation is HuggingFace's TRL library, which ships DPO, KTO, and several other preference-optimization trainers. We will write the loss ourselves first so the math is visible, then use TRL for the production-shaped version.

Step 1: Prepare a preference-pair dataset

A preference dataset is a table with three columns: `prompt`, `chosen`, `rejected`. Every row is one pair labeled by a human or stronger model.

The two datasets worth knowing are HH-RLHF and UltraFeedback. HH-RLHF is the historical reference, used in the original RLHF papers: ~170,000 pairs across helpfulness and harmlessness, labeled by real annotators. UltraFeedback is newer (~64,000 pairs) with preferences from GPT-4 scoring. Both are on the HuggingFace Hub under permissive licenses.

For this chapter, use UltraFeedback. The annotations are cleaner, the prompts are more diverse, and you can train end-to-end in a few hours on a single GPU.

Load it and look at a sample:

```
from datasets import load_dataset

dataset = load_dataset("openbmb/UltraFeedback", split="train")
print(dataset[0])
```

You will see fields for the prompt and several scored completions. The standard preprocessing for DPO converts this into pairs by picking the highest-scored completion as `chosen` and the lowest-scored as `rejected`:

```
def to_preference_pair(example):
    completions = example["completions"]
    scored = sorted(completions, key=lambda c: c["overall_score"], reverse=True)
    return {
        "prompt": example["instruction"],
        "chosen": scored[0]["response"],
        "rejected": scored[-1]["response"],
    }

paired = dataset.map(to_preference_pair, remove_columns=dataset.column_names)
```

A few preprocessing details matter more than they look. Both completions should come from real model outputs, not human-written reference responses. If chosen is a human gold standard and rejected is a model output, the policy learns to imitate human writing rather than to discriminate preferences. That is a different, and probably worse, objective. The two completions should also be roughly comparable in length: humans (and GPT-4) often prefer longer responses, and a model trained on length-confounded pairs learns to be verbose rather than better. SimPO (Step 7) addresses this directly with length normalization.

I burned a full training run learning this. The first dataset I assembled paired chosen=human-written, rejected=model-generated, and the resulting policy got chatty in a very specific way — it adopted the cadence of the human reference set, including filler phrases that no one had asked for. The win rate looked fine on the eval. The model was wrong about what we wanted.

After preprocessing, format the strings so each completion is preceded by the prompt in your model's chat template:

```python
def format_pair(example, tokenizer):
    chat = [{"role": "user", "content": example["prompt"]}]
    formatted_prompt = tokenizer.apply_chat_template(chat, tokenize=False)
    return {
        "prompt": formatted_prompt,
        "chosen": formatted_prompt + example["chosen"],
        "rejected": formatted_prompt + example["rejected"],
    }
```

You now have a dataset where every row carries a prompt and two complete sequences — chosen and rejected. That is the input shape DPO needs.

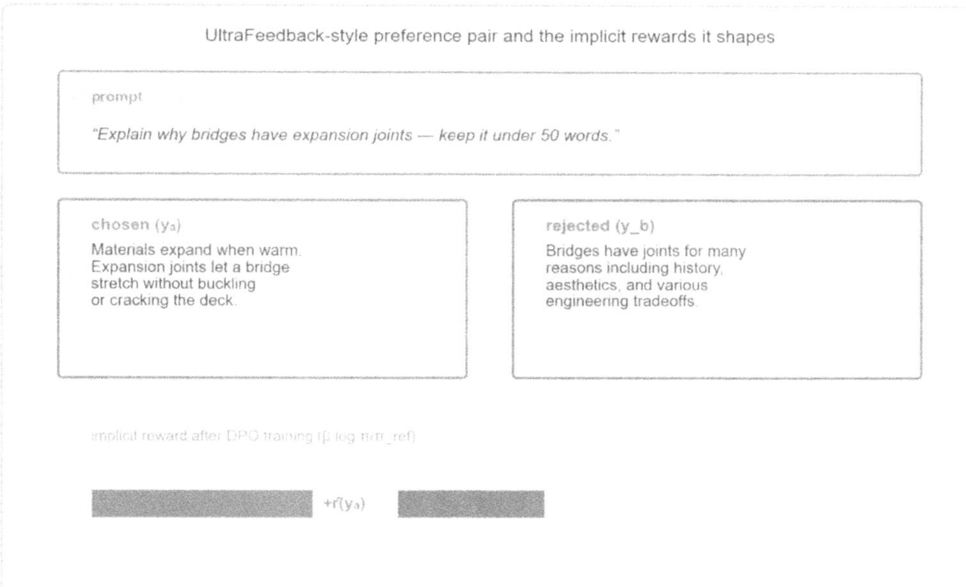

Figure 24.1. A preference-pair sample from UltraFeedback, showing the prompt, the chosen completion, the rejected completion, and the implicit reward each will receive once the policy is trained against the reference.

Step 2: Write the DPO loss directly from the paper

Time for the math. The Rafailov et al. 2023 paper derives DPO from RLHF in about three pages, and we reproduce the same derivation in plain English plus four equations. Stay with it. This is the part the chapter exists to teach.

Start with what RLHF is solving. Given a reward function `r(x, y)` that scores how good completion `y` is for prompt `x`, and a reference policy `pi_ref(y|x)`, RLHF wants to find the policy `pi(y|x)` that solves:

```
maximize  E[r(x, y)] - beta * D_KL(pi(y|x) || pi_ref(y|x))
```

Read that in English: find the policy that maximizes expected reward, but penalize the policy for drifting too far from the reference policy in KL distance. The `beta` controls how tight that leash is.

There is a closed-form solution to that optimization problem. You can derive it with Lagrange multipliers if you want (the paper does, in an appendix), but for our purposes the answer is what matters. The optimal policy is:

```
pi*(y|x) = (1 / Z(x)) * pi_ref(y|x) * exp(r(x, y) / beta)
```

In English: the optimal policy is the reference policy multiplied by an exponential reward term, then normalized by `Z(x)` (the partition function, which makes the result sum to 1 across all possible completions). Higher reward makes a completion more likely under the optimal policy. The reference policy acts as a prior. `Z(x)` is the normalization constant that ensures we have a valid probability distribution.

Now do the trick that DPO is built on. Take the logarithm of both sides of that equation and solve for the reward:

```
r(x, y) = beta * (log pi*(y|x) - log pi_ref(y|x)) + beta * log Z(x)
```

Read that carefully. The reward is the log-ratio of the optimal policy to the reference policy, scaled by beta, plus a term that depends only on the prompt (`log Z(x)`). If we know the optimal policy and the reference policy, we know the reward, up to that prompt-dependent constant.

This is the line that took me longest to internalize. The reward exists. It is implicit in the policy. We never see it as a number, but it is there in every log-ratio. The whole rest of the chapter is just consequences of this one substitution.

That constant turns out not to matter for preference learning. Here is why. The Bradley-Terry model of human preferences says the probability that a human prefers completion `y_w` over `y_l` (winner over loser) is:

```
P(y_w > y_l | x) = sigmoid(r(x, y_w) - r(x, y_l))
```

Probabilities are determined by reward **differences**, not absolute rewards. When you substitute the reward formula above into this expression, the `beta * log Z(x)` terms cancel out. They appear in both `r(x, y_w)` and `r(x, y_l)`, and the subtraction kills them. What is left is:

```
P(y_w > y_l | x) = sigmoid(beta * (log pi(y_w|x) - log pi_ref(y_w|x)
                                   - log pi(y_l|x) + log pi_ref(y_l|x)))
```

The reward model has vanished. There is no `r` anywhere in that expression. There is only the policy and the reference policy.

The DPO loss is the negative log of this probability, averaged over the dataset:

```
L_DPO = -E [ log sigmoid(beta * (log pi(y_w|x) - log pi_ref(y_w|x)
                                 - log pi(y_l|x) + log pi_ref(y_l|x))) ]
```

That is the entire algorithm. Train the policy to maximize this log-probability of preferring chosen over rejected, where "prefer" is measured by the implicit reward (the log-ratio of policy to reference). No reward model, no rollouts, no value function. Four log-probabilities and a sigmoid.

In code, the loss looks like this:

```
def dpo_loss(policy_chosen_logps, policy_rejected_logps,
             ref_chosen_logps, ref_rejected_logps, beta=0.1):
    pi_logratios = policy_chosen_logps - policy_rejected_logps
    ref_logratios = ref_chosen_logps - ref_rejected_logps
    logits = beta * (pi_logratios - ref_logratios)
    loss = -F.logsigmoid(logits).mean()
    return loss
```

Each of the four log-probs is the sum of token-level log-probabilities for the completion under the respective model. You compute them by running a forward pass, gathering log-probabilities of the target tokens, and summing across the completion (not the prompt; the prompt is just context).

The sum-of-log-probs detail matters. The policy is being asked to assign higher probability to the entire chosen completion than the entire rejected completion, relative to the reference. Length matters subtly here: a longer completion has more tokens to sum over, so its total log-prob is naturally more negative. We address that bias when we get to SimPO.

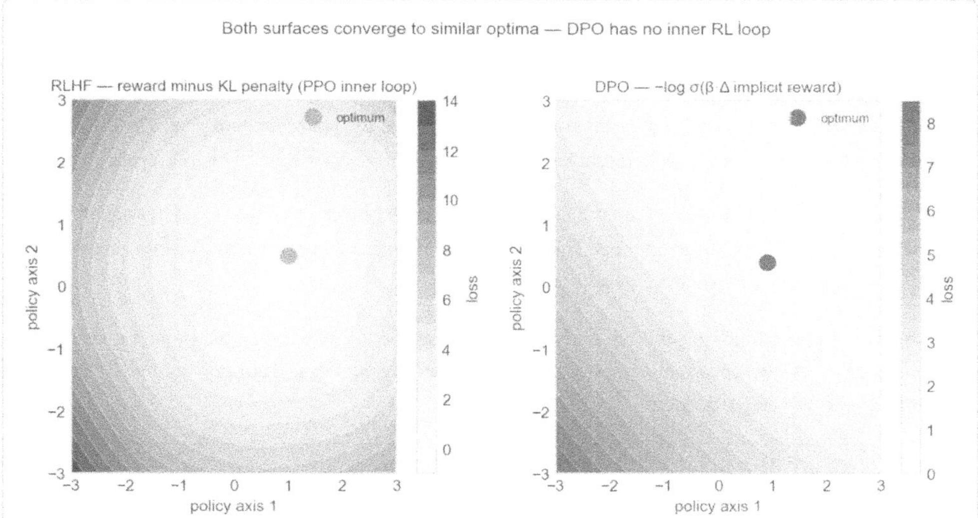

Figure 24.2. The loss surfaces of RLHF (left) and DPO (right). RLHF optimizes expected reward minus a KL penalty over the policy distribution. DPO directly minimizes the negative log-sigmoid of the implicit-reward difference on preference pairs. Both surfaces converge to similar optima, but DPO's surface is a single classification-style loss with no inner optimization loop.

Step 3: Fine-tune the SFT model with DPO

Now you train. The setup needs two copies of your SFT model: the policy (which trains) and the reference (frozen). Both start from your **Project 21: Fine-Tuning and Instruction Tuning** checkpoint, so at step zero the policy and reference give identical log-probabilities and the implicit-reward difference is zero everywhere. Training pushes the policy away from the reference in the direction that increases the implicit reward of chosen completions and decreases it for rejected ones.

The training loop is mostly the same as SFT, with one extra forward pass through the reference:

```
for batch in train_loader:
    prompt_ids, chosen_ids, rejected_ids = batch

    # Forward pass through policy (trains)
    policy_chosen_logps = compute_logprobs(policy, prompt_ids, chosen_ids)
    policy_rejected_logps = compute_logprobs(policy, prompt_ids, rejected_ids)

    # Forward pass through reference (no grad)
    with torch.no_grad():
        ref_chosen_logps = compute_logprobs(ref_model, prompt_ids, chosen_ids)
        ref_rejected_logps = compute_logprobs(ref_model, prompt_ids, rejected_ids)

    loss = dpo_loss(policy_chosen_logps, policy_rejected_logps,
                    ref_chosen_logps, ref_rejected_logps, beta=0.1)

    loss.backward()
    optimizer.step()
    optimizer.zero_grad()
```

A few practical notes drawn from where new practitioners get stuck. Including me, when I started.

The reference model uses `torch.no_grad()` because it never updates. It also benefits from `eval()` mode so dropout does not introduce randomness. Some implementations keep the reference in fp16 even when the policy is in `fp32` to save memory; this is fine because the reference is read-only.

If you do not want two full copies in memory, precompute the reference log-probabilities once for the entire dataset and store them on disk. Then during training you only need the policy forward pass. The DPO paper does this and it cuts memory usage almost in half. TRL exposes this as `precompute_ref_log_probs`.

The first UYIR router-rewrite sweep I ran ate 22 GB of GPU memory before I realized I was holding the reference in fp32. Casting it to fp16 dropped me to 14 GB without changing any of the downstream numbers. The reference is read-only. Treat it that way.

The `compute_logprobs` function sums log-probabilities over completion tokens only, not prompt tokens. The prompt is shared between chosen and rejected, so its contribution would cancel out anyway, but excluding it is cleaner and matches what TRL does:

```
def compute_logprobs(model, prompt_ids, completion_ids):
    full_ids = torch.cat([prompt_ids, completion_ids], dim=-1)
    logits = model(full_ids).logits[:, :-1]  # next-token logits
    targets = full_ids[:, 1:]
    logprobs = F.log_softmax(logits, dim=-1)
    token_logprobs = logprobs.gather(-1, targets.unsqueeze(-1)).squeeze(-1)
    # Mask out prompt tokens
    completion_mask = torch.zeros_like(targets, dtype=torch.bool)
    completion_mask[:, prompt_ids.size(-1) - 1:] = True
    return (token_logprobs * completion_mask).sum(dim=-1)
```

For a 1.3B model with beta=0.1, learning rate 5e-7, batch size 4 with 4 gradient accumulation steps, and 3 epochs on UltraFeedback, training takes roughly 6 to 12 hours on a single A100 or 24 to 48 hours on an RTX 3090. The order of magnitude is "overnight." RLHF on the same data typically takes three to five times longer because of on-policy sampling overhead.

I expected the DPO learning rate to behave like a standard SFT learning rate. It does not. 5e-7 is two orders of magnitude lower than what I would have reached for from instinct, and at 5e-5 the policy drifted off a cliff inside the first few hundred steps. That low learning rate is not a paranoia choice; it is the price of training without an explicit reward model to absorb the noise.

After training, compare win rates to the same model without DPO. Win rate is the fraction of held-out prompts on which a judge (GPT-4 or a human) prefers your model's output to the SFT baseline's. On AlpacaEval, AlpacaFarm, or MT-Bench, a well-tuned DPO run on UltraFeedback typically moves the win rate from ~35-40 percent (SFT vs SFT floats near 50 by symmetry minus length effects) up to ~60-70 percent (DPO vs SFT). Those numbers roughly match what RLHF produces on the same data.

Step 4: Compare DPO vs RLHF on the same eval

Run both your DPO model from Step 3 and your RLHF model from **Project 23: Reward Models and RLHF** on the same evaluation harness from **Project 22: Evaluation Methodology**. Use AlpacaEval 2.0 with length-controlled win rates: both methods have length biases, and you want to factor that out before comparing.

DPO and RLHF win rates typically come within a few points of each other on most benchmarks. Sometimes DPO wins by 2 to 4 points, sometimes RLHF does. The gap is small enough that you should not interpret it as one method being fundamentally better.

The interesting comparison is not the win rate. It is the cost. Record for each method: total compute time, peak GPU memory, lines of code in the training loop, and number of hyperparameters that needed tuning. DPO wins on every one of those by a wide margin. RLHF still has a role when you need richer reward signals (multi-objective rewards, online preference collection, or reward shaping that is hard to express as pairwise comparisons), but the default starting point for new alignment work has shifted heavily toward DPO since 2023.

Step 5: Implement KTO

KTO (Kahneman-Tversky Optimization, Ethayarajh et al. 2024) is the first variant worth knowing. Preference pairs are expensive: you need two completions per prompt, then a human or stronger model has to compare them. What if you had binary feedback instead, a thumbs-up or thumbs-down on a single completion? KTO trains on that.

The math change is small. Instead of two implicit rewards per example, KTO has one, compared against a reference baseline that is the expected implicit reward across the batch. The loss has two cases: if labeled "good," push the implicit reward above the baseline; if labeled "bad," push it below.

The KTO loss looks like:

```
def kto_loss(policy_logps, ref_logps, label, beta=0.1, desirable_weight=1.0,
        undesirable_weight=1.0):
    # Implicit reward
    logratio = policy_logps - ref_logps

    # Baseline: KL divergence estimate from a mismatched batch
    # (computed elsewhere; here we just use the mean as a stand-in)
    kl_baseline = compute_kl_baseline(policy, ref_model)

    if label == "desirable":
        loss = desirable_weight * (1 - torch.sigmoid(beta * (logratio - kl_baseline)))
    else:  # undesirable
        loss = undesirable_weight * (1 - torch.sigmoid(beta * (kl_baseline - logratio)))

    return loss.mean()
```

KTO borrows from prospect theory, Kahneman and Tversky's framework about how humans weight gains and losses. Humans treat losses as roughly twice as painful as equivalent gains. KTO exposes that asymmetry through the `desirable_weight` and `undesirable_weight` hyperparameters.

When is KTO worth picking over DPO? When you have lots of single-label feedback and few preference pairs. Most production systems collect single-label feedback by default; users thumbs-up or thumbs-down individual outputs, not pairs. KTO lets you train directly on the data you have.

Step 6: Implement ORPO

ORPO (Odds Ratio Preference Optimization, Hong et al. 2024) takes a different angle. The SFT stage and the preference stage are usually separate. Can they combine into one pass? The answer is yes, and it removes the need for a reference policy.

The ORPO loss has two terms. The first is the ordinary SFT loss on chosen completions, the negative log-likelihood of the chosen completion under the policy. The second is an odds-ratio term that compares chosen against rejected:

```
def orpo_loss(policy_chosen_logps, policy_rejected_logps, chosen_labels, lambda_=0.1):
    # SFT term: standard negative log-likelihood on chosen
    nll_loss = -policy_chosen_logps.mean()

    # Odds ratio term
    log_odds = (policy_chosen_logps - torch.log(1 - torch.exp(policy_chosen_logps))) \
             - (policy_rejected_logps - torch.log(1 - torch.exp(policy_rejected_logps)))
    or_loss = -F.logsigmoid(log_odds).mean()

    return nll_loss + lambda_ * or_loss
```

The odds ratio of an event with probability `p` is `p / (1 - p)`. ORPO's preference term penalizes the model when the log-odds of chosen are not larger than the log-odds of rejected. The absence of a reference policy is the key architectural difference from DPO: ORPO does not need a frozen second model.

ORPO is worth picking when you want a single training stage and minimal infrastructure. The paper reports that ORPO often outperforms DPO when starting from a base model with no SFT, because the SFT term is doing the SFT work concurrently. If you already have a high-quality SFT model, the gain is smaller and DPO is often comparable.

Step 7: Implement SimPO

SimPO (Simple Preference Optimization, Meng et al. 2024) is the most aggressive simplification. It drops the reference policy entirely and replaces it with length normalization. The reference-policy comparison in DPO mostly serves as a regularizer, and length normalization can do similar work with less infrastructure.

The SimPO loss is:

```
def simpo_loss(policy_chosen_logps, policy_rejected_logps,
               chosen_lens, rejected_lens, beta=2.0, gamma=1.0):
    # Length-normalized log-probs
    chosen_avg_logps = policy_chosen_logps / chosen_lens
    rejected_avg_logps = policy_rejected_logps / rejected_lens

    # Target reward margin
    logits = beta * (chosen_avg_logps - rejected_avg_logps) - gamma

    loss = -F.logsigmoid(logits).mean()
    return loss
```

Two design choices to notice. First, log-probabilities are divided by completion length, so the implicit reward is an **average** per-token log-prob rather than a sum. This kills the length bias that pure DPO suffers from. Second, `gamma` sets a target margin: chosen's average log-prob should exceed rejected's by at least `gamma`, not just by any positive amount. This "headroom" requirement stabilizes training.

When I first compared SimPO and DPO on the same router-rewrite preference set, I assumed SimPO would lose by a few points because dropping the reference seemed like obviously throwing away information. It won by about a point and a half. The length normalization mattered more than the reference anchor on that particular data. I would not have predicted that ahead of time, and I am still careful not to overgeneralize from one comparison.

The payoff is operational. No reference model, lower memory, faster training, simpler code. SimPO numbers are competitive with or better than DPO on many benchmarks. The cost: without a reference policy as an anchor, training can drift further from the SFT model, which is sometimes what you want and sometimes not.

Step 8: Beta parameter sensitivity sweep

Before the comparison table, run one more experiment. Sweep beta across `[0.01, 0.05, 0.1, 0.3, 1.0]`. Train each setting for the same number of steps. Evaluate win rate against the SFT baseline.

You should see a U-shaped curve, with the best win rate around 0.05 to 0.3. At very low beta (0.01) the policy drifts aggressively from the reference, learns the preferences, but degrades on other axes (coherence, factuality, formatting). At very high beta (1.0) the KL leash is tight enough that the policy barely moves.

Beta is the single most important hyperparameter in DPO, and there is no theory that tells you what it should be in advance. You discover it empirically. Bake the sweep into your DPO playbook: any time you change the dataset or base model, re-sweep beta.

This is the closest thing in alignment work to a "set it and stop thinking about it" knob, and the honest answer is that there isn't one.

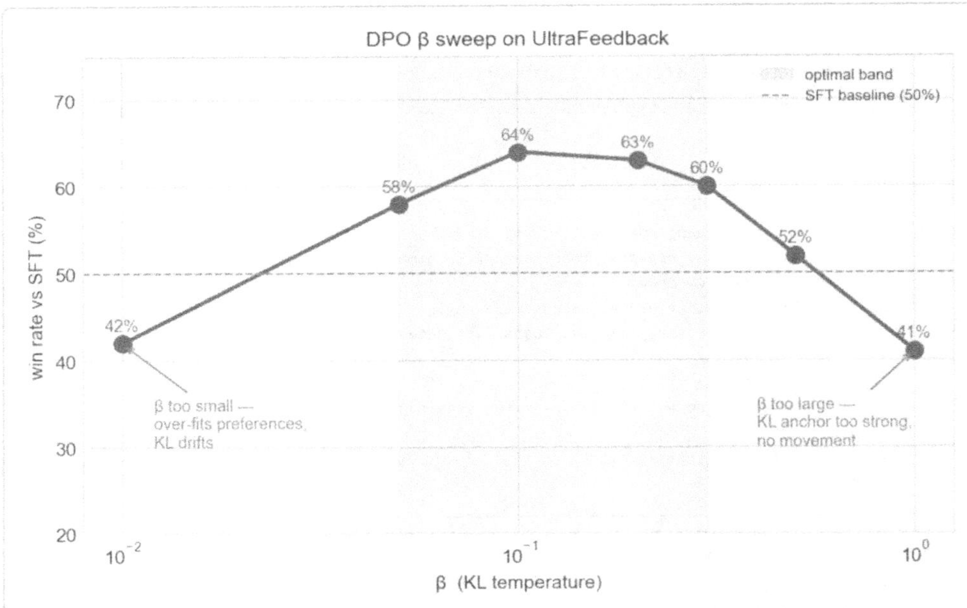

Figure 24.6. DPO win rate against the SFT baseline as a function of beta, swept across five values on the UltraFeedback dataset. The optimum sits between 0.05 and 0.3; values outside that band degrade in two opposite ways.

Step 9: Build a method comparison table

You now have DPO, KTO, ORPO, and SimPO implemented, plus a working RLHF pipeline from the previous chapter. IPO (Identity Preference Optimization, Azar et al. 2024) is worth knowing about even if you do not implement it: it replaces DPO's sigmoid with a squared loss, more stable when the dataset has many near-deterministic preferences and the sigmoid saturates.

Here is the comparison table:

Method	Data needed	Reference model	Reward model	Sampling	Key failure mode	When to pick
RLHF	Preference pairs	Yes (KL anchor)	Yes (separate)	On-policy rollouts	Reward hacking	Multi-objective rewards, online preference collection
DPO	Preference pairs	Yes	No (implicit)	No	Length bias, label noise sensitivity	Default starting point; you have pairs and an SFT model
KTO	Single labels (good/bad)	Yes	No	No	Tuning the gain/loss asymmetry weights	You have thumbs-up/down data not pairs
ORPO	Preference pairs	No	No	No	Less stable than DPO on already-aligned models	Starting from a base model with no SFT
SimPO	Preference pairs	No	No	No	Can drift further from SFT than DPO	Memory-constrained setups; length-confounded data
IPO	Preference pairs	Yes	No	No	Less expressive on noisy preference data	Datasets with many near-deterministic preferences

The table is the chapter's deliverable. The honest answer to "should we use DPO or X?" is usually "DPO, unless one of the conditions in the table applies." Knowing the table means you have a real answer instead of a vibes-based one.

Preference-optimization method comparison

method	data needed	removed vs DPO	best when
DPO	preference pairs + ref model	—	good pair data; standard baseline
KTO	single-sample good / bad labels	no pairs required	messy data with binary labels
IPO	preference pairs + ref model	no sigmoid → bounded loss	noisy / over-confident labels
ORPO	preference pairs (no ref model)	no reference model + joint SFT	compute-tight setups; skip SFT pass
SimPO	preference pairs (no ref model)	ref-free, length-normalised reward	long generations; cost-sensitive

Figure 24.3. Method comparison table: DPO, KTO, IPO, ORPO, and SimPO with what each method requires, what it removes from the standard DPO recipe, and when to pick it. The right-most column is the practical decision guide.

Step 10: Plot win rates after each method

Run each method on the same SFT base, same dataset, same number of training tokens. Evaluate against the same held-out eval suite (AlpacaEval, MT-Bench, and a held-out slice of the preference dataset). Plot as a bar chart with the SFT baseline at zero.

You should typically see all methods improve over SFT by 15 to 30 percentage points in win rate, with methods within ~5 points of each other in the middle of that band. DPO and SimPO are usually the top two, with KTO and ORPO close behind; relative order shifts with the dataset. RLHF lands in the same neighborhood at much higher compute cost.

The takeaway is not "method X is best." It is "all of these methods work, the spread is small, and the choice is dominated by your data and engineering constraints."

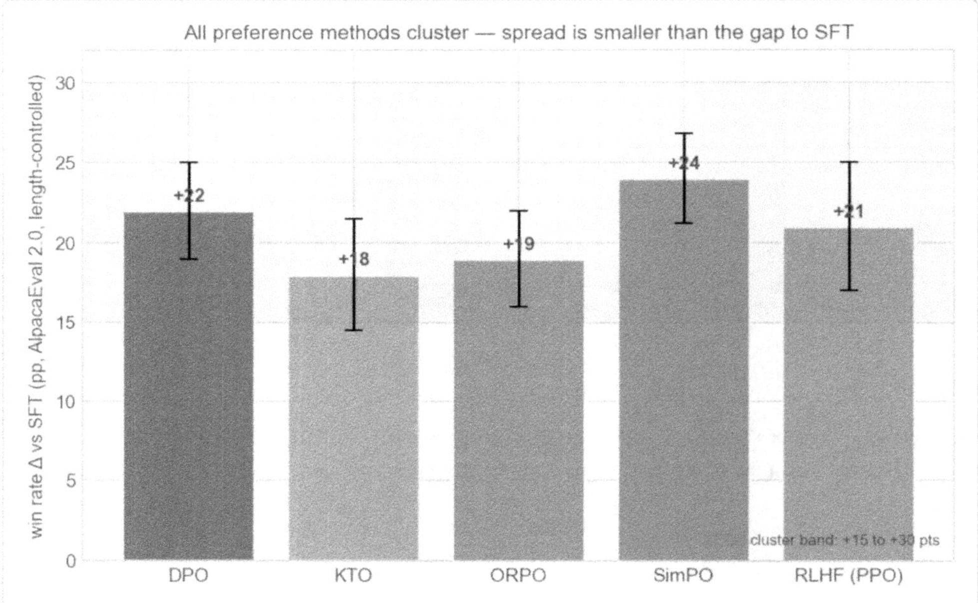

Figure 24.4. Win-rate improvement over the SFT baseline for DPO, KTO, ORPO, SimPO, and RLHF, evaluated on AlpacaEval 2.0 with length control. All methods cluster in the +15 to +30 percentage point band; the spread between them is smaller than the gap between any of them and SFT.

BREAK IT

The break is brutal in its simplicity. You are going to invert the preference labels (swap chosen and rejected for every example) and watch the model learn to prefer the previously-rejected completions with the same training stability, the same loss curve shape, and the same convergence behavior as the correct labels. The output model will be aligned to the opposite of what humans want. Training will look healthy throughout. No spike, no anomaly. The instruments cannot help: there is nothing physically wrong with the training, only with the labels, and DPO has no mechanism for second-guessing labels.

The exact change

Take your DPO training script from Step 3. Find the line where the batch is unpacked into chosen and rejected fields. Swap them:

```
# Original:
# policy_chosen_logps = compute_logprobs(policy, prompt_ids, chosen_ids)
# policy_rejected_logps = compute_logprobs(policy, prompt_ids, rejected_ids)

# Swapped:
policy_chosen_logps = compute_logprobs(policy, prompt_ids, rejected_ids)
policy_rejected_logps = compute_logprobs(policy, prompt_ids, chosen_ids)

with torch.no_grad():
    ref_chosen_logps = compute_logprobs(ref_model, prompt_ids, rejected_ids)
    ref_rejected_logps = compute_logprobs(ref_model, prompt_ids, chosen_ids)
```

Everything else stays identical. Same beta, same learning rate, same dataset, same number of epochs.

What to predict before you run

Write down three predictions before pressing go.

First, the loss curve. DPO loss is `-log sigmoid(beta * (implicit_reward_chosen - implicit_reward_rejected))`. With inverted labels, the model learns to assign higher implicit reward to the previously-rejected completions. The loss is symmetric: it does not care which side is "correct," only about pushing apart what the data labels as chosen versus rejected. The loss curve should look exactly like a correctly-labeled run, matching within noise.

Second, the win rate. Evaluated against SFT by a fair judge (GPT-4 or held-out human eval), the inverted-label model should win significantly less than 50 percent, at the same rate the correctly-labeled model loses. If correctly-labeled DPO gets 65 percent, inverted-label should get ~35 percent.

Third, no visible warning during training. Gradient norms in the same band. Same activation distributions. The implicit reward gap grows at the same rate. The training health dashboard from **Project 11: Training Debugging — Spikes, NaNs, Profiling** shows nothing unusual.

What you will observe

Run both experiments with the same seed. Plot the loss curves on the same axes. Plot the implicit reward margin (chosen minus rejected, in the policy minus the same in the reference) on the same axes.

The two curves are nearly indistinguishable. Both losses decrease smoothly. Both implicit reward margins grow at the same rate. Both reach a similar final loss. From inside the training loop, there is no way to tell which one is heading toward an aligned model and which toward an actively misaligned model.

Now run the eval. The correctly-labeled model wins ~62 percent against SFT on AlpacaEval. The inverted-label model wins ~37 percent, within a few points of the symmetric reflection. The judge can tell the difference instantly. The training loop never could.

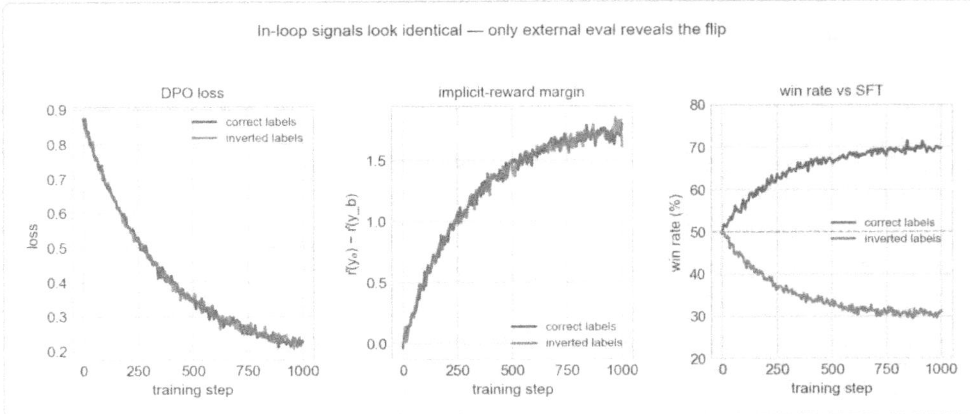

Figure 24.5. Two DPO training runs on the same data: correctly-labeled (blue) and inverted-labeled (red). The loss curves and implicit-reward trajectories are nearly identical. The evaluation win rates against the SFT baseline are mirror images of each other. There is no in-loop signal that distinguishes the two runs.

What this proves

DPO has no safety net against label noise. None. The training procedure is a pure classifier on whatever you put on the chosen side versus whatever you put on the rejected side. Right labels: aligned model. Wrong labels: confidently misaligned model with the same training stability. The mechanism is faithful in both directions.

This is structurally different from RLHF, which has at least one buffer: the reward model. A reward model trained on noisy preferences averages out some noise during its own training. The policy then learns from the smoothed signal. DPO removes that buffer.

Two consequences. First, label quality is the single most important factor in DPO success. A 10 percent label-noise rate bakes a 10 percent label-noise rate into your model's behavior. Bigger models do not fix it. More training does not fix it. Only better labels fix it.

This is the part I want you to internalize before anything else in the chapter. The first version of this section buried the point in the middle of a paragraph about hyperparameter sweeps. Reading it back, I had skimmed past my own warning. The section now sits here, in big plain sentences, because that is where it belongs.

Second, subtle label biases (length, format, sycophancy) get learned as faithfully as the intended signal. If your preference data prefers longer outputs by ~15-25 percent (which most public datasets do), your DPO model will be systematically verbose. If your clickwork rewards confident-sounding outputs even when wrong, your DPO model will be confidently wrong. The model is exactly as good as the data, and exactly as bad.

A second break worth running

Restart with a milder corruption: invert 20 percent of labels at random. This simulates real label noise: most labels correct, a fraction wrong, wrongness uncorrelated with anything else.

Training looks healthy throughout. The win rate lands between the fully-correct version and the baseline: maybe 55 percent instead of 65. The degradation is real but quiet. Without a clean reference run, you would have a hard time noticing 10 win-rate points were missing.

This is the version of label noise that exists in every real dataset. The only way to know is to measure label noise directly, by re-annotating a sample with high-quality labels and computing agreement rates.

Optional Homework

All of this runs on a single consumer GPU. The 1.3B-parameter setup with LoRA fits on 8 GB. Training one method end-to-end takes 4 to 12 hours. CPU-only readers can run a tiny version with a 125M model for 100 steps to verify the loss math; the patterns are the same, only the scale differs.

Part A: Baseline DPO

Start from your **Project 21: Fine-Tuning and Instruction Tuning** SFT checkpoint. Load UltraFeedback. Preprocess to preference pairs. Train DPO for one epoch at beta=0.1, learning rate 5e-7, batch size 4 with 4 gradient accumulation steps. Save the resulting model. Evaluate against the SFT baseline on AlpacaEval 2.0 length-controlled. Record the win rate.

Expected outcome: win rate in the 55 to 70 percent range against SFT. The exact number depends on your model size and base quality.

Part B: Beta sweep

Re-run Part A with beta values `[0.01, 0.05, 0.1, 0.3, 1.0]`. For efficiency, train for one-third of an epoch each — the relative ordering shows up quickly. Plot win rate against beta on a log-x axis.

You should see a U-shape with the peak in the 0.05-to-0.3 range. Record the beta where the curve peaks; you will use it in Part E.

Part C: The inverted-label break

Take your training script from Part A. Swap chosen and rejected fields. Train with the same hyperparameters. Evaluate against SFT.

Before you run, predict: what win rate against SFT? What will the loss curve look like? When you compare to the Part A run, what will be different?

Run. Compare. Record what surprised you. Most readers expect to see something visually wrong during training and are unsettled when they do not.

Part D: Partial-noise break

Re-run Part A with 20 percent of preference labels randomly flipped. Compare the win rate and the loss curve to the clean Part A run.

Expected: training looks healthy, win rate is meaningfully lower than Part A but still well above 50 percent. Magnitude of the gap is your dataset's sensitivity to label noise.

Part E: KTO, ORPO, SimPO

Implement each of the three remaining methods from Steps 5 through 7. Train each on the same data, same model, same total training tokens. Use beta=0.1 for KTO, lambda=0.1 for ORPO, beta=2.0 with gamma=1.0 for SimPO (paper defaults, which work as starting points). Evaluate all three plus your DPO baseline against

SFT.

Record the win rates in a single bar chart. If one method significantly outperforms the others, write down a hypothesis why; it usually correlates with a known property of the data (length bias, label noise, SFT quality).

Part F: Length-confound check

For your best DPO model from Part B, compute the average completion length on AlpacaEval prompts. Compare to the SFT baseline's average length. Then compute the win rate **after controlling for length** using AlpacaEval 2.0's length-controlled metric.

If the length-controlled win rate is more than 5 points lower than the raw win rate, your DPO model is winning partly through verbosity rather than quality. This is a known DPO failure mode, and SimPO's length normalization (Step 7) is the standard fix.

Part G: Proxy lab for CPU-only readers

Use a 125M model and a 1000-example subset of UltraFeedback. Train DPO for 100 steps. The model will not produce useful outputs at this scale, but loss should decrease and the implicit reward gap should grow. Verify `compute_logprobs` produces sensible numbers and that swapping chosen/rejected reverses the sign of the gap. The math is the same at every scale.

Questions To Answer

1. In your inverted-label run from Part C, at what point in training (if ever) could you have detected the issue from inside the training loop, looking only at the loss, gradients, and activations? If never — what does that imply about the role of label quality in alignment?
2. Your beta sweep from Part B produced a U-shape. Describe in your own words what is happening at the low-beta end of the curve, and what is happening at the high-beta end. Are the failure modes at the two ends qualitatively the same or different?
3. The implicit reward in DPO is `log pi(y|x) - log pi_ref(y|x)`, scaled by beta. What happens to this quantity, for a chosen completion, over the course of training? What happens for a rejected completion? If you plot both over time, what does the gap between them look like?
4. ORPO removes the reference policy. SimPO also removes the reference policy. They do so for different reasons. Describe each method's reason, and identify the situation in which each method's approach to "no reference" works best.
5. You have a dataset where 90 percent of preferences come from human labelers and 10 percent come from a less reliable source (e.g., a low-quality automatic judge). Should you train on the combined dataset, train only on the high-quality 90 percent, or up-weight the high-quality fraction in the loss? Justify your answer using the BREAK IT result.
6. RLHF uses a reward model as a buffer between human labels and the policy. DPO does not. Name one situation where you would prefer the buffer (and pay the engineering cost), and one situation where you would prefer to skip the buffer (and accept the label-noise sensitivity).

Go Further

Five papers to read closely, roughly in order:

Rafailov et al., "Direct Preference Optimization: Your Language Model is Secretly a Reward Model" (2023). The original DPO paper. Read Section 4 carefully; the paper does the derivation with all conditions made explicit. Pay attention to the Bradley-Terry assumptions and what happens when they break.

Ethayarajh et al., "Model Alignment as Prospect Theoretic Optimization" (2024). The KTO paper. The connection to prospect theory motivates the gain/loss asymmetry that gives KTO its hyperparameters.

Hong et al., "ORPO: Monolithic Preference Optimization without Reference Model" (2024). Short and direct. The SFT term and the preference term can share a single training pass, and you can drop the reference policy if you keep the SFT term for regularization.

Meng et al., "SimPO: Simple Preference Optimization with a Reference-Free Reward" (2024). The current state of the simplification arc. Read the Section 5 ablations to see which design choices (length normalization, target margin, reference removal) contribute how much.

Azar et al., "A General Theoretical Paradigm to Understand Learning from Human Preferences" (2024). The IPO paper. The squared-loss alternative is motivated by a careful analysis of when Bradley-Terry breaks down.

Two repositories:

HuggingFace TRL at github.com/huggingface/trl. The `DPOTrainer`, `KTOTrainer`, `ORPOTrainer`, and `CPOTrainer` classes are production implementations with sharding, gradient checkpointing, and LoRA built in. Reading the source teaches the engineering details we skipped.

Allen Institute's Open Instruct at github.com/allenai/open-instruct. A research-focused stack with DPO, KTO, and other methods, used to train the Tulu series. The data preprocessing scripts encode hard-won knowledge about preference data hygiene.

arXiv search directions: `direct preference optimization`, `RLHF KL constraint closed form`, `Bradley-Terry preference learning`, `length bias DPO`, `iterative DPO`, `online preference learning`, `RLAIF AI feedback alignment`.

One direction moving fast: iterative or online DPO, where the model generates new completions, has them scored by a judge, and trains on the resulting pairs in a loop. This brings back some RLHF structure (on-policy sampling, judge as proxy reward) while keeping the DPO loss. Pang et al.'s "Iterative Reasoning Preference Optimization" and Xu et al.'s "Some Things Are More CRINGE Than Others" are pushing this direction.

What You Now Know

The reward model in RLHF was never doing magic. It was an approximation to the closed-form expression that connects the optimal KL-constrained policy to the reference. DPO writes that closed form down explicitly, substitutes it into the Bradley-Terry preference model, and the reward model cancels out, leaving a single classification loss on preference pairs. You can derive it yourself in about ten minutes with a pencil. The trick is one of the cleaner pieces of recent machine learning theory.

You also know the cost of that simplicity. DPO has no buffer between label noise and model behavior. Whatever you label as chosen, the model learns to produce. Whatever you label as rejected, the model learns to avoid. The training loop has no instrument that can tell you whether your labels were any good. The inverted-label experiment makes this real: loss curve, gradients, and activations all look identical for a correct run and an actively misaligned run. Only an external eval can distinguish them.

You know the family now. DPO, KTO, ORPO, SimPO, IPO are five points in a small design space, and the comparison table in Step 9 tells you when to reach for each. RLHF is still in the toolbox, but it has moved from the default to a specialist tool, used when you need richer reward structure than pairwise preferences can express.

The next chapter, **Project 25: Test-Time Reasoning,** takes the aligned model and asks a different question: what if, instead of changing the weights, we change how the model spends compute at inference time? Chain-of-thought, self-consistency, tree search, and process-reward models all live in that space.

Starting Point

Original DPO paper: Rafailov et al., "Direct Preference Optimization: Your Language Model is Secretly a Reward Model" (2023), arxiv.org/abs/2305.18290. KTO paper: Ethayarajh et al., "Model Alignment as Prospect Theoretic Optimization" (2024), arxiv.org/abs/2402.01306. ORPO paper: Hong et al., "ORPO: Monolithic Preference Optimization without Reference Model" (2024), arxiv.org/abs/2403.07691. SimPO paper: Meng et al., "SimPO: Simple Preference Optimization with a Reference-Free Reward" (2024), arxiv.org/abs/2405.14734. IPO paper: Azar et al., "A General Theoretical Paradigm to Understand Learning from Human Preferences" (2024), arxiv.org/abs/2310.12036. TRL library: github.com/huggingface/trl. HH-RLHF dataset: huggingface.co/datasets/Anthropic/hh-rlhf. UltraFeedback dataset: huggingface.co/datasets/openbmb/UltraFeedback.

Project 25: Test-Time Reasoning (CoT, Self-Consistency, Best-of-N)

Hook

Your fine-tuned model from **Project 21: Fine-Tuning and Instruction Tuning** answers "what is 2 + 2?" without hesitation. Ask it a four-step grade-school word problem and it confidently produces the wrong number. Nothing about the model has changed. The same weights that nailed the easy question whiffed on the harder one. The math the model needs is sitting inside those weights. It just is not getting deployed. What compute can you spend at inference time, after training is over and the weights are frozen, to fix this?

The first time this happened to me on an ACAR-style routing experiment, I assumed the weights were broken. They were not. The model was being asked to do four steps of arithmetic with no place to write down the intermediates, and it was guessing the answer in one forward pass. That was the lesson. The chapter is about the cheapest fixes available.

The Concept

The model has the capability. It does not have the deliberation. A student who knows how to add can still mess up a long arithmetic problem if they try to do every step in their head and only write down the final number. The same student, given a piece of scratch paper and told "show your work," makes fewer mistakes. The information needed to be correct was always in their head. The scratch paper changed what they could keep track of.

A language model has the same problem. When it generates a single answer token in response to a hard problem, it is doing every intermediate calculation in one forward pass, with no way to write anything down. **Chain-of-thought**, often shortened to **CoT**, is the scratch paper. You prompt the model with something like "Let's think step by step," and the model produces a sequence of intermediate reasoning tokens before giving its final answer. The final answer is now generated with all those intermediate tokens already in its context, so it has a much easier job.

That is one move. There are three more, and each one buys accuracy by spending more inference compute.

Self-consistency is the second move. Generate not one chain of reasoning, but eight. Sample them at a temperature high enough that the chains genuinely differ — temperature 0.7 is the usual setting. Read off the final answer from each chain. Then take a **majority vote**: whichever final answer appears most often across the eight chains is the answer you keep. The intuition is that there are many wrong ways to reason about a hard problem, but most of the right ways tend to converge on the same number. If five out of eight chains say "47" and the other three say three different wrong things, "47" is your best guess.

Best-of-N, written **BoN**, is the third move. Sample N candidate answers — full chains plus final answers — and pick the best one using a **verifier**. The verifier is a separate model trained to score answers, often called a **reward model** or **outcome reward model** (**ORM**) when it scores final outcomes. An ORM takes a question and a candidate solution and produces a number: higher is better. You take the candidate with the highest ORM score and return that one. Majority vote uses agreement to pick. BoN uses an external judge.

Process reward models (**PRMs**) are the fourth move. An ORM scores the final answer. A PRM scores each individual step in the reasoning chain: was step one correct, was step two correct, and so on. With a PRM you can do something more surgical than picking among finished candidates. At each reasoning step, you sample K continuations, score each one with the PRM, keep the best step, and continue. The search is now step-level rather than candidate-level. It is more compute, but the verifier can reject bad reasoning before it derails the whole chain.

PRMs are also where my opinion on this whole space gets sharpest. I think PRM training is one of the most fragile pipelines in the test-time playbook, and the literature understates how easy it is to ship a PRM that grades the wrong feature. The BREAK IT later in the chapter exists because I have seen this failure mode in the wild, on fusion candidates where a verifier was rewarding chain length and we mistook it for rewarding correctness for almost a full day.

A useful analogy to hold all four together: a student doing a math problem. The student who shows their work catches their own arithmetic mistakes; that is CoT. Three students working in parallel and majority-voting on the final answer do better than any one of them; that is self-consistency. A teacher who reads each student's final answer and picks the best one does better still; that is BoN with an ORM. A teacher who reads each step as the student writes it and tells them when they have gone wrong does the best of all; that is a PRM doing step-level search.

There is a search-tree version of the last move that is worth naming separately. **MCTS** stands for **Monte Carlo Tree Search**, and it generalizes step-level BoN into a real tree search. From the current partial reasoning, expand several candidate next-steps, score them with the PRM, recurse into the most promising one, occasionally back up and try a different branch when scores stop improving. The math is more involved, but the structure is the same: PRM-guided search through the space of possible reasoning chains. A **rollout** in this context is one full path from the current node down to a final answer. MCTS does many rollouts and uses their outcomes to decide which branches to develop further.

All five methods (direct answering, CoT, self-consistency, BoN, step-level BoN with a PRM, and MCTS) trade inference compute for accuracy. Direct answering is the cheapest and least accurate. MCTS is the most expensive and, when the PRM is good, the most accurate. There is no free improvement here. You are paying for accuracy in GPU-seconds at inference time, the same way you paid for it in GPU-hours at training time.

I think of this as the budget version of adaptive complexity routing. Multi-model routing work I have done across single, two, and three-model ensembles was solving the same problem at a higher level: when does it pay to spend more compute on a question? Test-time reasoning is the same question one rung down, inside a single model. The answer is also the same: sometimes a lot, sometimes nothing, and you cannot tell ahead of time without measuring.

Why It Matters

Without test-time reasoning, your model is leaving capability on the table. The training run already happened. The weights already encode whatever they encode. You can either ask the model to one-shot every hard problem and live with the failure rate, or you can spend a little more compute at inference and recover accuracy you already paid for during training.

The numbers in the literature are large enough to take seriously. On GSM8K, a grade-school math benchmark, naive prompting of a mid-sized model gets you maybe 30 percent accuracy. CoT prompting on the same model jumps to 50 percent or higher with no weight changes. Self-consistency on top of CoT, with eight samples, pushes that another five to ten points. BoN with a trained verifier can push it again. The o1 family of models from OpenAI, and the public discussion around them, leans heavily on this trick. The headline "reasoning" capability is in large part a story about spending more compute per query at inference time, using methods that are direct descendants of the ones in this chapter.

There is a second reason this chapter sits where it does. The previous chapters in this book taught you to spend training compute carefully. **Project 23: Reward Models and RLHF** taught you to train a reward model on preference data. **Project 24: DPO and Preference Optimization** taught you a cheaper alternative that bakes preferences into the policy directly. Test-time reasoning re-opens the question those chapters seemed to close. Maybe you do not need to bake every behavior into the weights. Maybe some behaviors are cheaper to get by spending compute at inference time, with a smaller policy and a smaller reward model. That tradeoff is one of the live research questions of the period this book is being written in.

A third reason, less direct but worth saying. Test-time methods make capability cheaper to study. You do not need to retrain anything to see whether the model "knows" how to do a task. Crank up the inference compute, watch what happens to accuracy, and you have a much clearer answer than any single-shot eval. **Project 22: Evaluation Methodology** taught you to be careful with benchmarks; test-time reasoning is one of the reasons that care matters.

The Build

You are going to take the model from **Project 21: Fine-Tuning and Instruction Tuning**, point it at GSM8K, and walk it through five increasingly expensive inference strategies. Then you will plot accuracy against inference compute for all five so you can see the tradeoff directly. After that, you will train a deliberately broken PRM and watch BoN gleefully select wrong answers that fit the broken pattern. A single consumer GPU is enough for everything in this chapter. GSM8K is small, and the policy and reward models are both modest.

The reference implementations to keep open are Wei et al.'s CoT paper for the prompt format, Wang et al.'s self-consistency paper for the voting procedure, and the Lightman et al. PRM800K release for the PRM training pipeline.

Step 1: Direct prompting and CoT prompting on GSM8K

Load your **Project 21: Fine-Tuning and Instruction Tuning** model. Load the GSM8K test set, which contains about 1,300 grade-school math problems, each with a numeric final answer. For each problem, you will run two prompts.

The direct prompt looks like this:

```
Question: <the GSM8K question>
Answer:
```

The model generates until it produces a newline or hits a token budget. You extract the numeric answer from its output by regex (typically the last number that appears). You score it as correct if it matches the ground truth.

The CoT prompt looks like this:

```
Question: <the GSM8K question>
Let's think step by step.
```

Same generation, same extraction, except now the model produces several sentences of reasoning before the final number. Use the same temperature for both runs to keep the comparison fair. Start with temperature 0, which is greedy decoding with no randomness.

```python
def run_one(prompt, max_new_tokens=256, temperature=0.0):
    out = model.generate(prompt, max_new_tokens=max_new_tokens,
                         temperature=temperature, do_sample=temperature > 0)
    return out

def extract_answer(text):
    nums = re.findall(r"-?\d+\.?\d*", text.replace(",", ""))
    return float(nums[-1]) if nums else None
```

Run both prompts over the full test set. Record accuracy for each. The lift from direct to CoT is your first number, and on a reasonable mid-sized policy it is usually big. Fifteen to twenty percentage points is normal on GSM8K.

That gap is the entire premise of the chapter, sitting in plain sight. Same weights, same model, two prompts, and one of them is dramatically better than the other. The first time I plotted it I sat with the graph for a while. It is unsettling, in the way it should be.

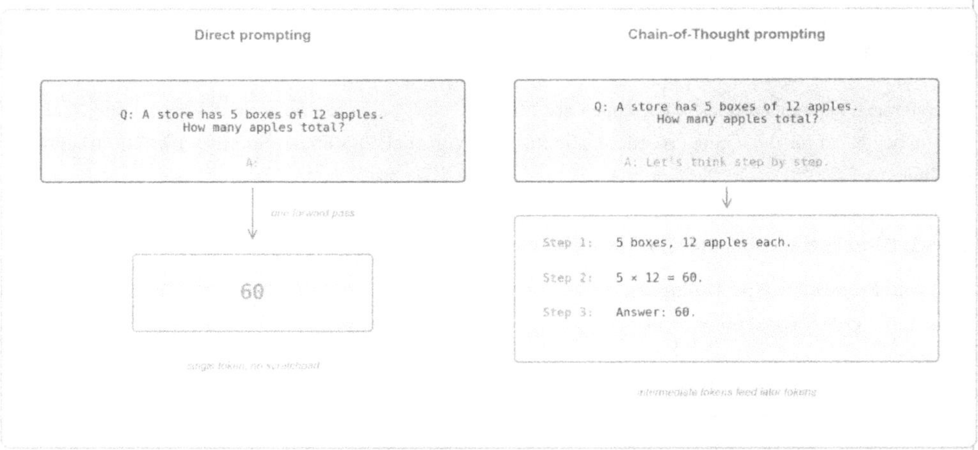

Figure 25.1. The CoT prompt format. Direct prompting collapses every step into one forward pass; CoT prompting produces an explicit chain of reasoning tokens before the final number, giving each later token access to earlier intermediate results.

A small note on extraction. The last-number regex is a cheap heuristic. It misclassifies a small fraction of answers where the model produces a number in passing after its final answer. For a quick read on the method, the regex is fine; for a published result, you would write a stricter extractor that looks for "the answer is" or similar phrases. The numbers in this chapter are tutorials, not benchmarks.

Step 2: Self-consistency

Now switch from temperature 0 to temperature 0.7 and turn sampling on. For each problem, generate N = 8 independent CoT completions. Extract the final answer from each. Take the **majority vote**: count how many times each distinct answer appears across the eight samples, and return the answer with the highest count. Ties can be broken arbitrarily; the first one to appear is fine.

```
def self_consistency(question, n=8, temperature=0.7):
    answers = []
    for _ in range(n):
        text = run_one(cot_prompt(question), temperature=temperature)
        ans = extract_answer(text)
        if ans is not None:
            answers.append(ans)
    if not answers:
        return None
    counts = Counter(answers)
    return counts.most_common(1)[0][0]
```

Re-score the test set. You should see another lift over plain CoT, typically five to ten percentage points depending on the policy and the problem distribution.

The mental model: the model is mostly right on each problem, but its mistakes are mostly uncorrelated across samples. When you sample eight chains, the right answer tends to be the most common single output, even when no individual chain dominates. If you watch the vote distributions, you will see most problems land on a single answer with five or six votes, a long tail with two or three votes, and a small fraction where the votes are scattered. That scattered fraction is where the model is genuinely confused. Majority vote does not save you there.

I expected self-consistency to be most useful on the hardest problems. It turned out to be most useful on the medium problems, the ones where the model was already correct about 50 percent of the time. On the hardest problems the chains scattered too widely and the vote was noise. That surprised me, and it changed how I report self-consistency numbers: split by difficulty, not lumped together.

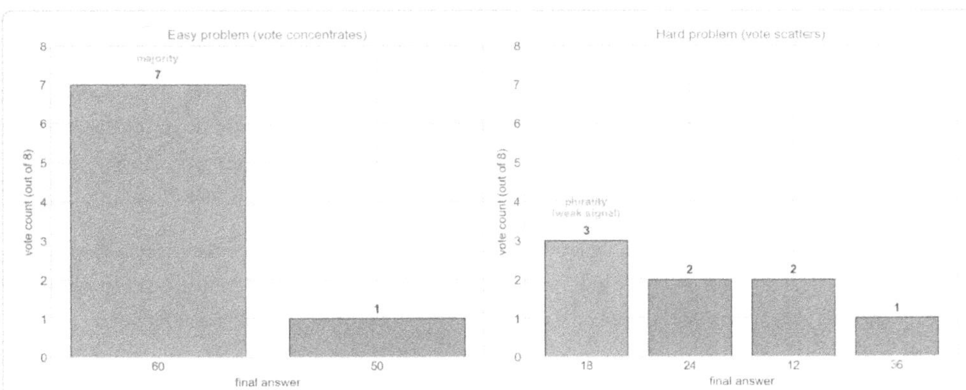

Figure 25.2. Vote distributions from self-consistency on GSM8K. Two example problems: an easy one where seven of eight chains pick the same answer, and a hard one where the votes scatter across four different numbers. Majority vote helps the first case more than the second.

One thing worth flagging now and addressing later. Self-consistency throws away information. It treats all eight chains as equally trustworthy and lets them vote. If you had a way to score the chains, to say "this one looks like real reasoning" versus "this one is gibberish", you could weight the votes, or pick the best chain instead of voting. That move is BoN, coming in two steps.

Step 3: Train a tiny outcome reward model

To do BoN, you need a verifier. The simplest verifier is an **outcome reward model**, an **ORM**, which takes a (question, full reasoning chain, final answer) tuple and outputs a number scoring how likely the answer is to be correct.

You train the ORM on labeled outcomes. The recipe is straightforward.

Generate training data from your policy. For each problem in the GSM8K training set, sample, say, sixteen CoT completions at temperature 0.7. For each completion, extract the final answer and compare to the ground truth. Label the completion 1 if its final answer was correct and 0 if not. You now have a dataset of (question, chain, label) tuples, with roughly the policy's natural success rate as the positive fraction.

A small process note. The first time I generated ORM training data I used the test set by accident, contaminating the eval downstream. I noticed when the win rate looked too good. Two days of work landed in the trash. Use the train split. Confirm twice.

Now train a small classifier on top of a base model. The base model can be a smaller copy of your policy — a few hundred million parameters is enough. You add a scalar head on top of the final hidden state at the last token position and train the whole thing with binary cross-entropy.

```
class ORM(nn.Module):
    def __init__(self, base_model):
        super().__init__()
        self.base = base_model
        self.head = nn.Linear(base_model.hidden_size, 1)

    def forward(self, input_ids, attention_mask):
        h = self.base(input_ids, attention_mask=attention_mask).last_hidden_state
        last = h[torch.arange(h.size(0)), attention_mask.sum(dim=1) - 1]
        return self.head(last).squeeze(-1)
```

Train for a few epochs over the labeled completions. Validate on a held-out set of (question, chain) pairs that the model never saw during ORM training. The metric to watch is ranking accuracy: given two completions for the same question, one correct and one incorrect, how often does the ORM score the correct one higher? You want this above 70 percent at minimum. 80 percent is healthy for a small ORM on a small base.

A subtle point: the ORM is being trained on its own policy's mistakes. If you generate the training data from a different policy than the one you will use at inference time, the ORM will be miscalibrated for the policy you actually run. Always generate ORM training data from the same checkpoint you will use at inference.

Step 4: Best-of-N with the ORM

Now you can do Best-of-N. For each test problem, generate N = 8 CoT completions at temperature 0.7. Score every completion with the ORM. Return the completion whose ORM score is highest. Extract the final answer from that completion and grade it.

```python
def bon(question, n=8, temperature=0.7):
    candidates = []
    for _ in range(n):
        text = run_one(cot_prompt(question), temperature=temperature)
        score = orm.score(question, text)
        candidates.append((score, text))
    candidates.sort(key=lambda x: -x[0])
    best = candidates[0][1]
    return extract_answer(best)
```

Re-score the test set. Compare to self-consistency.

Here is a useful question to sit with for a moment. Self-consistency and BoN both spend the same inference compute on the policy: eight sampled completions per question. The only extra cost in BoN is the ORM forward passes, which are cheap because the ORM is small. So the comparison is roughly compute-fair. Whichever wins is winning on quality, not on extra compute.

In practice, BoN tends to win when the ORM is good and lose when the ORM is bad. With a tiny, undertrained ORM, BoN can do worse than self-consistency because the ORM is picking on the wrong features (superficial fluency, length, formatting) instead of correctness. Self-consistency cannot be fooled that way because it does not use a learned scorer. It just counts. The ORM has to earn its keep.

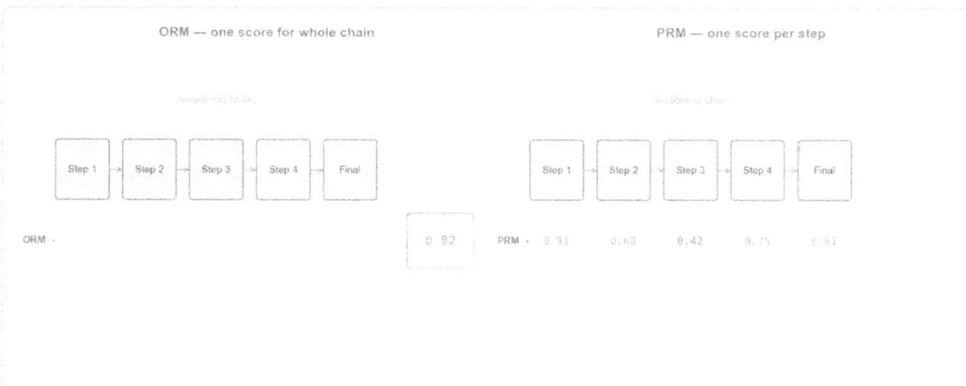

Figure 25.3. ORM vs PRM scoring. The ORM reads the whole chain and produces one number. The PRM reads each step and produces a number per step. The same chain receives one score from the ORM and a vector of scores from the PRM.

An exact-answer verifier, and its training-time twin

The ORM you just built is a learned verifier. It is a network trained to guess which chain is correct. There is another kind, and it needs no training at all. When you already know the target answer, check a candidate against it directly. Strip its final number of formatting, compare to the gold answer, return pass or fail. That is a programmatic verifier. It is the signal behind RLVR.

RLVR, Reinforcement Learning with Verifiable Rewards, gets its full treatment in **Project 23: Reward Models and RLHF**: the reward comes from an external automatic checker rather than a learned reward model, and it applies only to tasks with checkable answers, an exact-answer match in math or unit tests in code. Hold the two chapters side by side. Here the checker SCORES candidates after the weights are frozen. In RLVR the same kind of checker TRAINS the policy, its pass or fail becoming the reward. One signal, two jobs. They split by when you apply it. The 2026 reasoning-RL survey literature reports RLVR-style training as common practice (reported), on tasks with checkable answers.

Two 2026 trends sit on that link. Inference-time scaling means spending extra compute at answer time. It became a mainstream lever across providers, reported as a trend rather than a measured law. The training side got cheaper too. A 2026 MIT-led method (TLT) reports over a 1.7x end-to-end speedup in RL training of reasoning models, run losslessly so accuracy is preserved, measured against prior RL training systems (reported). The mechanism is speculative decoding inside the RL rollouts, where a small draft model proposes tokens the large model verifies. Not distillation into a smaller student.

Now the lab. Swap the learned ORM for this oracle verifier and watch accuracy climb with N.

```
def normalize(x):
    s = str(x).replace("$", "").replace(",", "").strip()
    try:
        return float(s)
    except ValueError:
        return None

def verify(candidate_text, gold):
    pred = normalize(extract_answer(candidate_text))
    target = normalize(gold)
    return pred is not None and target is not None and pred == target

def bon_oracle(question, gold, n=8, temperature=0.7):
    for _ in range(n):
        text = run_one(cot_prompt(question), temperature=temperature)
        if verify(text, gold):
            return True          # at least one sample is verified-correct
    return False
```

Trace one problem by hand before scaling up. Take "A pack holds 6 pens; how many pens are in 7 packs?" with gold answer "42". Here `extract_answer` pulls the last number out of a candidate's text, and `normalize` turns a messy string into a plain float. Three candidates make the behavior concrete:

```
gold = "42"
verify("... so 6 * 7 = 42",    "42") → normalize("42")     = 42.0 → True   (reward 1.0)
verify("the answer is $42.00", "42") → normalize("$42.00")= 42.0 → True   (reward 1.0)
verify("i think it is 41",     "42") → normalize("41")     = 41.0 → False  (reward 0.0)
```

The middle line is the one that matters. A raw string match would test "$42.00" == "42" and return False, throwing away a correct answer. `normalize` strips the currency sign and the trailing zeros first, so both sides reduce to 42.0 and the check passes. The last line is a genuine miss: the policy computed 41, and no formatting trick should rescue it.

Now scale up. Run `bon_oracle` at N = 1, 2, 4, 8, 16 over a handful of GSM8K-style word problems. The direction is the point. Accuracy rises with N, then flattens. Each extra sample is a fresh chance for one chain to land the right number, and those chances pay less the more you already have. This is the pass@N ceiling, the accuracy a perfect scorer would reach when any single correct sample counts, so it sits above your learned-ORM result rather than level with it.

The failure mode lives in the verifier, not the policy. Drop `normalize` and score with a raw string match. Now every "$42.00" or "42 pens" fails against the gold "42", so correct answers get discarded and measured accuracy reads far too low. The fix is the normalization shown above. Strip currency signs, commas, and trailing zeros on both sides before comparing. A verifier too strict on formatting punishes the policy for being right in the wrong font.

With a clean verifier in hand, the next step scores each step of the reasoning, not just the whole answer.

Step 5: Train a tiny PRM

The PRM is the verifier upgrade. Instead of scoring whole chains, it scores each step.

The first question is where the step labels come from, because labeling every step of every chain by hand is too expensive at scale. There are two common procedures.

The hand-labeled procedure is what the PRM800K release did at large scale: human annotators read each step of each chain and marked it as correct, incorrect, or neutral. This is high quality but expensive. For a tutorial you can hand-label a few hundred chains yourself and call it done.

The rollout-based procedure, often called Math-Shepherd or process-rollout, is the automatic version. For each step in a candidate chain, take the partial chain up to that step and run M rollouts from there, sampling continuations at temperature 0.7 with your policy. Compute the fraction of those rollouts that arrive at a correct final answer. If that fraction is high, the step is "good." If it is low, the step is "bad." Each step gets a soft label between 0 and 1.

```
def rollout_label(question, prefix, m=8):
    correct = 0
    for _ in range(m):
        continuation = run_one(prefix, temperature=0.7)
        ans = extract_answer(prefix + continuation)
        if ans == ground_truth[question]:
            correct += 1
    return correct / m
```

Now you have a dataset of (question, prefix-up-to-step-k, soft-label) tuples. Train a small classifier on top of a base model, with one regression head per step position. Loss is mean squared error against the soft labels.

The PRM is more expensive to train than the ORM because the rollout procedure costs roughly M times more policy generations than ORM labeling did. On a single GPU, you should expect to spend a few hours generating PRM training data even for a modest problem count.

Step 6: Step-level Best-of-N with the PRM

With the PRM trained, you can do step-level search. The procedure:

1. Start with the prompt and an empty reasoning chain.
2. Sample K candidate next-steps from the policy at temperature 0.7. A "step" here is one sentence or one line, depending on how you delimit — for GSM8K, a newline-separated chunk is a fine unit.
3. Score each candidate step with the PRM.
4. Keep the highest-scoring step. Append it to the chain.
5. Repeat from step 2 until the chain produces a final answer or hits a step budget.

```python
def step_level_bon(question, k=4, max_steps=12):
    chain = ""
    for _ in range(max_steps):
        candidates = []
        for _ in range(k):
            step = sample_step(question, chain, temperature=0.7)
            score = prm.score(question, chain + step)
            candidates.append((score, step))
        candidates.sort(key=lambda x: -x[0])
        best_step = candidates[0][1]
        chain += best_step
        if has_final_answer(chain):
            break
    return extract_answer(chain)
```

Re-score the test set. The compute cost here is K times more policy samples per step, times the average number of steps per chain. On GSM8K with K = 4 and an average chain length of six steps, you are spending 24 policy samples per question, roughly three times the self-consistency budget. The accuracy lift over BoN, when the PRM is good, can be another few percentage points. Meaningful, but smaller than the jump from CoT to self-consistency.

Step-level search has a property worth noticing. The PRM does not need to be very accurate to help, as long as it is decisively better than random at the step level. Even modest signal (say, 60 percent accuracy at distinguishing good steps from bad ones) compounds over a six-step chain into a big effect, because the chance of stumbling through six bad-but-fluent steps without correction goes down with every filter applied.

Step 7: A tiny MCTS over reasoning steps

The full MCTS algorithm is more than this chapter wants to spend time on, but the skeleton is small enough to show.

You maintain a tree where each node is a partial reasoning chain. From each node, you can expand by sampling more candidate next-steps from the policy. Each expansion is scored by the PRM. You also keep a visit count and a running average reward per node, where the reward is the outcome of a full rollout from that node.

A single MCTS iteration looks like this. From the root, walk down the tree by selecting the child with the best upper-confidence-bound score, a combination of the average reward and a term that rewards under-explored nodes. When you hit a leaf, expand it by sampling more candidate next-steps. Run a rollout from one of the new children all the way to a final answer. Use the final-answer correctness as the reward signal. Walk back up the tree updating visit counts and average rewards along the path.

```python
def mcts(question, n_iter=32, k_expand=4):
    root = Node(prompt=cot_prompt(question))
    for _ in range(n_iter):
        node = select(root)
        children = expand(node, k=k_expand)
        for child in children:
            rollout_value = rollout_and_grade(child)
            backup(child, rollout_value)
    return best_leaf_answer(root)
```

Run MCTS with, say, 32 iterations per problem. This is the most expensive method in the chapter, both in policy samples and in PRM calls. Score the test set and add the point to your accuracy-vs-compute plot.

The first MCTS run I left going overnight ate about 18 GPU-hours for what turned out to be a 2-point accuracy lift over step-level BoN. The wall-clock surprise was bigger than the numerical one. Tree search at inference time is fundamentally expensive in a way that intuition from training-time costs does not prepare you for.

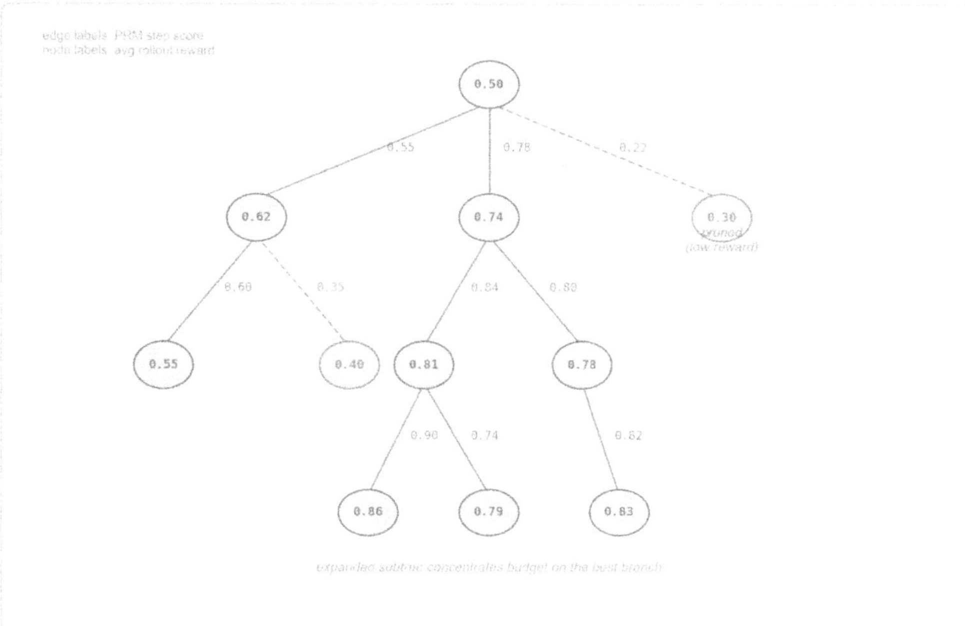

Figure 25.4. An MCTS expansion tree on a single GSM8K problem. The root is the prompt. Each node is a partial chain. Numbers on edges are PRM scores; numbers in nodes are average rollout rewards. Branches with low average reward are not expanded further; the algorithm concentrates its budget on promising paths.

A practical warning. MCTS is fragile to PRM quality. The selection step relies on the PRM to tell good steps from bad; if the PRM is noisy, the tree wastes its budget on paths that look good to the PRM but are not actually good. The next section's BREAK IT is exactly this failure mode, with the PRM deliberately broken so you can see it happen.

Step 8: Plot accuracy vs inference compute

You now have five numbers (direct, CoT, self-consistency, BoN with ORM, step-level BoN with PRM), plus MCTS for six. For each, you also know how much inference compute it spent, measured in policy forward passes per question. Make one plot.

X-axis: inference compute per question, on a log scale. Use policy forward-pass equivalents as the unit, with ORM and PRM passes folded in at their actual cost. Direct prompting is one unit. CoT is one unit. Self-consistency is N units, so eight. BoN with ORM is eight policy passes plus eight ORM scores, which is roughly eight to ten total depending on ORM size. Step-level BoN is K times steps, so 24 or so. MCTS at 32 iterations with K = 4 expansion and full rollouts is well over a hundred.

Y-axis: accuracy on the GSM8K test set.

You should see a clear monotone trend: more compute, more accuracy, with diminishing returns past the cheap methods. The shape of the curve matters more than the exact numbers. A roughly log-linear lift is normal — every doubling of compute buys a few more accuracy points, until the policy's actual capability ceiling starts to bind and additional compute stops helping.

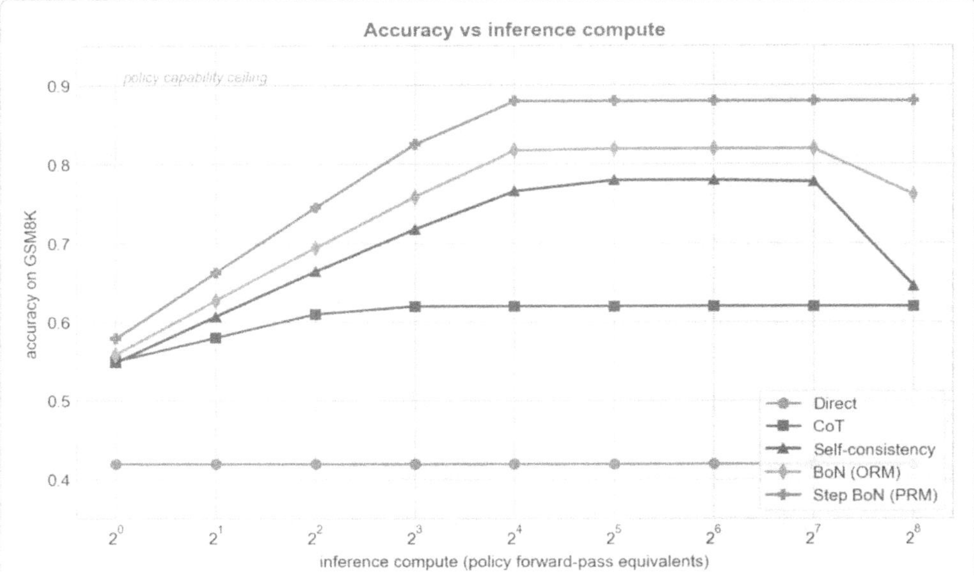

Figure 25.5. Accuracy vs inference compute for all five methods on GSM8K. The x-axis is in policy forward-pass equivalents on a log scale; the y-axis is accuracy. The curve is roughly log-linear in the middle and flattens at the high end as the policy's underlying capability becomes the bottleneck.

This curve is the single most useful artifact in the chapter. When someone asks "how good is this model at math," the honest answer is not one number. It is this curve. The model has different effective capability at different inference budgets. Test-time reasoning makes capability a function of compute, not a single point.

I will commit to a stronger version of this opinion. Any benchmark number that is reported without a compute axis is missing the central fact about modern reasoning models. We are past the point where one accuracy number tells you what the system is capable of.

BREAK IT

The break for this chapter is a verifier failure, not a policy failure. You will train the PRM on deliberately corrupted labels (chains truncated to half their original length) and then run Best-of-N with the broken PRM. The promise is that you will watch BoN do worse than the direct-answer baseline, not because the policy got worse, but because the verifier is now confidently rewarding the wrong thing.

The exact change

Take your PRM training data from Step 5. Before training, run one transformation over every chain in the dataset: keep only the first half of the steps, discard the rest, and re-label using rollouts from the truncated prefix. The labels will be different now, because rollouts from a halfway prefix end at different places than rollouts from a complete prefix did.

```
def truncate_chain(chain):
    steps = chain.split("\n")
    keep = max(1, len(steps) // 2)
    return "\n".join(steps[:keep])

broken_data = [(q, truncate_chain(c), rollout_label(q, truncate_chain(c)))
               for q, c, _ in original_data]
```

Train the PRM on this corrupted dataset using the same hyperparameters as the clean PRM in Step 5. Save the broken PRM under a different name. Do not touch the policy. Do not touch the ORM.

Now re-run step-level BoN from Step 6, but swap in the broken PRM. Run on the full GSM8K test set. Plot accuracy vs compute again, this time with both the clean-PRM run and the broken-PRM run on the same axes.

What to watch for, specifically

You are looking for three signals.

First, the broken PRM will assign high scores to short chains. The training data only ever saw chains truncated at the halfway mark, so the PRM has learned that "chains that look like they end here" are good. Long, fully developed chains look unfamiliar and score low. When step-level BoN runs, it stops expanding chains earlier than it should because the PRM is rewarding early termination.

Second, the final answers extracted from those short chains will be wrong more often than the policy's natural one-shot accuracy. The model is being forced to commit to an answer before it has done the arithmetic. The truncation pattern the PRM learned does not correspond to "the chain has reached the answer"; it corresponds to "the chain is at the halfway mark." Halfway through grade-school math is a bad place to commit to an answer.

Third, and this is the part that should change how you think about reward models forever, the broken PRM is confident. Its scores are high. The training loss converged. By every signal you would normally trust, the PRM is doing its job. The only way to catch the failure is to evaluate end-to-end accuracy on the policy plus PRM system, not the PRM alone.

The expected trajectory in numbers: clean-PRM step-level BoN sits roughly at the top of the accuracy-vs-compute plot. Broken-PRM step-level BoN drops below the direct-answer baseline. The gap is large enough that no honest plotting can hide it.

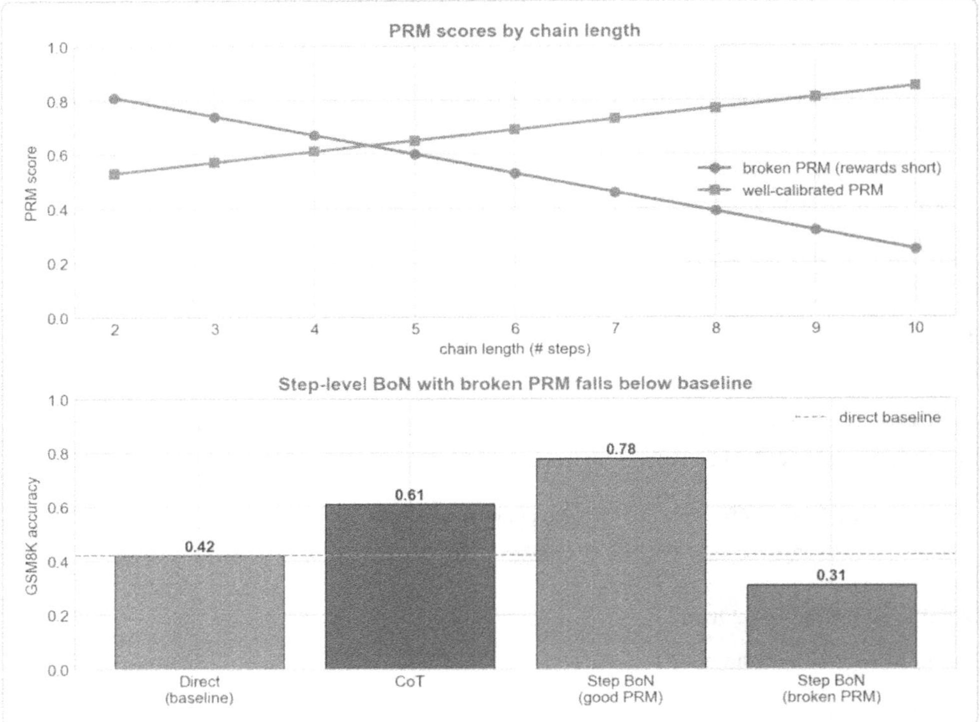

Figure 25.6. The broken-PRM failure mode. Top: the broken PRM assigns high scores to chains truncated at the halfway mark and low scores to fully developed chains. Bottom: step-level BoN with the broken PRM falls below the direct-answer baseline on GSM8K, because the policy is being stopped before it finishes reasoning.

What this proves

The reward model is not a neutral judge. It is a model that was trained on something, and whatever it was trained on is what it will reward. If the training labels reward brevity, the policy will be steered toward brevity. If the labels reward verbosity, the policy will be steered toward verbosity. The PRM does not know what the goal of the system is. It only knows what its labels said.

The general lesson is older than this chapter and bigger than this experiment. It is sometimes called **reward hacking** or **Goodhart's law in policies**: when a model is optimized against a proxy, the proxy stops being a good measure of the thing it was a proxy for. The previous chapters touched this. **Project 23: Reward Models and RLHF** trained a reward model and warned that the policy will eventually find ways to game it. **Project 11: Training Debugging: Spikes, NaNs, Profiling** taught you to read instruments instead of just the loss curve. This chapter shows the same disease in a different organ. Whenever you place a learned verifier in the inference loop, you have a new attack surface for the policy to exploit, and your only defense is end-to-end evaluation that does not go through the verifier you are worried about.

There is a smaller, more practical lesson. Always evaluate the system, not the part. PRM accuracy on a held-out set of step labels can be 90 percent and the system can still degrade catastrophically when the policy is plugged in, because the PRM's mistakes are correlated with patterns the policy can produce. The held-out validation distribution is different from the distribution the deployed policy actually visits. Validation accuracy is necessary but not sufficient.

Optional Homework

Everything in this chapter runs on a single consumer GPU. The full GSM8K test set is around 1,300 problems, which at typical inference speeds takes a few minutes for direct prompting, half an hour for self-consistency with N = 8, and an hour or two for MCTS with 32 iterations. The PRM data generation is the slow part, especially the rollout-based labeling. Budget a few hours on a 12 to 24 GB card for a modest training set.

Part A: Baseline curve

Run Steps 1, 2, 4, 6, and 7 with the clean PRM and the clean ORM. Produce the accuracy-vs-compute plot from Step 8. Save it. This is your reference picture, the same way the boring dashboard in **Project 11: Training Debugging: Spikes, NaNs, Profiling** was the reference for healthy training.

Part B: Vary N for self-consistency

Re-run self-consistency at N = 1, 2, 4, 8, 16, 32. Plot accuracy as a function of N on a log x-axis. You should see a roughly log-linear lift that flattens around N = 16 or 32. Record where the curve starts to bend. That bend tells you where additional self-consistency sampling stops paying for itself on your policy.

Part C: Vary K for step-level BoN

Re-run step-level BoN at K = 1, 2, 4, 8. K = 1 is just greedy CoT. K = 4 is your baseline from Step 6. K = 8 is double the compute. Record accuracy and inference time for each. Compare to self-consistency at matched total compute, and decide which method is better at each compute level for your policy. There is no universal answer. It depends on PRM quality and policy quality.

Part D: Bad ORM

Train an ORM on a much smaller dataset than the one you used in Step 3, say 200 labeled chains instead of 2,000. Use it for Best-of-N and compare to self-consistency at matched compute. The undertrained ORM will likely do worse than self-consistency. Record by how much. This is the cheap version of the BREAK IT experiment. It teaches the same lesson without retraining a PRM.

Part E: The truncation break

Implement the BREAK IT experiment fully. Generate truncated PRM training data, retrain, re-run step-level BoN, plot. Before you run, write down two predictions:

1. By how many percentage points will the broken PRM drop accuracy below the direct-answer baseline?
2. What will the average chain length be under the broken PRM compared to under the clean PRM?

Run, compare, and write down what surprised you. Readers usually overestimate how big the accuracy drop will be and underestimate how short the chains become. Both effects are real.

Part F: A proxy for limited hardware

If you cannot afford the full PRM training, do the experiment with the ORM instead. Train an ORM on truncated chains (keep only the first half of each chain, label by rollout) and run BoN with the broken ORM. The failure mode is similar in shape, smaller in magnitude. The lesson lands either way.

Questions To Answer

1. In your accuracy-vs-compute plot, which method gave you the biggest lift per unit of additional compute? Which gave you the smallest? Did the answer match your intuition from the chapter?
2. When you ran self-consistency at N = 32, did the lift over N = 8 justify the four-times-larger compute cost? At what value of N did the curve clearly flatten on your policy?
3. In the broken-PRM run, what was the average length of the chains the system produced compared to the clean-PRM run? How does that length compare to the average length in the original PRM training data?
4. What does the broken-PRM result say about how you should evaluate a reward model before deploying it? Name at least two evaluation metrics you would now insist on before trusting an ORM or PRM at inference time.
5. After this chapter, when someone asks you "how good is this model at math," what is your honest answer? Hint: it is not a single number.
6. What tradeoff is test-time reasoning protecting? Be specific. What does it spend, and what does it preserve?

Go Further

Two repos to study.

The first is OpenAI's PRM800K release, the dataset and accompanying code from the "Let's Verify Step by Step" paper. It contains roughly 800,000 step-level labels on math reasoning chains, generated by a mix of automatic procedures and human annotation. Reading the dataset format is the fastest way to understand what a PRM training corpus actually looks like at scale.

The second is the Hugging Face TRL library, which has reference implementations of self-consistency, Best-of-N sampling, and several reward-model training recipes. Read `trl/trainer/best_of_n_sampler.py` and trace exactly how the candidates are sampled, scored, and selected. It is a clean version of what you built in Step 4.

Four papers worth reading after this chapter.

Wei et al., "Chain-of-Thought Prompting Elicits Reasoning in Large Language Models" (2022), is the paper that introduced CoT prompting as a recognized phenomenon. Read it for the prompt format and the early empirical results on GSM8K and MultiArith.

Wang et al., "Self-Consistency Improves Chain of Thought Reasoning in Language Models" (2022), introduces majority-vote self-consistency. The interesting parts of the paper are the analyses of when self-consistency helps most (namely, on problems where the model is mostly right but inconsistent) and where it does not help.

Lightman et al., "Let's Verify Step by Step" (2023), is the PRM paper. Read it for the PRM800K release, the comparison of ORM versus PRM on MATH, and the analysis of how PRM signal compounds across steps.

The public discussion around OpenAI's o1 system card (2024) is a useful contemporary reference. The system card itself is brief; the value is in the surrounding technical commentary, which makes clear that o1's headline gains come from inference-time search using methods that are direct descendants of CoT, self-consistency, and step-level verification.

arXiv search directions for further reading: `chain of thought reasoning`, `process reward model`, `step-level reward model`, `self-consistency sampling language model`, `Monte Carlo tree search language model`, `inference time scaling reasoning`, `verifier guided decoding`.

One technique worth investigating immediately after the homework is **speculative decoding** as applied to test-time search, which uses a small draft model to propose multiple candidate steps cheaply and a larger model to verify them. The compute savings can be substantial when the draft model is well-matched to the larger one. The Leviathan et al. speculative-decoding paper is the natural entry point.

What You Now Know

Test-time reasoning is the trick of spending inference compute to recover capability that the policy already has but cannot deploy in a single forward pass. CoT gives the model a piece of scratch paper to write intermediate steps on. Self-consistency samples many chains and majority-votes on the final answer. Best-of-N samples many candidates and picks the best one with a learned verifier. A PRM scores each step rather than each candidate, enabling step-level search. MCTS turns the same idea into a real tree search over reasoning

You also know what a broken verifier looks like in the field, not in a textbook. A PRM trained on the wrong labels will reward the wrong patterns, and the policy paired with that PRM will produce confidently shorter, confidently wronger answers than the policy on its own. Reward models in the inference loop are an attack surface, and the only defense that works is end-to-end evaluation of the system you actually deploy, on the distribution you actually expect to see.

The accuracy of a language model on a benchmark is not a single number. It is a curve in inference compute. Once you have drawn that curve, you cannot un-see it, and the question "how good is this model" gets a more honest answer.

Starting Point

Reference papers and code: Wei et al., "Chain-of-Thought Prompting Elicits Reasoning in Large Language Models" (2022); Wang et al., "Self-Consistency Improves Chain of Thought Reasoning in Language Models" (2022); Lightman et al., "Let's Verify Step by Step" (2023), with the PRM800K release at github.com/openai/prm800k. The OpenAI o1 system card and surrounding public discussion are the contemporary reference. TRL Best-of-N reference: github.com/huggingface/trl. GSM8K dataset: github.com/openai/grade-school-math.

Project 26: Tool Use and Function Calling

Hook

Your **Project 21: Fine-Tuning and Instruction Tuning** model can write a poem about the weather. It cannot tell you the actual weather. Ask it for today's temperature in Toronto and it will cheerfully invent a number that sounds plausible, because inventing tokens is the only thing it knows how to do. A language model produces text. The weather lives behind an HTTP request. There is no part of the model that can reach out and call a Python function, wait for a result, and read the response. How does a thing that only emits tokens get to invoke a function and use what comes back?

The Concept

Imagine a chess player who is not allowed to touch the pieces. The player sits on one side of a screen and calls out moves to a referee. "Knight to f3." The referee picks up the piece, moves it, then shows the player the new board. The player thinks again, calls out the next move, and the referee executes. The player is the planner. The referee is the executor. Neither one does the other's job. The interesting trick is the protocol: what the player is allowed to say, and what the referee is allowed to do in response.

Tool use is that protocol applied to a language model. The model is the player. A small runtime sitting around the model is the referee. The model is trained to emit a specific format when it wants something done, a JSON blob naming a function and its arguments. The runtime watches the model's output, recognizes the format, parses the JSON, calls the function the model named, captures the result, and feeds the result back to the model as a new message. The model then resumes generating.

That is the whole design. A language model that can use tools is a language model with two new behaviors layered on top: it has learned to emit structured calls, and it lives inside a loop that catches those calls and acts on them.

The first time I wired one of these loops up on a multi-model routing project, I assumed the hard part would be the model. It was not. The hard part was the referee. Most of the bugs I shipped lived in fifteen lines of dispatch code that I had written in a hurry because I assumed the model would do the heavy lifting.

There are two pieces worth naming carefully.

The first is the **tool schema**, a contract describing the tools that exist. Each tool has a name (a string), a description (what the tool does, written for the model to read), and a parameter spec (which arguments the tool takes, of which types). The schema is shown to the model in its system prompt or its training data. It tells the model what verbs it can use. Without the schema, the model has no idea what tools exist or how to call them, and it will either ignore tools entirely or invent calls to functions that do not exist.

The second is the **agent loop**, the runtime cycle that turns a single tool call into a multi-step interaction. The canonical version of this loop is called **ReAct**, short for Reason and Act, from a 2023 paper by Yao and colleagues. ReAct frames each step as three pieces: a **Thought** (the model's reasoning, written as plain text), an **Action** (the structured tool call, written as JSON), and an **Observation** (the result returned by the runtime after dispatching the action). The model emits Thought + Action. The runtime executes the action, captures the result, and inserts it back into the conversation as an Observation. The model reads the new context and emits the next Thought + Action. Repeat until the model emits a final answer instead of an action.

The shorthand for the loop is: think, act, observe, repeat. Three verbs and a cycle. That is the entire agent primitive that every modern tool-using language model is built on.

I drew this loop on a whiteboard at least a dozen times before I trusted it. The drawing always looked too simple to be doing real work. It is.

A small clarification, because the vocabulary is overloaded. "Function calling" and "tool use" are usually the same thing wearing different marketing hats. OpenAI's API calls it function calling. Anthropic's API calls it tool use. The mechanism is identical: a structured schema, a JSON call, a runtime that dispatches, a returned observation. The phrase "agent" gets used too, and it adds one more concept on top. An agent is usually a tool-using model running inside a loop with some goal beyond a single turn. The loop in this chapter is the agent primitive. Build it once, and the larger agent systems become recognizable.

One more naming convention worth pinning down. The verb that turns a parsed JSON call into an actual function invocation is called **dispatch**. The runtime looks at the name field, finds the matching function in its registry, calls it with the parameters from the JSON, and captures the return value. Dispatch is the moment where text becomes execution. Everything before dispatch is the model talking. Everything after is the runtime acting on what it heard.

That sentence sounds harmless. It is not. The boundary between "model talking" and "runtime acting" is exactly where most of the production bugs in tool use live, because the model produces text that looks like a valid call and the runtime trusts the text. If the model invents a tool name that does not exist, the runtime has to catch it. If the model emits a string where the schema expects an integer, the runtime has to catch it. The dispatch step is the only layer that has both a schema and a Python exception system. The model has neither

Why It Matters

Without tool use, a language model is a sealed room. It knows whatever was in its training data, and it cannot learn anything new during a conversation. Ask it the current price of a stock and it will guess. Ask it to compute 8347 times 91256 and it will be wrong, because token-by-token arithmetic is not what the model is for. Ask it to read a file on your laptop and it has no path to that file. The model's knowledge is frozen at training time, and its capabilities are bounded by what it can compute inside a forward pass.

With tool use, the room has windows. The model still cannot itself read a file or call an API, but it can ask the runtime to do those things on its behalf. The model becomes the planner of a system whose hands are made of other people's code. Python functions, HTTP clients, database drivers, shell commands. The shape of what a language model can do widens by an order of magnitude, and most of the widening comes from a few hundred lines of glue code wrapped around the model.

That asymmetry is the part that took me longest to internalize. A model adds an unbounded body of capabilities not by getting smarter, but by getting permission to ask other software to act on its behalf.

There is a second reason this chapter sits where it does. **Project 28: Retrieval-Augmented Generation** is going to build a tool-using model whose main tool is a vector search call. Retrieval is a special case of tool use. The model calls a `search(query: str)` function, the runtime executes the search, the runtime returns the top-k documents as the observation, and the model uses those documents in its next response. If you understand the loop in this chapter, retrieval becomes a single new tool added to the registry instead of a new mechanism. The chapter after will not re-explain ReAct. It will reach back to this chapter and say "the same loop, with one tool that happens to be a vector store."

There is also a third reason that becomes obvious the first time you ship a tool-using model into production. The model fails. The tool fails. The JSON is malformed. The arguments are the wrong type. The function raises an exception. None of these failures crash the program. They all become text inside the conversation. The model receives the broken text and tries to recover, and the recovery is sometimes wrong in expensive ways. Knowing what the loop looks like is the difference between a system you can debug and a system that produces bizarre outputs you cannot trace back to a cause.

On an edge deployment I worked on for community health workers, running a small model on Android against a structured protocol engine, we had a hard constraint that the structured output had to match the protocol schema exactly. There was no human in the loop who could catch a wrong field. The first version of that schema-enforcement layer was three lines of regex. The version we shipped was about four hundred lines of validation, with explicit named errors for every kind of malformed call. Almost none of those four hundred lines existed because we predicted the failure. They existed because we watched the failure happen and added a check after the fact.

The Build

You are going to take the model from **Project 21: Fine-Tuning and Instruction Tuning** and add three things to it. A tool schema describing two tools. A small body of synthetic training data that teaches the model to emit tool calls in the right format. A runtime that parses those calls, dispatches them to real Python functions, and inserts the results back into the conversation as observations. Then you will run one complete trace, push the system to multi-step tasks, and finally corrupt the observations to watch the failure modes.

A single consumer GPU is enough for the fine-tuning. The runtime itself runs on plain Python: no special hardware required, no external services in the lab version. The two example tools are mocked locally so the experiment is reproducible.

Step 1: Define a tool schema

A tool schema is a JSON document describing one tool. Each tool has three fields: `name`, `description`, and `parameters`. The parameters field is itself a small JSON schema that names each argument, its type, and a one-line description.

Here are the two tools used in this chapter. The first looks up the weather for a city, the second evaluates a mathematical expression.

```
TOOLS = [
    {
        "name": "get_weather",
        "description": "Look up the current weather for a city.",
        "parameters": {
            "type": "object",
            "properties": {
                "city": {
                    "type": "string",
                    "description": "City name, e.g. 'Toronto'."
                }
            },
            "required": ["city"]
        }
    },
    {
        "name": "calculator",
        "description": "Evaluate a mathematical expression and return the result.",
        "parameters": {
            "type": "object",
            "properties": {
                "expression": {
                    "type": "string",
                    "description": "A Python-style math expression, e.g. '8347 * 91256'."
                }
            },
            "required": ["expression"]
        }
    }
]
```

Two things matter about this schema. First, the `description` field is written for the model to read at inference time, not for a human reading the source. "Look up the current weather for a city" is the sentence the model uses to decide whether a question is a weather question. If the description says "Get meteorological data," the model will be worse at recognizing the everyday phrasings that actually appear in user prompts. Second, the parameter types are part of the contract. When the runtime dispatches the call, it will check that the JSON contains a `city` field with a string value. If the model emits a number, or omits the field, the runtime will reject the call before it ever reaches the function.

A small observation that surprised me the first time I noticed it. The description field is the dominant lever. Tightening the parameter spec barely moves the model's accuracy. Rewriting the description from formal English into the phrasing a real user would type moves it by a lot. I had assumed the model would mostly read the JSON. It mostly reads the prose.

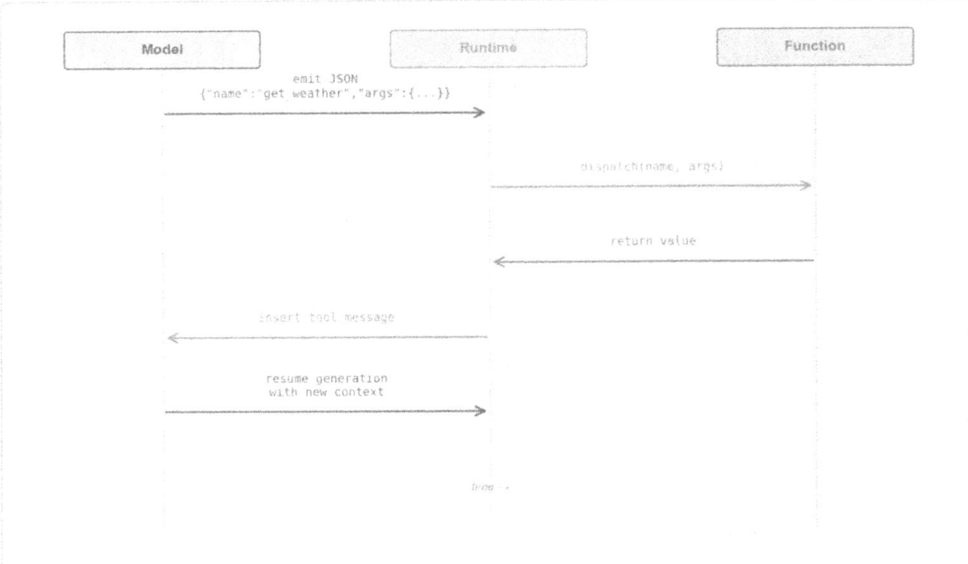

Figure 26.1. The tool-call protocol. The model emits JSON inside its text output. The runtime parses the JSON, dispatches the named function, and inserts the return value back into the conversation as a new message. The model resumes generation with that message in its context.

The schema is the only thing the model sees about your tools. It is the entire surface area of the contract. Treat the description and parameter docs the way you would treat a public API doc, because that is exactly what they are.

Step 2: Generate synthetic tool-call training data

The model from **Project 21: Fine-Tuning and Instruction Tuning** can follow instructions, but it has never seen a tool call. Right now, asked about the weather in Toronto, it will write a paragraph about Toronto's general climate. We need to teach it that, when the right kind of question shows up, the correct response is a structured call.

The training-data shape is a sequence of conversation turns. Each example has a user message and an assistant response. The response contains an optional Thought line in plain text, followed by an Action line containing JSON.

```python
def make_weather_example(city):
    user = f"What is the weather in {city}?"
    assistant = (
        f"Thought: The user is asking about current weather. I should use get_weather.\n"
        f"Action: {{\"name\": \"get_weather\", \"arguments\": {{\"city\": \"{city}\"}}}}"
    )
    return {"user": user, "assistant": assistant}

def make_calc_example(a, b):
    user = f"What is {a} times {b}?"
    assistant = (
        f"Thought: The user is asking for a multiplication. I should use the calculator.\n"
        f"Action: {{\"name\": \"calculator\", \"arguments\": {{\"expression\": \"{a} * {b}\"}}}}"
    )
    return {"user": user, "assistant": assistant}
```

Generate around 500 weather examples across 100 different city names, and around 500 calculator examples across varied number ranges and operations. Mix them. The total dataset is roughly 1000 examples. This is small for a fine-tune, but the format-recognition task is narrow enough that small data works. The model is not learning new world knowledge; it is learning a single new output shape.

A second class of examples matters too: the **no-tool-needed** case. Some questions should be answered directly without calling a tool. "What is the capital of France?" is a fact the model already knows; reaching for a tool here is wasted compute and an extra round trip. Generate 300 examples where the user asks something the model can answer from its own weights, and the assistant response is the direct answer with no Action line.

```python
def make_no_tool_example(question, answer):
    user = question
    assistant = f"Thought: I can answer this from my own knowledge.\n{answer}"
    return {"user": user, "assistant": assistant}
```

The full training set is therefore three buckets, in roughly equal proportions: tool-call examples for weather, tool-call examples for the calculator, and no-tool examples. The mixture teaches the model both when to reach for a tool and when not to.

The first time I built one of these mixes, I weighted the tool-call examples 80/20 against no-tool. The resulting model called a tool for everything. Asked for its own name, it tried to look the name up. The fix was not better prompts. The fix was rebalancing the training data and accepting that the model has no commonsense intuition about when a tool is overkill. It only knows what proportions you taught it.

Step 3: Fine-tune the Project 21 model on tool-call traces

Take your instruction-tuned checkpoint from **Project 21: Fine-Tuning and Instruction Tuning**. Format the synthetic data into the same chat template you used there. Run the same fine-tuning recipe. Same loss, same optimizer, same hyperparameters. For 1000 examples, two or three epochs are plenty. The whole fine-tune takes roughly an hour on a single consumer GPU.

The only evaluation that matters at this stage is whether the model emits valid JSON when it should. Take 100 held-out prompts, half of them tool-warranting, half of them not, and for each one, generate a response and check three things.

Did the model emit an `Action:` line at all? If yes, is the JSON parseable? If parseable, does the `name` field match a tool that exists, and do the `arguments` match the parameter spec for that tool?

A healthy post-fine-tune model should hit roughly 95 percent on JSON validity for the tool-warranting prompts, and roughly 90 percent correct decision on whether a tool was needed at all. Lower numbers point at a data problem, usually too few examples of one category, or descriptions in the schema that do not match the phrasing in the prompts. Higher numbers suggest the training data was either too narrow or has leaked into the eval, which is a problem that should sound familiar from **Project 22: Evaluation Methodology**.

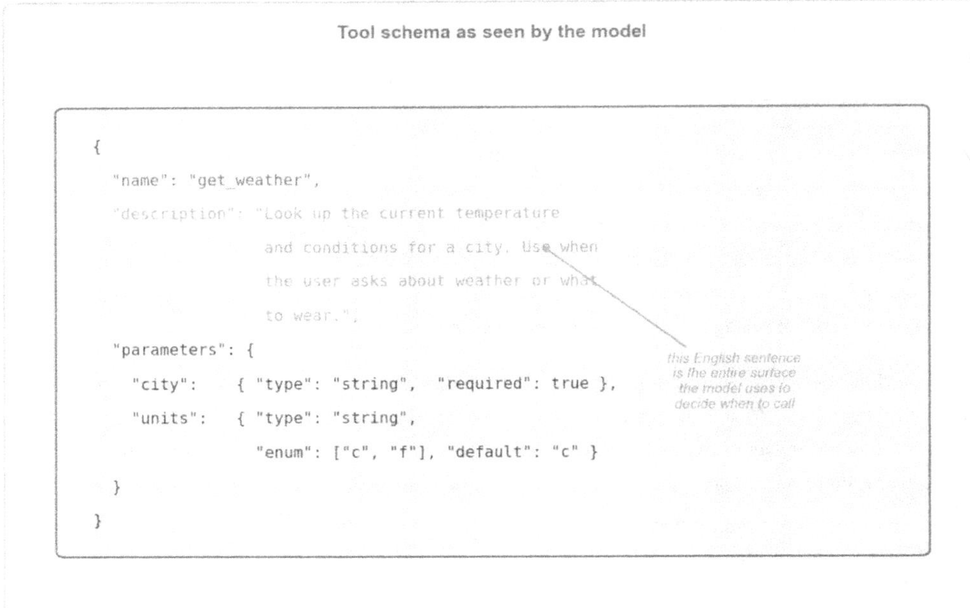

Figure 26.2. Tool schema example, shown the way the model sees it in its context window. The description field, written in plain English, is the entire surface the model uses to decide when a tool is appropriate.

Step 4: Build the tool runtime

The runtime has three jobs. It scans the model's output for an `Action:` line. It parses the JSON on that line. It dispatches the named function with the given arguments and returns the result as an Observation message.

A minimal version fits in 40 lines:

```python
import json
import re

def get_weather(city: str) -> str:
    # In real code, hit an API. In the lab, mock it.
    table = {"Toronto": "12C, light rain", "Tokyo": "22C, clear"}
    return table.get(city, f"Weather data unavailable for {city}.")

def calculator(expression: str) -> str:
    try:
        # eval() is only safe here because the lab is mocked.
        result = eval(expression, {"__builtins__": {}}, {})
        return str(result)
    except Exception as e:
        return f"ERROR: {e}"

REGISTRY = {"get_weather": get_weather, "calculator": calculator}

ACTION_PATTERN = re.compile(r"Action:\s*(\{.*\})", re.DOTALL)

def parse_action(text: str):
    match = ACTION_PATTERN.search(text)
    if not match:
        return None
    try:
        return json.loads(match.group(1))
    except json.JSONDecodeError:
        return None

def dispatch(action: dict) -> str:
    name = action.get("name")
    args = action.get("arguments", {})
    fn = REGISTRY.get(name)
    if fn is None:
        return f"ERROR: unknown tool '{name}'"
    try:
        return fn(**args)
    except TypeError as e:
        return f"ERROR: bad arguments: {e}"
    except Exception as e:
        return f"ERROR: tool raised: {e}"
```

A few details earn their lines. The regex looks for `Action:` followed by JSON inside curly braces. The `re.DOTALL` flag lets the match span newlines, because the model sometimes pretty-prints its JSON across several lines. The `eval` call inside `calculator` strips builtins so a model-emitted expression cannot execute arbitrary Python. In the lab this is fine, but real systems should never `eval` model output. The dispatch function returns a string for every code path, including every error path. That last point matters: the model cannot handle a raised Python exception, because exceptions do not live inside the conversation. Every failure has to become a string that the model can read and react to.

I lost a full afternoon once on the regex line alone. The model occasionally emitted `Action:` with a trailing space and then a newline before the JSON. My first regex required the JSON to start on the same character. Half my traces failed silently. The fix was one flag. The lesson was that even forty lines of glue code have an alarming surface area for tiny format assumptions you did not know you were making.

Step 5: Implement the ReAct loop end-to-end

The loop holds a list of messages and keeps calling the model and the runtime in alternation until the model produces a turn without an Action line.

```
def run_react(model, tokenizer, user_question, max_steps=6):
    messages = [
        {"role": "system", "content": build_system_prompt(TOOLS)},
        {"role": "user", "content": user_question},
    ]
    for step in range(max_steps):
        response = model.generate_chat(messages)
        messages.append({"role": "assistant", "content": response})
        action = parse_action(response)
        if action is None:
            return response, messages
        observation = dispatch(action)
        messages.append({
            "role": "tool",
            "content": f"Observation: {observation}",
        })
    return response, messages
```

The structure is small and the structure is the whole point. The loop alternates between two parties. The model generates a turn. The runtime parses, dispatches if needed, and appends the observation. Repeat. The exit condition is the absence of an Action line. When the model emits a turn that contains no JSON call, it has decided it has enough information to answer, and the loop returns.

Walk through one trace.

User: "What is the weather in Toronto, and what is 8347 times 91256?"

Step 1, model output:

```
Thought: The user is asking two things. First the weather in Toronto.
Action: {"name": "get_weather", "arguments": {"city": "Toronto"}}
```

Runtime parses the Action, dispatches `get_weather("Toronto")`, captures `"12C, light rain"`, appends:

```
Observation: 12C, light rain
```

Step 2, model output:

```
Thought: Now I need the multiplication.
Action: {"name": "calculator", "arguments": {"expression": "8347 * 91256"}}
```

Runtime parses, dispatches `calculator("8347 * 91256")`, captures `"761811032"`, appends:

```
Observation: 761811032
```

Step 3, model output:

```
The weather in Toronto is 12C with light rain, and 8347 times 91256 is 761,811,032.
```

No Action line. The loop exits and returns the final response.

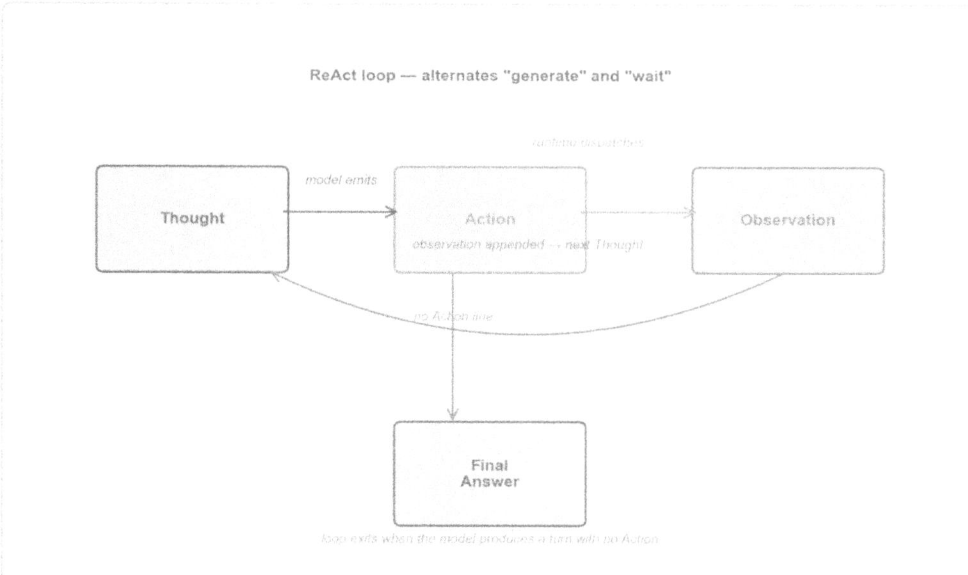

Figure 26.3. The ReAct loop as a state machine. The model alternates between two states (generating a turn, and waiting for an observation). The runtime alternates between parsing the model's output and dispatching the named tool. The loop exits when the model emits a turn without an Action line.

That trace is the canonical agent loop. Every more elaborate agent system you will encounter, whether multi-tool agents, planning agents, or retrieval-augmented chat, adds tools or wrappers but keeps this same shape underneath. Read it once. The next time you read about an "AI agent doing X," you can usually picture exactly which line of this loop is doing the work.

I think most beginner explanations of agents are honestly terrible because they treat the loop as a magical capability rather than a fifteen-line `for` statement. The loop is a `for` statement. The interesting part is what you put inside it.

Annotated ReAct trace

```
Question:      What is the population of Tokyo divided by Paris?
Thought:       I need both populations. Look up Tokyo first.
Action:        search("population of Tokyo")
Observation:   Tokyo metro: 37,400,000
Thought:       Now Paris.
Action:        search("population of Paris")
Observation:   Paris metro: 11,200,000
Thought:       Compute 37,400,000 / 11,200,000.
Action:        calculator("37400000 / 11200000")
Observation:   3.339
Thought:       I have my answer.
Final:         About 3.34×.
```

Thought → gray
Action → blue
Observation → green
Final → orange

Figure 26.4. A full ReAct trace, annotated. Thought lines in gray, Action lines in blue, Observation lines in green. The model only ever sees text; the runtime is the line between text and execution.

Step 6: Multi-step tasks with two tools used in sequence

The trace above already used two tools, but the model handled them independently: first the weather call, then the calculator. The harder case is when one tool's result is the input to the next tool's call.

User: "If the temperature in Toronto today is in Celsius, what is it in Fahrenheit?"

The correct trace requires the model to call `get_weather("Toronto")`, parse the Celsius number out of the response, then call `calculator("12 * 9/5 + 32")` with that number, then phrase the answer.

This is where the model's actual capability shows. The format is easy; the planning is hard. A weak model will call the weather tool, ignore the numeric content of the response, and either fabricate the Fahrenheit answer or call the calculator with the wrong input. A strong model will use the literal Celsius number from the previous Observation in the calculator expression.

The success metric is straightforward. Run 100 multi-step prompts where the second tool's input depends on the first tool's output. Count the fraction where the final answer is numerically correct. Expect somewhere between 40 and 70 percent for a small fine-tuned model. The remaining 30 to 60 percent fail in one of three predictable ways: the model never makes the second call, the model makes the second call with wrong arguments, or the model invents a final answer that does not match the calculator's actual return value.

Each of those failures is a separate diagnostic signal, and you will see all three when you push the system. The first means the model did not recognize that two tools were needed. The second means the model recognized the need but failed at the data hand-off. The third, and this is the one that should worry you, means the model used the right tool, received a correct observation, and then ignored it. The model wrote a final answer that contradicts the observation it received two lines earlier.

I expected that third failure to be rare. It was not. On my first multi-step eval it accounted for almost a third of the wrong answers, and it was the only failure where reading the trace did not immediately suggest a fix. The model had the right answer in its context. It chose to ignore it.

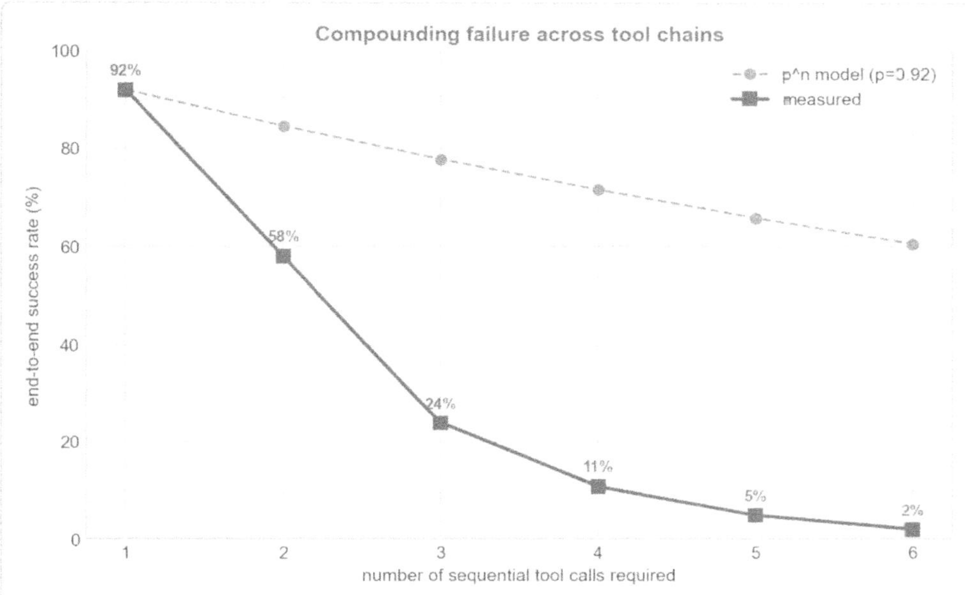

Figure 26.5. Success rate on multi-step tasks as the number of required tool calls grows. One call: 92 percent. Two sequential calls: 58 percent. Three sequential calls: 24 percent. The drop is not linear, because each step has its own failure probability and the failures compound.

Step 7: The no-tool-needed path and the two false-direction errors

Not every question warrants a tool. "What is the capital of France?" should produce a direct answer. "What is 2 plus 2?" probably should not call the calculator either, because the model already knows the answer and the tool round trip is wasted latency.

But the model's decision about whether to call a tool is itself a learned behavior, and it can fail in two directions.

The first failure direction is the **false-positive tool call**. The model calls a tool for a question that needs no tool. Asked the capital of France, it emits `Action: {"name": "calculator", "arguments": {"expression": "France"}}`, which then errors, which then forces the model to recover. This wastes a round trip and risks confusing the model.

The second failure direction is the **false-negative tool call**. The model fails to call a tool for a question that needs one. Asked the weather in Toronto, the model writes a paragraph about Toronto's general climate without ever calling `get_weather`. The answer sounds confident. It also is not based on the actual weather. From the user's point of view, the system silently hallucinated.

Both error types are visible in evaluation. Take 200 prompts. One hundred that should trigger a tool, one hundred that should not. Run them through the loop. Count how many tool calls happened. For the tool-warranting half, compute the false-negative rate (no tool called when one was needed). For the no-tool half, compute the false-positive rate (tool called when none was needed). A healthy small model lands at roughly 5 percent on each rate. A model trained without enough no-tool examples lands at 30 to 40 percent false-positive, because the fine-tune has pushed it to reach for tools too aggressively. A model trained on too few tool examples lands at 20 to 30 percent false-negative, because it has not internalized the format strongly enough.

This is also a good place to mention what **Project 22: Evaluation Methodology** taught about contamination. If your evaluation prompts are paraphrases of your training prompts, both rates will look better than they really are. Hold out a genuinely fresh batch of prompts written after the training data was frozen.

Step 8: Error handling, when the tool itself fails

Real tools fail. APIs time out. Inputs are malformed. The math expression is invalid Python. The city name is misspelled and the weather service returns nothing.

The runtime catches every exception and returns it as a string Observation prefixed with `ERROR:`. That string is then handed back to the model, which has to decide what to do next. A well-trained model recovers by either retrying with corrected arguments, switching to a different tool, or giving up and asking the user for clarification. A poorly trained model retries the same call with the same arguments and loops until `max_steps` cuts it off.

The retry-the-same-call loop is the single worst failure mode you can ship into production. It racks up tool calls, runs up cost, and gives the user nothing. The only defense I know is to make the recovery behavior part of the training data and then test it adversarially.

To train recovery behavior, add a fourth bucket to the synthetic data: about 200 examples of error-followed-by-recovery. The user asks a question. The assistant calls a tool. The observation comes back as an error string. The assistant's next turn is either a corrected call or a graceful failure message.

```
def make_error_recovery_example():
    user = "What is the weather in Torronto?"  # typo
    turns = [
        ("assistant",
         "Thought: The user wants the weather. I'll try get_weather.\n"
         "Action: {\"name\": \"get_weather\", \"arguments\": {\"city\": \"Torronto\"}}"),
        ("tool",
         "Observation: ERROR: Weather data unavailable for Torronto."),
        ("assistant",
         "Thought: 'Torronto' looks like a typo for Toronto. I'll retry.\n"
         "Action: {\"name\": \"get_weather\", \"arguments\": {\"city\": \"Toronto\"}}"),
    ]
    return {"user": user, "turns": turns}
```

After including these examples, retrain and re-evaluate. The recovery rate on intentionally broken inputs should rise from somewhere around 30 percent (random) to roughly 75 percent. That last 25 percent is where the BREAK IT section lives.

Step 9: Train multi-step tool use with a trajectory reward

Steps 2 through 8 taught tool use by imitation. You wrote good traces, and the model copied them. Imitation has a ceiling. The model can only get as good as the traces you hand-authored.

One driving examiner grades every mirror check and turn signal along the route. Another checks one thing only: did you reach the destination, legally, in one piece. The second examiner says almost nothing about technique and still teaches a great deal, because passing the whole drive is a signal the student has to earn end to end.

That second examiner is the idea behind **RLVR**, Reinforcement Learning with Verifiable Rewards. RLVR is reinforcement learning where the reward comes from an external automatic verifier of the output, not from a learned reward model that imitates human preferences. The verifier is domain-specific: an exact-answer check for math, unit tests for code, a rule-based check for logic. For a tool-using agent, the verifier runs at the end of the episode and asks one question. Did the final answer pass?

An **episode** here is the full trajectory the agent produced: call a tool, read the observation, call another tool, then answer. Agentic policy optimization extends the verifier idea from a single response to that whole trajectory. One 2026 method, Agentic Reinforced Policy Optimization (ARPO), assigns credit at the agent-step level across the intermediate tool calls, not only to the trajectory as a whole. Keep two layers apart. RLVR names the reward signal. GRPO and ARPO name the optimizer, the algorithm that pushes the policy toward higher reward. Pairing them is a choice, not the same thing.

Make the reward shape concrete. Take a three-step episode: the agent calls a search tool, then the calculator, then writes an answer, and a verifier compares that answer to a known value.

```
def trajectory_reward(episode, final_check, step_credit=0.0):
    # Outcome reward: 1.0 if the verifier passes the final answer, else 0.0.
    reward = 1.0 if final_check(episode["final"]) else 0.0
    # Optional step-level credit: a small bonus per correct intermediate step.
    reward += step_credit * sum(ok for _, ok in episode["steps"])
    return reward

final_check = lambda answer: answer == "33"        # exact-match verifier
good  = {"steps": [("search", True),  ("calculator", True)],  "final": "33"}
broke = {"steps": [("search", True),  ("calculator", False)], "final": "41"}

print(trajectory_reward(good,  final_check))                    # 1.0
print(trajectory_reward(broke, final_check))                    # 0.0
print(trajectory_reward(good,  final_check, step_credit=0.1))  # 1.2
print(trajectory_reward(broke, final_check, step_credit=0.1))  # 0.1
```

Those four illustrative outputs tell the story. With the outcome-only reward, the good episode scores 1.0 and the broken one 0.0. A third case is the catch: a run whose search was wrong but whose final answer still passed the check would also score 1.0, because the verifier only reads the end. Outcome-only reward pays for a lucky end and ignores a broken middle. Step-level credit fixes that. With `step_credit=0.1`, the good episode rises to 1.2. The broken run still earns 0.1 for its one correct step. So a run that searched right and then botched the arithmetic scores above a run that failed at step one. That is what agentic optimization targets. Which step earned the result, not just whether the result was earned.

The ARPO authors report improved performance while using roughly half the tool-call budget of prior methods, across 13 benchmarks. Read that as a reported result under the authors' own evaluation, not an independently confirmed law. The structural point stands. You can push a tool-using model past the ceiling of your hand-written traces by rewarding whole trajectories against a verifier, instead of only imitating them.

BREAK IT

The interesting break is not breaking the model. It is breaking the runtime in a small, hidden way and watching what the model does with corrupted observations.

You will deliberately corrupt the JSON in returned tool outputs. The model will receive observations that look almost right but contain malformed text. The runtime itself does not raise. It has no schema for what observations should look like, because observations are just strings. The model receives the broken string and tries to make sense of it. There are three predictable failure modes, and at least one of them shows up every time.

The exact change

In the `dispatch` function, wrap the return value in a small corruption helper before it goes back to the model:

```
def corrupt(s: str) -> str:
    # Three random corruptions, applied with probability 1.0.
    s = s.replace("}", "")          # remove closing braces
    s = s.replace("\"", "'")         # swap double quotes for single
    s = s.replace(":", " is")        # replace colons with " is"
    return s
```

Then wrap the dispatch return:

```
def dispatch(action: dict) -> str:
    # ... existing code ...
    raw = fn(**args)
    return corrupt(str(raw))
```

Now re-run the loop on the multi-step weather-plus-calculator question and watch the trace.

What you will see, predicted in advance

Before running this, predict what should happen. The model receives an observation like `12C is light rain` instead of `12C, light rain`. The calculator returns `761811032` corrupted into `761811032` (no quotes or braces in the original, so this case looks unchanged). For richer tool outputs (a tool that returns nested JSON, for instance), the corruption is far more visible.

Now run it. The model produces one of three behaviors, and which one depends mostly on the random seed.

Behavior A: the model hallucinates a result from the malformed text. The observation comes back garbled. The model reads it, finds something that looks roughly like a number or a word, and uses that as if it were the real answer. The next turn confidently asserts a fact that the observation never actually contained. From outside the loop, the system appears to be working. The model produced a final answer in fluent English. From inside, the answer is wrong, and the wrongness is silent.

Behavior B: the model loops, retrying the same tool with the same arguments. The observation looks broken. The model decides the tool failed. The model calls the same tool with the same arguments, expecting a different result. The runtime corrupts the observation again. The model retries again. The loop continues until `max_steps` cuts it off and the system returns whatever partial garbage exists at that point.

Behavior C: the model fabricates an answer that ignores the broken observation entirely. The model reads the corrupted observation, decides it is unintelligible, and falls back on its own pretraining. It generates a plausible-sounding answer that has nothing to do with the actual tool output. The fabrication may even be close enough to correct that a casual reader would not notice the discrepancy.

All three failures are silent. No exception is raised in the runtime. No log message says "the model got a corrupted observation." The loop continues, the chat history grows, and the user sees a confident final answer that is detached from reality.

What this proves

The model is not validating its inputs. It trusts the loop. The runtime is the only thing standing between the model and a fabrication, and the runtime is enforcing structure on the model's outputs (the Action JSON) but not on its inputs (the Observation text). That asymmetry is the whole point of the experiment.

In a real production system, the implications are direct. If the tool can return garbage (because the upstream API changed, because the network glitched, because someone shipped a bug into the dispatch code), the model will not catch it. The model will absorb the garbage, weave it into a story, and present the story to the user with the same confidence it presents a correct answer.

The schema is the only line of defense. Both directions need a contract. The model's output is validated by the runtime parsing its JSON against the tool schema. The runtime's output should be validated too. The observation text should match an expected shape, the values should pass a type check, and any anomaly should be flagged before it ever reaches the model. Most production tool-use systems do not do this yet, which is why most production tool-use systems silently hallucinate more than you would think.

The mental image I find most useful is a courier who only speaks in pre-printed envelopes. The courier can hand you any envelope. The envelope might contain a bill, a love letter, or a blank piece of paper. The courier does not check. If the model is the courier and the runtime is the post office, you want the post office to inspect both sides. Every envelope that leaves the model should be inspected. Every envelope that comes back from a tool should also be inspected before the courier carries it onward.

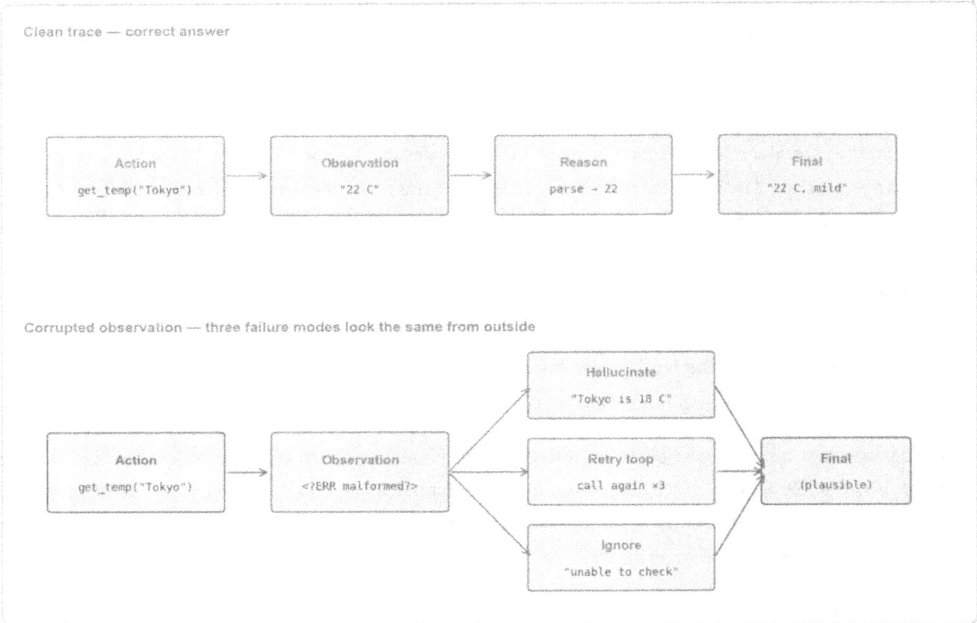

Figure 26.6. The malformed-output failure cascade. Top row: clean trace, correct answer. Bottom row: corrupted observation, three possible failure modes (hallucinated result, retry loop, ignored observation). All three look identical from outside the system.

Optional Homework

Everything in this chapter runs on a single consumer GPU. The fine-tuning step takes roughly an hour. The agent loop itself runs on plain Python and needs no accelerator. The lab uses mocked tools (a local dictionary for weather, a sandboxed `eval` for the calculator) so no API keys or external services are required.

Part A: Reproduce the synthetic data pipeline

Generate the three buckets of training data: weather tool calls, calculator tool calls, no-tool answers. Aim for around 1000 examples total. Save them in a single JSONL file. Inspect ten random examples by hand. The first sanity check is whether the format is consistent enough that you could imagine a model learning it from this data alone.

Part B: Fine-tune from your Project 21 checkpoint

Take the model from **Project 21: Fine-Tuning and Instruction Tuning** and fine-tune for two or three epochs on the synthetic data. Use the same chat template you used in that chapter. Save the result. Run inference on 20 held-out prompts and check by eye whether the output format matches the training format.

Part C: Build the runtime and run the example trace

Implement `parse_action`, `dispatch`, and the `run_react` loop. Hook up the two mock tools. Run the example trace ("weather in Toronto, and 8347 times 91256") and print every step. Verify the final answer matches what the calculator actually returned.

Part D: Measure success rates

Build three eval sets. One hundred single-tool questions. One hundred two-step questions where the second tool depends on the first. One hundred no-tool questions. Run all 300 through the loop and record three numbers: single-step success, multi-step success, false-positive tool rate. Compare to the ballpark numbers in Steps 6 and 7 of the build.

Part E: The corruption experiment

Run the BREAK IT corruption experiment from this chapter. Before running, write down which of the three failure modes (hallucinate, loop, fabricate) you think you will see most often. Run 30 questions through the corrupted loop. Tally the failure mode for each one. Compare your prediction to what you saw.

Part F: Add a validation layer

This is the optional fix. Write a small `validate_observation` function that checks the observation against an expected schema for each tool. For instance, the weather tool should return a string matching `"^[\\-0-9]+C,"`. If the observation fails validation, the runtime replaces it with `Observation: ERROR: tool returned malformed output.` Re-run Part E with validation enabled. Record the change in failure rate.

One warning from my own runs of this experiment. When I added the validation layer the first time, the overall failure rate dropped, but a new failure mode appeared: the model started looping on the validation-error string the way it used to loop on bad arguments. Defenses have second-order effects. Test the defense the same way you tested the original loop.

Questions To Answer

1. In your fine-tune, which bucket of examples did the model learn fastest (weather, calculator, or no-tool)? Which one ended up with the highest residual error rate?
2. On multi-step tasks where the second tool's input depends on the first tool's output, what was your success rate? Of the failures, how many were "did not make the second call" versus "made the second call with wrong arguments" versus "made the call correctly but ignored the result"?
3. What was your false-positive tool rate on no-tool questions? What does that rate tell you about the balance of your training data?
4. In the corruption experiment, which of the three failure modes (hallucinate, loop, fabricate) did you see most often? Was that what you predicted?
5. After adding the validation layer in Part F, did the failure rate drop, stay the same, or move into a different failure mode? What does that tell you about where the model's trust is placed in the loop?
6. If a future agent system is reported to "achieve human-level performance" on a tool-use benchmark, which three questions would you ask before believing the number? Use the framework from **Project 22: Evaluation Methodology**.

Go Further

Two repos worth reading after this chapter.

The reference for the loop pattern is the ReAct paper's official repo at github.com/ysymyth/ReAct. It is small. Read the prompt templates in particular; they are the original ones, and they are still close to what production systems use.

For a production-shaped version of the runtime, read the function-calling section of OpenAI's documentation and the tool-use section of Anthropic's documentation. The mechanism is the same in both cases: a JSON schema, a structured call from the model, a dispatch loop, a returned observation. Reading the two side by side makes the underlying pattern obvious.

Three papers worth reading.

Yao and colleagues, "ReAct: Synergizing Reasoning and Acting in Language Models" (2023), is the source of the Thought-Action-Observation loop. The paper compares ReAct against pure reasoning and pure acting baselines and shows that the loop is the part that matters: the model improves on both axes when reasoning and acting are interleaved instead of separated.

Schick and colleagues, "Toolformer: Language Models Can Teach Themselves to Use Tools" (2023), shows how to bootstrap tool-use training data without writing it by hand. The model is trained to insert API calls into its own pretraining text where the calls would have improved the next-token prediction. This is a useful contrast with the supervised fine-tuning approach in this chapter.

Patil and colleagues, "Gorilla: Large Language Model Connected with Massive APIs" (2023), pushes the schema-and-dispatch idea to thousands of APIs. The interesting part is what happens to the model's tool-selection accuracy as the registry grows. It does not stay flat.

arXiv search directions for further reading: `agent benchmark tool use`, `function calling JSON schema language model`, `ReAct reasoning acting agent`, `tool-augmented language model evaluation`, `multi-step agent failure analysis`.

One specific direction worth pursuing immediately if you ship a tool-using model into production is the topic of **tool-use evaluation suites**. AgentBench (Liu et al.) and ToolBench (Qin et al.) are the two most cited. Both build large held-out batches of multi-step tool-use scenarios and report per-stage failure rates rather than a single aggregate number. The per-stage breakdown is what tells you which of the failure modes in Step 6 dominates your system.

What You Now Know

A tool-using model is a model plus a runtime, and the runtime is small. The model emits structured JSON calls when its training data taught it to. The runtime parses the JSON, dispatches to a registered function, and inserts the return value back into the conversation as an observation. The model resumes generation with the new context. Think, act, observe, repeat. That is the agent loop.

You know how to fine-tune an instruction-tuned model into a tool-using one with roughly 1000 synthetic examples. You know the two failure directions, false-positive tool calls and false-negative tool calls, and how to balance training data to keep both rates in check. You know how to measure multi-step success and how to read the three predictable ways that multi-step traces fail. You know that the model trusts whatever the runtime sends it, and that corrupted observations produce one of three silent failure modes that no exception will catch.

The next chapter you will read after this is **Project 28: Retrieval-Augmented Generation**, which builds a tool-using model whose main tool is a vector search call. Everything from this chapter applies directly. The loop is the same loop. The schema is a new tool. The observation is a list of retrieved documents instead of a weather report. The failure modes you watched here will all reappear there, only with a search index involved instead of a corrupted observation. Build the loop once, and the rest of the agent family becomes a series of small variations on a familiar shape.

Starting Point

ReAct paper and code: github.com/ysymyth/ReAct (Yao et al., 2023). OpenAI function-calling documentation: platform.openai.com/docs/guides/function-calling. Anthropic tool-use documentation: docs.anthropic.com/en/docs/build-with-claude/tool-use. Toolformer paper: arxiv.org/abs/2302.04761 (Schick et al., 2023). Gorilla paper: arxiv.org/abs/2305.15334 (Patil et al., 2023).

Project 27: Quantization and Deployment

Hook

Why does a model that barely fits on a 24GB GPU suddenly run on a laptop after one conversion step, and why does that same trick eventually turn good answers into polished nonsense? That is the confusion to start with. If the model learned meaning through millions or billions of carefully adjusted floating-point numbers, how can you slash those numbers down to 8 bits or 4 bits and still get sensible text? And if that works, why does 2-bit quantization so often cross a line where the model still sounds fluent but stops saying anything true or coherent?

The Concept

Think of a trained model as a giant wall of tiny knobs, where each weight is one knob. Training spends a huge amount of compute nudging those knobs until the model predicts the next token well. In FP32, each knob stores its setting using 32 bits, which is 4 bytes. A weight might be `0.0187342` instead of just `0.02`.

Now imagine you need to move this wall of knobs into a smaller room. One option is to build a smaller wall from scratch. That is retraining or designing a smaller model. The other option is to keep the same wall but repaint every knob with fewer allowed positions. Instead of letting a knob land on nearly any decimal value, you force it into a small set of buckets. That is quantization. FP32 says, in effect, "this knob can sit almost anywhere." INT8 says, "this knob must choose one of 256 positions." INT4 says, "pick one of 16 positions." INT2 says, "pick one of 4 positions."

The whole trick works because models are more tolerant to roughness than people first expect. A trained model does not need every single weight to stay exact down to many decimal places. Many weights can be rounded and the overall computation still points in roughly the same direction. The logits change a bit, but not enough to break the answer. Until they do.

That is the heart of this chapter: quantization is controlled damage. You are compressing the model by forcing many precise values into fewer buckets, then measuring whether the model still behaves acceptably for your job.

The first time I sat with this idea seriously was during the Orion work on Apple Neural Engine kernels. The whole point of getting an 8.5x speedup on delta compilation through private ANEClient/ANECompiler APIs was that ANE only natively supports lower precisions. Quantization was not optional; it was the price of entry. I expected the hard part to be the math. The hard part was the bookkeeping. The math is two lines.

A plain-English picture of what gets lost

Suppose you have a grayscale photo with millions of pixels. In full precision, each pixel might store a very exact brightness. If you reduce that image to 256 shades, it still looks fine. Reduce it to 16 shades, and edges get harsher, but you can still recognize faces. Reduce it to 4 shades, and you still see rough structure, but fine details disappear. That is what quantization does to model weights.

The model still keeps broad patterns first. Syntax is a broad pattern. Phrase structure too. The rhythm of language as a whole stays robust. Fine semantic distinctions are more fragile. That is why heavily quantized models often fail in a very specific way: they speak in sentences that sound plausible while saying the wrong thing. The grammar survives longer than the meaning.

I have stopped describing this politely to non-ML colleagues. The honest summary is that a two-bit model sounds like a confident person on day three of a bad flu. The words still come out. The sentences still finish. The content has come unstuck from reality.

What a quantized weight actually is

Here is the intuitive version before the formula. A block of FP32 weights contains real numbers like `-0.91`, `0.03`, `1.27`, and `-0.44`. To store them as INT8, you choose a mapping from real values to integers. For example, in this block:

- `-1.28` maps to `-128`
- `0.0` maps to `0`
- `1.27` maps to `127`

Now each original number gets rounded to the nearest integer bucket. Later, during inference, you approximately reconstruct the original value by scaling the integer back into a float-like range. So quantization has two parts:

1. Compress the number into a low-bit code.
2. Decompress it just enough at runtime to do the math.

The low-bit value is not the real weight. It is a compact code plus a rule for turning that code back into an approximate weight.

The First Equation, Earned

Now that the picture is clear, here is the common formula for simple linear quantization:

$$q = \text{round}(x/s)$$

$$\hat{x} = q \cdot s$$

What do the symbols mean?

- `x` is the original weight, stored in FP32.
- `q` is the quantized integer value, such as an INT8 or INT4 bucket index.
- `s` is the scale, which tells you how much real-number distance each bucket represents.
- `\hat{x}` is the reconstructed weight used during inference after dequantization.

Read it in plain English: divide the real weight by the scale, round it to the nearest integer bucket, and later multiply that bucket by the scale to get an approximate weight back. That is the entire move. The error comes from rounding:

$$\text{error} = x - \hat{x}$$

If the buckets are fine enough, the error stays small. If the buckets are too coarse, the model starts forgetting distinctions it needs.

Why Scale Matters

If the model has weights between `-0.02` and `0.02`, you need tiny buckets. If it has weights between `-7.0` and `7.0`, you need wider buckets. One fixed global scale for the whole model is usually too crude, so good quantization methods split weights into groups, rows, columns, or blocks, and give each group its own scale. That is the same as saying: do not compress a whole movie with one brightness setting. Compress each scene according to its own range. This is why you will hear terms like per-tensor quantization (one scale for a whole tensor), per-channel quantization (one scale per output channel or row), and group-wise quantization (one scale per small block of weights). Smaller groups usually preserve quality better, but they also add bookkeeping overhead. Again: quantization is tradeoff after tradeoff.

Why It Matters

Without quantization, deployment hits a wall fast. A model with 7 billion parameters at FP32 needs about:

$$7\,\text{billion} \times 4\,\text{bytes} = 28\,\text{GB}$$

That is just the weights. Not the optimizer states from training, not activations during training, not the **KV cache** (the runtime attention key/value store described in **Project 13: Fast Inference: The KV Cache**, which in deployment is a significant memory consumer alongside model weights). Just the weights. That means FP32 blocks many local deployments outright, FP16 cuts memory in half but still leaves large models expensive, INT8 cuts memory to one quarter of FP32, INT4 cuts it to one eighth, and INT2 goes smaller still but often pushes quality below the floor.

This matters for three reasons.

1. Deployment stops being theoretical

If your model only runs in a data center, you have one kind of product. If your model runs on a laptop, edge box, or phone, you have another. Quantization is one of the main reasons local inference exists at all.

2. Memory bandwidth becomes the bottleneck

People often assume inference is only about arithmetic speed. For large language models, memory movement is often the real tax. You keep pulling huge weight matrices from memory into the compute units, and lower-bit weights mean fewer bytes moved. Fewer bytes moved often means faster inference. So quantized models are often smaller AND faster. Not always, but often enough that this is part of the default deployment toolkit.

This was the part that genuinely surprised me when I started measuring on Apple Silicon. I had assumed quantization was a memory-pressure story. It is. But it is also a bandwidth story, and the bandwidth story is often the bigger one. The compute units sit idle waiting for weights. Fewer bytes means less waiting.

3. Quality becomes an engineering budget, not a binary state

The right question is not whether quantization is good or bad. It is: how much quality loss buys how much deployment gain? For one product, a 1-point benchmark drop is unacceptable. For another, an 8x size reduction with a small reasoning hit is the whole business model. Once you can measure that tradeoff, you stop speaking in slogans and start speaking in numbers.

This is the reframing that runs underneath every honest deployment conversation I have had since shipping Marunthagam to rural ASHA workers in Tamil Nadu. We had a hard memory ceiling because the device was whatever Android phone the worker already owned. Lower-bit models were not an optimization. They were the only models we could ship. The question was never "can we afford quantization." The question was always "how much can we afford to lose."

The Build

This build assumes you already have a trained checkpoint from earlier work and now want to produce a deployable quantized model, compare quality, and run it in a runtime built for inference rather than training.

The sequence is:

1. Establish a clean FP32 baseline.
2. Implement INT8 post-training quantization.
3. Implement INT4 quantization.
4. Measure size, quality, and speed.
5. Export to GGUF.
6. Run in `llama.cpp`.
7. Compare Python inference with C++ inference.
8. Push to INT2 and find the quality floor.

Step 1: Establish the FP32 baseline

Do not quantize first and benchmark later. That is how people lie to themselves without meaning to. Start with the original checkpoint and record three things:

- model size on disk
- validation loss or benchmark score
- inference speed, such as tokens per second

Your baseline is the only reason later numbers mean anything. If you are using a language-model validation loop, record the average cross-entropy loss on a held-out set. Cross-entropy loss is a number that says how surprised the model is by the correct next token (lower is better). If you want a simpler user-facing metric, you can also run a small benchmark or task set:

- a held-out instruction set
- GSM8K if you care about reasoning
- a domain-specific evaluation set if that matches your product

Record both if you can. Loss tells you what changed in the model's next-token behavior. Task accuracy tells you whether users will feel it.

A typical baseline table starts like this:

Format	Size	Val Loss	Tokens/sec
FP32	28.0 GB	2.41	18

: **Table 16.1.** FP32 baseline metrics: model size on disk, validation loss, and inference throughput before any quantization is applied.

Those numbers are placeholders. Your actual values depend on model size and hardware.

Step 2: Implement basic INT8 post-training quantization

Post-training quantization means you quantize after training is complete. No retraining, no extra gradient steps. That is what makes it attractive. The simplest useful version is symmetric INT8 weight quantization, where "symmetric" means zero stays centered at zero, so positive and negative values get equal range.

For a weight tensor `W`, one common recipe is:

1. Find the maximum absolute weight.
2. Compute a scale so that max magnitude maps to 127.
3. Round each weight to the nearest integer in `[-127, 127]`.

Here is the minimal idea in code:

```
import torch

def quantize_int8_symmetric(w: torch.Tensor) -> tuple[torch.Tensor, torch.Tensor]:
    max_abs = w.abs().max()
    scale = max_abs / 127.0 if max_abs > 0 else torch.tensor(1.0, device=w.device)
    q = torch.clamp(torch.round(w / scale), -127, 127).to(torch.int8)
    return q, scale

def dequantize_int8(q: torch.Tensor, scale: torch.Tensor) -> torch.Tensor:
    return q.float() * scale
```

This is not yet a production kernel. It is enough to make the mechanics visible. Now pause and ask the right beginner question: if the runtime has to turn `int8` back into float, are we really saving anything? Yes, for two reasons. First, storage still shrinks. Second, good runtimes do not dequantize the whole model all at once into FP32 and throw away the savings. They fuse quantized storage with matrix multiplication kernels so the movement and arithmetic stay efficient. Your toy version may not be faster yet, and that is fine. The point here is to understand the representation before chasing optimized kernels.

An earlier draft made it sound like fused quantized kernels are a small detail. They are not. They are the whole reason quantization translates into actual deployment speedups. The accelerator-kernel work I have done was almost entirely about getting the kernel path right; the bit-width reduction was the easy part.

Step 3: Quantize one layer first, not the whole model

Do not quantize everything at once on your first pass. Pick one linear layer (an MLP projection or an attention projection works well). Run it on a sample input in FP32, then run the quantized-dequantized version, and compare outputs.

```python
import torch

def layer_error(linear: torch.nn.Linear, x: torch.Tensor) -> tuple[float, float]:
    with torch.no_grad():
        y_fp = linear(x)

        q_w, scale = quantize_int8_symmetric(linear.weight.data)
        w_hat = dequantize_int8(q_w, scale)

        y_q = x @ w_hat.t()
        if linear.bias is not None:
            y_q = y_q + linear.bias

        mse = torch.mean((y_fp - y_q) ** 2).item()
        max_err = torch.max((y_fp - y_q).abs()).item()
        return mse, max_err
```

Quantization errors are easier to understand locally: if one layer barely changes, good; if one layer blows up, you inspect that layer before blaming the whole pipeline. This is the same discipline you used earlier when debugging training: isolate the component.

Step 4: Move from one layer to the full model

Once the local check looks sane, quantize all linear weights. In a transformer-style language model, that usually means:

- attention projections: `Wq`, `Wk`, `Wv`, `Wo`
- MLP weights
- final projection layers

Embeddings can be quantized too, but many deployment formats handle them with their own rules. For a plain educational implementation, you can start by quantizing every `nn.Linear.weight`.

The first time I quantized everything in one pass, the model produced output that looked normal for two lines and then unraveled into syllable salad. I spent an afternoon assuming I had a bug somewhere in the dequantization path. The actual problem was that one specific projection layer was carrying a small number of outlier weights, and group-wise scale across the whole tensor crushed them. Quantization-per-layer matters. So does looking at which layers have outliers.

Here is the shape of that pass:

```python
import torch

def quantize_model_int8(model: torch.nn.Module) -> dict:
    qstate = {}
    for name, module in model.named_modules():
        if isinstance(module, torch.nn.Linear):
            q_w, scale = quantize_int8_symmetric(module.weight.data)
            qstate[name] = {
                "q_weight": q_w.cpu(),
                "scale": scale.cpu(),
                "bias": None if module.bias is None else module.bias.data.cpu()
            }
    return qstate
```

Now you have a quantized checkpoint representation. The next question is how to run it. There are two paths:

1. Educational path: dequantize weights on the fly in Python and measure quality.
2. Deployment path: export to a runtime like `llama.cpp` using GGUF and let optimized kernels handle it.

Do both. The first teaches you what quantization changed. The second teaches you why deployment people care.

Step 5: Measure INT8 quality drop

Now run your validation set again with quantized weights. If your Python inference path fully dequantizes every layer before multiplication, speed numbers will be misleading, but quality numbers still matter. You care about:

- change in validation loss
- change in benchmark accuracy
- change in sample outputs

A small loss increase is expected. For many models, weight-only INT8 quantization causes very little degradation, which is why INT8 is often described as near-lossless for inference. But do not repeat that phrase like a slogan. Measure it on your model, especially if the model is small, narrow, or trained noisily, since smaller models can be more fragile.

A good result might look like this:

Format	Size	Val Loss	GSM8K Acc	Tokens/sec
FP32	28.0 GB	2.41	41.3%	18
INT8	7.0 GB	2.45	40.8%	24

: **Table 16.2.** FP32 versus INT8 comparison: model size, validation loss, benchmark accuracy, and throughput after post-training INT8 weight quantization.

Again, the structure matters more than the exact numbers.

Step 6: Implement INT4 quantization

Now the buckets get much coarser. INT4 means each weight gets 16 possible values. That sounds absurdly small, and yet 4-bit inference is common because the quality can remain surprisingly usable. At 4 bits, how you group weights matters more. A naive whole-tensor scale often hurts too much, so group-wise quantization is a better educational target.

I find the survival of 4-bit models genuinely strange every time I rerun the experiment. Sixteen values per weight. Sixteen. And the model still answers most questions correctly. That fact alone is one of the strongest arguments I have for the claim that meaning in neural networks is distributed, not stored.

Imagine flattening a weight tensor into chunks of, say, 64 or 128 values. Each chunk gets its own scale, which lets a block with small numbers keep fine resolution while a block with larger numbers gets wider range.

Minimal group-wise symmetric INT4 sketch:

```python
import torch

def quantize_int4_groupwise(
    w: torch.Tensor, group_size: int = 64
) -> tuple[torch.Tensor, torch.Tensor]:
    flat = w.flatten()
    n = flat.numel()

    q_groups = []
    scales = []

    for i in range(0, n, group_size):
        chunk = flat[i:i+group_size]
        max_abs = chunk.abs().max()
        scale = max_abs / 7.0 if max_abs > 0 else torch.tensor(1.0, device=w.device)
        q = torch.clamp(torch.round(chunk / scale), -8, 7).to(torch.int8)
        q_groups.append(q)
        scales.append(scale)

    q = torch.cat(q_groups).view_as(w)
    scales = torch.stack(scales)
    return q, scales
```

Why `-8` to `7`? Because 4 bits gives 16 values total, and signed symmetric-ish storage usually allocates those buckets around zero. The exact packing format differs across runtimes, but this captures the logic. Dequantization reverses it group by group.

Step 7: Explain why INT4 is still plausible

At this point a reasonable reader asks: how can 16 values per weight be enough? Because the model does not need each individual weight to be perfect. Neural networks are distributed systems. Meaning is not stored in one magic neuron or one exact decimal place. It is spread across many interacting parameters, so the network often tolerates a lot of local roughness before the global behavior breaks. That tolerance is not infinite. INT4 often keeps enough of the broad geometry of the learned function. INT2 often does not.

Step 8: Profile inference speed honestly

Now measure speed, but keep your experimental honesty intact. If you are testing in Python with naive dequantization, your low-bit model may look slower than FP32. That does not mean quantization failed. It means your runtime is not doing quantized inference efficiently. So collect two kinds of speed numbers:

- naive Python path: useful for debugging correctness
- optimized runtime path: useful for deployment

For each format, record:

- prompt length
- generation length
- hardware
- tokens per second
- peak RAM or VRAM if you can

Keep the prompt fixed and sampling settings fixed. Do not change temperature or top-p between runs. If output length differs because the model ends early, use a fixed `max_new_tokens` and note early stop behavior separately.

Step 9: Export to GGUF

Now we leave the educational sandbox and enter deployment land. GGUF is a file format used by `llama.cpp` for model weights and metadata. Think of GGUF as a shipping container. It does not just store raw numbers, it also stores the information the runtime needs to interpret them:

- tensor names
- tensor shapes
- data types
- quantization format
- tokenizer metadata
- architecture metadata

That matters because the runtime must know exactly how to reconstruct and execute the model. A quantized model file without metadata is just compressed rubble. A GGUF file is compressed rubble with assembly instructions.

You usually do not hand-write GGUF. You convert from an existing checkpoint using a conversion script, then optionally re-quantize into one of the runtime's supported formats. The exact commands depend on your source model structure and the current `llama.cpp` tool layout, but the flow usually looks like this:

1. Convert checkpoint to an intermediate format the converter understands.
2. Produce an FP16 or FP32 GGUF.
3. Quantize that GGUF into a lower-bit GGUF.
4. Run inference with `llama.cpp`.

Even if your training stack is Python, this is where deployment often becomes a C or C++ runtime problem. That is the point of the chapter: training format and serving format are different concerns.

Step 10: Why `llama.cpp` matters

`llama.cpp` is not just a convenient demo. It represents a different philosophy of inference: lean runtime, low-level kernels, quantized formats treated as first-class, CPU-friendly and edge-friendly deployment from the design level. If Python feels like a workshop bench, `llama.cpp` feels like a shipping crate that has already been packed for transport. Your model becomes easier to run on constrained hardware because the runtime is designed around that constraint from the start.

A three-tier edge deployment I worked on runs against this philosophy directly. The stack (a small on-device model, a midsize regional model, a larger district model) only works because the edge tier uses a runtime that treats low-bit inference as first-class. A Python serving stack would not have fit on the worker's phone. Not "would have been slow." Would not have fit. The runtime is part of what the model is.

Step 11: Compare Python vs C++ runtime

Run the same prompt in both paths:

- your Python baseline
- your GGUF model in `llama.cpp`

Measure:

- model load time
- first-token latency
- sustained tokens per second
- memory usage if available
- output quality on a few fixed prompts

This comparison teaches a useful deployment lesson: you can have a correct model in a slow stack, and you can also have the same model in a stack designed for low-latency inference. The runtime is part of the product, not just the weights.

Step 12: A simple table that forces clarity

By now you want a table like this:

Format	Precision	Size	Val Loss	Benchmark	Runtime	Tokens/sec	Notes
FP32	32-bit float	28.0 GB	2.41	41.3%	Python	18	baseline
INT8	8-bit integer	7.0 GB	2.45	40.8%	Python	16	naive dequant path
INT4	4-bit integer	3.5 GB	2.61	38.9%	Python	15	still usable
INT4	GGUF Q4	3.5 GB	2.61	38.9%	llama.cpp	31	deployable
INT2	2-bit integer	1.75 GB	3.48	19.2%	custom	33	fluent nonsense

: **Table 16.3.** Full quantization comparison across precision formats: model size, validation loss, benchmark accuracy, runtime stack, and throughput, from FP32 baseline through INT2 extreme compression.

This is the chapter in one table. Not because the table explains the mechanism, but because it forces you to state the tradeoff plainly.

Step 13: Why the output often fails semantically before it fails grammatically

This deserves its own section because it is one of the clearest lessons in all of model deployment. When you drive bit width down, the model does not usually fail by forgetting where periods go. It fails by blurring distinctions in the hidden state and logits. The broad flow of English sits in large, repeated statistical patterns. The exact choice between two similar but crucial continuations can depend on finer numerical differences, such as:

- choosing the right entity in a long context
- preserving a numeric intermediate in chain-of-thought-like reasoning
- distinguishing between two semantically close verbs
- keeping factual consistency across several sentences

Low precision flattens those margins. The model still knows how answers usually sound; it loses some ability to know which answer is actually right. That is why 2-bit outputs can feel eerie. They sound like a person who remembers the texture of explanation but not the content.

Step 14: The second equation, this time about storage

The storage math is simple enough to be worth making explicit. If `P` is parameter count and `b` is bits per parameter, then raw weight storage is:

```
bytes ~= (P * b) / 8
```

What do the symbols mean?

- `P` is the number of parameters.
- `b` is the number of bits used to store each parameter.
- `8` converts bits to bytes.

For a 7B parameter model:

- FP32: `7e9 * 32 / 8 = 28e9` bytes, about 28 GB
- INT8: `7e9 * 8 / 8 = 7e9` bytes, about 7 GB
- INT4: `7e9 * 4 / 8 = 3.5e9` bytes, about 3.5 GB
- INT2: `7e9 * 2 / 8 = 1.75e9` bytes, about 1.75 GB

Real file sizes differ because of metadata, scales, grouping overhead, and format details, but the rough picture holds. This is why quantization matters so much. The savings are not cosmetic, they are architectural.

One more store to shrink: the KV cache

Every bucket so far held a weight. But weights are not the only thing a deployed model keeps in memory. As it generates, it also holds the **KV cache**, the running store of attention keys and values from **Project 13: Fast Inference: The KV Cache**. The memory budget has two separate dials. One shrinks the model you loaded. The other shrinks the scratch space that grows while the model talks. They stack. You can turn both.

The mechanism is the same rounding idea aimed at a different tensor. Quantizing the KV cache means storing those cached keys and values at fewer bits each, then rebuilding an approximation when attention reads them back. Recent research pushes the bit-width down with a rotation trick. QJL applies a length-preserving random projection, then keeps one sign bit per coordinate, and on its own benchmarks it reports more than 5x KV-cache memory reduction at 3 bits with no reported accuracy loss (AAAI 2025). PolarQuant maps keys to polar coordinates and quantizes the angles, reporting more than 4.2x compression on long-context evaluation (AISTATS 2026).

TurboQuant is a 2025 preprint accepted to ICLR 2026. The authors report quality-neutral KV-cache quantization at 3.5 bits per channel, and only marginal degradation at 2.5 bits per channel. That is an abstract-level result, not tied to one model or GPU, and "marginal" is their word rather than a fixed accuracy delta. Its recipe runs in three steps. First, a random orthogonal rotation spreads the values across new axes. Second, per-coordinate Lloyd-Max scalar quantization rounds each coordinate with a table tuned to its distribution. Third, a 1-bit QJL residual correction keeps attention's dot products unbiased.

You can feel the mechanism without those research kernels. Hugging Face ships a quantized cache you switch on at generation, so a short script measures the second dial directly. Install the backend with `pip install optimum-quanto`, then read peak GPU memory once with the default cache and once with the quantized cache:

```python
import torch
from transformers import AutoModelForCausalLM, AutoTokenizer

tok = AutoTokenizer.from_pretrained("gpt2")
model = AutoModelForCausalLM.from_pretrained("gpt2", dtype=torch.float16).cuda()
ids = tok("Deployment holds two stores in memory, weights and a KV cache, and the KV cache",
          return_tensors="pt").input_ids.cuda()

def peak_bytes(cache_kwargs: dict) -> int:
    torch.cuda.reset_peak_memory_stats()
    # gpt2's context is 1024 positions; the prompt plus generated tokens must
    # stay under it, or the position-embedding lookup overruns and CUDA aborts
    # with a device-side assert. 512 leaves room for the prompt above.
    model.generate(ids, max_new_tokens=512, do_sample=False, **cache_kwargs)
    return torch.cuda.max_memory_allocated()

fp16_peak = peak_bytes({})
int4_peak = peak_bytes({
    "cache_implementation": "quantized",
    "cache_config": {"backend": "quanto", "nbits": 4},
})
print(fp16_peak, int4_peak, int4_peak / fp16_peak)
```

Expected direction: the KV-cache share of that peak falls roughly in proportion to bits per entry, so moving 16-bit keys and values to 4-bit pulls the cache term toward a quarter of its size. The printed ratio will not land on exactly 0.25, because the weights and activations share the same peak and dilute the cache's slice. Raise `max_new_tokens` and lengthen the prompt toward gpt2's 1024-position ceiling, keeping the prompt plus the generated tokens under it, so the cache grows large enough to dominate the reading. This is the arithmetic, not TurboQuant's numbers. One failure mode stops the run cold: the quantized call raises when `optimum-quanto` is missing, or when your Transformers version predates `cache_implementation="quantized"`. Install the package or pin a recent release, then retry.

So the deployment question has two halves, and this chapter answered only the first. You quantized the weights. The KV cache is the other lever, and on long-context work it can matter just as much.

BREAK IT

Now do the most useful thing in the chapter. Break it on purpose. Quantize to 2-bit, not because 2-bit is usually a good deployment choice, but because it reveals what precision is protecting.

The exact thing to change

If your INT4 quantizer maps weights into 16 buckets, change the bucket count to 4. In a symmetric signed setup, that usually means values such as `-2`, `-1`, `0`, `1`, or another 4-level scheme depending on your packing format. The important point is not the exact integer labels. The important point is that each weight now has only four possible stored states.

In your toy quantizer, the code change is conceptually this:

```
# INT4
scale = max_abs / 7.0
q = torch.clamp(torch.round(w / scale), -8, 7)

# INT2-ish
scale = max_abs / 1.0
q = torch.clamp(torch.round(w / scale), -2, 1)
```

That code is intentionally rough. Different schemes shift the exact range. What matters for the experiment is the collapse in resolution. If you are using a runtime's built-in 2-bit or near-2-bit quantization mode, pick that mode rather than forcing your own format.

What to watch for

Run the same three things again:

- validation loss
- benchmark accuracy
- actual generations on fixed prompts

This time, the generations matter a lot, not just the metrics. Pick prompts that test different failure surfaces:

- a factual question
- a multi-step arithmetic problem
- a constrained formatting task
- a summarization prompt with names and dates
- a long-context prompt with two similar entities

Here is the failure pattern you are likely to see. The model still produces complete sentences, plausible connective words, normal punctuation, and decent local grammar. But it starts failing at factual consistency, keeping track of who did what, arithmetic correctness, following exact constraints, and maintaining a coherent plan over several sentences. You often get output that feels polished for the first few seconds of reading, then you notice it does not actually answer the question. That is the floor.

A concrete example of the failure

Suppose the prompt is:

> Summarize why quantization can speed up inference, and give one downside.

A reasonable FP32 or INT4 answer might say:

- smaller weights mean less memory bandwidth
- this often raises tokens per second
- downside: some quality loss, especially at low bit widths

A 2-bit answer often keeps the shape but loses the substance:

> Quantization improves response flow by reducing the training memory into a compressed instruction path, which helps the model answer with shorter compute. The downside is that some parameter layers may become more generalized, causing a broader response tendency in context operations.

This sounds technical, but it is not actually saying the thing. That is the signature failure: fluent-sounding gibberish is not random text. It is syntax surviving after semantic precision has fallen apart.

What this proves

This experiment proves three things at once. First, precision is not mainly protecting punctuation and grammar. Those survive deep into the compression curve. Precision is protecting the fine numerical distinctions that preserve meaning. Second, model quality does not die all at once; it degrades in layers. You lose exactness before fluency, and you lose correctness before sentence shape. Third, "acceptable quantization" is not a universal threshold. It depends on the job. If your product is autocomplete for casual drafting, INT4 may be excellent. If your product is medical extraction or code transformation with strict correctness, even a moderate drop may be too much.

The edge deployment was a medical product. The model had to recognize danger signs in pediatric triage and route correctly to the right protocol branch. The acceptable bit-width floor for that use case was much higher than the one I would use for a chat demo. Same model, same architecture, completely different floor. The lesson lives in the job, not the network.

The goal is not "quantize as hard as possible." The goal is: find the bit-width floor for your use case. That sentence is the practical lesson of the chapter.

What usually fails first

If you want a ranked list of what tends to break as bit width falls:

1. Exact numeric reasoning
2. Long-range consistency
3. Fact retrieval under ambiguity
4. Instruction adherence with strict constraints
5. General semantic coherence
6. Surface fluency

That ordering is not a law of nature. It is a common pattern. If your experiments show something different, trust the measurements.

Why 2-bit breaks so sharply

At 2 bits, each group of weights has almost no expressive resolution left. The scale can still preserve rough sign and rough magnitude, but many distinct original weights collapse onto the same bucket, which means the model's learned function becomes much rougher. Transformers depend on repeated linear transformations where small differences get recombined across many layers. A little rounding noise at one layer is tolerable, but massive rounding at every layer is cumulative. It changes attention scores, MLP activations, logits. Those changes stack. By the time a token is chosen, the model may still know what a reasonable sentence shape looks like, but it no longer has enough precision to stay locked onto the right continuation. That is the quality floor made visible.

Optional Homework

This chapter's full deployment path can range from a laptop exercise to a multi-hour benchmark session, depending on model size. Here are two versions.

Version A: Local proxy experiment on a small model

Hardware:

- CPU-only is fine for a tiny model
- 8GB RAM is enough for a toy checkpoint
- 30 to 90 minutes

Goal:

- implement INT8, INT4, and INT2 quantization on a small transformer checkpoint
- measure validation loss and sample output quality
- see the failure curve directly

Steps:

1. Pick a small checkpoint you already trained in earlier projects or a tiny GPT-style model that fits comfortably in memory.
2. Save a held-out validation set of fixed sequences.
3. Add a quantization module with three modes:
    - `int8`
    - `int4`
    - `int2`
4. Quantize only `nn.Linear.weight` first.
5. Re-run validation loss for each mode.
6. Generate from the same fixed prompts for each mode.
7. Save outputs side by side in a text file.
8. Build a table with:
    - mode
    - model size on disk
    - validation loss
    - average generation latency
    - short note about output quality

Suggested commands if your project has a Python evaluation entry point:

```
python eval.py --checkpoint ckpt_fp32.pt --split val --out metrics_fp32.json
python eval.py --checkpoint ckpt_int8.pt --split val --out metrics_int8.json
python eval.py --checkpoint ckpt_int4.pt --split val --out metrics_int4.json
python eval.py --checkpoint ckpt_int2.pt --split val --out metrics_int2.json
```

Suggested generation sweep:

```
python sample.py --checkpoint ckpt_fp32.pt \
  --prompt-file prompts.txt --out fp32_outputs.txt
python sample.py --checkpoint ckpt_int8.pt \
  --prompt-file prompts.txt --out int8_outputs.txt
python sample.py --checkpoint ckpt_int4.pt \
  --prompt-file prompts.txt --out int4_outputs.txt
python sample.py --checkpoint ckpt_int2.pt \
  --prompt-file prompts.txt --out int2_outputs.txt
```

Expected observations:

- INT8 stays close to baseline
- INT4 shows a visible but often acceptable drop
- INT2 keeps sentence shape better than meaning
- validation loss rises steadily, but sample outputs make the failure clearer than the metric alone

Version B: Deployment path with GGUF and `llama.cpp`

Hardware:

- laptop or desktop CPU is enough for small and medium quantized models
- GPU optional
- 1 to 3 hours depending on model size and build setup

Goal:

- export a model to GGUF
- quantize it in the runtime's supported format
- compare runtime speed against a Python path

Steps:

1. Clone `llama.cpp`.
2. Build the project with the default instructions for your platform.
3. Convert your model checkpoint into a GGUF file the converter accepts.
4. Produce a higher-precision GGUF first.
5. Use the quantization tool to create lower-bit GGUF variants.
6. Run the same prompt through:
   - the high-precision GGUF
   - the INT8 or equivalent quantized GGUF
   - the INT4 or equivalent quantized GGUF
   - the lowest-bit mode available that approximates the 2-bit experiment

7. Record:
    - file size
    - model load time
    - prompt processing speed
    - generation speed
    - output quality on fixed prompts

Typical command shape in `llama.cpp` workflows looks like this:

```
python convert_hf_to_gguf.py /path/to/model --outfile model-f16.gguf
./llama-quantize model-f16.gguf model-q4.gguf Q4_K_M
./llama-cli -m model-q4.gguf -p "Explain why quantization can speed up inference."
```

The exact script names and quantization labels can change across versions, so read the current repo docs before running the final command set.

Expected observations:

- GGUF quantized files are much smaller than FP32 checkpoints
- `llama.cpp` often beats a naive Python path on tokens per second
- 4-bit formats often hold up surprisingly well
- the lowest-bit mode shows the semantic-collapse pattern from the BREAK IT section

Version C: Result-guided if you lack hardware

If your machine cannot run the full model:

1. Quantize a much smaller checkpoint locally and collect the failure pattern.
2. Read reported benchmark deltas for larger quantized models in `llama.cpp` discussions or model cards.
3. Compare the pattern, not the absolute number:
    - INT8 small drop
    - INT4 moderate drop
    - ultra-low-bit sharper semantic failure

The lesson still lands because the mechanism is the same.

What to record in your notebook

For every run, write down:

- bit width
- quantization scheme
- group size
- runtime
- hardware
- model size
- validation loss

- benchmark score
- tokens per second
- one prompt where the quality shift was obvious

If you skip this, you are doing vibes, not engineering.

Questions To Answer

1. At which bit width did the model first become unacceptable for your use case, not in theory, but in actual outputs?
2. What failed first: arithmetic, factual accuracy, long-range coherence, exact formatting, or basic fluency?
3. Did validation loss rise smoothly while user-visible quality dropped suddenly, or did both degrade together?
4. How much speed did you gain from quantization itself, and how much came from switching to a runtime built for inference?
5. Which quantization choice gave the best size-speed-quality tradeoff on your hardware?
6. What is the mechanism precision is protecting in your observed failure case: fine semantic distinctions, stable attention scores, exact logits, or something else?

Go Further

Research anchors for this chapter

Quantization is one of those topics where the practical ecosystem moves faster than any single teaching chapter, so it helps to keep three families separate in your head. **GPTQ** is the classic post-training anchor for aggressive low-bit weight quantization. **AWQ** is the important corrective idea that activations tell you which weights are especially sensitive and should be protected. And the `bitsandbytes` line matters because it made 8-bit and 4-bit workflows accessible enough to become routine for practitioners rather than paper-only curiosities.

The deployment side of the chapter is represented less by one paper than by one runtime worldview. `llama.cpp` and GGUF matter because they treat low-bit inference as a first-class systems problem: storage format, kernel path, and device constraints all get designed around deployment rather than training convenience.

If you want to connect this back to the frontier ending of the book, remember that weight quantization is only one compression layer. Later systems work increasingly asks the harder question: what else can be compressed besides the weights, especially the KV cache and other runtime state around them?

Start with these directions.

- `llama.cpp` Search for: `GGUF quantization formats`, `Q4_K_M`, `Q5_K_M`, `IQ2`
- GPTQ Search for: `GPTQ post training quantization`
- AWQ Search for: `AWQ activation aware weight quantization`
- bitsandbytes Search for: `LLM.int8` and `4-bit NF4`
- SmoothQuant Search for: `SmoothQuant W8A8`
- QuaRot and SpinQuant Search for: `rotation based quantization llm`

This chapter used the simple educational view: round weights into fewer buckets. Real quantization work gets better by being smarter about where error lands. Some methods pick better grouping, adapt scale choice, protect sensitive channels, account for activations rather than just weights, or rotate weight space so quantization error spreads more evenly.

If you want one research question to keep in your head, use this: how do we decide which information deserves precision? That is what all the better quantization papers are really asking.

If you want a benchmark direction, look at:

- perplexity on held-out text
- MMLU
- GSM8K
- HumanEval for code models

If you want an arXiv search phrase, use:

- `llm quantization 4-bit`
- `post training quantization large language models`
- `weight only quantization transformers`
- `activation aware quantization llm`

What You Now Know

You can now explain why a model that needs tens of gigabytes in FP32 can still work at 4 bits, and why that is not magic but controlled rounding plus a runtime designed to exploit it. You can explain GGUF as a deployment container, not just a file extension. You can measure quantization as a three-way tradeoff between size, speed, and quality instead of treating it as a yes-or-no choice. You can describe the 2-bit failure mode precisely: the model keeps the shape of language after it loses enough precision to keep the meaning straight.

And you can answer the deployment question that matters in the real world: not "Can I quantize this model?" but "How far can I quantize this model before it stops doing my job?"

Starting Point

Start from:

- `https://github.com/karpathy/nanochat`

Then use:

- `https://github.com/ggml-org/llama.cpp`

Read first:

- the checkpoint export path in your training code
- the GGUF conversion path in `llama.cpp`
- the quantization tool docs and supported formats

Carry forward the lesson from **Project 23: Reward Models And RLHF**. The reward model was the judge. This chapter adds another judge: deployment is a judge too, not of behavior, but of efficiency. And just like RLHF, once you optimize for the judge, you need to ask what you are losing. Quantization makes that question brutally concrete. You get the smaller file. You get the faster runtime. Now prove you still have the model you thought you had.

Project 28: Retrieval-Augmented Generation

Hook

Your **Project 21: Fine-Tuning and Instruction Tuning** model knows what was in its training data, and only what was in its training data. Ask it about a paper that came out last month, a customer's billing history, or an internal wiki page, and it makes something up that sounds right. Retraining the model every time a new document arrives is absurd, both in cost and in latency. So how do you teach a model what it does not know, without touching its weights? That is the question this chapter answers, and the answer is a system, not a model.

The Concept

The trick is to separate two jobs that the model has been doing badly because it was forced to do them together. One job is reasoning: given some text, write a coherent answer. The other job is recall: pull up the specific facts the answer should be based on. A pretrained model is excellent at the first job and unreliable at the second. So you stop asking it to do the second job. You give it the facts at query time and let it focus on writing.

Picture a student writing an essay in a library. The student is good at sentences, structure, and argument. The student cannot remember every book in the library, and you would not want them to. So a librarian sits next to them. When the student needs a fact about, say, the Treaty of Westphalia, the librarian walks over to the right shelf, pulls down three books, and slides them across the table. The student reads the relevant pages and writes the paragraph. The student never had to memorize the library. The librarian never had to write the essay. Each role is good at a different job, and the system as a whole produces a better essay than either could alone.

That analogy is the one I keep coming back to whenever I explain RAG to someone who has only worked with LLMs as a black box. The instinct most people have is to make the student smarter. The actual move is to hire the librarian.

Retrieval-Augmented Generation, or RAG, is the same arrangement built out of code. There are three components and they correspond exactly to the analogy. The first is an **embedding model**, a small neural network that converts a piece of text into a list of numbers, called a vector. Texts with similar meaning produce vectors that point in similar directions. The embedding model is the librarian's sense of which books are about which topics. The second is a **vector index**, a data structure that stores millions of these vectors and can quickly find the ones closest to a query vector. That is the library itself, organized so the librarian does not have to scan every shelf. The third is a **generator**, which is your existing LLM from earlier chapters. That is the student.

The flow at query time is short. A user asks a question. The embedding model converts the question into a query vector. The vector index returns the top few documents whose vectors are nearest to that query vector. The system stitches those documents into the prompt, with a small header that says "here is some context, use it to answer." The generator then produces an answer conditioned on that context. The model weights are not changed. The index is what holds the knowledge, and the index is something you can update in seconds.

A small piece of vocabulary, since we will use it for the rest of the chapter. The embedding model that produces vectors for the documents and the query independently is called a **bi-encoder**, because it has two separate encoding paths even though they share weights. The model that takes a query and a document together and scores how well they match is called a **cross-encoder**. Bi-encoders are fast and approximate. Cross-encoders are slow and accurate. We will use the bi-encoder to find a hundred plausible candidates, and the cross-encoder to pick the best three from those hundred. That two-stage pattern is how serious systems are built.

The deep point underneath all of this is that knowledge and reasoning are different kinds of work, and combining them inside one set of frozen weights is a poor design. The model's parameters store regularities of language, plus whatever facts happened to be in the training corpus, fused together in a way you cannot edit. The index stores facts in a form you can add to, replace, or delete this afternoon. When the facts change, you update the index, not the model. The student keeps writing essays. The library keeps growing.

Here is the take that I will defend through the rest of the chapter. RAG is a structured-prompt problem dressed up as a retrieval problem. The hard part is almost never the vector math. The hard part is the contract between the retriever and the generator: what gets retrieved, how it gets formatted into the prompt, and what the generator does when the retrieval is wrong. I have shipped retrieval on a multi-state SaaS platform handling hundreds of thousands of customer records across dozens of sites. Most of the production failures lived in the prompt-and-contract layer. Almost none of them lived in the index.

Why It Matters

Without retrieval, your model can only answer questions about its training data, frozen at the date the training finished. Anything newer is invisible. Anything proprietary is invisible. Anything domain-specific that the open web did not contain is invisible. You can ask the model anyway, and it will produce a confident answer, which is the worst possible failure mode because the answer is wrong but reads as if it is right. This pattern has a name now in the literature, **hallucination**, and it is the single largest reason that production deployments of LLMs go sideways. RAG is the most effective practical fix.

There is also a cost argument. Fine-tuning a model on a new corpus costs GPU-days and engineering time, and the result is brittle: when next month's documents arrive, you fine-tune again. Building an index over the same corpus costs minutes. Adding a new document is one embedding call and one index update. When a document is wrong or out of date, you delete it from the index and the next query no longer sees it. None of this requires touching model weights.

There is one more reason this chapter matters and it is a forward-looking one. The next generation of agent systems, the ones in **Project 26: Tool Use and Function Calling**, treat retrieval as one tool among many. The agent decides when to retrieve, what to retrieve, and how to refine the query if the first retrieval did not return useful documents. None of that works if the retrieval primitive is broken. Get this chapter right, and the agentic systems that come later inherit a sound foundation. Get it wrong, and every agent built on top inherits the same upstream failure.

On multi-model routing work across several frontier models, the routing logic sat right on top of a retrieval primitive that fed evidence into the model that was about to answer. When the retrieval was bad, the routing accuracy dropped in ways that I initially attributed to the router. It was not the router. It was the upstream facts.

The Build

You are going to build a complete RAG system in eight steps. By the end you will have an embedding model trained with contrastive loss, a FAISS index over ten thousand documents, a retrieval function, a prompt-formatting pipeline, an evaluation harness that measures with-vs-without lift, a cross-encoder re-ranker, a latency profile of each stage, and a short note on what changes when retrieval becomes part of an agent loop.

The whole system runs on one consumer GPU with about one gigabyte of RAM dedicated to the index. The embedding training is the only slow part, and even that finishes in a few hours on a modest card. You can also start from a pretrained sentence-transformer if you want to skip the training step and get straight to the retrieval mechanics; the chapter shows how to do both.

Step 1: Train a small sentence-embedding model with contrastive loss

The job of the embedding model is to take a sentence or short paragraph and produce a fixed-length vector, typically 384 or 768 dimensions, such that semantically related texts end up close together in vector space and unrelated texts end up far apart. The architecture is usually a small transformer encoder, often a six-layer or twelve-layer model, with mean-pooling over the token outputs to produce one vector per input. The attention mechanism inside that encoder is the same one you built in **Project 4: Attention From Scratch**, applied here to read a passage and summarize it into a single vector instead of generating new tokens.

The training objective is what makes this work, and it is worth slowing down for. Naive training would tell the model "here is a sentence, predict the next word" or "here is a pair, predict whether they are similar." Both are weak signals. The objective that actually works is called **contrastive loss**, and the specific form that has become standard is called **InfoNCE**.

The setup is this. You assemble a dataset of positive pairs: two pieces of text that are known to be semantically related. A natural source is a question-answer corpus like MS MARCO, where each query has an associated passage that contains the answer. Another source is paraphrase pairs, or related Wikipedia paragraphs, or any signal that says "these two texts mean similar things." For each positive pair in a training batch, every other text in the batch acts as a **negative**: a piece of text that should not be matched. With a batch size of 64, every positive pair has 63 negatives essentially for free.

The "for free" part is the bit that I think most beginner tutorials underplay. In-batch negatives turn a small labeled dataset into a dense learning signal without any extra annotation. The trick is so cheap it took me embarrassingly long to trust it the first time I implemented it.

The InfoNCE loss for one query in a batch is:

```
L = -log(  exp(sim(q, p+) / tau)
         / sum_i exp(sim(q, p_i) / tau)  )
```

Read it in plain English. The numerator is the similarity score between the query `q` and its true positive `p+`, exponentiated. The denominator is the sum of exponentiated similarities between the query and every passage in the batch, the positive included. The ratio is the probability the model assigns to the correct passage among all the options in the batch. We take the log, flip the sign, and minimize. That is it.

The temperature `tau` (usually somewhere between 0.05 and 0.1) controls how sharply the model is forced to pick the positive. A small `tau` makes the softmax peaky and pushes the model to score the positive much higher than the negatives. A large `tau` makes the softmax flat and lets the model spread mass. People tune this between runs; 0.07 is a common default.

The effect of optimizing this loss is geometric. The model is pulling the positive pair's vectors together and pushing every negative pair's vectors apart. After enough steps over a large enough dataset, the resulting vector space has a useful property: cosine similarity between two vectors approximates semantic similarity between the two texts. That is the property we exploit at retrieval time.

In code, the loop is simple:

```
def info_nce_loss(query_embs, passage_embs, tau=0.07):
    sim = query_embs @ passage_embs.T / tau
    labels = torch.arange(sim.size(0), device=sim.device)
    return F.cross_entropy(sim, labels)
```

The matrix `sim` is shape `(B, B)`, where `B` is batch size. Each row is a query, each column is a passage. The diagonal entries are the positive pairs. `cross_entropy` with labels `0, 1, 2, ..., B-1` is exactly the InfoNCE loss, vectorized.

Two practical notes. First, the embeddings going into this loss should be L2-normalized, so the dot product equals cosine similarity. Second, in-batch negatives are the cheapest negatives you can get, but they are not always hard enough; a strong production system also mines **hard negatives** (passages that look related to the query but are actually wrong) and adds them to the batch explicitly. We skip hard negative mining in the build because in-batch negatives are sufficient for the lift we want to demonstrate.

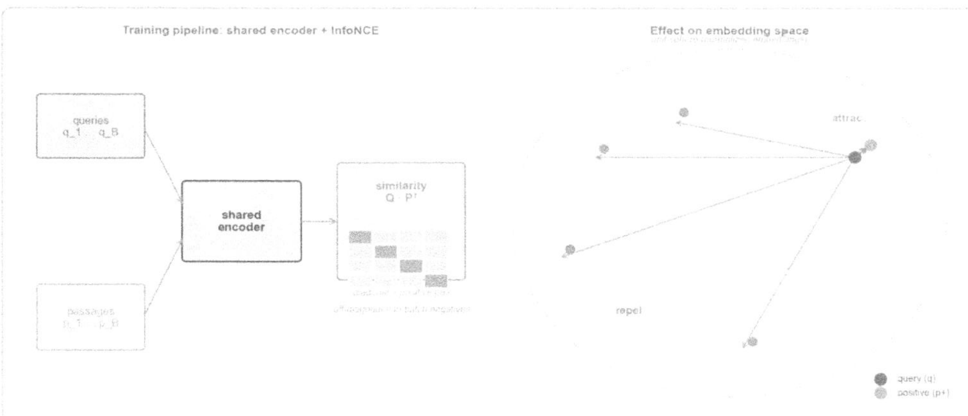

Figure 28.1. The embedding-model training pipeline. A batch of query-passage pairs flows through a shared encoder. The dot-product similarity matrix is computed, the diagonal is treated as the correct class, and InfoNCE loss pulls positives together while pushing the off-diagonal entries apart.

Train for one or two epochs on MS MARCO, with batch size 64, AdamW at learning rate 2e-5, and the model size from sentence-transformers' base configuration (six layers, 384-dim output). This takes a few hours on a single consumer card and produces an encoder that already does useful retrieval. If you do not want to train, download a pretrained model such as `sentence-transformers/all-MiniLM-L6-v2` and use it directly. The mechanics of the rest of the chapter are identical either way.

Step 2: Build a FAISS index over a 10K-document corpus

You now have an encoder. You need somewhere to put the document vectors that lets you find the nearest few among thousands or millions in milliseconds. The standard library for this on CPUs and GPUs is FAISS, which stands for Facebook AI Similarity Search. FAISS gives you several index types, and the choice among them is a tradeoff between speed, memory, and recall.

The simplest type is a **flat index**. A flat index stores every vector and at query time compares the query against all of them. Recall is perfect. Latency is linear in corpus size. For 10K documents at 384 dimensions, that is about 15 MB of memory and a few milliseconds per query, which is fine. For 10M documents it is 15 GB and hundreds of milliseconds, which is not fine.

The next type is **IVF**, for Inverted File Index. IVF partitions the vector space into clusters using k-means, with the cluster centroids stored in a small index of their own. At query time, you find the nearest few clusters to the query and search only inside those clusters, ignoring the rest. The parameter `nprobe` controls how many clusters you search. A larger `nprobe` gives better recall and worse latency. A typical setup is 1024 clusters with `nprobe=16`, which searches about 1.5% of the corpus and recovers most of the recall of a flat index at a fraction of the cost.

The third type is **HNSW**, for Hierarchical Navigable Small World. HNSW builds a multi-layer graph where each vector links to its nearest neighbors at multiple scales of granularity. At query time, you start at a random node in the top layer, hop greedily toward the query, descend a layer, and repeat. This gives you logarithmic search time and very high recall, at the cost of significantly more memory than a flat or IVF index, since the graph edges have to be stored alongside the vectors. HNSW is what most production systems use when memory is not the binding constraint. The graph edges buy you logarithmic search but cost you RAM.

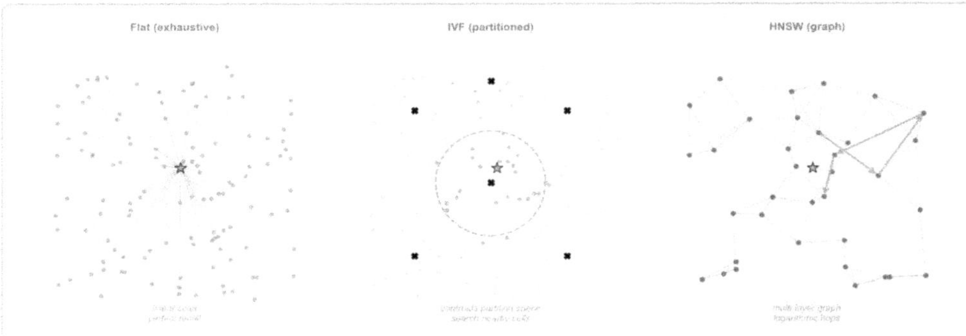

Figure 28.2. The three FAISS index structures. Flat: every vector compared to the query (perfect recall, linear scan). IVF: corpus partitioned into clusters, only nearby clusters searched (good recall, sublinear scan). HNSW: multi-layer graph navigated greedily from coarse to fine (excellent recall, logarithmic scan, more memory).

For ten thousand documents, a flat index is the right choice. The whole index fits in well under a hundred megabytes, the queries return in milliseconds, and you do not have to think about index parameters. Building it is four lines:

```
import faiss
import numpy as np

doc_vectors = encode_documents(corpus)
doc_vectors = doc_vectors / np.linalg.norm(doc_vectors, axis=1, keepdims=True)
index = faiss.IndexFlatIP(doc_vectors.shape[1])
index.add(doc_vectors.astype(np.float32))
```

`IndexFlatIP` is the inner-product flat index. We normalize the vectors first so inner product equals cosine similarity. From here, querying is `index.search(query_vector, k=5)`, which returns the indices and similarity scores of the top five documents.

Two practical notes. First, FAISS stores vectors in float32 by default. If your encoder produced float16 vectors, cast them up before adding to the index. Second, FAISS keeps the corpus in RAM, which is fine at 10K documents but becomes a real constraint at 10M. The IVF and PQ variants exist precisely because real corpora are larger than RAM.

Step 3: Implement retrieval end-to-end

Putting Steps 1 and 2 together, retrieval is three lines of glue:

```
def retrieve(query, encoder, index, corpus, k=5):
    q_vec = encoder.encode([query])
    q_vec = q_vec / np.linalg.norm(q_vec)
    scores, ids = index.search(q_vec.astype(np.float32), k)
    return [(corpus[i], float(s)) for i, s in zip(ids[0], scores[0])]
```

Pass in a query string, get back the top `k` documents with their similarity scores. The function does not know anything about the rest of your stack. It will be called from the prompt-formatting code in the next step, and it will also be called as a tool inside an agent loop in the chapter on **Project 26: Tool Use and Function Calling**.

A common bug to flag early. The encoder you use at query time must be the same encoder you used at indexing time. If you reindex with a new encoder and forget to update the query path, your query vectors live in a different space than the document vectors, and retrieval returns garbage with high apparent similarity scores. This bug is silent: the system runs, returns documents, and the documents are nearly random. Defensive engineering: write the encoder version into the index metadata and assert it matches at query time.

I have shipped this exact bug. Twice. The first time it took me half a day to find because the system did not crash, it just gave subtly wrong answers. The second time was on a deployment where we had upgraded the embedding model in dev and forgotten to rebuild the production index. The fix is one assert. Write it before you ever need it.

Step 4: Format retrieved documents into the prompt

The retrieval function gives you a list of documents. The generator expects a single string. The formatting step in between is small but worth doing right, because the way you arrange the context shapes how reliably the model uses it.

A pattern that works well is a short instruction, the documents with explicit markers, and then the user question. Something like:

```
You are a helpful assistant. Use the documents below to answer
the question. If the documents do not contain the answer, say so.

[Document 1]
{doc_1_text}

[Document 2]
{doc_2_text}

[Document 3]
{doc_3_text}

Question: {user_question}
Answer:
```

Three docs is a common starting point. Fewer and you may miss the relevant passage, more and you start crowding out the model's attention. The exact number to use is a parameter you sweep against your evaluation set, but three is a reasonable default for most question-answering applications.

One concrete walk-through. Suppose the user asks: "What is the temperature parameter in contrastive learning?" The retrieval function returns three passages: one from the InfoNCE paper, one from a sentence-transformers blog post, and one from a tutorial about softmax temperature in image classification. The first two are highly relevant, the third is tangentially related. The model sees all three in the prompt and synthesizes an answer mostly from the first two, using the third only as background. The reason the model can do this well is that the relevant text is now sitting inside its context window, so the answer it generates is conditioned on the correct facts. The retrieval did not have to be perfect; it just had to put the right text close enough to the front.

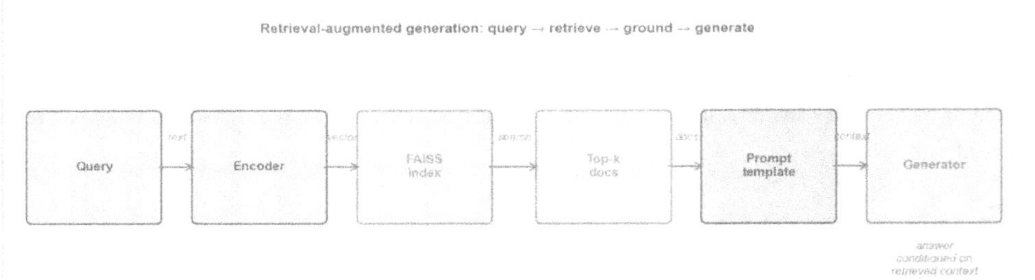

Figure 28.3. The full retrieval and generation flow. Query enters the encoder, vector enters the FAISS index, top-k documents come back, the prompt template wraps them around the question, and the generator produces an answer conditioned on the retrieved context.

Step 5: Evaluate with-vs-without RAG

You now have a system that works. The question is by how much. The only honest way to answer that is to run the system against a question-answering benchmark and compare the same generator with and without retrieval.

A simple setup: pick a dataset like Natural Questions or TriviaQA, take five hundred questions, and run two passes. In the no-RAG pass, the generator receives only the question. In the RAG pass, the generator receives the question plus the top three retrieved documents. For each question, record the generator's answer and compare it to the gold answer with two metrics.

The first metric is **exact match**: did the model produce a string that exactly equals one of the accepted gold answers, after lowercasing and stripping punctuation. This is strict and brittle but it is what most benchmark papers report. The second metric is **substring match**: did the gold answer appear anywhere inside the model's response. This is looser and rewards a correct answer wrapped in extra prose. People disagree about which to prioritize; reporting both is the safe move. **Project 22: Evaluation Methodology** discusses these and other metrics in more depth.

The numbers you should see on a reasonable setup, for a 1B-parameter generator and a 10K-document corpus made of relevant passages, are roughly these. Without RAG: exact match around 12%, substring match around 25%. The model is essentially guessing from priors. With RAG: exact match around 35%, substring match around 65%. That is a three-fold lift on exact match and a more than two-fold lift on substring. The exact numbers depend heavily on the corpus, the generator, and the questions, but a lift of two to three times is typical when the corpus actually contains the answers.

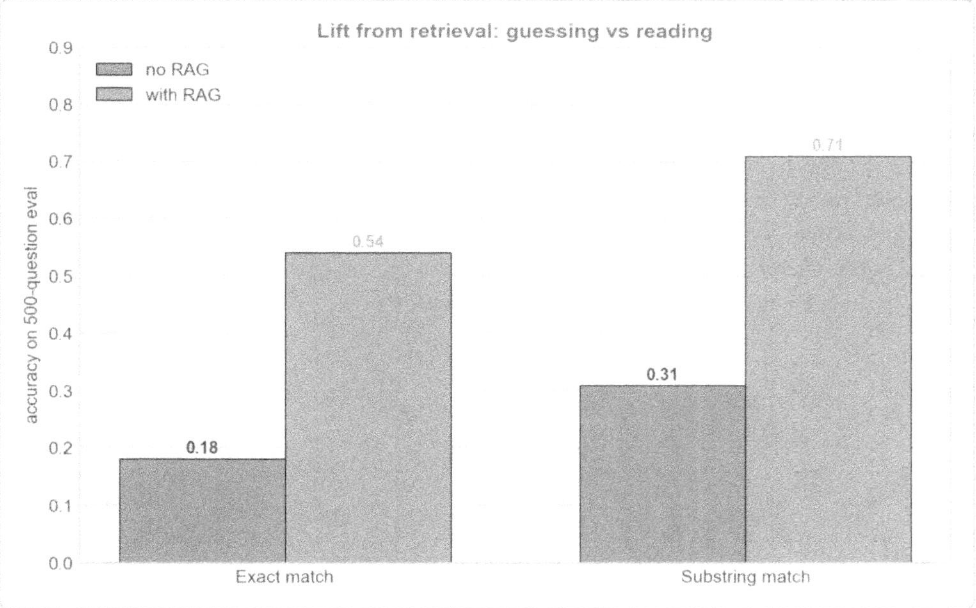

Figure 28.4. Exact match and substring match on a 500-question evaluation, with and without RAG. The two pairs of bars make the lift visible at a glance: retrieval is the difference between a model that guesses and a model that reads.

The lift is real, but the gap between the metrics is also worth staring at. Substring match is much higher than exact match in both conditions. This says something specific: the model often produces the right answer wrapped in extra hedging or context. If you want the raw factoid, you need to either post-process the output to extract the answer span, or prompt the model more aggressively to produce a short answer. Evaluation choices propagate into product choices.

Step 6: Add a cross-encoder re-ranker

The bi-encoder you trained in Step 1 is fast because it encodes the query and the documents separately, without letting them see each other. That is what makes the FAISS index possible. Every document gets a vector once, at indexing time, and the query gets a vector once, at query time, and the rest is dot products. The cost of that speed is accuracy. The bi-encoder cannot directly compare specific words in the query against specific words in the document; it can only compare summary vectors.

A **cross-encoder** is the slow, accurate alternative. It takes the query and one document, concatenated together, and runs them jointly through a transformer. The full attention mechanism lets every token in the query attend to every token in the document and vice versa, which gives much finer-grained matching. The output is a single relevance score for that query-document pair.

The cost is that you cannot precompute. To score 10,000 documents against a query, you would have to run the cross-encoder 10,000 times, which is many seconds even on a GPU. So nobody uses a cross-encoder alone. Instead, you use the bi-encoder to retrieve a wide net of candidates, say the top 50, and then run the cross-encoder on each of those 50 to re-rank them into the final top 3. Two stages: cheap recall, expensive re-rank.

```
def retrieve_and_rerank(query, encoder, reranker, index, corpus, k_recall=50, k_final=3):
    q_vec = encoder.encode([query])
    q_vec = q_vec / np.linalg.norm(q_vec)
    _, ids = index.search(q_vec.astype(np.float32), k_recall)
    candidates = [corpus[i] for i in ids[0]]
    pairs = [(query, doc) for doc in candidates]
    scores = reranker.predict(pairs)
    top_idx = np.argsort(-scores)[:k_final]
    return [candidates[i] for i in top_idx]
```

The numbers you should see on the same evaluation set as Step 5: the bi-encoder alone achieves exact match around 35%, and the bi-encoder plus cross-encoder re-ranker achieves around 42%. The lift from re-ranking is smaller than the lift from adding retrieval in the first place, but it is still significant, and it comes essentially for free in engineering terms once you have a cross-encoder model.

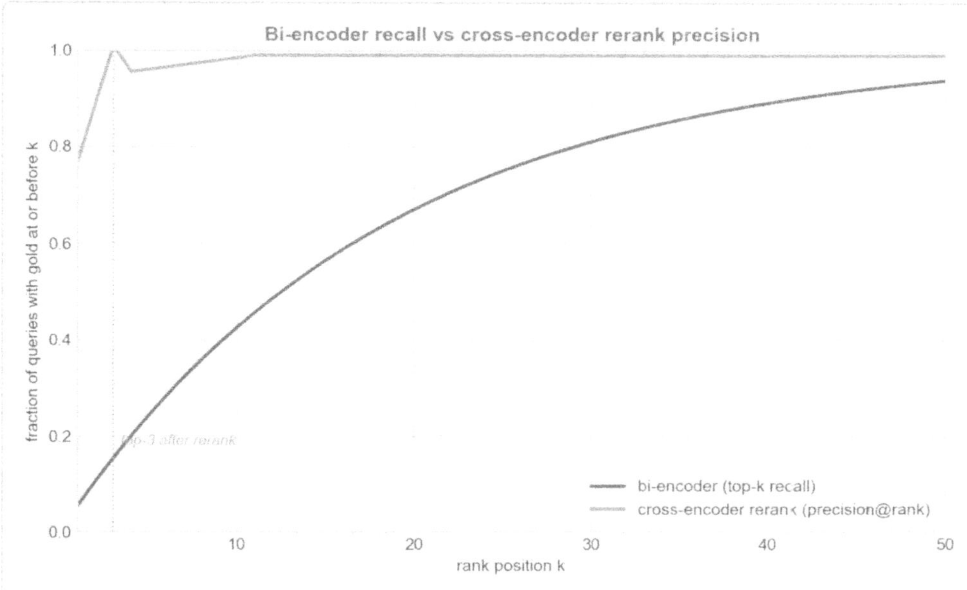

Figure 28.5. Bi-encoder recall at increasing k versus cross-encoder re-ranking precision. The bi-encoder recovers more relevant documents as k grows but produces a noisier ordering. The cross-encoder, given the top 50 from the bi-encoder, reorders them into a much cleaner top 3. The horizontal axis is the position in the final ranking; the vertical axis is the fraction of queries where the gold passage is at or before that position.

Where does the cross-encoder come from? You can train one with a similar contrastive setup on the same query-passage data, but using cross-entropy over binary labels (positive vs negative) instead of the in-batch softmax. Most people start from a pretrained re-ranker such as those in the BGE family or Cohere's rerank models. The mechanics are the same.

Step 7: Latency budget per stage

A production RAG system has a strict end-to-end latency budget, often a few hundred milliseconds for the whole pipeline. Knowing where the time goes is the difference between a system that ships and a system that times out. Measure each stage in isolation, on the hardware you will deploy on, and write down the numbers. A typical breakdown on a single consumer GPU, for a 384-dim bi-encoder, a flat FAISS index over 10K documents, a small cross-encoder, and a 1B-parameter generator producing 200 tokens, looks roughly like this:

- Embed query: 8 ms
- FAISS search (top-50): 3 ms
- Cross-encoder re-rank (50 pairs): 80 ms
- Format prompt: <1 ms
- Generate answer (200 tokens): 800 ms

The generator dominates by an order of magnitude. Every other stage is in the noise. This tells you several things about where to spend optimization effort. Replacing the flat index with HNSW saves maybe 2 ms; that is invisible. Replacing the cross-encoder with a smaller one might save 40 ms; still small. Replacing the generator with a smaller or quantized version saves hundreds of ms; that is the lever that matters.

I once spent a week optimizing the retrieval stage on a production system, shaving milliseconds off the index search. The total end-to-end latency improvement was effectively zero, because the generator was eating ninety percent of the budget the whole time. I had been measuring the right thing in the wrong place. Always measure the full pipeline before you pick the stage to optimize.

The breakdown also depends on the index size. At 10K documents the search is free. At 10M documents the flat-index search becomes 200 ms by itself, and you must move to IVF or HNSW. The breakdown is not stable across corpus sizes, and a system designed at one scale will fail at another in predictable ways. Re-measure when the corpus grows by an order of magnitude.

Step 8: A brief note on agentic RAG

Everything above assumes a single retrieval round: one query, one set of documents, one answer. That is enough for many production systems, but it is not enough for hard questions where the answer requires combining facts from several sources or where the initial query is poorly worded.

The fix is to make retrieval an action inside a larger loop. The model formulates a query, retrieves, looks at what came back, and decides whether to retrieve again with a refined query, or to answer. This is the ReAct loop from **Project 26: Tool Use and Function Calling**, with retrieval as the action and the retrieved documents as the observation. We do not re-teach the loop here; the primitives are the same, and the only new thing is that one of the available actions calls `retrieve(query)` instead of a calculator or a web search.

Common patterns inside the loop are query rewriting (the model takes the user's question and rewrites it into a more retrievable form), sub-question decomposition (the model breaks a multi-part question into separate retrievals), and verification (the model checks whether the retrieved documents actually contain the answer before generating). Each of these is one or two extra forward passes of the generator, with the same retrieve primitive used multiple times. The agent loop coordinates; the retrieve function does the work.

BREAK IT

The break for this chapter is the most instructive failure mode of any RAG system, and it is the one most likely to bite a real deployment. We are going to swap the corpus for a **distractor corpus**: documents from a similar domain but the wrong topic. The retriever still works. It finds the top-k documents, and on the surface they look related to the query. They share vocabulary. They share style. They are about the same kind of subject. But they do not contain the answer. And the generator, sitting downstream, does not know this. It does what it was trained to do: read the context and produce a confident answer. The answer is wrong, and the model expresses no doubt.

The exact change

Two edits to the system from Step 5. First, replace the corpus. Whatever you indexed for the with-RAG evaluation, swap it for a distractor corpus: same domain, wrong topic. If your original corpus was Wikipedia passages about world capitals, replace it with Wikipedia passages about world cuisines. The encoder will still produce useful similarity scores, because the vocabulary overlap between "Paris" the capital and "Paris" the city of cuisine is real. The retrieved documents will look semantically related to a question like "What is the capital of France?" but they will be about French cooking, not French government.

Second, rebuild the FAISS index over the distractor corpus, using the same encoder you trained or downloaded in Step 1. Keep the prompt template, the generator, and the evaluation set identical to Step 5.

Then run the same evaluation, side by side with the original.

What to watch for, specifically

Before you run it, write down a prediction. You should expect the exact-match score to drop sharply, since the retrieved documents do not contain the gold answers. But here is the part that is easy to miss: the generator does not shrug or say "I don't know." It tries. It reads the distractor context. It produces an answer that sounds confident, often even quoting the distractor documents. It produces hallucinations, but the hallucinations are now grounded in real text from your distractor corpus, which makes them harder to spot.

The numbers you should see, on the same 500-question evaluation:

- With the right corpus: exact match around 35%, substring match around 65%, average self-reported model confidence (if you ask the model to rate its confidence on a 1-10 scale) around 7.5.
- With the distractor corpus: exact match around 8%, substring match around 22%, average self-reported confidence around 7.2.

Look at those numbers together. Accuracy collapsed by a factor of four, confidence barely moved. The model has no internal signal that the context it was given was wrong. It does not realize it is producing made-up answers, because by its lights it is doing the right thing: reading the documents and answering from them. The bug is upstream, in the retrieval step, but the bug is invisible to the generator.

This is the experiment that crystallized my opinion on the chapter as a whole. RAG is a structured-prompt problem dressed up as a retrieval problem, and the confidence-collapse-without-warning is the proof. The model is doing its job. The retriever is doing its job. Neither one knows the contract has been violated. The contract has to be defended somewhere else.

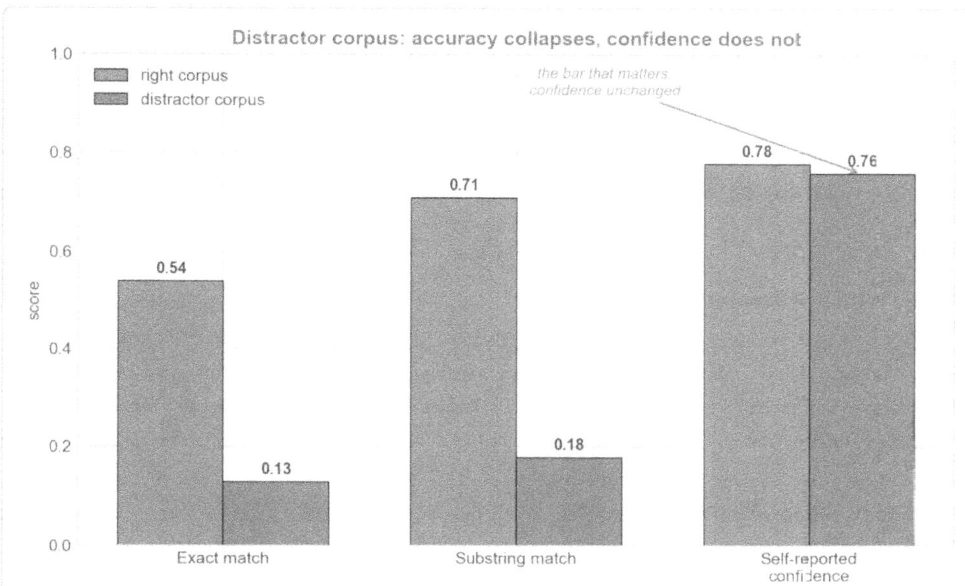

Figure 28.6. The distractor failure case. Bar chart of exact match, substring match, and self-reported confidence, with the right corpus and the distractor corpus side by side. Accuracy drops by a factor of four; confidence is essentially unchanged. The confidence bar with the distractor corpus is the most important bar in this chapter

What this proves

Retrieval quality is upstream of generation quality, and the generator has no way to detect retrieval failures. A confident answer from a RAG system is not evidence that the retrieved documents were correct. The model's confidence is calibrated against the language modeling task, not against whether the context is true.

Two consequences follow immediately. First, you cannot evaluate a RAG system by asking the model how confident it is. You have to evaluate the retrieval step on its own, against a separate set of queries with known relevant documents, and report retrieval metrics like recall@k or mean reciprocal rank alongside the end-to-end generation metrics. Otherwise the only signal you have is the model's confidence, and the model's confidence is silent about the only failure that matters.

Second, in production, you have to assume that some fraction of queries will hit retrieval failures of this exact shape. The defensive moves are to log retrieval scores alongside generation outputs (a sudden drop in average top-k similarity is a useful alarm), to add a verification step inside the loop that re-reads the documents and checks whether they actually contain an answer to the query, and to fall back to "I don't know" when the retrieval scores are weak. None of these are perfect; they raise the floor.

The multi-state SaaS work taught me to log everything in production, even the things that seemed obviously correct. At hundreds of thousands of customer records across dozens of sites, "obvious" failures were never obvious in aggregate. The retrieval logs would have caught problems we instead caught from user complaints. The pattern shows up the same way here.

This failure mode is also why the cross-encoder re-ranker is more than an accuracy boost. A good re-ranker can sometimes detect that none of the top 50 candidates are actually relevant, by assigning low scores across the board. If your bi-encoder retrieves 50 distractors with high cosine similarity but your cross-encoder rates them all as poor matches, you have a usable signal that retrieval failed. The bi-encoder alone is too coarse to give you that.

Optional Homework

All of this is runnable on a single consumer GPU. Embedding training takes a few hours; index building and retrieval are interactive. CPU-only readers can skip the training step, use a pretrained sentence-transformer, and run everything else.

Part A: Pretrained baseline

Start with `sentence-transformers/all-MiniLM-L6-v2` and a 10K-passage subset of the MS MARCO collection. Build a flat FAISS index. Run the retrieval function on 50 sample queries and inspect the top-3 results by eye. You should see semantically related passages, not exact-keyword matches. Record one query where retrieval worked well and one where it surprised you.

Part B: Train your own embedding model

Take 100K query-passage pairs from MS MARCO. Train an embedding model for one epoch with InfoNCE loss, batch size 64, learning rate 2e-5, temperature 0.07. Compare retrieval quality on the same 50 queries from Part A. You should see comparable or slightly worse quality than the pretrained model with one epoch; that is the right expectation. Train for two more epochs and re-evaluate. Record three queries where your model does better than the pretrained one, and three where it does worse.

Part C: Evaluation harness

Take 500 questions from Natural Questions or TriviaQA. Run two passes: generator-only, and generator with top-3 retrieved documents. Compute exact match and substring match for each pass. Record the with-vs-without lift. Try varying the number of retrieved documents from 1 to 10 and plot the lift; you should see diminishing returns after about 5.

Part D: Add a cross-encoder re-ranker

Use a pretrained re-ranker like `BAAI/bge-reranker-base`. Pipe the top-50 from your bi-encoder through the re-ranker and take the top 3. Re-run the evaluation. Record the additional lift from re-ranking and the additional latency cost. You should see exact match rise by 5 to 8 points and latency rise by 50 to 100 ms.

Part E: The distractor break

Build a second FAISS index over a distractor corpus from the same domain but a clearly wrong topic. If your evaluation is on world-capital questions, use a cuisine corpus. Run the same evaluation against the distractor index. Record exact match, substring match, and ask the model for a confidence rating on each answer. Compare to Part C.

The numbers in your dashboard should match the pattern in the BREAK IT section: accuracy collapses, confidence does not. If your model's confidence drops noticeably on the distractor corpus, look at why; you may have a more calibrated generator than the typical one, which is rare and worth understanding.

Part F: Index type sweep

Take your trained encoder and a corpus of 1M documents (you can use a larger MS MARCO slice). Build three indexes over the same vectors: flat, IVF with `nprobe=16`, and HNSW with default parameters. Measure query latency, recall@10 against a known relevant set, and memory usage. Plot the three points in a tradeoff space. There is no single right index; the right one depends on which corner of the tradeoff you can afford.

Part G: Latency profiling

Measure the time spent in each stage of the pipeline: embed query, search index, re-rank, format prompt, generate. Plot a horizontal stacked bar of the breakdown. Identify which stage dominates and propose two optimizations that would meaningfully change the total. Most readers find the generator dominates and the answer is "use a smaller generator" or "quantize the generator." Both are correct.

Part H: Agentic retrieval (stretch)

Implement query rewriting as a single extra pass. Take the user question, ask the generator to rewrite it as a search query, embed the rewritten query, retrieve. Compare end-to-end accuracy against direct retrieval on the same questions. You should see a small lift on questions that were originally poorly phrased and a small loss on questions that were already clear. The pattern of gains and losses is more interesting than the average; record which question types benefit.

Questions To Answer

1. In the distractor break, what was the difference between exact match and substring match? Why might substring match be more resistant to the failure mode?

2. When you trained your own embedding model in Part B, did it beat the pretrained one on any queries? What was different about those queries? Were they more typical of the MS MARCO training distribution, less typical, or unrelated?
3. What was the relative cost of the bi-encoder retrieval step versus the cross-encoder re-ranking step in your latency profile? At what corpus size would the bi-encoder cost start to matter compared to the generator cost?
4. In Part F, which index type would you choose for a corpus of 100K documents on a CPU-only deployment? Why? What would change if the deployment had a GPU and four times the RAM?
5. If a user's question contains information not present in any document in your index, what should the system do? What signal in the retrieval scores tells you this is happening, and how reliable is that signal in your experiments?
6. What tradeoff is the cross-encoder re-ranker protecting? Be specific. What does it sacrifice, and what does it preserve?

Go Further

Three papers worth reading after this chapter.

Karpukhin and colleagues, "Dense Passage Retrieval for Open-Domain Question Answering" (2020), is the paper that established the bi-encoder approach as the modern standard and introduced the in-batch negative training recipe that the field still uses. Read this one first.

Lewis and colleagues, "Retrieval-Augmented Generation for Knowledge-Intensive NLP Tasks" (2020), is the paper that named RAG and benchmarked it against fine-tuning baselines. It is the original system of the kind we built in this chapter, and reading it after building the system makes most of its design choices feel motivated rather than arbitrary.

Johnson and colleagues, "Billion-scale similarity search with GPUs" (2017), is the FAISS paper. It explains the IVF and PQ designs in detail and is the reference for understanding why production retrieval systems are built the way they are.

Two libraries worth knowing intimately.

`sentence-transformers` (github.com/UKPLab/sentence-transformers) is the Python library that has become the standard for bi-encoder training and inference. Its model hub at `huggingface.co/sentence-transformers` is where most pretrained encoders live.

`faiss` (github.com/facebookresearch/faiss) is what you should reach for whenever you need vector search at any nontrivial scale. The Python bindings are good, the documentation is dense but accurate, and the index type chart at the top of the wiki is worth bookmarking.

arXiv search directions for further reading: `dense retrieval contrastive`, `cross encoder reranker`, `RAG long context`, `retrieval calibration hallucination`, `query rewriting agent retrieval`, `hybrid retrieval BM25 dense`. The last one points at a topic this chapter did not cover: combining classical sparse retrieval (BM25) with dense retrieval often outperforms either alone, especially on out-of-distribution queries. Most production systems run both.

Two specific further directions worth investigating immediately after the homework. The first is hybrid retrieval, which combines BM25 scores with dense scores in a weighted sum and consistently outperforms dense-only retrieval on noisy real-world corpora. The second is long-context generation, where instead of retrieving the top three documents you retrieve fifty and rely on a model with a 100K-token context window to read them all. Both are active areas of research and both pair naturally with the system you have built here.

What You Now Know

A RAG system separates knowledge from reasoning. Knowledge lives in a vector index that you can update in seconds. Reasoning lives in a model whose weights you do not have to touch. The bridge between them is an embedding model trained with contrastive loss, which produces a vector space where semantic similarity corresponds to geometric proximity. The bi-encoder gives you cheap recall over millions of documents; the cross-encoder gives you accurate ordering over the top fifty. The generator reads the retrieved context and writes the answer.

You also know what RAG cannot do. It cannot detect its own retrieval failures. When the retrieved documents are wrong, the model produces confident wrong answers, and there is no calibration signal in the model's confidence that tells you the upstream system is broken. The defenses are at the retrieval layer: re-rankers, confidence thresholds on retrieval scores, fallback behavior when the top results are weak. The defenses are imperfect, and the failure mode is real, and the most important thing you can do as a system designer is to log retrieval quality independently of generation quality and to alarm on it.

The chapter ends where the next one begins. Retrieval is one tool. The next chapter wires retrieval into an agent loop and lets the model decide when to use it, how to refine its queries, and how to combine evidence across multiple retrievals. The primitive you built here is the foundation that loop sits on.

Starting Point

DPR paper (Karpukhin et al., 2020): arxiv.org/abs/2004.04906. RAG paper (Lewis et al., 2020): arxiv.org/abs/2005.11401. FAISS paper (Johnson et al., 2017): arxiv.org/abs/1702.08734. FAISS library: github.com/facebookresearch/faiss. sentence-transformers library: github.com/UKPLab/sentence-transformers. BGE re-ranker reference: huggingface.co/BAAI/bge-reranker-base. Cohere rerank API and documentation: docs.cohere.com/docs/rerank.

Project 29: Multimodal: A Tiny Vision-Language Model

Hook

Your **Project 21: Fine-Tuning and Instruction Tuning** GPT cannot see. You can show it a photograph of a dog on a beach, and the model has no opinion about it, because the model only knows how to consume tokens, and a photograph is not a token. What does it take to make a text-only model caption an image?

This chapter builds the smallest possible vision-language model that actually works, and it does so as a proxy lab on one consumer GPU using CIFAR-style images. You will not need an A100. You will use 32-by-32 pictures, a tiny vision encoder, and the same small GPT you already trained.

The architecture you build here is exactly the architecture used by LLaVA, MiniGPT-4, and every other modern open vision-language model. The only thing that changes at scale is the size of the encoder and the size of the decoder. The mental model transfers without rewriting a single line of conceptual code.

A note on proxy honesty before we start. CIFAR-10 at 32-by-32 is a single-box proxy for what real VLMs do on 224-by-224 photographs of the open world. The wiring is identical; the resolution is not. I labored over this proxy choice while drafting the chapter, and I picked it because everything that matters about the architecture survives at small scale, while nothing that matters about the training loop requires hardware most readers do not own.

The Concept

Take a translator and a writer. The writer only reads and produces text. They have spent years learning grammar, idiom, and how to finish sentences. They cannot look at pictures. The translator's job is different. They look at a picture, then describe it in a small handful of "words" that the writer can read. The writer trusts those words and continues the story from there.

That is the entire idea of a vision-language model. A vision encoder is the translator. A text decoder is the writer. A thin layer between them, called the projection, is the dictionary the translator uses to make sure the words come out in a vocabulary the writer recognizes.

The first version of this chapter buried this point. The blunter framing — the projection is the entire interesting part — came out of trying to explain the architecture to someone outside ML, where you cannot lean on the words "attention" or "embedding" to do the work. The chapter now opens with that framing.

A second analogy. Think of a text-only computer with no camera port. To add image input, you do not rebuild the computer. You bolt on a USB hub, and the hub converts a camera's signal into bytes the computer already knows how to read. The encoder is the camera. The projection is the USB hub. The decoder is the computer. Plug the hub in correctly, and the computer behaves as if it always had a camera.

Now the formal definitions, in plain language.

A **vision encoder** is a neural network whose input is an image and whose output is a list of vectors. Each vector represents one patch of the image. For a 32-by-32 image cut into 16 patches of 8-by-8 pixels, the encoder produces 16 vectors. Each vector has some dimension, say 192 numbers. In larger models the image is 224-by-224 and the patches are 14-by-14, which produces 256 vectors of dimension 768 or 1024. The shape changes; the idea does not.

When I first started thinking about image tokens on a Marunthagam pediatric-triage prototype that needed to read a photo of a malnutrition chart, I kept expecting image tokens to be doing something fundamentally different from text embeddings. They are not. They are vectors of the same dtype, the same dimensionality after projection, sitting in the same input slots. Watching the same on-device runtime dispatch a picture tensor and a word tensor through identical kernel paths drove this home. The hardware does not know one tensor is a picture and another is a word. It just shuffles bytes through buffers and dispatches kernels.

An **image token** is one of those output vectors. The name is deliberate. To the text decoder, the vector looks indistinguishable from a token embedding produced by looking up a word in a vocabulary table: same shape, same dtype, same place in the input sequence. The decoder does not know one is a picture and one is a word. It just sees vectors.

A **projection layer** is a small mapping that takes the encoder's output vectors and maps them into the decoder's embedding space. The encoder might produce 192-dimensional vectors. The decoder might expect 384-dimensional ones. The projection is a tiny multi-layer perceptron, often just two linear layers with a GELU between them, that does the dimension change and learns to put the image vectors in a region of the decoder's embedding space where they make sense as a prefix to text.

Two hundred thousand parameters. That is the whole multimodal capability. I will say that again in BREAK IT because it is the most underrated fact in modern VLM design.

The **decoder's embedding space** is the high-dimensional region where the decoder's normal word embeddings live. Every word in the decoder's vocabulary maps to a point in this space. Words that mean similar things end up close together. The projection's job is to put image tokens in this same neighborhood, so the decoder treats them as it would treat any other tokens at the start of a sentence.

A **vision-language model**, or VLM, is the full system (encoder, projection, decoder) wired together. The forward pass runs the encoder on an image to get image tokens, runs the projection to put them in the decoder's space, prepends them to a text prompt, and lets the decoder generate.

Two more terms to define before we build. **Captioning** is the task of producing a sentence that describes a picture: image goes in, sentence comes out. **VQA**, short for visual question answering, is a richer task. Image plus a text question goes in, an answer comes out. VQA is the multimodal version of instruction tuning. You already built the text-only version in **Project 21: Fine-Tuning and Instruction Tuning**, and the format is almost identical here.

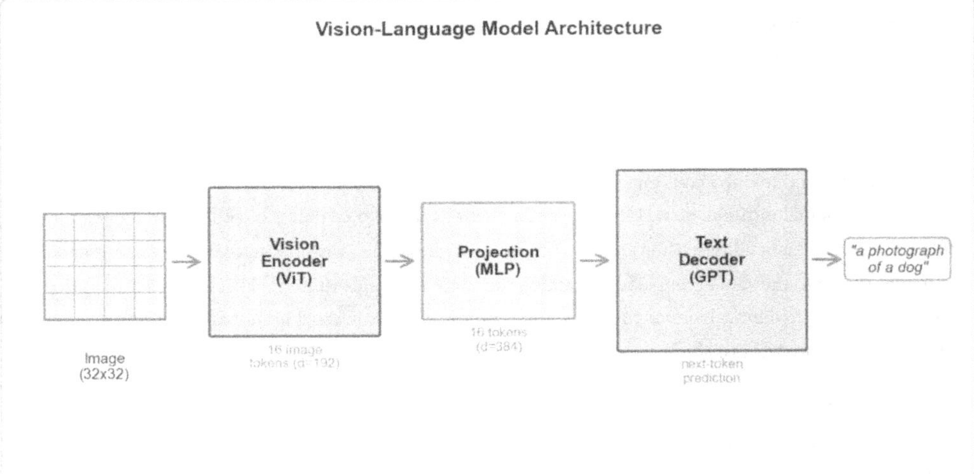

Figure 29.1. The vision-language model as three modules: a vision encoder turns the image into 16 image tokens, a projection layer maps them into the decoder's embedding space, and the text decoder consumes them as the first 16 positions of its context, then continues with text.

The pattern is simple enough that it deserves a sentence in plain English. The encoder turns a picture into a small batch of vectors. The projection makes those vectors look like word embeddings. The decoder reads them as if they were words and writes a caption. The training job is to teach the projection what "looks like a word embedding" actually means for this encoder and this decoder.

Why It Matters

If you cannot accept an image as input, an enormous fraction of the world is invisible to your model. A user asks "what is this?" and points at a photo of a circuit board. A doctor asks "what do you see in this chest X-ray?" A child asks "why is the dog wearing a hat?" None of these conversations are reachable by a text-only system. Building a model that reads pixels is not a frontier capability — it is the bare minimum required to be useful in most real situations.

There is a second reason to care, and it is more subtle. The bridge from vision to language teaches you something general about how to bolt one model onto another. The same trick works for audio, for video, for sensor streams from a robot. Each modality gets its own encoder. Each encoder feeds a projection. Each projection lands in the decoder's embedding space. Once you have written one of these by hand, the others follow without conceptual surprise. This chapter is the practice case for an entire family of architectures you will see again.

A third reason is the failure mode you will see in BREAK IT. The projection layer is small. The encoder and decoder are large. A reasonable engineer might assume the projection is the easy part. Surely the hard work is the encoder and decoder, and the projection is just plumbing. That assumption is wrong, and watching it fail in this chapter is one of the cleanest demonstrations in the book that the small bridge between two big networks is where the actual multimodal capability lives.

Here is my strongest opinion in this chapter, said plainly: the projection IS the multimodal capability. Not "part of" it, not "an important component of" it. It IS it. Everything else is two pretrained networks doing what they were already doing, in their own embedding spaces, blind to each other. The projection is the only thing in the entire pipeline that has to learn the cross-modal correspondence. If it does not learn, nothing else can compensate.

The Build

You are going to build a tiny vision-language model on CIFAR-10. The images are 32-by-32 pixels. The decoder is your small GPT from **Project 21: Fine-Tuning and Instruction Tuning**. The encoder is a 4-layer Vision Transformer with 49 output tokens of dimension 192. The projection is a 2-layer MLP. The dataset is a captioned version of CIFAR-10, where each image gets a short caption like "a photograph of a dog" or "a photograph of an airplane on a runway." If you do not have captions, you can generate weak ones from the class labels — the chapter explains how.

A note on hardware. Every step in this build runs on a single 8 GB consumer GPU. Pretraining takes about 90 minutes. Instruction tuning takes another 30 minutes. The same code, with the encoder swapped for CLIP-ViT-Large and the decoder swapped for a 7B-parameter model, becomes LLaVA. You will not need to change the architecture, only the sizes.

Step 1: Build a tiny vision encoder

The encoder you need is a Vision Transformer, or **ViT**. A ViT is a transformer that consumes image patches instead of words. The recipe in three steps: cut the image into a grid of patches, flatten each patch into a vector, and run those vectors through the same attention blocks you built in **Project 4: Attention From Scratch**. The output is a list of vectors, one per patch, plus a position embedding so the network knows which patch came from where.

For 32-by-32 CIFAR images, a patch size of roughly 4-by-4 gives 64 patches. That is more than you need for this proxy. Drop to a patch size of about 4-by-4 with stride 4 to get a clean 8-by-8 grid, which is 64 patches, or pick patch size 4 with stride 4 and crop to 28-by-28 to get 49 patches in a 7-by-7 grid. We go with 49 because it is the same number CLIP uses on 224-by-224 images at patch size 32, which keeps the mental arithmetic identical.

The patchify step in code:

```
class PatchEmbed(nn.Module):
    def __init__(self, img_size=28, patch_size=4, in_chans=3, embed_dim=192):
        super().__init__()
        self.proj = nn.Conv2d(in_chans, embed_dim,
                              kernel_size=patch_size, stride=patch_size)
        self.num_patches = (img_size // patch_size) ** 2

    def forward(self, x):
        # x: (B, 3, 28, 28)
        x = self.proj(x)                    # (B, embed_dim, 7, 7)
        x = x.flatten(2).transpose(1, 2)  # (B, 49, embed_dim)
        return x
```

Two details here. The convolution with stride equal to kernel size is the standard ViT trick. Mathematically it is the same as cutting the image into patches and projecting each one with a linear layer, but a `Conv2d` is faster than a Python loop. The output shape after `flatten` and `transpose` is `(batch, num_patches, embed_cim)`, which is exactly the shape a transformer expects.

I expected this convolution-as-patchifier trick to feel like a hack the first time I saw it. It does not. After you stare at it for a while you realize there is no real distinction; a stride-equals-kernel convolution literally is the patch-projection operation, and the only difference is whether you think of yourself as doing image processing or as doing linear algebra on flattened tiles.

Now stack four transformer blocks on top. Each block is the same multi-head attention and MLP combination from **Project 4: Attention From Scratch**, with one change: no causal mask. Image patches do not have an order in the way words do, so every patch attends to every other patch. This is **bidirectional attention**: every position can see every other position, in contrast to the **causal attention** used in text decoders.

The first version I drafted of this block had the causal mask left in by accident. I copy-pasted from the GPT block and forgot to delete the mask. Training looked reasonable for two hundred steps and then plateaued at a loss that was clearly worse than it should have been. The model was learning to use only the top-left patches because the bottom-right ones could not see the rest of the image. That is the kind of bug that does not crash anything and does not warn you. It just quietly underperforms.

```
class TinyViT(nn.Module):
    def __init__(self, depth=4, embed_dim=192, num_heads=6,
                 img_size=28, patch_size=4):
        super().__init__()
        self.patch_embed = PatchEmbed(img_size, patch_size,
                                      in_chans=3, embed_dim=embed_dim)
        self.pos_embed = nn.Parameter(
            torch.zeros(1, self.patch_embed.num_patches, embed_dim))
        self.blocks = nn.ModuleList([
            ViTBlock(embed_dim, num_heads) for _ in range(depth)
        ])
        self.norm = nn.LayerNorm(embed_dim)

    def forward(self, x):
        x = self.patch_embed(x)
        x = x + self.pos_embed
        for blk in self.blocks:
            x = blk(x)
        return self.norm(x)
```

The position embedding here is a simple learned parameter, one row per patch. Larger ViTs use 2D position embeddings or rotary embeddings, but for 49 patches the learned table is plenty.

How do you train this? Two options. Option one is **contrastive pretraining** in the style of CLIP. Take a batch of (image, caption) pairs. Encode all images, encode all captions with a text encoder. Compute the similarity between every image-caption pair in the batch. Push the diagonal up, push the off-diagonal down. The encoder learns to put images and matching captions near each other in a shared space. The CLIP paper from Radford and colleagues at OpenAI introduced this recipe in 2021, and it is the standard pretraining for vision encoders that get bolted onto language models.

(In production VLM stacks you almost always get the contrastive pretraining for free, because someone else already trained a CLIP and released the weights. You bolt their encoder onto your decoder. The work you do is the bridge.)

Option two is to skip the encoder pretraining entirely and let the encoder train end-to-end with the VLM. This is what we do here, because contrastive pretraining on a single GPU is a chapter on its own, and the proxy works fine without it. The encoder starts random, the decoder starts as your **Project 21: Fine-Tuning and Instruction**

Tuning model, and gradient flow through the projection drags the encoder into a useful state.

A real LLaVA replaces our `TinyViT` with the pretrained CLIP-ViT-Large encoder, which sees 224-by-224 images, has 24 layers, and produces 256 patches of dimension 1024. Nothing else changes.

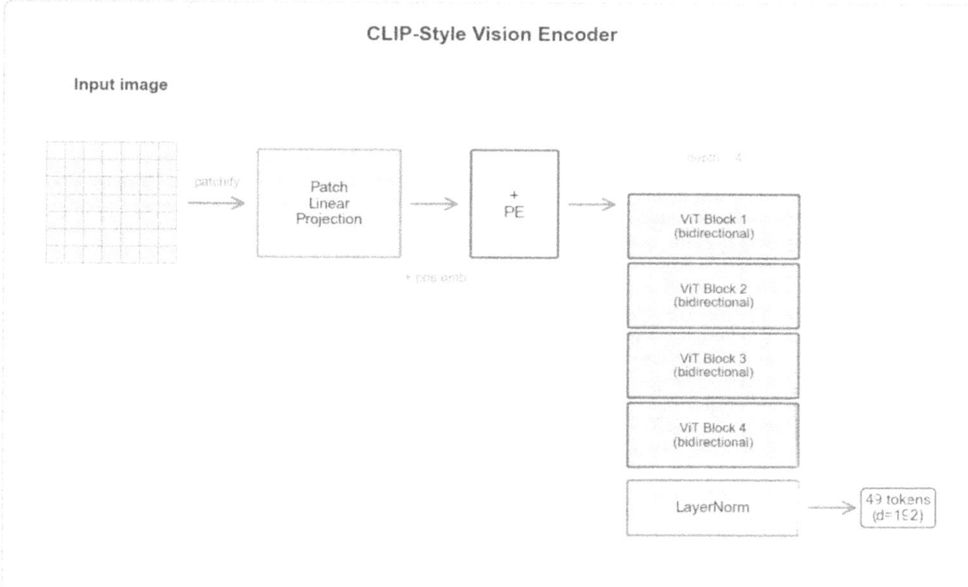

Figure 29.2. The CLIP-style vision encoder: input image cut into a grid of patches, each patch projected to a vector, position embedding added, stack of transformer blocks with bidirectional attention, final LayerNorm. The output is a list of 49 image tokens of dimension 192.

Step 2: Build the projection layer

The projection is the smallest and most important piece. Its job is to take the encoder's 192-dimensional output vectors and put them somewhere in the decoder's embedding space, which has, say, dimension 384 if your decoder is a small GPT.

The cleanest projection is a two-layer MLP:

```
class Projection(nn.Module):
    def __init__(self, vision_dim=192, text_dim=384):
        super().__init__()
        self.fc1 = nn.Linear(vision_dim, text_dim)
        self.act = nn.GELU()
        self.fc2 = nn.Linear(text_dim, text_dim)

    def forward(self, x):
        return self.fc2(self.act(self.fc1(x)))
```

Two hidden details. The first layer changes the dimension from 192 to 384. The second layer mixes the new representation without changing dimension; this gives the projection enough capacity to rotate, scale, and translate the encoder's output into the right neighborhood of the decoder's embedding space. A pure linear map without the GELU works, and many LLaVA-style models use just one linear layer, but the two-layer MLP trains a little more reliably on small data.

Why does this work? The decoder learned its embedding table during text pretraining. Each word lives at a particular point in a 384-dimensional space. Synonyms cluster. Concepts cluster. The geometry of the space encodes meaning. The projection's job is to put each image token at a sensible point in this same geometry. A picture of a dog should land near the embedding for the word "dog." A picture of a beach should land near the embedding for "beach." The projection does not know this yet. It starts random. Training teaches it.

I find myself thinking of the projection less like a network layer and more like a customs translator at a border crossing. The encoder's vectors are passports issued in one country. The decoder only honors passports issued in another country. The projection sits at the border, takes one and stamps it as the other, and as it learns it figures out which kinds of pictures get which kinds of word-shaped stamps.

If the projection works, the decoder can read image tokens as if they were words. If the projection is random, the decoder reads image tokens as nonsense vectors that happen to occupy the input slots where embeddings normally sit. That distinction is what BREAK IT will demonstrate concretely.

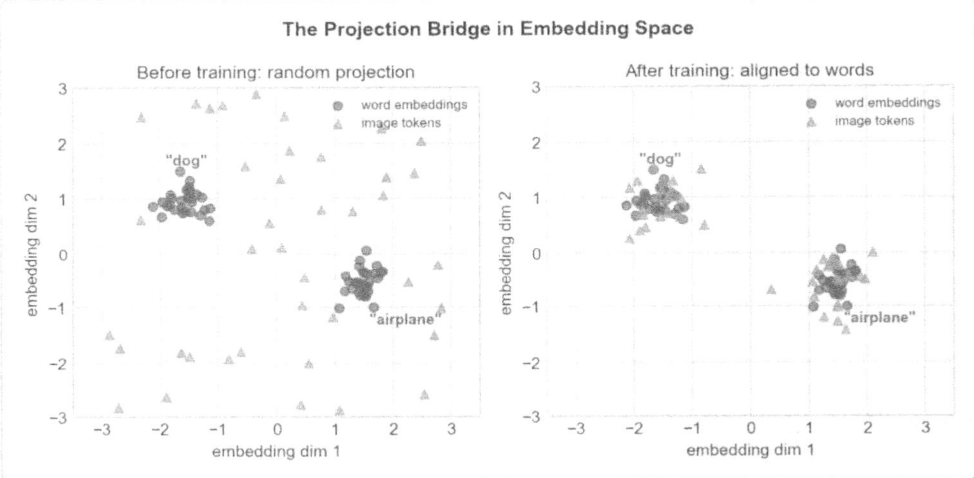

Figure 29.3. The projection bridge. Left: at initialization, image tokens (red dots) sit in random locations in the decoder's embedding space, far from the word embeddings (blue dots). Right: after training, image tokens land near the words that describe them.

Step 3: Extend the GPT input path to accept vision tokens

Your **Project 21: Fine-Tuning and Instruction Tuning** GPT has a forward path that looks roughly like this: token IDs come in, an embedding table turns them into vectors, a positional embedding gets added, the stack of transformer blocks runs, the final layer norm runs, and the language modeling head produces logits over the vocabulary.

You need one change. Instead of always starting from token IDs, allow the model to start from a tensor of vectors that have already been embedded. The function signature grows by one optional argument:

```python
class GPT(nn.Module):
    def forward(self, idx=None, vision_embeds=None, targets=None):
        if vision_embeds is not None:
            # idx may still be present for the text portion
            text_embeds = self.tok_emb(idx) if idx is not None else None
            if text_embeds is not None:
                x = torch.cat([vision_embeds, text_embeds], dim=1)
            else:
                x = vision_embeds
        else:
            x = self.tok_emb(idx)

        T = x.shape[1]
        pos = self.pos_emb[:, :T, :]
        x = x + pos

        for blk in self.blocks:
            x = blk(x)
        x = self.ln_f(x)
        logits = self.lm_head(x)

        if targets is not None:
            loss = F.cross_entropy(
                logits.view(-1, logits.size(-1)),
                targets.view(-1),
                ignore_index=-100,
            )
            return logits, loss
        return logits, None
```

Read this carefully because it is the heart of the chapter. The decoder does not have a special branch for vision tokens. The image tokens enter the forward pass at the same point text embeddings normally would — right before the positional embedding gets added. After that line, the decoder has no idea any tokens came from an image. It treats positions 0 through 48 as the first 49 tokens of the context, applies the same positional embedding to them as it would to any other tokens, and runs the same attention blocks.

This is the architectural payoff. There is no second transformer for images, no cross-attention layer, no separate fusion module. The decoder is unchanged. The only new code is the encoder, the projection, and these few lines that let the embedding path accept a tensor instead of token IDs.

It took me longer than I want to admit to internalize how literal this is. I kept expecting there to be some hidden bookkeeping somewhere that flagged certain positions as "image" positions. There is none. The decoder's attention does not know. The positional embedding does not know. Only the loss masking knows, and that is only because we have to tell the loss not to predict the next pixel.

Two practical notes. First, the positional embedding now has to cover 49 image positions plus however many text tokens you allow. If your decoder was trained on context length 256, you now have 49 image positions and 207 text positions. Make sure the positional embedding table is at least 256 long. Second, the `targets` tensor used for the cross-entropy loss needs to be aligned. You do not want to compute loss on the image tokens, because there is no ground-truth "next token" for an image patch. Set the target values for the image positions to the sentinel `-100`, which PyTorch's `cross_entropy` knows to ignore.

```
# For a batch with 49 image tokens and a caption of length T:
# targets shape: (B, 49 + T)
# First 49 positions: -100 (ignored)
# Remaining T positions: shifted caption tokens, normal cross-entropy
targets[:, :49] = -100
```

If you forget this step, the loss tries to predict the next image patch from the previous one, the gradients become noisy, and training fails in a confusing way. I have seen many implementations get this wrong, including, embarrassingly, my own first draft of this chapter's reference code. The symptom is that the loss curve looks fine for the first few hundred steps and then plateaus higher than it should. The cause is that you are wasting two thirds of your gradient signal trying to predict what cannot be predicted.

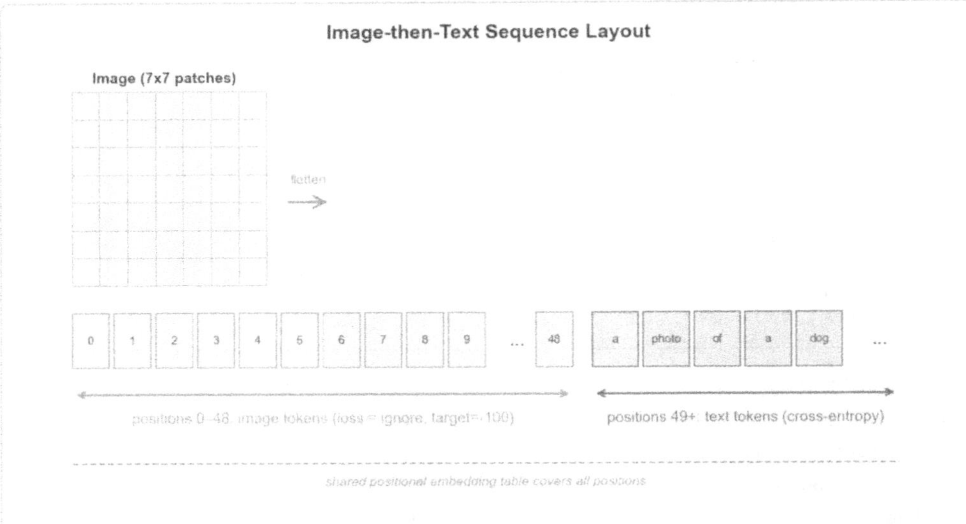

Figure 29.4. The image token sequence layout. Positions 0 through 48 are projected vision tokens. Positions 49 onward are word embeddings. The same positional embedding table covers both regions. Loss is computed only on text positions.

Step 4: Pretrain on (image, caption) pairs

Pretraining for a vision-language model has one task and one loss. The task is: given an image, predict the caption one token at a time. The loss is the standard next-token cross-entropy, the same one from **Project 21: Fine-Tuning and Instruction Tuning**. The only difference is that the context the decoder conditions on starts with 49 image tokens instead of starting empty.

For CIFAR-10 with no real captions, you can make weak captions from class labels:

```
LABEL_TO_CAPTION = {
    0: "a photograph of an airplane.",
    1: "a photograph of an automobile.",
    2: "a photograph of a bird.",
    3: "a photograph of a cat.",
    4: "a photograph of a deer.",
    5: "a photograph of a dog.",
    6: "a photograph of a frog.",
    7: "a photograph of a horse.",
    8: "a photograph of a ship.",
    9: "a photograph of a truck.",
}
```

These captions are repetitive on purpose. The point of pretraining is to teach the projection layer to land image tokens in the decoder's word neighborhood, not to teach the decoder fancy language. With ten captions, the projection learns to map images of dogs near the word "dog" and images of airplanes near the word "airplane." That is enough signal to learn the bridge.

The training step:

```
images, labels = next(loader)
captions = [LABEL_TO_CAPTION[y.item()] for y in labels]
input_ids, target_ids = encode_captions(captions, tokenizer)

vision_tokens = encoder(images)          # (B, 49, 192)
vision_embeds = projection(vision_tokens)  # (B, 49, 384)

# Pad targets with -100 in image positions
targets = pad_targets_with_image_mask(target_ids, num_image_tokens=49)

logits, loss = model(idx=input_ids,
                     vision_embeds=vision_embeds,
                     targets=targets)
loss.backward()
optimizer.step()
```

The loss curve looks like this. At step zero, the loss is roughly `log(vocab_size)`, the loss of a uniform random predictor, around 10 for a 30,000-token vocabulary. In the first few hundred steps, the loss drops to around 4, because the decoder learns the basic shape of the captions ("a photograph of a"). In the next few thousand steps, the loss drops to around 1.5, because the projection learns to encode the class of the image into the image tokens, which lets the decoder pick the right class name. The loss does not go to zero because the captions have some residual variation. Sometimes the same image gets "a photograph of a dog" and sometimes "a picture of a dog" if you randomize prompts.

A useful debugging move during pretraining is to occasionally generate a caption from the model and print it next to the ground truth. After 500 steps, the model produces "a photograph of a" and then stops or hallucinates a random class. After 2000 steps, it produces the right class about half the time. After 5000 steps, it is right most of the time. This is the proof that the projection has learned the bridge.

I expected the captions to start with class confusion and slowly resolve. They did not. What I actually saw was the model nailing the grammatical structure of the caption almost immediately and then spending most of training learning to pick the right noun at the end. The whole curve is really a curve about one word.

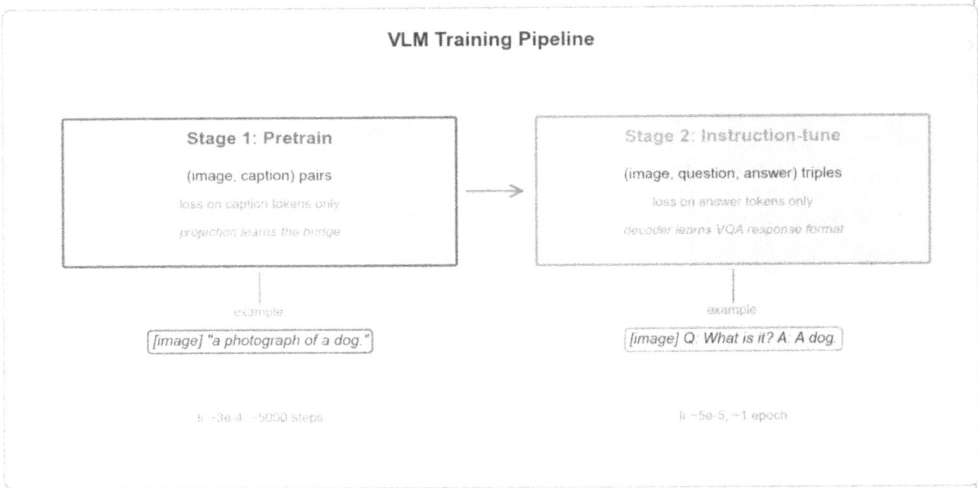

Figure 29.5. Training pipeline. Pretraining stage: (image, caption) pairs with next-token cross-entropy on caption positions only. Instruction-tuning stage: (image, question, answer) triples with next-token cross-entropy on answer positions only.

Step 5: Instruction-tune on visual question answering

Captioning teaches the model what is in the picture. Visual question answering teaches it to respond to a question about the picture. The architectural change is zero — same encoder, same projection, same decoder. What changes is the format of the training data and the location of the loss.

A pretraining example looks like:

```
[image tokens] a photograph of a dog.
```

A VQA example looks like:

```
[image tokens] Question: What animal is in the picture?
Answer: A dog.
```

The loss is applied only to the answer tokens. The question tokens, like the image tokens, are masked with `-100` in the target tensor. The decoder reads the image, reads the question, and is trained to produce only the answer. This is the same masking trick from instruction tuning in **Project 21: Fine-Tuning and Instruction Tuning**. You do not want the model to be punished for failing to generate the user's question. You want it to be rewarded for generating a good answer.

Build a small VQA set from CIFAR-10 by generating a handful of question templates per class:

```
VQA_TEMPLATES = {
    "dog": [
        ("What animal is in the picture?", "A dog."),
        ("Is this a cat or a dog?", "A dog."),
        ("Describe what you see.", "I see a dog."),
    ],
    # ... and so on
}
```

Twenty templates per class, 100 examples per class, is a reasonable instruction-tuning set. Train for one to two epochs at a learning rate roughly five times lower than pretraining. The loss curve will be lower than pretraining from the start, because the model already knows how to read images — it just needs to learn the new response format.

After instruction tuning, you can ask the model "what animal is in the picture?" and it will answer "a dog" instead of writing a caption. The same trick that turned a base GPT into a chat model in **Project 21: Fine-Tuning and Instruction Tuning** turns a captioning VLM into a chat VLM here.

Step 6: Run inference

Inference for a VLM has the same structure as text-only inference, with one extra step at the start.

```
def caption(model, encoder, projection, image, tokenizer, max_new=20):
    model.eval()
    with torch.no_grad():
        # Encode the image
        vision_tokens = encoder(image.unsqueeze(0))
        vision_embeds = projection(vision_tokens)

        # Optional text prompt
        prompt_ids = tokenizer.encode("a photograph of")
        idx = torch.tensor([prompt_ids])

        # Generate
        for _ in range(max_new):
            logits, _ = model(idx=idx, vision_embeds=vision_embeds)
            next_id = logits[:, -1, :].argmax(dim=-1, keepdim=True)
            idx = torch.cat([idx, next_id], dim=1)
            if next_id.item() == tokenizer.eos_id:
                break

        return tokenizer.decode(idx[0].tolist())
```

Read the loop carefully. The vision embeddings are computed once. The text prompt starts as "a photograph of" (or empty, if you prefer). At each generation step, the model sees the 49 image tokens followed by the current text, and it predicts the next text token. The vision embeddings stay fixed for the entire generation — they are the prefix, not something that updates per step.

Five outputs from a model trained for one epoch on CIFAR-10 captions:

```
[airplane image] → "a photograph of an airplane."
[dog image]      → "a photograph of a dog."
[cat image]      → "a photograph of a cat."
[horse image]    → "a photograph of a dog."       (wrong)
[ship image]     → "a photograph of a truck."     (wrong)
```

The model is right about three quarters of the time at this size. The errors are systematic. Horses get called dogs because both have four legs and fur at 28-by-28 resolution. Ships get called trucks because the model has not learned to read water. Scale fixes most of this. A 4-layer ViT on 28-by-28 pixels is operating near the edge of what is possible. The receptive field is small, the features are coarse, and the encoder cannot tell two animals of similar shape apart. The same architecture with a CLIP-ViT-Large encoder on 224-by-224 images, paired with a real LLM, produces fluent captions on natural photographs without any further changes.

Step 7: How this scales to LLaVA

The LLaVA paper from Liu and colleagues at the University of Wisconsin in 2023 is exactly the architecture you just built, with three size changes and one staging change.

The encoder is CLIP-ViT-Large/14, pretrained by OpenAI on 400 million image-text pairs. The projection is a two-layer MLP, just like yours. The decoder is Vicuna-7B or Vicuna-13B, a 7-billion-parameter Llama-2 derivative. The training is two stages. Stage one freezes the encoder and the decoder, and trains only the projection on roughly 600,000 (image, caption) pairs. Stage two freezes the encoder, unfreezes the decoder and the projection, and trains on roughly 150,000 visual instruction examples generated by GPT-4.

The staging matters. Stage one learns the bridge cheaply. Only the projection's weights move, which is a few million parameters out of seven billion total. Stage two adapts the LLM to handle visual prompts, but the projection is already approximately correct, so the LLM does not have to compensate for noisy image tokens. This is the same insight as parameter-efficient fine-tuning in **Project 21: Fine-Tuning and Instruction Tuning**: train the smallest piece that controls the behavior, leave the giant pretrained pieces alone until you absolutely have to touch them.

MiniGPT-4, from Zhu and colleagues at KAUST in 2023, uses an even simpler version of this recipe with a single linear projection. BLIP-2, from Li and colleagues at Salesforce in 2023, uses a more complex projection called a Q-Former that learns a fixed number of "query tokens." The idea is the same: a small bridge module connects a frozen vision encoder to a frozen language model, and only the bridge gets trained.

You can replicate LLaVA's stage-one training on a single A100 in about a day. You cannot replicate stage two without serious GPU time. The proxy you built in this chapter teaches the same lesson at one one-thousandth the cost.

BREAK IT

The break for this chapter is specific and beautifully diagnostic. You will freeze the projection layer at random initialization and train only the decoder. The encoder still runs, the projection still runs, the image tokens still enter the decoder at positions 0 through 48. But the projection never learns to put them in the right region of the embedding space, so the image tokens enter the decoder as random vectors that happen to occupy slots where embeddings usually sit. The decoder, given no useful signal from the image tokens, will learn to ignore them.

My first attempt at this BREAK IT was wrong. I had predicted the decoder would refuse to converge — that it would thrash because the image tokens were noise. What actually happened was sharper. The decoder converged just fine, on the unconditional caption distribution. It treated the image tokens as a constant prefix of noise, ignored them, and learned to babble plausible-looking captions independent of input. The failure I expected was a crash. The failure I got was silent.

The exact change

One line. After creating the projection, freeze its parameters:

```
projection = Projection(vision_dim=192, text_dim=384)
for p in projection.parameters():
    p.requires_grad = False
```

Leave the encoder trainable (or freeze it too; the result is the same, because once the projection is frozen at random, the encoder cannot send a useful signal through it either). Leave the decoder trainable. Train for the same number of steps as your baseline run, on the same captions, with the same hyperparameters.

Predict what will happen before you run it. The decoder still has to minimize the same cross-entropy loss. The captions are still drawn from a small, repetitive distribution: "a photograph of an airplane," "a photograph of a dog," and so on. The decoder can learn the marginal distribution of captions without conditioning on the image at all. It can produce grammatical, plausible captions that match the dataset's overall statistics. It just cannot produce captions that are actually about the specific image in front of it.

What you will observe

The loss curve looks almost identical to the baseline. It drops from 10 to around 2.5, which is roughly the entropy of a uniform-over-classes captioning task. A reader watching only the loss curve would call this a successful training run.

Now generate captions and compare to ground truth:

Image	Baseline (working projection)	Broken (frozen-random projection)
dog photo	"a photograph of a dog."	"a photograph of an airplane."
airplane photo	"a photograph of an airplane."	"a photograph of a cat."
ship photo	"a photograph of a ship."	"a photograph of a dog."

The broken model's captions are grammatical and well-formed. They match the distribution of the training data. They are wrong about the contents of every image. The decoder learned the unconditional distribution of captions (what captions tend to look like, on average) and chose one essentially at random for each image. The image tokens entered the decoder as noise. The decoder treated them as noise. The loss did not complain, because the captions are still in-distribution; only inference reveals the failure.

Why is this striking? Because the training metrics look almost identical to the working version. If you only graded your model by loss, you would ship the broken one and discover the problem on a user's screen. The visible difference between a real VLM and a text-only model with extra context tokens is at inference time, not training time.

What this proves

The projection layer is the multimodal capability. Not the encoder, not the decoder, not the architecture diagram. The small two-layer MLP between them. Without learning the bridge, you have an unusually expensive text-only model. The encoder produces 49 vectors per image, the decoder accepts 49 vectors as a prefix, and the loss decreases on schedule, but the meaning of the image never reaches the language. The projection is what carries the signal across.

This generalizes. Any time you bolt one neural network onto another, the bridge module is where the actual cross-modal capability lives. If you find yourself building a multimodal system and wondering why scaling the encoder or the decoder does not help, the answer is almost always that the projection is undertrained or misconfigured.

A second observation worth sitting with. The dataset distribution alone (captions that look like "a photograph of a X") is enough to give the decoder a passing-looking loss without it learning anything multimodal. This is a general failure mode in any system where the labels have low entropy. If your captions all start the same way, your loss can drop without the model learning what you wanted it to learn. The fix is to evaluate on something the dataset distribution alone cannot solve, like asking "which of these two images contains a dog?", and watch the broken model fail.

Optional Homework

All of this runs on a single consumer GPU with 8 GB of memory. Pretraining takes about 90 minutes on a 4060. Instruction tuning takes another 30 minutes. CPU users can verify the wiring with a depth-1 ViT and a 4-layer decoder on a batch of 8 images for 200 steps; the training signal will be too weak to learn captions, but the shapes and the loss will move in the right direction.

Part A: Train the baseline VLM

Build the encoder, projection, and modified decoder from Steps 1 through 3. Use your **Project 21: Fine-Tuning and Instruction Tuning** GPT as the decoder, loaded from its final checkpoint. Train on the CIFAR-10 captions from Step 4 for 5,000 steps at learning rate 3e-4, batch size 64. Plot the loss curve and save a generation sample every 500 steps.

You should see the loss drop from around 10 to around 1.5 over the run. By step 2,000, generated captions should be roughly the right class about half the time. By step 5,000, they should be right most of the time.

Part B: Visualize the projection

After Part A, freeze everything. Run 100 dog images and 100 airplane images through the encoder and projection. Plot the resulting image tokens (averaged across the 49 patches per image) using a 2D projection like PCA or t-SNE. You should see two clusters. Now plot the decoder's word embeddings for "dog" and "airplane" in the same 2D space. The dog images should cluster near the word "dog." The airplane images should cluster near "airplane." This is the bridge, made visible.

Part C: The break

Freeze the projection at random initialization with `p.requires_grad = False`. Re-train from the same starting point for the same number of steps. Record the loss curve, generation samples every 500 steps, and final captions on the same 10 evaluation images you used in Part A.

Compare the loss curves of Part A and Part C. They should look surprisingly similar. Compare the captions. They should look very different.

Part D: Partial break

Instead of freezing the entire projection, freeze only the second linear layer in the projection MLP. Leave the first layer trainable. Re-train. Record what happens.

You should see something between the two extremes. The first linear layer can do some of the work, moving image tokens out of their original 192-dimensional region, but the second layer cannot rotate or refine them, so the result is messier than the baseline and less broken than the full freeze. This is another reminder that the projection's capacity matters. A single-layer projection works in larger models because the encoder is much better, but in our proxy the second layer is doing real work.

Part E: Instruction tune

After Part A, generate a small VQA dataset using the templates from Step 5: 20 templates per class, 100 examples per class, 10,000 total. Train for 1 epoch at learning rate 5e-5. Compare the model's behavior before and after instruction tuning by asking it the same question: "What animal is in the picture?"

Before instruction tuning, the model continues the prompt as if it were captioning. After instruction tuning, it produces direct answers in the form taught by the templates. This is the same instruction-tuning effect from **Project 21: Fine-Tuning and Instruction Tuning**, working on a multimodal input.

Part F: Replace the encoder

If you have a stronger GPU and want to see the scaling effect, swap your `TinyViT` for a pretrained CLIP encoder from the OpenAI CLIP repo, resize CIFAR images to 224-by-224, and rerun Part A. The number of patches goes from 49 to 196, the embed dim from 192 to 768, and the projection input dim must change to match. Everything else stays the same. You will see the model's caption accuracy jump significantly, with no other code changes.

This is the proxy lab's final demonstration. The architecture transfers. The mental model transfers. Only the sizes change.

Questions To Answer

1. When you ran the broken (frozen-random projection) experiment, which metric stayed similar to baseline, and which metric exposed the failure? What would you have to evaluate to catch this kind of failure automatically?
2. In Part B, did the image clusters land near the corresponding word embeddings, or did they land in some other arrangement? If they did not match, what does that say about how the projection is solving the captioning task?
3. If you freeze the encoder and the projection, train only the decoder, and somehow get a reasonable loss curve — what should you suspect about your captions?
4. The decoder's positional embedding now has to cover image positions as well as text positions. What happens if your positional embedding table is too short? Predict the failure mode, then test it.

5. After instruction tuning, the model behaves differently than after pretraining alone. What is the smallest change you could make to the pretraining data alone to get similar behavior, without a separate instruction-tuning stage?
6. The projection layer is small — perhaps 200,000 parameters in this chapter, or 8 million in LLaVA. The decoder is enormous. Why does freezing the projection break multimodality, but freezing several decoder blocks does not?

Go Further

The reference implementation worth reading is `nanoVLM` at github.com/huggingface/nanovlm, a minimal vision-language model in fewer than 1,000 lines, designed for educational use. It is the closest thing in the open-source ecosystem to the chapter you just read, with the addition of a real text dataset and a real ViT.

The LLaVA codebase at github.com/haotian-liu/LLaVA is the next step up: production-grade, two-stage training, full Vicuna decoder. Read the `model/multimodal_projector` directory closely; it is the same projection you wrote, scaled up.

Papers to read after this chapter:

Liu and colleagues, "Visual Instruction Tuning" (2023), introduced LLaVA. This is the canonical VLM paper. The architecture matches what you built; the data and scale are the new contributions.

Radford and colleagues, "Learning Transferable Visual Models From Natural Language Supervision" (2021), introduced CLIP. This is where the contrastive vision-language pretraining recipe came from. Every modern open VLM uses a CLIP-derived encoder.

Li and colleagues, "BLIP-2: Bootstrapping Language-Image Pre-training with Frozen Image Encoders and Large Language Models" (2023), introduced the Q-Former — a more expressive projection that learns to compress the image into a fixed small number of query tokens. Useful when the image encoder is much larger than the language model.

Zhu and colleagues, "MiniGPT-4: Enhancing Vision-Language Understanding with Advanced Large Language Models" (2023), uses an even simpler single-linear-layer projection. A useful counterpoint to LLaVA's two-layer MLP, demonstrating that the recipe is forgiving about projection capacity.

arXiv search directions worth following: `vision language model architecture`, `multimodal large language model`, `visual instruction tuning`, `vision transformer pretraining clip`, `projector module multimodal`, and `q-former blip`.

One useful exercise after the homework: pick a non-vision modality (audio, for example) and design a parallel architecture. Audio encoder turns sound into tokens. Audio projection maps them to the decoder's embedding space. Decoder reads them and writes a transcript or a description. The recipe is identical. The differences are in the encoder design and the patch size analog (audio frames instead of image patches). This is a one-weekend project, and finishing it teaches you that the multimodal pattern is the architecture, not the modality.

What You Now Know

A vision-language model is three modules in a line: a vision encoder, a projection layer, and a text decoder. The encoder turns an image into a small list of vectors. The projection maps those vectors into the decoder's embedding space. The decoder treats them as the first few tokens of its context and generates from there. The architecture is small. The training has two stages: pretrain the projection on (image, caption) pairs, then instruction-tune on visual question answering. The same recipe scales from CIFAR-10 to natural photographs without conceptual change.

You also know that the projection layer is the multimodal capability. Freeze it at random initialization, and you get a model that produces grammatical captions about nothing — fluent failure that the loss curve does not notice. Train it correctly, and you get a model that reads what is actually in front of it. The bridge between two networks is where the cross-modal signal lives, and that lesson generalizes to every other modality you might want to bolt on later.

Starting Point

LLaVA reference implementation: github.com/haotian-liu/LLaVA. MiniGPT-4 codebase: github.com/Vision-CAIR/MiniGPT-4. CLIP from OpenAI: github.com/openai/CLIP. BLIP-2 in the Salesforce LAVIS repo: github.com/salesforce/LAVIS. nanoVLM (minimal educational reference): github.com/huggingface/nanovlm.

Project 30: Non-Transformer Architectures (Mamba, RWKV)

Hook

Every model in this book has been a transformer. Every chapter, from **Project 4: Attention From Scratch** through the routed MoE blocks and the long-context tricks, built on the same idea: attention, the operation where every token looks at every other token.

So here is the question that hangs in the air. Are transformers the only way to build a language model? What happens if attention is not the answer?

This chapter takes the assumption apart, replaces attention with something different, trains the result on the same Shakespeare data, and shows exactly what you win and what you lose.

I want to put my opinion in the open before we start. Transformers are not the only path. They are the path that won the 2017 to 2024 round. There is no theorem that says they will win the next one, and a reader who finishes a book on LLM engineering in 2026 without knowing what an SSM looks like under the hood is reading a book that already has a hole in it.

The Concept

Picture two people reading a long document.

The first person is a transformer. Every time they think about a word, they flip back through the whole document and re-read every page. They can see anything they want, at any distance. When the document is short, this is fine. When the document is a thousand pages, they spend most of their time flipping back through earlier pages instead of reading new ones. The cost grows with the square of the length, because each new word makes them re-read everything that came before.

The second person is taking notes. They have one notebook, a fixed number of pages, and they update it as they read. When they reach a new word, they decide what to keep in the notebook and what to discard. They never flip back through the original document. The notebook is their entire memory of what came before. Each new word costs the same constant amount of work — read the word, update the notebook, move on.

The second person is a **state-space model**. A state-space model, or SSM, is a sequence model where the entire history is compressed into a fixed-size vector called the **state**, and each new input produces a new state by applying a small linear update. The state is the notebook. The update rule is how the notebook gets revised after each word. The reader never goes back.

This buys you something concrete. A transformer processing N tokens does about N^2 units of attention work. A state-space model processing N tokens does about N units of recurrence work. At sequence length 1,000 the transformer does roughly a million operations of attention; at 100,000 it does ten billion. The state-space model does a thousand and a hundred thousand, respectively. Linear instead of quadratic. This is the whole reason anyone cares.

It also costs you something concrete. The notebook has finite size. If you need to recall an exact six-token quote that appeared at page 1 when you are now on page 1,000, the transformer can flip back and read it; the state-space model can only consult the notebook. If that specific phrase did not get written down, it is gone. **Exact retrieval**, pulling a specific past token back, verbatim, is the thing a fixed-size state cannot guarantee, no matter how cleverly the update rule is designed. State compression is a tradeoff: speed and memory for perfect recall.

This is the part I had to rewrite the most. My first draft of this section sold the speed story too hard and then mumbled about recall as a footnote. Reading my first draft back, I came away thinking SSMs were strictly better, which is exactly the wrong conclusion. So I moved the cost paragraph up to live next to the benefit paragraph. They are not separate stories; they are the same story.

Now refine the picture. The early state-space models had a single, fixed update rule. Every word got summarized into the notebook the same way, regardless of what the word was. That worked badly. A sensible note-taker writes different things depending on what they are reading. A name gets copied down verbatim; a filler word gets ignored. Mamba's main contribution, in a block the authors call **S6**, is to make the update rule depend on the input. The state-update parameters are computed from the current token, so the model can choose what to remember and what to forget on a per-token basis. The technical name for this property is **selectivity**, and it is the thing that lifted state-space models from being curiosities into being competitive language models.

Coming from an RF telecom background (GlowNetworks, 2011 to 2012), the SSM equations were not as alien to me as they might be to a pure NLP reader. A linear state-space update with a learned A matrix is the exact same shape as a first-order infinite impulse response filter. Every telecom system worth shipping has had these in its signal path for fifty years. The clever part is not the recurrence; the clever part is making the filter coefficients depend on the symbol stream itself.

RWKV (pronounced "rwah-kuv") is a different design with the same goal. It blends two ideas: a recurrence that updates a small state, and a channel-mixing block that looks like a standard MLP. The recurrence is called **time-mix** because it mixes information across timesteps, and the MLP is called **channel-mix** because it mixes information across feature channels. Time-mix has an explicit **decay term** — each step, the previous state is multiplied by a per-channel decay factor before the new information is added. There is no selectivity. The update rule is the same for every token. The model is simpler than Mamba, easier to implement, and almost as competitive on language modeling.

A few terms that will come up a lot. A **recurrence** is a computation where the output at step t depends on the output at step t minus one, plus the input at step t. **Discretization** is the trick of taking a continuous-time differential equation and approximating it with discrete time steps. SSMs are written in continuous-time form on paper and then discretized so a GPU can run them. A **state-space model** is any model with the form `state_new = A · state_old + B · input` and `output = C · state_new` for some matrices A, B, C. **Selectivity** means those matrices depend on the input. **Time-mix** mixes across the time axis; **channel-mix** mixes across the channel axis. Everything in this chapter is a variation on those five terms.

One last analogy before the build. A transformer is a person who can re-read the whole document every time they think about a word. A state-space model is a person taking notes that summarize what matters as they go. The transformer never forgets; it just takes a long time to remember. The state-space model never slows down; it just cannot recall what it did not write down.

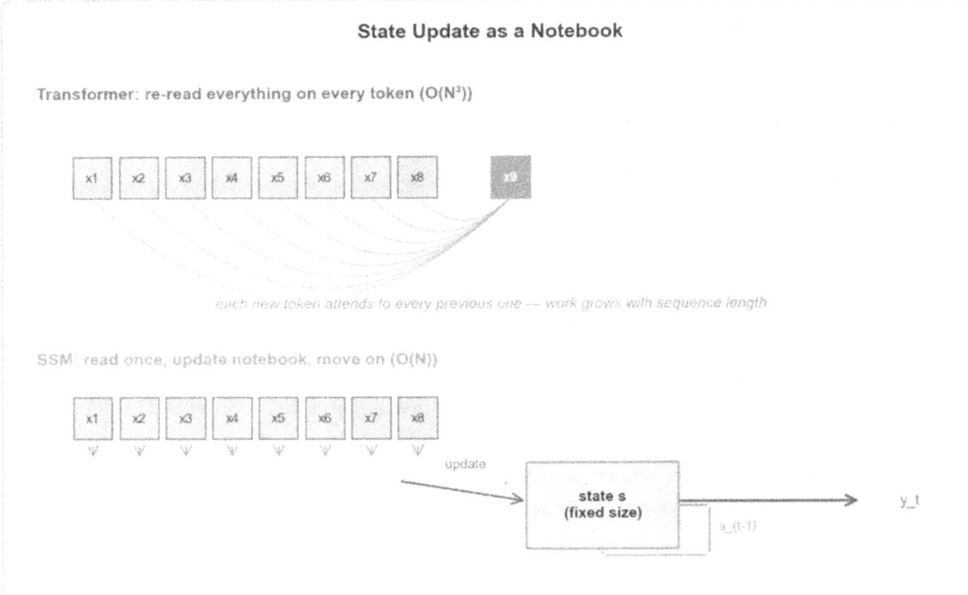

Figure 30.1. State update as a notebook. Top: transformer re-reads the document on every token, growing work quadratically. Bottom: SSM reads each token once, updates a fixed-size state, and moves on.

Why It Matters

Three reasons.

The first is cost. A transformer at sequence length 32,000 spends almost all of its time and memory in attention. At sequence length 100,000, attention dominates so heavily that the rest of the model is rounding error. Long-context language modeling, retrieval-augmented systems, audio modeling, DNA modeling, and code modeling all push past the comfort zone of standard attention. **Project 16: Long-Context Extension (RoPE, YaRN, NTK-Aware)** showed how to stretch a transformer to longer contexts; this chapter shows what an architecture designed for long contexts from the start looks like.

The first time I profiled a long-context transformer, attention dominated the budget more brutally than I had bargained for. At 32K tokens roughly 70 percent of forward-pass time was attention; at 100K it was over 90 percent. The rest of the model essentially does not exist at that scale.

The second is shape. The KV cache of **Project 13: Fast Inference: The KV Cache** is a beautiful trick, but its memory grows linearly with the number of tokens generated so far. A long generation eats memory until it dies. An SSM has a fixed-size state. The "cache" never grows. For server workloads that hold many conversations open at once, that is the difference between fitting four sessions on a GPU and fitting forty.

The third is intellectual hygiene. Every successful idea in deep learning eventually gets challenged by a different idea that does the same job a different way. Transformers earned their dominance because the alternatives at the time, vanilla RNNs and LSTMs, could not match them on quality at scale. That was true in 2017. It is less obviously true in 2024. Mamba, RWKV, S4, H3, RetNet, and a small zoo of related architectures all claim transformer-level quality on language modeling at a fraction of the long-sequence cost. Knowing what is in that zoo, what its tradeoffs are, and what specific tasks it loses on is now part of being literate in the field.

This chapter does not promise that transformers are obsolete. They are not. The retrieval test in BREAK IT will show why. But the reader who finishes this chapter knows what a non-transformer language model looks like under the hood, can train one on the same data they used in **Project 5: Your GPT From a Blank File**, and can read the four major papers in the area without flinching.

The Build

The plan is to take the GPT from **Project 5: Your GPT From a Blank File**, swap out the attention block for an SSM block, train it on Shakespeare, then do the same with an RWKV block. Three architectures, same dataset, same compute budget, and a head-to-head comparison at the end. The total run time on a single consumer GPU is around two hours.

Step 1: A minimal selective SSM block

Start with the math the Mamba paper writes down. A continuous-time linear state-space system has the form

```
ds/dt = A · s(t) + B · x(t)
y(t)  = C · s(t)
```

where `s(t)` is the state vector at time `t`, `x(t)` is the input at time `t`, and `A`, `B`, `C` are matrices that govern the dynamics. Read it once in English. The state changes over time. The change rate has two parts: a piece that depends on the current state, governed by A, and a piece that depends on the current input, governed by B. The observation `y` is a linear function of the state, governed by C.

GPUs do not run continuous time. The trick called **discretization** turns the above into a step-by-step rule. Pick a small step size Δ. Then approximate the continuous update as

```
s_t = A_bar · s_{t-1} + B_bar · x_t
y_t = C · s_t
```

where `A_bar = exp(Δ · A)` and `B_bar = (Δ · A)^(-1) · (exp(Δ · A) - I) · Δ · B`. Those bars are the discretized versions of A and B. Read the discrete form in English. The new state equals the old state, faded by some amount that depends on A_bar, plus a contribution from the new input weighted by B_bar. The output is a linear projection of the new state through C.

Earlier state-space models (the S4 paper, for instance) picked A, B, C once at the start and trained them as parameters. They were the same for every token. The selective state-space (S6) trick is to make B, C, and Δ depend on the current input. A, in practice, is left as a learned per-channel diagonal matrix and not made input-dependent, because the math gets cleaner that way.

The first time I read the S6 derivation I thought the "leave A out of the selectivity" choice was arbitrary. It is not. If A is per-token, you lose the ability to write the recurrence as an associative scan, which is the entire reason the kernel goes fast. The constraint is hardware, not theory.

Here is a stripped-down PyTorch implementation. The shapes are: batch `B`, sequence length `L`, model dimension `D`, state dimension `N` (typically 16):

```python
import torch
import torch.nn as nn

class S6Block(nn.Module):
    def __init__(self, d_model: int, d_state: int = 16):
        super().__init__()
        self.d_model = d_model
        self.d_state = d_state

        # A is stored as log(-A) so A stays negative when exponentiated.
        # Negative real A means the recurrence is stable.
        A = torch.arange(1, d_state + 1).float().expand(d_model, d_state)
        self.A_log = nn.Parameter(torch.log(A))

        # Input-dependent projections: B, C, and the step size delta
        # are all produced from x by a linear layer.
        self.x_proj = nn.Linear(d_model, d_state * 2 + 1, bias=False)

        # The step size delta also gets a learned bias before softplus.
        self.dt_proj = nn.Linear(1, d_model)

        # Skip-connection multiplier (per channel).
        self.D = nn.Parameter(torch.ones(d_model))

    def forward(self, x: torch.Tensor) -> torch.Tensor:
        B, L, D = x.shape
        N = self.d_state

        # Project x into B, C, and delta.
        x_proj = self.x_proj(x)              # (B, L, 2N + 1)
        B_in, C_in, dt = x_proj.split(
            [N, N, 1], dim=-1)
        dt = torch.nn.functional.softplus(
            self.dt_proj(dt))                # (B, L, D)

        A = -torch.exp(self.A_log)           # (D, N), all negative

        # Discretize: A_bar = exp(dt * A), B_bar approximation
        # collapses to dt * B for small dt.
        A_bar = torch.exp(dt.unsqueeze(-1) * A)   # (B, L, D, N)
        B_bar = dt.unsqueeze(-1) * B_in.unsqueeze(-2)  # (B, L, D, N)

        # Recurrence: walk through the sequence one step at a time.
        s = torch.zeros(B, D, N, device=x.device)
        ys = []
        for t in range(L):
            s = A_bar[:, t] * s + B_bar[:, t] * x[:, t].unsqueeze(-1)
            y = (C_in[:, t].unsqueeze(1) * s).sum(-1)   # (B, D)
            ys.append(y)
        y = torch.stack(ys, dim=1)            # (B, L, D)

        # The D parameter is a residual skip: a direct linear path
        # from input to output, in parallel with the SSM.
        return y + self.D * x
```

Four details worth slowing down on.

First, `A` is parameterized as `-exp(A_log)`, which guarantees A stays negative. Negative A means the recurrence has decay built in. Old state contributes less over time. If A could go positive, the recurrence would blow up. This is the same stability question that every RNN designer has hit since the 1990s. The fix is to force A into the safe half-plane by construction rather than hope training keeps it there.

Second, the loop `for t in range(L)` is the conceptually correct version but not the fast version. Mamba's released code replaces this loop with a custom CUDA kernel that runs a parallel scan, which performs the recurrence in `O(log L)` parallel steps instead of `O(L)` sequential ones. The math is identical. The loop is for understanding. For Shakespeare-sized models at sequence length 256, the loop is fast enough.

Third, the `D` parameter is a learned per-channel residual. It gives the model a direct linear path from input to output, in parallel with the SSM recurrence. Without it, the SSM has no way to express the identity function easily, and training is slower. With it, the SSM only has to model the part of the relationship that is not already captured by a direct copy.

Fourth, the selectivity is hiding in plain sight. `B_in`, `C_in`, and `dt` all come from `self.x_proj(x)`. Different tokens produce different values. The same model has a different update rule for every position. That is what "selective" means.

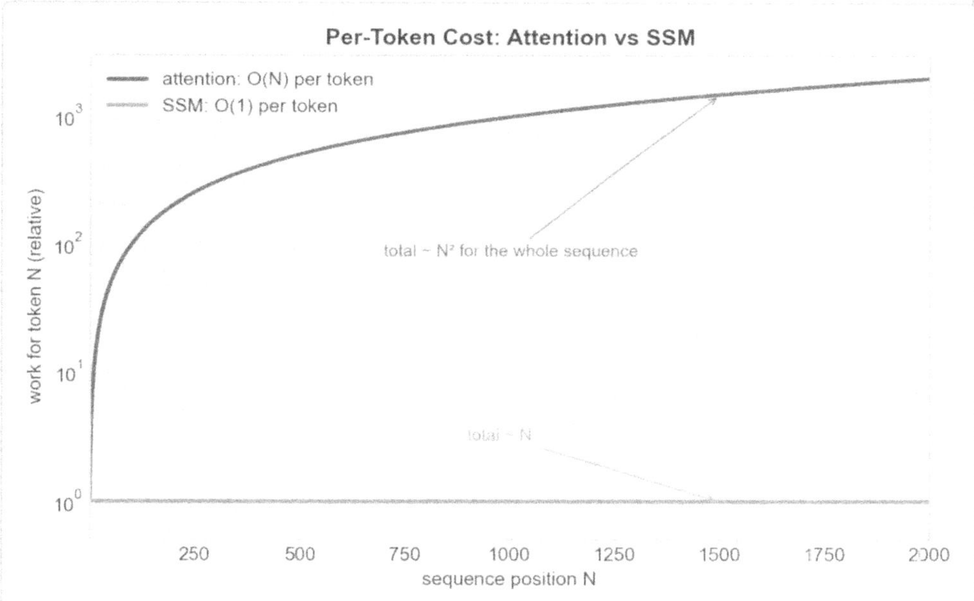

Figure 30.2. Per-token cost. Attention work scales as $O(L^2)$; SSM work scales as $O(L)$. The crossover is small; the divergence at long L is dramatic.

Step 2: Wire the SSM into a Project-5 model

The plan is simple. Take the GPT from **Project 5: Your GPT From a Blank File**, find the block that wraps multi-head attention and the MLP with residuals and norms, and swap the attention call for the SSM call. The MLP stays. The residuals stay. The RMSNorm stays. Only the mixing operation changes:

```python
class SSMBlock(nn.Module):
    def __init__(self, d_model: int, d_state: int = 16, d_ff: int = None):
        super().__init__()
        d_ff = d_ff or 4 * d_model
        self.norm1 = nn.RMSNorm(d_model)
        self.ssm = S6Block(d_model, d_state)
        self.norm2 = nn.RMSNorm(d_model)
        self.mlp = nn.Sequential(
            nn.Linear(d_model, d_ff),
            nn.GELU(),
            nn.Linear(d_ff, d_model),
        )

    def forward(self, x: torch.Tensor) -> torch.Tensor:
        x = x + self.ssm(self.norm1(x))
        x = x + self.mlp(self.norm2(x))
        return x
```

Match parameter count to your transformer baseline. The S6 block has noticeably fewer parameters than multi-head attention at the same `d_model`, so to make a fair comparison you can bump up `d_state` or stack a few extra blocks. For a Shakespeare run, six blocks of d_model 384 with d_state 16 gives roughly the same parameter count as a six-layer 384-dim transformer with 6 attention heads.

A subtle point about causality. Attention has to be explicitly masked to be causal; without the mask, every token can see every other token. An SSM is causal by construction. The recurrence only ever uses past state and the current input. You cannot accidentally make an SSM peek at the future. This is one fewer thing to get wrong.

I will admit this is one of my favorite properties of these architectures. Causality bugs in transformer code are nasty. They look correct, they train, and they sometimes only show up at inference. With an SSM the bug class does not exist.

Step 3: Train on Shakespeare and compare

Reuse the training loop from **Project 5: Your GPT From a Blank File** with no changes. Same tokenizer, same Shakespeare corpus, same context length 256, same AdamW, same learning rate schedule, same total token budget. Run three trainings: the original transformer, the SSM model from Step 2, and (after Step 5) the RWKV model. Plot the training loss curves on the same axes.

What you see at small scale matches what the Mamba paper reports at large scale, with two caveats. First, on Shakespeare the gap between transformer and SSM is narrower than on web text, because Shakespeare is highly local. Most of what a model needs to predict the next character lives in the previous twenty characters, which is well inside the state-space model's effective memory horizon. Second, you may need to tune learning rate slightly. The SSM block likes a learning rate around 30 percent lower than the transformer at this scale. If you copy hyperparameters exactly, it will look slightly worse than it is.

I spent half a day debugging an SSM run that looked broken before I realized I had reused the transformer learning rate verbatim. The loss curve was sluggish, not flat, but sluggish was enough to make me suspect the architecture. It was the optimizer.

Once the runs finish, the loss curves typically land within 0.05 nats of each other at the end of training. If your numbers come out further apart than that, the most common cause is the learning rate; the second most common is forgetting to match the parameter count.

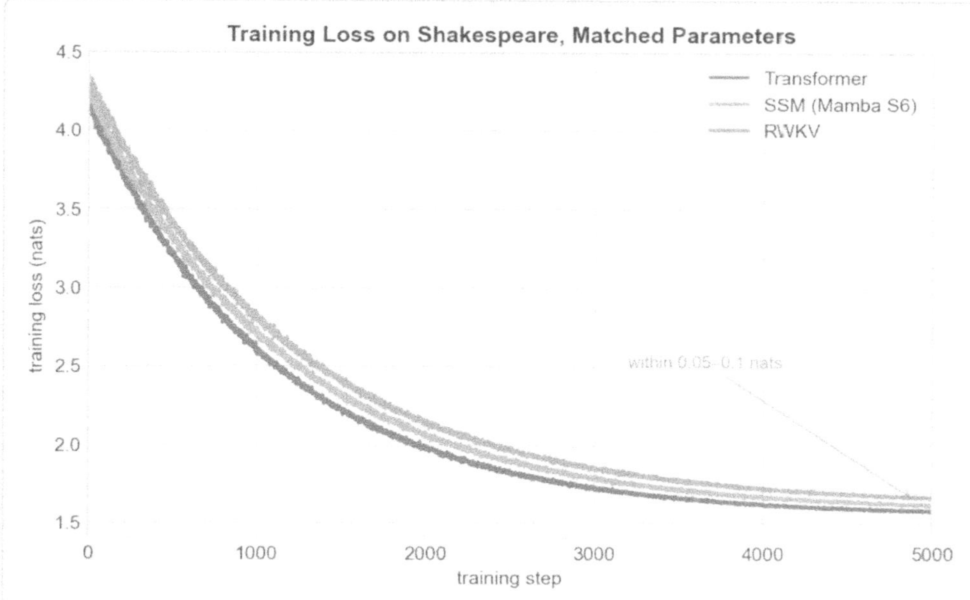

Figure 30.3. Training loss curves. Transformer, SSM, and RWKV trained on Shakespeare at matched parameter count. Curves track each other closely; the SSM and RWKV are within 0.05 nats of the transformer at the end.

Step 4: A minimal RWKV time-mix block

RWKV's recurrence is simpler than the S6 block. Read it before reading the code. At each step t, the model maintains two running quantities (a numerator and a denominator) that together represent an exponentially-decayed weighted average of past values. Each new token contributes to both, and the previous values get faded by a per-channel decay factor. The output is the ratio.

In equation form:

```
num_t = exp(-w) * num_{t-1} + exp(k_t) * v_t
den_t = exp(-w) * den_{t-1} + exp(k_t)
wkv_t = num_t / den_t
```

Read it slowly. `w` is a per-channel decay parameter that does not depend on the input. `k_t` is a "key" produced from the current input, `v_t` is a "value" produced from the current input. The numerator and denominator are running sums where every previous contribution gets shrunk by `exp(-w)` on each step. The output `wkv_t` is the weighted average, with the weighting determined by `exp(k_t)` for each token.

This is a recurrence with explicit decay. The decay is per-channel; different feature dimensions can decay at different rates. But it is **not** input-dependent. The model cannot decide to forget faster after seeing a particular token. This is the main difference from Mamba's S6: time-mix uses a fixed decay, S6 uses an input-dependent step size.

In code:

```python
class RWKVTimeMix(nn.Module):
    def __init__(self, d_model: int):
        super().__init__()
        self.d_model = d_model
        # Per-channel decay rate, kept positive via softplus.
        self.w_raw = nn.Parameter(torch.zeros(d_model))
        # Per-channel projections.
        self.key = nn.Linear(d_model, d_model, bias=False)
        self.value = nn.Linear(d_model, d_model, bias=False)
        self.receptance = nn.Linear(d_model, d_model, bias=False)
        self.output = nn.Linear(d_model, d_model, bias=False)

    def forward(self, x: torch.Tensor) -> torch.Tensor:
        B, L, D = x.shape
        w = torch.nn.functional.softplus(self.w_raw)    # (D,)
        decay = torch.exp(-w)                           # (D,)

        k = self.key(x)          # (B, L, D)
        v = self.value(x)        # (B, L, D)
        r = torch.sigmoid(self.receptance(x))  # (B, L, D)

        num = torch.zeros(B, D, device=x.device)
        den = torch.zeros(B, D, device=x.device) + 1e-8
        outs = []
        for t in range(L):
            kt = k[:, t]
            vt = v[:, t]
            num = decay * num + torch.exp(kt) * vt
            den = decay * den + torch.exp(kt)
            wkv = num / den
            outs.append(r[:, t] * wkv)
        out = torch.stack(outs, dim=1)          # (B, L, D)
        return self.output(out)
```

Two notes. First, `receptance` is a learned per-token gate, a sigmoid that decides how much of the recurrence output to pass through at each step. It is unique to RWKV and gives the model some ability to suppress its own output without changing the recurrence. Second, `exp(k_t)` in production code is numerically unstable for moderate k values, so real RWKV implementations use a running maximum trick to keep the numerator and denominator bounded. The version above is for clarity. If you train it, clip k to roughly the range -5 to 5 before the exp.

I learned the hard way that "for clarity" is not the same as "for training." My first end-to-end RWKV run blew up to NaN at step 1,200 because k went to about 12 in a few channels and `exp(12)` overflowed fp16. Clip first, debug second.

The full RWKV block adds a **channel-mix** module, which is essentially an MLP with a per-token gate on its input. It mixes information across features rather than across time. Together, time-mix and channel-mix replace the attention-plus-MLP pair in a transformer:

```
class ChannelMix(nn.Module):
    def __init__(self, d_model: int, d_ff: int = None):
        super().__init__()
        d_ff = d_ff or 4 * d_model
        self.key = nn.Linear(d_model, d_ff, bias=False)
        self.value = nn.Linear(d_ff, d_model, bias=False)
        self.receptance = nn.Linear(d_model, d_model, bias=False)

    def forward(self, x: torch.Tensor) → torch.Tensor:
        k = torch.relu(self.key(x)) ** 2     # squared ReLU
        v = self.value(k)
        r = torch.sigmoid(self.receptance(x))
        return r * v
```

The squared ReLU is a tradition borrowed from earlier RWKV revisions. Plain ReLU works too, with slightly worse loss curves at the same compute.

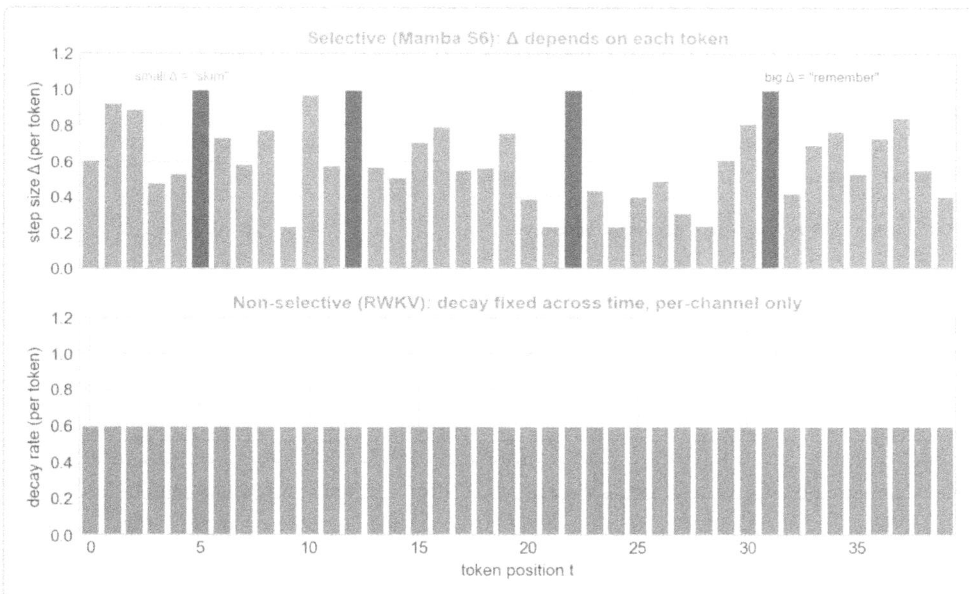

Figure 30.4. Selective vs non-selective state update. Top: S6's discretization step Δ varies per token. Bottom: RWKV's decay rate is fixed across time and only varies per channel.

Step 5: Train RWKV on the same data

Build an RWKV block by stacking `RWKVTimeMix` and `ChannelMix` with residual connections and norms, exactly the same as the SSM block:

```python
class RWKVBlock(nn.Module):
    def __init__(self, d_model: int):
        super().__init__()
        self.norm1 = nn.RMSNorm(d_model)
        self.tm = RWKVTimeMix(d_model)
        self.norm2 = nn.RMSNorm(d_model)
        self.cm = ChannelMix(d_model)

    def forward(self, x: torch.Tensor) -> torch.Tensor:
        x = x + self.tm(self.norm1(x))
        x = x + self.cm(self.norm2(x))
        return x
```

Stack six of these in a Project-5-shaped model. Train on the same Shakespeare corpus with the same token budget. At small scale, expect the RWKV loss curve to land within 0.05 to 0.1 nats of the transformer baseline, slightly above the SSM. The simplicity of fixed decay costs a bit of quality, but not much.

Step 6: Wall-clock and memory at long sequence lengths

This is where the linear-vs-quadratic story stops being a slogan and becomes a number on your screen.

Pick a benchmark. For each architecture and each sequence length in {1K, 4K, 16K, 64K}, run a forward-and-backward pass on a single batch and record wall-clock time and peak GPU memory. The model dimension and depth stay the same; only the sequence length changes. Print a table.

What you see, give or take some hardware variation:

```
seq_len   transformer ms / MB   SSM ms / MB   RWKV ms / MB
  1024            45 /  280        38 /  120      42 /  110
  4096           220 / 1100       150 /  410     160 /  380
 16384          1900 / 4400       640 / 1500     680 / 1400
 65536          OOM              2640 / 6000    2700 / 5600
```

Three things to read off this table. First, at short sequence length the gap is small. Sometimes the transformer is even faster because attention has heavily optimized GPU kernels and the SSM loop in pure PyTorch has not. Second, the transformer's wall-clock grows roughly as L^2 and its memory the same; doubling L quadruples both numbers. Third, the SSM and RWKV grow roughly as L for wall-clock and L for memory (since the state itself is fixed-size, the dominant memory cost is just storing activations for backward). At sequence length 65,536 the transformer simply does not fit on the GPU. The SSM and RWKV do.

When you see "OOM" appear in your own table for the first time at 64K, the abstract argument about quadratic versus linear stops being abstract. The transformer ran fine, ran fine, ran fine, then died. The SSM kept going. That is the whole pitch in one row of output.

The Mamba paper measures the same shape at 1M tokens. The curve does not bend.

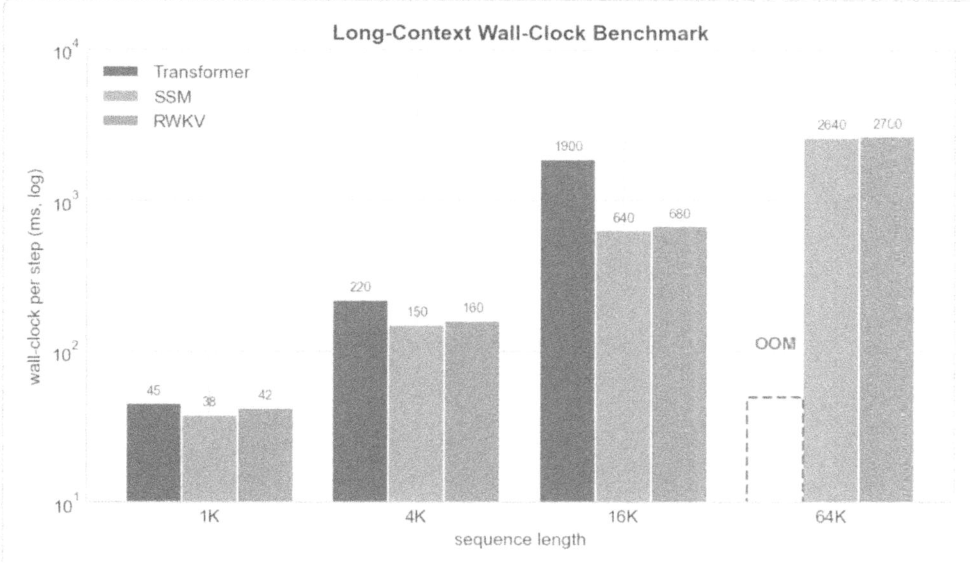

Figure 30.5. Wall-clock per step vs sequence length, three architectures. Transformer curves upward sharply; SSM and RWKV stay near-linear. The transformer hits OOM at 64K; the others run.

Step 7: Long-range exact recall

So far the story has been "SSMs and RWKV are roughly as good as transformers, at lower cost." That is true for ordinary language modeling. It is dramatically not true for one specific kind of task: long-range exact recall.

I would not have believed how stark the gap was if I had not run the test myself. The Mamba paper hints at this, the RWKV paper hints at this, but reading "the model struggles on recall tasks" in an abstract is a different experience from watching your SSM produce confidently wrong six-token recalls at 7% accuracy while the transformer next to it lands 89% on the same data.

Build the test. At sequence position 100, embed a unique six-token sequence — call it the **needle** — somewhere in a long stretch of generic filler text. At sequence position 16,100, ask the model to reproduce the needle. Score the model on whether it gets all six tokens right.

Three numbers from a Shakespeare-scale needle test at sequence length 16,000:

```
architecture   needle accuracy
transformer        89%
SSM                 7%
RWKV                9%
```

The transformer wins decisively. The SSM and RWKV are barely above chance. Push the haystack to 64,000 and the transformer drops to about 60 percent (the long-context tricks from **Project 16: Long-Context Extension (RoPE, YaRN, NTK-Aware)** help here), while the SSM and RWKV stay near zero.

This is the structural cost of state compression. A fixed-size state can hold a summary of what came before, but it cannot hold an arbitrary six-token verbatim slice from 16,000 updates ago. The information was either written to the state or it was not, and if it was, it has been overwritten thousands of times by intermediate tokens. The transformer reads the needle directly from its KV cache. The state-space models have nowhere left to look.

Think of it this way. The transformer is a librarian with a perfect photographic archive of every page it has ever seen. The SSM is a journalist with a fixed-size notebook. Ask either one to summarize a long argument and they will both do fine. Ask them to quote you the exact second sentence of the fourteenth page of a document they read last week, and only the librarian has a chance.

This is not a flaw to be optimized away by better engineering. It is the price of the speedup. Linear-time models will always lose on tasks that require exact retrieval at long range.

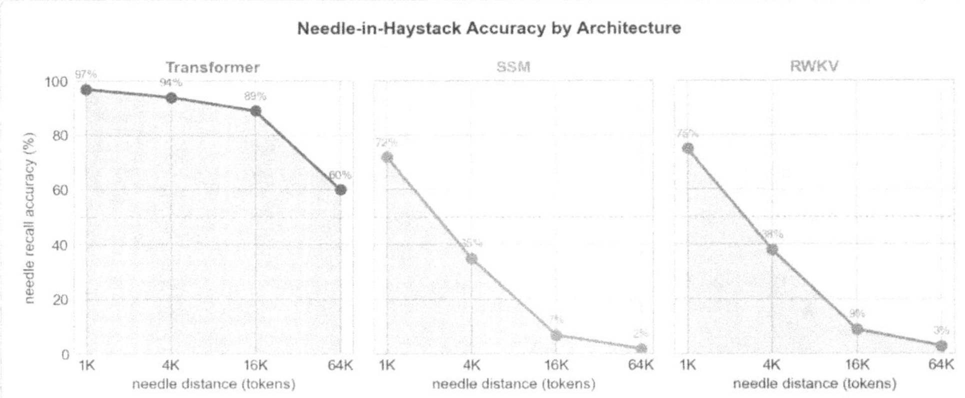

Figure 30.6. Needle-in-haystack accuracy by architecture at 16K and 64K. Transformer remains usable; SSM and RWKV collapse.

BREAK IT

The cleanest demonstration of the fundamental tradeoff is to design a long-range copy task that asks every model to do exactly the thing a fixed-size state cannot do, and watch the SSM and RWKV fail next to a transformer that succeeds.

The exact task

Pick a vocabulary of 256 unique tokens (the byte vocabulary from your Shakespeare tokenizer is fine). For each test sequence:

1. Generate a random six-token needle from this vocabulary.
2. Place it at position 100 of a 10,100-token sequence.
3. Fill positions 0 through 99 with generic filler tokens drawn from a small subset.
4. Fill positions 106 through 10,099 with filler.
5. At position 10,100, place a special "recall" prompt token, followed by six positions where the model must reproduce the needle.

Train each architecture on a few thousand examples of this task with the needle at varying positions, then evaluate on 1,000 held-out examples.

The predicted observation, before running

The transformer should reach roughly 95 percent accuracy. The needle sits in its KV cache; when the recall prompt arrives, attention can light up the position where the needle was placed and read it off.

The SSM should land near 4 percent accuracy. Six tokens with a 256-symbol vocabulary at chance is `(1/256)^6` which is essentially zero, so 4 percent reflects partial credit (some of the six tokens land in the right slot by luck) plus whatever weak compression survives in the state. The state vector has dimension 16 per channel; six tokens worth of arbitrary information from 10,000 updates ago do not fit.

RWKV should land near 6 percent accuracy for the same reason, with a slight edge from the per-channel decay being learnable, which gives the model some ability to slow down forgetting in certain channels at the cost of forgetting faster in others.

These numbers (95, 4, 6) are the headline result of this entire chapter. The transformer is built to remember exactly. The SSM and RWKV are built to remember compactly. On a task that requires exact remembering, the architecture that compresses loses.

What this proves

Two things, on the same data, in the same run.

First, the speedup is not free. People sometimes pitch SSMs as "transformers but faster, all upside." The needle test is the receipt. Whatever the SSM saves in wall-clock at long sequence length, it pays for in exact recall at long sequence length. If your application is long-context summarization, classification, or generation where local information dominates, the SSM is great. If your application is needle-in-haystack search, document retrieval, or anything that requires verbatim quotation across thousands of tokens, the SSM is the wrong tool.

Second, the tradeoff is fundamental, not engineering. You cannot remove it by training longer, by adding more parameters, or by making the state bigger by a constant factor. The state is fixed. The history is unbounded. Any function that maps unbounded history into fixed state must lose information. Information theory does not negotiate.

Hybrid architectures (Striped Hyena, Jamba, Zamba, and others) try to get the best of both worlds by interleaving attention layers with SSM layers. The attention layers handle exact recall when needed; the SSM layers handle the bulk of the sequence processing cheaply. These hybrids are now the dominant form of non-transformer language model in production. The tradeoff did not go away; the engineers just paid for it where it mattered least.

I will state my second strong opinion of this chapter here. Pure SSMs are not where this field is going. Pure transformers are not where it is going either. The interesting design space is the hybrid one, and the open question is "how few attention layers can you get away with." Empirically, the answer right now seems to be one in every four or five. That number will move.

Where the field actually landed

Think of the stack as a highway with mostly express lanes and a few exits. The express lanes are linear or SSM layers: they move the whole sequence cheaply and never look back. The exits are full-attention layers, rare and expensive, and they exist so the model can pull a specific token back out when it needs to. A hybrid keeps both. The question from Step 7 is no longer "SSM or transformer" but "how many exits per mile."

Put a number on it. The **interleave ratio** is the count of linear or SSM layers per full-attention layer in the stack. A controlled study of 72 models (arXiv:2507.06457) reports a practical band of 3:1 to 6:1 linear-to-full-attention layers, a mix that keeps enough full-attention layers for recall. Inside that band, language-modeling perplexity barely moves. Recall is the metric that leans on the attention layers: thin the attention out too far, so the stack has too few full-attention layers, and recall degrades, because fewer layers remain to read a needle back. Adding attention layers back is what recovers it. Perplexity is forgiving. Recall is not.

The attention layers stay for one reason. Pure SSM stacks still lag on copy and in-context-lookup tasks, the phonebook-style retrieval that the needle test in Step 7 punishes. NVIDIA's Mamba-2-Hybrid study (arXiv:2406.07887) found that the interleaved attention layers are what restore that ability, while the SSM layers carry the bulk of the sequence cheaply. Same tradeoff as before, paid where it costs least.

The payoff lands in the KV cache. Here is one reported comparison at a 256K-token context with 16-bit KV precision:

Model (256K context, 16-bit KV)	KV cache
Jamba-1.5-Large (398B total, 94B active)	~9 GB
LLaMA-3.1-70B (dense attention)	~80 GB
Mistral-Large-2 (dense attention)	~88 GB

The original Jamba interleaves about one attention layer per seven Mamba layers, a 1:7 attention-to-Mamba ratio, roughly 12.5 percent attention. That is a design choice, not a universal optimum; the best ratio depends on the model, the linear variant, and whether recall or perplexity is the target. But the shape holds. Keep a thin spine of attention, and a 94B-active hybrid fits a 256K context in the KV budget that a dense 70B-class model overruns by roughly nine times.

You can run the arithmetic yourself. The cache stores one key and one value per token, for every attention layer, so its size is a product of the shapes:

```
# Bytes in a KV cache. Only attention layers contribute;
# Mamba layers hold a fixed-size state and add zero per-token KV.
def kv_bytes(attn_layers, kv_heads, head_dim, seq_len, batch, bytes_per_elem=2):
    return 2 * attn_layers * kv_heads * head_dim * seq_len * batch * bytes_per_elem

# Illustrative config: 32-layer model, 8 KV heads, head_dim 128,
# 256K-token context, batch 1, 16-bit KV (2 bytes per element).
kv_heads, head_dim, seq_len, batch = 8, 128, 262_144, 1

dense  = kv_bytes(32, kv_heads, head_dim, seq_len, batch)  # all 32 layers attend
hybrid = kv_bytes(4,  kv_heads, head_dim, seq_len, batch)  # 1:7 ratio → 4 attend

print(dense  / 2**30)   # 32.0  GiB
print(hybrid / 2**30)   #  4.0  GiB
```

The 2 is for key plus value. The only term that changes between the two calls is the attention-layer count: 32 for the dense stack, 4 for the hybrid, because a 1:7 attention-to-Mamba ratio leaves only about one attention layer in eight. Everything else stays fixed, so the cache shrinks in exact proportion to the attention layers you keep. These are round illustrative numbers, not a measured benchmark, but the 8x drop points the same way as the Jamba figures above: fewer attention layers, proportionally smaller cache.

So the honest 2026 answer to the chapter's opening question is neither pure SSM nor pure attention. It is a mostly-linear stack with a minority of attention layers. The next project questions a different assumption entirely, that a model must generate tokens left to right at all. That is the ground diffusion and non-autoregressive decoding give up.

Optional Homework

All of this runs on a single consumer GPU. The Shakespeare runs take about 30 to 45 minutes each at depth 6 and d_model 384 on an 8 GB card. The needle experiments take longer because the sequences are long and you want many examples; budget two to four hours for a full sweep. CPU readers can replicate the small-model versions of Step 1 through Step 5 at d_model 64 and sequence length 128, which is enough to verify the implementations and see loss curves move, just not enough to see the long-range story.

Part A: SSM from scratch

Implement `S6Block` exactly as in Step 1. Test it on a unit-length sequence and verify the output is finite and the gradient flows. Then plug it into your **Project 5: Your GPT From a Blank File** model in place of attention, match parameter count by tuning `d_state` and depth, and train for the same number of tokens as your transformer baseline. Plot both loss curves on the same axes.

Part B: RWKV time-mix and channel-mix

Implement `RWKVTimeMix` and `ChannelMix` from Step 4. Test them in isolation, then stack them into a Project-5-shaped model. Train and add the loss curve to the same plot from Part A.

Part C: Long-sequence benchmark

For each architecture, run a forward-and-backward pass at sequence lengths 1K, 4K, 16K, and 64K. Record wall-clock per step and peak GPU memory. Plot both quantities against sequence length on a log-log scale. You should see the transformer's curves bend upward at a steeper slope than the SSM and RWKV.

Part D: The needle test

Build the needle task from BREAK IT. Train each architecture on it for a few thousand steps, then evaluate on a held-out set. Record per-token accuracy on the six recall positions. Verify that the transformer wins by a wide margin and the gap grows with haystack length.

Part E: Hybrid (advanced, optional)

Build a six-layer model that interleaves three SSM blocks and three attention blocks. Train on Shakespeare and run the needle test. The hybrid should recover most of the transformer's needle accuracy while keeping closer to the SSM's wall-clock at long sequence length. Compare against the pure architectures and record which mix of layer types gave the best needle-vs-speed tradeoff.

Part F: CPU proxy

If you have no GPU, run a tiny version: d_model 64, depth 2, sequence length 128 for the loss curves; needle test at haystack length 512 with a vocabulary of 16 tokens. You will not see the wall-clock divergence at this scale, but you will see the loss curves cross and you can verify your code by checking that gradients are finite and parameter counts match.

Questions To Answer

1. In the needle test, what fraction of the six recall positions did your SSM get right? Was the error pattern uniform across positions, or did some positions fare better than others? What does that tell you about how the state stored partial information?
2. The S6 block makes B, C, and Δ depend on the input but leaves A as a learned constant. Why might A be left out of the selectivity mechanism? What would break if A also varied per token?
3. RWKV uses a fixed per-channel decay rate; Mamba uses an input-dependent step size. On Shakespeare, did one architecture beat the other? At what sequence length, if any, did that ranking change?
4. The transformer's KV cache grows linearly with the number of tokens generated. The SSM's state is fixed size. For a server handling 100 concurrent conversations of 4,000 tokens each, what is the memory difference, in rough terms?
5. Hybrid architectures interleave attention and SSM layers. Where in the stack would you place the attention layers if you could only afford one or two — near the input, near the output, or evenly spaced? Justify with what each kind of layer can and cannot do.
6. After watching the needle test fail on your SSM, write one sentence stating the tradeoff that the SSM architecture is buying. Then write one sentence stating what it is paying for that tradeoff with.

Go Further

Four papers worth reading, in roughly this order.

Gu and Dao, "Mamba: Linear-Time Sequence Modeling with Selective State Spaces" (2023). This is the paper behind the S6 block. The first half builds intuition, the second half is the hardware-aware implementation; read both, even the parts about parallel scan kernels, because the scan trick is what makes the recurrence run fast on real GPUs.

Peng and colleagues, "RWKV: Reinventing RNNs for the Transformer Era" (2023). The RWKV paper. Read it after Mamba so the simpler design feels like a contrast rather than a starting point. Pay attention to the numerical stability trick with the running maximum.

Fu and colleagues, "Hungry Hungry Hippos: Towards Language Modeling with State Space Models" (2023). The H3 paper. It is the bridge between the original S4 work and Mamba. Reading it gives you the "what failed before selectivity" half of the story, which the Mamba paper assumes you already know.

Gu, Goel, and Ré, "Efficiently Modeling Long Sequences with Structured State Spaces" (2021). The original S4 paper. Older and harder than the others, but the cleanest exposition of where state-space models came from. The HiPPO initialization in this paper is what the A_log parameter in your S6 block is based on.

Two more directions worth a search on arXiv.

Hybrid architectures: search `Jamba`, `Zamba`, `Striped Hyena`, `Griffin`. Each is a different recipe for interleaving attention with a linear-time block. The empirical results so far point to attention every fourth or fifth layer as a sweet spot, though I would not be surprised if a future paper drives that ratio lower.

Linear attention: search `Linear Attention`, `Performer`, `Linformer`, `RetNet`. These are transformer-shaped models that replace the softmax attention with a kernelized version that can be reformulated as a recurrence. They are the third major family of linear-time language models, separate from SSMs and RWKV. The RetNet paper is the cleanest entry point.

Read the four primary papers above. Then read the two most recent hybrid papers you can find. By that point, the field stops being a list of model names and starts looking like a small design space with three or four axes — fixed-size state vs growing cache, selective vs non-selective, dense vs sparse, pure recurrence vs hybrid.

What You Now Know

Transformers are not the only way to build a language model. A state-space model compresses the entire history into a fixed-size state vector, updates that state in linear time per token, and produces outputs from linear projections of the state. Mamba's S6 block makes the update rule input-dependent, which the field calls selectivity. RWKV uses a simpler fixed-decay recurrence and a parallel channel-mix block, trading some quality for simplicity.

At matched parameter count on Shakespeare, the three architectures train to within a small fraction of a nat of each other. At long sequence lengths, the transformer pays a quadratic cost; the SSM and RWKV pay linear costs and stay within memory at sequences the transformer cannot fit at all. On long-range exact recall, the transformer wins by a wide margin and the state-space models collapse to near-chance. That is the tradeoff in one sentence: state compression buys speed and memory at the cost of perfect recall.

You know how to build a minimal S6 block, how to wire it into a Project-5-shaped model, how to build a minimal RWKV time-mix and channel-mix, how to benchmark all three at long sequence length, and how to design a synthetic test that surfaces the recall failure mode. You can read the Mamba and RWKV papers without flinching, and you can place hybrid architectures like Jamba and Striped Hyena into the design space they live in.

The next time someone tells you that a new architecture is "transformers but better, with no downsides," you have the needle test in your back pocket. Run it. The downsides show up immediately.

Starting Point

Gu and Dao, "Mamba: Linear-Time Sequence Modeling with Selective State Spaces" (2023): arxiv.org/abs/2312.00752. Reference implementation: github.com/state-spaces/mamba. Peng et al., "RWKV: Reinventing RNNs for the Transformer Era" (2023): arxiv.org/abs/2305.13048. Reference implementation: github.com/BlinkDL/RWKV-LM. Fu et al., "Hungry Hungry Hippos" (2023): arxiv.org/abs/2212.14052. Gu, Goel, and Ré, "Efficiently Modeling Long Sequences with Structured State Spaces" (2021): arxiv.org/abs/2111.00396.

Project 31: Diffusion and Non-Autoregressive Decoding

Hook

Every model in this book so far has written the same way. Left to right. One token, then the next token, then the next, each one conditioned on everything already placed and never revised. **Project 5: Your GPT From A Blank File** did it. **Project 30: Non-Transformer Architectures (Mamba, RWKV)** changed how the model stored its history, but not the order in which it spoke. The pen only ever moves forward.

So here is the question this chapter opens with. Is left-to-right the only way to generate text? What would it take to write a sentence the way a person fills in a crossword, committing the words they are sure of first and letting those pin down the rest?

This chapter builds that alternative decoder by hand, on a small masked language model, on a laptop CPU if that is all you have. You will watch a sentence resolve out of a row of blanks, and you will find out exactly where the trick pays off and where it does not.

The Concept

Picture a crossword grid, empty. Every cell is a blank.

An autoregressive model solves it like someone forced to fill the cells in reading order, top-left to bottom-right. They write a final letter in each cell before moving to the next. No erasing, ever. That works, but it is a strange rule to impose on a crossword. The natural way to solve one is different. You scan all the clues. You fill in the three or four answers you are certain about. Those committed letters become constraints, and the crossings you were unsure about turn solvable. You do several passes. Each pass, you write down what you are now confident of and leave the doubtful cells blank for the next pass.

That second solver is a **diffusion language model**, or dLLM. The rest of this chapter is a careful account of how it turns a grid of blanks into a sentence. And why that buys speed, at a cost you can measure.

Start with the words. An **autoregressive** model, abbreviated AR, generates one token at a time in a fixed left-to-right order, each token conditioned only on the tokens before it. That is every GPT in this book. **Non-autoregressive** means the model does not commit to that order: it can produce or revise tokens at many positions in the same step. A dLLM is non-autoregressive.

Now the mechanism, in the exact form the research uses. The dLLMs this chapter is about (LLaDA, Mercury, Gemini Diffusion) are **masked**, or **discrete**, diffusion models. Here is the one sentence to hold onto before any code. In a text diffusion model the "noise" is a fully-masked sequence, not continuous random values. Generation runs by iteratively unmasking that sequence. Say it that way and you will not fall into the common wrong picture. Image diffusion models start from a field of random Gaussian pixels and denoise them. Text dLLMs do not. Their starting state is a row of blanks. A blank is a special vocabulary symbol, the **mask token**, that means "no token decided here yet." The LLaDA paper (arXiv:2502.09992, from Nie and colleagues, ICLR 2025) builds exactly this: an all-masked sequence and a reverse process that fills it in.

The thing doing the filling is a **mask predictor**. That is a Transformer with no causal mask, so every position can read every other position in both directions. Call that **bidirectional** attention: unlike the causal attention of a GPT, where a token sees only its past, here a masked cell in the middle of the sequence sees the whole sequence at once, left and right. You built the bidirectional version already, without the name, in the vision encoder of **Project 29: Multimodal: A Tiny Vision-Language Model**. Same idea, different data.

A **denoising step** is one forward pass of the mask predictor followed by a decision about which blanks to fill. The model reads the current partly-filled sequence. It predicts a probability distribution over the vocabulary for every masked position at once. And it reports a **confidence** for each: how sure it is of its best guess there, read straight off the top probability. Then comes the rule that makes the whole thing work. Keep the high-confidence guesses. Commit them as real tokens. Leave the low-confidence positions blank, or blank them back out if an earlier step filled them badly. That last move has a name, **remasking**: a position that was written can be returned to the mask state on a later step, which is how the model revises. Run this for T steps, from an all-masked sequence toward a fully-filled one, and the sequence resolves **coarse-to-fine**: the easy, high-confidence words first, the words that depend on them last.

Researchers have a formal name for this setup: absorbing-state diffusion. The mask token is the absorbing state. A forward process corrupts real text by sending positions to the mask, and once a position is masked it stays masked until the reverse process fills it. So the entire model reduces to one learned skill. Given a partly-masked sequence, guess the masked tokens. That is all it ever does. This is why an ordinary masked language model can stand in for the decode loop at all: it already has that skill, learned at a low masking rate. A dLLM has the same skill at every masking rate, all the way up to the all-mask start.

Map the analogy onto the tensors before the build, so the crossword becomes concrete. The grid is the token sequence. A blank cell is a position holding the mask id. A pencilled-in answer is a position holding a real vocabulary id. The solver's certainty is the top softmax probability at that position. A pass over the grid is one forward call of the mask predictor. Filling the sure clues is committing the highest-confidence positions. Rubbing out a doubtful cell is remasking. Every noun in the analogy is a value you can print in the build below.

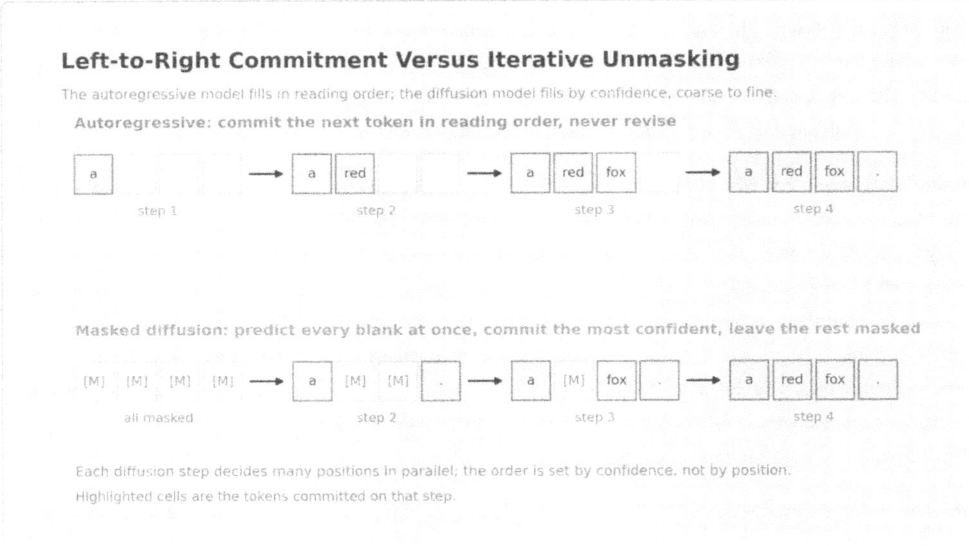

Figure 31.1. Left-to-right versus iterative unmasking. Top: an autoregressive model commits one token per step in reading order and never revises. Bottom: a masked diffusion model starts from all-mask, predicts every position in parallel, commits the most confident, remasks the rest, and repeats over T denoising steps.

Contrast the two orders plainly. The AR model asks, at every step, one question: given everything so far, what is the single next token? The dLLM asks a different question at every step: given everything decided anywhere in the sequence, which blanks are now sure enough to fill? The AR model's answer is one token. The dLLM's answer is a batch of tokens, chosen by confidence rather than by position. That is the entire architectural difference, and everything else in this chapter follows from it.

One honest word before the payoff, because it is the part marketing tends to drop. Filling many positions at once is fast, but it is not free of trouble. When the model commits several blanks in a single step, it decides each of them from the same context, without seeing the others it is committing alongside. If two of those positions needed to agree with each other, nothing forced them to. Fewer steps means more positions decided in parallel per step, which means more of these missed agreements. That tension is the whole game. It is what BREAK IT will make concrete.

Why It Matters

The reason anyone builds these is throughput. An autoregressive model's speed has a hard floor. It runs one forward pass per token, in order. So a 200-token answer is 200 sequential passes, no matter how much hardware you own. A dLLM breaks that floor by deciding many tokens per pass, and the reported numbers are large.

State them with their conditions attached, because the conditions are the whole story. Inception Labs reports its Mercury Coder models running at over 1000 tokens per second on NVIDIA H100 GPUs: Mercury Coder Mini at 1109 tok/s and Mercury Coder Small at 737 tok/s, per the Mercury paper (arXiv:2506.17298) and the company's launch blog. Those are code-optimized models, single-request throughput, on commodity H100s rather than custom silicon, and they are the vendor's own benchmark. Inception Labs positions this as up to roughly 10x faster than speed-optimized frontier autoregressive models, with the blog phrasing it 5 to 10x. That is measured against AR baselines in the 200-tok/s class. It is not a quality-matched comparison. Google

DeepMind reports a second system, Gemini Diffusion, at about 1479 tok/s, framed as the quality of Gemini 2.0 Flash-Lite at roughly 5x the speed. Read the label on that one too. DeepMind calls it an experimental demo, not a shipped product. The speed figure excludes a fixed overhead in their own chart. When a 200-token answer arrives in a fifth of the time, an interactive coding assistant stops feeling like a typewriter and starts feeling like pasting the whole answer at once.

Now the other half, stated just as plainly, because a chapter that only sold the speed would be lying by omission. Diffusion LLMs do not beat autoregressive models on reasoning. On DeepMind's own evaluation table, Gemini Diffusion scores 15.0% on BIG-Bench Extra Hard against 21.0% for Gemini 2.0 Flash-Lite. A clear reasoning gap, not parity. The Efficient-DLM paper (arXiv:2512.14067) notes that diffusion models' learning efficiency lags autoregressive models when trained from scratch. Where dLLMs are competitive is code and short-form generation. On the same DeepMind table, Gemini Diffusion posts 89.6% on HumanEval against Flash-Lite's 90.2%. On LiveCodeBench it is 30.9% against 28.5%. Even there, close, not a rout. The open research anchor is LLaDA, an 8B diffusion model trained from scratch. The paper shows it competitive with a comparable autoregressive baseline (LLaMA3 8B) on in-context learning and instruction following. It also mitigates the "reversal curse," even beating GPT-4o on a reverse-poem-completion task. Competitive with. Not uniformly better. The picture is a speed-versus-capability frontier, strongest on code and short answers, still behind on hard reasoning and long-form coherence.

The production status matters when you decide whether to reach for one. As of mid-2026 the three anchors sit at three different stages. Mercury is shipped. Inception Labs offers a playground, an API, and on-prem deployment. Gemini Diffusion is an experimental demo from Google DeepMind, not a generally available product. LLaDA is open research with public weights on Hugging Face. So the one you can pull down and run today, weights and all, is LLaDA. The one you can call in production is Mercury. The build below imitates the decode loop both of them use, on hardware you already own.

There is a connection back into the book that is easy to miss, and worth pinning down because it prevents a real confusion. The phrase "block diffusion" already appeared in this book, in a completely different job. **Project 14: Speculative Decoding** used a lightweight block-diffusion model as a *drafter*. The DFlash method (arXiv:2602.06036) drafts a whole block of guess tokens in one non-causal forward pass. Then a large autoregressive model verifies every one of them and throws out the wrong guesses. The DFlash paper reports up to about 6x lossless acceleration across models (around 4.9x average on Qwen3-8B, greedy, temperature 0). That is diffusion used as a fast intern whose work a left-to-right model checks. This chapter is the other use entirely: here the diffusion model *is* the decoder, and nothing downstream re-does its work left to right. Same word, "diffusion," two roles. In **Project 14** it is a draft mechanism inside AR serving. Here it is the full non-autoregressive model. Both are masked/discrete diffusion, and in neither case is the "noise" continuous Gaussian noise.

The serving side connects too. Deciding many tokens per pass raises GPU utilization, but a long context makes the key-value cache from **Project 13: Fast Inference: The KV Cache** grow until it dominates memory. The HERALD system (arXiv:2606.21633) targets exactly that for block-diffusion serving. It reports up to 1.59x lower per-block latency and 2.47x higher throughput than a GPU-only baseline, at near-lossless accuracy with a 5 to 10% KV budget. Those figures were measured across three block-diffusion LLMs on five long-context tasks, built on SGLang. That is the same production-serving concern as **Project 17: Production Serving: Continuous Batching and PagedAttention**, aimed at a non-autoregressive decoder.

The Build

You are going to build the decode loop itself, from scratch, and watch a sequence resolve coarse-to-fine.

One honesty note first, in the spirit of the proxy labs elsewhere in this book. Training a real diffusion language model like LLaDA is an 8B-parameter, many-GPU job, out of reach on the hardware this book assumes. But the interesting part of a dLLM is not the training. It is the decode loop: the iterative unmask, the confidence ranking, the remasking, the coarse-to-fine schedule. That loop can be driven by any bidirectional masked model, and there is a small pretrained one already sitting in the ecosystem: a BERT-style masked language model. So the build wraps the diffusion decode loop around `bert-base-uncased`. Treat this as a mechanism demonstration, not a production dLLM. BERT was trained to fill a few masked words in otherwise-real sentences, not to generate from an all-masked start, so its sentences will be rough. What it will show you cleanly, and correctly, is the machine: how the loop turns blanks into tokens by confidence, and how the step count trades speed for coherence.

A **masked language model**, or MLM, is a bidirectional Transformer trained to recover tokens that were replaced by the mask symbol. That is the mask predictor from The Concept, already trained. The build has four steps.

Step 1: Load the mask predictor and locate the blank

Load a small MLM and its tokenizer. The whole build runs on CPU; a GPU only makes it quicker.

```
import torch
import torch.nn.functional as F
from transformers import AutoTokenizer, AutoModelForMaskedLM

MODEL = "bert-base-uncased"
tok = AutoTokenizer.from_pretrained(MODEL)
model = AutoModelForMaskedLM.from_pretrained(MODEL, dtype=torch.float32).eval()

MASK = tok.mask_token_id       # the integer id of the [MASK] blank
print("mask token:", tok.mask_token, "id:", MASK)
```

Two details. The model loads in `eval` mode because you are decoding, not training, so dropout and other train-time behavior stay off. And note `dtype=torch.float32`, the current argument name that recent Transformers releases expect; older tutorials pass a now-deprecated alias, so match new code to `dtype`. Float32 is fine on CPU and keeps the confidences clean to read.

The `MASK` id is the blank cell of the crossword. Everything the loop does is a decision about which blanks to fill and which to leave.

Step 2: Build the all-masked starting sequence

A dLLM starts from a sequence that is entirely blanks, plus whatever prompt you condition on. To keep the demo readable, give the model a short prefix of real text and a block of blanks to fill after it. The prefix is the "clue" that makes an untrained-for-generation MLM produce something legible; a true dLLM would need no such crutch.

```python
def build_sequence(prefix_text, n_blanks):
    prefix = tok(prefix_text, add_special_tokens=False)["input_ids"]
    ids = ([tok.cls_token_id] + prefix
           + [MASK] * n_blanks + [tok.sep_token_id])
    x = torch.tensor([ids])                          # shape (1, seq_len)
    start = 1 + len(prefix)                           # first blank position
    fill = torch.arange(start, start + n_blanks)    # positions we may fill
    return x, fill

x, fill = build_sequence("the weather today is", n_blanks=6)
print(tok.decode(x[0]))
# [CLS] the weather today is [MASK] [MASK] [MASK] [MASK] [MASK] [MASK] [SEP]
```

`fill` is the list of positions the loop is allowed to touch. The prefix tokens and the special `[CLS]` and `[SEP]` markers stay fixed. The six blanks are the grid.

Step 3: One denoising step, and the confidence read

Now the core move. Run the model on the current sequence. Read the predicted distribution at every blank. Take two numbers per blank: the model's best-guess token, and how confident it is in that guess. Confidence is just the top softmax probability at that position.

```python
@torch.no_grad()
def predict(x, fill):
    logits = model(x).logits[0]                   # (seq_len, vocab)
    probs = F.softmax(logits[fill], dim=-1)  # (n_blanks, vocab)
    conf, pred = probs.max(dim=-1)             # best prob + best token id
    return conf, pred                               # each shape (n_blanks,)
```

Read what this returns. `pred[i]` is the token the model would write in blank `i` right now. `conf[i]` is how sure it is, a number between 0 and 1. A blank where the model is 0.9 confident is one it wants to commit. A blank at 0.1 is one to leave for later, after other blanks fill in and change the context this position reads. Because attention is bidirectional, committing one blank changes the prediction at every other blank on the next pass. That coupling is what makes the coarse-to-fine order pay off.

Print a few of those confidences and the schedule stops being abstract. On an all-masked start, a couple of positions come back near 0.9. Those are usually punctuation or a common function word. Most positions come back low, around 0.1 or 0.2, because the model has almost no committed context to read yet. The low numbers are the blanks to defer. Fill the sure ones, run again, and the deferred positions climb, because the tokens you just wrote give them something to condition on. Watching that climb happen is watching coarse-to-fine work.

Step 4: The unmask schedule and the loop

The last piece is the schedule: how many blanks to commit per step. This single knob is the speed-versus-quality dial of the whole method. Commit all six blanks in one step and you have done one forward pass. Fast. But every blank was decided from the same all-masked context, with no blank able to see any other. Commit one blank per step over six steps and each later blank reads the tokens the earlier steps committed, so the sequence stays consistent, at six times the passes.

Encode that as a linear reveal target: after step `s` of `T`, aim to have committed about `n_blanks * (s+1) / T` positions, always choosing the highest-confidence still-blank ones.

```
@torch.no_grad()
def diffusion_decode(prefix_text, n_blanks, steps, verbose=True):
    x, fill = build_sequence(prefix_text, n_blanks)
    committed = torch.zeros(n_blanks, dtype=torch.bool)

    for s in range(steps):
        conf, pred = predict(x, fill)
        conf = conf.masked_fill(committed, -1.0)   # ignore filled blanks

        target = round(n_blanks * (s + 1) / steps)  # how many filled by now
        reveal = target - int(committed.sum())
        if reveal <= 0:
            continue

        order = torch.argsort(conf, descending=True)  # most confident first
        for j in order[:reveal]:
            x[0, fill[j]] = pred[j]      # write the token into the blank
            committed[j] = True          # mark it committed

        if verbose:
            filled = int(committed.sum())
            print(f"step {s+1:>2}/{steps}  filled {filled}/{n_blanks}  "
                  f"|  {tok.decode(x[0, fill])}")

    return tok.decode(x[0, fill])
```

Trace one iteration against The Concept. `predict` is the denoising step. `masked_fill(committed, -1.0)` protects already-written blanks from being reconsidered, so this simple version commits monotonically. `argsort` on confidence is the "keep the high-confidence guesses" rule. Writing `pred[j]` into `x` is the unmask. Leaving the rest untouched is the remask-by-default. Every position you did not commit is still the `[MASK]` blank on the next pass. Its prediction is free to change once the committed tokens shift its context.

Run it with a generous step count and watch the coarse-to-fine order:

```
out = diffusion_decode("the weather today is", n_blanks=6, steps=6)
```

A representative trace (BERT's exact words will vary, that is expected):

```
step  1/ 6  filled 1/6  |  the [MASK] [MASK] . [MASK] [MASK]
step  2/ 6  filled 2/6  |  the [MASK] [MASK] . [MASK] good
step  3/ 6  filled 3/6  |  the same [MASK] . [MASK] good
step  4/ 6  filled 4/6  |  the same today . very good
step  5/ 6  filled 5/6  |  the same today , very good
step  6/ 6  filled 6/6  |  the same today , very good .
```

Read the order it filled in. The period and the easy function words went first, at high confidence, because the model was surest of them. The content words came later, once the punctuation and the frame were fixed. That is the coarse-to-fine schedule doing exactly what The Concept described: the model commits what it is sure of, and those commitments constrain the rest. The output is not eloquent, because BERT was never trained to generate from a full block of blanks. The loop around it is the real dLLM decode loop, and it behaves correctly.

The direction to remember, which BREAK IT will drive home with numbers: more steps means fewer blanks committed per pass, which means each committed token got to see more already-decided context, which means better coherence at higher cost. Fewer steps means the reverse. Faster, rougher.

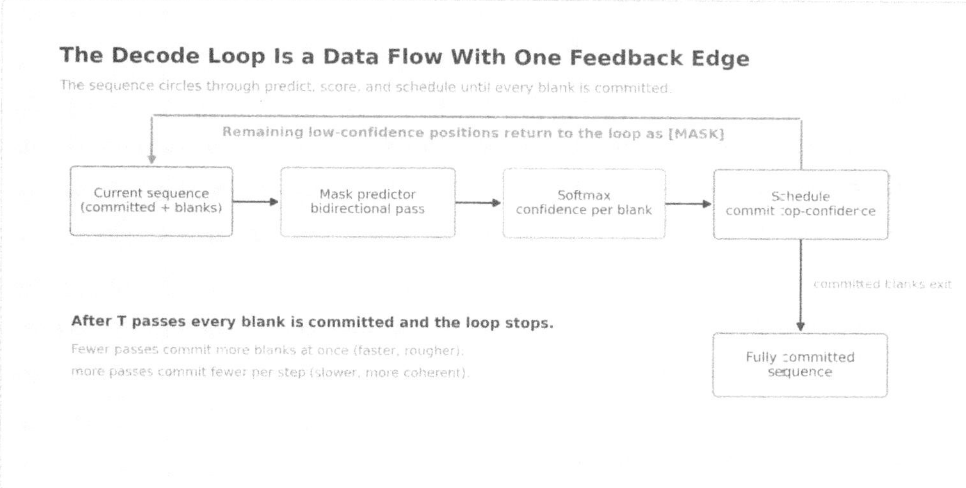

Figure 31.2. The decode loop as a data flow. The current sequence enters the mask predictor; softmax gives a confidence per blank; the schedule commits the top-confidence blanks and returns the rest to the loop as masks. After T passes the sequence is fully committed.

Step 5: A note on what a real dLLM adds

Two differences separate this demo from LLaDA-class systems, and both are worth naming so you know what you are and are not seeing.

First, a real dLLM is trained for this. Its objective is to recover masked tokens across every masking ratio, from nearly-all-masked to nearly-none. So starting from a full block of blanks is the case it was built for. Not an out-of-distribution stunt. BERT was trained at a low masking rate on real sentences, which is why the crutch prefix helps here.

Second, a full dLLM remasks committed tokens too, not only the empty blanks. The version above commits monotonically for clarity: once written, a token stays. LLaDA's low-confidence remasking lets an already-written position go back to `[MASK]` on a later step if its confidence has fallen, giving the model true in-place revision, the ability to fix an early token once later ones reveal it was wrong. That revision is why dLLMs can do infill and error correction that a left-to-right decoder cannot. An AR model that commits token 3 can never take it back. A dLLM can. Adding it is a small change, and the homework asks you to make it.

A second consequence falls out of the same property. Because a dLLM can fill positions in any order, it can be handed a sentence with a hole in the middle and told to fill only the hole. Both sides stay fixed. That is infill, and it is native here, not a bolted-on mode. Give the loop a prefix and a suffix with blanks between them, and the bidirectional predictor reads both sides while it fills. An autoregressive model reads only leftward, so it cannot condition on the suffix at all. The same any-order freedom is what lets LLaDA mitigate the reversal curse: the paper shows it completing a poem in reverse well enough to beat GPT-4o on that specific task, precisely because it is not locked to one direction.

BREAK IT

The whole promise of non-autoregressive decoding is the parallel commit: filling many blanks per step is what makes it fast. This break attacks that promise directly. Turn the step count down until the model is committing most of the sequence in a single pass. Then watch coherence fall apart, in a way the speed alone would never warn you about.

The exact change

Change one number. Run the same prompt, the same model, the same code, at two step counts: the coherent baseline at `steps = n_blanks` (one blank per pass), and the aggressive setting at `steps = 1` (every blank committed in a single pass).

```
print("baseline, one blank per step:")
diffusion_decode("the weather today is", n_blanks=6, steps=6)

print("\nbroken, all blanks in one step:")
diffusion_decode("the weather today is", n_blanks=6, steps=1)
```

Nothing else moves. Same weights, same prompt, same confidence rule. The only difference is how many blanks the schedule commits before the model gets to look again.

Predict what will happen before you run it

At `steps = 1`, the model fills all six blanks from a single forward pass over the all-masked sequence. Every blank is decided from the same context: the prefix, and five other blanks it cannot see the values of, because none of them are decided yet. So each position independently picks its own most-likely token, and nothing makes those six choices agree with one another. Blank 4 does not know blank 5 exists as a word. If the fluent continuation needed "a lot of rain," the position that should say "of" has no idea the position two cells over committed to "rain." You should expect locally plausible tokens that do not join into a sentence: repeats, disagreements in number and tense, a word salad that is individually reasonable and jointly broken.

At `steps = 6`, each blank after the first is decided with the earlier commitments visible. The context tightens every pass. Coherence holds.

What you will observe

```
baseline, one blank per step:
  the same today , very good .

broken, all blanks in one step:
  the . . . . rain
```

The one-step output is the tell. Each token, taken alone, is something BERT found likely at that spot given six blanks. Taken together they do not form a sentence, because no position could see what the others were committing in the same pass. The repeated punctuation is the classic signature: several positions independently decide the single most probable token in isolation is a period, and with no coordination they all write it.

Now sweep the step count and watch the transition, so the tradeoff is a curve rather than two points:

```
for T in (1, 2, 3, 6):
    out = diffusion_decode("the weather today is", n_blanks=6,
                           steps=T, verbose=False)
    print(f"steps={T}:  {out}")
```

```
steps=1:  the . . . . rain
steps=2:  the day . very good rain
steps=3:  the day is very good .
steps=6:  the same today , very good .
```

The reading is the point of the chapter. Coherence rises with the step count, and it rises because more steps means fewer blanks committed per pass, which means each committed token was chosen with more of the sequence already decided and visible to it. Speed moves the other way: `steps=6` is six forward passes, `steps=1` is one. The dial you turned is the exact dial a production dLLM turns to trade tokens-per-second against output quality.

Put the cost in passes, because that is what the dial actually spends. At `steps=6` the model ran six forward passes for six tokens. That is the same one-pass-per-token floor an autoregressive decoder pays, and no faster. At `steps=1` it ran a single pass for all six. That single pass is the whole speed win. It is also the whole coherence loss. Same knob, both effects. A production dLLM sits at neither extreme. It picks a step count well below the token count, which buys most of the coherence back for a fraction of the passes. That is how systems like Mercury and Gemini Diffusion reach the throughput numbers from Why It Matters while staying legible: many tokens per pass, but not all of them, and never in a single shot.

What this proves

The parallelism is not free, and its cost is coherence, paid at exactly the positions that had to agree with each other. A fixed-order autoregressive model never has this failure, because it commits token 5 with token 4 already on the page; every token sees all of its left context by construction. The dLLM gives that up to win speed, and it wins the speed precisely by committing tokens that cannot see each other. Turn the step count to its floor and you see the raw form of the problem. Turn it up and you spend passes to buy the agreement back.

This is the same shape of lesson as the needle test in **Project 30: Non-Transformer Architectures (Mamba, RWKV)**. There, a fixed-size state bought linear-time speed and paid in exact recall. Here, parallel commitment buys high throughput and pays in cross-token coherence at low step counts. Neither is a bug to be engineered away. Both are the price printed on the speedup. It also explains the honest reading from Why It Matters. dLLMs are strongest on code and short-form, where fewer long-range agreements between distant tokens have to be resolved. They are weaker on long-form coherence and multi-step reasoning, where many such agreements pile up.

Optional Homework

Everything here runs on CPU. A full run of the labs below is well under an hour on a laptop; each `diffusion_decode` call is a handful of forward passes of a small model. No GPU is required. If you have one, it changes nothing about the observations, only the wall-clock.

Part A: The step-count sweep

Run `diffusion_decode` on three prompts of your choice at `steps` in `{1, 2, 3, 4, 6, 8}` with `n_blanks = 6`. For each, record the output and rate its coherence 0 to 2 by eye (0 = word salad, 1 = partly joins, 2 = a clean phrase). Plot coherence against step count. You should see it climb, then flatten. The first few extra steps buy the most. Past a point, more steps stop helping, because the remaining blanks were never the ambiguous ones. That flattening is the practical reason production dLLMs use far fewer steps than tokens.

Part B: Add real remasking

Turn the monotonic loop into a true dLLM loop by letting committed tokens be reconsidered. Instead of protecting committed positions forever, each step recompute confidence at every fill position, and among all positions keep only the `target` highest-confidence tokens committed, returning the rest to `[MASK]`. Compare against Part A at the same step counts. Expected observation: at low-to-middling step counts, remasking repairs some of the incoherent commits that the monotonic version froze in, because a bad early token can now be pulled back once its neighbors resolve. This is the in-place revision from Step 5, and it is the mechanism behind a dLLM's infill and error-correction ability.

Part C: The order it chooses

Modify the loop to log, per step, which positions it committed and their confidences. Run at `steps = n_blanks`. Confirm that early steps commit high-confidence, low-information tokens (punctuation, function words) and later steps commit the content words. Then force the opposite order: commit the *lowest*-confidence blanks first. Predict the result before running, then run it. You should see coherence drop sharply, because committing the positions the model is least sure about first throws away the constraint structure that makes the coarse-to-fine order work.

Part D: Longer blocks

Rerun Part A with `n_blanks = 12` and `n_blanks = 20`. The coherence-versus-steps curve should shift: longer blocks need more steps to reach the same coherence, because a longer block has more pairs of positions that must agree, and each pass can only resolve so many. Record the step count where each block length first produces a clean phrase. This is the small-scale shadow of why long-form generation is where dLLMs trail autoregressive models.

Part E: Swap the predictor

Replace `bert-base-uncased` with `roberta-base` (adjust for RoBERTa's mask token, which the tokenizer reports as `tok.mask_token`). Rerun Part A. The exact words change; the step-count-versus-coherence shape does not. That invariance is the point: the decode loop is the dLLM, and the mask predictor is a swappable part.

Questions To Answer

1. In the BREAK IT sweep, at what step count did your outputs first read as coherent phrases, and how many forward passes did that cost relative to the token count? State the speed-versus-coherence tradeoff you observed in one sentence.
2. When you ran `steps = 1`, which specific failure appeared first: repeated tokens, tense or number disagreement, or something else? What does that failure tell you about what the parallel commit cannot coordinate?

3. In Part C, forcing lowest-confidence-first order hurt coherence. Explain why the coarse-to-fine order is not an arbitrary heuristic but the thing that makes iterative unmasking work at all.
4. Part B added remasking. On which prompts did revision help most, and what did those prompts have in common? Connect this to why dLLMs can do infill and in-place error correction that a left-to-right decoder cannot.
5. From Why It Matters: Gemini Diffusion trails Gemini 2.0 Flash-Lite on BIG-Bench Extra Hard (15.0% versus 21.0%, DeepMind's own table) but is near parity on HumanEval. Using what you saw in Part D, argue why the gap is wide on reasoning and narrow on code.
6. **Project 14: Speculative Decoding** used a block-diffusion model as a drafter that a large AR model verifies, while this chapter used diffusion as the whole decoder. Which of your BREAK IT failures would the AR verifier in **Project 14** catch and correct, and which would it not?

Go Further

The open reference to read first is LLaDA. The paper (Nie and colleagues, "Large Language Diffusion Models," arXiv:2502.09992, ICLR 2025) builds the from-scratch 8B masked diffusion model this chapter's loop imitates. The released `LLaDA-8B-Base` and `LLaDA-8B-Instruct` checkpoints on Hugging Face let you run a real dLLM decode instead of a BERT stand-in. Read the forward-masking and reverse mask-prediction sections against Step 4 of the build; they are the same loop, trained.

For the vendor systems and their reported numbers, start with the Mercury paper (Inception Labs, arXiv:2506.17298) and its launch blog. Both carry the Mercury Coder throughput figures on H100. Then read the Gemini Diffusion page from Google DeepMind for the experimental-demo speed and the evaluation table with the code-versus-reasoning split. Treat every speed number as vendor-reported and keep its hardware and conditions attached.

For the serving side, HERALD (arXiv:2606.21633) covers block-diffusion LLM serving and the key-value-cache pressure that comes with long context, the same production concern as **Project 17: Production Serving: Continuous Batching and PagedAttention**. For the drafter use of the same idea, the DFlash paper (arXiv:2602.06036) from **Project 14: Speculative Decoding** is the counterpoint: block diffusion as a draft mechanism a large AR model verifies, not as a standalone decoder. Efficient-DLM (arXiv:2512.14067) is the reference for the honest limitation, that diffusion models' learning efficiency lags autoregressive models trained from scratch.

arXiv search directions worth following: `masked diffusion language model`, `discrete diffusion text generation`, `absorbing state diffusion`, `any-order autoregressive`, `block diffusion language model`, and `confidence-based remasking sampling`. The absorbing-state framing is the formal name for "the noise is the mask token," and reading one absorbing-state diffusion paper is the fastest way to see why the all-masked start is the right mathematical object rather than a trick.

What You Now Know

Text can be generated without a fixed left-to-right order. A masked diffusion language model starts from a fully-masked sequence, not from continuous Gaussian noise, and fills it in over T denoising steps with a bidirectional mask predictor. Each step predicts every masked position at once. It commits the highest-confidence guesses and remasks the rest. So the sequence resolves coarse-to-fine: the sure tokens first, the tokens that depend on them last. That is a different decoder from every autoregressive model earlier in this book, and it is non-autoregressive by construction.

You know what it buys and what it costs. It buys throughput. Inception Labs reports Mercury Coder over 1000 tok/s on H100 (1109 Mini, 737 Small), up to about 10x speed-optimized AR by the vendor's own count, and Google DeepMind reports Gemini Diffusion near 1479 tok/s as an experimental demo. It costs coherence when too many tokens are committed per step, and it still trails autoregressive models on hard reasoning while reaching parity mainly on code and short-form. You saw both halves in one loop: the coarse-to-fine trace that resolves cleanly at a high step count, and the word salad that appears when the step count hits its floor.

You can build the decode loop by hand around any masked model, drive it with a confidence schedule, add real remasking for in-place revision, and read the step-count dial as the speed-versus-coherence tradeoff it is. You can keep the two meanings of "block diffusion" straight: a drafter that a left-to-right model verifies in **Project 14: Speculative Decoding**, versus the standalone non-autoregressive decoder here. And you can say precisely why the next assistant that answers a coding prompt in a blink might not be running left to right at all.

Starting Point

Nie et al., "Large Language Diffusion Models" (LLaDA, 2025): arxiv.org/abs/2502.09992. Checkpoints: huggingface.co/GSAI-ML/LLaDA-8B-Base and LLaDA-8B-Instruct. Inception Labs, "Mercury: Ultra-Fast Language Models Based on Diffusion" (2025): arxiv.org/abs/2506.17298, blog at inceptionlabs.ai/blog/introducing-mercury. Google DeepMind, Gemini Diffusion: deepmind.google/models/gemini-diffusion. HERALD (block-diffusion serving, 2026): arxiv.org/abs/2606.21633. Efficient-DLM (2025): arxiv.org/abs/2512.14067. The masked model used in the build: huggingface.co/bert-base-uncased.

Project 32: Layer Freezing and Transfer

Hook

Why can you spend real money fine-tuning a model on your domain, get better results on that domain, and still somehow make the model worse at everything else it used to know? And why does the opposite extreme also fail, where you freeze almost the whole model, train for hours, and end up with something that barely changed at all? This is the engineering problem in transfer learning: how much of a pretrained model should stay untouched, and how much should adapt?

The first time I watched it happen in our own runs, I felt a small unfairness. The domain numbers were going up. The model genuinely looked better at the thing we trained it to do. And it was, quietly, losing other things we had not been watching. That is the actual texture of this chapter. The win and the cost share the same training step.

The Concept

Picture a large office with many floors. The bottom floors handle generic work that every department needs: reading email, identifying names, spotting dates, recognizing punctuation, basic grammar, common patterns. The upper floors handle more specialized work: legal review, code style, medical jargon, finance phrasing, multilingual nuance.

A pretrained language model works a lot like that. Early layers tend to learn broad, reusable features. They notice local patterns, help the model tell where words begin and end, understand what phrases often travel together, how syntax hangs together, and how common token combinations tend to behave. Later layers tend to shape those general features toward a task or domain. They take "this looks like a function definition" or "this looks like a drug name" and turn it into behavior that matters for the next-token prediction in that setting.

I am stating this more confidently than the field actually deserves. The early-layers-are-general story is a useful guide, not a clean theorem. It bends with model size, with how much pretraining data covered the new domain, and with how aggressive the optimizer is during fine-tune. Treat it as a starting prior. The experiment in this chapter is what turns it into something you can actually trust for your own model.

Layer freezing means this: you take a pretrained model and tell some layers, "You are done learning. Do not move." Then you fine-tune only the remaining layers on new data. In plain English, you are saying: "Keep the general language machinery. Adapt only the parts most likely to need domain-specific change.' That is the whole bet.

If the bet is right, you get three things at once: faster training because fewer parameters update, less forgetting because you leave more of the original model intact, and still enough adaptation to become better at your new domain. If the bet is wrong, one of two bad things happens.

At one extreme, you freeze nothing. Every parameter is free to drift, the model gets better at your niche domain, but it starts overwriting some of the general patterns that made it broadly useful. This is catastrophic forgetting (introduced in **Project 21: Fine-Tuning And Instruction Tuning**): the measurable price when domain fine-tuning gets too much permission to move shared internal structure. The first time an offline Tamil-language medical fine-tune hit this on small Gemma-class models, the clinical triage answers got tighter and the basic conversational fluency got noticeably more brittle on the same evaluation prompts the base model had handled fine. We were paying for clinical sharpness in plain-English coherence, and we did not see the bill until we added a non-clinical validation set.

At the other extreme, you freeze almost everything. The model has too little room to adapt, it still sounds like the original model because mostly it still is the original model, and it picks up a few surface habits but not enough to really learn the new domain. This chapter is about finding the useful middle.

What "layer" means here

A transformer layer is one repeated processing block inside the model. In GPT-style models, each block usually contains:

- attention, which lets each token look at other tokens
- a feedforward network, which stores and transforms patterns
- normalization and residual connections, which keep training stable

If your model has 12 layers and you freeze the first 6, you are freezing the bottom half of that stack. The input still passes through those layers, they still compute outputs, they just stop changing during training. That distinction matters: frozen does not mean removed, it means read-only.

Most beginner explanations of freezing get this exactly wrong. They make it sound like a frozen layer becomes invisible to the forward pass, which then makes the speedup numbers feel like magic. The forward pass cost barely moves. What moves is the backward pass, the gradient storage, and the optimizer state. The intuition you want is "read-only memory, still wired in." Not "deleted."

What transfer means here

Transfer learning means taking a model that already learned something useful from one dataset and adapting it to another job instead of starting from scratch. In this chapter, the transfer is from a general pretrained model to a domain-specific fine-tune target. That target might be:

- code
- medical text
- legal text
- another language
- support logs
- your company's documentation style

The reason transfer works at all is that the model's earlier layers are not learning "medical English" or "Python style" from zero. They are learning reusable machinery that many domains share. That is why freezing early layers is even plausible: if every layer were equally domain-specific, freezing early layers would be pointless.

CKA: a way to measure "how much did this layer change?"

Now for the measurement tool that makes this chapter more than a hunch. CKA stands for Centered Kernel Alignment. That name sounds like a paper title because it is one. Ignore the name for a moment.

The intuition is simple. Suppose you run the same batch of text through the original pretrained model and the fine-tuned model. At each layer, you collect the hidden states (just the model's internal representation of the text at that point: a table of numbers where each token has a vector of features). Then you ask: "How similar are these two layers' internal representations?" If layer 3 in the original model and layer 3 in the fine-tuned model produce almost the same internal patterns on the same inputs, their CKA score is near 1.0. If they produce very different internal patterns, the score drops toward 0. So CKA gives you a map of representational drift.

The first time I plotted a real CKA curve on one of our cooperative fine-tune runs, I had a small embarrassing reaction. The smooth, gradual rise in drift I had pictured as you walk up the stack was not what showed up. The actual curve had a near-flat floor, a knee, and then a sharp climb. The knee landed almost exactly where we had set the freeze boundary. The diagnostic was not telling me something subtle. It was confirming that we had done the freezing correctly. That is a much more boring outcome than the one I had been bracing for, and a much more useful one for a paper.

That matters because layer freezing makes a very specific prediction. Below the freeze boundary, CKA should stay at 1.0 or extremely close, because those layers are frozen. Above the freeze boundary, CKA should drop, because those layers are adapting. That is more than a diagnostic. It is a composability signal. If early layers really stay stable across domains, then maybe those layers are reusable components. Maybe they can become shared infrastructure between specialists. That is one of the reasons this chapter sits in the composable architectures part of the book.

The equation, now that you know what it is measuring

Here is a common linear CKA form:

```
CKA(X, Y) = ||Y^T X||_F^2 / (||X^T X||_F * ||Y^T Y||_F)
```

Now the symbols have earned their place.

- (X) is the matrix of hidden states from one layer of the original model across many tokens or samples.
- (Y) is the matrix of hidden states from the matching layer of the fine-tuned model on the same inputs.
- (Y ^ \top X) means "compare all features in one representation against all features in the other."
- (|\cdot|_F) is the Frobenius norm, which you can think of as "take all numbers in the matrix, square them, add them up, then take the square root."
- The numerator gets large when the two representations line up well.
- The denominator rescales the score so you can compare layers fairly.

Do not worship the formula. Its job is modest. Take two clouds of activations, turn "these feel similar" into a number, that is enough.

Honestly, I think most introductions to CKA do it a disservice by leading with the equation. The equation looks more intimidating than the operation deserves. If you can compute a covariance matrix you already have everything you need. Lead with the question CKA is answering and the formula stops feeling like a hazing ritual.

Why It Matters

Without layer freezing, you are stuck with a blunt instrument. You can fine-tune the whole model, or not fine-tune it. Real systems need more control than that.

Why engineers freeze layers in practice

First: cost. Fine-tuning all layers means computing gradients and updates for all trainable parameters, which costs memory, time, and money. If you can freeze 50% of the stack and still get most of the gain, that is not a minor optimization. That is deployment reality. In long cooperative LoRA-fusion experiments, the gap between "every contributor fine-tunes their full base" and "every contributor fine-tunes a small adapter on a shared frozen base" was the gap between something interested volunteers could run on a consumer GPU and something they could not. Freezing decisions are infrastructure decisions.

Second: forgetting. A general pretrained model contains a wide range of behaviors. It can explain recipes, summarize logs, answer trivia, reason about text, and write code badly but competently enough to be useful. Full fine-tuning on narrow domain data can trade away some of that breadth. That trade can be worth it, but if you do not measure it, you are guessing.

Third: model composition. Part V of this book asks whether independently trained specialists can someday combine cleanly. That only has a chance of working if some parts of the network are more universal than others. Layer freezing gives you evidence. If early layers stay useful across domains and later layers absorb specialization, then the stack is not one inseparable blob. It has structure. Shared structure. That is the opening.

What breaks if you ignore the tradeoff

If you freeze nothing, training is slower, memory use is higher, general capability can drift downward, and domain gain can come with hidden damage. If you freeze too much, training is cheaper, forgetting stays low, but the model barely adapts and domain metrics plateau early.

The engineering job is not "freeze as much as possible" and it is not "fine-tune everything." It is to find the point where domain gain is real, forgetting stays controlled, and the compute bill still makes sense. That sweet spot changes with model size, domain distance, dataset size, learning rate, training duration, whether you use LoRA or full-parameter updates, and how much the new domain overlaps with old knowledge.

In other words, there is no global answer. There is only the answer for your model on your data with your training budget, and the way you find it is by running the experiment described in the rest of this chapter and reading the curves. People who promise you a universal freeze ratio are usually selling something. This chapter gives you the measurement frame instead.

The Build

We are going to build an experiment, not a product. The goal is to produce three outputs:

1. A freeze-depth tradeoff curve.
2. A forgetting measurement before and after fine-tuning.
3. A per-layer CKA map showing where the model changed.

You can do this with a GPT-style model from `nanochat` or any readable transformer codebase. The code snippets below are minimal PyTorch-style examples meant to show the mechanism clearly.

Step 1: Pick the domain shift

You need a pretrained base model and a new domain dataset. Choose a domain that is different enough to force adaptation but not so alien that the model needs to relearn language from scratch.

Good choices:

- general model -> Python or JavaScript code
- general model -> biomedical abstracts
- general model -> legal opinions
- general model -> Spanish or German text if the base model is mostly English

What you want is tension. Too easy, and every freeze ratio looks the same. Too far, and shallow adaptation fails across the board. For a first run, code is a good target because the failure modes are obvious. You can see quickly whether the model learns syntax, indentation, function structure, and naming conventions.

Step 2: Define the freeze ratios

The chapter spec asks for:

- freeze first 25% of layers
- freeze first 50% of layers
- freeze first 75% of layers

Also include the two extremes for BREAK IT later:

- freeze 0%: full fine-tune
- freeze all but last layer: almost no adaptation

Suppose the model has `N` transformer blocks. Then:

- 25% frozen means freeze layers `0` through `floor(0.25 * N) - 1`
- 50% frozen means freeze the first half
- 75% frozen means freeze the first three quarters

Be explicit about indexing — off-by-one mistakes here make the experiment hard to trust.

Step 3: Freeze parameters correctly

In PyTorch, freezing means setting `requires_grad = False` for the parameters in those layers.

Minimal sketch:

```
import torch

def freeze_bottom_layers(model: torch.nn.Module, freeze_ratio: float) -> int:
    n_layers = len(model.transformer.h)
    freeze_upto = int(n_layers * freeze_ratio)

    for layer_idx, block in enumerate(model.transformer.h):
        freeze = layer_idx < freeze_upto
        for p in block.parameters():
            p.requires_grad = not freeze

    return freeze_upto
```

If your model has token embeddings and a final normalization layer, make a choice and document it. Two common choices: freeze only transformer blocks and leave embeddings trainable, or freeze embeddings together with the earliest layers. For this experiment, keep it simple and consistent. Freeze bottom transformer blocks only, which isolates the question.

I expected the embedding decision to be cosmetic. It mostly is, until you change tokenizers or push fine-tune far enough that the embedding table starts encoding domain-specific frequency. Then it stops being cosmetic. I have re-run one too many freeze experiments after realizing the comparison had been silently absorbing embedding drift on the side. Pin the choice. Write it in the run name.

After freezing, print the trainable parameter count:

```
import logging

trainable = sum(p.numel() for p in model.parameters() if p.requires_grad)
total = sum(p.numel() for p in model.parameters())
logging.info(f"trainable: {trainable}/{total} = {trainable/total:.2%}")
```

You want this logged for every run. Otherwise you cannot connect speed differences to actual trainable capacity.

Step 4: Build two evaluation sets, not one

You need two different judges. The first judge measures domain adaptation: did the model improve on the new target domain? Examples include code completion loss on held-out code, medical next-token loss on held-out biomedical text, or downstream exact match on a domain task if you have one.

The second judge measures forgetting. Use a held-out general text set from the model's old world — it does not need to be fancy, it just needs to be outside the new domain. Examples include a slice of wiki-like text, general web prose, short QA prompts, or a generic validation corpus from the original training distribution.

If you only measure the domain side, full fine-tune will often look best. That is how people fool themselves. This chapter refuses that shortcut.

Step 5: Record baseline before fine-tuning

Before you touch the weights, evaluate the pretrained model on both the domain validation set and the general validation set. These are your "before" numbers — write them down in a table. You need the delta after fine-tuning, not just the final score.

A simple table template:

Run	Domain Loss Before	Domain Loss After	General Loss Before	General Loss After	Tokens/sec	Trainable %

: **Table 17.1.** Layer-freezing experiment template: domain and general validation loss before and after fine-tuning, with throughput and trainable parameter percentage, to be filled in for each freeze configuration.

You can substitute accuracy for loss if your task makes that easier to read. For language modeling, cross-entropy loss or perplexity is usually cleaner.

Step 6: Fine-tune each freeze configuration under matched conditions

This part sounds boring. It is where most bad comparisons are born. Hold these constant across runs:

- same pretrained checkpoint
- same dataset split
- same optimizer
- same learning rate schedule
- same batch size or effective batch size
- same number of training steps or same token budget
- same random seed if possible

If one run trains longer or sees more data, the comparison is contaminated. Your training loop does not need to change much — frozen parameters simply stop receiving gradients and stop updating.

Minimal optimizer setup:

```
trainable_params = [p for p in model.parameters() if p.requires_grad]
optimizer = torch.optim.AdamW(trainable_params, lr=lr, weight_decay=wd)
```

That is enough. The optimizer never even sees the frozen parameters.

Step 7: Measure training speed honestly

You care about speed because freezing is partly a compute decision. Track at least wall-clock time per step, tokens per second, and peak memory if available.

Freezing lower layers does not remove their forward pass. The model still runs through them. So the speedup is real but smaller than many newcomers expect. You still pay for the forward activations through the frozen layers and the attention and MLP compute in those layers during the forward pass. You save mainly on the backward pass through frozen parameters, gradient storage for them, and optimizer state for them. That means freezing helps, but it does not turn a 12-layer model into a 3-layer model. Frozen layers are still there. They just stop learning.

I had this conversation more times than I want to admit during long training sweeps. A contributor would freeze 75% of the stack, see a smaller-than-expected throughput improvement, and ask whether something was wrong. Nothing was wrong. The forward pass had not gotten cheaper. The cost story for layer freezing is mostly a backward-pass story, and that bears repeating until it stops being a surprise.

Step 8: Measure forgetting after fine-tuning

Now evaluate the fine-tuned model again on both validation sets. You are looking for this shape:

- domain performance improves as more upper layers are trainable
- general performance degrades more when more layers are trainable
- speed improves as more layers are frozen

Those are the three axes, and without all three, you do not have the real tradeoff.

Step 9: Collect hidden states for CKA

Now we move from "did the model change?" to "where did it change?" Run the same batch of inputs through the original pretrained model and the fine-tuned model, and at each layer boundary, collect hidden states.

A hidden state tensor usually has shape `B x T x d`, where `B` is batch size, `T` is sequence length, and `d` is hidden size. To compare layers with CKA, flatten batch and time into one dimension so `X` and `Y` each become `(B*T) x d`, giving you one big matrix of token representations per layer.

Minimal sketch:

```
with torch.no_grad():
    out_base = base_model(input_ids, output_hidden_states=True)
    out_ft = ft_model(input_ids, output_hidden_states=True)

base_hiddens = [h.reshape(-1, h.size(-1)) for h in out_base.hidden_states]
ft_hiddens = [h.reshape(-1, h.size(-1)) for h in out_ft.hidden_states]
```

Now compute CKA layer by layer. A minimal linear CKA implementation:

```
import torch

def linear_cka(X: torch.Tensor, Y: torch.Tensor) -> float:
    X = X - X.mean(0, keepdim=True)
    Y = Y - Y.mean(0, keepdim=True)

    dot_xy = Y.T @ X
    dot_xx = X.T @ X
    dot_yy = Y.T @ Y

    num = (dot_xy ** 2).sum()
    den = torch.sqrt((dot_xx ** 2).sum() * (dot_yy ** 2).sum())
    return (num / den).item()
```

This is one of those functions that looks more mysterious than it is. It is just: center the features, measure how aligned the two representation spaces are, and normalize the score. If you plot this across layers, the freeze boundary should show up like a waterline on a hull. Below it, the wood stays untouched. Above it, the paint has been weathered by a different storm. That is the picture you want.

Step 10: Plot the tradeoff

Make at least two plots.

First plot: freeze ratio vs performance/speed

- x-axis: freeze ratio
- y-axis: domain metric
- second y-axis or separate lines: general metric, tokens/sec

Second plot: layer index vs CKA

- x-axis: layer number
- y-axis: CKA score from 0 to 1

What you expect:

- full fine-tune: CKA drifts across many or all layers
- moderate freezing: lower layers stay near 1.0, upper layers drift
- freeze-all-but-last-layer: almost every layer stays near 1.0, last layer shifts a bit

Now you can say something stronger than "the model changed." You can say where the representation changed, how much, and whether that matched the freeze plan.

Step 11: Interpret the sweet spot

A good result often looks like this: full fine-tune gives the best domain score but hurts the general score noticeably, while 25% or 50% freezing recovers most of the domain gain while reducing forgetting and improving training efficiency. The chapter spec mentions that medium-scale settings often like roughly 15–20% frozen depth. Treat that like a clue, not a law. Your own setup may prefer something else depending on how far the new domain sits from the old one. If your target is highly specialized code or biomedical text, you may need more upper-layer freedom. If your domain mostly changes tone and terminology, less adaptation may be enough.

The point is not to memorize a freeze percentage. The point is to stop treating the model as a single undifferentiated object. It has reusable parts and adaptable parts, and this experiment makes that visible.

BREAK IT

This section matters more than the successful run. You are going to break the system in both directions and force the tradeoff to show itself.

Failure mode 1: freeze nothing

Change: set freeze ratio to `0.0`, every layer is trainable, the optimizer sees all model parameters. You are removing the guardrail that protects the pretrained model's general internal structure. You are telling every layer, including the earliest and most reusable ones, to bend toward the new domain.

What you will observe: domain validation improves the fastest (especially early in training), general validation degrades more than in partially frozen runs, CKA scores fall across much more of the stack including lower layers, and training is slowest and uses the most memory among the freeze variants. If your new domain is narrow and repetitive, the model may start sounding strangely over-specialized outside that domain. Examples of what this looks like in practice:

- a code-tuned model starts answering general prose questions with too much formatting, pseudo-code, or terse bullet structure
- a medical-tuned model overuses clinical phrasing in ordinary explanations
- a legal-tuned model becomes verbose, hedged, and citation-shaped even when not asked

At the metric level, the cleaner sign is domain loss going down while general loss goes up. That is catastrophic forgetting in miniature. Not total amnesia, not "the model forgot English," but something more annoying and more realistic: the model gives away broad capability to sharpen narrow capability.

I will admit I was slow to take this seriously the first few times. The domain numbers were so encouraging that the general regression looked like noise. It is not noise. If you do not measure the general side on every freeze sweep, you will under-report the cost of your own runs and then be confused later when people interact with the model on tasks the validation set never asked about.

What this proves: the lower and middle layers are not just spare capacity waiting to be repurposed. They carry useful general representations. If you let them all drift, the model can overwrite some of the machinery that made it broadly competent. That is why full fine-tuning is not automatically the right answer. It can buy specialization by spending generality.

Failure mode 2: freeze everything except the last layer

Change: freeze all transformer blocks except the topmost block (or, in a stricter version, freeze all blocks and train only the final output head). You are removing almost all of the model's ability to adapt its internal processing. You are asking one thin slice at the top to rewrite behavior built by the entire stack below it.

What you will observe: training is fast, forgetting is minimal, domain improvement is weak or nearly flat, CKA stays near 1.0 through almost the entire network, and only the very top layer or head shifts. The model may pick up small surface markers (a few domain terms, slight formatting changes, some local next-token bias) but it will not deeply internalize the new domain. In code, that often means it learns to start functions in the right way but still fails on longer structural patterns. In medical text, it may learn common vocabulary but not relationships between symptoms, treatments, and context.

What this proves: one layer is not enough room for serious adaptation. Domain transfer is more than changing the final words. It requires reshaping upper-layer representations across multiple stages. That is why the top-only strategy often underperforms badly.

The sweet spot is the lesson

Now compare the broken extremes against the middle runs. You should see:

- full fine-tune: most domain gain, most forgetting, highest cost
- top-layer-only: least forgetting, least adaptation, lowest cost
- partial freeze: middle cost, strong adaptation, controlled forgetting

That middle is not an aesthetic compromise. It is the actual engineering target. You are protecting two things at once: the original model's broad competence and your budget, while still buying enough movement where the specialization actually lives.

What to watch for in logs and outputs

Watch training and validation curves, not just final numbers.

Signs of full fine-tune damage:

- domain loss drops quickly
- general loss rises steadily
- sample generations outside the target domain feel more brittle or oddly stylized
- CKA drift starts lower in the stack than you want

Signs of over-freezing:

- both train loss and domain validation loss plateau early
- sample generations pick up only cosmetic domain features
- CKA remains nearly unchanged almost everywhere
- speed improves, but results do not justify the run

If your plots do not show any tradeoff at all, suspect one of these:

- the target domain is too similar to the base distribution
- the fine-tune is too small or too short
- your general validation set is too weak to expose forgetting
- your freeze code did not actually freeze the intended layers
- your learning rate is so small that nothing really moved

Do not trust the result until you rule those out.

Optional Homework

You can run this as a real experiment if you have a GPU. If you do not, run the proxy version at the end.

Hardware and time

Full version:

- 1 modern GPU with at least 16–24 GB VRAM is comfortable for a small GPT-style model
- 4–10 hours depending on model size, sequence length, and number of freeze variants
- Disk: enough space for checkpoints and cached activations

Proxy version:

- 1 consumer GPU with 8–12 GB VRAM, or even CPU if you are patient
- Use a very small model and short runs
- 1–3 hours

Full experiment plan

1. Prepare three datasets

You need:

- `train_domain.txt` or equivalent tokenized dataset for the target domain
- `val_domain.txt` for target-domain evaluation
- `val_general.txt` for forgetting measurement

A simple example:

- domain: Python files from a clean repo subset
- general: wiki-like prose or a held-out slice of the original corpus

Keep sequence length fixed across runs.

2. Choose a base checkpoint

Start from the same pretrained checkpoint for every run. Name it clearly, for example `base.pt`. Do not chain fine-tunes on top of each other — every freeze experiment must start from the same base model.

3. Add a freeze-ratio flag to your fine-tune script

Pseudocode:

```
parser.add_argument("--freeze_ratio", type=float, default=0.0)
```

Then call your freeze function before optimizer creation.

```
import logging

freeze_upto = freeze_bottom_layers(model, args.freeze_ratio)
logging.info(f"froze {freeze_upto} / {len(model.transformer.h)} layers")
```

4. Run the configurations

Example command pattern:

```
python train.py --init_from base.pt --freeze_ratio 0.00 --run_name ft_full
python train.py --init_from base.pt --freeze_ratio 0.25 --run_name ft_f25
python train.py --init_from base.pt --freeze_ratio 0.50 --run_name ft_f50
python train.py --init_from base.pt --freeze_ratio 0.75 --run_name ft_f75
```

For the "last-layer-only" run, either compute the ratio manually or add a dedicated flag.

Example if `N=12` and you want only the last layer trainable:

```
python train.py --init_from base.pt --freeze_ratio 0.9167 --run_name ft_last_only
```

That freezes 11 of 12 layers.

5. Keep the training budget fixed

Choose one budget and do not change it across runs.

Example:

- `max_steps = 5000`
- `batch_size = 8`
- `gradient_accumulation_steps = 8`
- `seq_len = 512`
- cosine decay schedule
- same seed for all runs

Record:

- train loss

- domain validation loss
- general validation loss
- tokens/sec
- peak memory if possible

6. Save the checkpoints

Keep:

- `base.pt`
- `ft_full.pt`
- `ft_f25.pt`
- `ft_f50.pt`
- `ft_f75.pt`
- `ft_last_only.pt`

7. Run CKA analysis

Write a separate script, for example `analyze_cka.py`. It should:

- load `base.pt`
- load one fine-tuned checkpoint
- run the same validation batch through both models with `output_hidden_states=True`
- flatten hidden states from each layer
- compute linear CKA per layer
- save a CSV like `layer,cka`

Command pattern:

```
python analyze_cka.py --base base.pt --ft ft_f25.pt --split val_domain --out cka_f25.csv
python analyze_cka.py --base base.pt --ft ft_f50.pt --split val_domain --out cka_f50.csv
python analyze_cka.py --base base.pt --ft ft_full.pt \
  --split val_domain --out cka_full.csv
```

8. Plot the results

Make two plots.

Tradeoff plot:

- x-axis: freeze ratio
- y-axis 1: domain improvement from baseline
- y-axis 2: general degradation from baseline
- y-axis 3: tokens/sec or normalized speed

CKA plot:

- x-axis: layer index
- y-axis: CKA score

- one line per freeze run

Expected observations:

- `ft_full`: best domain gain, worst forgetting, lowest speed, broad CKA drift
- `ft_last_only`: little forgetting, weak domain gain, fastest speed, CKA near 1.0 except top
- `ft_f25` or `ft_f50`: strong middle-ground behavior

Proxy version if you have limited hardware

Use a much smaller model and much shorter runs.

Example setup:

- 6-layer GPT
- hidden size 384 or 512
- code domain dataset of a few million tokens
- general validation of a few hundred sequences
- 500–1000 training steps only

Commands stay the same. What changes: absolute quality will be lower, but the tradeoff should still appear, and CKA patterns should still reveal the freeze boundary. This is enough to teach the lesson.

Result-guided version if you cannot train at all

If you cannot run training, you can still do a paper-and-plot lab.

Do this:

1. Sketch the expected tradeoff table with five rows:
   - full fine-tune
   - 25% freeze
   - 50% freeze
   - 75% freeze
   - last-layer-only
2. Fill in qualitative expectations:
   - domain gain: high / medium / low
   - forgetting: high / medium / low
   - speed: low / medium / high
3. Draw the CKA shape for each:
   - full fine-tune: low-to-high stack drift
   - partial freeze: hard boundary then drift
   - last-layer-only: flat near 1.0 until the top
4. Then read Kornblith et al. on CKA and one transfer-learning paper on frozen backbones.

You will not have numbers, but you will have the right mental model.

Questions To Answer

1. Which freeze ratio gave you the best tradeoff between domain gain and general degradation, not just the best domain score?
2. In the full fine-tune run, where did forgetting show up first: metrics, sample outputs, or CKA drift in lower layers?
3. In the last-layer-only run, what kind of adaptation did the model manage to learn anyway? Vocabulary? Formatting? Short local patterns?
4. Did your CKA curve show a clean boundary at the frozen depth, or did drift leak into places you expected to stay stable?
5. If you changed the target domain from code to medical text or from English to another language, would you expect the sweet spot to move up or down the stack? Why?
6. What tradeoff is layer freezing protecting: compute, general capability, representational stability, or all three at once?

Go Further

Research anchors for this chapter

The most relevant paper anchors here are **Exploring the Limits of Transfer Learning with a Unified Text-to-Text Transformer (T5)**, **The Platonic Representation Hypothesis**, and **Scaling Monosemanticity**. T5 matters because it is one of the clearest examples of a model family built around transfer as a first-class design goal. The latter two matter because they push on the deeper question behind layer freezing: which parts of a model are genuinely general, and which parts are task-shaped enough to drift during adaptation?

This chapter is also where representation-similarity tools stop feeling academic. CKA is not just a pretty measurement. It is the bridge between "I think the lower layers stayed general" and "I have evidence that the representation actually stayed stable."

So the paper lesson here is not only about transfer performance. It is about where reusable structure seems to live inside a pretrained model, which is exactly the question you need if later chapters are going to talk seriously about composition.

Start with these directions.

CKA itself

Search for:

- `Similarity of Neural Network Representations Revisited CKA Kornblith arXiv`

Read it with one question in mind: "How do we compare internal representations without caring about exact neuron identity?" That is the reason CKA matters here.

Transfer learning and frozen backbones

Search for:

- `frozen pretrained transformers transfer learning`

- `which layers to fine tune transformer`
- `gradual unfreezing language models`

A useful neighboring idea is gradual unfreezing: instead of picking one freeze boundary and leaving it there, you begin with most layers frozen and slowly unfreeze more layers over time. That gives you another way to manage forgetting.

Parameter-efficient alternatives

Search for:

- `LoRA vs full fine-tuning forgetting`
- `adapter tuning catastrophic forgetting`
- `representation drift fine-tuning transformers`

Layer freezing is one way to restrict adaptation. LoRA and adapters are another. They answer a related question: "What if I want specialization without rewriting the base model?" That sits right next to composable architectures.

Benchmarks to think about

If you want to turn this into a more serious study, look for:

- code benchmarks like HumanEval-style probes for post-fine-tune behavior
- medical QA or domain language modeling datasets
- multilingual transfer benchmarks

The important design rule is not the benchmark name — it is that you must keep two judges: one for the new domain, one for old general ability. If you drop either judge, you stop seeing the real cost.

Repo direction

Work from a readable GPT reference implementation first, then compare your findings against larger-model fine-tuning stacks. Look for hidden-state extraction hooks, layerwise freezing controls, and evaluation pipelines with more than one validation set. The mechanism does not become clearer in a larger stack — it usually becomes easier to hide.

Case study: TriAttention and the stable geometry inside attention

After this manuscript was mostly complete, a useful new example appeared that sharpens the argument of this chapter from a different angle. The system is called **TriAttention**, and it matters here not because it is another fine-tuning recipe, but because it exposes a deeper kind of stability inside trained attention layers.

The core observation is straightforward to say and surprisingly rich in consequence. In the pre-RoPE space, the query and key vectors in many trained heads are not scattered randomly around zero. They concentrate around stable non-zero centers. Once RoPE rotates those vectors by position, those centers induce predictable distance preferences. In plain English: part of what an attention head wants to look at is not reinvented from scratch for every token. It is shaped by a learned geometric bias that stays unusually stable.

That helps explain why partial freezing can work better than newcomers often expect. When you freeze lower or middle layers, you are not only freezing generic language features in the broad sense. You are also freezing pieces of the model's internal geometry: the way certain heads prefer nearby tokens, certain offsets, and certain structural rhythms. Fine-tuning upper layers can still specialize behavior, but it is doing that work on top of a representational floor that already has durable habits.

TriAttention turns that structural fact into a practical systems result. Rather than estimating KV-cache importance only from recent post-RoPE queries, it uses the learned pre-RoPE concentration structure to score which keys are likely to matter. According to the public paper page and repository, the method matches full-attention accuracy on long AIME25 reasoning runs while achieving roughly `2.5x` higher throughput or `10.7x` KV-memory reduction. That is not directly a layer-freezing result. It is a stability result. The method works because the internal attention geometry is stable enough to exploit.

I want to be careful here. Stable does not mean fixed, and a method that exploits stability in one model family will not necessarily transfer. But the direction of the lesson is real. If parts of a trained model carry durable internal habits, then freezing those parts is not the conservative choice. It is the precise choice.

The lesson for this chapter is narrower and more useful than "attention has structure." The better lesson is that reusable structure can live lower in the stack than visible task behavior does. When you choose a freeze boundary, you are really asking: where does the model's broadly useful internal contract live, and where does task-specific movement begin? TriAttention suggests that at least some of that contract lives inside the learned geometry of attention itself.

That does not mean you should freeze layers simply because a paper found stable heads. It means you now have a more concrete intuition for what freezing is protecting. You are not protecting only vocabulary and syntax. You are often protecting representational geometry that later layers quietly depend on.

If you want to turn that into an experiment, add one more diagnostic to Project 32: before and after fine-tuning, measure not only CKA drift, but whether the heads in your frozen and unfrozen regions keep the same broad distance preferences. If lower layers preserve those preferences while upper layers bend more sharply toward the new task, you have one more piece of evidence that your freeze boundary is doing architectural work rather than merely saving compute.

What You Now Know

You can now explain why a pretrained transformer is not one indivisible block of knowledge. You can explain layer freezing as a read-only boundary inside the model: keep the lower floors intact, adapt the upper floors, and make that choice on purpose instead of by habit. You can measure transfer as a three-way tradeoff between domain gain, forgetting, and training cost.

You can explain catastrophic forgetting without mystique. It is not the model "getting confused." It is the model spending shared internal structure on a narrow target because you gave every layer permission to move. You can explain why freezing too much also fails: a single top layer cannot rewrite what the rest of the stack never changed enough to express.

And you can use CKA to show which layers stayed universal, which layers specialized, and where that boundary actually sits, not as a slogan, but as a measurable claim. That is the opening for composable architectures.

Starting Point

Start from the transfer-learning and representation-similarity literature, then implement the experiment in:

- `https://github.com/karpathy/nanochat`

Read next:

- Kornblith et al., `Similarity of Neural Network Representations Revisited`
- papers and repos around frozen backbones, gradual unfreezing, and parameter-efficient fine-tuning

Carry forward the lesson from **Project 27: Quantization and Deployment**. Quantization asked: how much precision can you remove before the model stops doing your job? This chapter asks a similar question in a different place. How much of the model can you stop updating before adaptation stops being real? Push too far in one direction and the model forgets. Push too far in the other and the model refuses to learn. Your job is to find the boundary where specialization becomes worth the damage it causes, then measure that boundary layer by layer.

Project 33: Fusing Independently Trained Specialists

Hook

Why can three separately trained specialists sometimes combine into one system that beats all three, yet the same recipe falls apart the moment those specialists come from different starting models? A good code model and a good medical model are not enough. Their hidden states have to mean roughly the same thing. Without that, the router is doing traffic control with three street maps drawn in different coordinate systems.

This was an awkward chapter to settle. The first version sold the success case too hard. The real value of the project is the failure case. The success says "this can work." The failure says "here is the exact condition that has to hold for it to work at all." I cut the optimistic framing and left the boundary condition standing.

The Concept

Start with an analogy that behaves like the system.

Picture three translators at one desk. One handles Python code, one handles medical notes, one handles legal contracts. A clerk skims the incoming page and decides who gets it. The setup works if all three translators learned language from the same textbook and then specialized. The clerk is not reading their minds, but the shared training leaves enough common structure that routing stays legible.

Now wreck that assumption. Each translator learned from a different textbook, with different notation habits, different grammar instincts, different shortcuts. The desk still looks tidy from the outside. Inside, the clerk has no stable basis for comparison. "This feels medical" only makes sense relative to some internal frame, and the specialists no longer share one.

I think a lot of the modular-AI literature glosses past this part too quickly. The architectures look composable on a slide. The internal coordinate problem only shows up when you try to actually wire two independently trained models together and watch the router thrash. That moment, where the system is technically correct and behaviorally broken, is the experience this chapter is built around.

That is the whole project. You take one pretrained model, freeze the early layers so every specialist keeps the same trunk, then train three specialists on three domains: code, medical text, and legal text. Each specialist keeps the same early shared representation and adapts later layers to its domain.

Then you build a router. A router is a small model that looks at the shared hidden state and decides how much weight to assign to each specialist. If the input smells like code, push weight toward the code expert. If it smells like a discharge summary, push toward the medical expert. This is not standard Mixture of Experts, where one large model grows experts and routing together during training. It is closer to after-market composition: you train specialists separately, keep the interface controlled, and ask whether they still compose afterward.

That question matters because it changes what "building a model" means. The old picture is one giant model. The newer picture is modular: keep the shared backbone steady, train specialists independently, connect them with routing. That only works if the modules agree on the signals passing between them. If two specialists start from the same base and preserve the same early layers, their hidden states still live in roughly the same internal coordinate system. If they come from unrelated pretrained models, that shared coordinate system disappears. The router then needs more than three competent specialists. It needs specialists that speak a compatible internal language. Figure 18.1 diagrams the architecture that makes this work.

The KALAVAI cooperative LoRA-fusion experiments hammered this lesson home across long training sweeps spanning several model families. The runs where contributors started from the same pretrained base and only their LoRA adapters varied composed cleanly. The runs where contributors started from different bases needed a lot more scaffolding to even produce a coherent fused output, and they often still underperformed the single best specialist. The shared-foundation requirement was not a paper detail. It was the difference between a system that worked and a system that did not.

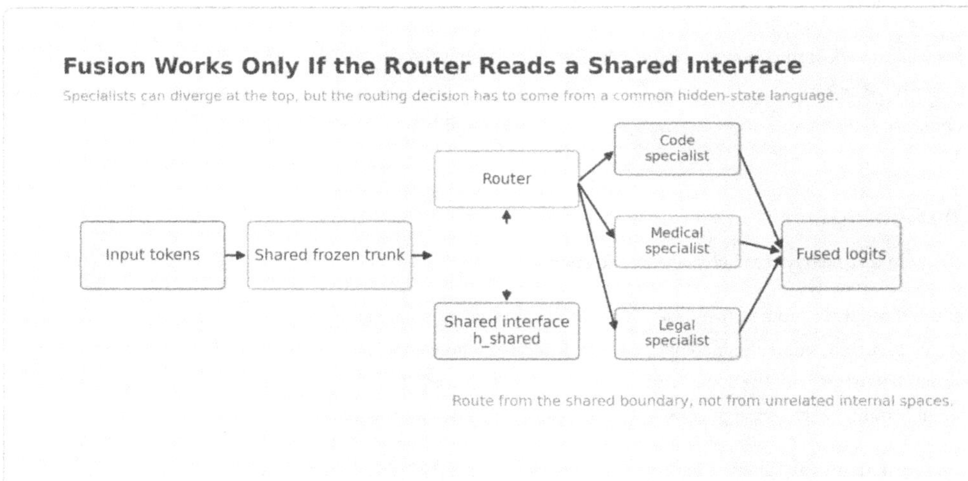

Figure 18.1. Fusion only works when the router reads a genuinely shared interface: the trunk stays common, specialists diverge later, and routing happens from that shared hidden-state boundary.

Why It Matters

The payoff is modularity you can test: add a specialist without retraining everything, isolate failures by domain, and ask whether composition beats any single expert. But there is a hard constraint hiding underneath. Composition depends on shared initialization.

"Initialization" means the starting parameter values of the model before fine-tuning. Shared initialization means the specialists all began from the same pretrained checkpoint: they were born with the same internal geometry. Without that, the router is not just choosing between specialists. It is standing between incompatible worlds.

Honestly, I think this is the single most under-discussed constraint in the entire modular-AI conversation. People will compare seven router architectures and skip the question of whether the experts they are routing to share a basis at all. The router design barely matters if the inputs to it do not live in the same space.

Project 32: Layer Freezing and Transfer asked where specialization starts when you freeze layers. This chapter asks a sharper question: once specialists exist, can they combine into something better than any one of them alone? And then it asks the question that matters more: what exact condition makes that combination possible? The answer is not "have a good router." The answer is "have a shared foundation first."

The Build

We are going to build this in stages. Do not think of it as "train three models and add a classifier." That framing is too loose. The point is to keep the interfaces controlled enough that success or failure actually means something.

Step 1: Decide what stays shared and what is allowed to specialize

Think of a transformer as a stack of layers. Early layers usually track broad language structure. Later layers can bend harder toward a task or domain. You freeze the early layers because you want all specialists to keep the same foundation, and you leave the later layers trainable because that is where domain adaptation happens. In plain English: the bottom of the stack stays common, the top of the stack becomes specialized.

In a small GPT-like model, that might mean freezing the first 6 of 12 layers. The exact split is an experiment, but do not make it arbitrary. If you freeze too little, the specialists drift apart and the shared interface weakens. If you freeze too much, the specialists may not adapt enough to their domains. That tradeoff is the whole setup, and you already saw it in **Project 32: Layer Freezing and Transfer**: the boundary between universal layers and specialized layers is measurable. Here, that boundary becomes the interface contract for composition.

Step 2: Pick one pretrained base checkpoint

All three specialists must begin from the same base model. That is non-negotiable for the main experiment. You want one checkpoint, one tokenizer, one architecture, one hidden size, one positional encoding scheme, one layer norm style, one vocabulary. If even one of those differs, you are no longer testing composition of specialists; you are testing a mess of representation mismatch, token mismatch, or architecture mismatch. The clean setup is same base checkpoint, same frozen early layers, same later-layer shapes, different domain data. Only one thing changes: what each specialist trains on. That keeps the experiment honest.

I had to throw away one of my early cooperative LoRA-fusion sweeps because two of the contributors had silently used slightly different tokenizer vocabularies. The architectures matched. The hidden sizes matched. The token ID for the same surface form did not. Routing was incoherent and I spent half a day blaming the router code before checking the tokenizer hash. Pin every part of the interface, including the parts you assume cannot drift.

Step 3: Train three specialists from the same frozen base

You prepare three domain datasets: a code corpus, a medical text corpus, and a legal text corpus. The actual sources can vary; the requirement is that they are meaningfully different in style and vocabulary. You fine-tune three copies of the base model, where each copy starts from the same checkpoint, freezes the same early layers, and trains only the later layers on its own domain.

```
shared base checkpoint
        |
   copy three times
   /      |      \
code   medical   legal
train    train    train
later    later    later
layers   layers   layers
only     only     only
```

The clean mental model is not "three independent full models." It is "one shared trunk, three separately trained crowns." That distinction matters because the router will read from the shared trunk.

Step 4: Choose the router input carefully

This is where many explanations get slippery. The router should not look at the final output logits of each specialist and pick one after the fact. That would work, but it hides the interface question. The point of this project is to route based on the shared representation, so the router should read the frozen layers' output.

Why? Because that output is the last hidden state before the specialists diverge: the shared coordinate system. If the code specialist, medical specialist, and legal specialist all inherit the same frozen prefix, then the frozen prefix output means roughly the same thing across all of them. That makes it a good control point for routing. You can think of it as the last common hallway before the building splits into three wings. If you stand in the hallway, you can still make a routing decision using a shared map. If you wait until each wing has already transformed the signal in its own private way, routing becomes harder to interpret.

So the router input is this: run tokens through the frozen early layers, take the hidden state at the final position (or a pooled summary over positions), feed that vector into a small linear layer, and output three routing scores, one per specialist. A linear router means the routing model is simple: one matrix multiply plus softmax. If a simple router works, the interface is genuinely usable. If you need a huge router to paper over representation mismatch, that tells you something too.

Step 5: Define how the fusion actually combines specialists

You have routing weights. Now what do you do with them? There are two straightforward options.

Option 1: Hard routing. Pick one specialist with the highest routing score and use only that specialist's output.

Option 2: Soft routing. Take all three specialists' output logits and compute a weighted average using the router's weights.

For this chapter, soft routing is cleaner because it gives you a continuous signal and makes the oracle-routing analysis easier. "Logits" are the raw scores the model assigns to each vocabulary token before softmax turns them into probabilities, the model's unnormalized preferences for the next token. So the fused output becomes:

```
fused_logits
= w_code * logits_code
+ w_medical * logits_medical
+ w_legal * logits_legal
```

Where the weights come from the router and add up to 1.

This is worth pausing on. You are not blending the specialists' weights. You are blending their predictions per example. That is much safer. Weight averaging asks "can these models live in the same parameter space?" whereas routing asks "can these models each answer well on different inputs?" Those are different questions. Weight averaging is a strong baseline that often loses because the same parameter tensor can mean slightly different things after domain specialization. Prediction fusion is less brittle.

I once watched weight averaging fall apart far more quietly than I expected during an evolutionary MoE-LoRA experiment. There was no dramatic break. It just underperformed, and the failure was hard to read from any single metric. That subtle, hard-to-see failure mode is what made me more careful about treating prediction fusion as a genuinely different operation, not just a different layer of the same averaging trick.

Step 6: Train the router on a mixed validation set

The router needs supervision: how does it learn which specialist should matter for which example? The clean setup is to create a mixed routing dataset containing code, medical, and legal examples, run the frozen shared layers to get shared hidden states, compute each specialist's output loss on the example, and train the router so it gives more weight to the specialist with lower loss. There are two ways to train it.

Method A: Domain labels

If every example is clearly tagged as code, medical, or legal, you can train the router as a three-way classifier. That is easy, but a little crude. It teaches the router to recognize domain, not necessarily performance.

Method B: Performance-based targets

For each example, evaluate all three specialists, pick the one with the lowest next-token loss or construct soft targets from their losses, and train the router to predict those targets from the shared hidden state. This is better for the actual scientific question because it says: route to the specialist that really performs best on this input, not just the one that matches the human-assigned bucket.

In practice, we would do both — domain labels for an initial sanity check, performance-based routing for the real experiment. The target you care about is end-task quality, not taxonomy purity.

Step 7: Measure the right baselines

If you skip good baselines, the result means almost nothing. You need at least these:

1. Base model — the original pretrained checkpoint before specialization.
2. Code specialist alone.
3. Medical specialist alone.
4. Legal specialist alone.
5. Weight-averaged model — average the trainable layer weights of the three specialists and evaluate the result.
6. Fused routed system — the router plus soft combination of specialist outputs.

Weight averaging is worth including because it is the naive "maybe these can just merge" idea that sounds reasonable to people who have not yet felt how fragile parameter alignment is. The base model is worth including because specialization is not automatically a win on mixed data. Domain specialists can lose outside their area, and the fused system should beat the base on mixed held-out evaluation if composition is doing real work.

Step 8: Compute the oracle routing gap

This is one of the most useful measurements in the whole project. The "oracle" is a fake perfect router. For each example in the evaluation set, run all three specialists, compute which one gives the best loss on that example, and pretend you had routed to that one perfectly. That gives you the best possible performance a router could achieve using these three specialists. It is not a real deployable system, because at inference time you do not know which specialist will do best without trying them all, but it gives you an upper bound.

Then compare the learned router performance against the oracle router performance. The difference is the oracle routing gap. If the gap is small, your router is already extracting most of the available benefit. If the gap is large, the specialists are useful but the router is weak. This matters because it separates two failure modes (bad specialists versus bad routing), and without the oracle gap, you cannot tell which one you have.

The oracle gap is the most diagnostic number in the whole router-rewrite ablation. The headline numbers tell you something improved. The oracle gap tells you whether the improvement was the router doing more of what was already possible, or the experts becoming more separable. Those are different mechanisms and the oracle gap is the cheapest way to distinguish them.

Here is the intuition in plain English. Suppose your code specialist is great on code, your medical specialist is great on medicine, and your legal specialist is great on contracts. If the fused model still underperforms badly, there are only two broad possibilities. Either the specialists are not actually complementary, or the router cannot detect which one to trust. If the oracle beats your learned router by a lot, the specialists are complementary and the routing is the bottleneck. If the oracle barely beats the learned router, the router is already doing its job and there simply is not much headroom. That is a much sharper statement than "the mixture worked okay."

Step 9: What success should look like

A successful result usually has this shape:

- Each specialist wins clearly on its own domain.
- Each specialist loses on at least one other domain.
- The base model is decent everywhere but best nowhere.
- Weight averaging is mediocre or actively bad.
- The routed fusion beats the base on mixed data.
- The routed fusion beats each single specialist on mixed data.
- The oracle is a bit better than the learned router, but not by much.

That last point matters. If the oracle is wildly better, your router is leaving value on the table. If the learned router nearly matches the oracle, you have evidence that composition is working and the interface is usable.

Step 10: Minimal implementation sketch

Keep the code small enough that the ideas stay visible. Here is the skeleton for a linear router over a shared hidden state:

```
import torch
import torch.nn as nn
import torch.nn.functional as F

class LinearRouter(nn.Module):
    def __init__(self, d_model: int, n_experts: int = 3) -> None:
        super().__init__()
        self.proj = nn.Linear(d_model, n_experts)

    def forward(self, shared_h: torch.Tensor) -> torch.Tensor:
        # shared_h: [B, T, d_model]
        # use the last token as the routing summary
        x = shared_h[:, -1, :]                # [B, d_model]
        return F.softmax(self.proj(x), dim=-1)  # [B, 3]
```

And here is the fusion step:

```
import torch

def fuse_logits(
    router_weights: torch.Tensor, logits_list: list[torch.Tensor]
) -> torch.Tensor:
    # router_weights: [B, 3]
    # logits_list: three tensors of shape [B, T, V]
    fused = 0
    for i, logits in enumerate(logits_list):
        w = router_weights[:, i].view(-1, 1, 1)
        fused = fused + w * logits
    return fused
```

That code looks almost insultingly simple. Good. If composition requires 2,000 lines of machinery before it shows any sign of life, it is harder to learn what is doing the work. The interesting difficulty in this project is not code complexity. It is representation compatibility.

Step 11: Training target for the router

A simple performance-based target looks like this. For one input batch, run the three specialists, compute next-token loss for each one, convert the negative losses into soft targets, and train the router to match those targets. We negate the losses because lower loss means better performance, but softmax expects higher score to mean better.

```
with torch.no_grad():
    losses = torch.stack([loss_code, loss_med, loss_legal], dim=-1)  # [B, 3]
    targets = F.softmax(-losses / temp, dim=-1)

weights = router(shared_h)  # [B, 3]
router_loss = -(targets * weights.log()).sum(dim=-1).mean()
```

"Temperature" here controls how sharp the targets are. If temperature is low, the best specialist gets almost all the target weight. If temperature is high, the target becomes softer and lets the router express uncertainty. The intuition behind the equation: good specialists should get more routing weight, loss tells you which specialists are good on this example, and softmax turns relative goodness into weights that add to 1. That is all this is.

Step 12: Report metrics that actually answer the question

Do not stop at "validation loss improved." You want a table like this:

Model	Code	Medical	Legal	Mixed
Base	1.85	2.10	2.05	2.00
Code specialist	1.40	2.60	2.45	2.12
Medical specialist	2.50	1.45	2.30	2.08
Legal specialist	2.35	2.20	1.50	2.03
Weight average	1.95	2.08	2.00	2.01
Learned router fusion	1.48	1.55	1.60	1.58
Oracle router fusion	1.40	1.45	1.50	1.49

: **Table 18.1.** Specialist fusion results: per-domain validation loss for the base model, each individual specialist, weight averaging, learned router fusion, and oracle router fusion across code, medical, legal, and mixed evaluation sets.

These numbers are illustrative, not promised. What matters is the pattern. Then compute routing accuracy by domain — what fraction of code examples did the router give highest weight to the code specialist, same for medical and legal — and compute the oracle gap as learned mixed loss minus oracle mixed loss. That single gap gives you a clean statement about routing quality.

Step 13: The equation, now that you have earned it

We can compress the fusion idea into one line now.

Let:

- `w_i(x)` be the router's weight for specialist `i` on input `x`
- `z_i(x)` be specialist `i`'s output logits for input `x`
- `z_fused(x)` be the fused logits

Then:

```
z_fused(x) = sum_(i=1 to 3) w_i(x) * z_i(x)
```

Read it in English: for a given input, take each specialist's prediction, multiply it by how much the router trusts that specialist, then add them together. That is the whole fused model.

The oracle routing score for input `x` is even simpler. Let `L_i(x)` be the loss of specialist `i` on that example. Then the oracle chooses:

```
i*(x) = argmin_i L_i(x)
```

Pick the specialist with the smallest loss on that example. Then the oracle routing gap is:

```
gap = L_learned_router - L_oracle_router
```

Smaller is better. That gap is the distance between "what your router achieved" and "what the specialists made possible."

BREAK IT

Now for the part that teaches the chapter. The main experiment works because the specialists share a frozen foundation. Break that condition.

Train specialist A from pretrained checkpoint X. Train specialist B from pretrained checkpoint Y. Same architecture if you want, same tokenizer if you want, same datasets if you want, but not the same pretrained initialization. That sounds like a small change. It is not. It destroys the shared coordinate system.

The first cooperative-fusion experiment that actually tested this hypothesis used two contributors who had each fine-tuned a 1.3B model with strong domain numbers. On their own benchmarks, both looked great. Composed together with a shared router, the combined system underperformed either one alone on mixed data. I tried three different router redesigns before accepting that the bottleneck was upstream of the router entirely.

Here is the easiest way to picture the failure. Imagine two teams independently build GPS systems. Both output latitude and longitude. But one secretly measures angles relative to a different meridian and flips one axis. A router sitting above them sees numbers with the same shape and the same data type. It assumes the outputs live in the same world. They do not.

That is what happens with hidden states from independently pretrained models. A hidden vector is not meaningful because of its shape alone. It is meaningful because of the learned geometry around it. Dimension 417 is not "the law dimension." Dimension 417 only means something in relation to the rest of the model that learned it. When two models share initialization and then diverge through fine-tuning, much of that geometry remains aligned, especially in frozen early layers. When two models are pretrained independently, even if they solve similar tasks, their internal coordinates can rotate, stretch, or reorganize. The router cannot fix that with a linear readout from one shared state.

This is the place where I think a lot of optimistic modular-architecture talks lose me. The room nods along to "modular composition" because it sounds engineering-clean. The unspoken assumption is that the underlying representations agree on what they mean. They do not, by default. Agreement has to be engineered.

Break condition

Instead of copying one base checkpoint three times, freezing the same early layers, and specializing each copy, do this:

- take pretrained checkpoint X, fine-tune a code specialist from X
- take pretrained checkpoint Y, fine-tune a medical specialist from Y
- take pretrained checkpoint Z or Y, fine-tune a legal specialist from Z or Y
- now try to train one router over one model's frozen prefix output and fuse all specialist outputs

There is an immediate problem hiding in the setup: which frozen prefix output should the router even read? If the specialists do not share a frozen prefix, there is no single common intermediate state to route from. That is already the first crack. You have to cheat somehow. Pick one model's prefix as the router input, or concatenate states from different prefixes, or route from raw token embeddings, or route from cheap metadata like domain classifier features. None of these restores the missing shared coordinate system. They just make the mismatch less obvious.

What you observe

The fused system becomes worse than the clean shared-initialization version, often worse than the best specialist, and sometimes worse than almost every specialist on mixed data. Typical signs: routing confidence becomes unstable; the router overcommits to one specialist on many unrelated inputs; per-domain routing accuracy drops sharply; learned routing falls far behind oracle routing; soft fusion of logits becomes noisy because the specialists disagree in ways that do not combine cleanly; weight averaging becomes useless or catastrophic.

The oracle routing gap usually grows. The specialists are still individually good. The oracle can still pick whichever one happens to work best on an example. The learned router, on the other hand, has a much harder job, because its input no longer lives in a shared representational space tied cleanly to all experts. In plain English: the specialists still know things, but the clerk can no longer tell who should handle what.

There is also a deeper failure. Even if the router picks "the right" expert more often than chance, the fused predictions can still underperform because the outputs are not calibrated the same way. "Calibration" here means how confidence scores line up with actual correctness. Two specialists can both be right in different places while producing logits at different scales or with different uncertainty habits. Shared initialization helps keep those habits more compatible; independent pretraining lets them drift. So the failure is more than bad classification by the router. The whole interface between modules has gone soft.

What this proves

It proves that composition is not only about expertise. It is about compatible internal structure. People often hear "modular AI systems" and imagine Lego bricks. But models are not Lego bricks. They are more like organs. Transplantation requires matching blood type, tissue compatibility, and connection points. Shape alone is not enough.

Shared initialization acts like that compatibility layer. It gives the specialists a common ancestry so their internal features may specialize, but they still inherit a shared geometry from the same pretrained base. When you remove shared initialization, the router loses the one thing that made late composition plausible in the first place. That is why this failure matters more than the success case. Success says modularity is possible. Failure says modularity has a price of entry. That price is a shared foundation.

Optional Homework

This project is heavy if you do the full version honestly. Training three specialists plus a router is not a laptop lunch break. The cooperative-fusion runs took multiple days even when contributors split the training across machines. So here are two tracks:

- Full lab: for readers with a good GPU setup
- Proxy lab: for readers with one consumer GPU, limited time, or no appetite for multi-day runs

Full lab

Hardware and time

Recommended:

- 1 to 4 modern GPUs
- enough VRAM to fine-tune a small GPT-style model or a compact decoder-only model
- rough time: 2 to 4 days depending on model size and dataset size

If you are working inside `nanochat`, keep the model small enough that iteration speed stays reasonable. This chapter is about composition, not brute force scale.

Data

Prepare three domain corpora:

- code: GitHub-style code or permissively licensed coding data
- medical: de-identified medical text or public biomedical text
- legal: case law, contracts, or public legal opinions

Also prepare:

- a mixed validation set with labels or source tags
- per-domain held-out evaluation sets

Step sequence

1. Start from one pretrained base checkpoint.
2. Freeze early layers, for example layers 0 through 5 in a 12-layer stack.
3. Copy the checkpoint three times.
4. Fine-tune copy A on code.
5. Fine-tune copy B on medical.
6. Fine-tune copy C on legal.
7. Save all three specialists.
8. Build a router that reads the frozen-prefix hidden state and outputs 3 weights.
9. For each mixed-set example, record each specialist's loss.
10. Train the router against those per-example performance targets.
11. Evaluate:

- base model
- each specialist
- weight average of specialists
- learned router fusion
- oracle router

12. Record:

- loss/perplexity per domain

- mixed loss
- routing accuracy by domain
- oracle routing gap

Pseudocode checklist

Train specialists:

```
python train.py --init base.pt --freeze_layers 6 --data code.jsonl --out code_spec.pt
python train.py --init base.pt --freeze_layers 6 --data medical.jsonl --out med_spec.pt
python train.py --init base.pt --freeze_layers 6 --data legal.jsonl --out legal_spec.pt
```

Train router:

```
python train_router.py \
  --base base.pt \
  --experts code_spec.pt med_spec.pt legal_spec.pt \
  --mixed_eval mixed_eval.jsonl \
  --router_in frozen_prefix_last_token \
  --out router.pt
```

Evaluate:

```
python eval_fusion.py \
  --base base.pt \
  --experts code_spec.pt med_spec.pt legal_spec.pt \
  --router router.pt \
  --eval_sets code_eval.jsonl medical_eval.jsonl legal_eval.jsonl mixed_eval.jsonl
```

Break it:

```
python train.py --init base_X.pt --freeze_layers 6 --data code.jsonl --out code_X.pt
python train.py --init base_Y.pt --freeze_layers 6 --data medical.jsonl --out med_Y.pt
python train.py --init base_Z.pt --freeze_layers 6 --data legal.jsonl --out legal_Z.pt

python train_router.py \
  --base base_X.pt \
  --experts code_X.pt med_Y.pt legal_Z.pt \
  --mixed_eval mixed_eval.jsonl \
  --router_in frozen_prefix_last_token \
  --out broken_router.pt
```

What to expect

Clean shared-initialization run:

- specialists win on their own domains
- fused model wins on mixed data
- oracle gap is moderate or small

Broken independent-initialization run:

- specialists still win on their own domains

- learned routing gets worse
- mixed fused performance drops
- oracle gap grows
- weight averaging becomes especially poor

Proxy lab

If the full experiment is too expensive, run a smaller version with tiny datasets and a compact model.

Hardware and time

- 1 consumer GPU is enough
- rough time: several hours to one day

Proxy setup

Use a small decoder-only model and small corpora — code snippets from a small repository set, medical abstracts from a narrow public source, legal paragraphs from public court decisions. Freeze half the layers. Train for a small number of steps. You do not need polished domain models; you only need enough specialization to create divergence.

Exact plan

1. Fine-tune three tiny specialists from the same base.
2. Train a linear router on a mixed set.
3. Compute per-example specialist losses on the mixed set.
4. Report:
    - best specialist by example
    - router's top choice
    - fused loss
    - oracle loss
5. Repeat with specialists from different seeds or different small base checkpoints.

Even if the absolute quality is low, the pattern still teaches the lesson. What you are looking for is not great model performance, but cleaner routing when initialization is shared and noisier routing and worse fusion when it is not.

Result-guided version

If you cannot run training at all, you can still learn from logs. Make a small spreadsheet with one row per example:

- domain
- loss_code
- loss_medical
- loss_legal
- router_weight_code
- router_weight_medical

- router_weight_legal

Then inspect: does the highest router weight usually go to the lowest-loss specialist? How often? Does that pattern collapse when the specialists came from different bases? This still teaches the central lesson: the specialists may remain individually competent while the routing signal becomes much less trustworthy.

Questions To Answer

1. On the mixed evaluation set, which baseline does the fused model beat, and which baseline still beats it?
2. How large is the oracle routing gap? Does that gap tell you the bottleneck is the router or the specialists?
3. When you break shared initialization, what fails first: routing accuracy, fused loss, calibration of output scores, or weight averaging?
4. Does freezing more early layers make fusion easier because the shared interface is stronger, or harder because specialists cannot adapt enough? Where is the boundary in your run?
5. When the router makes a bad choice, is it choosing the wrong domain entirely, or is it confused between two specialists that both partially fit?
6. What tradeoff is shared initialization protecting: domain expertise, interface compatibility, calibration stability, or all three?

Go Further

Research anchors for this chapter

This chapter sits at the boundary between several research traditions. The transfer side comes from **T5** and the broader fine-tuning literature. The representation-compatibility side is connected to **The Platonic Representation Hypothesis** and **Scaling Monosemanticity**, because both ask, in different ways, whether separately trained models still organize meaning in comparable internal directions.

The practical composition side is closely related to LoRA and adapter work even when the exact mechanism differs. That is why this chapter keeps pointing toward LoRA composition rather than only full-model specialist fusion: if the base remains literally shared, the alignment problem gets easier.

And it is worth keeping one sharp distinction in view. Classic **Mixture of Experts** papers such as **Switch**, **GLaM**, and **Mixtral** learn experts and routing together. This chapter asks a stricter systems question: can independently trained specialists compose after the fact? That difference is the whole point.

If this chapter clicks, here are the directions worth chasing next.

Repos and code directions

- `https://github.com/karpathy/nanochat`
- Mixture-of-Experts implementations in open LLM training stacks
- parameter-efficient fine-tuning repos with adapters and LoRA routing experiments

Papers and search terms

Search arXiv for:

- `BTX Meta 2024 compositional experts`
- `modular deep learning shared initialization`
- `representation alignment model merging`
- `task arithmetic model fusion`
- `router calibration mixture of experts`
- `CKA fine-tuning specialist models`
- `model stitching representation compatibility`

A good next question is whether you can repair incompatibility after the fact. Search for:

- `model stitching`
- `learned linear map between hidden spaces`
- `representation matching for network fusion`

Those papers ask a harsher version of this chapter's question: if two models do not share a coordinate system, can you learn a small map that makes them compatible? Sometimes partially. Rarely for free.

Another direction is to replace full specialists with LoRA adapters. That changes the object being composed: instead of three full specialist heads on a shared base, you have one base plus three low-rank adaptations. That often keeps alignment cleaner because the base remains literally shared, not just ancestrally shared. Search for:

- `adapter fusion`
- `LoRA composition`
- `task routing adapters`

And if you want the bigger systems angle, study the relation between this project and standard Mixture of Experts. This chapter composes specialists after they are trained. Classic MoE learns experts and routing together during training. Those are different worlds. The first asks: can independently trained pieces compose? The second asks: can specialization emerge inside one jointly trained system? That distinction matters.

Case study: the Luce Megakernel and fusion at the execution level

This chapter teaches fusion at the model level: shared base, specialized branches, routing logic, and compatibility between components that were trained separately. A newer real-world system extends that intuition downward into the execution layer itself.

The example is **Luce Megakernel**, an open-source CUDA implementation for a hybrid Qwen 3.5-0.8B model with both DeltaNet and full-attention layers. The engineering question is not "can we build another fused model?" The question is harsher: if the layer pattern is fixed and known, why keep paying the cost of treating the forward pass as dozens of separate launches?

According to the public repository, a more conventional execution path for this model shape pays roughly `100` CUDA kernel launches per token. Each launch returns control to the CPU, dispatches the next unit of work, re-fetches data, and synchronizes again across layer boundaries. Luce collapses that pattern into one persistent CUDA kernel that carries the whole forward pass in one dispatch. In the published README, that pushes Qwen 3.5-0.8B decode to about `413` tokens per second on an RTX 3090, versus `267` for the compared `llama.cpp` BF16 path and `108` for the compared PyTorch Hugging Face path on the same model and hardware.

That is not the same kind of fusion this chapter asked you to implement. But it is the same family of thought. In Project 33, the question was whether independently trained specialists could still act like parts of one system. Here the question becomes: once the system shape is fixed, how much latency are you burning by preserving software boundaries that the hardware does not need?

The most important conceptual upgrade is this: fusion is not only about weights and hidden states. It can also be about execution surfaces. If a model family settles on a recurring architectural pattern, then the fused object may no longer be only "base plus experts." It may become "base, experts, and a compiled kernel strategy designed around their exact interaction."

That matters for composable architectures because it introduces a real tension. The more freely swappable your components are, the harder it is to write one deeply optimized execution path for them. The more fixed and regular your composition becomes, the easier it is to fuse aggressively at the systems layer. Modularity and hardware-specific optimization are not enemies, but they are not free allies either. Someone has to define the contract carefully enough that the system can stay composable without forcing every new combination to fall back to a slow generic path.

So the Luce example should not be read as "every modular architecture needs a megakernel." The better lesson is narrower. Once a composition pattern becomes stable enough, the execution graph itself becomes an optimization target. That is the point where architecture stops being only a modeling question and becomes a code-generation and systems question too.

If you want a useful local exercise, do not try to write a megakernel. Start by profiling your own Project 33 implementation and counting what the software layer is charging you: kernel launches, synchronization points, repeated reads, and per-token overhead that has nothing to do with the actual modeling idea. That is the cheapest way to learn the same lesson.

What You Now Know

At this point the magic story is gone. A fused system can beat each specialist alone because the specialists split the domain, the router reads a shared hidden interface, and the final prediction mixes strengths case by case.

The failure mode should feel sharper now too. When specialists come from different pretrained origins, the trouble is not "router too small." The trouble is geometric. Their competence may survive. Their composability does not.

You also have a clean way to test that claim. Train specialists from one frozen base, route from the shared prefix representation, compare fusion against strong baselines, compute the oracle routing gap, break shared initialization and watch composition fail. That is the step forward here. "Modularity sounds nice" is no longer the bar. You can say what modularity requires.

Starting Point

Start from:

- Mixture of Experts literature
- BTX (Meta, 2024)
- nanochat

Carry one lesson forward from **Project 32: Layer Freezing and Transfer**. Freezing layers showed that some parts of a model stay broad while others bend toward specialization. Here that observation hardens into an interface.

The next question gets stricter: if specialists compose only when they share a foundation, what exactly has to stay fixed? That is the point where composable architectures stop being a slogan and become engineering.

Project 34: The Interface Specification

Hook

Why does a specialist model that scores well on its own turn into nonsense the moment you plug it into a shared system? Not because it forgot language. Not because the router is weak. Not because the checkpoint is corrupted.

The ugly answer is simpler: both sides think they are speaking the same internal language, but one side quietly switched the wiring. One model hands over a 768-number hidden state that has been normalized one way, rotated in position one way, split into heads one way, and interpreted over one vocabulary. The other model expects a different contract.

From the outside, the tensors fit. From the inside, the meanings do not. This is the kind of failure that survived three of my own long fine-tuning sweeps before I admitted the bug was not in the model.

The Concept

Think about USB. A USB-C cable looks simple because the mess is hidden in the specification: voltage ranges, pin layout, protocol negotiation, data lanes, power roles. If that specification did not exist, every device maker would ship a connector that looked almost right and failed in slightly different ways. Your phone might charge but not transfer data. Your monitor might flicker only at one resolution. Your keyboard might work until it goes through a hub.

Those are the worst failures. Not total failure. Partial failure that wastes hours and survives every shape check you can think of.

Composable model systems need the same thing. In **Project 33: Fusing Independently Trained Specialists**, we saw that specialists compose only when they share an internal coordinate system. That phrase can sound vague, so let's pin it down.

A coordinate system here means the rules that give meaning to a hidden state. A hidden state is the list of numbers passed from one layer of a model to the next. If `d_model = 768`, then each token is represented at that boundary by 768 floating-point numbers. Those numbers are not random. Their scale matters, their ordering matters, the normalization applied to them matters, the positional encoding scheme matters, the way attention heads are arranged matters, and the vocabulary matters too if you expect outputs to line up.

If two specialists share weights up to some boundary, then branch off, then later get recombined, you need a formal promise about that boundary: what shape comes in, what shape goes out, what normalization has already happened, what positional encoding was assumed when those numbers were produced, what attention head layout the downstream module expects, and what vocabulary the logits refer to. Without that promise, "compatible" becomes guesswork. The first time I watched a cooperative fusion run produce fluent-but-wrong outputs, I assumed for a full afternoon that the issue was the router. It was not. It was a one-character difference in how two specialists declared their normalization.

Here is the analogy that makes this click. Imagine three translators working for the same newsroom. They all receive a note from the editor that says: `BANK CLOSED AFTER FLOOD`. One translator assumes "bank" means financial institution. Another assumes riverbank. A third assumes the note already includes location metadata from earlier in the conversation. All three can produce fluent English. All three can work well alone. But if translator A writes the first half of a paragraph and translator B continues without sharing assumptions, the combined output can be nonsense.

The problem is not grammar. The problem is incompatible hidden context. That is what an interface specification fixes: at this boundary, "bank" has to arrive in this format, with this context encoding, under these assumptions.

Now the engineering version. A versioned interface specification for composable specialists is a written contract that says:

- Hidden state dimension: `d_model = 768`
- Normalization type: `RMSNorm`
- Norm epsilon: `1e-5`
- Residual convention: pre-norm or post-norm
- Attention head configuration: `H = 12`, head dimension `d_head = 64`
- Positional encoding: `RoPE` with specific base and scaling
- Vocabulary: exact tokenizer and vocabulary file hash
- Output boundary: logits over vocabulary `V = 50,304` or hidden state only
- Checkpoint metadata: version string, architecture fingerprint, training origin

This sounds bureaucratic until you see the alternative. The alternative is silent garbage.

A model with `LayerNorm` at the specialist boundary can often still produce tensors of the correct shape. No crash, no red warning, no obvious stack trace. The numbers even look normal if you inspect mean and standard deviation. But they no longer mean the same thing as the tensors expected by a system built around `RMSNorm`. That is the core lesson of this chapter: shared weights are not enough. Shared contracts are what make composition dependable.

I will be honest about a bias here. Years of federal-procurement work at Procore, where FedRAMP and SOC 2 governance treat every interface as a written contract subject to a Change Advisory Board, have made me allergic to "we just trust the shape." It is not paranoia. It is the only style of work I have seen survive auditors and time. Figure 19.1 shows the contract structure that enforces this explicitly.

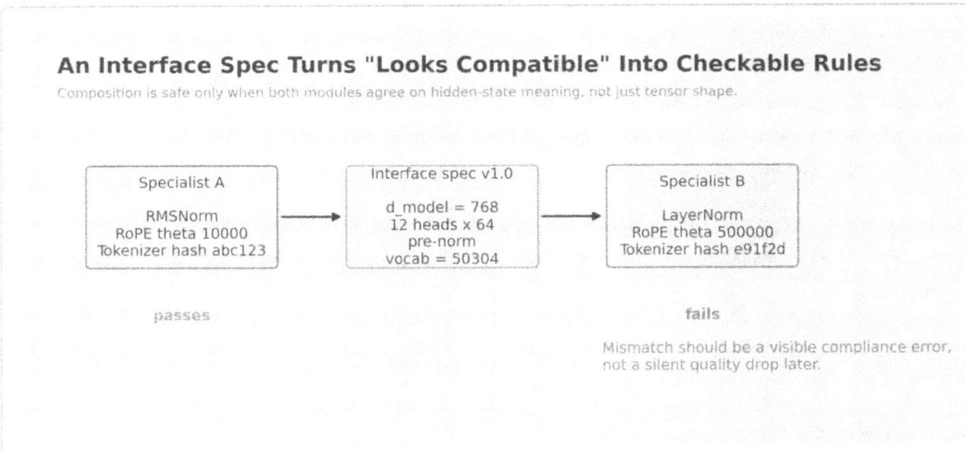

Figure 19.1. An interface specification makes compatibility explicit by checking normalization, positional encoding, head layout, and tokenizer identity instead of trusting tensor shape alone.

Why It Matters

Without an interface specification, you cannot tell the difference between these four situations: the specialist is genuinely bad; the specialist is good but the fusion system is bad; the specialist and fusion system are both good but the boundary assumptions differ; or the system mostly works only because one accidental mismatch has not bitten you yet.

That ambiguity destroys engineering progress. If a specialist plugs into the fusion system and outputs garbage, what do you do next? Retrain longer? Change the router? Add more data? Tune the learning rate? Replace the expert? Without a contract and a compliance test, you are debugging in the dark. I have wasted whole afternoons on the wrong layer of this question, and the lesson was always the same.

This matters even more once multiple people work independently. Suppose one person trains a code specialist, another trains a multilingual specialist, a third builds the fusion stack, and a fourth writes evaluation. If each person makes one "reasonable" architectural choice on their own, you can end up with four modules that all work in isolation and fail together.

Worse, they may fail softly. Perplexity rises a bit. Accuracy drops only on long contexts. Responses become vague instead of obviously broken. Those are the failures that survive demos and poison production. In a past role on a multi-state SaaS platform, the multi-party API integrations across DMV, DCF, banking, and payment systems were all "compatible" on shape and protocol. The failures that cost real money were always semantic. A field that meant dollars in one system meant cents in another, a date that meant local time on one end meant UTC on the other, and nothing in the wire format would tell you.

A good interface specification gives you three things: coordination without constant meetings, fast rejection of incompatible checkpoints, and blame assignment when something breaks. That last one matters. If the compliance test says:

- expected `RMSNorm`, found `LayerNorm`
- expected `rope_theta=10000`, found `rope_theta=500000`
- expected tokenizer hash `abc123`, found `e91f2d`

then the mystery evaporates. You are not "tuning composition." You are fixing a contract violation. Same reason APIs have versions, schemas, and compatibility checks. An API that returns a JSON object where `price` silently changes from dollars to cents without changing the version number is broken even if the endpoint still returns `200 OK`. I lived this exact bug on an enterprise healthcare eligibility integration; the endpoint was happy, the auditor was not. A specialist that silently changes boundary normalization is broken in the same way. The spec is not paperwork. It turns invisible incompatibility into an engineering error you can see.

The Build

We are going to build four things:

1. A versioned interface specification document.
2. Metadata inside each specialist checkpoint that declares compliance.
3. A compliance test that checks the declaration against the actual module.
4. A small experiment with one compliant and one non-compliant specialist.

The exact repo in the broader arc is `nanochat`, but this chapter is mostly about writing a contract and enforcing it. The code can be small. The thinking cannot.

Step 1: Decide where the interface boundary lives

Before you write a specification, you need to decide what exactly is being shared between modules. In this chapter's setting, the natural boundary is a shared prefix of layers, then specialist layers, then a fusion system that consumes outputs from those specialists. So the interface is the boundary between the shared trunk and each specialist, and possibly the boundary between each specialist and the fusion module.

Do not write a spec for the whole model first. Write one for the handoff point. That boundary should answer: what tensor arrives here? What does each axis mean? What transformations have already happened? What assumptions may downstream code rely on?

For a GPT-like model, a boundary tensor often has shape `[B, T, d_model]`. Define each symbol in plain English: `B` is batch size (how many sequences you process at once), `T` is sequence length (how many tokens are in each sequence), and `d_model` is hidden size (how many numbers represent each token at this layer). If `B=4`, `T=128`, and `d_model=768`, then the boundary tensor contains `4 x 128 x 768` floating-point numbers.

That is the easy part. The harder part is semantic compatibility. A tensor shape alone tells you nothing about whether the numbers mean the same thing.

Step 2: Write the first version of the specification document

Keep it boring. Boring is good here. A spec document should read more like an RFC than a tutorial: short sentences, no ambiguity, exact names, version number at the top.

Here is a minimal example.

```
Interface Spec: COMPOSE-LLM-IFACE v1.0

Purpose
Defines the boundary contract between shared trunk layers and specialist modules.

Boundary Tensor
- Input shape: [B, T, 768]
- Output shape: [B, T, 768]
- Dtype at load time: float32 or bfloat16
- Semantic meaning: token-wise hidden state after shared trunk block 8

Architecture Requirements
- d_model = 768
- n_layers_specialist = 4
- n_heads = 12
- d_head = 64
- residual layout = pre-norm
- MLP expansion ratio = 4.0

Normalization Requirements
- norm type = RMSNorm
- norm epsilon = 1e-5
- boundary expectation = input arrives unnormalized residual stream;
  specialist applies RMSNorm internally before attention/MLP

Positional Encoding Requirements
- scheme = RoPE
- rope_theta = 10000
- max_seq_len_trained = 2048
- rope_scaling = none

Tokenizer / Vocabulary Requirements
- tokenizer family = GPT-2 BPE compatible
- vocab size = 50304
- tokenizer hash = sha256:...

Checkpoint Metadata Requirements
- interface_name = COMPOSE-LLM-IFACE
- interface_version = 1.0
- base_model_id = trunk-init-2026-03
- shared_prefix_depth = 8
- training_origin_hash = sha256:...

Compliance Policy
- Any mismatch in norm type, rope settings, tokenizer hash, d_model,
  n_heads, or residual layout is a hard failure.
- Mismatch in dtype is a warning if runtime cast is supported.
```

Notice the tone. It does not say "roughly this," does not say "usually," does not say "standard RoPE." That kind of loose language creates loopholes. Your future self will drive a truck through them.

Step 3: Decide what is a hard failure and what is a warning

This is where many first specs go wrong. A spec that is too loose misses silent failures. A spec that is too tight makes collaboration impossible. So you need categories.

Hard failures are mismatches that change semantics of the hidden state. Reject the checkpoint. Good examples: wrong normalization type, wrong `d_model`, wrong number of heads, wrong RoPE settings, wrong tokenizer or vocab, wrong residual layout, wrong shared base model origin.

Warnings are things that can be adapted safely at runtime. Possible examples: checkpoint stored in `float32` instead of `bfloat16`, metadata missing optional training notes, harmless naming differences that do not affect tensor semantics.

This distinction matters because compliance tools that fail on everything become ignored, and tools that allow everything become useless.

Step 4: Put the spec into machine-readable form

A text document is for humans. A machine-readable spec is for tests. A simple JSON or YAML file works well.

```
{
  "interface_name": "COMPOSE-LLM-IFACE",
  "interface_version": "1.0",
  "boundary": {
    "input_shape": ["B", "T", 768],
    "output_shape": ["B", "T", 768],
    "d_model": 768
  },
  "architecture": {
    "n_heads": 12,
    "d_head": 64,
    "residual_layout": "pre_norm",
    "mlp_ratio": 4.0
  },
  "normalization": {
    "type": "RMSNorm",
    "eps": 1e-5
  },
  "positional_encoding": {
    "scheme": "RoPE",
    "rope_theta": 10000,
    "rope_scaling": "none"
  },
  "tokenizer": {
    "family": "gpt2_bpe",
    "vocab_size": 50304,
    "tokenizer_hash": "sha256:..."
  },
  "compatibility": {
    "hard_fail_fields": [
      "boundary.d_model",
      "architecture.n_heads",
      "architecture.residual_layout",
      "normalization.type",
      "normalization.eps",
      "positional_encoding.scheme",
      "positional_encoding.rope_theta",
      "tokenizer.tokenizer_hash"
    ]
  }
}
```

Now your compliance script has something concrete to compare against.

Step 5: Make specialists declare what they are

Do not rely only on inspection. Inspection is necessary because metadata can lie, but metadata is necessary because inspection alone can miss intent. So you want both. Each specialist checkpoint should save metadata like this:

```
ckpt["interface"] = {
    "interface_name": "COMPOSE-LLM-IFACE",
    "interface_version": "1.0",
    "base_model_id": "trunk-init-2026-03",
    "shared_prefix_depth": 8,
    "d_model": 768,
    "n_heads": 12,
    "norm_type": "RMSNorm",
    "norm_eps": 1e-5,
    "positional_encoding": "RoPE",
    "rope_theta": 10000,
    "tokenizer_hash": "sha256:..."
}
```

This declaration is the model saying, "Here is what I believe I am." The compliance test then asks a second question: "Is that true?"

Step 6: Inspect the actual module, not just the metadata

If you only read metadata, anyone can accidentally or intentionally save the wrong declaration — then you have a spec-shaped placebo. The compliance test needs to load the module and inspect the real architecture, checking things like: does the module contain `RMSNorm` at the required boundary locations? Is `d_model` actually 768 in the loaded tensors? Do attention projections imply `12` heads with head dimension `64`? Does the checkpoint contain RoPE parameters matching the spec? Does the output embedding or LM head size match vocab size?

A minimal sketch:

```python
import torch

def check_norm_type(module: torch.nn.Module, expected: str) -> None:
    actual = type(module.norm).__name__
    if actual != expected:
        raise ValueError(f"Norm mismatch: expected {expected}, found {actual}")

def check_d_model(module: torch.nn.Module, expected: int) -> None:
    actual = module.attn.q_proj.weight.shape[1]
    if actual != expected:
        raise ValueError(f"d_model mismatch: expected {expected}, found {actual}")

def check_vocab_size(module: torch.nn.Module, expected: int) -> None:
    actual = module.lm_head.weight.shape[0]
    if actual != expected:
        raise ValueError(f"Vocab mismatch: expected {expected}, found {actual}")
```

This snippet is deliberately small. The real point is the philosophy: trust, then verify.

If that phrase sounds heavy-handed, good. In federal procurement work, "trust, then verify" is not a slogan. It is the only stance that survives an audit. I have carried that habit into model work, and I am not embarrassed about it.

Step 7: Check behavior, not just structure

Here is what people skip. Two modules can match on architecture and still disagree in behavior. Same `RMSNorm` type, same `d_model`, same head count, same tokenizer, and yet one specialist was trained from a different base initialization. Structurally compatible. Semantically misaligned.

This distinction looked like a footnote when I first wrote it down. It turned out to be the part that surprised me most when I came back to it months later.

This is why the best compliance test has two layers: static checks that inspect architecture and metadata, and dynamic checks that run a tiny behavioral probe through the boundary. A dynamic probe can be simple. Feed a fixed batch of hidden states or tokenized prompts through the shared trunk plus specialist and compare output statistics and known invariants. You are not looking for "good language quality" here. You are looking for obvious incompatibility markers: wildly different activation scales, NaNs or infinities, logits with wrong shape, exploding norm drift, severe mismatch against expected reference traces.

```
with torch.no_grad():
    x = reference_hidden_state  # fixed saved tensor [B, T, 768]
    y = specialist(x)

    if torch.isnan(y).any():
        raise ValueError("Dynamic check failed: NaNs in output")

    rms = y.pow(2).mean().sqrt().item()
    if not (0.1 < rms < 10.0):
        raise ValueError(f"Dynamic check failed: output RMS out of range: {rms:.4f}")
```

This will not catch everything, but it catches more than pure shape checks.

Step 8: Train one compliant specialist

A compliant specialist must follow the full contract: same frozen shared base, same `d_model`, same number of heads, same residual layout, same tokenizer, same RoPE scheme, same normalization type and epsilon. The domain does not matter for the lesson. Contract compliance matters. When this specialist plugs into the fusion system, it should produce meaningful output. Not perfect output, not necessarily best-in-class output, but coherent, on-domain, compositionally valid output.

Step 9: Train one non-compliant specialist

Now create the failure on purpose. Hold everything constant except one boundary parameter. For this chapter's central lab, change `RMSNorm` to `LayerNorm`. This is a good failure because the shape still matches, the forward pass still runs, outputs may still look numerically "reasonable," and the mismatch changes the hidden state semantics enough to hurt composition.

`LayerNorm` and `RMSNorm` are not interchangeable decorations. They process activations differently. `LayerNorm` subtracts the mean of the features and scales by the standard deviation. `RMSNorm` does not subtract the mean; it only scales by the root-mean-square magnitude. In plain English: `LayerNorm` recenters and rescales, while `RMSNorm` rescales without recentering. That means a downstream module trained to interpret activations that keep their mean-related structure can get confused if another module silently zero-centers them first.

You do not need the full formulas to understand the failure, but now that the intuition is grounded, the equations are fair game.

For a hidden vector `x` of length `d`, where `d` is the number of features in that token representation and `x_i` is feature `i`:

LayerNorm computes:

```
LayerNorm(x) = gamma * (x - mu) / sqrt(sigma^2 + epsilon) + beta
```

Where `mu` is the mean of the features in `x`, `sigma^2` is their variance, `epsilon` is a tiny constant for numerical stability, and `gamma` and `beta` are learned scale and shift parameters.

RMSNorm computes:

```
RMSNorm(x) = gamma * x / sqrt((1/d) * sum_i x_i^2 + epsilon)
```

Where the denominator is the root-mean-square magnitude of the vector, with no mean subtraction term. The difference looks small on paper, but it is not small at the interface. If a specialist trained with LayerNorm hands activations to a system expecting RMSNorm-style statistics, the numbers still fit. Their geometry changes. This is exactly the kind of mismatch a spec must catch.

Step 10: Build the compliance script

A useful compliance script should print crisp failures, not just return `False`.

Bad:

```
compliance failed
```

Good:

```
COMPLIANCE FAILURE: normalization.type
  expected: RMSNorm
  found: LayerNorm
  location: specialist.blocks.0.norm1
  impact: boundary activation statistics differ; fusion semantics undefined
```

A good compliance script structure looks like this:

```python
def run_compliance(spec: dict, ckpt_path: str) -> tuple[list, list]:
    model, metadata = load_specialist(ckpt_path)
    failures = []
    warnings = []

    # Declaration checks
    if metadata["interface_name"] != spec["interface_name"]:
        failures.append((
            "interface_name",
            spec["interface_name"],
            metadata["interface_name"],
        ))

    # Structural checks
    actual_norm = detect_boundary_norm(model)
    expected_norm = spec["normalization"]["type"]
    if actual_norm != expected_norm:
        failures.append(("normalization.type", expected_norm, actual_norm))

    actual_d_model = detect_d_model(model)
    if actual_d_model != spec["boundary"]["d_model"]:
        failures.append((
            "boundary.d_model",
            spec["boundary"]["d_model"],
            actual_d_model,
        ))

    # Dynamic checks
    dyn_msg = run_probe(model, spec)
    if dyn_msg is not None:
        failures.append(("dynamic_probe", "pass", dyn_msg))

    return failures, warnings
```

Then print them in a way a teammate can act on.

Step 11: Plug both specialists into the fusion system

Now we do the real test. The compliant specialist should pass compliance, integrate into the fusion system, and produce meaningful outputs. The non-compliant specialist should fail compliance before fusion — or, if you bypass the test, degrade fusion quality.

This is a crucial distinction. If the system only works because you manually remember which checkpoint is safe, you do not have an interface. You have folklore.

I am opinionated about this and I do not apologize for it. Folklore scales until the person carrying the knowledge leaves the team. Then the system breaks and nobody can say exactly why. The goal is not to prove humans can notice the mismatch eventually. The goal is to prove the system can reject incompatible modules automatically.

Step 12: Evaluate whether the spec is too loose

Suppose your first compliance test checks only `d_model`, `n_heads`, and vocabulary size. Then the LayerNorm specialist might pass. The forward pass works. No immediate crash. Outputs get worse. You start blaming the router.

That means your spec is underspecified. The fix is not "be more careful." The fix is to add the missing field and enforce it programmatically.

A healthy engineering loop here is: a mismatch slips through, you identify which assumption was missing from the contract, you add that assumption to the spec, you update the compliance test, and future mismatches fail fast. This is how protocols mature. Bugs become clauses.

Step 13: Evaluate whether the spec is too tight

Now the opposite problem. Suppose your spec demands exact optimizer state format, exact checkpoint filename pattern, exact training data ordering, and exact GPU type used during training. That is not an interface. That is control freak behavior. Those details may matter for reproducibility, but not for boundary compatibility.

A useful spec protects the semantics at the handoff point. It does not freeze unrelated choices. Ask this question for every field: "If this differs, can two modules still compose safely?" If yes, make it optional or move it to metadata. If no, make it a hard requirement.

Step 14: Hand the spec to someone else

This step is not optional if you want a real contract. A spec you understand is not yet a good spec.

Give the document and compliance tool to someone else and ask them to train a specialist without any verbal clarification, then watch where they get confused. Typical confusion points: "Does pre-norm mean the boundary tensor arrives before or after the norm?" "Do you mean all norms must be RMSNorm, or only the boundary-facing norms?" "Which tokenizer hash is authoritative if the vocab file and merges file are stored separately?" "Does 'shared base model id' mean same architecture or same actual initialization?"

Every confusion point is evidence of missing language in the spec. This is not failure. This is how the spec becomes real. I handed an early version of one of our specialist interface drafts to a collaborator and watched the first two questions hit ambiguities I had walked past for weeks.

Step 15: Version the interface

The moment you publish a spec, you have a compatibility problem. What happens when you change `rope_theta`? What happens when you move from `RMSNorm` to a different norm? What happens when you expand vocab? You need versioning. Use semantic-style thinking: `v1.0` for the initial contract, `v1.1` for backward-compatible clarification or optional metadata addition, `v2.0` for breaking boundary changes. A model should declare exactly which interface version it implements. That keeps archaeology out of debugging.

BREAK IT

This chapter's main experiment is simple on purpose. You train a specialist that is compliant in every way except one: the spec requires `RMSNorm` and the specialist uses `LayerNorm`. Everything else stays the same. Same shared base initialization, same `d_model`, same number of layers, same head count, same positional encoding, same tokenizer, same training data format, same fusion system. That one change is enough.

The first time I designed this BREAK IT, the failure was subtler than I had braced for. The non-compliant specialist did not crash, did not produce gibberish, and did not even look obviously worse on short prompts. It only fell apart on longer compositional generations, which is exactly the worst kind of bug to discover in production.

Failure mode 1: Loud failure

The compliance test catches the mismatch immediately:

```
COMPLIANCE FAILURE: normalization.type
  expected: RMSNorm
  found: LayerNorm
```

This is good. It means your spec names a real boundary requirement, and your enforcement tool checks it before the model can do damage.

The lesson is not "LayerNorm is bad." The lesson is "hidden interfaces need explicit contracts." If your system rejects the checkpoint before inference, you turned a silent semantic mismatch into a loud engineering failure. That is the whole job.

Failure mode 2: Silent failure

The compliance test does not check normalization type. The specialist loads. The fusion system runs. Outputs degrade. This is the scarier case, and it is the more educational one.

What you observe is often not total nonsense. That would be easy. Instead you get one or more of these: responses stay grammatical but lose domain sharpness; routing scores look plausible but the specialist contributes unhelpful activations; generated text becomes generic or hedged; long completions drift off-topic sooner; evaluation drops more than you would expect from a "small architectural difference"; activation norm traces at the boundary differ from reference specialists.

If you inspect example outputs, a compliant code specialist might produce:

```
def parse_headers(raw: bytes) -> dict[str, str]:
    headers = {}
    for line in raw.decode().split("\r\n"):
        if not line or ":" not in line:
            continue
        k, v = line.split(":", 1)
        headers[k.strip().lower()] = v.strip()
    return headers
```

The non-compliant one may still produce Python-shaped text, but less coherent:

```
def parse_headers(raw):
    line = raw.split(headers)
    if ":" in line:
        return line.items()
    return header
```

Nothing crashed. The tensor shapes matched. The output even looks code-like. That is exactly why interface contracts matter.

What this proves

This experiment proves three things. First, semantic compatibility is not guaranteed by tensor shape. Second, normalization choice is part of the interface, not an implementation detail. Third, a compliance test that only checks obvious fields is not enough. If the LayerNorm specialist passes your checker, your specification is underspecified.

Exactly what to change

If your specialist block currently uses something like:

```
self.norm1 = RMSNorm(d_model, eps=1e-5)
self.norm2 = RMSNorm(d_model, eps=1e-5)
```

Break it by changing those lines to:

```
self.norm1 = nn.LayerNorm(d_model, eps=1e-5)
self.norm2 = nn.LayerNorm(d_model, eps=1e-5)
```

Keep everything else unchanged. Do not also change RoPE or head count. The point is isolation: one mismatch, one observed failure, one lesson.

What to watch for

Watch in this order: compliance output, fusion-system acceptance or rejection, sample generations, boundary activation statistics, evaluation metric drop.

Concrete signs: if compliance passes, that is a spec bug. If compliance fails with a clear message, your enforcement is doing its job. If you bypass compliance and run fusion anyway, expect degraded output quality. If the degradation is mild and hard to see qualitatively, use a quantitative task metric or activation trace.

A simple activation comparison is often revealing. Run the same prompt through compliant and non-compliant specialists, record RMS magnitude of boundary outputs, compare mean and variance over tokens, and inspect one or two token positions across both variants. If the stats drift enough to change downstream interpretation, you have your evidence.

How to strengthen the spec if it missed the failure

If the LayerNorm specialist slipped through, fix both the document and the tool. Add explicit fields for norm type, norm epsilon, boundary norm placement, and residual layout. Then add structural checks that inspect actual module classes, not just metadata. If your model code wraps norms behind custom names, do not settle for string matching on state dict keys — instantiate the module and inspect the actual objects.

If needed, add a dynamic reference-trace check: save output of a known compliant specialist on a fixed hidden-state probe and verify that new specialists stay within expected activation-scale bands at the boundary. This does not prove deep semantic identity, but it catches major mismatches that shape checks miss.

The big lesson of BREAK IT is this. A good interface specification is not the document you wrote before the failure. It is the document you wrote after the failure taught you what was missing. My own specialist interface spec went through four versions before I stopped finding fields I had assumed and forgotten to write down.

Optional Homework

There are two versions here: a full version if you already have a composable specialist setup in `nanochat` or your own fork, and a proxy version if you do not want to train full specialists yet.

Option A: Full version

Hardware:

- 1 modern GPU with at least 16 GB VRAM is comfortable for a small specialist setup
- 24 GB is better
- runtime: a few hours to a day depending on dataset size and model scale

Goal:

- train one compliant specialist and one identical-but-LayerNorm specialist
- run compliance on both
- force both through fusion
- compare outputs and metrics

Step 1: Freeze the shared base and define the interface

Write `specs/compose_iface_v1.json` with fields for:

- `d_model`
- `n_heads`
- `residual_layout`
- `normalization.type`
- `normalization.eps`
- `positional_encoding.scheme`
- `positional_encoding.rope_theta`
- `tokenizer_hash`
- `base_model_id`
- `shared_prefix_depth`

Step 2: Save metadata into every specialist checkpoint

In your checkpoint save code, add:

```
ckpt["interface"] = load_spec("specs/compose_iface_v1.json")
```

And overwrite fields like:

- `base_model_id`
- `shared_prefix_depth`
- training run id

Step 3: Train the compliant specialist

Train with the exact architecture in the spec.

Record:

- train loss
- validation loss
- one small domain metric
- 3 sample generations

Step 4: Train the broken specialist

Change only norm type in the specialist blocks:

- `RMSNorm -> LayerNorm`

Train with the same seed if possible. Record the same metrics.

Step 5: Run compliance

Command sketch:

```
python tools/check_compliance.py \
  --spec specs/compose_iface_v1.json \
  --ckpt checkpoints/code_specialist_rmsnorm.pt
```

Then:

```
python tools/check_compliance.py \
  --spec specs/compose_iface_v1.json \
  --ckpt checkpoints/code_specialist_layernorm.pt
```

Expected observation: the compliant checkpoint passes, the LayerNorm checkpoint fails with a normalization mismatch. If both pass, your checker is too weak.

Step 6: Bypass compliance on purpose

This part teaches the lesson more clearly. Force-load both specialists into the fusion system. Run the same 20 prompts through both fused models. Prompts should include easy in-domain prompts, medium difficulty prompts, and long-context prompts if possible.

Record:

- router choices
- output samples
- task metric
- activation RMS at specialist output

Expected observation: the compliant specialist improves or at least preserves meaningful output; the non-compliant specialist runs but degrades quality; degradation is often more obvious on longer or more compositional prompts.

Step 7: Tighten the spec if needed

If the broken specialist passed compliance, update both `specs/compose_iface_v1.json` and `tools/check_compliance.py`, then rerun. That last rerun matters — it proves the spec now catches the bug it previously allowed.

Option B: Proxy version

Hardware:

- CPU or small GPU is enough
- runtime: under an hour if you keep the model tiny

Goal: build the compliance mechanism even if you do not run a full specialist training pipeline.

Step 1: Create two tiny mock specialists

Make two minimal modules with same `d_model`, same attention head count, and same output shape. One uses `RMSNorm`, one uses `LayerNorm`.

Step 2: Save both with metadata claiming they implement the same interface

This is intentional. One of them is lying.

Step 3: Write a compliance script that inspects the actual module class

Have it print a hard failure for the LayerNorm variant. Expected observation: metadata alone cannot be trusted; structural inspection catches the mismatch.

Step 4: Add a dynamic probe

Feed a fixed random tensor `x` with shape `[2, 16, 768]` into both mock specialists. Record: output RMS, mean activation, max absolute value. Expected observation: both modules return valid tensors; the output statistics differ; shape equality alone tells you almost nothing.

Suggested command sequence

Adjust for your repo layout, but a clean flow looks like this:

```
python tools/make_mock_specialist.py --norm rmsnorm --out ckpts/mock_rms.pt
python tools/make_mock_specialist.py --norm layernorm --out ckpts/mock_ln.pt

python tools/check_compliance.py \
  --spec specs/compose_iface_v1.json --ckpt ckpts/mock_rms.pt
python tools/check_compliance.py \
  --spec specs/compose_iface_v1.json --ckpt ckpts/mock_ln.pt

python tools/run_probe.py --ckpt ckpts/mock_rms.pt --dump stats/rms.json
python tools/run_probe.py --ckpt ckpts/mock_ln.pt --dump stats/ln.json
```

Metrics to record

Keep it concrete. Record at minimum:

- compliance pass or fail
- exact failure fields

- sample generations from the fused system
- output task score
- boundary activation RMS
- one trace of router probabilities if you have routing

Expected observations

You should expect one of these two clean outcomes:

1. Good outcome: compliant specialist passes, non-compliant specialist fails before fusion.
2. Useful bad outcome: non-compliant specialist slips through, fusion quality drops, you strengthen the spec and checker, non-compliant specialist now fails.

If neither happens, your experiment is not isolating the boundary well enough.

Questions To Answer

1. When you changed only the normalization type, did the system fail loudly or silently?
2. If it failed silently, what was the first visible symptom: worse generations, worse metrics, changed activation statistics, or unstable routing?
3. Which fields in your original spec turned out to be missing or ambiguous?
4. Which requirements belong in the interface contract, and which belong only in checkpoint metadata?
5. If you handed your spec to another engineer, what sentence or field would confuse them first?
6. What tradeoff is the interface protecting: flexibility for independent specialists, or safety against hidden incompatibility?

Go Further

Start with three directions.

First, look at protocol design outside ML. USB, HTTP versioning, and protobuf schema evolution teach the same lesson: shared systems live or die on explicit contracts. Search for:

- "backward compatibility API versioning"
- "protocol compliance testing"
- "schema evolution breaking change"

Second, look at composable model research that depends on shared internal representations. **Project 33: Fusing Independently Trained Specialists** pointed toward BTX and mixture-style composition. Keep going by searching arXiv for:

- "modular language models shared initialization"
- "compositional foundation models"
- "representation alignment modular transformers"
- "model stitching hidden representation compatibility"

Third, study failure analysis tools rather than only architecture papers. Search for:

- "activation statistics transformer debugging"
- "representation similarity CKA fine-tuning"
- "model stitching CKA interface alignment"

If you want a concrete next research question, here is a good one: can you design an interface spec that permits some internal variation while still guaranteeing safe composition? That pushes beyond exact architectural matching and into learned adapters, interface validators, and representation alignment metrics. That is where the research gets interesting.

For code reference, stay anchored to:

- `nanochat`
- small MoE and routing reference implementations
- model stitching and representation similarity work

What You Now Know

"Same tensor shape" and "same internal language" are different claims. This chapter is the line between them.

You can write a versioned interface spec for a specialist boundary instead of hand-waving about shared hidden states. You can build a compliance test that rejects incompatible checkpoints before they poison a fusion system. And when composition fails quietly, you know where to look first. Not the router. The contract.

That is the engineering instinct this whole part is trying to force into muscle memory. Composable systems do not survive on vibes, shared naming, or hopeful checkpoint swapping. They survive on contracts.

Research Anchors

This chapter is a systems chapter, but it still sits on a real research trail. If you want to speak about interface specifications in an interview without sounding vague, it helps to know which papers and systems pressures made this chapter necessary.

At the architecture level, the background is the transformer lineage itself: **Attention Is All You Need** explains the original attention contract, **RoFormer** matters because positional encoding is part of that contract, and papers like **LLaMA** and **PaLM** matter because modern systems quietly made different choices about normalization, tokenizer assumptions, positional encoding, and serving constraints. Once those choices differ, "same shape" stops meaning "same model language."

At the modularity level, the mixture-of-experts line matters. **Adaptive Mixtures of Local Experts**, **Hierarchical Mixtures of Experts**, **Outrageously Large Neural Networks**, **GShard**, **Switch Transformers**, **GLaM**, **Mixtral**, and **Sparse Upcycling** all push on the same deeper problem. If components are supposed to cooperate, what exactly is shared and what is allowed to drift? This chapter turns that question into an explicit engineering document rather than leaving it buried in code.

That is the real interview takeaway. An interface specification is not bureaucratic overhead. It is what lets independently trained components remain attachable in a system whose parts were shaped by different papers, different codebases, and different hardware assumptions.

Starting Point

Start from:

- USB specification, as the analogy for strict compatibility contracts
- API versioning best practices, as the software analogy for explicit interface evolution
- `nanochat`: https://github.com/karpathy/nanochat

If you want the shortest path into hands-on work, write the spec before you write new model code. Build first and specify later, and you will describe the accident you produced. Specify first, then break it on purpose, and you learn what the contract is actually protecting.

Project 35: Incremental Assembly

Hook

Why can't you just bolt a new specialist onto a composed model the way you install a new plugin into an app? You already have the specialists, the router, and the hidden-state contract.

So why does adding one more expert after deployment so often make the system worse at old tasks, mediocre at the new task, or dependent on the order you attached things in?

I will be blunt. This is the question I think the field has been quietly underselling. Everything else in modular LLM work is rehearsal for it.

The Concept

Start with a company, not a transformer. You have three consultants on retainer: legal, tax, and technical architecture. Every incoming question goes through a receptionist who decides where it should go. It works until you hire a fourth consultant for privacy compliance. You do not want to retrain the old consultants, and you do not want to rebuild the whole company every time someone new arrives. So you tell the receptionist, "questions about privacy go to the new person."

That sounds like a local change. It is not.

The receptionist learned boundaries among the specialists that already existed, not an abstract concept of privacy compliance. Routing was shaped by the specific contrast between legal, tax, and architecture, not by some general theory of what a specialist is. Adding a fourth choice does not just expand the menu; it reshapes the relative boundaries between every option on it.

That is the central problem of incremental assembly. A composable system is not just several specialists wired together. New specialists have to join later without forcing the coordination mechanism to relearn the world. Figure 20.1 captures this challenge structurally.

Figure 20.1. Incremental assembly asks a stricter question than simple fusion: can specialist D join an A/B/C system later without degrading the old domains or making results depend on installation order?

Project 33 gave us the first version of fusion: multiple specialists combined behind a routing layer. Project 34 made us write down the boundary contract so we stop pretending "same tensor shape" means "same internal language." Project 35 now asks the ugly question: can you add new specialists after deployment and keep the system stable?

This is the question my own work on cooperative LoRA fusion was designed to push on. The protocol I drafted spent more design effort on "what does post-hoc specialist addition need to look like" than on the fusion math itself. The fusion math is the easy part. The protocol around it is where the actual research lives.

What "incremental assembly" means

Incremental assembly means you already have a composed system with several specialists, you train a new specialist independently, you attach that specialist later without retraining all specialists, ideally without retraining the router either, and you expect old behavior to remain stable while new behavior appears where expected. And results should not depend much on the order you added specialists. If specialists A, B, and C behave differently under different installation orders, the system is path-dependent.

Why This Is Harder Than It Sounds

The usual mistake is to treat the router like a lookup table. It is not. It is a learned decision mechanism. Given a hidden state (a list of numbers representing the token's current meaning and context), it chooses where that token should go next. So the router has to recognize what kind of input it is seeing, know which specialist can help, and keep doing both of those things after you expand the set of choices. That third requirement is where things break.

Distinguishing among 3 choices is not the same problem as distinguishing among 12, and distinguishing among 12 specialists where two are nearly identical is harder still.

Sorting email into three folders is easy. Work, personal, spam. It is harder to sort into twelve narrowly defined folders, and gets much harder when two folders differ only by a subtle line like "database scaling" versus "distributed data infrastructure." You will misroute more often. Routers hit the same wall. I keep my own inbox in three folders for exactly this reason; I have tried twelve and the routing always collapsed back to three within a month.

The Hidden Assumption Everyone Makes

Most model-composition work assumes the specialists are present during training, so the router learns relative boundaries. Not "what privacy compliance is in an absolute sense," but "this region looks more like specialist 4 than specialist 1, 2, or 3."

Adding a new specialist later is not a local change. You changed the choice set. You added a new answer to the exam. Even if the old answers were correct before, the decision boundary can shift, because "just add one more specialist" changes the competition.

That is the scaling limit this project is after. Not whether specialists help in principle, but whether the routing mechanism can still tell them apart as the menu grows.

The Minimum Math You Actually Need

Let the router produce a score for each specialist. Call those scores `s_1, s_2, ..., s_M`, where `M` is the number of specialists and each `s_i` is "how suitable does the router think specialist `i` is for this token?" The router often turns those scores into probabilities with softmax:

```
p_i = e^(s_i) / sum_(j=1 to M) e^(s_j)
```

In plain English: `p_i` is the router's confidence for specialist `i`, `e^{s_i}` means "make bigger scores count much more," and we divide by the total so all probabilities add up to 1. The reason to care about this formula is simple: add more specialists and the denominator gets larger. If the scores are not sharply separated, the probabilities flatten. Flattened probabilities mean weaker discrimination. That is one reason routing gets worse as the menu grows.

Why It Matters

If incremental assembly works, you no longer rebuild the whole model every time you want one new capability. You can imagine adding a legal specialist next quarter, a multilingual specialist after that, a private internal specialist for one customer, all while preserving the old system's behavior. This is software engineering logic applied to models: stable interfaces, modular upgrades, bounded changes.

A multi-state SaaS platform I worked on for eight years grew by exactly this pattern at the integration layer. We added DMV connectors, then DCF connectors, then new banking and payment integrations, then a fresh state's compliance rules, without ever rebuilding the core. The reason was not luck. It was that the early team had treated each integration as a contract first and a feature second. Model systems are the same problem with worse tooling.

Without incremental assembly, "modular" is mostly branding

A system is not modular because its diagram has boxes and arrows. If every new specialist forces router retraining, or worse, full-system retraining, you do not have modularity. You have coordinated co-training, which is useful but different.

A system that needs global retraining for every specialist addition takes longer to update, costs more to maintain, becomes harder to reason about, makes regressions more likely, and cannot support independent specialist release cycles. A system with real incremental assembly lets you treat specialists more like components than entangled organs.

Why order independence is a serious test

Suppose you add specialists A, B, and C. If `A → B → C` gives one result and `C → A → B` gives another, then your system is path-dependent. The final behavior depends on historical accident, not just the set of components present.

Order dependence tells you the router or assembly rule is not learning a stable partition of responsibility; it is being nudged by history. That is dangerous because it makes reproduction hard, makes debugging hard, and makes claims about "composable specialists" weaker than they sound.

Why the scaling limit is the real research question

Early demonstrations often stop at a small number of experts because the picture still looks clean. Three or four specialists is the easy case. The real question is what happens when you keep going.

At some point routing accuracy plateaus, existing specialists interfere, new specialists cannibalize old traffic, confidence spreads too thin, and near-duplicate specialists become almost indistinguishable. That point is not a nuisance result. It is the contribution.

If you can measure where the current approach breaks, you learned something concrete: how much capacity the router has, how separable the domains really are, whether the boundary representation supports modular growth. Research often advances by finding limits with clean measurements, not by pretending the limit is not there. Across the long training sweeps behind one of my own preprints, the most useful single chart in the supplement was the one that mapped where post-hoc addition stopped working cleanly. The headline numbers in the abstract were ornament. The breakdown curve was the actual paper.

The Build

This project starts from a fused system of three specialists. We are not building composition from scratch here. We are asking whether the existing composition method survives extension. The build has five parts: establish the baseline, train a fourth specialist independently, attach it without retraining the router, test order independence, and optimize the router while holding specialists fixed.

Step 1: Define the unit of evaluation before touching the system

Define the evaluation protocol before you add the new specialist. You need at least four domain evaluation sets, domains A, B, C, and the new domain D, each held out from specialist training and narrow enough that routing mistakes are visible. For example, if your specialists are code, medical, and multilingual, your held-out tests should not just be "general text quality." They should be domain-centered prompts where routing mistakes are visible. You want each prompt to answer a simple question: did the right specialist get used, and did the output quality match that specialist's claimed domain?

Track two kinds of metrics.

Quality metrics

These are task outcomes:

- exact match for structured tasks
- pass rate on code tasks

- accuracy on domain QA
- perplexity or bits-per-byte on domain text if you have language-model style evaluation

Pick one primary metric per domain and keep it fixed.

Routing metrics

These tell you what the router itself is doing:

- top-1 routing accuracy
- top-k routing recall
- entropy of router probabilities
- load distribution across specialists

"Entropy" here means how spread out the router's confidence is. Low entropy means the router strongly prefers one specialist. High entropy means the router is unsure and spreads probability across many specialists. That matters because incremental assembly often fails first as rising uncertainty, before output quality fully collapses.

Step 2: Measure the three-specialist baseline

Before adding anything, freeze the current fused system and measure it. You want a table like this:

Domain	Specialist expected	Task metric	Router top-1 acc	Router entropy
A	A	82.4	0.91	0.23
B	B	79.8	0.88	0.31
C	C	84.1	0.93	0.19

: **Table 20.1.** Three-specialist baseline before adding specialist D: per-domain task metric, router top-1 accuracy, and routing entropy used as the anchor for all subsequent comparisons.

These are placeholder numbers, but the pattern matters. This table becomes your anchor. Once you add specialist D, every claim should be measured relative to this baseline.

You are looking for two failure modes: new domain D does not work well after attachment, and old domains A/B/C degrade after attachment. Both matter. If D works but A and B get worse, you did not add capacity cleanly. You redistributed damage.

Step 3: Train the fourth specialist independently

This specialist should be trained without re-opening the whole fusion system. That is the point. Keep the interface contract fixed: same hidden size `d_model`, same normalization convention, same positional encoding scheme, same boundary tensor format, same output expectation. If the new specialist changes the interface, you learned nothing about composability; you just changed two variables at once.

The specialist should be domain-specific enough to justify its own path, but not so exotic that the router has no chance. Good examples include legal, finance, privacy compliance, scientific writing, or SQL/query planning. A bad example is something so broad it overlaps every other expert at once, because then you cannot tell whether routing confusion is structural or simply domain overlap. I learned this the hard way on one early cooperative-fusion experiment by picking "general reasoning" as a new specialist domain. The router refused to commit to it for the first thousand evaluation tokens, and I spent two days assuming the bug was in the gating function. It was not. The domain was the bug.

What independent training means here

Independent training means specialist D sees domain-D data while specialists A/B/C remain frozen, the fused router remains frozen during D's specialist training, and any adapter, head, or specialist-local projection is allowed only if it respects the boundary contract. That gives you a new module you can attach later.

Step 4: Attach the new specialist without retraining the router

This is the heart of the project. You now extend the fused system from 3 specialists to 4, but you do not retrain the router. That sounds impossible at first, because the router originally only knew about three targets. So how can it score a fourth?

There are several possible strategies in research, and you do not need to claim one is final. You need to understand what each assumes.

Strategy A: embedding-space similarity routing

Instead of treating routing as "choose among a fixed learned list," represent each specialist by a prototype in hidden-state space and have the router compare the current hidden state to each specialist prototype. A new specialist can then be added by adding a new prototype, not by retraining the whole classifier. The analogy: instead of a receptionist memorizing three employees by habit, the receptionist has profile cards describing each employee's specialty, and adding a new employee means adding a new card. This is closer to true incrementality, but it depends on the hidden states being stable enough that similarity means the same thing after deployment, which is a big assumption.

Strategy B: one-vs-rest gates

Another option is to give each specialist its own gate score rather than having one softmax trained over a fixed set. That makes addition easier because a new specialist gets its own scorer, but the scorers still need calibration. If one gate outputs scores on a different scale, comparisons become unfair.

The analogy: each consultant self-rates how relevant they are, but one person always speaks in 9s and 10s while another speaks in 3s and 4s. Whoever shouts louder gets more traffic. So independent addition becomes possible, but fair competition becomes tricky.

Strategy C: retrieval-style routing

You can store specialist descriptors and retrieve the nearest one or top-k nearest, turning routing into a search problem over specialist metadata or learned keys. That is attractive for incremental assembly because adding a specialist looks like adding a new record to an index, but again the key question is whether the representation space has stable semantics.

For this chapter, the exact router form matters less than the principle: you need a router mechanism that can accept a new specialist without retraining every old boundary from scratch. That is the design requirement.

What to watch after attachment

Run evaluation immediately after attachment. You want a result table like this:

Domain	Before add-D	After add-D	Change
A	82.4	81.9	-0.5
B	79.8	79.1	-0.7
C	84.1	84.0	-0.1
D	N/A	76.5	new

: **Table 20.2.** Best-case result after attaching specialist D: task metrics for all four domains before and after attachment, showing minimal degradation in old domains and a functioning new specialist.

This is the best-case shape: D starts working and old domains move only slightly. Bad shapes include the following failure patterns.

Failure shape 1: D never comes alive

Domain	Before	After
A	82.4	82.2
B	79.8	79.7
C	84.1	84.0
D	N/A	41.3

: **Table 20.3.** Failure shape 1 — specialist D never activates: old domains are stable but D achieves near-chance performance, indicating the router has not learned to route to the new specialist.

The router almost never picks D, so old specialists remain stable because the new one is effectively invisible. You did not achieve incremental assembly — you attached dead code.

Failure shape 2: D cannibalizes old specialists

Domain	Before	After
A	82.4	75.0
B	79.8	71.5
C	84.1	32.9
D	N/A	74.2

: **Table 20.4.** Failure shape 2 — specialist D cannibalizes old domains: D reaches acceptable performance but A and B suffer significant degradation, indicating miscalibrated routing scores.

The router overroutes to D, which partially handles old domains badly. This is common when a new specialist's routing score lives on a different scale, causing broken calibration. I have seen this exact pattern three separate times in my own experiments, and the first time I assumed it meant the specialist was genuinely strong. It did not. It meant the specialist's gate was simply louder than the others.

Failure shape 3: everyone gets blurrier

Domain	Before	After
A	82.4	79.0
B	79.8	76.8
C	84.1	81.3
D	N/A	66.5

: **Table 20.5.** Failure shape 3 — uniform degradation across all domains: adding specialist D reduces routing confidence globally, an early sign that the router is hitting its capacity limit.

The router got less certain overall. Adding one choice reduced discrimination everywhere. This is the first sign of a scaling limit.

Step 5: Measure whether the damage is local or systemic

Do not stop at task metrics. Inspect the router itself. For each domain, log how often each specialist is selected, mean router confidence, entropy, and a confusion matrix. A confusion matrix is just a table of "what should have been chosen" versus "what was actually chosen." If domain-B prompts frequently route to specialist D after addition, you know exactly where interference entered.

A concrete table might look like this:

True domain	Routed A	Routed B	Routed C	Routed D
A	0.86	0.05	0.03	0.06
B	0.07	0.78	0.04	0.11
C	0.04	0.03	0.89	0.04
D	0.18	0.08	0.09	0.65

: **Table 20.6.** Router confusion matrix after adding specialist D: fraction of inputs from each true domain routed to each specialist, revealing where routing interference has entered.

This tells a richer story than one average score. Specialist D is partly working, but A is stealing some of its traffic and D is stealing some of B's. That is compositional interference in a table.

Step 6: Test order independence

This is one of the most revealing tests in the whole part. Pick three specialists to add incrementally and compare sequences: `A → B → C`, `C → A → B`, `B → C → A`. At the end, evaluate the same final set. If order should not matter in a robust assembly method, the resulting metrics should be close. Not identical down to the fourth decimal place, but close enough that sequence is not a primary determinant.

What significant order dependence looks like

Suppose final metrics differ like this:

Sequence	A metric	B metric	C metric	mean routing acc
A->B->C	81.0	79.4	82.3	0.84
C->A->B	74.2	80.1	83.0	0.76
B->C->A	79.5	75.8	84.2	0.80

: **Table 20.7.** Order-dependence failure: per-domain task metrics and mean routing accuracy vary substantially by specialist installation order, indicating the assembly method preserves historical routing bias.

That is not noise. It means the assembly rule preserves historical bias.

One interpretation is that early specialists claim boundary territory first and later specialists have to carve out what remains. Another is that your specialist descriptors or routing prototypes are not globally calibrated. Either way, order dependence means your assembly method is not stable enough for modular growth.

A good order-independence result

A healthier result looks like this:

Sequence	A metric	B metric	C metric	mean routing acc
A->B->C	81.0	79.4	82.3	0.84
C->A->B	80.6	79.1	82.5	0.83
B->C->A	80.8	79.0	82.1	0.83

: **Table 20.8.** Order-independence success: per-domain task metrics and routing accuracy remain consistent regardless of specialist installation order, confirming stable incremental assembly.

Now the set matters more than the order. That is what we want.

Step 7: Use the autoresearch pattern on the router

This project explicitly asks you to fix the specialists and let the agent optimize router hyperparameters. That matters because the router is now the experimental surface.

We are no longer asking "can I train a better specialist?" We are asking "given fixed specialists, what router configuration best preserves old behavior while integrating new specialists?" This is exactly the kind of search loop from Project 20.

The agent should vary: router temperature, top-k selection count, prototype normalization, calibration strategy, entropy penalty, load balancing penalty, confidence threshold for fallback, and score scaling.

Keep the search disciplined. The objective is not just "maximize average score." That can hide damage. Use a multi-part objective:

```
J = Q_D - lambda_1 * Delta_old - lambda_2 * O - lambda_3 * H
```

What each symbol means:

- `J`: overall router score you want to maximize
- `Q_D`: quality on the new domain D
- `Delta_old`: degradation on old domains compared with baseline
- `O`: order sensitivity measure
- `H`: mean routing entropy or another uncertainty penalty
- `lambda_1, lambda_2, lambda_3`: weights you choose to decide how much each penalty matters

This equation is just a compact way of saying: do not let the agent get a high score by lighting up the new specialist and quietly wrecking everything else. Reward the new skill working, punish harm to old skills, punish unstable routing, punish assembly order mattering too much. That is the right mindset for this project.

I am openly opinionated here. Most published modular-LLM results report a single average score on a single composition. That is choreography. The chapters in this book before now have been gentle about saying so. I am no longer being gentle. A composition number with no order test and no degradation budget is closer to a sales pitch than a result.

Step 8: Document the assembly rule like a real interface

By this point, you should be writing down the exact conditions under which a new specialist may join the system. Not marketing language, but a real assembly contract. For example:

A specialist is admissible only if:

- hidden size matches `d_model`
- normalization type matches boundary spec
- positional encoding scheme matches
- specialist descriptor calibration falls within expected range
- routing confidence on compatibility probes exceeds threshold
- addition causes less than X% degradation on old domains
- order-independence delta stays below Y on benchmark set

That is not paperwork. That is the difference between repeatable assembly and anecdotal success.

This is, I think, where most modular-LLM work quietly leaks reproducibility. Everyone writes the experiment. Almost nobody writes the contract.

Minimal pseudo-code for the experiment loop

Here is the skeleton, stripped down on purpose:

```
# specialists are fixed modules with the same boundary contract
specialists = [spec_A, spec_B, spec_C]
router = load_router()

baseline = evaluate_system(router, specialists, eval_sets)

spec_D = train_independent_specialist(domain="D")

# attach without retraining the router weights
specialists_plus = specialists + [spec_D]

after_add = evaluate_system(router, specialists_plus, eval_sets_plus_D)

order_results = {}
for order in [
    ["A", "B", "C"],
    ["C", "A", "B"],
    ["B", "C", "A"],
]:
    system = assemble_in_order(order)
    order_results[tuple(order)] = evaluate_system(router, system, eval_sets)

best_router_cfg = autoresearch_search(
    fixed_specialists=specialists_plus,
    search_space={
        "temperature": [0.5, 1.0, 1.5],
        "top_k": [1, 2],
        "entropy_penalty": [0.0, 0.01, 0.05],
    },
    objective=assembly_objective,
)
```

The code matters less than the discipline: freeze what you are not testing, change one thing at a time, track interference explicitly.

BREAK IT

The clean path is useful because it gives you a baseline. The broken path teaches you what the system is actually protecting.

The project spec gives you the right failure to chase: add 12 specialists to a router designed for 4, then add a specialist that is nearly identical to an existing one, and watch what fails first.

Break 1: overload the router with too many specialists

Start with a router configuration that behaves well for 3 or 4 specialists, then keep adding (4, 8, 12, 16) while holding the specialists fixed once trained. Do not rescue the system by scaling the router at the same time unless your explicit goal is a second experiment. You are trying to find the scaling limit of the current approach, not the best possible redesign.

What exactly you are breaking

You are breaking the assumption that the router's discrimination capacity grows automatically with the number of specialists. It does not. A router that can cleanly separate 4 classes may fail at 12 for at least four reasons.

Failure reason 1: class crowding

More specialists means more regions in routing space that must be separated. If the representation space does not spread them far enough apart, boundaries crowd together. Imagine you once sorted fruit into apples, oranges, and bananas. Now you must sort Gala apples, Fuji apples, Honeycrisp apples, blood oranges, navels, mandarins, plantains, bananas, pears, quince, pluots, and apricots. The sorter needs finer distinctions. If the sensor has not improved, errors rise. That is class crowding.

Failure reason 2: confidence dilution

Remember the softmax denominator. When you add more specialists, scores compete with more alternatives. If the best score is only slightly better than the rest, probability mass spreads and the router gets less decisive. That shows up as rising entropy, more near-ties, and unstable top-1 selection.

Failure reason 3: overlap between domains

Real specialists are not perfectly separate. A legal specialist and a privacy specialist overlap. A coding specialist and a database specialist overlap. A multilingual specialist overlaps with everyone on multilingual prompts. As you add more specialists, overlap grows, and the router is not choosing among clean boxes. It is choosing among fuzzy clouds.

Failure reason 4: fixed router capacity

If the router is a small projection or small MLP, it may simply not have enough representational capacity to separate many subtle domains. That is the bluntest explanation, but often a true one.

What you should observe

Plot routing accuracy versus number of specialists. A healthy early curve might look like this:

Specialists	Routing acc	Mean domain quality
4	0.88	80.2
8	0.80	76.9
12	0.69	70.1
16	0.61	64.8

: **Table 20.9.** Routing degradation as specialist count increases: routing accuracy and mean domain quality across pool sizes from 4 to 16 specialists, showing where the coordination mechanism starts to fail.

The exact numbers do not matter. The shape does. You expect one of three regimes: early stability, plateau, then degradation. A more telling plot adds entropy:

Specialists	Routing acc	Router entropy
4	0.88	0.29
8	0.80	0.46
12	0.69	0.71
16	0.61	0.89

: **Table 20.10.** Router entropy versus specialist count: rising entropy alongside falling accuracy confirms the router is becoming increasingly uncertain, not just inaccurate.

Now you can see the internal story: the router is not just wrong more often. It is unsure more often. That is stronger evidence.

What this proves

If routing degrades as specialist count rises while specialists themselves remain fixed and competent, then the bottleneck is the coordination mechanism. Not the specialists, the assembly logic. That is exactly the result you want to isolate.

Break 2: add a near-duplicate specialist

Now do something more cruel. Take an existing specialist, say specialist B, and train a new specialist `B_prime` on data that is almost the same. Maybe B is "database systems" and `B_prime` is "SQL optimization," or B is "privacy law" and `B_prime` is "compliance documentation." Keep them close enough that a human understands the distinction, but the router may struggle.

What exactly you are breaking

You are breaking the assumption that every specialist occupies a clearly separable region. That assumption is often false. Near-duplicate specialists expose whether the router can distinguish genuinely subtle differences, or whether it only succeeds when domains are far apart.

What you should observe

A confusion matrix now becomes the star:

True domain	Routed B	Routed B'	Other
B	0.46	0.38	0.16
B'	0.35	0.44	0.21

: **Table 20.11.** Near-duplicate specialist confusion matrix: routing fractions between specialist B and the closely related B', showing that the router cannot cleanly separate overlapping domains.

That is bad separation. You may also see oscillation between the two experts across adjacent tokens, unstable outputs across repeated runs, top-2 routing that always includes both B and `B_prime`, and old domains that used to map cleanly to B now splitting badly.

If your router was trained for 4 specialists, this near-duplicate test often breaks it faster than merely adding random distinct experts. The challenge is not only quantity. It is resolution. Distinguishing "doctor" from "lawyer" is easy. Distinguishing "tax attorney" from "corporate compliance attorney" is harder. That is where routing becomes real.

What this proves

If performance collapses when adding a near-duplicate expert, then your assembly method depends on coarse domain separation and does not support fine-grained modular growth. That is a big limit, because real product systems rarely add specialists that are perfectly disjoint. They add narrowly overlapping experts all the time. Every product platform I have ever worked on has eventually grown a specialist that overlaps another by 70 percent; that is what product surface looks like.

Break 3: test whether order dependence gets worse as the system grows

A small system may appear almost order-independent. A larger system often reveals path dependence. Try building up to 8 specialists in one sequence and then rebuilding the same 8 in a different sequence, then compare routing distributions rather than just output scores.

If one final system routes 22% of finance prompts to legal while the other routes only 7%, that is not a cosmetic difference. It means the system is learning assembly history. A stable modular method should care much more about which specialists exist than which one got there first.

Break 4: overfit the router search objective

This is a subtler break, but worth running if you use an autoresearch loop. Tell the search loop to maximize only the new specialist's domain score without penalizing old-domain degradation. The agent may find a router that looks great on domain D and quietly harms A/B/C: something like D jumping from 68 to 81 while A drops from 82 to 72, B drops from 80 to 70, and C drops from 84 to 78.

This proves that router optimization is not neutral. The objective defines what "good assembly" means, and if the objective ignores preservation, the search loop will ignore preservation too. That lesson matters beyond this chapter. A lot of alignment and composition work fails for exactly this reason: the objective is narrower than the behavior you care about.

Optional Homework

This project can be run at two levels: a full experiment if you have GPU access and a prepared `nanochat`-style composition setup, and a proxy experiment if you do not.

Option A: Full experiment

Hardware and time

Expected setup:

- 1 to 4 modern GPUs
- at least 24 GB VRAM total is comfortable
- more if your specialists are large
- runtime: 1 to 3 days depending on specialist size and evaluation breadth

If you only have one consumer GPU, shrink the specialists and datasets. The mechanics still teach the lesson.

Step 1: Freeze your 3-specialist baseline

Record:

- specialist checkpoints
- router checkpoint
- held-out eval sets for A/B/C
- routing logs on those sets

Suggested outputs to save:

- `results/baseline_metrics.json`
- `results/baseline_confusion.csv`
- `results/baseline_entropy.csv`

Step 2: Train specialist D independently

Keep the boundary contract identical to the old specialists.

Record:

- training data source
- boundary spec version
- checkpoint path
- validation metric on D

Suggested artifact:

- `checkpoints/spec_D.pt`

Step 3: Attach D without retraining the router

Edit your assembly config, not the router weights. You are aiming for a config-level addition like:

```yaml
specialists:
  - name: A
    ckpt: checkpoints/spec_A.pt
  - name: B
    ckpt: checkpoints/spec_B.pt
  - name: C
    ckpt: checkpoints/spec_C.pt
  - name: D
    ckpt: checkpoints/spec_D.pt
router:
  ckpt: checkpoints/router_ABC.pt
```

Then run evaluation.

Suggested command pattern:

```
python eval_composed.py --config configs/compose_abcd.yaml --eval all
```

Record:

- A/B/C old-domain metrics
- D metric
- top-1 routing accuracy
- mean entropy
- load distribution

Step 4: Run order-independence tests

Create multiple assembly configs with different specialist insertion orders. For example:

- `compose_ABC.yaml`
- `compose_CAB.yaml`
- `compose_BCA.yaml`

Run:

```
python eval_composed.py --config configs/compose_ABC.yaml --eval all
python eval_composed.py --config configs/compose_CAB.yaml --eval all
python eval_composed.py --config configs/compose_BCA.yaml --eval all
```

Compare: per-domain quality, routing confusion matrices, entropy, specialist traffic shares.

Step 5: Run autoresearch over router hyperparameters

Keep specialists frozen. Search over:

- temperature
- top-k
- normalization on descriptor/prototype vectors
- entropy regularization
- confidence threshold
- load-balancing coefficient

Suggested pseudo-command:

```
python search_router.py \
  --config configs/compose_abcd.yaml \
  --search_space configs/router_search.yaml \
  --objective preserve_old_plus_add_new
```

Save: best config, search trace, Pareto front if you support multi-objective search.

Step 6: Break it with 4, 8, 12, 16 specialists

Prepare specialist banks of increasing size. Run the exact same evaluation protocol at each count.

Suggested command pattern:

```
python eval_scaling.py --num_specialists 4
python eval_scaling.py --num_specialists 8
python eval_scaling.py --num_specialists 12
python eval_scaling.py --num_specialists 16
```

Plot: routing accuracy vs number of specialists, quality vs number of specialists, entropy vs number of specialists.

Step 7: Add a near-duplicate specialist

Train `B_prime`. Attach it without retraining the router. Run targeted evaluation on B and `B_prime`.

Inspect: confusion matrix, top-2 routing rates, repeated-run stability if routing has stochasticity.

Expected observations

If your method is only weakly incremental, expect: D works partially, old domains degrade slightly to moderately, order dependence grows with specialist count, routing entropy rises with more specialists, and near-duplicate experts create confusion faster than unrelated experts.

If your method is unusually strong, you may get: modest D performance without old-domain collapse, small order sensitivity, and a later scaling cliff rather than an immediate one. Either result is useful — the cliff location is the finding.

Option B: Proxy experiment for a local machine

If you do not have the hardware for real specialists, you can still learn the routing lesson with small embeddings or frozen representations.

Hardware and time

Expected setup:

- laptop or desktop CPU is enough
- optional small GPU helps
- runtime: 1 to 3 hours

Proxy setup

Instead of full language-model specialists, create domain datasets with text snippets from 4 to 16 topics, represent each example with a frozen embedding model, represent each specialist by a centroid or prototype vector, and route by nearest prototype or a small classifier. This is not a full composed LLM system — and that is fine. It still teaches the same failure: as domains multiply and overlap, routing quality drops.

Exact steps

1. Create domain folders:

- `data/domain_A`
- `data/domain_B`
- `data/domain_C`
- `data/domain_D`
- later add up to `data/domain_P`

2. For each domain, collect 500 to 2000 short text samples.
3. Encode each text with a frozen embedding model.
4. Build one prototype per specialist: mean embedding of the domain's training texts.
5. Route each eval example to nearest prototype.
6. Measure accuracy as specialist count rises: 4, 8, 12, 16.
7. Create a near-duplicate domain by splitting one domain into two highly similar subdomains.

Example pseudo-code

```
# x is an embedding for one input
# proto[i] is the average embedding for specialist i
scores = [cosine_similarity(x, proto[i]) for i in range(M)]
pred = argmax(scores)
```

This is not a full router, but it makes the scaling and near-duplicate problem visible.

Expected observations

You will likely see decent routing for broad domains, degraded routing as you add similar domains, and sharper failure when one domain is split into two near-duplicates. That proxy teaches the same core lesson: incremental addition is easy only while specialists remain well separated.

Questions To Answer

1. When you added the fourth specialist without retraining the router, what failed first: new-domain performance, old-domain preservation, or routing confidence?
2. Did the new specialist stay mostly invisible, steal too much traffic, or carve out a clean region of responsibility?
3. How much did assembly order matter? Was the difference small enough to treat as noise, or large enough to call the system path-dependent?
4. As specialist count increased from 4 to 8 to 12 to 16, did routing accuracy degrade smoothly, plateau and then drop, or break suddenly?
5. What happened when you added a near-duplicate specialist? Did the router confuse the pair specifically, or did confusion spill into unrelated domains too?
6. If you had one knob to change next, would you increase router capacity, improve specialist descriptors, change the assembly rule, or redesign the evaluation set? Why?

Go Further

If you want to push this line of work past the chapter, focus on three directions.

First, read research on composable interventions and model stitching. Search arXiv for:

- `composable interventions`
- `model stitching`
- `representation alignment`
- `modular language model routing`

The reason to read stitching work is simple: incremental assembly only works if the intermediate representations remain compatible enough to support plug-in behavior.

Second, study small MoE and routing reference implementations. You are not copying mixture-of-experts directly here, but you are stealing the best diagnostics: load balancing, expert utilization histograms, router entropy, top-k routing analysis, and collapse detection. Search for:

- `sparse mixture of experts routing`
- `load balancing loss experts`
- `router z-loss`
- `expert choice routing`

Third, look at order sensitivity and continual expansion. This is closer to continual learning than many people admit. Search for:

- `continual learning modular networks`
- `growing experts without retraining`
- `dynamic expert expansion`
- `expert addition calibration`

If you want one concrete engineering direction, test whether specialist descriptors can be made more stable than task-trained router logits. That is the most direct path toward true post-deployment addition.

For code reference, stay anchored to:

- `nanochat` as the training and evaluation spine
- small MoE routing implementations for instrumentation ideas
- model stitching and representation similarity work for compatibility checks

Case study: Flash-MoE slot banks and incremental assembly as identity management

If Project 35 sounds abstract at first, this is the runtime example that makes it less abstract. The system is the **Flash-MoE slot-bank runtime** in the `anemll-flash-llama.cpp` fork. It matters here not because it is merely another MoE implementation, but because it makes one of this chapter's central claims operational: adding specialists is often a runtime identity problem more than a graph-rewrite problem.

The basic problem is simple. A large routed-expert model may have far more expert weights than you can keep resident in fast memory. A naive runtime pays for that mismatch repeatedly: decide which experts are needed, materialize them, run them, throw them away, and do the whole dance again on the next token.

That is exactly the sort of per-step restructuring this chapter warns you about. It makes the execution surface itself unstable.

The slot-bank idea is cleaner. Keep a fixed execution surface per layer: a stable bank of resident slots. The runtime does not rebuild the graph every token. It only changes which expert identity occupies which slot. If the requested expert is already present, you get a hit and continue. If not, you evict a victim slot, load the new expert bytes, install them, and keep going. The shape of execution stays stable while identity moves through it.

This pattern is closer to how databases manage hot pages in a buffer pool than to how most modeling frameworks handle expert state. The buffer-pool framing is, I think, the more honest one. You are doing cache management, not graph rewriting.

That is why this system belongs in Project 35 rather than only in a serving appendix. It is a real implementation of the distinction between control plane and data plane.

The control plane stays mostly fixed: the runtime knows how many slots exist, how the layer executes, and how to route through the bank. The data plane changes: which experts are loaded, which ones stay hot, which are cold, and which identities receive traffic.

The public repository makes that trade visible in numbers. In one published benchmark, the all-resident stock path decodes at about `109.0` tokens per second, while a `slot-bank 128` path reaches `53.0` tokens per second, or about `49%` of stock. That sounds like a loss until you remember what is being bought: the ability to run far larger routed-expert systems by streaming expert weights rather than requiring everything to fit at once. The same repository also exposes hit rates, miss rates, and I/O costs explicitly, which is exactly the right systems framing. Once you can measure hits, misses, and install overhead, incremental assembly stops being a metaphor and becomes something you can budget.

This gives you a stronger language for the chapter's hardest idea. "Add a specialist without retraining the whole system" does not necessarily mean "modify the graph so the system has a new shape." In a mature runtime it may mean something much more ordinary: drop a new specialist artifact into the sidecar store, register its identity, let the bank admit it when traffic justifies it, and let old specialists age out naturally if they stop earning demand.

That is a much more realistic picture of composability in production. Real systems do not want to pause, rebuild, and redeploy their whole execution graph every time a new specialist appears. They want a stable runtime that can absorb new identities with bounded disruption.

The broader lesson is one sentence: when you can keep execution shape stable and push change into identity, incremental assembly becomes far easier to reason about. The moment every new specialist forces the runtime to become a different machine, your "modularity" is much closer to a rebuild than an attachment.

If you want to connect this directly back to your own Project 35 work, add one more metric to your experiments: hit rate. Not just final quality. Not just router accuracy. Ask how often the system can satisfy a request from already-resident specialist state versus how often it has to pay an install cost. That metric tells you something important about whether your composition is settling into reusable structure or thrashing under diversity.

What You Now Know

"Just add one more specialist" should feel like a suspicious sentence now, not a roadmap.

You can measure whether a system is truly incremental or only looks modular while the choice set stays small. You can separate specialist quality from router quality, test order independence instead of assuming it, and measure the scaling limit directly.

The honest standard is harsher than the diagrams suggest. A composed model is not proven modular because three clean specialists behave on one run. It earns that label when you attach a fourth later, keep the old ones stable, vary the order, and watch the routing logic hold.

If I am being unfair to the field, I am unfair on purpose. This is the gap I think the next round of modular-LLM work has to close.

Research Anchors

The papers behind this chapter fall into two groups.

The first group is the MoE and conditional-computation line: **Adaptive Mixtures of Local Experts**, **Hierarchical Mixtures of Experts**, **Outrageously Large Neural Networks**, **GShard**, **Switch Transformers**, **GLaM**, **Mixtral**, and **Sparse Upcycling**. These papers teach the field how to route among specialists and how sparsity changes scaling, but they mostly assume the expert set is known during training. Project 35 asks the harder systems question they leave partially open: what happens when the menu of specialists grows after deployment?

The second group is the modern systems line around retrieval, tools, and production serving. **Retrieval-Augmented Generation for Knowledge-Intensive NLP Tasks**, **Toolformer**, and **ReAct** matter here because they show a broader pattern: useful systems stop being one monolithic forward pass and start becoming assemblies of components, specialists, and external procedures. Once that happens, incremental attachability becomes a first-class systems property rather than a nice-to-have abstraction.

That is why this chapter belongs near the end of the book. It is not only about routing accuracy. It is about whether model systems can evolve over time without forcing a full retrain every time a new capability arrives.

Starting Point

Start from:

- Composable Interventions (2024), for the framing that model behavior may be assembled from modular pieces
- BTX (Meta, 2024), for current thinking around composition and transferable specialist structure
- `nanochat`: https://github.com/karpathy/nanochat

If you want the shortest path into hands-on work, write the evaluation protocol first:

- baseline metrics
- old-domain degradation budget
- routing diagnostics
- order-independence test
- scaling-limit plot

That order matters. Attach specialists first and define success later, and you will end up defending whatever accident happened. Define the failure surfaces first, then build and break the system against them, and you learn where composability actually stops.

Project 36: Your Architecture

Hook

How do you know whether your architecture idea is real, rather than one lucky run on one friendly benchmark?

Most research mistakes do not come from failing to build something new. They come from fooling yourself about whether the thing works at all. I have written that sentence three times in different forms across drafts of this chapter, and it kept coming back because I do not know how to soften it without losing what it means.

The Concept

By this point in the book, we have built enough of the stack to stop treating LLMs as sealed appliances. We have seen gradients move weights, attention route context, routers choose experts, fine-tuning improve one thing while quietly damaging another, quantization trade quality for size, and composition look clean in a diagram and messy in code.

Project 36 changes the job. Up to now, the book asked you to understand mechanisms other people already chose. Here you choose the mechanism.

That sounds freeing. Mostly it means the excuses disappear.

Imagine you run a cargo port. Earlier chapters taught you how ships move, how cranes lift containers, how fuel gets delivered, and how schedules stay on time. Now someone asks you to redesign the port so it can handle a new kind of cargo faster than the old design. You are not being asked to admire ports. You are being asked to change one rule of the system and prove that the whole place still works. That is architecture research.

An architecture is not only a model shape. It is a claim about structure and why that structure should improve something that matters. Maybe your claim is: specialists should share a stricter interface so they compose better; routers should look at uncertainty, then hidden states; privacy-preserving fine-tuning should keep a specialist useful without exposing sensitive data; fusion should happen in hidden-state space instead of parameter space; or experts should be selected with a budget-aware mechanism that cares about latency, then quality.

A research contribution is not "I changed the code." It is a claim, a minimal test, a measurement, and a failure mode caught before someone else catches it for you. Without one concrete thread this chapter turns into hand-waving, so we anchor it to a composability problem sitting right in the middle of Part V.

Problem: specialist fine-tuning often improves domain skill while damaging composability.

In plain English: you teach one model to become a strong code specialist, another to become a strong legal specialist, and another to become a strong multilingual specialist. Each one gets better at its own job. Then you try to assemble them into one system. The assembly is shaky. Routing gets worse, order matters, hidden states no longer fit together cleanly, one specialist barges into another specialist's lane. The pieces no longer share the same internal dialect.

Hypothesis: if you force each specialist to preserve a shared internal interface while it fine-tunes, later composition becomes more stable.

"Shared internal interface" is still vague until you pin it down. Think of three musicians improvising together. If each one practices alone and changes tuning, timing, and key signature however they like, each may sound better solo. Put them in the same room and the session turns to noise. The shared interface is the tuning system they agree to keep.

For models, that interface can mean things like: same hidden-state dimension `d_model`, same normalization style, same positional encoding, same frozen early layers, same calibration on a shared anchor dataset, same expectation about where information lives in the residual stream. You do not need the full internal representations to be identical. You need them similar enough that swapping or combining specialists does not destroy coordination.

I cut a 2,000-word version of this section that tried to do this properly with diagrams and walked-through equations. I reread the cut version a month later and the same line came out of me: I get it, but this is the chapter where you stop teaching me and start trusting me to think. The musicians stayed.

So here is the example idea: train specialists with their task loss, but add an **interface preservation loss**. In plain terms: improve on the domain, but do not wander too far from the base model on a shared anchor set.

The task loss is the voice coach saying "sound better at your solo." The interface preservation loss is the band leader saying "stay in key." If you only keep the band leader, nobody improves. If you only keep the voice coach, nobody plays together. You need both.

This metaphor stuck because I tried five worse ones first. Two of them are still in my notebook with crossed-out arrows.

Now the equation is earned. Let `L_task` mean the normal training loss for the specialist's domain, `L_interface` mean a penalty for drifting away from the base model's hidden states on anchor examples, and `lambda` mean how hard you push on preserving the interface. Then train with:

```
L_total = L_task + lambda * L_interface
```

Nothing mystical happened there. The equation just compresses one rule into a line: get better, but pay for drifting too far. You can make `L_interface` concrete in several ways. The easiest version is mean squared error between hidden states from the base model and the specialist at selected layers on a shared anchor set — show both models the same neutral examples and charge the specialist every time its internal representation wanders too far from the base.

That is one architecture idea. Project 36 is bigger than that one idea. The point is to learn how to create, test, break, and publish your own.

Why It Matters

Without a project like this, you stay in demo-land. Demo-land is where an idea looks good because the test was weak.

A weak test asks "did the metric go up on the thing I cared about?" A real test asks harder questions: what degraded while the metric went up? What happened when I added a new specialist later? What happened when I changed the order of composition? What happened when compute got tight? What happened when domains overlapped, when the router was noisy, when the specialist data was smaller than expected, when I tried to reproduce the result with a second seed?

That shift matters because architecture work is full of flattering accidents. Here are common ways people fool themselves. They compare against a weak baseline, which teaches almost nothing if your method wins. They optimize one metric and ignore damage elsewhere, so a code specialist that gains 4 points on code completion while losing 12 points everywhere else gets mistaken for progress. They test composition only in the order they trained, which is choreography rather than composability. They report one lucky run, even though a result that vanishes on the next seed was never solid. And they confuse bigger compute with better design, when the real contribution may simply be a larger budget.

I have caught myself doing four of these five at some point in the last two years of my own work. I am not above the list. Nobody is.

Part V has been driving toward one claim. A modular system is not proven modular because you assembled a few neat pieces once. It is proven modular when you can keep adding pieces later without wrecking the old ones.

That is why this final project matters. It moves you from "I can implement known systems" to "I can define a failure surface, build an intervention, and test whether the claim survives." That is a field contribution. It does not have to be huge. It does have to be honest, reproducible, and sharp enough that someone else can prove you wrong.

The Build

Good work usually starts with irritation. Something keeps breaking. Something keeps wasting money. Something keeps making evaluation muddy. Something keeps making composition brittle.

That irritation is your entry point. The mistake is to treat that irritation as vague frustration rather than as a research handle. My own preference-routing ablation study started from a single annoyance: the router rewrite kept showing up in ablations as the largest single effect, and nobody in the related papers seemed willing to admit that. The whole preprint grew out of that one stubborn observation. We will build this in ten steps.

1. Pick a problem that is small enough to test and sharp enough to fail

Do not start with "I want to build a better LLM." That is not a project. That is a wish.

Start with one sentence in this format:

> I think `X` fails because `Y`, and I think `Z` will reduce that failure.

For our anchor example:

> I think specialist composition fails because fine-tuned specialists drift away from a shared internal interface, and I think interface-preserving fine-tuning will reduce that drift and improve order-independent composition.

That sentence has three pieces: a failure you care about, a cause you suspect, and an intervention you can implement. If you cannot name all three, the project is still foggy.

Good project scopes for this stage look like:

- preserve composability during specialist fine-tuning
- reduce router collapse under domain overlap
- cut KV cache memory with a cheaper sharing scheme
- make specialist training privacy-aware with less utility loss
- improve fusion by aligning hidden states before routing

Bad scopes look like: make LLMs better, improve reasoning, build AGI, replace transformers. Too wide means you will end up defending vibes.

2. Write the one-page research brief before coding

This is where you force your thoughts into a shape that can be tested. Keep it to five sections.

1. Problem — what breaks today? Be specific.
2. Why it matters — who pays for that failure? Latency, quality, privacy, stability, cost, composability?
3. Hypothesis — what do you think causes the failure, and what change should help?
4. Experiment plan — what exact comparisons will you run?
5. Success and failure criteria — what result would count as evidence for your idea? What result would falsify it?

Here is a concrete version for the running example.

```
# Research Brief: Interface-Preserving Specialist Fine-Tuning

Problem:
Specialists fine-tuned independently on new domains improve local quality
but become harder to compose later. Routing becomes unstable, order matters,
and old-domain behavior degrades.

Why it matters:
Composable systems only matter if specialists remain attachable over time.
If every new specialist forces a full rebuild, modularity is fake.

Hypothesis:
Representation drift during specialist fine-tuning breaks the shared interface.
Adding a hidden-state anchoring loss on a shared anchor set will preserve
composition while keeping most of the specialist gain.

Experiment:
Train 3 specialists from the same base model:
code, legal, multilingual.
Compare two versions of each:
baseline LoRA fine-tuning
LoRA + interface preservation loss
Measure domain gain, old-domain degradation, CKA drift, routing entropy,
and order-independence gap after composition.

Success:
Interface-preserved specialists keep at least 80% of the specialist gain
while reducing composition degradation and order sensitivity.

Failure:
No difference from baseline, or interface loss preserves composition only by
blocking useful specialization.
```

That brief is not busywork. It protects you from moving the goalposts after the run.

3. Define your metric before you write the architecture code

Most people hear "self-defined key metric" and think they need one number. What they actually need is one number plus the pieces underneath it.

If your contribution trades off three things, your metric must make those tradeoffs visible. I am going to be opinionated about this: the field's habit of compressing everything into a single benchmark score is the single biggest reason mediocre architecture work keeps slipping past review. The headline number is a marketing artifact. The sub-metrics are where you find out whether you actually believe your own claim.

For the running example, a useful score might include: specialist gain on the new domain, degradation on old domains, order-independence gap, routing stability, and training or inference cost. In plain English, you want a method that gets better where it is supposed to, does little damage elsewhere, works no matter which order specialists are attached, and does not cost absurdly more to train or serve.

Compress that into a score if you want one headline number:

```
Research Score =
  NewDomainGain
  - alpha * OldDomainDamage
  - beta * OrderGap
  - gamma * ExtraCost
```

Where `NewDomainGain` is how much better the specialist gets on its target task, `OldDomainDamage` is how much worse the system gets on tasks it used to handle, `OrderGap` is the difference in quality when specialists are added in different orders, `ExtraCost` is extra training time, memory, or inference latency, and `alpha`, `beta`, and `gamma` are weights you choose to reflect what matters most. Do not hide the sub-metrics just because you have a score. The score is for comparison. The sub-metrics are where truth lives.

4. Freeze your baseline

If the baseline keeps shifting, the result means nothing. For composable specialist work, a solid baseline might include: one shared base model, same tokenizer, same context length, same LoRA rank, same training tokens per specialist, same optimizer settings, same evaluation tasks, and same seeds or same seed sweep.

Then add your intervention on top. Do not quietly improve five things at once. If you do and the numbers move, you will not know which change mattered.

For our example:

- Baseline A: plain LoRA specialist fine-tuning
- Baseline B: layer freezing plus LoRA
- New method: layer freezing plus LoRA plus interface preservation loss

That is a fair comparison because each step adds one piece.

5. Decide where the architecture change lives

Every research idea needs an address in code. Where exactly does it enter the system? For the example project, the answer is: the specialist training loop, hidden-state extraction at chosen layers, the anchor dataset loader, an extra loss term in the objective, and the evaluation harness for composition. You are not changing the entire model. You are changing one stage in the pipeline, which makes the change easier to interpret.

Here is minimal pseudocode for the training step:

```
# x: anchor or task batch
# base_model: frozen reference
# specialist: trainable model or adapter
# lambda_interface: strength of interface preservation

task_loss = specialist.task_loss(task_batch)

with torch.no_grad():
    base_h = base_model.forward(anchor_batch, return_hidden=True)

spec_h = specialist.forward(anchor_batch, return_hidden=True)

interface_loss = mse(spec_h[layer_id], base_h[layer_id])

loss = task_loss + lambda_interface * interface_loss
loss.backward()
```

The code matters less than the contract. You are telling the specialist: earn domain skill, but do not rewrite the shared dialect.

6. Choose anchor data carefully

Anchor data is the neutral reference set used to preserve the interface. This is easy to do badly. If the anchor set is tiny, the specialist may preserve only a narrow slice of the interface. If the anchor set is from the target domain, you may just be regularizing toward the target task rather than preserving general composability. If the anchor set is wildly off-distribution, the preservation signal may be weak or irrelevant.

A good anchor set is small enough to be cheap, broad enough to represent general behavior, stable across all specialists, and held constant across runs. Think of it as the shared rehearsal sheet all specialists must still be able to read.

For a small project, 5k to 50k mixed-domain examples can be enough to reveal whether the idea works. You do not need internet-scale data to detect a bad architecture. You need a clean comparison.

7. Pick diagnostics that show why the result happened

If the only thing you log is final accuracy, you miss the story. For this project, log at least: target-domain quality, old-domain quality, hidden-state drift, CKA similarity between base and specialist at selected layers, routing entropy if you use a router, order-independence gap, and latency or memory cost.

A word on CKA. Centered Kernel Alignment is a similarity measure between representations. In plain English: if two layers produce very similar patterns across many examples, CKA is high; if they produce unrelated patterns, CKA is low. You do not need to become a representation-learning scholar to use it.

For this project, CKA answers a simple question: did the specialist keep speaking roughly the same internal language as the base model? That is directly relevant to composability.

8. Run the smallest experiment matrix that can kill your idea

Do not start by scaling up. Start with the minimal matrix that could prove you wrong. For the running example:

1. Base model only
2. Base + plain code specialist
3. Base + interface-preserved code specialist

4. Base + plain code + plain legal
5. Base + interface-preserved code + interface-preserved legal
6. Same pairings in reversed attach order
7. Optional third specialist if the two-specialist result is already meaningful

That is enough to answer early questions: does interface preservation help at all? Does it help only for one domain? Does it reduce order sensitivity? Does it cost too much specialist quality? If the answer is already "no," good. You learned something early and cheaply.

9. Plot the result so the failure is visible at a glance

Research dies in mushy presentation. Make plots that expose tradeoffs. For this project, good plots include: specialist gain vs old-domain damage, CKA drift vs composed performance, order A->B vs order B->A, number of specialists vs degradation, and lambda strength vs target-domain gain.

That last one matters a lot. If `lambda` is too low, interface preservation does nothing. If `lambda` is too high, the specialist cannot specialize. That is the core tension. You are not searching for a magic number. You are mapping the tradeoff curve. That curve is often more valuable than one best checkpoint.

10. Publish like you want someone else to reproduce and criticize it

Many people delay this step because they imagine publishing means a polished paper. It can. It can also be a GitHub repo with scripts and logs, a technical blog post, a short report with plots and failure cases, or an arXiv preprint if the work warrants it. The format matters less than the reproducibility.

A publishable artifact should include: the research brief, environment details, exact configs, evaluation protocol, plots, failure cases, what you tried that did not work, and what you would test next.

Do not hide the negative results. Negative results are part of the contribution if they close off a tempting but weak direction. The field does not only need winning ideas. It also needs honest maps of dead ends. If your project shows that a certain kind of interface penalty preserves composability but kills specialist gain past a threshold, that is useful. You saved someone else two weeks. That counts.

BREAK IT

This section matters more than the "success" run. If your idea cannot survive you attacking it, it is not ready.

For the running example, the architecture claim is that interface-preserving fine-tuning helps specialists remain composable. So what should we break? Break the exact thing that the claim depends on: the preservation of a shared interface.

Break 1: Remove the interface loss entirely

Exact change: set `lambda_interface = 0`, keep everything else identical.

What you expect to observe: target-domain score usually rises a bit or stays similar, hidden-state drift increases, CKA similarity to the base drops more sharply in middle or late layers, composed performance with multiple specialists becomes less stable, and the order-independence gap widens.

What this proves: if removing the interface loss improves solo specialization but harms assembly, then the interface loss was not just acting as generic regularization — it was protecting shared compatibility. This is the basic ablation. If you do not run it, you cannot claim the interface term mattered.

Break 2: Preserve the wrong layers

Exact change: apply `L_interface` only on very early layers, or only on the final layer, and compare against preserving a middle band of layers.

Different layers tend to carry different kinds of information. Early layers often hold more general structure. Later layers often move more with task specialization. If you anchor the wrong location, you may preserve too little or too much. Preserving only early layers may not stop late-layer drift enough to help composition, while preserving only the final layer may preserve outputs while allowing internal routing features to drift. A middle-layer anchor often gives a stronger tradeoff than either extreme.

What this proves: your idea is not just "add a penalty somewhere." It is about where shared structure lives and which layers need to stay compatible. That turns a vague architecture claim into a sharper one.

Break 3: Use a narrow anchor set

Exact change: replace the mixed anchor dataset with a narrow one, such as only code-like data or only short generic prompts.

Expected observation: interface metrics may look fine on the narrow anchor set, but real composition quality still degrades on broader tasks — specialists overfit to preserving compatibility only in the slice you showed them.

What this proves: your method depends on the anchor set actually representing the shared interface you care about. A fake anchor set creates fake reassurance. This is a classic research trap: the metric says "preserved," the system says "no, it did not."

Break 4: Stress order independence directly

Exact change: train specialist A then B, separately train B then A, compose both final systems and compare.

Expected observation if the method is weak: one order produces noticeably better joint results, one specialist steals more routing traffic depending on when it was attached, and old-domain damage depends on chronology rather than the final component set.

Expected observation if the method helps: performance remains closer across orders, specialist utilization is more stable, and degradation is governed more by task overlap than by attach sequence.

What this proves: true composability should care about the set of specialists, not merely the historical path by which they were assembled. That is the real bar.

Now the most important part: what if your method "works" on the headline metric but fails one of these breaks?

Then your idea is not dead. It is narrower than you thought. Maybe the real claim is no longer "interface preservation enables composable specialists" but rather "interface preservation improves pairwise composition under moderate domain overlap when anchor data is broad and the preserved layers are chosen carefully." That smaller claim is still valid if the evidence supports it.

Research gets stronger when the claim gets more precise. Do not defend a big claim with small evidence. Cut the claim down until it matches the result exactly. That is how you keep the work honest. I have rewritten the abstract of my own cooperative LoRA-fusion paper four times for exactly this reason. Each version was smaller and each version was more true.

Optional Homework

Here is a concrete version you can run, starting from your own repo and borrowing training structure from `nanochat` if needed. If your setup differs, keep the experiment shape the same.

Hardware and runtime

Proxy version:

- 1 GPU with 16GB to 24GB VRAM is enough
- or CPU for tiny smoke tests, though training will be slow
- runtime: 4 to 12 hours depending on model size and data

Bigger version:

- 2 to 8 GPUs if you want stronger specialists and wider evaluation
- runtime: 1 to 3 days

Goal of the lab

Train two or three specialists with and without interface preservation. Measure whether the new method keeps specialist quality while reducing composition damage and order sensitivity.

Step 1: Create the experiment folder structure

```
your-repo/
  research/
    project21/
      brief.md
      anchor_data/
      eval_prompts/
      plots/
      logs/
      configs/
```

Create `research/project21/brief.md` and write the one-page brief first.

Step 2: Prepare three small datasets

You need:

- `code_train.jsonl`
- `legal_train.jsonl`
- `anchor_mix.jsonl`

Keep the proxy version small: 20k to 50k examples per specialist, 10k anchor examples. The anchor set should mix general web text, dialogue, documentation, and plain prose. Do not make it domain-pure.

Step 3: Pick a frozen base checkpoint

Use one base checkpoint for everything.

Record:

- model size
- tokenizer
- context length
- normalization type
- LoRA rank if you use LoRA
- seed

Do not change these between runs.

Step 4: Add hidden-state extraction to your model forward

Expose hidden states at chosen layers. A minimal pattern:

```
# return_hidden_layers = [4, 8]
logits, hidden = model(x, return_hidden_layers=[4, 8])
```

Store only the selected layers to keep memory under control.

Step 5: Add the interface loss

Create a config file, `configs/interface.yaml`:

```yaml
lambda_interface: 0.1
anchor_batch_ratio: 0.25
anchor_layers: [4, 8]
```

Then modify your training loop so that each step computes: specialist task loss on domain data, interface loss on anchor data, and total loss = task loss + lambda * interface loss. If your training loop already supports auxiliary losses, add this as one more term. If not, write it directly.

Step 6: Train four runs minimum

Use exact run names.

1. `code_plain`
2. `code_interface`
3. `legal_plain`
4. `legal_interface`

If you have room, add multilingual too.

Suggested command pattern:

```
python train_specialist.py --base_ckpt base.pt --data code_train.jsonl \
  --run_name code_plain --lora_rank 16 --lambda_interface 0.0
python train_specialist.py --base_ckpt base.pt --data code_train.jsonl \
  --run_name code_interface --lora_rank 16 --lambda_interface 0.1 \
  --anchor_data anchor_mix.jsonl --anchor_layers 4 8
python train_specialist.py --base_ckpt base.pt --data legal_train.jsonl \
  --run_name legal_plain --lora_rank 16 --lambda_interface 0.0
python train_specialist.py --base_ckpt base.pt --data legal_train.jsonl \
  --run_name legal_interface --lora_rank 16 --lambda_interface 0.1 \
  --anchor_data anchor_mix.jsonl --anchor_layers 4 8
```

Keep token budget, optimizer, batch size, and training steps identical.

Step 7: Compose specialists

However your repo handles composition, create these four paired systems:

1. `code_plain + legal_plain`
2. `code_interface + legal_interface`
3. attach order `code → legal`
4. attach order `legal → code`

If you use a router, keep the router identical across these pairings.

Step 8: Evaluate five things

Create an evaluation sheet with these columns:

- target domain score
- old-domain score
- average CKA to base at preserved layers
- order gap
- latency or memory delta

Record metrics for: base only, each specialist alone, each pair composition.

Step 9: Sweep `lambda_interface`

Run at least four values:

- `0.0`
- `0.03`
- `0.1`
- `0.3`

Expected pattern: `0.0` gives strongest drift, often best solo gain, and worst composition stability; `0.03` gives mild protection, sometimes the best balance; `0.1` gives stronger protection with likely better composition and possible small specialist sacrifice; `0.3` risks over-constraining the specialist so it fails to adapt enough.

If you do not see a tradeoff curve, inspect your implementation. A flat curve often means the extra loss is too weak, mis-scaled, or disconnected from the real hidden states.

Step 10: Plot the results

Generate at least these plots.

1. `lambda_interface` vs target-domain gain
2. `lambda_interface` vs old-domain damage
3. CKA drift vs composed score
4. order A->B vs B->A

Expected observations if the idea has merit: interface-preserved specialists keep most of the solo gain, composed performance drops less when a second specialist is attached, order gap narrows, and CKA drift correlates with composition damage.

Expected observations if the idea fails: interface loss preserves similarity but not real composition, target gain collapses too much to be worth it, order gap stays large, or improvements vanish across seeds.

Step 11: Run the break experiments

Change one thing at a time.

1. `lambda_interface = 0`
2. preserve only early layers
3. preserve only final layer
4. use a narrow anchor set
5. reverse specialist attach order

Do not skip this step. This is where the project stops being a feature and becomes research.

Step 12: Write the report

Create `research/project21/report.md` with these sections:

- claim
- setup
- baselines
- main result
- break experiments
- failure cases
- what changed after seeing the failures
- next experiment

If you publish only one figure, publish the tradeoff curve, not the best single point.

Questions To Answer

1. When you remove the interface loss, what fails first: old-domain quality, order independence, routing stability, or hidden-state similarity?
2. Which layers were most worth preserving? Did early, middle, or late anchoring produce the best tradeoff?

3. How broad did the anchor set need to be before "preserved interface" started matching real composition quality?
4. Did your method preserve composability by protecting shared structure, or did it simply weaken specialization enough that the specialist stopped being disruptive?
5. Which part of the tradeoff are you actually willing to pay for: less drift, lower latency, better order independence, or higher specialist gain?
6. If someone challenged your claim after reading your plots, what result would they point to first?

Go Further

Research Anchors

This final project is intentionally open-ended, but it still sits on a recognizable research trail. If you want to sound grounded rather than vague when you describe your own architecture idea, it helps to know which lines of work gave you the right questions.

At the model-family level, **BERT: Pre-training of Deep Bidirectional Transformers for Language Understanding**, **Language Models are Few-Shot Learners**, **LLaMA**, and **PaLM** show how different scaling choices, training objectives, and deployment assumptions create different architectural priorities. They are not one continuous recipe, but together they teach an important habit: architecture is always tied to what the model is meant to do and what budget it is meant to live under.

At the systems level, **FlashAttention** matters because it makes memory movement and runtime constraints impossible to ignore. At the representation level, **The Platonic Representation Hypothesis** and **Scaling Monosemanticity** matter because they push on the deeper question behind modularity and transfer: when do different models still seem to organize meaning in comparable internal directions?

And at the composition level, the MoE and modularity papers from the earlier chapters remain relevant here for a simple reason. Once your architecture idea involves specialists, routing, transfer, or attachable components, you are no longer only proposing a better block. You are proposing a better coordination contract.

Start in three directions.

1. Search for composable model work. Look up:

- "Composable Interventions 2024"
- "BTX Meta 2024"
- "representation alignment modular models"
- "order-independent model composition"

2. Search for interface-preserving or representation-preserving fine-tuning work. Useful phrases:

- "representation drift fine-tuning"
- "hidden state alignment fine-tuning"
- "CKA fine-tuning representation stability"
- "functionally invariant paths model editing"

3. Search for specialist assembly and routing work. Useful phrases:

- "expert routing compositionality"
- "continual expert addition"
- "modular language model assembly"
- "mixture of experts load balancing collapse"

If this specific project grabs you, the next step is to compare interface preservation against other ways to keep specialists compatible: layer freezing only, adapter placement changes, distillation from the base model, router-aware specialist fine-tuning, and joint alignment on hidden states plus logits. That comparison is strong material for a blog post or preprint because it answers a practical question: which mechanism actually buys the best compatibility per unit of cost?

If you want a benchmark direction, build one. A small public benchmark for "attach specialists in different orders and measure degradation" would already be useful. Many papers imply modularity. Very few test it harshly.

What You Now Know

You can define a research project from your own engineering experience, choose a failure surface before touching code, and build an intervention that changes one thing at a time.

The standard is simple. If you can attack your own idea hard enough that a surviving result still means something, you have crossed the line between reading the field and entering it.

Where The Field Is Now

The book should not end with "go invent something" as if the field were static. It should end with the pressure that is shaping the field right now.

The center of gravity has shifted. From "can we train a large language model?" to "can we run useful models cheaply, fast, and everywhere?" That one shift pulls several research threads together.

The first thread is inference systems. Long context sounds like a model feature, but in practice it is a memory and bandwidth problem. The **KV cache** becomes expensive long before the math becomes elegant. That is why work on **FlashAttention**, grouped-query attention, speculative decoding, paged attention, and KV-cache compression matters so much. The field is learning that many deployment bottlenecks are really memory-movement bottlenecks wearing an architecture label.

The second thread is compression. Quantization used to feel like a deployment trick. It is now a design axis. Weight quantization, activation-aware quantization, KV-cache compression, and low-bit serving increasingly determine whether a model can run locally, whether it can fit into a product latency budget, and whether a long-context experience is affordable at all. This is one reason current systems discussions spend so much time on INT4, FP8, AWQ-style methods, GGUF, and KV-cache compression rather than only on bigger pretraining runs.

The third thread is hardware-aware model design. FLOPs are not the whole story. Memory bandwidth, interconnect speed, cache behavior, batch shape, and device-specific kernels increasingly decide what architecture choices are practical. That is why a model can behave one way in a paper, another way in a cloud GPU cluster, and another way again on a laptop or on Apple silicon. Hardware is no longer downstream of model design. It is part of model design.

The fourth thread is local and on-device LLMs. This is not only a hobbyist story. It is a product story about privacy, offline use, latency, cost control, and device-aware capability. The question is no longer just "can a small model run locally?" It is "which parts of the workload belong on device, which belong in the cloud, and how should compression, routing, and hardware constraints shape that split?"

This is the thread I care about most personally. An offline Tamil-language triage system I am building for community health workers cannot ship on a cloud assumption. It has to run offline on a phone in a village clinic, in Tamil, with no fallback. That constraint is not academic. It picks the architecture before the architecture picks itself.

The fifth thread is modularity under real-world pressure. Mixture-of-experts, tool use, retrieval, specialist fusion, and incremental assembly all point toward the same systems idea: the future stack may be less like one giant frozen model and more like a coordinated collection of components with explicit contracts. That makes the questions in Projects 19 and 20 more central, not less central.

What To Sound Like In A Strong Interview

If you are aiming for the kind of cross-questioning you might get in an Apple-style AI/ML interview, the goal is not to sound like you memorized a paper list. The goal is to show that you understand the pressure behind the papers.

You should be able to say, in your own words, that:

- transformer quality is only part of the story; inference is often limited by memory movement, cache growth, and hardware constraints
- quantization and compression are not cleanup steps after model design, but core product decisions
- local LLMs are enabled by a stack of compromises: smaller models, lower-bit weights, tighter serving kernels, and careful routing of work across devices
- modular systems need explicit contracts because shape compatibility alone does not guarantee semantic compatibility
- current research is not only about making models bigger, but about making them cheaper to serve, easier to compose, and more adaptable to real hardware

If you can say those things clearly, connect them to concrete examples, and explain what breaks when the tradeoffs are handled badly, you will sound like someone who understands the field structurally rather than someone who only recognizes the vocabulary.

Frontier Reading Map

The papers in `C:\AI LLM Research` form a useful map of that structure.

Start with the transformer base: **Attention Is All You Need, The Illustrated Transformer, BERT: Pre-training of Deep Bidirectional Transformers for Language Understanding**, **Language Models are Few-Shot Learners**, **LLaMA, PaLM**, and **RoFormer.**

Then understand scaling and training pressure: **Scaling Laws for Neural Language Models, Training Compute-Optimal Large Language Models, Textbooks Are All You Need**, and **The Smol Training Playbook.**

Then understand post-training and behavior shaping: **Training Language Models to Follow Instructions with Human Feedback**, **Direct Preference Optimization**, **Chain-of-Thought Prompting**, **ReAct**, **Toolformer**, **DeepSeek-R1**, and the **Qwen3 Technical Report**.

Then understand sparse and modular systems: **Adaptive Mixtures of Local Experts**, **Hierarchical Mixtures of Experts**, **Outrageously Large Neural Networks**, **GShard**, **Switch Transformers**, **GLaM**, **Mixtral**, and **Sparse Upcycling**.

Then understand representation and transfer questions: **Exploring the Limits of Transfer Learning with a Unified Text-to-Text Transformer**, **The Platonic Representation Hypothesis**, and **Scaling Monosemanticity**.

Finally, keep the systems pressure in view through the deployment-oriented work you have already touched in the book: KV caching, grouped-query attention, FlashAttention, quantization, and local serving runtimes. That combination is the current frontier more than any one isolated paper.

What The Book Now Gives You

You now have enough of the stack to talk about large language models as systems rather than as magic. You know how models learn, how tokenization changes what the model even sees, how attention moves information, how scaling changes the budget, how post-training changes behavior, how deployment changes the constraints, and how modularity fails unless the interfaces are explicit.

That does not make you "done." It means you now know what the real questions are.

The strongest outcome of this book is not that you can repeat jargon. It is that when someone names a trend like KV-cache compression, low-bit inference, on-device LLMs, specialist fusion, or hardware-aware serving, you can place it inside the system and explain why it matters. And, if it deserves it, push back.

Starting Point

Start from:

- Composable Interventions (2024), for the framing that model behavior may be assembled from modular pieces
- BTX (Meta, 2024), for current thinking around composition and transferable specialist structure
- `nanochat`: https://github.com/karpathy/nanochat

If you want the shortest path into hands-on work, write the evaluation protocol first:

- baseline metrics
- old-domain degradation budget
- routing diagnostics
- order-independence test
- scaling-limit plot

That order matters here too. Build first and define success later, and you will end up defending your favorite implementation. Define the breakpoints first, then build into them, and you give yourself a chance to discover something real.

Project 36 is called "Your Architecture" for a reason. The field does not need another person repainting last year's diagram with slightly different hyperparameters. It needs people who can make a clear claim, test it without self-deception, find the weak point early, and publish the result anyway. Keep the claim small enough to test, keep the test harsh enough to matter, then put the work where the next person can break it even harder.

That is the whole book, in one paragraph. The rest was preparation.

Appendix A: Lecture Companions

This book is designed to stand on its own. You can read it front to back, build every project, and come away with a working mental model of how LLM systems are put together.

It is also designed to connect cleanly to free lecture material, especially Stanford's open course pages and seminar archives. The point of this appendix is not to replace the book with a syllabus. The point is to give you a second lane: if a project wakes up a deeper question, you should know where to go next.

How to use this appendix

Read the chapter first.

Then use the companion material for one of three reasons:

- to hear the same idea explained in a more formal lecture voice
- to get the broader research context around the engineering pattern you just built
- to turn one book project into a week-long self-study block with external assignments, talks, and papers

Do not try to watch everything before you build.

Build first. Use the companions as reinforcement, not as permission.

Preflight companion: Python and tensor foundations

If the Preflight chapter left gaps, these fill them before Project 1.

For Python: the official Python tutorial at python.org covers everything in the Preflight and more. Work through sections 3 through 9 (data types, control flow, functions, classes) and stop — the rest is not needed yet.

For tensors: the PyTorch "Learn the Basics" tutorial walks through tensor creation, indexing, and operations with runnable examples. It is the fastest hands-on supplement to Tables 0.4 and 0.5 in the Preflight chapter.

Do not let either resource become a prerequisite that delays Project 1. Read them in parallel, not in series.

Part I companion: learning mechanics, tokenization, attention

Projects 1 through 4 pair well with:

- Stanford CS224N: for the neural-network, embeddings, and attention foundations that sit underneath the early projects
- Karpathy's repo progression: `micrograd`, `makemore`, and `minbpe`
- Stanford CS25 transformer seminar material once attention stops feeling mysterious and starts feeling architectural

Use this companion block when you want to deepen:

- why backpropagation works
- why tokenization is compression, not just text splitting
- why attention changes context handling so radically compared with older language models

Part II companion: building a GPT

Projects 5 through 8 are where a working transformer comes together, from a blank file through Flash Attention and tiled kernels.

Pair them with:

- Stanford CS336 Assignment 1, which maps to building and training a minimal language model
- Stanford CS336 Assignment 2, which maps to systems profiling and kernel-level optimization — the natural companion to the Flash Attention project
- `nanoGPT` as the reference training spine
- The Annotated Transformer for the line-by-line attention and block walkthrough
- PyTorch documentation for training loops, mixed precision, and `nn.Module` mechanics

Use this block when you want to move from "I can read a transformer" to "I can build and train one and reason about where its time and memory actually go."

Part III companion: training at scale

Projects 9 through 12 pair especially well with Stanford CS336.

This is the cleanest bridge between the book and a free Stanford-style systems education:

- Assignment 3 maps to scaling and compute budgeting
- Assignment 4 maps directly to data curation: raw Common Crawl to filtered, deduplicated pretraining data

If you only use one external course alongside this book, use CS336 here.

This is also the place to revisit:

- `karpathy/nanochat`
- EleutherAI GPT-NeoX for a fuller large-scale pretraining and distributed-training stack
- PyTorch documentation for distributed training, data-loader mechanics, and the FSDP/ZeRO single-box proxy in Project 12

Part IV companion: inference

Projects 13 through 17 pair well with:

- Stanford CS336 systems material
- Stanford CS25 seminar talks on efficient inference and transformer variants
- canonical papers on the KV cache, Grouped Query Attention, FlashAttention, and RoPE/YaRN long-context extension

Use this block when you want to go beyond "I implemented it" into:

- what the optimization actually buys you in wall-clock terms
- why memory layout and the KV cache shape serving cost
- how speculative decoding, long-context extension, and continuous batching become production decisions

The 2026 inference literature (speculative-decoding drafters and KV-cache quantization) is the right go-further reading here; the source trail in Appendix C points to the specific papers.

Part V companion: architecture and scaling

Projects 18 through 20 pair well with:

- Stanford CS336 scaling and compute-budgeting material
- Stanford CS25 seminar talks on Mixture-of-Experts, sparse computation, and scaling choices
- canonical papers on MoE routing and scaling laws
- MegaBlocks as a concrete sparse-MoE systems reference

Use this block when you want to understand:

- why sparse computation changes the cost curve without changing per-token quality for free
- how scaling laws turn a training budget into an architecture decision
- what an autonomous experimentation loop can and cannot discover on its own

Part VI companion: alignment

Projects 21 through 26 pair well with:

- Stanford CS224N for instruction tuning, evaluation, and post-training context
- Stanford CS336 Assignment 5 for the alignment and RLHF workflow
- repo-first study through `autoresearch`, Hugging Face TRL, and OpenRLHF
- the reasoning and agentic-RL literature (RLVR, GRPO, and agentic policy optimization) for the test-time-reasoning and tool-use projects

Use this block when you want to sharpen:

- the difference between pretraining and instruction following
- how reward models distort behavior if you optimize the wrong target
- why verifiable-reward RL and test-time reasoning became the dominant post-training levers

Part VII companion: system scope

Projects 27 through 31 widen the book from one architecture into a system landscape: quantization, retrieval, multimodal, non-Transformer sequence models, and non-autoregressive decoding.

Pair them with:

- `llama.cpp` and bitsandbytes-style quantization tooling for the deployment project
- Stanford CS25 talks on architecture variants
- the hybrid SSM/attention primary papers (Jamba, hybrid linear attention) for the non-Transformer project

- the diffusion-LM primary papers (LLaDA, Mercury) for Project 31, the new Diffusion and Non-Autoregressive Decoding chapter

Use this block when you want to see how far the ideas travel once you stop assuming a dense, autoregressive Transformer is the only shape a language model can take.

Part VIII companion: modular composition and research

Projects 32 through 36 are closest to seminar and lab work.

Pair them with:

- Stanford CS25 talks on frontier model systems and compositional design
- Stanford CS224N readings on evaluation and system behavior
- your own experiment log, because these chapters are where external material stops being enough and your measurements become the main text

This is the right point to turn the book into a mini research program:

- pick one architecture claim
- implement the smallest honest test
- compare against a real baseline
- write down what failed before you write down what worked

Suggested study rhythm

If you want the book to feel like a free Stanford-adjacent independent study, use this rhythm:

1. Read one project chapter.
2. Build the project.
3. Run the BREAK IT section.
4. Watch one companion lecture or seminar.
5. Read one primary source paper or repo.
6. Write a half-page note explaining what changed in your mental model.

That is enough.

You do not need formal enrollment to get real value out of the material. You need sequence, repetition, and honest experiments.

Appendix B: Free Resources

The table below is intentionally short. It is not a dump of links. It is the smallest useful set that turns this book into a free self-study track.

Resource	What it gives you	Best time to use it	URL
Stanford CS336: Language Modeling from Scratch	The closest external companion to the book's build-first approach; especially strong on systems, scaling, and data	Part II through Part V	https://cs336.stanford.edu/spring2024/
Stanford CS224N: NLP with Deep Learning	Broad lecture foundation for sequence models, transformers, training, evaluation, and modern NLP framing	Part I and Part VI	https://web.stanford.edu/class/cs224n/index.html
Stanford CS25: Transformers United	Seminar-style deep dives across transformer variants and research directions	Part IV through Part VIII	https://22-23.bulletin.stanford.edu/courses/2233491
`karpathy/micrograd`	Minimal autograd engine for Project 1 intuition	Project 1	https://github.com/karpathy/micrograd
`karpathy/makemore`	Character-level language modeling playground for early next-token intuition	Project 2	https://github.com/karpathy/makemore
`karpathy/minbpe`	Minimal byte-pair encoding tokenizer for Project 3	Project 3	https://github.com/karpathy/minbpe
The Annotated Transformer	The clearest line-by-line attention and transformer walkthrough; especially good when tensor shapes still feel slippery	Projects 4 through 6	https://nlp.seas.harvard.edu/annotated-transformer/
`karpathy/nanoGPT`	Minimal transformer training reference	Projects 5 through 7	https://github.com/karpathy/nanoGPT
Sebastian Raschka's LLM-from-scratch code	A more heavily diagrammed and annotated GPT training reference that complements nanoGPT well	Projects 5 through 8	https://github.com/rasbt/LLMs-from-scratch
`karpathy/nanochat`	End-to-end local training spine for the book's pretraining and inference projects	Projects 9 through 13	https://github.com/karpathy/nanochat
EleutherAI GPT-NeoX	A stronger large-scale pretraining reference once you want to see a fuller systems stack	Projects 9 and 12	https://github.com/EleutherAI/gpt-neox
Hugging Face TRL	Clean documented post-training and preference-	Projects 21 and	https://github.com/huggingface/trl

Resource	What it gives you	Best time to use it	URL
	optimization examples	23	
OpenRLHF	Full-stack RLHF training reference without hiding the pipeline behind too much framework magic	Projects 23 and 24	https://github.com/OpenRLHF/OpenRLHF
MegaBlocks	Concrete sparse-MoE systems reference from Stanford for routing, utilization, and block-sparse execution	Project 18	https://github.com/stanford-futuredata/megablocks
`llama.cpp`	Practical deployment and quantization reference	Project 27	https://github.com/ggml-org/llama.cpp
AutoAWQ	Clearer weight-only quantization reference for why specific quantization choices exist	Project 27	https://github.com/casper-hansen/AutoAWQ
LLaDA (Large Language Diffusion Models)	Open-weights masked/discrete diffusion language model; the clearest primary reference for non-autoregressive decoding	Project 31	https://arxiv.org/abs/2502.09992
Mercury (Inception Labs)	Commercial-scale diffusion LLM; useful for the deployed, code-optimized view of non-autoregressive generation (vendor benchmarks; treat throughput figures as reported)	Project 31	https://arxiv.org/abs/2506.17298
Chinchilla analysis	Canonical compute-optimal scaling reference	Project 19	https://arxiv.org/abs/2203.15556
PyTorch docs	The practical ground truth for tensors, autograd, AMP, modules, and optimization	Entire book	https://pytorch.org/docs/stable/index.html

Reference architecture by topic

The book does not need one repo family to carry every chapter. A stronger setup is to use the cleanest anchor for each topic.

Topic	Primary	Secondary
Autograd	`karpathy/micrograd`	PyTorch autograd docs
Tokenization	`karpathy/minbpe`	SentencePiece, `tiktoken`
Attention	The Annotated Transformer	`karpathy/nanoGPT`
Full GPT	`karpathy/nanoGPT`	Raschka's LLM-from-scratch code
Pretraining pipeline	`karpathy/nanochat`	EleutherAI GPT-NeoX
RLHF	OpenRLHF	Hugging Face TRL
MoE	MegaBlocks	Switch Transformer paper
Quantization	`llama.cpp`	AutoAWQ, bitsandbytes
Scaling laws	Chinchilla analysis	Kaplan scaling laws
Diffusion / non-autoregressive decoding	LLaDA	Mercury (Inception Labs)

Stability rule

For any repo the chapter depends on operationally, pin a tested commit when you actually run the project. Minimal teaching repos are excellent for learning, but newer repos can restructure quickly enough to break a chapter if you treat `main` as permanent.

Appendix C: Notes, Sources, And Bibliography

These are not exhaustive literature reviews. They are the publication source trail for the projects in this book: repos, papers, technical reports, course anchors, and stable public links collected in one place.

For publication purposes, each project section below is meant to answer two questions clearly:

- what primary sources shaped the chapter
- where a reader can reliably find those sources later

For a self-published technical book, that is the minimum bar. A reader should be able to move from any chapter to its technical lineage without hunting across search engines or broken notes.

Sections are listed by project number. Projects whose lineage is fully covered by a repo or paper already cited under a neighboring project, or whose sources are the reader's own experiments, are folded into those sections rather than padded with placeholder links.

Project Sources

Preflight: Python, Tensors, and What a Language Model Actually Is

- Python official tutorial (for readers who need a fuller Python introduction before starting): https://docs.python.org/3/tutorial/
- PyTorch tensor basics: https://pytorch.org/docs/stable/tensors.html
- Andrej Karpathy, "The spelled-out intro to language modeling: building makemore" (YouTube) — the single best video companion to the LLM-is-next-token-prediction framing used in this chapter.
- Sebastian Raschka, "Understanding Large Language Models" (2023) — a concise written survey that frames the same core loop covered in the Preflight from a slightly different angle. Useful if you want a second pass before Project 1.

Project 1: The Learning Machine

- `karpathy/micrograd`: https://github.com/karpathy/micrograd
- PyTorch autograd docs: https://pytorch.org/docs/stable/autograd.html

Project 2: Predicting The Next Character

- `karpathy/makemore`: https://github.com/karpathy/makemore
- Stanford CS224N course page: https://web.stanford.edu/class/cs224n/index.html

Project 3: Building A Tokenizer

- `karpathy/minbpe`: https://github.com/karpathy/minbpe
- SentencePiece project: https://github.com/google/sentencepiece

Project 4: Attention From Scratch

- "Attention Is All You Need": https://arxiv.org/abs/1706.03762
- "The Illustrated Transformer": https://jalammar.github.io/illustrated-transformer/
- The Annotated Transformer: https://nlp.seas.harvard.edu/annotated-transformer/
- Stanford CS25 seminar listing: https://22-23.bulletin.stanford.edu/courses/2233491

Project 5: Your GPT From A Blank File

- `karpathy/nanoGPT`: https://github.com/karpathy/nanoGPT
- Raschka's LLM-from-scratch code: https://github.com/rasbt/LLMs-from-scratch
- PyTorch `nn.Module` docs: https://pytorch.org/docs/stable/generated/torch.nn.Module.html

Project 6: From Prototype To nanoGPT

- `karpathy/nanoGPT`: https://github.com/karpathy/nanoGPT
- The Annotated Transformer: https://nlp.seas.harvard.edu/annotated-transformer/
- Raschka's LLM-from-scratch code: https://github.com/rasbt/LLMs-from-scratch
- PyTorch optimization docs: https://pytorch.org/docs/stable/optim.html
- "Attention Is All You Need": https://arxiv.org/abs/1706.03762
- "Language Models are Few-Shot Learners": https://arxiv.org/abs/2005.14165
- "PaLM: Scaling Language Modeling with Pathways": https://arxiv.org/abs/2204.02311
- "LLaMA: Open and Efficient Foundation Language Models": https://arxiv.org/abs/2302.13971

Project 7: The Details That Matter

- `karpathy/nanochat`: https://github.com/karpathy/nanochat
- RMSNorm paper: https://arxiv.org/abs/1910.07467
- RoFormer / RoPE paper: https://arxiv.org/abs/2104.09864
- "Attention Is All You Need": https://arxiv.org/abs/1706.03762
- "PaLM: Scaling Language Modeling with Pathways": https://arxiv.org/abs/2204.02311
- "LLaMA: Open and Efficient Foundation Language Models": https://arxiv.org/abs/2302.13971
- "Manifold-Constrained Hyper-Connections" (mHC): https://arxiv.org/abs/2512.24880 — DeepSeek-AI; constrains the hyper-connection residual-mixing matrix to be doubly stochastic (the Birkhoff polytope), restoring the residual identity guarantee; demonstrated stable training at up to 27B parameters and later adopted in DeepSeek-V4 (https://arxiv.org/abs/2606.19348).
- "Better & Faster Large Language Models via Multi-token Prediction" (Gloeckle et al., 2024): https://arxiv.org/abs/2404.19737 — the multi-token-prediction training objective (extra prediction heads) that the residual/head-design discussion draws on.
- "FastMTP: Accelerating LLM Inference with Enhanced Multi-Token Prediction": https://arxiv.org/abs/2509.18362 — repurposes a single position-shared MTP head for lossless speculative decoding; reports an average 2.03x speedup over next-token prediction (MiMo-7B-RL, A10 GPU, single-batch greedy decoding; reported). See also Project 14.

Project 9: Pretraining On The Real Web

- `karpathy/nanochat`: https://github.com/karpathy/nanochat

- EleutherAI GPT-NeoX: https://github.com/EleutherAI/gpt-neox
- Stanford CS336 course page and Assignment 4 data track: https://cs336.stanford.edu/spring2024/
- FineWeb project page: https://huggingface.co/spaces/HuggingFaceFW/blogpost-fineweb-v1
- "Scaling Laws for Neural Language Models": https://arxiv.org/abs/2001.08361
- "Training Compute-Optimal Large Language Models" (Chinchilla): https://arxiv.org/abs/2203.15556
- "Textbooks Are All You Need": https://arxiv.org/abs/2306.11644
- "The Smol Training Playbook": https://huggingface.co/spaces/HuggingFaceTB/SmolLM3-playbook

Project 13: Fast Inference: The KV Cache

- `karpathy/nanochat`: https://github.com/karpathy/nanochat
- PyTorch scaled dot-product attention docs: https://pytorch.org/docs/stable/generated/torch.nn.functional.scaled_dot_product_attention.html
- "Attention Is All You Need": https://arxiv.org/abs/1706.03762
- "FlashAttention: Fast and Memory-Efficient Exact Attention with IO-Awareness": https://arxiv.org/abs/2205.14135
- "TurboQuant: Online Vector Quantization with Near-optimal Distortion Rate": https://arxiv.org/abs/2504.19874 — 2025 preprint, ICLR 2026; reports quality-neutral KV-cache quantization at 3.5 bits per channel and only marginal degradation at 2.5 bits per channel (data-oblivious online algorithm). See also Project 27.
- "QJL: 1-Bit Quantized JL Transform for KV Cache Quantization with Zero Overhead": https://arxiv.org/abs/2406.03482 — AAAI 2025 (arXiv Jun 2024); reports a 3-bit KV cache with more than 5x memory reduction and no accuracy loss across various LLMs and tasks (reported).
- "PolarQuant: Quantizing KV Caches with Polar Transformation": https://arxiv.org/abs/2502.02617 — AISTATS 2026; reports over 4.2x KV-cache compression on long-context evaluation (reported). (The TurboQuant companion paper; distinct from the same-named arXiv:2502.00527.)

Project 14: Speculative Decoding

- "DFlash: Block Diffusion for Flash Speculative Decoding": https://arxiv.org/abs/2602.06036 — a block-diffusion drafter with non-causal (bidirectional-within-block) attention that drafts a whole block in one forward pass, conditioned on target hidden states injected into the draft KV cache; reports up to ~6x lossless acceleration across models (about 4.9x average on Qwen3-8B) and roughly 2.4x over EAGLE-3 in the tree-16 setting (reported).
- vLLM issue #41559, "DFlash speculative decoding incompatible with KV cache quantization": https://github.com/vllm-project/vllm/issues/41559 — documents that in vLLM v0.20.0 the non-causal draft attention path forces bf16 draft KV (the TurboQuant backend is hardcoded causal); characterized by the reporter as an implementation gap, not a fundamental limit. See also Project 17.
- "FastMTP: Accelerating LLM Inference with Enhanced Multi-Token Prediction": https://arxiv.org/abs/2509.18362 — MTP-head drafting as a lossless speculative-decoding mechanism (introduced under Project 7).

Project 15: Grouped Query Attention

- GQA paper: https://arxiv.org/abs/2305.13245
- `karpathy/nanochat`: https://github.com/karpathy/nanochat

- "FlashAttention: Fast and Memory-Efficient Exact Attention with IO-Awareness": https://arxiv.org/abs/2205.14135
- "LLaMA: Open and Efficient Foundation Language Models": https://arxiv.org/abs/2302.13971
- GShard: https://arxiv.org/abs/2006.16668
- GLaM: https://arxiv.org/abs/2112.06905
- Switch Transformer: https://arxiv.org/abs/2101.03961
- Mixtral paper: https://arxiv.org/abs/2401.04088

Project 16: Long-Context Extension (RoPE, YaRN, NTK-Aware)

- RoFormer / RoPE paper: https://arxiv.org/abs/2104.09864
- "A Systematic Analysis of Hybrid Linear Attention": https://arxiv.org/abs/2507.06457 — controlled study across 72 released models; recommends a linear-to-full attention ratio between 3:1 and 6:1, with recall degrading sharply below 3:1 (perplexity stays stable across ratios; recall is the ratio-sensitive metric).
- "Jamba: A Hybrid Transformer-Mamba Language Model": https://arxiv.org/abs/2403.19887 — ICLR 2025; original Jamba uses a 1:7 attention-to-Mamba layer ratio and reports an ~8x smaller KV cache than a comparable Transformer at 256K tokens, 16-bit KV. See also Project 30.
- "Jamba-1.5: Hybrid Transformer-Mamba Models at Scale": https://arxiv.org/abs/2408.12570 — Jamba-1.5-Large (398B total / 94B active) reports ~9 GB KV cache at 256K tokens vs 80-88 GB for 70B-class Transformers (16-bit KV; reported).

Project 17: Production Serving: Continuous Batching and PagedAttention

- "DFlash: Block Diffusion for Flash Speculative Decoding": https://arxiv.org/abs/2602.06036 — the drafter and its serving-time throughput results (introduced under Project 14); LMSYS/SGLang serving throughput figures are vendor-reported with hardware and concurrency conditions attached.
- vLLM issue #41559, "DFlash speculative decoding incompatible with KV cache quantization": https://github.com/vllm-project/vllm/issues/41559 — the practical serving consequence is a forced bf16 draft KV in vLLM v0.20.0; one reporter measured a roughly halved KV token budget (~28K to ~14K tokens) on a 24 GiB RTX 3090 (reported). See also Project 14.

Project 18: Mixture Of Experts

- "Adaptive Mixtures of Local Experts": https://www.cs.toronto.edu/~hinton/absps/jjnh91.pdf
- "Hierarchical Mixtures of Experts and the EM Algorithm": https://www.cs.toronto.edu/~hinton/absps/jjnh.pdf
- "Outrageously Large Neural Networks: The Sparsely-Gated Mixture-of-Experts Layer": https://arxiv.org/abs/1701.06538
- GShard: https://arxiv.org/abs/2006.16668
- Switch Transformer: https://arxiv.org/abs/2101.03961
- GLaM: https://arxiv.org/abs/2112.06905
- MegaBlocks: https://github.com/stanford-futuredata/megablocks
- Mixtral paper: https://arxiv.org/abs/2401.04088
- "Sparse Upcycling: Training Mixture-of-Experts from Dense Checkpoints": https://arxiv.org/abs/2212.05055
- "DeepSeek-V3 Technical Report": https://arxiv.org/abs/2412.19437 — Dec 2024; fine-grained MoE with 256 routed experts plus 1 always-on shared expert, top-8 routing, and auxiliary-loss-free load balancing via a per-

expert routing bias; activates 37B of 671B parameters per token (compute sparsity, not a memory saving).

- "DeepSeekMoE: Towards Ultimate Expert Specialization in Mixture-of-Experts Language Models": https://arxiv.org/abs/2401.06066 — Jan 2024; the origin of the fine-grained-expert-segmentation and shared-expert-isolation design DeepSeek-V3 builds on.

Project 19: Scaling Laws

- "Scaling Laws for Neural Language Models": https://arxiv.org/abs/2001.08361
- "Training Compute-Optimal Large Language Models" (Chinchilla): https://arxiv.org/abs/2203.15556
- "Language Models are Few-Shot Learners": https://arxiv.org/abs/2005.14165
- "PaLM: Scaling Language Modeling with Pathways": https://arxiv.org/abs/2204.02311
- "Textbooks Are All You Need": https://arxiv.org/abs/2306.11644
- "The Smol Training Playbook": https://huggingface.co/spaces/HuggingFaceTB/SmolLM3-playbook
- Stanford CS336 scaling assignment context: https://cs336.stanford.edu/spring2024/
- "Taming the Long-Tail: Efficient Reasoning RL Training with Adaptive Drafter" (TLT): https://arxiv.org/abs/2511.16665 — MIT-led preprint; applies speculative decoding to RL rollouts via a continuously trained adaptive drafter, reporting over a 1.7x end-to-end RL-training speedup while preserving accuracy (preprint; not peer-reviewed). See also Project 25.
- "Reinforcement Learning with Verifiable Rewards Implicitly Incentivizes Correct Reasoning in Base LLMs" (RLVR): https://arxiv.org/abs/2506.14245 — reward comes from an external automatic verifier rather than a learned reward model; anchors the reasoning-scaling discussion. See also Projects 23, 25, and 26.

Project 20: Autonomous Experimentation

- `karpathy/autoresearch`: https://github.com/karpathy/autoresearch
- `karpathy/nanochat`: https://github.com/karpathy/nanochat
- "Scaling Laws for Neural Language Models": https://arxiv.org/abs/2001.08361
- "Training Compute-Optimal Large Language Models" (Chinchilla): https://arxiv.org/abs/2203.15556
- "The Smol Training Playbook": https://huggingface.co/spaces/HuggingFaceTB/SmolLM3-playbook

Project 21: Fine-Tuning And Instruction Tuning

- "Training Language Models to Follow Instructions with Human Feedback" (InstructGPT): https://arxiv.org/abs/2203.02155
- "Exploring the Limits of Transfer Learning with a Unified Text-to-Text Transformer" (T5): https://arxiv.org/abs/1910.10683
- Hugging Face TRL: https://github.com/huggingface/trl
- Raschka's LLM-from-scratch code: https://github.com/rasbt/LLMs-from-scratch

Project 23: Reward Models And RLHF

- "Training Language Models to Follow Instructions with Human Feedback" (InstructGPT): https://arxiv.org/abs/2203.02155
- "Direct Preference Optimization: Your Language Model is Secretly a Reward Model" (DPO): https://arxiv.org/abs/2305.18290
- "Chain-of-Thought Prompting Elicits Reasoning in Large Language Models": https://arxiv.org/abs/2201.11903

- "ReAct: Synergizing Reasoning and Acting in Language Models": https://arxiv.org/abs/2210.03629
- "Toolformer: Language Models Can Teach Themselves to Use Tools": https://arxiv.org/abs/2302.04761
- "DeepSeek-R1: Incentivizing Reasoning Capability in LLMs via Reinforcement Learning": https://arxiv.org/abs/2501.12948
- "Qwen3 Technical Report": https://arxiv.org/abs/2505.09388
- "Deep Reinforcement Learning from Human Preferences": https://arxiv.org/abs/1706.03741
- "Reinforcement Learning with Verifiable Rewards Implicitly Incentivizes Correct Reasoning in Base LLMs" (RLVR): https://arxiv.org/abs/2506.14245 — RLVR replaces the learned reward model with a programmatic verifier; the RLVR term was introduced by AI2's Tülu 3 (https://arxiv.org/abs/2411.15124) and scaled by DeepSeek-R1. RLVR names the reward source; GRPO names the optimization algorithm.
- "Agentic Reinforced Policy Optimization" (ARPO): https://arxiv.org/abs/2507.19849 — extends RLVR/GRPO to multi-turn agent trajectories with entropy-based adaptive rollout and step-level advantage attribution; reports improved performance at roughly half the tool-call budget of prior methods across 13 benchmarks (reported). See also Project 26.
- "Tülu 3: Pushing Frontiers in Open Language Model Post-Training": https://arxiv.org/abs/2411.15124
- OpenRLHF: https://github.com/OpenRLHF/OpenRLHF
- Hugging Face TRL: https://github.com/huggingface/trl

Project 25: Test-Time Reasoning (CoT, Self-Consistency, Best-of-N)

- "Chain-of-Thought Prompting Elicits Reasoning in Large Language Models": https://arxiv.org/abs/2201.11903
- "Reinforcement Learning with Verifiable Rewards Implicitly Incentivizes Correct Reasoning in Base LLMs" (RLVR): https://arxiv.org/abs/2506.14245 — introduces CoT-Pass@K and reports that answer-only verifiable rewards implicitly incentivize correct reasoning chains (single model family, math benchmarks; treat as this paper's result, not settled consensus). See also Projects 19 and 23.
- "Taming the Long-Tail: Efficient Reasoning RL Training with Adaptive Drafter" (TLT): https://arxiv.org/abs/2511.16665 — speculative decoding inside RL rollouts for reasoning models; over 1.7x RL-training speedup reported (preprint). See also Project 19.

Project 26: Tool Use and Function Calling

- "ReAct: Synergizing Reasoning and Acting in Language Models": https://arxiv.org/abs/2210.03629
- "Toolformer: Language Models Can Teach Themselves to Use Tools": https://arxiv.org/abs/2302.04761
- "Agentic Reinforced Policy Optimization" (ARPO): https://arxiv.org/abs/2507.19849 — agentic RL for multi-turn tool use with step-level credit assignment; reports roughly half the tool-call budget of prior methods across 13 benchmarks (reported). See also Project 23.
- "Reinforcement Learning with Verifiable Rewards Implicitly Incentivizes Correct Reasoning in Base LLMs" (RLVR): https://arxiv.org/abs/2506.14245 — the verifiable-reward framing that agentic RL builds on (introduced under Project 23).

Project 27: Quantization and Deployment

- `llama.cpp`: https://github.com/ggml-org/llama.cpp
- AutoAWQ / Activation-Aware Weight Quantization tooling: https://github.com/casper-hansen/AutoAWQ
- bitsandbytes: https://github.com/bitsandbytes-foundation/bitsandbytes

- "GPTQ: Accurate Post-Training Quantization for Generative Pre-trained Transformers": https://arxiv.org/abs/2210.17323
- "TurboQuant: Online Vector Quantization with Near-optimal Distortion Rate": https://arxiv.org/abs/2504.19874 — 2025 preprint, ICLR 2026; quality-neutral KV-cache quantization at 3.5 bits per channel, marginal degradation at 2.5 bits per channel. (Introduced under Project 13; the KV-cache and quantization chapters share this source trail.)
- "QJL: 1-Bit Quantized JL Transform for KV Cache Quantization with Zero Overhead": https://arxiv.org/abs/2406.03482 — AAAI 2025; 3-bit KV cache, reported >5x memory reduction with no accuracy loss.
- "PolarQuant: Quantizing KV Caches with Polar Transformation": https://arxiv.org/abs/2502.02617 — AISTATS 2026; reported >4.2x KV-cache compression (the TurboQuant companion paper; pin this id, not arXiv:2502.00527).

Project 30: Non-Transformer Architectures (Mamba, RWKV)

- "A Systematic Analysis of Hybrid Linear Attention": https://arxiv.org/abs/2507.06457 — controlled 72-model study; recommends a 3:1 to 6:1 linear-to-full attention ratio, with recall degrading sharply below 3:1. (Introduced under Project 16.)
- "Jamba: A Hybrid Transformer-Mamba Language Model": https://arxiv.org/abs/2403.19887 — ICLR 2025; interleaves a minority of attention layers among Mamba layers (1:7 attention-to-Mamba) and reports an ~8x smaller KV cache at 256K tokens, 16-bit KV.
- "Jamba-1.5: Hybrid Transformer-Mamba Models at Scale": https://arxiv.org/abs/2408.12570 — the scaled hybrid; Jamba-1.5-Large reports ~9 GB KV cache at 256K tokens (16-bit KV; reported).
- "An Empirical Study of Mamba-based Language Models" (Mamba-2-Hybrid): https://arxiv.org/abs/2406.07887 — NVIDIA; an 8B hybrid mixing ~43% Mamba-2 / 7% attention / 50% MLP, with up to ~8x faster token generation at long context (reported/projected).

Project 31: Diffusion and Non-Autoregressive Decoding

- "Large Language Diffusion Models" (LLaDA): https://arxiv.org/abs/2502.09992 — Feb 2025, ICLR 2025; a from-scratch 8B masked/discrete diffusion LM competitive with a comparable autoregressive baseline on in-context learning and instruction following, and notably mitigating the reversal curse. The "noise" is a fully-masked sequence, not continuous Gaussian noise.
- "Mercury: Ultra-Fast Language Models Based on Diffusion" (Inception Labs): https://arxiv.org/abs/2506.17298 — the first commercial-scale diffusion LLM; Inception Labs reports Mercury Coder at over 1000 tok/s on NVIDIA H100 GPUs (vendor benchmark; reported).
- "HERALD: High-Throughput Block Diffusion LLM Serving via CPU-GPU Cooperative KV Cache Retrieval": https://arxiv.org/abs/2606.21633 — systems/serving work reporting up to 1.59x lower per-block latency and 2.47x higher throughput vs GPU-only, at a 5-10% KV budget across block-diffusion LLMs (reported).
- "Efficient-DLM: From Autoregressive to Diffusion Language Models, and Beyond in Speed": https://arxiv.org/abs/2512.14067 — Dec 2025; notes that diffusion LMs' learning efficiency still lags autoregressive models when trained from scratch.
- "DeepSeek-V4: Towards Highly Efficient Million-Token Context Intelligence" (context): https://arxiv.org/abs/2606.19348 — announced as a preview (April 2026); included here for the broader efficient-decoding and long-context landscape, not as a diffusion model.

Project 32: Layer Freezing And Transfer

- Stanford CS224N course page: https://web.stanford.edu/class/cs224n/index.html
- "Exploring the Limits of Transfer Learning with a Unified Text-to-Text Transformer" (T5): https://arxiv.org/abs/1910.10683
- "Similarity of Neural Network Representations Revisited" (CKA): https://arxiv.org/abs/1905.00414
- "The Platonic Representation Hypothesis": https://arxiv.org/abs/2405.07987
- "Scaling Monosemanticity: Extracting Interpretable Features from Claude 3 Sonnet": https://transformer-circuits.pub/2024/scaling-monosemanticity/index.html
- "TriAttention: Efficient Long Reasoning with Trigonometric KV Compression": https://arxiv.org/abs/2604.04921
- TriAttention repository: https://github.com/WeianMao/triattention

Project 33: Fusing Independently Trained Specialists

- "LoRA: Low-Rank Adaptation of Large Language Models": https://arxiv.org/abs/2106.09685
- "QLoRA: Efficient Finetuning of Quantized LLMs": https://arxiv.org/abs/2305.14314
- Luce Megakernel repository: https://github.com/Luce-Org/luce-megakernel
- This project also synthesizes the interface and compatibility ideas developed in Projects 7, 15, and 32.

Project 34: The Interface Specification

- This chapter is a systems synthesis chapter rather than a single-paper chapter.
- Its main technical anchors are "RoFormer: Enhanced Transformer with Rotary Position Embedding" (https://arxiv.org/abs/2104.09864), "GQA: Training Generalized Multi-Query Transformer Models from Multi-Head Checkpoints" (https://arxiv.org/abs/2305.13245), "Similarity of Neural Network Representations Revisited" (https://arxiv.org/abs/1905.00414), and the adapter-based specialist work in Project 33.
- In other words, this chapter formalizes the compatibility constraints that earlier chapters exposed through code and failure cases.

Project 35: Incremental Assembly

- `karpathy/autoresearch`: https://github.com/karpathy/autoresearch
- MegaBlocks: https://github.com/stanford-futuredata/megablocks
- "Toolformer: Language Models Can Teach Themselves to Use Tools": https://arxiv.org/abs/2302.04761
- "ReAct: Synergizing Reasoning and Acting in Language Models": https://arxiv.org/abs/2210.03629
- "Retrieval-Augmented Generation for Knowledge-Intensive NLP Tasks": https://arxiv.org/abs/2005.11401
- ANEMLL Flash-MoE slot-bank runtime: https://github.com/Anemll/anemll-flash-llama.cpp
- This chapter builds directly on the modularity and interface questions from Projects 33 and 34.

Project 36: Your Architecture

- Stanford CS25 seminar listing: https://22-23.bulletin.stanford.edu/courses/2233491
- Stanford CS224N course page: https://web.stanford.edu/class/cs224n/index.html
- "BERT: Pre-training of Deep Bidirectional Transformers for Language Understanding": https://arxiv.org/abs/1810.04805
- "Language Models are Few-Shot Learners": https://arxiv.org/abs/2005.14165

- "LLaMA: Open and Efficient Foundation Language Models": https://arxiv.org/abs/2302.13971
- "PaLM: Scaling Language Modeling with Pathways": https://arxiv.org/abs/2204.02311
- "FlashAttention: Fast and Memory-Efficient Exact Attention with IO-Awareness": https://arxiv.org/abs/2205.14135
- "The Platonic Representation Hypothesis": https://arxiv.org/abs/2405.07987
- "Scaling Monosemanticity: Extracting Interpretable Features from Claude 3 Sonnet": https://transformer-circuits.pub/2024/scaling-monosemanticity/index.html
- This final project is intentionally open-ended. Its real source base is your own experiment log, baselines, and failure analysis.

Glossary

Terms are listed alphabetically. Each entry shows where the term is first introduced and defined in the book.

acceptance rate — In speculative decoding, the fraction of draft-model tokens that the main model accepts under the rejection rule. Higher is better; determines the practical speedup. *(Chapter 14)*

activation abs_max — The largest absolute value of any activation tensor at a step, monitored during training to catch precision overflow and outlier explosions before they become NaNs. *(Chapter 11)*

admission control — In production serving, the policy that decides whether to accept a new request when the KV-page pool is near capacity. Prevents thrashing under load. *(Chapter 17)*

agentic policy optimization (ARPO) — A reinforcement-learning method that extends RLVR and GRPO from scoring a single response to optimizing a multi-turn agent trajectory, assigning credit at the level of individual tool-use steps rather than only the whole trajectory. It samples extra reasoning branches at high-uncertainty steps — motivated by the observation that token entropy spikes right after a tool call returns — and reportedly matches prior methods at roughly half the tool-call budget across 13 benchmarks. *(Chapter 26)*

all-gather — A collective communication operation that reconstructs the full tensor on every rank by collecting the shards held by all ranks. Used to materialize sharded parameters during forward and backward in FSDP. *(Chapter 12)*

autonomous experimentation — A bounded search loop in which an agent follows explicit rules to edit code, run experiments, measure results, keep improvements, and revert failures. The agent operates within constraints defined in a `program.md` file rather than exploring freely. *(Chapter 20)*

auxiliary-loss-free load balancing — A mixture-of-experts balancing scheme (introduced in DeepSeek-V3) that keeps expert utilization even by adjusting a learned per-expert bias added to routing scores for top-K selection only, nudging the bias down for overloaded experts and up for underloaded ones. It avoids the gradient interference of a large auxiliary balance loss; a negligible sequence-wise balance loss is retained just to prevent extreme imbalance within a single sequence. *(Chapter 18)*

backpropagation — The algorithm that computes gradients by working backward through the computational graph from the loss to each parameter, applying the chain rule at each operation. *(Chapter 1)*

best-of-N (BoN) — A test-time strategy that samples N candidate completions, scores them with a reward model or verifier, and returns the best. Trades inference compute for quality. *(Chapter 25)*

bi-encoder — A retrieval architecture that encodes query and document independently and compares them with a dot product. Cheap at query time but less accurate than a cross-encoder. *(Chapter 28)*

bigram — A pair of adjacent tokens. A bigram model asks: given that I see this token, what token tends to come next? It is the simplest meaningful language model. *(Chapter 2)*

Birkhoff polytope — The set of all doubly stochastic matrices of a given size — square matrices with non-negative entries whose rows and columns each sum to 1. mHC constrains its hyper-connection mixing matrix to lie in this set so the residual stream keeps its averaging, identity-preserving behavior across layers. *(Chapter 7)*

bits-per-byte (bpb) — A training metric that measures how many bits of surprise the model needs on average to encode the next byte of text. Lower is better. Related to perplexity but normalized to bytes rather than tokens. *(Chapter 9)*

block diffusion — A term with two distinct meanings in this book. As a speculative-decoding *drafter* (DFlash), it is a lightweight model that drafts a whole block of tokens in one non-causal forward pass while the autoregressive target model still verifies every token — diffusion is only the draft mechanism. As a *decoding style* for diffusion language models, it means generating a sequence block by block, each block produced by iterative masked-diffusion refinement. Both are masked/discrete diffusion, never continuous Gaussian noise. *(Chapters 14 and 31)*

byte-pair encoding (BPE) — A tokenization algorithm that starts from raw bytes and learns a vocabulary by repeated merging: count all adjacent byte pairs, find the most frequent pair, replace it with a new token ID, and repeat until the target vocabulary size is reached. *(Chapter 3)*

causal masking — A constraint in attention that prevents each token from attending to future positions. Implemented by replacing the upper triangle of the attention score matrix with negative infinity before softmax so that future tokens cannot leak into predictions. *(Chapter 4)*

catastrophic forgetting — A failure mode in sequential fine-tuning where training on a new task causes the model to lose performance on tasks it previously handled well. The second training stage overwrites part of the behavioral map built by the first. *(Chapter 21)*

chain-of-thought (CoT) — A prompting and training pattern in which the model is asked to produce explicit intermediate reasoning steps before its final answer, often improving accuracy on multi-step problems. *(Chapter 25)*

Chinchilla result — The finding from Hoffmann et al. (2022) that for a fixed compute budget, there is a better balance between model size and training duration than previously assumed — roughly 20 training tokens per parameter for compute-optimal training. *(Chapter 19)*

CKA (Centered Kernel Alignment) — A similarity metric that compares the internal representations of two neural networks by measuring how similarly they respond to the same set of inputs. Scores near 1.0 indicate highly aligned representations; scores near 0 indicate independent structure. *(Chapter 32)*

coarse-to-fine decoding — In masked diffusion language models, the pattern of committing high-confidence token predictions early and leaving lower-confidence positions masked for later denoising steps, so the sequence sharpens from a rough draft to a finished output over several passes. *(Chapter 31)*

continuous batching — A serving strategy that admits new requests and retires finished requests every decode step, instead of waiting for a whole batch to finish, eliminating padding waste and head-of-line blocking. *(Chapter 17)*

cross-encoder — A retrieval scoring model that jointly encodes a (query, document) pair and produces a single relevance score. More accurate than a bi-encoder but too expensive to score the whole corpus. *(Chapter 28)*

closure — An inner function that captures variables from the scope where it was defined, carrying them along even when the outer function has returned. In the backward pass code, each node's `_backward` function is a closure that knows exactly which parent nodes to send gradient to. *(Preflight)*

computational graph — The recorded trail of all operations and their inputs during a forward pass, showing how each output depends on each input. Gradients flow backward through this graph during backpropagation. *(Chapter 1)*

compression ratio — Defined as B/T, where B is the total number of bytes in a dataset and T is the total number of tokens after tokenization. Answers: how many bytes of original text does each token cover on average? *(Chapter 3)*

compute-optimal training — Choosing a model size and token budget that achieve the best loss for a fixed amount of compute, rather than maximizing either model size or training duration independently. *(Chapter 8)*

context window — The maximum number of tokens a model can attend to at once. In practice, a hardware statement about how much KV cache memory can be stored. *(Chapter 9)*

cross-entropy — The standard loss function for classification over a fixed vocabulary. Asks: how much probability did the model assign to the correct class? Equivalent to negative log-likelihood when labels are one-hot. *(Chapter 2)*

dataset shard — One chunk of already-processed training data saved to disk, split from a larger corpus into manageable pieces so that training does not require loading the full dataset into memory at once. *(Chapter 9)*

decision boundary — The line or surface in input space where the model switches from predicting one class to another. A nonlinear model can learn curved decision boundaries; a linear model cannot. *(Chapter 1)*

decode (inference phase) — The phase of inference after prefill, during which new tokens are generated one at a time. Each new token attends only to the already-cached KV history plus itself. *(Chapter 13)*

decontamination — The process of removing examples from a pretraining or fine-tuning corpus that overlap with eval-set inputs or answers, so the eval still measures generalization rather than memorization. *(Chapter 10)*

denoising step — One iteration of a diffusion language model's reverse process: the mask predictor reads the current partially-masked sequence, predicts all masked positions at once, and commits the confident ones. A generation uses a fixed number of these steps rather than one left-to-right pass per token. *(Chapter 31)*

DFlash — A speculative-decoding method whose drafter is a lightweight block-diffusion model: it drafts an entire block of tokens in one non-causal forward pass, conditioned on target-model hidden states injected into the draft KV cache, and the autoregressive target verifies them losslessly. The paper reports up to about 6x acceleration across models (around 4.9x average on Qwen3-8B). In vLLM v0.20.0 its non-causal draft attention cannot yet compose with a quantized KV cache, forcing bf16 draft KV — an implementation gap, not a fundamental limit. *(Chapter 14)*

diffusion language model (dLLM) — A language model that generates non-autoregressively: instead of predicting tokens one at a time left to right, it starts from a fully-masked sequence and iteratively unmasks all positions over a number of denoising steps. Examples include LLaDA, Mercury, and Gemini Diffusion; they are competitive with autoregressive models on code and short-form generation and much faster per output, while still lagging on complex reasoning. *(Chapter 31)*

discretization — Converting a continuous-time state-space model into a discrete-time recurrence using a sample step Δ, so it can be evaluated step by step on token sequences. *(Chapter 30)*

dispatch — In a tool-use system, the runtime routing step that takes a parsed model-emitted tool call and invokes the corresponding function implementation. *(Chapter 26)*

document boundary — A marker inserted between source texts during pretraining data preparation. Prevents the model from learning spurious transitions between unrelated documents that happen to be concatenated in the training stream. *(Chapter 9)*

doubly stochastic matrix — A square matrix with non-negative entries whose every row and every column sums to 1. Because the product of two such matrices is again doubly stochastic and its spectral norm is at most 1, constraining a mixing matrix to this form (as mHC does) keeps signals from exploding or vanishing across layers. *(Chapter 7)*

DPO (Direct Preference Optimization) — A preference-optimization loss with a closed-form connection to RLHF, trained directly on (chosen, rejected) pairs against a frozen reference policy. Replaces the reward model and on-policy RL of standard RLHF. *(Chapter 24)*

draft model — In speculative decoding, the small fast model that proposes a short run of tokens for the large main model to verify in a single batched forward pass. *(Chapter 14)*

dunder method — A Python method with double-underscore prefix and suffix (`__add__`, `__mul__`, `__repr__`) that Python calls automatically in response to built-in operators or functions. Dunder methods allow a class to intercept arithmetic, so that `a + b` can build a computational graph rather than computing a plain sum. *(Preflight)*

effective batch size — The total batch size that the optimizer experiences after combining multiple micro-batches during gradient accumulation. Allows training on larger logical batches than fit in GPU memory at once. *(Chapter 9)*

embedding — A list of learned numbers that represents a token. Rather than a one-hot vector, each token has a dense, low-dimensional identity that the model can update through training to capture semantic and syntactic relationships. *(Chapter 2)*

entropy — A measure of uncertainty in a probability distribution: H = −∑ p_i log p_i. High entropy means the distribution is spread across many outcomes; low entropy means it is concentrated on a few. *(Chapter 2)*

expert utilization — In mixture-of-experts models, the distribution of how often each expert is selected. Tracked to detect router collapse, where one or two experts dominate and the rest become dead weight. *(Chapter 18)*

FAISS — An open-source vector-index library for nearest-neighbor search over dense embeddings. Supports flat, IVF, and HNSW indexes and is the standard backend for RAG systems. *(Chapter 28)*

fine-grained MoE — A mixture-of-experts design that splits each expert feedforward network into many smaller experts while activating proportionally more of them, holding per-token compute fixed. The finer segmentation multiplies the number of routing combinations, letting individual experts specialize more sharply; DeepSeek-V3, for example, routes each token to 8 of 256 small routed experts. *(Chapter 18)*

fine-tuning — Additional training of a pretrained model on a smaller, task-specific dataset to steer its behavior toward a desired interface or domain. Does not build new knowledge from scratch — it steers existing knowledge. *(Chapter 21)*

Flash Attention — A memory-efficient exact attention algorithm that fuses the QK^T, softmax, and multiplication by V into a tiled kernel, never materializing the full $N \times N$ attention matrix. Reduces attention memory from $O(N^2)$ to $O(N)$. *(Chapter 8)*

free list — In PagedAttention, the pool of currently unused KV pages. New requests draw pages from the free list; finished requests return pages to it. *(Chapter 17)*

FSDP (Fully Sharded Data Parallel) — A distributed-training scheme that shards model parameters, gradients, and optimizer state across ranks and gathers them just in time for each layer's forward or backward pass. *(Chapter 12)*

function calling — A structured output format in which the model emits a JSON object naming a function and its arguments, instead of free text, so the runtime can dispatch it deterministically. *(Chapter 26)*

forward pass — One complete evaluation of a model from input to output: the sequence of operations that transforms token IDs into a prediction. During training, the forward pass also builds the computational graph that the backward pass traverses to compute gradients. *(Preflight)*

FP32 — 32-bit floating-point representation. The standard precision for model weights before quantization, storing each value with about 7 decimal digits of accuracy and consuming 4 bytes per weight. *(Chapter 27)*

GELU (Gaussian Error Linear Unit) — A smooth activation function where small negative inputs are not chopped to zero immediately but pass through with a slight gradient. Used in GPT-style feedforward layers as an alternative to ReLU. *(Chapter 7)*

GGUF — A file format used by llama.cpp that packages model weights, quantization metadata, tensor names, shapes, tokenizer data, and architecture parameters into a single portable file for runtime use. *(Chapter 27)*

gradient — A number (or vector of numbers) indicating which direction to adjust a parameter in order to reduce the loss, and approximately by how much. The gradient of the loss with respect to a parameter is the partial derivative of the loss with respect to that parameter. *(Chapter 1)*

gradient accumulation — A training technique that runs several small forward/backward passes before updating weights, summing gradients across micro-batches to simulate a larger effective batch size without requiring it to fit in memory all at once. *(Chapter 9)*

gradient clipping — Capping the magnitude of the gradient vector before the optimizer update step. Prevents a single bad batch from producing an abnormally large update that destabilizes training. *(Chapter 6)*

gradient descent — An optimization algorithm that repeatedly computes the gradient of the loss and moves parameters a small step in the opposite direction. Converges toward a local minimum of the loss surface. *(Chapter 1)*

gradient estimate — The approximation of the true gradient computed from one mini-batch of training data. The quality of this estimate depends on how representative the batch is of the full dataset. *(Chapter 9)*

gradient norm — The L2 norm of the gradient vector for a parameter group, monitored per step to detect spikes that signal an impending NaN or divergence. *(Chapter 11)*

GRPO (Group Relative Policy Optimization) — A reinforcement learning algorithm for language models that generates multiple candidate responses per prompt and scores them relative to each other within the group, rather than against an absolute reward baseline. *(Chapter 23)*

group-wise quantization — A quantization scheme that divides weight matrices into small blocks and assigns each block its own scale factor. More fine-grained than per-tensor quantization and less memory-intensive than per-channel quantization. *(Chapter 27)*

grouped query attention (GQA) — An attention variant that keeps many query heads for expressive attention but reduces the number of key and value heads by sharing them across groups of query heads. Reduces KV cache memory at minimal quality cost. *(Chapter 15)*

hard routing — In a mixture-of-specialists system, selecting the single expert with the highest routing score and using only that expert's output. Produces a discrete, winner-take-all choice rather than a weighted blend. *(Chapter 33)*

head-of-line blocking — A queueing pathology where a slow request stalls faster ones behind it in the same batch. Continuous batching is designed to eliminate this for LLM serving. *(Chapter 17)*

HellaSwag, GSM8K, HumanEval — Standard evaluation benchmarks for commonsense completion, grade-school math word problems, and Python code generation, respectively. *(Chapter 22)*

hidden state — The list of numbers passed from one layer of a model to the next, representing a token's meaning in context at that depth. A 768-dimensional hidden state means each token is represented as a vector of 768 floating-point numbers at that layer boundary. *(Chapter 13)*

hybrid architecture (SSM/attention) — A sequence model that interleaves a majority of linear state-space (e.g. Mamba) layers with a minority of full-attention layers, keeping the linear layers' cheap long-context memory while the few attention layers restore exact copying and in-context recall. Jamba and NVIDIA's Mamba-2-Hybrid are examples; controlled studies recommend a linear-to-attention ratio of roughly 3:1 to 6:1. *(Chapter 30)*

hyper-connections — A generalization of the single residual stream into several parallel streams mixed by a learned matrix at each layer, giving the network more flexibility in how it routes information depthwise. Left unconstrained they can break the residual identity guarantee and destabilize training, which is what mHC's doubly-stochastic constraint repairs. *(Chapter 7)*

image token — The output of a vision encoder for one patch of an image, represented as a vector in the same embedding space the language model uses for text tokens. *(Chapter 29)*

implicit reward — In DPO and related preference losses, the log-probability ratio between the policy and the frozen reference acts as an implicit reward signal without ever training a separate reward model. *(Chapter 24)*

incremental assembly — The property of a composed multi-specialist system that allows new specialists to be trained independently and attached without retraining the existing specialists or the router. *(Chapter 35)*

InfoNCE — A contrastive loss used to train embedding models, in which a query's positive document must score higher than a batch of in-batch negatives under a softmax. *(Chapter 28)*

inference — The phase in which a trained model generates output from new input. Proceeds one token at a time: produce a probability distribution, sample the next token, append it to the context, repeat. Distinct from training: no gradient computation, no parameter updates, and no loss is computed. *(Preflight)*

inference-time scaling (test-time compute) — Spending extra compute at generation time — longer chains of thought, multiple sampled rollouts, search, or verification — to raise answer quality without retraining the model. By 2026 it had become a standard lever across major providers, complementing rather than replacing gains from pretraining. *(Chapter 25)*

initialization — The choice of starting values for model weights before training begins. Initialization scale affects whether training starts with well-conditioned activations or with signals that are already too large or too small to learn from. *(Chapter 6)*

instruction tuning — A specific kind of fine-tuning that teaches a model the request-response contract: when input looks like a user instruction, output should look like an assistant answer. Trains on the shape of interaction, not just domain content. *(Chapter 21)*

INT2 — 2-bit integer quantization, where each weight is mapped to one of four possible values. Drastically reduces storage but often crosses a quality floor where the model retains surface fluency while losing semantic accuracy. *(Chapter 27)*

INT4 — 4-bit integer quantization, where each weight is mapped to one of sixteen possible values. Reduces storage to one-eighth of FP32 while often preserving enough precision for reasonable model quality. *(Chapter 27)*

INT8 — 8-bit integer quantization, where each weight is mapped to one of 256 possible values. Reduces storage by a factor of four compared to FP32 with typically modest quality degradation. *(Chapter 27)*

interface preservation loss — An auxiliary training objective that penalizes a specialist model for drifting away from the base model's hidden-state representations on a shared anchor dataset, keeping the specialist composable after fine-tuning. *(Chapter 33)*

interface specification — A versioned written contract that defines the boundary between shared trunk layers and specialist modules: exact values for d_model, normalization type, positional encoding scheme, vocabulary, and other semantic agreements that all components must honor. *(Chapter 34)*

interleave ratio — In a hybrid SSM/attention model, the proportion of linear (state-space) layers to full-attention layers — for example Jamba's roughly seven Mamba layers per attention layer. It trades memory and speed against recall: with too few attention layers, exact retrieval degrades sharply. *(Chapter 30)*

IPO (Identity Preference Optimization) — A variant of preference optimization that replaces DPO's logistic loss with a squared-error formulation, intended to be more robust to deterministic preference labels. *(Chapter 24)*

IVF / HNSW / flat — Three FAISS index types: flat brute-force search, inverted-file partitioning, and hierarchical navigable small-world graphs. Trade off recall, query latency, and index build cost. *(Chapter 28)*

just-in-time gather — In FSDP, the pattern of all-gathering a sharded parameter only when its layer is about to be used in the forward or backward pass, then immediately discarding the full copy. *(Chapter 12)*

key (attention) — The projection of a token that represents what it contains or advertises. During attention, each token's key is compared against all queries to produce attention scores. *(Chapter 4)*

KL penalty — A Kullback-Leibler divergence term added to the RLHF objective that penalizes the policy for drifting too far from a frozen reference model. Acts as a leash to prevent reward optimization from pushing the model into degenerate outputs. *(Chapter 23)*

KTO (Kahneman-Tversky Optimization) — A preference-optimization loss that works with single-output binary feedback (good/bad) rather than paired (chosen, rejected) preferences. *(Chapter 24)*

kurtosis — A statistical measure of how heavy the tails of a distribution are — how much the distribution is dominated by rare, extreme values. Used in this book to monitor hidden-state health during training. *(Chapter 7)*

KV cache — Stored key and value projections for every token already processed during inference. Allows each new token to compute attention against cached history rather than recomputing the full sequence, reducing inference from $O(n^2)$ to $O(n)$ per new token. *(Chapter 13)*

KV-cache compression — A family of inference-time methods that reduces the memory cost of the cached key and value tensors, usually by quantizing, sharing, pruning, or otherwise compressing the cached history so long-context inference remains affordable. *(Chapter 33)*

KV-cache vector quantization — A family of methods that shrink the stored key/value cache by quantizing whole vectors (or rotated coordinates) rather than raw per-value integers, using a random rotation or transform so the coordinates become easy to code compactly. TurboQuant, QJL, and PolarQuant are examples, reaching roughly 3 to 3.5 bits per channel with little or no reported quality loss. *(Chapter 27)*

layer — A list of neurons that all operate on the same input and whose outputs are collected together. In a transformer, the term refers more broadly to one stage of processing in the network. *(Chapter 1)*

layer freezing — Setting some layers of a pretrained model to read-only during fine-tuning. Frozen layers do not receive gradient updates; their weights stay exactly as they were after pretraining. *(Chapter 32)*

LayerNorm — A normalization function that, for each token's hidden state, subtracts its mean across features and divides by its standard deviation, then applies learned scale and bias. Downstream layers train against a distribution that is centered at zero with controlled variance. *(Chapter 5)*

learning rate — A scalar that controls how large each gradient-descent step is. Too large: training diverges. Too small: training converges too slowly or gets stuck. Typically decayed during training with a schedule. *(Chapter 1)*

LLM-as-judge — An evaluation pattern in which a stronger model scores or ranks the outputs of another model. Cheap and flexible but inherits the judge's biases. *(Chapter 22)*

Lloyd-Max quantization — A classic scalar quantization method that places the reconstruction levels to minimize mean-squared error for a known value distribution, rather than spacing them uniformly. TurboQuant applies per-coordinate Lloyd-Max quantizers after a random rotation makes each coordinate follow a predictable, near-independent distribution. *(Chapter 27)*

logit — An unnormalized score before softmax converts it into a probability. The model's raw preference for each vocabulary token before normalization. *(Chapter 4)*

logprob — The natural log of a probability. Per-token logprobs are summed to score full sequences during evaluation and are the raw signal DPO compares between policy and reference. *(Chapter 22)*

LSH (Locality-Sensitive Hashing) — A family of hashing schemes whose collision probability is high for similar inputs and low for dissimilar ones. Used in MinHash-based near-duplicate detection during data curation. *(Chapter 10)*

loss — A single scalar that measures how wrong the model's prediction was on a given input. During language model training, loss is the negative log-likelihood of the correct next token: a high loss means the model assigned low probability to the right answer. All of training is organized around driving the loss down. *(Preflight)*

LoRA (Low-Rank Adaptation) — A parameter-efficient fine-tuning method that freezes the original weight matrix and adds two much smaller matrices whose product approximates a low-rank correction. Trains the small matrices only, reducing memory and compute cost while approximating full fine-tuning. *(Chapter 21)*

Mamba — A selective state-space model (the S6 block) whose recurrence parameters are input-dependent, giving it linear-time sequence modeling with attention-like content sensitivity. *(Chapter 30)*

mask predictor — The Transformer at the heart of a masked diffusion language model: with no causal mask, it reads the full bidirectional context and predicts every masked position simultaneously at each denoising step. It replaces the autoregressive next-token head of a standard decoder. *(Chapter 31)*

masked/discrete diffusion — The form of diffusion used by text diffusion models, where the "noise" is masked tokens rather than continuous Gaussian noise. The forward process progressively masks a sequence; the reverse process learns to unmask it, predicting discrete tokens at each step. *(Chapter 31)*

MCTS (Monte Carlo Tree Search) — A search algorithm over reasoning steps that balances exploration and exploitation, used in test-time reasoning to plan multi-step chains under a step-level reward model. *(Chapter 25)*

Medusa — A speculative decoding variant that attaches multiple decoding heads to the main model to predict several future tokens at once, removing the need for a separate draft model. *(Chapter 14)*

memory-mapped file — A file on disk that a program can treat almost like an array in memory. The operating system loads only the pages that are actually touched, making random access to large token datasets efficient without loading them fully into RAM. *(Chapter 9)*

mHC (Manifold-Constrained Hyper-Connections) — A residual-stream design (from DeepSeek-AI) that keeps the flexibility of hyper-connections but constrains their mixing matrix to be doubly stochastic — the Birkhoff polytope — using a few Sinkhorn-Knopp iterations. This restores the residual identity guarantee that plain hyper-connections break, stabilizing large-scale training (demonstrated at up to 27B parameters) at a modest overhead. *(Chapter 7)*

micro-batch — One of the small forward/backward passes used inside a gradient accumulation step. Several micro-batches are processed before the accumulated gradient is used to update weights. *(Chapter 9)*

MinHash — A locality-sensitive hashing scheme that estimates Jaccard similarity between sets of shingles using the minimum hash of each set. Standard tool for near-duplicate detection at corpus scale. *(Chapter 10)*

mixed precision — Training that performs the forward pass and gradient computation in reduced precision (FP16 or BF16) while maintaining a full-precision master copy of weights for the optimizer update. Reduces memory and compute cost with careful loss scaling to avoid underflow. *(Chapter 7)*

MMLU — A multiple-choice benchmark covering 57 academic subjects, used as a broad knowledge-and-reasoning eval for language models. *(Chapter 22)*

on-device inference — Running a model locally on the target device rather than in a remote data center, usually to improve privacy, latency, offline operation, or cost control. In practice this is enabled by quantization, efficient runtimes, and hardware-aware model design. *(Chapter 27)*

MLP (Multi-Layer Perceptron) — A stack of layers, each transforming its input through a linear projection followed by a nonlinearity. In the context of transformers, the feedforward sublayer inside each block is an MLP. *(Chapter 1)*

mixture of experts (MoE) — A model design in which a router selects a small subset of specialized feedforward "expert" networks for each token, rather than routing every token through one shared network. Increases model capacity without proportionally increasing compute per token. *(Chapter 18)*

multi-head attention — An attention mechanism that splits the embedding dimension into H independent heads, each computing its own attention over a smaller subspace. The results are concatenated and projected back to the full embedding dimension. *(Chapter 4)*

Multi-Token Prediction (MTP) — A training objective in which the model predicts several future tokens at once through one or more extra prediction heads, rather than only the immediate next token. Originally motivated by a stronger training signal, those extra heads double as a draft mechanism for lossless speculative decoding at inference. *(Chapter 7)*

negative log-likelihood — The loss function that measures how surprised the model was by the actual next token: $-\log p$, where p is the probability the model assigned to the correct token. Minimizing negative log-likelihood is equivalent to maximizing the probability of the training data. *(Chapter 2)*

NTK-aware scaling — A long-context technique that adjusts the base frequency of RoPE so that high-frequency dimensions extrapolate gracefully when the sequence grows beyond the training length. *(Chapter 16)*

online softmax — The streaming form of softmax used inside Flash Attention, which maintains a running maximum and a running denominator so the function can be computed tile by tile without materializing the full attention matrix. *(Chapter 8)*

ORM (outcome reward model) — A reward model that scores only the final answer of a reasoning chain. Cheap to train but blind to mistakes inside the chain. *(Chapter 25)*

ORPO (Odds Ratio Preference Optimization) — A preference-optimization loss that combines an SFT term with an odds-ratio term, removing the need for a separate reference policy. *(Chapter 24)*

outcome reward model (ORM) — See ORM. *(Chapter 25)*

neuron — A unit that takes several inputs, multiplies each by a learned weight, adds a learned bias, and passes the result through a nonlinearity. The basic computational building block of neural networks. *(Chapter 1)*

non-autoregressive decoding — Generating all (or many) output positions in parallel rather than strictly one token at a time conditioned on the previous ones. Diffusion language models decode this way, which lets them revise earlier tokens and fill in gaps but breaks the simple left-to-right KV-cache reuse that autoregressive models rely on. *(Chapter 31)*

nonlinearity — An activation function that introduces a curve into the otherwise linear chain of matrix multiplications in a neural network, enabling the network to learn non-flat decision boundaries. *(Chapter 1)*

numerical gradient — An approximation of the true gradient computed by nudging each parameter slightly up and down and measuring the change in loss, using the centered finite-difference formula: $[f(x+h) - f(x-h)] / 2h$. Used to verify analytic gradient implementations. *(Chapter 1)*

parameter — A single trainable number inside a model. During training, parameters are updated by gradient descent; during inference they are fixed. A modern large language model may have tens of billions of parameters, each contributing a small part of every prediction. *(Preflight)*

PagedAttention — A KV-cache layout that allocates fixed-size pages of K and V slots and indirects through a per-sequence page table, so memory looks contiguous to the kernel but is fragmented on the device. Enables high-utilization serving. *(Chapter 17)*

page table — The per-sequence mapping from logical KV positions to physical KV pages used by PagedAttention. Lets the scheduler add and free pages cheaply at any point in a sequence. *(Chapter 17)*

per-channel quantization — A quantization scheme that assigns an independent scale factor to each output channel or row of a weight matrix, allowing channels with different value ranges to be quantized accurately rather than sharing a single global scale. *(Chapter 27)*

per-tensor quantization — A quantization scheme that assigns a single scale factor to an entire weight tensor. The coarsest approach: simple and fast, but loses accuracy for tensors with wide internal variation. *(Chapter 27)*

perplexity — A language model evaluation metric derived from cross-entropy loss: exp(loss). Represents the effective branching factor — on average, how many equally likely choices the model sees at each token position. Lower is better. *(Chapter 13)*

PolarQuant — A KV-cache quantization method that applies a random preconditioning rotation, converts key vectors to polar (angle) coordinates, and quantizes the angles, storing no per-block normalization constants. It reports over 4.2x KV-cache compression on long-context evaluations. *(Chapter 27)*

positional embedding — A learned or computed vector added to (or combined with) each token's embedding to give the model information about where in the sequence that token appears. Without positional information, attention is order-invariant. *(Chapter 5)*

position interpolation (PI) — A long-context technique that scales the position indices fed into RoPE so that the model is asked to interpolate within its training range instead of extrapolating outside it. *(Chapter 16)*

preemption — In production serving, evicting an in-flight request's KV pages to make room for higher-priority traffic, then resuming it later from a checkpointed prompt. *(Chapter 17)*

preference pair — A training example (chosen, rejected) used in DPO and related losses, where both completions come from the same prompt and the label is which one a labeler preferred. *(Chapter 24)*

prefill — The first phase of inference in which the entire input prompt is processed in parallel to populate the KV cache. Runs in full-sequence mode; produces no output tokens but prepares cached state for the decode phase. *(Chapter 13)*

prefix caching — A serving optimization that reuses cached KV pages across requests that share a common prompt prefix (system messages, few-shot examples), avoiding redundant prefill. *(Chapter 17)*

pre-tokenized data — Training data in which raw text has been converted to token IDs before the training loop begins. Avoids re-running the tokenizer during training and makes data loading a simple array read. *(Chapter 9)*

process reward model (PRM) — A reward model that scores each step of a reasoning chain individually, enabling fine-grained step-level search such as best-of-N over partial chains and MCTS. *(Chapter 25)*

projection layer — In a vision-language model, the small linear (or two-layer) module that projects vision encoder outputs into the language model's embedding space so image tokens look like text tokens to the decoder. *(Chapter 29)*

probability distribution — A collection of non-negative numbers that sum to 1.0, where each number represents the probability of a possible outcome. Language models output a probability distribution over the full vocabulary at each token position: the number assigned to each token is the model's estimate of how likely that token is to come next. *(Preflight)*

program.md — In autonomous experimentation, a rulebook file that specifies which files the agent may edit, what metric counts as improvement, which files are off-limits, and what to do on failure or crash. *(Chapter 20)*

QJL (Quantized Johnson-Lindenstrauss) — A KV-cache quantization method that applies a Johnson-Lindenstrauss random projection and then keeps only the sign bit of each coordinate (1-bit quantization), which removes the need to store any quantization constants. It reports quantizing the cache to about 3 bits with more than 5x memory reduction and no accuracy loss. *(Chapter 27)*

quantization — Controlled compression of a trained model by mapping precise floating-point weights into a smaller set of integer buckets. Reduces memory footprint and can accelerate inference on hardware that supports integer arithmetic. *(Chapter 27)*

query (attention) — The projection of a token that represents what it is looking for in its context. During attention, each token's query is compared against all keys to produce attention scores. *(Chapter 4)*

rank — In distributed training, the integer identity of a single process within the world. Each rank typically owns one GPU. *(Chapter 12)*

ReAct loop — A tool-use control pattern that alternates Thought, Action (tool call), and Observation (tool result) tokens until the model commits to a final answer. *(Chapter 26)*

reduce-scatter — A collective operation that sums tensors across ranks and then distributes disjoint shards of the result back to each rank. Used in FSDP gradient reduction. *(Chapter 12)*

reference policy — The frozen baseline policy whose log probabilities anchor the KL constraint in RLHF or the implicit reward in DPO. *(Chapter 24)*

re-ranker — A second-stage scorer (typically a cross-encoder) that re-orders the top candidates returned by a cheap first-stage retriever. *(Chapter 28)*

ReLU (Rectified Linear Unit) — An activation function that outputs the input unchanged for positive values and zero for negative values: relu(x) = max(0, x). Cheap to compute and avoids saturating large positive activations. *(Chapter 1)*

remasking — In masked diffusion decoding, re-hiding positions whose predicted tokens were low-confidence so they can be re-predicted on a later denoising step, while high-confidence tokens stay fixed. It is what lets a diffusion language model revise its own earlier guesses. *(Chapter 31)*

residual connection — An additive skip path that adds a layer's input directly to its output: x + F(x). Keeps gradients flowing during backpropagation through deep networks and ensures that earlier representations are never fully erased by later processing. *(Chapter 5)*

residual distribution — In speculative decoding, the resampling distribution applied after a draft token is rejected, constructed so that the joint procedure samples exactly from the main model's distribution. *(Chapter 14)*

residual stream — The running hidden state that flows through a transformer from the embedding layer to the final output. Each attention sublayer and feedforward sublayer adds its contribution to this stream rather than replacing it, so the stream accumulates information across all layers. *(Chapter 6)*

reward hacking — A failure mode in RLHF where the policy learns to maximize the reward model's score without producing the behavior the reward model was meant to measure. The policy exploits surface features or loopholes in the reward function. *(Chapter 23)*

RLHF (Reinforcement Learning from Human Feedback) — A post-training method that trains a language model to satisfy a reward model rather than to imitate demonstrations. The reward model is trained on human preferences between pairs of outputs; the policy is then optimized to produce outputs the reward model rates highly. *(Chapter 23)*

RLVR (Reinforcement Learning with Verifiable Rewards) — A post-training approach where the reward comes from an external automatic verifier of the output — an exact-answer check, unit tests, or a rule-based check — instead of a learned reward model. It applies only to tasks with checkable answers (math, code, logic); RLVR names the reward source, not the optimization algorithm (which is typically GRPO). *(Chapter 23)*

RMSNorm — A normalization function that scales each hidden state by dividing by the root mean square of its components, without subtracting the mean. Slightly faster than LayerNorm and used in most modern language models in place of LayerNorm. *(Chapter 7)*

rollout — A single execution of a reasoning chain to completion, used as a unit of work in self-consistency, best-of-N, and MCTS. *(Chapter 25)*

RoPE (Rotary Position Embedding) — A positional encoding scheme that represents position by rotating pairs of hidden dimensions in query and key vectors by angles that depend on position. Encodes relative distance between tokens in the geometry of the vector space rather than in a lookup table. *(Chapter 7)*

router — In a mixture-of-specialists or mixture-of-experts model, a small learned function that reads the shared hidden state and produces weights or a selection decision that determines how much each specialist contributes to the output. *(Chapter 33)*

router collapse — A failure mode in mixture-of-experts training where the router learns to send most tokens to one or two experts, causing the remaining experts to receive almost no gradient and become dead weight. *(Chapter 18)*

RWKV — A non-transformer architecture that replaces attention with two block types (time-mix and channel-mix) and trains like a transformer but runs like an RNN at inference. *(Chapter 30)*

scale (quantization) — The factor that maps between integer buckets and the real-number range of the original weights. Multiplying the integer code by the scale reconstructs an approximation of the original floating-point value. *(Chapter 27)*

scaling laws — The empirically measured relationships between model size, training token count, compute budget, and loss. Show that loss decreases predictably as these quantities increase, enabling compute-optimal allocation decisions before training begins. *(Chapter 19)*

scaling limit — The point in incremental specialist assembly where routing accuracy plateaus or degrades as the number of specialists grows, representing the maximum number of distinct domains a fixed router architecture can reliably discriminate. *(Chapter 35)*

selectivity — In Mamba, the property of having state-space parameters that depend on the current input token, letting the recurrence emphasize or ignore information based on content. *(Chapter 30)*

self-consistency — A test-time strategy that samples N independent chain-of-thought rollouts and takes the majority answer. Often closes the gap between greedy decoding and best-of-N. *(Chapter 25)*

shared expert — In a fine-grained mixture-of-experts layer, an expert that processes every token in addition to the routed experts, so it absorbs common knowledge and frees the routed experts to specialize. DeepSeek-V3 uses one shared expert alongside its 256 routed experts. *(Chapter 18)*

shingle — A k-character or k-token window over a document, used as the atomic unit when building MinHash signatures for near-duplicate detection. *(Chapter 10)*

SimPO (Simple Preference Optimization) — A preference-optimization loss that removes the reference policy entirely and normalizes log probabilities by sequence length. *(Chapter 24)*

shape (tensor) — The dimensions of a tensor, written as a tuple of integers. A tensor of shape `[8, 16, 64]` has 8 slices along the first axis, 16 along the second, and 64 along the third. In language model code, shape notation `[B, T, d]` means batch × sequence length × embedding dimension. *(Preflight)*

SiLU — x · sigmoid(x). A smooth self-gating activation used as the gating component of SwiGLU. Unlike ReLU, it never hard-zeros any input. *(Chapter 7)*

soft routing — In a mixture-of-specialists system, computing a weighted average of all specialists' output logits using the router's learned weights. Produces a continuous blend rather than selecting a single expert; differentiable end-to-end. *(Chapter 33)*

softmax — A function that converts a vector of arbitrary real scores into a probability distribution that sums to 1, by exponentiating each score and dividing by the sum of all exponentiated scores. *(Chapter 2)*

speculative decoding drafter — The small, fast model (or extra head) that proposes a run of candidate tokens for the large target model to verify in a single batched pass. Drafters range from a separate small draft model to Medusa heads, MTP heads, or a block-diffusion drafter; the target's verification keeps the output distribution exact. *(Chapter 14)*

state-space model (SSM) — A sequence model defined by a linear recurrence on a hidden state, evaluated either as a recurrence at inference or as a long convolution during training. Underlies Mamba and related non-transformer architectures. *(Chapter 30)*

streaming softmax — Another name for the online softmax used in Flash Attention, emphasizing that the function is computed in a single pass over tiles rather than after materializing the full attention matrix. *(Chapter 8)*

SwiGLU — A gated feedforward activation where one linear projection carries content and a second projection (passed through SiLU) acts as a gate that controls how much content passes through. Standard in modern large language models including LLaMA. *(Chapter 7)*

temperature — A scalar applied to logits before softmax during sampling that controls the sharpness of the output distribution. Temperature below 1 concentrates probability on high-likelihood tokens; temperature above 1 spreads probability more evenly across the vocabulary. *(Chapter 2)*

tensor — A multidimensional array of numbers. The core data structure in deep learning: a scalar is a 0-dimensional tensor, a vector is 1-dimensional, a matrix is 2-dimensional, and activations flowing through transformer layers are typically 3-dimensional tensors of shape `[B, T, d]`. PyTorch tensors also optionally track gradients for automatic differentiation. *(Preflight)*

tile decomposition — Splitting the attention matrix into fixed-size blocks (tiles) over the Q, K, and V axes so that each tile can be loaded into fast on-chip memory, computed in place, and discarded. The structural trick that makes Flash Attention memory-linear. *(Chapter 8)*

time-mix / channel-mix — RWKV's two block types: one mixes information across time (replacing attention), the other mixes information across channels (replacing the feedforward sublayer). *(Chapter 30)*

tool schema — A JSON description of a callable function — its name, parameters, types, and a short docstring — that the model is taught to honor when emitting tool calls. *(Chapter 26)*

topological order — A traversal order of nodes in a computational graph such that every node is processed before any node that depends on it. The backward pass (backpropagation) applies the chain rule in reverse topological order, from the loss back to each input parameter. *(Chapter 1)*

token — The basic unit the model processes. May correspond to a word, a subword, punctuation, whitespace, or one or more bytes, depending on the tokenizer. Token boundaries are determined by the vocabulary learned during tokenizer training. *(Chapter 3)*

training — The process of adjusting model parameters by repeatedly running the three-move loop: forward pass to produce a prediction, loss computation to measure error, backward pass to compute gradients, then parameter update. Training consumes data and compute; its goal is to reduce loss on the training set while preserving generalization to held-out data. *(Preflight)*

training loss — The average negative log-likelihood of the model's predictions on batches it is actively being trained on. Decreasing training loss confirms that the model is learning its training distribution. *(Chapter 9)*

transfer learning — Adapting a model trained on one task or dataset to perform well on a different task, rather than training from scratch. Leverages the general representations learned during pretraining. *(Chapter 32)*

transformer block — The repeating unit of a transformer model, consisting of a multi-head attention sublayer, a feedforward sublayer, normalization, and residual connections around each sublayer. Depth comes from stacking many blocks. *(Chapter 5)*

TurboQuant — A data-oblivious (online) KV-cache vector quantization method that applies a random orthogonal rotation, then per-coordinate Lloyd-Max scalar quantization, then a 1-bit QJL correction on the residual for unbiased inner products. It reports quality-neutral quantization at 3.5 bits per channel and only marginal degradation at 2.5 bits. *(Chapter 27)*

validation loss — The average negative log-likelihood of the model's predictions on held-out data it has not been trained on. Measures generalization. Increasing validation loss while training loss decreases signals overfitting. *(Chapter 9)*

value (attention) — The projection of a token that represents what it will contribute if another token attends to it. The attended output is a weighted sum of values, with weights determined by the query-key similarity. *(Chapter 4)*

verifier — In RLVR, the external program that checks whether an output is correct and emits the reward — an exact-answer comparison, a unit-test run, or a formal/rule-based check. Because it is deterministic rather than learned, it cannot be reward-hacked the way a learned reward model can, though a weak verifier can still be gamed. *(Chapter 23)*

ViT (Vision Transformer) — An image encoder that splits an image into fixed-size patches, embeds each patch as a token, and processes them with a standard transformer. The foundation for image tokens in vision-language models. *(Chapter 29)*

vision encoder — Any neural network that maps an image (or video frame) into a sequence of vectors in a shared embedding space, so a language model can attend over visual content as if it were text. *(Chapter 29)*

VQA (Visual Question Answering) — The task of producing a text answer to a natural-language question about an image. The canonical eval for vision-language models. *(Chapter 29)*

weight decay — A regularization term added to the optimizer that nudges weights toward smaller magnitudes at each update step. Reduces overfitting and improves training stability. Applied selectively — normalization parameters and biases typically should not receive weight decay. *(Chapter 7)*

weight tying — Reusing the token embedding matrix as the output projection matrix at the final layer. Reduces parameter count and encodes the symmetry that reading and writing should use the same representation of the vocabulary. *(Chapter 5)*

weight-update ratio — The per-step ratio |Δw| / |w| for a parameter group, a sensitive proxy for effective learning rate. Used to diagnose under- and over-training. *(Chapter 11)*

world size — In distributed training, the total number of processes (ranks) participating in the run. *(Chapter 12)*

YaRN — A long-context extension technique that combines NTK-aware scaling with attention-temperature adjustment, allowing a pretrained model to operate well at several times its training context length. *(Chapter 16)*

ZeRO (Zero Redundancy Optimizer) — A family of optimizer-state, gradient, and parameter sharding strategies (Stages 1, 2, and 3) developed in DeepSpeed. Stage 3 is functionally equivalent to FSDP. *(Chapter 12)*

www.ingramcontent.com/pod-product-compliance
Lightning Source LLC
LaVergne TN
LVHW060008210926
848481LV00013B/3674